Lecture Notes in Computer Science 13108

More information about this subseries at https://link.springer.com/bookseries/7407

Teddy Mantoro · Minho Lee ·
Media Anugerah Ayu · Kok Wai Wong ·
Achmad Nizar Hidayanto (Eds.)

Neural Information Processing

28th International Conference, ICONIP 2021
Sanur, Bali, Indonesia, December 8–12, 2021
Proceedings, Part I

Springer

Editors
Teddy Mantoro ⓘ
Sampoerna University
Jakarta, Indonesia

Media Anugerah Ayu ⓘ
Sampoerna University
Jakarta, Indonesia

Achmad Nizar Hidayanto ⓘ
Universitas Indonesia
Depok, Indonesia

Minho Lee ⓘ
Kyungpook National University
Daegu, Korea (Republic of)

Kok Wai Wong ⓘ
Murdoch University
Murdoch, WA, Australia

ISSN 0302-9743   ISSN 1611-3349   (electronic)
Lecture Notes in Computer Science
ISBN 978-3-030-92184-2   ISBN 978-3-030-92185-9   (eBook)
https://doi.org/10.1007/978-3-030-92185-9

LNCS Sublibrary: SL1 – Theoretical Computer Science and General Issues

This Springer imprint is published by the registered company Springer Nature Switzerland AG
The registered company address is: Gewerbestrasse 11, 6330 Cham, Switzerland

Preface

Welcome to the proceedings of the 28th International Conference on Neural Information Processing (ICONIP 2021) of the Asia-Pacific Neural Network Society (APNNS), held virtually from Indonesia during December 8–12, 2021.

The mission of the Asia-Pacific Neural Network Society is to promote active interactions among researchers, scientists, and industry professionals who are working in neural networks and related fields in the Asia-Pacific region. APNNS has Governing Board Members from 13 countries/regions – Australia, China, Hong Kong, India, Japan, Malaysia, New Zealand, Singapore, South Korea, Qatar, Taiwan, Thailand, and Turkey. The society's flagship annual conference is the International Conference of Neural Information Processing (ICONIP).

The ICONIP conference aims to provide a leading international forum for researchers, scientists, and industry professionals who are working in neuroscience, neural networks, deep learning, and related fields to share their new ideas, progress, and achievements. Due to the current COVID-19 pandemic, ICONIP 2021, which was planned to be held in Bali, Indonesia, was organized as a fully virtual conference.

The proceedings of ICONIP 2021 consists of a four-volume set, LNCS 13108–13111, which includes 226 papers selected from 1093 submissions, representing an acceptance rate of 20.86% and reflecting the increasingly high quality of research in neural networks and related areas in the Asia-Pacific. The conference had four main themes, i.e., "Theory and Algorithms," "Cognitive Neurosciences," "Human Centred Computing," and "Applications."

The four volumes are organized in topical sections which comprise the four main themes mentioned previously and the topics covered in three special sessions. Another topic is from a workshop on Artificial Intelligence and Cyber Security which was held in conjunction with ICONIP 2021. Thus, in total, eight different topics were accommodated at the conference. The topics were also the names of the 20-minute presentation sessions at ICONIP 2021. The eight topics in the conference were: Theory and Algorithms; Cognitive Neurosciences; Human Centred Computing; Applications; Artificial Intelligence and Cybersecurity; Advances in Deep and Shallow Machine Learning Algorithms for Biomedical Data and Imaging; Reliable, Robust, and Secure Machine Learning Algorithms; and Theory and Applications of Natural Computing Paradigms.

Our great appreciation goes to the Program Committee members and the reviewers who devoted their time and effort to our rigorous peer-review process. Their insightful reviews and timely feedback ensured the high quality of the papers accepted for publication. Finally, thank you to all the authors of papers, presenters, and participants at the conference. Your support and engagement made it all worthwhile.

December 2021

Teddy Mantoro
Minho Lee
Media A. Ayu
Kok Wai Wong
Achmad Nizar Hidayanto

Organization

Honorary Chairs

Jonathan Chan   King Mongkut's University of Technology Thonburi, Thailand
Lance Fung   Murdoch University, Australia

General Chairs

Teddy Mantoro   Sampoerna University, Indonesia
Minho Lee   Kyungpook National University, South Korea

Program Chairs

Media A. Ayu   Sampoerna University, Indonesia
Kok Wai Wong   Murdoch University, Australia
Achmad Nizar   Universitas Indonesia, Indonesia

Local Arrangements Chairs

Linawati   Universitas Udayana, Indonesia
W. G. Ariastina   Universitas Udayana, Indonesia

Finance Chairs

Kurnianingsih   Politeknik Negeri Semarang, Indonesia
Kazushi Ikeda   Nara Institute of Science and Technology, Japan

Special Sessions Chairs

Sunu Wibirama   Universitas Gadjah Mada, Indonesia
Paul Pang   Federation University Australia, Australia
Noor Akhmad Setiawan   Universitas Gadjah Mada, Indonesia

Tutorial Chairs

Suryono   Universitas Diponegoro, Indonesia
Muhammad Agni Catur Bhakti   Sampoerna University, Indonesia

Proceedings Chairs

Adi Wibowo   Universitas Diponegoro, Indonesia
Sung Bae Cho   Yonsei University, South Korea

Publicity Chairs

Dwiza Riana   Universitas Nusa Mandiri, Indonesia
M. Tanveer   Indian Institute of Technology, Indore, India

Program Committee

Abdulrazak Alhababi   Universiti Malaysia Sarawak, Malaysia
Abhijit Adhikary   Australian National University, Australia
Achmad Nizar Hidayanto   University of Indonesia, Indonesia
Adamu Abubakar Ibrahim   International Islamic University Malaysia, Malaysia
Adi Wibowo   Diponegoro University, Indonesia
Adnan Mahmood   Macquarie University, Australia
Afiyati Amaluddin   Mercu Buana University, Indonesia
Ahmed Alharbi   RMIT University, Australia
Akeem Olowolayemo   International Islamic University Malaysia, Malaysia
Akira Hirose   University of Tokyo, Japan
Aleksandra Nowak   Jagiellonian University, Poland
Ali Haidar   University of New South Wales, Australia
Ali Mehrabi   Western Sydney University, Australia
Al-Jadir   Murdoch University, Australia
Ana Flavia Reis   Federal Technological University of Paraná, Brazil
Anaissi Ali   University of Sydney, Australia
Andrew Beng Jin Teoh   Yonsei University, South Korea
Andrew Chiou   Central Queensland University, Australia
Aneesh Chivukula   University of Technology Sydney, Australia
Aneesh Krishna   Curtin University, Australia
Anna Zhu   Wuhan University of Technology, China
Anto Satriyo Nugroho   Agency for Assessment and Application of
                             Technology, Indonesia
Anupiya Nugaliyadde   Sri Lanka Institute of Information Technology,
                             Sri Lanka
Anwesha Law   Indian Statistical Institute, India
Aprinaldi Mantau   Kyushu Institute of Technology, Japan
Ari Wibisono   Universitas Indonesia, Indonesia
Arief Ramadhan   Bina Nusantara University, Indonesia
Arit Thammano   King Mongkut's Institute of Technology Ladkrabang,
                             Thailand
Arpit Garg   University of Adelaide, Australia
Aryal Sunil   Deakin University, Australia
Ashkan Farhangi   University of Central Florida, USA

Atul Negi	University of Hyderabad, India
Barawi Mohamad Hardyman	Universiti Malaysia Sarawak, Malaysia
Bayu Distiawan	Universitas Indonesia, Indonesia
Bharat Richhariya	IISc Bangalore, India
Bin Pan	Nankai University, China
Bingshu Wang	Northwestern Polytechnical University, Taicang, China
Bonaventure C. Molokwu	University of Windsor, Canada
Bo-Qun Ma	Ant Financial
Bunthit Watanapa	King Mongkut's University of Technology Thonburi, Thailand
Chang-Dong Wang	Sun Yat-sen University, China
Chattrakul Sombattheera	Mahasarakham University, Thailand
Chee Siong Teh	Universiti Malaysia Sarawak, Malaysia
Chen Wei Chén	Chongqing Jiaotong University, China
Chengwei Wu	Harbin Institute of Technology, China
Chern Hong Lim	Monash University, Australia
Chih-Chieh Hung	National Chung Hsing University, Taiwan
Chiranjibi Sitaula	Deakin University, Australia
Chi-Sing Leung	City University of Hong Kong, Hong Kong
Choo Jun Tan	Wawasan Open University, Malaysia
Christoph Bergmeir	Monash University, Australia
Christophe Guyeux	University of Franche-Comté, France
Chuan Chen	Sun Yat-sen University, China
Chuanqi Tan	BIT, China
Chu-Kiong Loo	University of Malaya, Malaysia
Chun Che Fung	Murdoch University, Australia
Colin Samplawski	University of Massachusetts Amherst, USA
Congbo Ma	University of Adelaide, Australia
Cuiyun Gao	Chinese University of Hong Kong, Hong Kong
Cutifa Safitri	Universiti Teknologi Malaysia, Malaysia
Daisuke Miyamoto	University of Tokyo, Japan
Dan Popescu	Politehnica University of Bucharest
David Bong	Universiti Malaysia Sarawak, Malaysia
David Iclanzan	Sapientia Hungarian Science University of Transylvania, Romania
Debasmit Das	IIT Roorkee, India
Dengya Zhu	Curtin University, Australia
Derwin Suhartono	Bina Nusantara University, Indonesia
Devi Fitrianah	Universitas Mercu Buana, Indonesia
Deyu Zhou	Southeast University, China
Dhimas Arief Dharmawan	Universitas Indonesia, Indonesia
Dianhui Wang	La Trobe University, Australia
Dini Handayani	Taylors University, Malaysia
Dipanjyoti Paul	Indian Institute of Technology, Patna, India
Dong Chen	Wuhan University, China

He Chen	Nankai University, China
He Huang	Soochow University, China
Hea Choon Ngo	Universiti Teknikal Malaysia Melaka, Malaysia
Heba El-Fiqi	UNSW Canberra, Australia
Heru Praptono	Bank Indonesia/Universitas Indonesia, Indonesia
Hideitsu Hino	Institute of Statistical Mathematics, Japan
Hidemasa Takao	University of Tokyo, Japan
Hiroaki Inoue	Kobe University, Japan
Hiroaki Kudo	Nagoya University, Japan
Hiromu Monai	Ochanomizu University, Japan
Hiroshi Sakamoto	Kyushu Institute of Technology, Japan
Hisashi Koga	University of Electro-Communications, Japan
Hiu-Hin Tam	City University of Hong Kong, Hong Kong
Hongbing Xia	Beijing Normal University, China
Hongtao Liu	Tianjin University, China
Hongtao Lu	Shanghai Jiao Tong University, China
Hua Zuo	University of Technology Sydney, Australia
Hualou Liang	Drexel University, USA
Huang Chaoran	University of New South Wales, Australia
Huang Shudong	Sichuan University, China
Huawen Liu	University of Texas at San Antonio, USA
Hui Xue	Southeast University, China
Hui Yan	Shanghai Jiao Tong University, China
Hyeyoung Park	Kyungpook National University, South Korea
Hyun-Chul Kim	Kyungpook National University, South Korea
Iksoo Shin	University of Science and Technology, South Korea
Indrabayu Indrabayu	Universitas Hasanuddin, Indonesia
Iqbal Gondal	RMIT University, Australia
Iuliana Georgescu	University of Bucharest, Romania
Iwan Syarif	PENS, Indonesia
J. Kokila	Indian Institute of Information Technology, Allahabad, India
J. Manuel Moreno	Universitat Politècnica de Catalunya, Spain
Jagdish C. Patra	Swinburne University of Technology, Australia
Jean-Francois Couchot	University of Franche-Comté, France
Jelita Asian	STKIP Surya, Indonesia
Jennifer C. Dela Cruz	Mapua University, Philippines
Jérémie Sublime	ISEP, France
Jiahuan Lei	Meituan, China
Jialiang Zhang	Alibaba, China
Jiaming Xu	Institute of Automation, Chinese Academy of Sciences
Jianbo Ning	University of Science and Technology Beijing, China
Jianyi Yang	Nankai University, China
Jiasen Wang	City University of Hong Kong, Hong Kong
Jiawei Fan	Australian National University, Australia
Jiawei Li	Tsinghua University, China

Jiaxin Li	Guangdong University of Technology, China
Jiaxuan Xie	Shanghai Jiao Tong University, China
Jichuan Zeng	Bytedance, China
Jie Shao	University of Science and Technology of China, China
Jie Zhang	Newcastle University, UK
Jiecong Lin	City University of Hong Kong, Hong Kong
Jin Hu	Chongqing Jiaotong University, China
Jin Kyu Kim	Facebook, USA
Jin Ren	Beijing University of Technology, China
Jin Shi	Nanjing University, China
Jinfu Yang	Beijing University of Technology, China
Jing Peng	South China Normal University, China
Jinghui Zhong	South China University of Technology, China
Jin-Tsong Jeng	National Formosa University, Taiwan
Jiri Sima	Institute of Computer Science, Czech Academy of Sciences, Czech Republic
Jo Plested	Australian National University, Australia
Joel Dabrowski	CSIRO, Australia
John Sum	National Chung Hsing University, China
Jolfaei Alireza	Federation University Australia, Australia
Jonathan Chan	King Mongkut's University of Technology Thonburi, Thailand
Jonathan Mojoo	Hiroshima University, Japan
Jose Alfredo Ferreira Costa	Federal University of Rio Grande do Norte, Brazil
Ju Lu	Shandong University, China
Jumana Abu-Khalaf	Edith Cowan University, Australia
Jun Li	Nanjing Normal University, China
Jun Shi	Guangzhou University, China
Junae Kim	DST Group, Australia
Junbin Gao	University of Sydney, Australia
Junjie Chen	Inner Mongolia Agricultural University, China
Junya Chen	Fudan University, China
Junyi Chen	City University of Hong Kong, Hong Kong
Junying Chen	South China University of Technology, China
Junyu Xuan	University of Technology, Sydney
Kah Ong Michael Goh	Multimedia University, Malaysia
Kaizhu Huang	Xi'an Jiaotong-Liverpool University, China
Kam Meng Goh	Tunku Abdul Rahman University College, Malaysia
Katsuhiro Honda	Osaka Prefecture University, Japan
Katsuyuki Hagiwara	Mie University, Japan
Kazushi Ikeda	Nara Institute of Science and Technology, Japan
Kazuteru Miyazaki	National Institution for Academic Degrees and Quality Enhancement of Higher Education, Japan
Kenji Doya	OIST, Japan
Kenji Watanabe	National Institute of Advanced Industrial Science and Technology, Japan

Kok Wai Wong	Murdoch University, Australia
Kitsuchart Pasupa	King Mongkut's Institute of Technology Ladkrabang, Thailand
Kittichai Lavangnananda	King Mongkut's University of Technology Thonburi, Thailand
Koutsakis Polychronis	Murdoch University, Australia
Kui Ding	Nanjing Normal University, China
Kun Zhang	Carnegie Mellon University, USA
Kuntpong Woraratpanya	King Mongkut's Institute of Technology Ladkrabang, Thailand
Kurnianingsih Kurnianingsih	Politeknik Negeri Semarang, Indonesia
Kusrini	Universitas AMIKOM Yogyakarta, Indonesia
Kyle Harrison	UNSW Canberra, Australia
Laga Hamid	Murdoch University, Australia
Lei Wang	Beihang University, China
Leonardo Franco	Universidad de Málaga, Spain
Li Guo	University of Macau, China
Li Yun	Nanjing University of Posts and Telecommunications, China
Libo Wang	Xiamen University of Technology, China
Lie Meng Pang	Southern University of Science and Technology, China
Liew Alan Wee-Chung	Griffith University, Australia
Lingzhi Hu	Beijing University of Technology, China
Linjing Liu	City University of Hong Kong, Hong Kong
Lisi Chen	Hong Kong Baptist University, Hong Kong
Long Cheng	Institute of Automation, Chinese Academy of Sciences, China
Lukman Hakim	Hiroshima University, Japan
M. Tanveer	Indian Institute of Technology, Indore, India
Ma Wanli	University of Canberra, Australia
Man Fai Leung	Hong Kong Metropolitan University, Hong Kong
Maram Mahmoud A. Monshi	Beijing Institute of Technology, China
Marcin Wozniak	Silesian University of Technology, Poland
Marco Anisetti	Università degli Studi di Milano, Italy
Maria Susan Anggreainy	Bina Nusantara University, Indonesia
Mark Abernethy	Murdoch University, Australia
Mark Elshaw	Coventry University, UK
Maruno Yuki	Kyoto Women's University, Japan
Masafumi Hagiwara	Keio University, Japan
Masataka Kawai	NRI SecureTechnologies, Ltd., Japan
Media Ayu	Sampoerna University, Indonesia
Mehdi Neshat	University of Adelaide, Australia
Meng Wang	Southeast University, China
Mengmeng Li	Zhengzhou University, China

Miaohua Zhang	Griffith University, Australia
Mingbo Zhao	Donghua University, China
Mingcong Deng	Tokyo University of Agriculture and Technology, Japan
Minghao Yang	Institute of Automation, Chinese Academy of Sciences, China
Minho Lee	Kyungpook National University, South Korea
Mofei Song	Southeast University, China
Mohammad Faizal Ahmad Fauzi	Multimedia University, Malaysia
Mohsen Marjani	Taylor's University, Malaysia
Mubasher Baig	National University of Computer and Emerging Sciences, Lahore, Pakistan
Muhammad Anwar Ma'Sum	Universitas Indonesia, Indonesia
Muhammad Asim Ali	Shaheed Zulfikar Ali Bhutto Institute of Science and Technology, Pakistan
Muhammad Fawad Akbar Khan	University of Engineering and Technology Peshawar, Pakistan
Muhammad Febrian Rachmadi	Universitas Indonesia, Indonesia
Muhammad Haris	Universitas Nusa Mandiri, Indonesia
Muhammad Haroon Shakeel	Lahore University of Management Sciences, Pakistan
Muhammad Hilman	Universitas Indonesia, Indonesia
Muhammad Ramzan	Saudi Electronic University, Saudi Arabia
Muideen Adegoke	City University of Hong Kong, Hong Kong
Mulin Chen	Northwestern Polytechnical University, China
Murtaza Taj	Lahore University of Management Sciences, Pakistan
Mutsumi Kimura	Ryukoku University, Japan
Naoki Masuyama	Osaka Prefecture University, Japan
Naoyuki Sato	Future University Hakodate, Japan
Nat Dilokthanakul	Vidyasirimedhi Institute of Science and Technology, Thailand
Nguyen Dang	University of Canberra, Australia
Nhi N. Y. Vo	University of Technology Sydney, Australia
Nick Nikzad	Griffith University, Australia
Ning Boda	Swinburne University of Technology, Australia
Nobuhiko Wagatsuma	Tokyo Denki University, Japan
Nobuhiko Yamaguchi	Saga University, Japan
Noor Akhmad Setiawan	Universitas Gadjah Mada, Indonesia
Norbert Jankowski	Nicolaus Copernicus University, Poland
Norikazu Takahashi	Okayama University, Japan
Noriyasu Homma	Tohoku University, Japan
Normaziah A. Aziz	International Islamic University Malaysia, Malaysia
Olarik Surinta	Mahasarakham University, Thailand

Olutomilayo Olayemi Petinrin	Kings University, Nigeria
Ooi Shih Yin	Multimedia University, Malaysia
Osamu Araki	Tokyo University of Science, Japan
Ozlem Faydasicok	Istanbul University, Turkey
Parisa Rastin	University of Lorraine, France
Paul S. Pang	Federation University Australia, Australia
Pedro Antonio Gutierrez	Universidad de Cordoba, Spain
Pengyu Sun	Microsoft
Piotr Duda	Institute of Computational Intelligence/Czestochowa University of Technology, Poland
Prabath Abeysekara	RMIT University, Australia
Pui Huang Leong	Tunku Abdul Rahman University College, Malaysia
Qian Li	Chinese Academy of Sciences, China
Qiang Xiao	Huazhong University of Science and Technology, China
Qiangfu Zhao	University of Aizu, Japan
Qianli Ma	South China University of Technology, China
Qing Xu	Tianjin University, China
Qing Zhang	Meituan, China
Qinglai Wei	Institute of Automation, Chinese Academy of Sciences, China
Qingrong Cheng	Fudan University, China
Qiufeng Wang	Xi'an Jiaotong-Liverpool University, China
Qiulei Dong	Institute of Automation, Chinese Academy of Sciences, China
Qiuye Wu	Guangdong University of Technology, China
Rafal Scherer	Częstochowa University of Technology, Poland
Rahmadya Handayanto	Universitas Islam 45 Bekasi, Indonesia
Rahmat Budiarto	Albaha University, Saudi Arabia
Raja Kumar	Taylor's University, Malaysia
Rammohan Mallipeddi	Kyungpook National University, South Korea
Rana Md Mashud	CSIRO, Australia
Rapeeporn Chamchong	Mahasarakham University, Thailand
Raphael Couturier	Université Bourgogne Franche-Comté, France
Ratchakoon Pruengkarn	Dhurakij Pundit University, Thailand
Reem Mohamed	Mansoura University, Egypt
Rhee Man Kil	Sungkyunkwan University, South Korea
Rim Haidar	University of Sydney, Australia
Rizal Fathoni Aji	Universitas Indonesia, Indonesia
Rukshima Dabare	Murdoch University, Australia
Ruting Cheng	University of Science and Technology Beijing, China
Ruxandra Liana Costea	Polytechnic University of Bucharest, Romania
Saaveethya Sivakumar	Curtin University Malaysia, Malaysia
Sabrina Fariza	Central Queensland University, Australia
Sahand Vahidnia	University of New South Wales, Australia

Saifur Rahaman	City University of Hong Kong, Hong Kong
Sajib Mistry	Curtin University, Australia
Sajib Saha	CSIRO, Australia
Sajid Anwar	Institute of Management Sciences Peshawar, Pakistan
Sakchai Muangsrinoon	Walailak University, Thailand
Salomon Michel	Université Bourgogne Franche-Comté, France
Sandeep Parameswaran	Myntra Designs Pvt. Ltd., India
Sangtae Ahn	Kyungpook National University, South Korea
Sang-Woo Ban	Dongguk University, South Korea
Sangwook Kim	Kobe University, Japan
Sanparith Marukatat	NECTEC, Thailand
Saptakatha Adak	Indian Institute of Technology, Madras, India
Seiichi Ozawa	Kobe University, Japan
Selvarajah Thuseethan	Sabaragamuwa University of Sri Lanka, Sri Lanka
Seong-Bae Park	Kyung Hee University, South Korea
Shan Zhong	Changshu Institute of Technology, China
Shankai Yan	National Institutes of Health, USA
Sheeraz Akram	University of Pittsburgh, USA
Shenglan Liu	Dalian University of Technology, China
Shenglin Zhao	Zhejiang University, China
Shing Chiang Tan	Multimedia University, Malaysia
Shixiong Zhang	Xidian University, China
Shreya Chawla	Australian National University, Australia
Shri Rai	Murdoch University, Australia
Shuchao Pang	Jilin University, China/Macquarie University, Australia
Shuichi Kurogi	Kyushu Institute of Technology, Japan
Siddharth Sachan	Australian National University, Australia
Sirui Li	Murdoch University, Australia
Sonali Agarwal	Indian Institute of Information Technology, Allahabad, India
Sonya Coleman	University of Ulster, UK
Stavros Ntalampiras	University of Milan, Italy
Su Lei	University of Science and Technology Beijing, China
Sung-Bae Cho	Yonsei University, South Korea
Sunu Wibirama	Universitas Gadjah Mada, Indonesia
Susumu Kuroyanagi	Nagoya Institute of Technology, Japan
Sutharshan Rajasegarar	Deakin University, Australia
Takako Hashimoto	Chiba University of Commerce, Japan
Takashi Omori	Tamagawa University, Japan
Tao Ban	National Institute of Information and Communications Technology, Japan
Tao Li	Peking University, China
Tao Xiang	Chongqing University, China
Teddy Mantoro	Sampoerna University, Indonesia
Tedjo Darmanto	STMIK AMIK Bandung, Indonesia
Teijiro Isokawa	University of Hyogo, Japan

Thanh Tam Nguyen	Leibniz University Hannover, Germany
Thanh Tung Khuat	University of Technology Sydney, Australia
Thaweesak Khongtuk	Rajamangala University of Technology Suvarnabhumi, Thailand
Tianlin Zhang	University of Chinese Academy of Sciences, China
Timothy McIntosh	Massey University, New Zealand
Toan Nguyen Thanh	Ho Chi Minh City University of Technology, Vietnam
Todsanai Chumwatana	Murdoch University, Australia
Tom Gedeon	Australian National University, Australia
Tomas Maul	University of Nottingham, Malaysia
Tomohiro Shibata	Kyushu Institute of Technology, Japan
Tomoyuki Kaneko	University of Tokyo, Japan
Toshiaki Omori	Kobe University, Japan
Toshiyuki Yamane	IBM, Japan
Uday Kiran	University of Tokyo, Japan
Udom Silparcha	King Mongkut's University of Technology Thonburi, Thailand
Umar Aditiawarman	Universitas Nusa Putra, Indonesia
Upeka Somaratne	Murdoch University, Australia
Usman Naseem	University of Sydney, Australia
Ven Jyn Kok	National University of Malaysia, Malaysia
Wachira Yangyuen	Rajamangala University of Technology Srivijaya, Thailand
Wai-Keung Fung	Robert Gordon University, UK
Wang Yaqing	Baidu Research, Hong Kong
Wang Yu-Kai	University of Technology Sydney, Australia
Wei Jin	Michigan State University, USA
Wei Yanling	TU Berlin, Germany
Weibin Wu	City University of Hong Kong, Hong Kong
Weifeng Liu	China University of Petroleum, China
Weijie Xiang	University of Science and Technology Beijing, China
Wei-Long Zheng	Massachusetts General Hospital, Harvard Medical School, USA
Weiqun Wang	Institute of Automation, Chinese Academy of Sciences, China
Wen Luo	Nanjing Normal University, China
Wen Yu	Cinvestav, Mexico
Weng Kin Lai	Tunku Abdul Rahman University College, Malaysia
Wenqiang Liu	Southwest Jiaotong University, China
Wentao Wang	Michigan State University, USA
Wenwei Gu	Chinese University of Hong Kong, Hong Kong
Wenxin Yu	Southwest University of Science and Technology, China
Widodo Budiharto	Bina Nusantara University, Indonesia
Wisnu Ananta Kusuma	Institut Pertanian Bogor, Indonesia
Worapat Paireekreng	Dhurakij Pundit University, Thailand

Xiang Chen	George Mason University, USA
Xiao Jian Tan	Tunku Abdul Rahman University College, Malaysia
Xiao Liang	Nankai University, China
Xiaocong Chen	University of New South Wales, Australia
Xiaodong Yue	Shanghai University, China
Xiaoqing Lyu	Peking University, China
Xiaoyang Liu	Huazhong University of Science and Technology, China
Xiaoyang Tan	Nanjing University of Aeronautics and Astronautics, China
Xiao-Yu Tang	Zhejiang University, China
Xin Liu	Huaqiao University, China
Xin Wang	Southwest University, China
Xin Xu	Beijing University of Technology, China
Xingjian Chen	City University of Hong Kong, Hong Kong
Xinyi Le	Shanghai Jiao Tong University, China
Xinyu Shi	University of Science and Technology Beijing, China
Xiwen Bao	Chongqing Jiaotong University, China
Xu Bin	Northwestern Polytechnical University, China
Xu Chen	Shanghai Jiao Tong University, China
Xuan-Son Vu	Umeå University, Sweden
Xuanying Zhu	Australian National University, Australia
Yanling Zhang	University of Science and Technology Beijing, China
Yang Li	East China Normal University, China
Yantao Li	Chongqing University, China
Yanyan Hu	University of Science and Technology Beijing, China
Yao Lu	Beijing Institute of Technology, China
Yasuharu Koike	Tokyo Institute of Technology, Japan
Ya-Wen Teng	Academia Sinica, Taiwan
Yaxin Li	Michigan State University, USA
Yifan Xu	Huazhong University of Science and Technology, China
Yihsin Ho	Takushoku University, Japan
Yilun Jin	Hong Kong University of Science and Technology, Hong Kong
Yiming Li	Tsinghua University, China
Ying Xiao	University of Birmingham, UK
Yingjiang Zhou	Nanjing University of Posts and Telecommunications, China
Yong Peng	Hangzhou Dianzi University, China
Yonghao Ma	University of Science and Technology Beijing, China
Yoshikazu Washizawa	University of Electro-Communications, Japan
Yoshimitsu Kuroki	Kurume National College of Technology, Japan
Young Ju Rho	Korea Polytechnic University, South Korea
Youngjoo Seo	Ecole Polytechnique Fédérale de Lausanne, Switzerland

Yu Sang   PetroChina, China
Yu Xiaohan   Griffith University, Australia
Yu Zhou   Chongqing University, China
Yuan Ye   Xi'an Jiaotong University, China
Yuangang Pan   University of Technology Sydney, Australia
Yuchun Fang   Shanghai University, China
Yuhua Song   University of Science and Technology Beijing
Yunjun Gao   Zhejiang University, China
Zeyuan Wang   University of Sydney, Australia
Zhen Wang   University of Sydney, Australia
Zhengyang Feng   Shanghai Jiao Tong University, China
Zhenhua Wang   Zhejiang University of Technology, China
Zhenqian Wu   University of Electronic Science and Technology
                          of China, China
Zhenyu Cui   University of Chinese Academy of Sciences, China
Zhenyue Qin   Australian National University, Australia
Zheyang Shen   Aalto University, Finland
Zhihong Cui   Shandong University, China
Zhijie Fang   Chinese Academy of Sciences, China
Zhipeng Li   Tsinghua University, China
Zhiri Tang   City University of Hong Kong, Hong Kong
Zhuangbin Chen   Chinese University of Hong Kong, Hong Kong
Zongying Liu   University of Malaya, Malaysia

Contents – Part I

Theory and Algorithms

Metric Learning Based Vision Transformer for Product Matching 3
 Lei Huang, Wei Shao, Fuzhou Wang, Weidun Xie, and Ka-Chun Wong

Stochastic Recurrent Neural Network for Multistep Time Series
Forecasting. 14
 Zexuan Yin and Paolo Barucca

Speaker Verification with Disentangled Self-attention 27
 Junjie Guo, Zhiyuan Ma, Haodong Zhao, Gongshen Liu,
 and Xiaoyong Li

Multi Modal Normalization 40
 Neeraj Kumar, Ankur Narang, Brejesh lall, and Srishti Goel

A Focally Discriminative Loss for Unsupervised Domain Adaptation. 54
 Dongting Sun, Mengzhu Wang, Xurui Ma, Tianming Zhang, Nan Yin,
 Wei Yu, and Zhigang Luo

Automatic Drum Transcription with Label Augmentation Using
Convolutional Neural Networks 65
 Tianyu Xu, Pei Dai, Baoyin He, Mei Zhang, and Jinhui Zhu

Adaptive Curriculum Learning for Semi-supervised Segmentation
of 3D CT-Scans 77
 Obed Tettey Nartey, Guowu Yang, Dorothy Araba Yakoba Agyapong,
 JinZhao Wu, Asare K. Sarpong, and Lady Nadia Frempong

Genetic Algorithm and Distinctiveness Pruning in the Shallow Networks
for VehicleX 91
 Linwei Zhang and Yeu-Shin Fu

Stack Multiple Shallow Autoencoders into a Strong One: A New
Reconstruction-Based Method to Detect Anomaly. 103
 Hanqi Wang, Xing Hu, Liang Song, Guanhua Zhang, Yang Liu,
 Jing Liu, and Linhua Jiang

Learning Discriminative Representation with Attention and Diversity
for Large-Scale Face Recognition 116
 Zhong Zheng, Manli Zhang, and Guixia Kang

Multi-task Perceptual Occlusion Face Detection with Semantic
Attention Network.   129
   Lian Shen, Jia-Xiang Lin, and Chang-Ying Wang

RAIDU-Net: Image Inpainting via Residual Attention Fusion and Gated
Information Distillation   141
   Penghao He, Ying Yu, Chaoyue Xu, and Hao Yang

Sentence Rewriting with Few-Shot Learning for Document-Level Event
Coreference Resolution   152
   Xinyu Chen, Sheng Xu, Peifeng Li, and Qiaoming Zhu

A Novel Metric Learning Framework for Semi-supervised Domain
Adaptation   165
   Rakesh Kumar Sanodiya, Chinmay Sharma, Sai Satwik, Aravind Challa,
   Sathwik Rao, and Leehter Yao

Generating Adversarial Examples by Distributed Upsampling   177
   Shunkai Zhou, Yueling Zhang, Guitao Cao, and Jiangtao Wang

CPSAM: Channel and Position Squeeze Attention Module.   190
   Yuchen Gong, Zhihao Gu, Zhenghao Zhang, and Lizhuang Ma

A Multi-Channel Graph Attention Network for Chinese NER   203
   Yichun Zhao, Kui Meng, and Gongshen Liu

GSNESR: A Global Social Network Embedding Approach
for Social Recommendation   215
   Bing-biao Xiao and Jian-wei Liu

Classification Models for Medical Data with Interpretative Rules   227
   Xinyue Xu, Xiang Ding, Zhenyue Qin, and Yang Liu

Contrastive Goal Grouping for Policy Generalization in Goal-Conditioned
Reinforcement Learning.   240
   Qiming Zou and Einoshin Suzuki

Global Fusion Capsule Network with Pairwise-Relation Attention
Graph Routing   254
   Xinyi Li, Song Wu, and Guoqiang Xiao

MA-GAN: A Method Based on Generative Adversarial Network
for Calligraphy Morphing.   266
   Jiaxin Zhao, Yang Zhang, Xiaohu Ma, Dongdong Yang, Yao Shen,
   and Hualiang Jiang

One-Stage Open Set Object Detection with Prototype Learning.   279
   Yongyu Xiong, Peipei Yang, and Cheng-Lin Liu

Aesthetic-Aware Recommender System for Online Fashion Products   292
 Bei Zhou, Basem Suleiman, and Waheeb Yaqub

DAFD: Domain Adaptation Framework for Fake News Detection   305
 Yinqiu Huang, Min Gao, Jia Wang, and Kai Shu

Document Image Classification Method Based on Graph Convolutional
Network   317
 Yangyang Xiong, Zhongjian Dai, Yan Liu, and Xiaotian Ding

Continual Learning of 3D Point Cloud Generators   330
 Michał Sadowski, Karol J. Piczak, Przemysław Spurek,
 and Tomasz Trzciński

Attention-Based 3D ResNet for Detection of Alzheimer's Disease Process . . .   342
 Mingjin Liu, Jialiang Tang, Wenxin Yu, and Ning Jiang

Generation of a Large-Scale Line Image Dataset with Ground Truth Texts
from Page-Level Autograph Documents   354
 Ayumu Nagai

DAP-BERT: Differentiable Architecture Pruning of BERT   367
 Chung-Yiu Yau, Haoli Bai, Irwin King, and Michael R. Lyu

Trash Detection on Water Channels   379
 Mohbat Tharani, Abdul Wahab Amin, Fezan Rasool, Mohammad Maaz,
 Murtaza Taj, and Abubakar Muhammad

Tri-Transformer Hawkes Process: Three Heads are Better Than One   390
 Zhi-yan Song, Jian-wei Liu, Lu-ning Zhang, and Ya-nan Han

PhenoDeep: A Deep Learning-Based Approach for Detecting Reproductive
Organs from Digitized Herbarium Specimen Images   402
 Abdelaziz Triki, Bassem Bouaziz, Jitendra Gaikwad, and Walid Mahdi

Document-Level Event Factuality Identification Using Negation
and Speculation Scope   414
 Heng Zhang, Zhong Qian, Xiaoxu Zhu, and Peifeng Li

Dynamic Network Embedding by Time-Relaxed Temporal Random Walk . . .   426
 Yifan Song, Darong Lai, Zhihong Chong, and Zeyuan Pan

Dual-Band Maritime Ship Classification Based on Multi-layer
Convolutional Features and Bayesian Decision   438
 Zhaoqing Wu, Yancheng Cai, Xiaohua Qiu, Min Li, Yujie He, Yu Song,
 and Weidong Du

Context-Based Anomaly Detection via Spatial Attributed Graphs
in Human Monitoring   450
  Kang Zhang, Muhammad Fikko Fadjrimiratno, and Einoshin Suzuki

Domain-Adaptation Person Re-Identification via Style Translation
and Clustering   464
  Peiyi Wei, Canlong Zhang, Zhixin Li, Yanping Tang, and Zhiwen Wang

Multimodal Named Entity Recognition via Co-attention-Based Method
with Dynamic Visual Concept Expansion.   476
  Xiaoyu Zhao and Buzhou Tang

Ego Networks.   488
  Andrei Marin, Traian Rebedea, and Ionel Hosu

Cross-Modal Based Person Re-identification via Channel Exchange
and Adversarial Learning   500
  Xiaohui Xu, Song Wu, Shan Liu, and Guoqiang Xiao

SPBERT: an Efficient Pre-training BERT on SPARQL Queries
for Question Answering over Knowledge Graphs   512
  Hieu Tran, Long Phan, James Anibal, Binh T. Nguyen,
  and Truong-Son Nguyen

Deep Neuroevolution: Training Neural Networks Using a Matrix-Free
Evolution Strategy.   524
  Dariusz Jagodziński, Łukasz Neumann, and Paweł Zawistowski

Weighted P-Rank: a Weighted Article Ranking Algorithm Based
on a Heterogeneous Scholarly Network   537
  Jian Zhou, Shenglan Liu, Lin Feng, Jie Yang, and Ning Cai

Clustering Friendly Dictionary Learning.   549
  Anurag Goel and Angshul Majumdar

Understanding Test-Time Augmentation.   558
  Masanari Kimura

SphereCF: Sphere Embedding for Collaborative Filtering   570
  Haozhuang Liu, Mingchao Li, Yang Wang, Wang Chen,
  and Hai-Tao Zheng

Concordant Contrastive Learning for Semi-supervised Node Classification
on Graph   584
  Daqing Wu, Xiao Luo, Xiangyang Guo, Chong Chen, Minghua Deng,
  and Jinwen Ma

Improving Shallow Neural Networks via Local and Global Normalization . . .   596
  Ning Jiang, Jialiang Tang, Xiaoyan Yang, Wenxin Yu, and Peng Zhang

Underwater Acoustic Target Recognition with Fusion Feature.   609
  Pengyuan Qi, Jianguo Sun, Yunfei Long, Liguo Zhang, and Tianye

Evaluating Data Characterization Measures for Clustering Problems
in Meta-learning   621
  *Luiz Henrique dos S. Fernandes, Marcilio C. P. de Souto,
  and Ana C. Lorena*

ShallowNet: An Efficient Lightweight Text Detection Network Based
on Instance Count-Aware Supervision Information   633
  Xingfei Hu, Deyang Wu, Haiyan Li, Fei Jiang, and Hongtao Lu

Image Periodization for Convolutional Neural Networks   645
  Kailai Zhang, Zheng Cao, and Ji Wu

BCN-GCN: A Novel Brain Connectivity Network Classification Method
via Graph Convolution Neural Network for Alzheimer's Disease   657
  Peiyi Gu, Xiaowen Xu, Ye Luo, Peijun Wang, and Jianwei Lu

Triplet Mapping for Continuously Knowledge Distillation   669
  Xiaoyan Yang, Jialiang Tang, Ning Jiang, Wenxin Yu, and Peng Zhang

A Prediction-Augmented AutoEncoder for Multivariate Time Series
Anomaly Detection   681
  Sawenbo Gong, Zhihao Wu, Yunxiao Liu, Youfang Lin, and Jing Wang

Author Index   693

Theory and Algorithms

Metric Learning Based Vision Transformer for Product Matching

Lei Huang[1], Wei Shao[1], Fuzhou Wang[1], Weidun Xie[1],
and Ka-Chun Wong[1,2(✉)]

[1] Department of Computer Science, City University of Hong Kong,
Hong Kong (SAR), China
{lhuang93-c,weishao4-c,wang.fuzhou,weidunxie2-c}@my.cityu.edu.hk,
kc.w@cityu.edu.hk
[2] Hong Kong Institute for Data Science, City University of Hong Kong,
Hong Kong (SAR), China

Abstract. Product image matching is essential on e-commerce platforms since the target customers have to distinguish the same products organized in varied manners by different companies. Machine learning methods can be applied for such kind of task, which treat it as a classification problem. However, it is intrinsically intractable to enforce the model to expand to new products due to its invisibility to unknown categories. Metric learning has emerged as a promising approach to address this issue, which is designed to measure the distances between data points. In this paper, a metric learning based vision transformer (ML-VIT) is proposed to learn the embeddings of the product image data and match the images with small Euclidean distances to the same products. The proposed ML-VIT adopts Arcface loss to achieve intra-class compactness and inter-class dispersion. Compared with Siamese neural network and other pre-trained models in terms of F1 score and accuracy, ML-VIT is proved to yield modest embeddings for product image matching.

Keywords: Metric learning · Vision transformer · Product matching · Arcface

1 Introduction

Product matching aims to retrieve the same product among a large amount of product data on e-commerce platforms, which could improve the customer experience on the purchasing side while speeding up the listing process of the merchant on the selling side. The processing of the mapping data of products including images and descriptions to the specific products grows critical [20]. The same product has a high probability of being organized in varied manners by different online shops. For example, sellers only present the product itself while others exhibit the package in one image. Previous work focus on extracting the attributes of the product to feed into the matching algorithm to construct the

© Springer Nature Switzerland AG 2021
T. Mantoro et al. (Eds.): ICONIP 2021, LNCS 13108, pp. 3–13, 2021.
https://doi.org/10.1007/978-3-030-92185-9_1

product inventory [7,15,17]. The attributes of the products have been explored involve the descriptions [20], titles [1,14] and reviews [22]. However, the direction towards the product image matching is less considered.

Recently, deep neural network models have become mainstream methods for product matching [12,16,18,19]. Neural networks including CNN and RNN can be applied to extract the features of the information automatically and accurately. Specifically, CNN based neural networks can extract the image features while RNN based models are able to learn the semantic meaning of the text data. Previous research treat product matching as a simplex classification task [20] or binary matching distinguish [12,16]. It is indeed suitable for specific products with massive information in small classes. Nevertheless, enforcing the deep learning model to expand to new products is intrinsically intractable due to its inherent invisibility to unknown categories. Besides, the model trained on the binary matching (i.e., pairwise input return match or not) renders a combinatorial explosion in the number of products especially in a large dataset, leading to a significant increase in computing [4]. Metric learning has emerged as a promising methodology to address this issue, which is designed to measure the distances between data points [3,9,10]. By focusing on the distance function tuned to a specific task, it has been revealed to be robust through combining with nearest neighbor methods that depend on the distances or similarities [11]. Metric learning is widely applied in image classification especially face recognition [6] which is similar to the product matching task and a series of Softmax-loss based functions [4,13,24] are proposed to enforce the model to be more discriminative.

Inspired by the success of metric learning in feature embedding and vision transformer [5] in most benchmark tasks, a metric learning based vision transformer(ML-VIT) is proposed in this paper. ML-VIT incorporates the pretrained vision transformer with resnet to obtain the feature embeddings of the input image and then adopt Arcface loss to achieve the intra-class compactness and the inter-class dispersion. Arcface aims at adjusting the angles of the feature embeddings depending on their classes, ML-VIT thereby learns distributions of the data based on the classification task. In order to demonstrate the high performance of ML-VIT, experiments of Siamese neural network and other pre-trained models including vgg [21], resnet [8] and efficientnet [23] are conducted to compare the capabilities. The results illustrate ML-VIT can yield modest embeddings for product image matching.

2 Method

2.1 ML-VIT

An overview of ML-VIT is outlined in Fig. 1. In general, ML-VIT is trained on the image data in classification mode. Transformer requires 1D sequence-like tokens as input while the image data is 2D. To transfer the input size to feed it into the vision transformer, we follow the previous work [5] to reshape the feature $x \in \mathbb{R}^{H \times W \times C}$ into the flattened 2D patches $x \in \mathbb{R}^{N \times (P^2 \times C)}$, where H

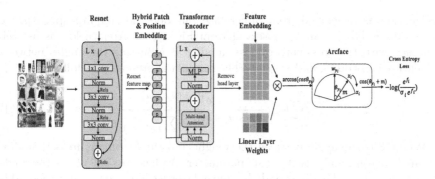

Fig. 1. Overview of ML-VIT

is the height size and W is the width size (i.e., resolution of the image), C is the number of channels, P is the resolution of patches and N is the number of patches, which can be regarded as the length of the sequence. Specifically, ML-VIT adopts resnet to extract the feature map as the patch embedding in conjunction with the position embedding, which is called hybrid architecture in previous work [5].

The feature map of resnet that is the embedding projection E and 1D position embedding are added together (Eq. 1) as the input of transformer encoder. The embeddings then go through several blocks which contain Layernorm (LN), multi-head attention (MHA) and multiple layer perceptrons (MLP). It is noted that we remove the head layer of vision transformer since it is only applied as the backbone model to get the feature embeddings and followed by subsequent classify layers.

$$x_{emb} = [x_p^1 E; x_p^2 E; \ldots x_p^N E] + E_{pos}, \quad E \in \mathbb{R}^{(P^2 \times C) \times D}, E_{pos} \in \mathbb{R}^{(N+1) \times D} \quad (1)$$

Classifier layers normally incorporate Softmax function to map the logits into the probabilities. If Softmax activation function is combined with the loss function, it becomes Softmax loss (Eq. 3), where x_i is the feature embedding of the ith sample, N is the number of samples, n is the batch size, W is the weight and b is the bias term. Softmax loss is a special form of cross-entropy loss (Eq. 2), where C is the number of classes, p_j is the probability and is calculated by Softmax function.

$$L_1 = -\sum_{j=1}^{C} y_j \log p_j \quad (2)$$

$$L_2 = -\frac{1}{N} \sum_{i=1}^{N} \log \frac{e^{W_{y_i}^T x_i + b_{y_i}}}{\sum_{j=1}^{n} e^{W_j^T x_i + b_j}} \quad (3)$$

Sphereface loss firstly introduces the angularly discriminative feature to the softmax loss, which transforms the Euclidean distance to angular margin [13].

Arcface loss is developed from Sphereface but does not require approximate computations [4]. It can be observed from Fig. 1 that Arcface loss is as well combined with the cross-entropy loss, aiming at adjusting the angles between the feature embedding and weights. The formula is presented as follows,

$$L_3 = -\frac{1}{N} \sum_{i=1}^{N} \log \frac{e^{s\left(\cos(\theta_{y_i}+m)\right)}}{e^{s\left(\cos(\theta_{y_i}+m)\right)} + \sum_{j=1,j\neq y_i}^{n} e^{s\cos\theta_j}} \tag{4}$$

Where θ is the angle between the weight and the feature x, m is the additive angle margin and s is the scale parameter. Arcface loss calculates the angle between feature embeddings and the linear layer weights firstly, and then adds the margin to increase the inter-class angle. In general, ML-VIT works in classification mode during the training stage.

2.2 Inference for Matching

The traditional machine learning methods that require a pair of inputs treat the matching as binary classification (i.e., 1 denotes matching and 0 denotes not). It is admitted that this kind of method can uniform the training stage and inference stage. However, a pair of inputs double even triple the dataset, leading to huge computations. Furthermore, the construction of input pairs is quite tricky to sample for effective model training.

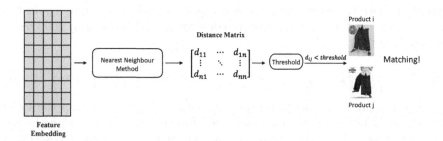

Fig. 2. The inference process of ML-VIT

Different from the aforementioned methods, ML-VIT accepts the same single image data input as the classification task. In the inference stage, ML-VIT is utilized as the feature extractor. The process of the inference is depicted in Fig. 2. It can be observed that ML-VIT is in conjunction with the nearest neighbour method to obtain the distance matrix. There are various distance metrics such as Euclid distance, cosine similarity, Hamming distance and so on. ML-VIT adopts standard Euclid distance as the metric, since other distances are not observed with significant performance gains. After chalking up the distance matrix, a threshold is set up to decide whether a pair of images are matched, that is, if d_{ij} is less than the threshold, image i and image j belong to the same product.

3   Experiment

In this section, we evaluate the matching ability of ML-VIT on the Shopee Price Match Guarantee dataset[1]. Because ML-VIT utilizes hybrid patch embedding extracted by resnet, we denote this model "ML-VIT_resnet". In order to illustrate the excellent performance of VIT, state-of-art(sota) models are evaluated on the same dataset by the same metrics (F1 score and accuracy).

3.1   Implementation

Dataset. The Shoppe Price Match Guarantee dataset is employed in this paper to investigate model scalability. It contains 34250 images, belonging to 11014 products. Most of the products have two or three different images that require matching. The dataset is divided into three parts, 80% of the dataset is the training dataset to train the machine learning models, 10% is the development dataset to select the appropriate parameters and 10% is to test the performance of models.

Experiments Settings. Table 1 illustrates the parameters selection of ML-VIT_resnet. The parameters are selected according to the model performance on the development dataset. It should be noted that we adopt an adaptive schedule to adjust the learning rate dynamically. The start learning rate is $1e^{-5}$ and the range of the value is from $1e^{-6}$ to $1e^{-5} \times$ batch size. Parameter selection of other sota models is achieved by the same method for a fair comparison.

Table 1. Parameters selection of ML-VIT-resnet.

Parameters	Value
Batch size	4
Learning rate	$1e^{-6}$–$1e^{-5} \times$ batch size
Epoch	10
Scale s of	64
Margin m of Arcface	0.17

Metrics. We benchmark ML-VIT on the test dataset to compare the performance of ML-VIT with other sota models. It is evaluated by F1 score and accuracy which is also adopted by the sota models.

---

[1] https://www.kaggle.com/c/shopee-product-matching/data.

3.2   Comparison with State-of-Art Models

In this section, Siamese neural network and other pre-trained models with the same architecture as ML-VIT's are implemented to obtain a fair comparison with proposed ML-VIT. The Siamese neural network has two inputs and feeds the two inputs into two neural networks, which respectively map the inputs to the new space to form the representation. Siamese neural networks learn the feature embedding by judging a pair of images matching instead of classifying. We adopt vgg16, resnet, efficientnet_b0, efficientnet_b4, base vision transformer that contains relatively small parameters compared with "large" and "huge" version, and vision transformer with resnet hybrid patch embedding as the backbone for Siamese based structure and Arcface & classification based structure. The former models are denoted by "Siamese" prefix while the latter are denoted by "ML" prefix.

Table 2. Comparison with state-of-art models.

Models	F1 score (%)	Accuracy (%)
Siamese-vgg16	50.56	36.67
Siamese-resnet50	56.33	44.07
Siamese-efficientnet_b0	50.89	37.54
Siamese-efficientnet_b4	52.74	39.89
Siamese-VIT_base	58.05	46.40
Siamese-VIT_restnet	59.10	47.43
ML-vgg16	64.17	52.64
ML-restnet50	69.78	62.10
ML-efficientnet_b0	75.00	73.00
ML-efficientnet_b4	74.91	70.91
ML-VIT_base	76.15	73.89
ML-VIT_restnet	**78.02**	**75.52**

The results are presented in Table 2. It can be observed that the proposed ML-VIT_restnet achieves the best performance whose F1 score is 78.02% and accuracy is 75.52%. The models that incorporate vision transformer outperform other pre-trained models in both structures, proving it is also suitable for feature extraction as well as classification. Another enlightening point is that all the Arcface & classification based metric learning models gain a huge increase on the performance than Siamese based metric learning models. The reason hidden behind the disappointing performance of Siamese models is that the negative samples are hard to select and the outputs collapsing to a constant without stop-gradient strategies [2]. Besides, the training time of Arcface & classification based models is also less than that of Siamese based models since Siamese neural network doubles the data. Detailly, ML-VIT_resnet takes half an hour to train one epoch while Siamese-VIT_resnet requires two hours and a half. Therefore, ML-VIT does not only achieves high-level performance but is also more efficient.

3.3   Inspection of ML-VIT

In order to get the intuition of the learned embedding by ML-VIT, we project the embeddings into 2D and 3D space by t-SNE descending dimension method. t-SNE can preserve the distance which is as well called similarity, it is thereby qualified to discover similar items. Firstly, the original image vectors are depicted in Fig. 3. Although some of the points in the same class are very close together, most of the points are scattered around the feature space in an irregular manner. The distribution of original data points is disorganized, which means it is hard for the nearest neighbor method to compare the similarity of two points accurately.

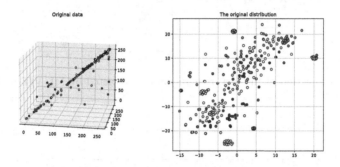

Fig. 3. The projection of the feature embeddings of original images

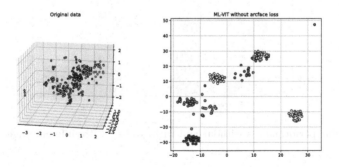

Fig. 4. The projection of the feature embeddings extracted by ML-VIT without Arcface loss (Color figure online)

When the original image features go through ML-VIT which does not incorporate Arcface loss, the output feature embeddings become compact in the same class while those in different classes are dispersive. As illustrated in Fig. 4, it is clear to discriminate the points belonging to different classes. Although there still exist some points that are not close enough to each other in the same classes (i.e., the bluegreen and violet data points), it demonstrates the ability of ML-VIT to yield modest feature embeddings.

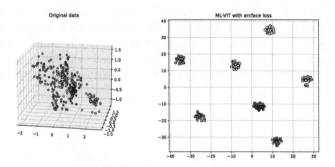

Fig. 5. The projection of the feature embeddings extracted by ML-VIT with Arcface loss

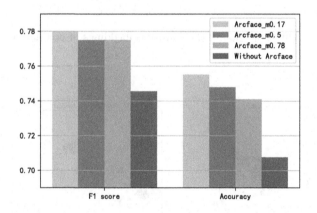

Fig. 6. The performances without and with Arcface with different angle margin

If Arcface loss is combined with the ML-VIT feature extractor, the result is more satisfactory. The distribution of the feature embeddings that are fed into ML-VIT with Arcface loss is presented in Fig. 5. It can be observed that the data points are further away from other points in different classes while the points of the same classes are closer to each other compared with Fig. 4. The distances between points can be regarded as the correspondent geodesic distances on the hypersphere. In other words, the distance implies the angle between the feature embedding and the correspondent weights. Angle margin is crucial in Arcface since it has the same effect as the geodesic distance margin. By analysing the addictive angle margin of Arcface loss numerically, it can be found that the margin enforces the intra-class angles to shrink and inter-class angles to increase because the margin is only added into the numerator and on of the terms in the denominator (Eq. 4). Actually, the performance with Arcface is better than the model without Arcface (Fig. 6). Moreover, the small additive margin shows a stronger ability to discriminate the embeddings (Fig. 6).

4 Conclusion

In this paper, ML-VIT is proposed to match the product images and achieves the best performance compared with Siamese models and other sota pre-trained models. Unlike the traditional classification machine learning methods, ML-VIT utilizes a metric learning structure to learn the similarities of the feature embeddings, which jumps out of the fixed class limitation. By combining with the nearest neighbor method, ML-VIT enables a fast and scalable manner to find similar items on e-commerce platforms. The feature extractor can be replaced by other pre-trained models, but ML-VIT illustrates that vision transformer combined with resnet patching embedding is more suitable for product matching task since it does not introduce any image-specific inductive biases. In addition to the backbone model, ML-VIT incorporates Arcface loss which shows high performance on face recognition tasks to obtain more discriminative feature embeddings, which indeed benefits the performance for subsequent image matching.

In general, ML-VIT can be employed to match the same products quickly and accurately, preventing the customers from getting lost in the massive product information. With regard to future work, more research about the loss function can be carried out to enable the model structure more suitable for matching tasks.

Acknowledgements. The work described in this paper was substantially supported by the grant from the Research Grants Council of the Hong Kong Special Administrative Region [CityU 11200218], one grant from the Health and Medical Research Fund, the Food and Health Bureau, The Government of the Hong Kong Special Administrative Region [07181426], and the funding from Hong Kong Institute for Data Science (HKIDS) at City University of Hong Kong. The work described in this paper was partially supported by two grants from City University of Hong Kong (CityU 11202219, CityU 11203520). This research is also supported by the National Natural Science Foundation of China under Grant No. 32000464.

References

1. Akritidis, L., Bozanis, P.: Effective unsupervised matching of product titles with k-combinations and permutations. In: 2018 Innovations in Intelligent Systems and Applications (INISTA), pp. 1–10 (2018). https://doi.org/10.1109/INISTA.2018.8466294
2. Chen, X., He, K.: Exploring simple Siamese representation learning. arXiv preprint arXiv:2011.10566 (2020)
3. Cheng, G., Yang, C., Yao, X., Guo, L., Han, J.: When deep learning meets metric learning: remote sensing image scene classification via learning discriminative CNNs. IEEE Trans. Geosci. Remote Sens. **56**(5), 2811–2821 (2018)
4. Deng, J., Guo, J., Xue, N., Zafeiriou, S.: ArcFace: additive angular margin loss for deep face recognition. In: Proceedings of the IEEE/CVF Conference on Computer Vision and Pattern Recognition, pp. 4690–4699 (2019)
5. Dosovitskiy, A., et al.: An image is worth 16 × 16 words: transformers for image recognition at scale. arXiv preprint arXiv:2010.11929 (2020)
6. Guillaumin, M., Verbeek, J., Schmid, C.: Is that you? Metric learning approaches for face identification. In: 2009 IEEE 12th International Conference on Computer Vision, pp. 498–505 (2009). https://doi.org/10.1109/ICCV.2009.5459197
7. Guy, I., Radinsky, K.: Structuring the unstructured: from startup to making sense of eBay's huge eCommerce inventory. In: Proceedings of the 40th International ACM SIGIR Conference on Research and Development in Information Retrieval, SIGIR 2017, p. 1351. Association for Computing Machinery, New York (2017). https://doi.org/10.1145/3077136.3096469
8. He, K., Zhang, X., Ren, S., Sun, J.: Deep residual learning for image recognition. In: Proceedings of the IEEE Conference on Computer Vision and Pattern Recognition, pp. 770–778 (2016)
9. Kaya, M., Bilge, H.Ş: Deep metric learning: a survey. Symmetry **11**(9), 1066 (2019)
10. Kim, W., Goyal, B., Chawla, K., Lee, J., Kwon, K.: Attention-based ensemble for deep metric learning. In: Ferrari, V., Hebert, M., Sminchisescu, C., Weiss, Y. (eds.) ECCV 2018. LNCS, vol. 11205, pp. 760–777. Springer, Cham (2018). https://doi.org/10.1007/978-3-030-01246-5_45
11. Kulis, B., et al.: Metric learning: a survey. Found. Trends Mach. Learn. **5**(4), 287–364 (2012)
12. Li, J., Dou, Z., Zhu, Y., Zuo, X., Wen, J.-R.: Deep cross-platform product matching in e-commerce. Inf. Retr. J. **23**(2), 136–158 (2019). https://doi.org/10.1007/s10791-019-09360-1
13. Liu, W., Wen, Y., Yu, Z., Li, M., Raj, B., Song, L.: SphereFace: deep hypersphere embedding for face recognition. In: Proceedings of the IEEE Conference on Computer Vision and Pattern Recognition, pp. 212–220 (2017)
14. Majumder, B.P., Subramanian, A., Krishnan, A., Gandhi, S., More, A.: Deep recurrent neural networks for product attribute extraction in ecommerce. arXiv preprint arXiv:1803.11284 (2018)
15. Mauge, K., Rohanimanesh, K., Ruvini, J.D.: Structuring e-commerce inventory. In: Proceedings of the 50th Annual Meeting of the Association for Computational Linguistics (Volume 1: Long Papers), pp. 805–814 (2012)
16. Peeters, R., Bizer, C., Glavaš, G.: Intermediate training of Bert for product matching. Small **745**(722), 2–112 (2020)
17. Putthividhya, D., Hu, J.: Bootstrapped named entity recognition for product attribute extraction. In: Proceedings of the 2011 Conference on Empirical Methods in Natural Language Processing, pp. 1557–1567 (2011)

18. Ristoski, P., Petrovski, P., Mika, P., Paulheim, H.: A machine learning approach for product matching and categorization. Semant. Web 9(5), 707–728 (2018)
19. Rivas-Sánchez, M., De La Paz Guerrero-Lebrero, M., Guerrero, E., Bárcena-Gonzalez, G., Martel, J., Galindo, P.L.: Using deep learning for image similarity in product matching. In: Rojas, I., Joya, G., Catala, A. (eds.) IWANN 2017. LNCS, vol. 10305, pp. 281–290. Springer, Cham (2017). https://doi.org/10.1007/978-3-319-59153-7_25
20. Shah, K., Kopru, S., Ruvini, J.D.: Neural network based extreme classification and similarity models for product matching. In: Proceedings of the 2018 Conference of the North American Chapter of the Association for Computational Linguistics: Human Language Technologies, vol. 3 (Industry Papers), pp. 8–15 (2018)
21. Simonyan, K., Zisserman, A.: Very deep convolutional networks for large-scale image recognition. arXiv preprint arXiv:1409.1556 (2014)
22. Sun, L.: Research on product attribute extraction and classification method for online review. In: 2017 International Conference on Industrial Informatics-Computing Technology, Intelligent Technology, Industrial Information Integration (ICIICII), pp. 117–121. IEEE (2017)
23. Tan, M., Le, Q.V.: EfficientNet: rethinking model scaling for convolutional neural networks (2019). https://arxiv.org/abs/1905.11946
24. Wang, H., et al.: CosFace: large margin cosine loss for deep face recognition. In: Proceedings of the IEEE Conference on Computer Vision and Pattern Recognition, pp. 5265–5274 (2018)

Stochastic Recurrent Neural Network for Multistep Time Series Forecasting

Zexuan Yin$^{(\boxtimes)}$ and Paolo Barucca

Department of Computer Science, University College London,
London WC1E 7JE, UK
{zexuan.yin.20,p.barucca}@ucl.ac.uk

Abstract. Time series forecasting based on deep architectures has been gaining popularity in recent years due to their ability to model complex non-linear temporal dynamics. The recurrent neural network is one such model capable of handling variable-length input and output. In this paper, we leverage recent advances in deep generative models and the concept of state space models to propose a stochastic adaptation of the recurrent neural network for multistep-ahead time series forecasting, which is trained with stochastic gradient variational Bayes. To capture the stochasticity in time series temporal dynamics, we incorporate a latent random variable into the recurrent neural network to make its transition function stochastic. Our model preserves the architectural workings of a recurrent neural network for which all relevant information is encapsulated in its hidden states, and this flexibility allows our model to be easily integrated into any deep architecture for sequential modelling. We test our model on a wide range of datasets from finance to healthcare; results show that the stochastic recurrent neural network consistently outperforms its deterministic counterpart.

Keywords: State space models · Deep generative models · Variational inference

1 Introduction

Time series forecasting is an important task in industry and academia, with applications in fields such as retail demand forecasting [1], finance [2–4], and traffic flow prediction [5]. Traditionally, time series forecasting was dominated by linear models such as the autoregressive integrated moving average model (ARIMA), which required prior knowledge about time series structures such as seasonality and trend. With an increasing abundance of data and computational power however, deep learning models have gained much research interest due to their ability to learn complex temporal relationships with a purely data-driven approach; thus requiring minimal human intervention and expertise in the subject matter. In this work, we combine deep learning with state space models (SSM) for sequential modelling. Our work follows recent trend that combines

© Springer Nature Switzerland AG 2021
T. Mantoro et al. (Eds.): ICONIP 2021, LNCS 13108, pp. 14–26, 2021.
https://doi.org/10.1007/978-3-030-92185-9_2

the powerful modelling capabilities of deep learning models with well understood theoretical frameworks such as SSMs.

Recurrent neural networks (RNN) are a popular class of neural networks for sequential modelling. There exists a great abundance of literature on time series modelling with RNNs across different domains [6–16]. However, vanilla RNNs have deterministic transition functions, which may limit their expressive power at modelling sequences with high variability and complexity [18]. There is recent evidence that the performance of RNNs on complex sequential data such as speech, music, and videos can be improved when uncertainty is incorporated in the modelling process [19–24]. This approach makes an RNN more expressive, as instead of outputting a single deterministic hidden state at every time step, it now considers many possible future paths before making a prediction. Inspired by this, we propose an RNN cell with stochastic hidden states for time series forecasting, which is achieved by inserting a latent random variable into the RNN update function. Our approach corresponds to a state space formulation of time series modelling where the RNN transition function defines the latent state equation, and another neural network defines the observation equation given the RNN hidden state. The main contributions of our paper are as follows:

1. we propose a novel deep stochastic recurrent architecture for multistep-ahead time series forecasting which leverages the ability of regular RNNs to model long-term dynamics and the stochastic framework of state space models.
2. we conduct experiments using publicly available datasets in the fields of finance, traffic flow prediction, air quality forecasting, and disease transmission. Results demonstrate that our stochastic RNN consistently outperforms its deterministic counterpart, and is capable of generating probabilistic forecasts

2 Related Works

2.1 Recurrent Neural Networks

The recurrent neural network (RNN) is a deep architecture specifically designed to handle sequential data, and has delivered state-of-the-art performance in areas such as natural language processing [25]. The structure of the RNN is such that at each time step t, the hidden state of the network - which learns a representation of the raw inputs - is updated using the external input for time t as well as network outputs from the previous step $t-1$. The weights of the network are shared across all time steps and the model is trained using back-propagation. When used to model long sequences of data, the RNN is subject to the vanishing/exploding gradient problem [26]. Variants of the RNN such as the LSTM [27] and the GRU [28] were proposed to address this issue. These variants use gated mechanisms to regulate the flow of information. The GRU is a simplification of the LSTM without a memory cell, which is more computationally efficient to train and offers comparable performance to the LSTM [29].

2.2 Stochastic Gradient Variational Bayes

The authors in [21] proposed the idea of combining an RNN with a variational auto-encoder (VAE) to leverage the RNN's ability to capture time dependencies and the VAE's role as a generative model. The proposed structure consists of an encoder that learns a mapping from data to a distribution over latent variables, and a decoder that maps latent representations to data. The model can be efficiently trained with Stochastic Gradient Variational Bayes (SGVB) [30] and enables efficient, large-scale unsupervised variational learning on sequential data. Consider input x of arbitrary size, we wish to model the data distribution $p(x)$ given some unobserved latent variable z (again, of arbitrary dimension). The aim is maximise the marginal likelihood function $p(x) = \int p(x|z)p(z)\,dz$, which is often intractable when the likelihood $p(x|z)$ is expressed by a neural network with non-linear layers. Instead we apply variational inference and maximise the evidence lower-bound (ELBO):

$$\log p(x) = \log \int p(x|z)p(z)\,dz = \log \int p(x|z)p(z)\frac{q(z)}{q(z)}\,dz$$
$$\geq \mathbb{E}_{z \sim q(z|x)}[\log p(x|z)] - KL(q(z|x)||p(z)) = ELBO, \quad (1)$$

where $q(z|x)$ is the variational approximation to true the posterior distribution $p(z|x)$ and KL is the Kullback-Leibler divergence. For the rest of this paper we refer to $p(x|z)$ as the decoding distribution and $q(z|x)$ as the encoding distribution. The relationship between the marginal likelihood $p(x)$ and the $ELBO$ is given by

$$\log p(x) = \mathbb{E}_{z \sim q(z|x)}[\log p(x|z)] - KL(q(z|x)||p(z))$$
$$+ KL(q(z|x)||(p(z|x)), \quad (2)$$

where the third KL term specifies the tightness of the lower bound. The expectation $\mathbb{E}_{z \sim q(z|x)}[\log p(x|z)]$ can be interpreted as an expected negative reconstructed error, and $KL(q(z|x)||p(z))$ serves as a regulariser.

2.3 State Space Models

State space models provide a unified framework for time series modelling; they refer to probabilistic graphical models that describe relationships between observations and the underlying latent variable [35]. Exact inference is feasible only for hidden Markov models (HMM) and linear Gaussian state space models (LGSS) and both are not suitable for long-term prediction [31]. SSMs can be viewed a probabilistic extension of RNNs. Inside an RNN, the evolution of the hidden states h is governed by a non-linear transition function f: $h_{t+1} = f(h_t, x_{t+1})$ where x is the input vector. For an SSM however, the hidden states are assumed to be random variables. It is therefore intuitive to combine the non-linear gated mechanisms of the RNN with the stochastic transitions of the SSM; this creates a sequential generative model that is more expressive than the RNN and better capable of modelling long-term dynamics than the SSM. There are many recent works that draw connections between SSM and VAE using an RNN. The authors in [18] and [19] propose a sequential VAE with nonlinear state transitions in the latent space, in [32] the authors investigate various inference schemes for variational RNNs, in [22] the authors propose to stack a stochastic SSM layer on top of a deterministic RNN layer, in [23] the authors propose a latent transition scheme that is stochastic conditioned on some inferable parameters, the authors in [33] propose a deep Kalman filter with exogenous inputs, the authors in [34] propose a stochastic variant of the Bi-LSTM, and in [37] the authors use an RNN to parameterise a LGSS.

3 Stochastic Recurrent Neural Network

3.1 Problem Statement

For a multivariate dataset comprised of $N+1$ time series, the covariates $\boldsymbol{x}_{1:T+\tau} = \{\boldsymbol{x}_1, \boldsymbol{x}_2, ...\boldsymbol{x}_{T+\tau}\} \in \mathbb{R}^{N \times (T+\tau)}$ and the target variable $y_{1:T} \in \mathbb{R}^{1 \times T}$. We refer to the period $\{T+1, T+2, ...T+\tau\}$ as the prediction period, where $\tau \in \mathbb{Z}^+$ is the number of prediction steps and we wish to model the conditional distribution

$$P(y_{T+1:T+\tau}|y_{1:T}, \boldsymbol{x}_{1:T+\tau}). \tag{3}$$

3.2 Stochastic GRU Cell

Here we introduce the update equations of our stochastic GRU, which forms the backbone of our temporal model:

$$\boldsymbol{u}_t = \sigma(\boldsymbol{W}_u \cdot \boldsymbol{x}_t + \boldsymbol{C}_u \cdot \boldsymbol{z}_t + \boldsymbol{M}_u \cdot \boldsymbol{h}_{t-1} + \boldsymbol{b}_u) \tag{4}$$

$$\boldsymbol{r}_t = \sigma(\boldsymbol{W}_r \cdot \boldsymbol{x}_t + \boldsymbol{C}_r \cdot \boldsymbol{z}_t + \boldsymbol{M}_r \cdot \boldsymbol{h}_{t-1} + \boldsymbol{b}_r) \tag{5}$$

$$\tilde{\boldsymbol{h}}_t = tanh(\boldsymbol{W}_h \cdot \boldsymbol{x}_t + \boldsymbol{C}_h \cdot \boldsymbol{z}_t + \boldsymbol{r}_t \odot \boldsymbol{M}_h \cdot \boldsymbol{h}_{t-1} + \boldsymbol{b}_h) \tag{6}$$

$$\boldsymbol{h}_t = \boldsymbol{u}_t \odot \boldsymbol{h}_{t-1} + (1 - \boldsymbol{u}_t) \odot \tilde{\boldsymbol{h}}_t, \tag{7}$$

where σ is the sigmoid activation function, $\boldsymbol{z}_t$ is a latent random variable which captures the stochasticity of the temporal process, $\boldsymbol{u}_t$ and $\boldsymbol{r}_t$ represent the update and reset gates, $\boldsymbol{W}, \boldsymbol{C}$ and $\boldsymbol{M}$ are weight matrices, $\boldsymbol{b}$ is the bias matrix, $\boldsymbol{h}_t$ is the GRU hidden state and $\odot$ is the element-wise Hadamard product. Our stochastic adaptation can be seen as a generalisation of the regular GRU, i.e. when $\boldsymbol{C} = 0$, we have a regular GRU cell [28].

3.3 Generative Model

The role of the generative model is to establish probabilistic relationships between the target variable y_t, the intermediate variables of interest $(\boldsymbol{h}_t, \boldsymbol{z}_t)$, and the input $\boldsymbol{x}_t$. Our model uses neural networks to describe the non-linear transition and emission processes, and we preserve the architectural workings of an RNN - relevant information is encoded within the hidden states that evolve with time, and the hidden states contain all necessary information required to estimate the target variable at each time step. A graphical representation of the generative model is shown in Fig. 1a, the RNN transitions are now stochastic, faciliated by the random variable z_t. The joint probability distribution of the generative model can be factorised as follows:

$$p_\theta(y_{2:T}, \boldsymbol{z}_{2:T}, \boldsymbol{h}_{2:T}|\boldsymbol{x}_{1:T}) = \prod_{t=2}^{T} p_{\theta_1}(y_t|\boldsymbol{h}_t)p_{\theta_2}(\boldsymbol{h}_t|\boldsymbol{h}_{t-1}, \boldsymbol{z}_t, \boldsymbol{x}_t)p_{\theta_3}(\boldsymbol{z}_t|\boldsymbol{h}_{t-1}) \quad (8)$$

where

$$p_{\theta_3}(\boldsymbol{z}_t|\boldsymbol{h}_{t-1}) = N(\boldsymbol{\mu}(\boldsymbol{h}_{t-1}), \boldsymbol{\sigma}^2(\boldsymbol{h}_{t-1})I) \quad (9)$$

$$\boldsymbol{h}_t = GRU(\boldsymbol{h}_{t-1}, \boldsymbol{z}_t, \boldsymbol{x}_t) \quad (10)$$

$$y_t \sim p_{\theta_1}(y_t|\boldsymbol{h}_t) = N(\mu(\boldsymbol{h}_t), \sigma^2(\boldsymbol{h}_t)), \quad (11)$$

where GRU is the stochastic GRU update function given by (4)–(7). (9) defines the prior distribution of z_t, which we assume to have an isotropic Gaussian prior (covariance matrix is diagonal) parameterised by a multi-layer perceptron (MLP). When conditioning on past time series for prediction, we use (9), (10) and the last available hidden state $\boldsymbol{h}_{last}$ to calculate $\boldsymbol{h}_1$ for the next sequence, otherwise we initialise them to $\boldsymbol{0}$. We refer to the collection of parameters of the generative model as θ, i.e. $\theta = \{\theta_1, \theta_2, \theta_3\}$. We refer to (11) as our generative distribution, which is parameterised by an MLP.

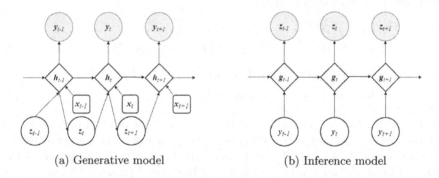

(a) Generative model   (b) Inference model

Fig. 1. Proposed generative and inference models

3.4   Inference Model

We wish to maximise the marginal log-likelihood function $\log p_\theta(y_{2:T}|\boldsymbol{x}_{2:T})$, however the random variable z_t of the non-linear SSM cannot be analytically integrated out. We instead maximise the variational lower bound (ELBO) with respect to the generative model parameters θ and some inference model parameter which we call ϕ [36]. The variational approximation of the true posterior $p(\boldsymbol{z}_{2:T}, \boldsymbol{h}_{2:T} | y_{1:T}, \boldsymbol{x}_{1:T})$ can be factorised as follows:

$$q_\phi(\boldsymbol{z}_{2:T}, \boldsymbol{h}_{2:T} | y_{1:T}, \boldsymbol{x}_{1:T}) = \prod_{t=2}^{T} q_\phi(\boldsymbol{z}_t | y_{1:T}) q_\phi(\boldsymbol{h}_t | \boldsymbol{h}_{t-1}, \boldsymbol{z}_t, \boldsymbol{x}_t) \qquad (12)$$

and

$$q_\phi(\boldsymbol{h}_t | \boldsymbol{h}_{t-1}, \boldsymbol{z}_t, \boldsymbol{x}_t) = p_{\theta_2}(\boldsymbol{h}_t | \boldsymbol{h}_{t-1}, \boldsymbol{z}_t, \boldsymbol{x}_t), \qquad (13)$$

where p_{θ_2} is the same as in (8); this is due to the fact that the GRU transition function is fully deterministic conditioned on knowing $\boldsymbol{z}_t$ and hence p_{θ_2} is just a delta distribution centered at the GRU output value given by (4)–(7). The graphical model of the inference network is given in Fig. 1a. Since the purpose of the inference model is to infer the filtering distribution $q_\phi(\boldsymbol{z}_t | y_{1:t})$, and that an RNN hidden state contains a representation of current and past inputs, we use a second GRU model with hidden states g_t as our inference model, which takes the observed target values y_t and previous hidden state g_{t-1} as inputs and maps g_t to the inferred value of $\boldsymbol{z}_t$:

$$g_t = GRU(g_{t-1}, \boldsymbol{y}_t) \qquad (14)$$

$$\boldsymbol{z}_t \sim q_\phi(\boldsymbol{z}_t | \boldsymbol{y}_{1:t}) = N(\boldsymbol{\mu}(g_t), \boldsymbol{\sigma}^2(g_t)I). \qquad (15)$$

3.5   Model Training

The objective function of our stochastic RNN is the ELBO $L(\theta, \phi)$ given by:

$$L(\theta, \phi) = \int \int q_\phi \log \frac{p_\theta}{q_\phi} d\boldsymbol{z}_{2:T} d\boldsymbol{h}_{2:T}$$

$$= \sum_{n=2}^{T} \mathbb{E}_{q_\phi}[\log p_\theta(y_t | \boldsymbol{h}_t)] - KL(q_\phi(\boldsymbol{z}_t | \boldsymbol{y}_{1:t}) \| p_\theta(\boldsymbol{z}_t | \boldsymbol{h}_{t-1})), \quad (16)$$

where p_θ and q_ϕ are the generative and inference distributions given by (8) and (12) respectively. During training, we use the posterior network (15) to infer the latent variable $\boldsymbol{z}_t$ used for reconstruction. During testing we use the prior network (9) to forecast 1-step-ahead $\boldsymbol{z}_t$, which has been trained using the KL term in the ELBO function. We seek to optimise the ELBO with respect to decoder parameters θ and encoder parameters ϕ jointly, i.e. we wish to find:

$$(\theta^*, \phi^*) = \underset{\theta, \phi}{\operatorname{argmax}} L(\theta, \phi). \qquad (17)$$

Since we do not back-propagate through a sampling operation, we apply the reparameterisation trick [30] to write

$$\boldsymbol{z} = \boldsymbol{\mu} + \boldsymbol{\sigma} \odot \boldsymbol{\epsilon}, \qquad (18)$$

where $\boldsymbol{\epsilon} \sim \boldsymbol{N}(0, I)$ and we sample from $\boldsymbol{\epsilon}$ instead. The KL divergence term in (16) can be analytically computed since we assume the prior and posterior of $\boldsymbol{z}_t$ to be normally distributed.

3.6   Model Prediction

Given the last available GRU hidden state h_{last}, prediction window τ and covariates $x_{T+1:T+\tau}$, we generate predicted target values in an autoregressive manner, assuming that at every time step the hidden state of the GRU h_t contains all relevant information up to time t. The prediction algorithm of our stochastic GRU is given by Algorithm 1.

Input: $\tau, h_{last}, x_{T+1:T+\tau}$
Output: $y_{T+1:T+\tau}$
for $t \leftarrow 1$ *to* τ **do**
$\quad$ $z_t \sim p_{\theta_3}(z_t|h_{last})$
$\quad$ $h_t \leftarrow GRU(h_{last}, z_t, x_t)$
$\quad$ $y_t \sim p_{\theta_1}(y_t|h_t)$
$\quad$ $h_{last} \leftarrow h_t$
end

Algorithm 1: Prediction algorithm for stochastic GRU

4   Experiments

We highlight the model performance on 6 publicly available datasets:

1. Equity options trading price time series available from the Chicago Board Options Exchange (CBOE) datashop. This dataset describes the minute-level traded prices of an option throughout the day. We study 3 options with Microsoft and Amazon stocks as underlyings where x_t = underlying stock price and y_t = traded option price
2. The Beijing PM2.5 multivariate dataset describes hourly PM2.5 (a type of air pollution) concentrations of the US Embassy in Beijing, and is freely available from the UCI Machine Learning Repository. The covariates we use are x_t = temperature, pressure, cumulated wind speed, Dew point, cumulated hours of rainfall and cumulated hours of snow, and y_t = PM2.5 concentration. We use data from 01/11/2014 onwards
3. The Metro Interstate Traffic Volume dataset describes the hourly interstate 94 Westbound traffic volume for MN DoT ATR station 301, roughly midway between Minneapolis and ST Paul, MN. This dataset is available on the UCI Machine Learning Repository. The covariates we use in this experiment are x_t = temperature, mm of rainfall in the hour, mm of snow in the hour, and percentage of cloud cover, and y_t = hourly traffic volume. We use data from 02/10/2012 9AM onwards
4. The Hungarian Chickenpox dataset describes weekly chickenpox cases (childhood disease) in different Hungarian counties. This dataset is also available on the UCI Machine Learning Repository. For this experiment, y_t = number of chickenpox cases in the Hungarian capital city Budapest, x_t = number of chickenpox cases in Pest, Bacs, Komarom and Heves, which are 4 nearby counties. We use data from 03/01/2005 onwards

We generate probabilistic forecasts using 500 Monte-Carlo simulations and we take the mean predictions as our point forecasts to compute the error metrics. We tested the number of simulations from 100 to 1000 and found that above 500, the differences in performance were small, and with fewer than 500 we could not obtain realistic confidence intervals for some time series. We provide graphical illustrations of the prediction results in Fig. 2a–2f. We compare our model performance against an AR(1) model assuming the prediction is the same as the last observed value ($y_{T+\tau} = y_T$), a standard LSTM model and a standard GRU model. For the performance metric, we normalise the root-mean-squared-error (rmse) to enable comparison between time series:

$$nrmse = \frac{\sqrt{\frac{\sum_{i=1}^{N}(y_i - \hat{y}_i)^2}{N}}}{\bar{y}}, \tag{19}$$

where $\bar{y} = mean(y)$, $\hat{y}_i$ is the mean predicted value of y_i, and N is the prediction size. For replication purposes, in Table 1 we provide (in order): number of training, validation and conditioning steps, (non-overlapping) sequence lengths used for training, number of prediction steps, dimensions of z_t, h_t and g_t, details about the MLPs corresponding to (9) (z_t prior) and (15) (z_t post) in the form of (n layers, n hidden units per layer), and lastly the size of the hidden states of the benchmark RNNs (LSTM and GRU). we use the ADAM optimiser with a learning rate of 0.001. In Table 2, 3, 4 and 5 we observe that the nrmse of the stochastic GRU is lower than its deterministic counterpart for all datasets investigated and across all prediction steps. This shows that our proposed method can better capture both long and short-term dynamics of the time series. With respect to multistep time series forecasting, it is often difficult to accurately model the long-term dynamics. Our approach provides an additional degree of freedom facilitated by the latent random variable which needs to be inferred using the inference network; we believe this allows the stochastic GRU to better capture the stochasticity of the time series at every time step. In Fig. 2e for example, we observe that our model captures well the long-term cyclicity of the traffic volume, and in Fig. 2d where the time series is much more erratic, our model can still accurately predict the general shape of the time series in the prediction period.

Table 1. Model and training parameters

Dataset	Train	Val	Cond	Seq length	Pred	z_t	h_t	g_t	z_t prior	z_t post	RNN hid
Options	300	30	10	10	30	50	64	64	(4,64)	(4,64)	64
PM2.5	1200	200	10	10	30	50	64	64	(4,64)	(4,64)	64
Traffic volume	1000	200	20	20	30	30	128	128	(4,128)	(4,128)	128
Hungarian chickenpox	300	150	10	10	30	50	128	128	(4,128)	(4,128)	128

Table 2. nrmse for 30 steps-ahead options price predictions

Option	Description	Ours	AR(1)	LSTM	GRU
MSFT call	strike 190, expiry 17/09/2021	**0.0010**	0.0109	0.0015	0.0015
MSFT put	strike 315, expiry 16/07/2021	**0.0004**	0.0049	0.0006	0.0007
AMZN put	strike 3345, expiry 22/01/2021	**0.0032**	0.0120	0.0038	0.0038

Table 3. nrmse for 30 steps-ahead PM2.5 concentration predictions

Steps	5	10	15	20	25	30
Ours	**0.1879**	**0.2474**	**0.4238**	**0.4588**	**0.6373**	**0.6523**
AR(1)	0.3092	1.0957	0.7330	0.6846	1.0045	1.1289
LSTM	0.4797	0.6579	0.4728	0.4638	0.8324	0.8318
GRU	0.4846	0.5553	0.4789	0.4919	0.6872	0.6902

Table 4. nrmse for 30 steps-ahead traffic volume predictions

Steps	5	10	15	20	25	30
Ours	**0.4284**	**0.2444**	**0.2262**	**0.2508**	**0.2867**	**0.2605**
AR(1)	1.2039	1.0541	1.0194	1.0283	1.1179	1.0910
LSTM	0.8649	0.5936	0.4416	0.4362	0.5591	0.5446
GRU	0.8425	0.5872	0.4457	0.4376	0.5510	0.5519

Table 5. nrmse for 30 steps-ahead Hungarian chickenpox predictions

Steps	5	10	15	20	25	30
Ours	**0.6585**	**0.6213**	**0.5795**	**0.5905**	0.6548	**0.5541**
AR(1)	0.7366	0.7108	0.9126	0.9809	1.0494	1.0315
LSTM	0.7215	0.6687	0.9057	1.0717	0.8471	0.7757
GRU	0.6795	0.6379	0.6825	0.6196	**0.6355**	0.6739

Table 6. nrmse of MLP benchmark and our proposed model for 30 steps-ahead forecasts

	MSFT call	MSFT put	AMZN put	PM2.5	Metro	Chickenpox
Ours	**0.0010**	**0.0004**	**0.0032**	**0.6523**	**0.2605**	**0.5541**
MLP	0.0024	0.0005	0.0141	0.7058	0.6059	0.5746

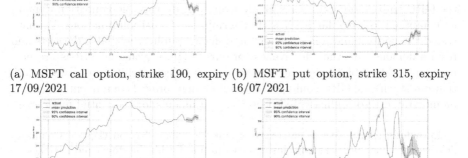

(a) MSFT call option, strike 190, expiry 17/09/2021
(b) MSFT put option, strike 315, expiry 16/07/2021

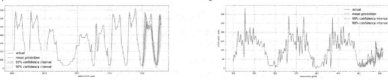

(c) AMZN put option, strike 3345, expiry 22/01/2021
(d) PM2.5 concentration forecasts up to 30 steps ahead

(e) Traffic volume forecasts up to 30 steps ahead
(f) Hungarian chickenpox cases forecasts up to 30 steps ahead

Fig. 2. Model prediction results on different datasets

To investigate the effectiveness of our temporal model, we compare our prediction errors against a model without a temporal component, which is constructed using a 3-layer MLP with 5 hidden nodes and ReLU activation functions. Since we are using covariates in the prediction period (3), we would like to verify that our model can outperform a simple regression-type benchmark which approximates a function of the form $y_t = f_\psi(x_t)$; we use the MLP to parameterise the function f_ψ. We observe in Table 6 that our proposed model outperforms a regression-type benchmark for all the experiments, which shows the effectiveness of our temporal model. It is also worth noting that in our experiments we use the actual values of the future covariates. In a real forecasting setting, the future covariates themselves could be outputs of other mathematical models, or they could be estimated using expert judgement.

5   Conclusion

In this paper we have presented a stochastic adaptation of the Gated Recurrent Unit which is trained with stochastic gradient variational Bayes. Our model design preserves the architectural workings of an RNN, which encapsulates all relevant information into the hidden state, however our adaptation takes inspiration from the stochastic transition functions of state space models by injecting a latent random variable into the update functions of the GRU, which allows the GRU to be more expressive at modelling highly variable transition dynamics compared to a regular RNN with deterministic transition functions. We have tested the performance of our model on different publicly available datasets and results demonstrate the effectiveness of our design. Given that GRUs are now popular building blocks for much more complex deep architectures, we believe that our stochastic GRU could prove useful as an improved component which can be integrated into sophisticated deep learning models for sequential modelling.

Acknowledgments. We would like to thank Dr Fabio Caccioli (Dpt of Computer Science, UCL) for proofreading this manuscript and for his questions and feedback.

References

1. Bandara, K., Shi, P., Bergmeir, C., Hewamalage, H, Tran, Q., Seaman, B.: Sales demand forecast in E-commerce using a long short-term memory neural network methodology. In: ICONIP, Sydney, NSW, Australia, pp. 462–474 (2019)
2. McNally, S., Roche, J., Caton, S.: Predicting the price of Bitcoin using machine learning. In: PDP, Cambridge, UK, pp. 339–343 (2018)
3. Hu, Z., Zhao, Y., Khushi, M.: A Survey of Forex and stock price prediction using deep learning. Appl. Syst. Innov. **4**(9) (2021). https://doi.org/10.3390/asi4010009
4. Zhang, R., Yuan, Z., Shao, X.: A new combined CNN-RNN model for sector stock price analysis. In: COMPSAC, Tokyo, Japan (2018)
5. Lv, Y., Duan, Y., Kang, W.: Traffic flow prediction with big data: a deep learning approach. IEEE Trans. Intell. Transp. **16**(2), 865–873 (2014)
6. Dolatabadi, A., Abdeltawab, H., Mohamed, Y.: Hybrid deep learning-based model for wind speed forecasting based on DWPT and bidirectional LSTM Network. IEEE Access. **8**, 229219–229232 (2020). https://doi.org/10.1109/ACCESS.2020. 3047077
7. Alazab, M., Khan, S., Krishnan, S., Pham, Q., Reddy, M., Gadekallu, T.: A multidirectional LSTM model for predicting the stability of a smart grid. IEEE Access. **8**, 85454–85463 (2020). https://doi.org/10.1109/ACCESS.2020.2991067
8. Liu, T., Wu, T., Wang, M, Fu, M, Kang, J., Zhang, H.: Recurrent neural networks based on LSTM for predicting geomagnetic field. In: ICARES, Bali, Indonesia (2018)
9. Lai, G., Chang, W., Yang, Y., Liu, H.: Modelling long-and short-term temporal patterns with deep neural networks. In: SIGIR (2018)
10. Apaydin, H., Feizi, H., Sattari, M.T., Cloak, M.S., Shamshirband, S., Chau, K.W.: Comparative analysis of recurrent neural network architectures for reservoir inflow forecasting: Water. vol. 12 (2020). https://doi.org/10.3390/w12051500
11. Di Persio, L., Honchar, O.: Analysis of recurrent neural networks for short-term energy load forecasting. In: AIP (2017)
12. Meng, X., Wang, R., Zhang, X., Wang, M., Ma, H., Wang, Z.: Hybrid neural network based on GRU with uncertain factors for forecasting ultra-short term wind power. In: IAI (2020)
13. Khaldi, R., El Afia, A., Chiheb, R.: Impact of multistep forecasting strategies on recurrent neural networks performance for short and long horizons. In: BDIoT (2019)

14. Mattos, C.L.C., Barreto, G.A.: A stochastic variational framework for recurrent gaussian process models. Neural Netw. **112**, 54–72 (2019). https://doi.org/10.1016/j.neunet.2019.01.005

15. Seleznev, A., Mukhin, D., Gavrilov, A., Loskutov, E., Feigin, A.: Bayesian framework for simulation of dynamical systems from multidimensional data using recurrent neural network. Chaos, vol. 29 (2019). https://doi.org/10.1063/1.5128372

16. Alhussein, M., Aurangzeb, K., Haider, S.: Hybrid CNN-LSTM model for short-term individual household load forecasting. IEEE Access. **8**, 180544–180557 (2020). https://doi.org/10.1109/ACCESS.2020.3028281

17. Sezer, O., Gudelek, M., Ozbayoglu, A.: Financial time series forecasting with deep learning: a systematic literature review: 2005–2019. Appl. Soft Comput. **90** (2020). https://doi.org/10.1016/j.asoc.2020.106181

18. Chung, J., Kastner, K., Dinh, L., Goel, K., Courville, A., Bengio, Y.: A recurrent latent variable model for sequential data. In: NIPS, Montreal, Canada (2015)

19. Bayer, J., Osendorfer, C.: Learning stochastic recurrent networks. arXiv preprint. arXiv: 1411.7610 (2014)

20. Goyal, A., Sordoni, A., Cote, M., Ke, N., Bengio, Y.: Z-forcing: training stochastic recurrent networks. In: NIPS, California, USA, pp. 6716–6726 (2017)

21. Fabius, O., Amersfoort, J.R.: Variational recurrent auto-encoders. arXiv preprint. arXiv **1412**, 6581 (2017)

22. Fraccaro, M., Sønderby, S., Paquet, U., Winther, O.: Sequential neural models with stochastic layers. In: NIPS, Barcelona, Spain (2016)

23. Karl, M., Soelch, M., Bayer, J., Smagt, P.: Deep Variational Bayes Filters: Unsupervised learning of state space models from raw data. In: ICLR, Toulon, France (2017)

24. Franceschi, J., Delasalles, E., Chen, M., Lamprier, S., Gallinari, P.: Stochastic latent residual video prediction. In: ICML, pp. 3233–3246 (2020)

25. Young, T., Hazarika, D., Poria, S., Cambria, E.: Recent trends in deep learning based natural language processing. IEEE Comput. Intell. Mag. **13**(3), 55–75 (2018). https://doi.org/10.1109/MCI.2018.2840738

26. Pascanu, R., Mikolov, T., Bengio, Y.: On the difficulty of training recurrent neural networks. In: ICML, Atlanta, GA, USA, pp. 1310–1318 (2013)

27. Hochreiter, S., Schmidhuber, J.: Long short-term memory. Neural Comput. **9**(8), 1735–1780 (1997)

28. Cho, K., Merrienboer, B., Bahdanau, D., Bengio, Y.: On the properties of neural machine translation: encoder-decoder approaches. In: SSST, Doha, Qatar, pp. 103–111 (2014)

29. Chung, J., Gulcehre, C., Cho, K., Bengio, Y.: Empiral evaluation of gated recurrent neural networks on sequence modelling. arXiv preprint. arXiv: 1412.3555 (2014)

30. Kingma, D., Welling, M.: Auto-encoding variational Bayes. In: ICLR, Banff, Canada (2014)

31. Liitiainen, E., Lendasse, A.: Long-term prediction of time series using state-space models. In: ICANN, Athens, Greece (2006)

32. Krishnan, R., Shalit, U., Sontag, D.: Structured inference networks for nonlinear state space models. In: AAAI, California, USA, pp. 2101–2109 (2017)

33. Krishnan, R., Shalit, U., Sontag, D.: Deep Kalman filters. arXiv preprint. arXiv **1511**, 05121 (2015)

34. Shabanian, S., Arpit, D., Trischler, A., Bengio, Y.: Variational bi-LSTMs. arXiv preprint. arXiv **1711**, 05717 (2017)

35. Durbin, J., Koopman, S.: Time Series Analysis by State Space Methods, vol. 38. Oxford University Press, Oxford (2012)

36. Jordan, M., Ghahramani, Z., Jaakkola, T., Saul, L.: An introduction to variational methods for graphical models. Mach. Learn. **37**(2), 183–233 (1999)
37. Rangapuram, S., Seeger, M., Gasthaus, J., Stella, L., Yang, Y., Janushowski, T.: Deep state space models for time series forecasting. In: NIPS, Montreal, Canada (2018)

Speaker Verification with Disentangled Self-attention

Junjie Guo, Zhiyuan Ma, Haodong Zhao, Gongshen Liu$^{(\boxtimes)}$, and Xiaoyong Li

School of Electronic Information and Electrical Engineering,
Shanghai Jiao Tong University, Shanghai, China
{jason1998,presurpro,zhaohaodong,lgshen,xiaoyongli}@sjtu.edu.cn

Abstract. In prior works of speaker verification, self-attention networks attract remarkable interests among end-to-end models and achieve great results. In this paper, we have proposed an integrated framework which is used for speaker verification (SV), disentangled self-attention network (DSAN), which focuses on the self-attention in depth. Based on Transformer, attention computation in DSAN is divided into two parts, a pairwise part which learns the relationship between two frames and a unary part which learns the importance of each frame. The original self-attention mechanism trains these two parts together which hinders the learning of each part. We show the effectiveness of this modification on speaker verification task. The proposed model trained on TIMIT, AISHELL-1 and VoxCeleb shows significant performance improvement over LSTM and traditional self-attention network. And we improve the interpretability of the model. Our best result yields an equal error rate (EER) result of 0.91% on TIMIT and 2.11% on VoxCeleb-E.

Keywords: Speaker verification · Self-attention · Disentangled network

1 Introduction

Speaker verification (SV) is the process of verifying, which is based on a speaker's enrolled utterances, whether an input speech utterance is spoken by a claimed speaker. According to whether the speech content have constraints, speaker verification task can be divided into two categories: text-dependent speaker verification (TD-SV) and text-independent speaker verification (TI-SV).

Traditionally, speaker embedding is extracted using i-vector [6] with a probabilistic linear discriminant analysis (PLDA) [11] backend classifier, which gains improvement over the GMM-UBMs [20]. GMM-SVM [3] is also common technique for speaker verification.

For better learning results, a lot of studies combine deep neural network (DNN) with speaker verification task. E. Variani [27] uses DNN to capture long-term speaker characteristics and gets better performance. x-vector [23] and its modified models [24,29], is proposed in the literature, which achieve superior

© Springer Nature Switzerland AG 2021
T. Mantoro et al. (Eds.): ICONIP 2021, LNCS 13108, pp. 27–39, 2021.
https://doi.org/10.1007/978-3-030-92185-9_3

performance than i-vector models. Time Delay Neural Network (TDNN) [26] uses the x-vector as frontend and uses a stack of 1-D dilated CNNs to capture local connection. End-to-End SV systems have attracted much attention because of its convenient training procedure. They use different models (e.g., recurrent neural network [9], multi-channel speech enhancement [16], multi-feature Integration [13]), and introduce various kinds of loss functions (e.g., triplet loss [33], GE2E [30]) to discriminate whether input utterance is from the claimed person.

However, how to find the most relevant part of the input still needs improvement. Attention mechanisms have been shown effective to this problem, which contributes a lot to sequence-to-sequence models in Neural Machine Translation (NMT) [28] or Automatic Speech Recognition (ASR) [4]. And self-attention is a specific type of attention that applies attention mechanism to every position of input.

For speaker verification, attention mechanism has been applied in different parts of network. K. Okabe [18] shows attention in pooling layer got better performance than traditional average and statistics pooling. Lee [12] uses mutual-attention which takes a pair of input utterances to generate an utterance-level embedding. In [34], they uses multi-attention mechanism to capture different kinds of information from the encoder features. Attention mechanism is implemented in the process of aggregating frame-level features with learned attention weights in [1]. And P. Safari [21] introduces tandem self-attention encoding and pooling (SAEP) mechanism to obtain a discriminative speaker embedding. These researches have achieved good performance, but they still can't answer why attention mechanism brings improvement or dig into attention calculation.

So in this paper, we present an integrated framework for speaker verification based on Transformer [28] and disentangled Non-Local Neural Networks [32]. We disentangle the calculation in Transformer into pairwise term and unary term, which helps learning each term better. This modification helps capturing relationship between different frames, finding frames which have global influence. After assessed on two datasets, our model has achieved better performance in creating discriminative representation for speakers. In the meanwhile, we put forward an explanation on what unary term and pairwise term learn and why unary term contributes more than pairwise term. This also explains which features attention mechanism extracts from input frames.

The rest of this paper is organized as follows. In Sect. 2, we review the disentangled Non-local Network and introduce our disentangled self-attention Network (DSAN). The proposed model is introduced in Sect. 3. Our experimental setup and results is showed in Sect. 4. And we draw a conclusion in Sect. 5.

2   Disentangled Self-attention

In this section, we start with the Disentangled Non-local Network (DNL) [32] in CV area, then explain the design philosophy in disentangled self-attention network.

2.1   Disentangled Non-local Network

The Non-local Network [31] models long-range dependency between pixels. It computes pairwise relations between features of two positions to capture long-range dependencies as:

$$y_i = \sum_{j \in \Omega} \omega(x_i, x_j) g(x_j),$$

$$\omega(x_i, x_j) = \sigma(q_i k_j^T) \tag{1}$$

where Ω denotes the set of all pixels on feature map, x_i and x_j represent the input features, $\omega(x_i, x_j$ is the embedded similarity function from pixel j (referred to as a **key** pixel) to pixel i (referred to as a **query** pixel),. At first glance, $\omega(x_i, x_j)$ appears to represent only a *pairwise* relationship in the non-local block, through a dot product operation. [32] proposed that $\omega(x_i, x_j)$ contained both *pairwise* relationship and *unary* information. Unary term represented the global impact on all *query* vectors. They proved $\omega(x_i, x_j)$ can be changed to 2. μ_q and μ_k represent the mean value of q k vector.

$$\omega(x_i, x_j) = \sigma(q_i k_j^T) = \sigma(\underbrace{(q_i - \mu_q)(k_j - \mu_k)^T}_{\text{pairwise}} + \underbrace{\mu_q k_j^T}_{\text{unary}}) \tag{2}$$

2.2   Disentangled Self-attention Network

Inspired by their work, we make improvement to self-attention in Transformer [28] and apply it on speaker verification task. Scaled Dot-Product Attention in Transformer computes attention function on a set of queries simultaneously, packed together into a matrix $Q = [q_1, ..., q_n]$. The keys and values are also packed together into matrices $K = [k_1, ..., k_n]$ and $V = [v_1, ..., v_n]$. $\sqrt{d_k}$ is the scaling factor.

$$O = Softmax(\frac{QK^T}{\sqrt{d_k}})V \tag{3}$$

With single input Q_i, K_j, V_k, the Eq. 3 can be expressed as:

$$o = Softmax(\frac{q_i k_j^T}{\sqrt{d_k}})v_k = Softmax(\frac{(q_i - \mu_q)(k_j - \mu_k)^T + \mu_q k_j^T}{\sqrt{d_k}})v_k \tag{4}$$

where μ_q and μ_k denote the average value of q_i and k_j respectively. Equation 4 proof sketch will be showed in the next part.

According to Eq. 4, $q_i k_j^T$ can be divided into *pairwise* term and *unary* term.

$$\omega_{pair}(q_i, k_j) = \sigma((q_i - \mu_q)(k_j - \mu_k)^T) \tag{5}$$

$$\omega_{unary}(q_i, k_j) = \sigma(\mu_q k_j^T) \tag{6}$$

Equation 5 represents the mutual influence between *query* and *key* for both q_i and k_j participate in the calculation. Equation 6 represents the impact of *key* on all *query* for only k_j and mean value of q_i are involved. We conduct experiments on the above two modifications separately, and then multiply them up to get the total attention output.

2.3 Proof of Disentangled Function

The disentangled function is used at dot products between query q and key k in Scaled Dot-Product attention. It can be expanded as:

$$q_i k_j^T = (q_i - \mu_q)(k_j - \mu_k)^T + \mu_q k_j^T + q_i \mu_k^T - \mu_q \mu_k^T \tag{7}$$

Note that the last two terms ($q_i \mu_k^T$ and $\mu_q \mu_k^T$) are factors in common with both the numerator and denominator so that they can be eliminated:

$$
\begin{aligned}
\sigma(q_i k_j^T) &= \frac{exp(q_i k_j^T)}{\sum_{t=1}^N exp(q_i k_j^T)} \\
&= \frac{exp((q_i - \mu_q)(k_j - \mu_k)^T + \mu_q k_j^T)}{\sum_{t=1}^N exp((q_i - \mu_q)(k_J - \mu_k)^T + \mu_q k_j^T)} \times \frac{exp(q_i \mu_k^T + \mu_q \mu_k^T)}{exp(q_i \mu_k^T + \mu_q \mu_k^T)} \\
&= \frac{exp((q_i - \mu_q)(k_j - \mu_k)^T + \mu_q k_j^T)}{\sum_{t=1}^N exp((q_i - \mu_q)(k_j - \mu_k)^T + \mu_q k_j^T)} \\
&= \sigma(\underbrace{(q_i - \mu_q)^T (k_j - \mu_k)}_{\text{pairwise}} + \underbrace{\mu_q^T k_j}_{\text{unary}}) \\
&= \omega_{pair}(q_i, k_j) \cdot \omega_{unary}(q_i, k_j)
\end{aligned}
$$

Then we try to explain why traditional self-attention doesn't learn these relationship well. Consider the back-propagation of loss L to the pairwise and unary terms:

$$
\begin{aligned}
\frac{\partial L}{\sigma(\omega_p)} &= \frac{\partial L}{\sigma(\omega)} \cdot \frac{\partial \omega}{\sigma(\omega_p)} = \frac{\partial L}{\sigma(\omega)} \cdot \sigma(\omega_u) \\
\frac{\partial L}{\sigma(\omega_u)} &= \frac{\partial L}{\sigma(\omega)} \cdot \frac{\partial \omega}{\sigma(\omega_u)} = \frac{\partial L}{\sigma(\omega)} \cdot \sigma(\omega_p)
\end{aligned}
\tag{8}
$$

It can be found that both gradients depend on the input value of the other term. We can imagine that when we calculate the gradient of ω_u, it relies on ω_p. So when ω_p gets close to 0, the gradient of ω_u also becomes small, which may causes gradient vanishment. The same situation also occurs in the calculation of gradient ω_p.

3 The Proposed Model

The architecture of the disentangled self-attention network (DSAN) is showed in Fig. 1. The model mainly consists of three parts: front-end feature extraction layer, disentangled self-attention layer and utterance level embedding layer.

Given input acoustic frame vectors, the proposed model generates utterance-level representation. During the training phase, we use generalized end-to-end (GE2E) loss [30]. The internal structure of each part will be described in the following subsections.

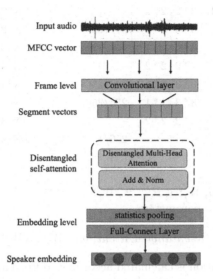

Fig. 1. Disentangled self-attention network overview. The audio goes through front-end feature extraction layer, disentangled self-attention layer, statistic pooling layer and becomes speaker embedding.

3.1 Front-End Feature Extraction Layer

The front-end feature extraction layer is modified based on VGG [10]. It includes three convolution blocks, which sense local features in time dimension and reduce feature dimension. Each convolution block has two concatenated conventional layers which is followed by a max pooling with 2×2 stride. For input data with N frames, we get output $h' \in \mathbb{R}^{M \times N/8 \times D'}$. M is the feature map numbers which increased gradually. At time dimension, these feature maps are concatenated into a new vector sequence $h \in \mathbb{R}^{N/8 \times D}$, where $D = MD'$.

3.2 Disentangled Self-attention Layer

After feature extraction, we get input sequences $h = [h_1, ..., h_N]i \in (1, ..., N)$ where $h_i \in \mathbb{R}^D$. They are sent to the multi-head attention with disentangled self-attention illustrated in Sect. 2. Multi-head attention is calculated as:

$$MultiHead(Q, K, V)$$
$$= Concat(head_1, ..., head_h)W^O$$

Table 1. The front-end feature extraction layer

Layer	Size	Stride	Input/output dim.	Feature size
Conv11	3 * 3	1 * 1	1/64	N * 40
Conv12	3 * 3	1 * 1	64/64	N * 40
Maxpool1	2 * 2	2 * 2	–	N/2 * 20
Conv21	3 * 3	1 * 1	64/128	N/2 * 20
Conv22	3 * 3	1 * 1	128/128	N/2 * 20
Maxpool2	2 * 2	2 * 2	–	N/4 * 10
Conv31	3 * 3	1 * 1	128/256	N/4 * 10
Conv32	3 * 3	1 * 1	256/256	N/4 * 10
Maxpool3	2 * 2	2 * 2	–	N/8 * 5
Flatten	–	–	256/1	N/8 * 1280

And for each *head*, $W_i^Q \in \mathbb{R}^{N \times d_k}$, $W_i^K \in \mathbb{R}^{N \times d_k}$, $W_i^V \in \mathbb{R}^{N \times d_v}$ and $W^O \in \mathbb{R}^{hd_v \times N}$ are projection parameters.

$$head_i = DisentangledAttention(QW_i^Q, KW_i^K, VW_i^V) \tag{9}$$

In Eq. 9, the attention is calculated according to Eq. 5 and Eq. 6. The output of disentangled attention mechanism is sent to the position-wise feed-forward network. It contains two linear transformations with a ReLU activation as:

$$FFN(h) = max(0, hW_1 + b_1)W_2 + b_2 \tag{10}$$

where h is the input, $W_1 \in \mathbb{R}^{d_m \times d_{ff}}$ and $W_2 \in \mathbb{R}^{d_{ff} \times d_m}$ are trainable parameters.

3.3 Utterance Embedding Layer

After the attention layer, a statistics pooling is applied to the output matrix $A_{t,i}$ following [25], where attention value over t is represented by mean (μ_t) and standard deviation (σ_t). By concatenating the mean vector (μ_t) and std (σ_t) vector, the utterance level vector is then generated.

$$\mu_U = mean(A_{t,})$$
$$\sigma_U = std(A_{t,}) \tag{11}$$
$$V_U = concatenate(\mu_U, \sigma_U)$$

The final speaker embeddings ($embedding_U$) are obtained after three fully connected layers with V_U as input.

4 Experiments

4.1 Experimental Setting

Dataset Description. We conduct experiments on TIMIT [7], AISHELL-1 [2] and VoxCeleb [5,17] dataset. The **TIMIT** data corpus contains 6,300 sentences from 630 speakers of 8 major dialects of American English. For each speaker, 10 utterances are released. The **AISHELL-1** contains 141,600 utterances from 400 speakers. For each speaker, around 360 utterances are released. The **VoxCeleb** contains two data sets. VoxCeleb1 includes 148642 voices of 1211 people in the train set, and 4874 voices of 40 people in the test set; VoxCeleb2 includes 145569 voices of 5994 people in the train set and 36237 voices of 118 people in the test set. In experiment process, we use whole VoxCeleb2 for training, and then use the VoxCeleb1-E data provided by VoxCeleb for testing. VoxCeleb1-E contains 552536 data pairs selected from VoxCeleb1's training set and test set, and we select the first 200000 of them for testing.

Experiment Setup. We use librosa toolkit [15] to preprocess data for our system. For each frame, we extract 40-dimensional log-mel-filterbank energies based on Mel-Frequency Cepstral Coefficients (MFCCs) with the standard 25ms window size and 10ms shift. As to computer resource, we used one GTX 1080Ti for training phase and evaluation phase.

Model Architecture. Parameters in front-end feature extraction layer shows in Table 1. It turns 40-dimensional input MFCC features to a 1280-dimensional vector. For the attention layer configuration, we use multi-head attention with disentangled self-attention. We employ $h = 6$ parallel heads which gets better results than others. For each head, we use $d_k = d_v = 30$, and position-wise feedforward dimensions are $d_{ff} = 1280$. Statistics pooling layer concatenates the mean vector (μ_t) and std (σ_t) vector. The dimensions of fully connected layers are equal to 1280, and we get 512-dimensional speaker embeddings. The loss function is GE2E loss introduced in [30].

Evaluation. In the evaluation stage, we use Equal Error Rate (EER), accuracy and minDCF to measure models' performance. EER is defined as the threshold when false acceptance rate (FAR) equals to false rejection rate (FRR). The lower equal error rate value is, the higher accuracy of the verification system is, and the better speaker verification system performs. minDCF specifies different weights for false acceptance rate and false rejection rate, and the calculation formula is:

$$DCF = C_{FR} * E_{FR} * P_{target} + C_{FA} * E_{FA} * (1 - P_{target}) \qquad (12)$$

where $C_{FR} = 10$, C_{FA} and $P_{target} = 0.01$.

4.2 Experimental Results

As shown in Table 2, we involve two modifications for disentangled self-attention network. We use LSTM model and traditional self-attention network as baselines.

For TIMIT dataset, traditional self-attention network outperforms the baselines. DSAN with pairwise term achieves 1.62% EER and 98.38% accuracy while DSAN with unary term achieves 1.37% EER and 98.63% accuracy. DSAN achieves 0.91% EER and 99.09% accuracy, which reduces 51.85% in EER compared with traditional self-attention network.

Table 2. Equal Error Rate (EER) and accuracy on TIMIT and AISHELL-1 dataset

Dataset	TIMIT			AISHELL-1			VoxCeleb1-E		
Method	EER	Acc.	minDCF	EER	Acc.	minDCF	EER	Acc.	minDCF
LSTM	3.77%	96.10%	0.00879	4.39%	95.61%	0.01976	6.24%	87.52%	0.02496
Self-Attention	1.89%	98.11%	0.00382	3.99%	96.01%	0.00916	2.99%	97.01%	0.00634
Pairwise DSAN	1.62%	98.38%	0.00327	3.45%	96.56%	0.00805	2.42%	97.58%	0.00521
Unary DSAN	1.37%	98.63%	0.00328	3.24%	96.76%	0.00735	2.21%	97.78%	0.00522
DSAN	**0.91%**	99.09%	0.00200	**2.10%**	97.90%	0.00548	**2.11%**	97.89%	0.00503

 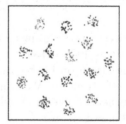

Fig. 2. Embedding visualization using t-SNE. One speaker is represented by one color, and one utterance is indicated by one point. The left picture is from traditional self-attention model, the right picture is from disentangled self-attention network.

The experiments on AISHELL-1 show similar results. Traditional self-attention network has 3.99% EER and 96.01% accuracy. DSAN with pairwise term, DSAN with unary term, DSAN outperforms the baseline by a relative EER decrease of 47.37%.

On VoxCeleb-E, traditional self-attention network result is 2.99% EER and 97.01% accuracy. DSAN with pairwise term achieves 2.42% EER and 97.58% accuracy while DSAN with unary term achieves 2.21% EER and 97.78% accuracy. DSAN achieves 2.11% EER and 97.89% accuracy, which reduces 29.43% in EER compared with traditional self-attention network.

From upper experimental results, we get the conclusion that self-attention mechanism helps a lot in information extraction, while traditional self-attention calculation hinders the learning process of pairwise and unary terms. With disentangling them, self-attention mechanism can have a better performance.

To further test the quality of embedding vectors, t-SNE [14] is used to visualize the distribution of embedding vectors by projecting these high-dimensional vectors into a 2D space. We use 15 random speakers and 50 random utterances for each speaker. In Fig. 2, each speaker is represented by one color and each utterance is represented by a point. The left picture is from Traditional Self-Attention Network and the right picture is from DSAN. In the left picture, some clusters get too close to be distinguished with each other. However, in the right picture, the embedding performs better separation property with DSAN.

4.3 Ablation Study

We discuss the effect of pairwise term and unary term in this section. The designs which use pairwise term and unary term separately achieve 1.37% EER and 1.62% EER on TIMIT dataset, 3.45% EER and 3.24% EER on AISHELL-1 dataset, 3.85%EER and 3.56% EER on VoxCeleb-E. So the final results benefit from modifications. Comparing EER between pairwise term and unary term, we find that unary term gains an additional 0.25% EER improvement on TIMIT, 0.21% on AISHELL-1 and on 0.29% VoxCeleb-E. This phenomenon indicates that traditional self-attention hinders the effect of unary term, and unary term contributes more in speaker verification task.

Explanation of Unary Term. There exists some frequencies which have bigger influence in the attention learning stage. In the calculation, These *key* frequencies k have the same dominant impact on all *query* frequencies q. For listener, the individual partials are not heard separately but are blended together by the ear into a single tone, which means unique timbre is from the mixture of a series of frequencies. And as mentioned in [8], speaker fundamental frequency (SFF) varies widely because this feature is depended on individual physiological dispositions.

The visualization of attention matrices supports our conclusion. We calculate attention matrices with the same input utterance in Fig. 3. In Fig. 3, the color represents the attention value of corresponding position, and white color means the attention value is high while dark color means the attention value is low. High attention value represents high influence in the calculation.

In each head, the distribution of high attention value is not uniform. Compared with traditional self-attention, high attention value is centralized at some parts. We recognize several white horizontal lines and vertical lines. This linear distribution shows some input vectors have higher attention value than all other vectors, which means the corresponding frames are indeed vital to all frames. This phenomenon corresponds with our conjecture.

4.4   Comparison with Other Methods

Table 3 shows comparison results of DSAN with state-of-the-art systems on TIMIT and VoxCeleb datasets. X-vector, which used to be the best model for speaker verification [25], had 4.92% EER on TIMIT and 5.40% EER on VoxCeleb-E. In [22], c-vector used multi-head self-attention mechanism and achieved 2.86% EER. M. Qi [19] brought SECNN which introduced squeeze-and-excitation (SE) components with a simplified residual convolutional neural network (ResNet) and got results with 2.58% EER. M.India [10] used self-attention encoding and pooling layer and achieved 3.18% EER on VoxCeleb-E. J.Lee [12] took a pair of input utterances and generates an utterance-level embedding using attention, this model got 2.49% EER on VoxCeleb1 test set.

Our proposed model achieves best result with 0.91% EER on TIMIT and 2.11% EER on VoxCeleb-E, which indicated that our modifications do improve discrimination ability for SV task.

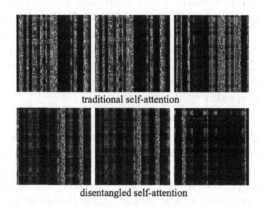

traditional self-attention

disentangled self-attention

Fig. 3. Visualization of attention matrices. Upper row shows attention matrices from 3 different heads from traditional self-attention network and the lower row shows the corresponding matrices for same speaker from DSAN.

Table 3. Performances of different models on TIMIT and VoxCeleb-E dataset. * Dual Attention Net is evaluated on Voxceleb1 test set.

Architecture	Equal Error Rate/%	
	TIMIT	VoxCeleb-E
x-vector	4.92	5.40
c-vector with attention [22]	2.86	–
SECNN [19]	2.58	–
Double MHA [10]	–	3.18
Dual Attention Net [12]*	–	2.49
DSAN	0.91	2.11

5 Conclusions

In this paper, the disentangled self-attention network (DSAN) is proposed to extract speaker embeddings from utterances. We demonstrate that frames and relationship between frames both contribute to attention calculation. However, they are coupled in traditional self-attention which hinders the learning of each term. With DSAN, our model obtains better performance on speaker verification task than traditional self-attention model with 0.91% EER on TIMIT and 2.11% on VoxCeleb-E. And unary term makes more contributions to the final result. Based on acoustic knowledge, we give an explanation to this phenomenon which we think can further guide our work in SV.

Acknowledgement. This research work has been funded by the National Natural Science Foundation of China (Grant No. 61772337, U1736207).

References

1. Bhattacharya, G., Alam, M.J., Kenny, P.: Deep speaker embeddings for short-duration speaker verification. In: Interspeech, pp. 1517–1521 (2017)
2. Bu, H., Du, J., Na, X., Wu, B., Zheng, H.: AISHELL-1: an open-source mandarin speech corpus and a speech recognition baseline. In: 2017 20th Conference of the Oriental Chapter of the International Coordinating Committee on Speech Databases and Speech I/O Systems and Assessment (O-COCOSDA), pp. 1–5. IEEE (2017)
3. Campbell, W.M., Sturim, D.E., Reynolds, D.A.: Support vector machines using GMM supervectors for speaker verification. IEEE Signal Process. Lett. **13**(5), 308–311 (2006)
4. Chan, W., Jaitly, N., Le, Q., Vinyals, O.: Listen, attend and spell: a neural network for large vocabulary conversational speech recognition. In: 2016 IEEE International Conference on Acoustics, Speech and Signal Processing (ICASSP), pp. 4960–4964. IEEE (2016)
5. Chung, J.S., Nagrani, A., Zisserman, A.: VoxCeleb2: deep speaker recognition. arXiv preprint arXiv:1806.05622 (2018)
6. Dehak, N., Kenny, P.J., Dehak, R., Dumouchel, P., Ouellet, P.: Front-end factor analysis for speaker verification. IEEE Trans. Audio Speech Lang. Process. **19**(4), 788–798 (2010)
7. Garofalo, J.S., Lamel, L.F., Fisher, W.M., Fiscus, J.G., Pallett, D.S.: DARPA TIMIT acoustic-phonetic continuous speech corpus CD-ROM. NIST speech disc 1-1.1. STIN 93, 27403 (1993)
8. Grawunder, S., Bose, I.: Average speaking pitch vs. average speaker fundamental frequency-reliability, homogeneity, and self report of listener groups. In: Proceedings of the International Conference Speech Prosody, pp. 763–766 (2008)
9. Heigold, G., Moreno, I., Bengio, S., Shazeer, N.: End-to-end text-dependent speaker verification. In: 2016 IEEE International Conference on Acoustics, Speech and Signal Processing (ICASSP), pp. 5115–5119. IEEE (2016)
10. India, M., Safari, P., Hernando, J.: Double multi-head attention for speaker verification. In: ICASSP 2021–2021 IEEE International Conference on Acoustics, Speech and Signal Processing (ICASSP), pp. 6144–6148. IEEE (2021)

11. Ioffe, S.: Probabilistic linear discriminant analysis. In: Leonardis, A., Bischof, H., Pinz, A. (eds.) ECCV 2006. LNCS, vol. 3954, pp. 531–542. Springer, Heidelberg (2006). https://doi.org/10.1007/11744085_41

12. Li, J., Lee, T.: Text-independent speaker verification with dual attention network. In: Proceedings of the Interspeech 2020, pp. 956–960 (2020)

13. Li, Z., Zhao, M., Li, J., Li, L., Hong, Q.: On the usage of multi-feature integration for speaker verification and language identification. In: Proceedings of the Interspeech 2020, pp. 457–461 (2020)

14. Maaten, L.v.d., Hinton, G.: Visualizing data using t-SNE. J. Mach. Learn. Res. **9**, 2579–2605 (2008)

15. McFee, B., et al.: librosa/librosa: 0.6. 3 (2019). https://doi.org/10.5281/zenodo.2564164

16. Mošner, L., Matějka, P., Novotný, O., Černocký, J.H.: Dereverberation and beamforming in far-field speaker recognition. In: 2018 IEEE International Conference on Acoustics, Speech and Signal Processing (ICASSP), pp. 5254–5258. IEEE (2018)

17. Nagrani, A., Chung, J.S., Zisserman, A.: VoxCeleb: a large-scale speaker identification dataset. arXiv preprint arXiv:1706.08612 (2017)

18. Okabe, K., Koshinaka, T., Shinoda, K.: Attentive statistics pooling for deep speaker embedding. In: Proceedings of the Interspeech 2018, pp. 2252–2256 (2018)

19. Qi, M., Yu, Y., Tang, Y., Deng, Q., Mai, F., Zhaxi, N.: Deep CNN with se block for speaker recognition. In: 2020 Information Communication Technologies Conference (ICTC), pp. 240–244. IEEE (2020)

20. Reynolds, D.A., Quatieri, T.F., Dunn, R.B.: Speaker verification using adapted gaussian mixture models. Digit. Signal Process. **10**(1–3), 19–41 (2000)

21. Safari, P., Hernando, J.: Self-attention encoding and pooling for speaker recognition. In: Proceedings of the Interspeech 2020, pp. 941–945 (2020)

22. Sankala, S., Rafi, B.S.M., Kodukula, S.R.M.: Self attentive context dependent speaker embedding for speaker verification. In: 2020 National Conference on Communications (NCC), pp. 1–5. IEEE (2020)

23. Snyder, D., Garcia-Romero, D., Povey, D., Khudanpur, S.: Deep neural network embeddings for text-independent speaker verification. In: Interspeech, pp. 999–1003 (2017)

24. Snyder, D., Garcia-Romero, D., Sell, G., McCree, A., Povey, D., Khudanpur, S.: Speaker recognition for multi-speaker conversations using x-vectors. In: ICASSP 2019–2019 IEEE International Conference on Acoustics, Speech and Signal Processing (ICASSP), pp. 5796–5800. IEEE (2019)

25. Snyder, D., Garcia-Romero, D., Sell, G., Povey, D., Khudanpur, S.: X-vectors: robust DNN embeddings for speaker recognition. In: 2018 IEEE International Conference on Acoustics, Speech and Signal Processing (ICASSP), pp. 5329–5333. IEEE (2018)

26. Snyder, D., Ghahremani, P., Povey, D., Garcia-Romero, D., Carmiel, Y., Khudanpur, S.: Deep neural network-based speaker embeddings for end-to-end speaker verification. In: 2016 IEEE Spoken Language Technology Workshop (SLT), pp. 165–170. IEEE (2016)

27. Variani, E., Lei, X., McDermott, E., Moreno, I.L., Gonzalez-Dominguez, J.: Deep neural networks for small footprint text-dependent speaker verification. In: 2014 IEEE International Conference on Acoustics, Speech and Signal Processing (ICASSP), pp. 4052–4056. IEEE (2014)

28. Vaswani, A., et al.: Attention is all you need. In: Advances in Neural Information Processing Systems, pp. 5998–6008 (2017)

29. Villalba, J., et al.: State-of-the-art speaker recognition for telephone and video speech: the JHU-MIT submission for NIST SRE18. In: Interspeech, pp. 1488–1492 (2019)
30. Wan, L., Wang, Q., Papir, A., Moreno, I.L.: Generalized end-to-end loss for speaker verification. In: 2018 IEEE International Conference on Acoustics, Speech and Signal Processing (ICASSP), pp. 4879–4883. IEEE (2018)
31. Wang, X., Girshick, R., Gupta, A., He, K.: Non-local neural networks. In: Proceedings of the IEEE Conference on Computer Vision and Pattern Recognition, pp. 7794–7803 (2018)
32. Yin, M., et al.: Disentangled non-local neural networks. In: Vedaldi, A., Bischof, H., Brox, T., Frahm, J.-M. (eds.) ECCV 2020. LNCS, vol. 12360, pp. 191–207. Springer, Cham (2020). https://doi.org/10.1007/978-3-030-58555-6_12
33. Zhang, C., Koishida, K.: End-to-end text-independent speaker verification with triplet loss on short utterances. In: Interspeech, pp. 1487–1491 (2017)
34. Zhu, Y., Ko, T., Snyder, D., Mak, B., Povey, D.: Self-attentive speaker embeddings for text-independent speaker verification. In: Proceedings of the Interspeech 2018, pp. 3573–3577 (2018)

Multi Modal Normalization

Neeraj Kumar[1,2]($\boxtimes$), Ankur Narang[1], Brejesh lall[2], and Srishti Goel[1]

[1] Hike Private Limited, Delhi, India
{neerajku,ankur,srishti}@hike.in
[2] Indian Institute of Technology, Delhi, India
brejesh@ee.iitd.ac.in

Abstract. In this paper, we propose a novel normalization framework, multi-modal normalization(MultiNorm) that learns the multiple modalities through affine transformations involved in the normalization architecture. We have shown its effectiveness in speech-driven facial video generation and video emotion detection which are complex problems due to its multi-modal aspects in audio and video domain.

The multi-modal normalization uses various features such as mel spectrogram, pitch, energy from audio signals and predicted keypoint heatmap/ optical flow and a single image to learn the respective affine parameters to generate highly expressive video. The incorporation of multimodal normalization has given superior performances in the proposed video synthesis from an audio and a single image against previous methods on various metrics such as SSIM (structural similarity index), PSNR (peak signal to noise ratio), CPBD (image sharpness), WER (word error rate), blinks/sec and LMD (landmark distance). In video emotion detection, we have leveraged the mel-spectrogram and optical flow in the multi modal normalization to learn the affine parameters that helps in improving the accuracy.

Keywords: Normalization · Gaussian mixture model · Audio to video synthesis · Video emotion detection

1 Introduction

There are various normalization techniques being proposed to incorporate various high and low level information such as shape, style, texture, etc. Instance Normalization (IN) [31] is being proposed to remove the instance specific contrast information from the image. Inspired by this, Adaptive Instance Normalization [14] shows that by simply incorporating the image features through the affine parameters of normalization, one can control the style of the generated image. IN incorporates the spatial information diluting the global statistics information of feature responses which can be undesirable depending on the task. For this, Batch-Instance Normalization(BIN) [24] was introduced which adaptively normalizes the styles to the task and controls the amount of style information being propagated through each channel using a learnable gate parameter. U-GAT-IT

© Springer Nature Switzerland AG 2021
T. Mantoro et al. (Eds.): ICONIP 2021, LNCS 13108, pp. 40–53, 2021.
https://doi.org/10.1007/978-3-030-92185-9_4

[17] has used instance(IN) and layer normalization [2] (LN) to dynamically capture the ratio of IN and LN in the network for smoother style transfer from one domain to another.

All the above normalization techniques have worked on capturing the styles of image and no work is done to capture the styles of audio and its mutual dependence on images in multi-modal applications through normalization. In this paper, we propose multi-modal normalization in the proposed architectures to learn the complex multi model tasks such as speech driven facial video generation and video emotion detection. We have built the architecture based on [19] to show how multi-modal normalization helps in generating highly expressive videos using the audio and person's image as input. The video emotion detection architecture has leveraged the 3D CNN to detect the emotions in the video.

Our main contributions are:

- We have shown that multi modal normalization helps in learning the multi modal data distribution with the mixture of multivariate Gaussian distribution through normalization (Sect. 3).
- Multi-Modal normalization opens the non-trivial path to capture the mutual dependence among various domains (Fig. 1). Generally, various encoder architectures [32] are used to convert the various features of multiple domains into latent vectors, and then the concatenated vectors are fed to the decoder to model the mutual dependence and generate the required output. The proposed architecture uses the normalization architecture to incorporate the multi-modal mutual dependence into the architecture.
- Incorporation of multi-modal normalization for video synthesis and video emotion detection have shown superior performance on multiple qualitative and quantitative metrics such as SSIM(structural similarity index), PSNR(peak signal to noise ratio), CPBD (image sharpness), WER(word error rate), blinks/sec, LMD(landmark distance) and accuracy.

2   Related Work

Normalization techniques are the active area of research in deep learning. The first normalization technique proposed was Batch Normalization(BN) [15]. It speeds up the convergence, helps in faster training of the network and makes the loss landscape smoother. It was proposed in discriminative models but its advantages have been extended to generative models as well.

Instance Normalization(IN) [31] makes the network contrast agnostic of the original image which is useful in style transfer. Adaptive instance normalization (AdaIN) [14] is the one step advancement in the IN which calculates the style image based affine parameters of normalization.

Style GAN [16] uses the mapping network to map the input into latent space which then controls the generator using AdaIN [14]. Each feature map of AdaIN is normalized separately and then scaled and biased using the corresponding scalar components from style features coming from the latent space. Spatially Adaptive Normalization proposed in SPADE [25] takes seMultiNormtic maps as an input to make learnable parameters(γ(y) and β(y)) spatially adaptive to it. In U-GAT-IT [17], the architecture proposes Adaptive and Layer Instance Normalization which uses both normalization to learn the parameters of the standardization process. [22] uses the optical flow and structure for motion images as an input to generate the affine parameters(γ' s and β' s) which are multiplied with normalized feature map and fed to the generator to create world consistent videos.

We have used the features of speech and video domains to generate the affine parameter γ's and β's in our multi-modal normalization. The proportion of different parameters are controlled by learnable parameters(ρ' s) which is fed into softmax function to make the sum equal to 1. We have applied this multi-modal normalization in the generation of realistic videos given and audio and a static image.

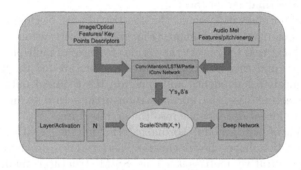

Fig. 1. Higher level architecture of multi modal normalization

3   Multi-Modal Normalization - Theory

The $\widehat{x}_i \in N(0,1)$ is the normalized feature map with zero mean and unit covariance (Eq. 2) [13]. Need of normalization occurs at local level as the distribution invariance assumption is not satisfied. The scale(γ) and bias(β) are used so that it gives flexibility to work with normalized input and also with scaled normalized input, if there is a need, thus increasing the representation power.

$$\mu_B = \frac{1}{m}\sum_{i=1}^{m} x_i \quad \sigma_B^2 = \frac{1}{m}\sum_{i=1}^{m}(x_i - \mu_B)^2 \tag{1}$$

$$\widehat{x}_i = \frac{x_i - \mu_B}{\sqrt{\sigma_B^2 + \epsilon}} \quad y_i = \gamma \cdot \widehat{x}_i + \beta \tag{2}$$

In our approach, the γ and β can be viewed as mean(μ) and co-variance(Σ) (Eq. 3) which is learned from the data distribution($p(x)$) $\in N(\mu(x), \Sigma(x))$ due to the fact that first-order networks inherently exploit Gaussian distributions. This can be explained by the Central Limit Theorem that the outputs of convolution, linear layers follow a multivariate Gaussian distribution for their feature representations [13]. Figure 1 depicts the higher level diagram of MultiNorm. The means and covariances are learnt through various deep learning architectures namely 2D convolution, LSTM, partial convolution, etc.

The y_i learns the data distribution $(p(x))$ through affine parameters of normalization network given in below equation:

$$y_i = \Sigma^{\frac{1}{2}} \cdot \widehat{x}_i + \mu \tag{3}$$

In multi modal setting, the multi modal data is learned from the Gaussian mixture of data distribution (Eq. 4) and ρ follows the Multinomial distribution.

$$p(x) = \sum_{j=1}^{K} \rho_j \cdot \phi(x; \mu, \Sigma) \qquad \sum_{j=1}^{K} \rho_j = 1 \tag{4}$$

This is a mixture of K components multivariate Gaussian distributions, $p_{\mu_j, \Sigma_j}(x) := \frac{1}{\sqrt{|2\pi\Sigma_j|}} e^{-\frac{1}{2}(x-\mu)^T \Sigma_j^{-1}(x-\mu_j)}$ with unknown parameters(μ_j, Σ_j).

In our setting , we have used K = 5 and the data distribution is learnt from different multivariate Gaussian distribution corresponding to multi modal features (Eq. 5).

$$y_i = \rho_1 \cdot (\Sigma_1^{\frac{1}{2}} \cdot \widehat{x}_i + \mu_1) + \rho_2 \cdot (\Sigma_2^{\frac{1}{2}} \cdot \widehat{x}_i + \mu_2)$$
$$+ \rho_3 \cdot (\Sigma_3^{\frac{1}{2}} \cdot \widehat{x}_i + \mu_3) + \rho_4 \cdot (\Sigma_4^{\frac{1}{2}} \cdot \widehat{x}_i + \mu_4) + \rho_5 \cdot (\Sigma_5^{\frac{1}{2}} \cdot \widehat{x}_i + \mu_5) \tag{5}$$

$$\sum_{j=1}^{5} \rho_j = 1 \tag{6}$$

The above setting of multi modal normalization helps in learning the various modes of multi modal complex problem by learning the distribution through mixture of Gaussian. It helps in preventing the model from getting mode collapse [20, 26, 27] which is usually the case of learning the data distribution from multivariate Gaussian distribution.

4   Multi Modal Normalization in Audio to Video Generation

In this section, we will design an architecture which generates speech synchronized realistic video on the target face using an audio and image of the target person as an input.

4.1   Architectural Design

Figure 2 shows the proposed architecture which is GAN-based which consists of a generator and a discriminator which uses multi-modal normalization technique to generate the realistic expressive videos.

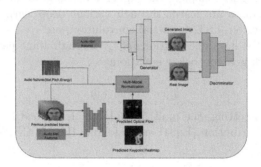

Fig. 2. Proposed architecture for audio to video synthesis

Generator. The generator consists of convolution layers with several layers having MultiNorm residual block (refer Fig. 3). 13 dimensional mel spectrogram features goes to the initial layers of generator and multi-modal normalization is used in the subsequent layers to generate expressive videos.The optical flow predictor/keypoint predictor is used to predict the optical flow/keypoint heatmap which is fed to normalization along with pitch, frequency and image to learn the affine transformations involved in the multi modal normalization. The third last layers incorporates the class activation map(CAM) [36] to focus the generator to learn distinctive features to create expressive videos. The generator uses various architectures namely multi-modal normalization, CAM and optical flow predictor/keypoint predictor.

KeyPoint Heatmap Predictor. The predictor model is based on hourglass architecture [29] that estimates K heatmaps $H_K \in [0,1]HW$, one for each keypoint, from the input image, each of which represents the contour of a specific facial part, e.g., upper left eyelid and nose bridge. It captures the spatial configuration of all landmarks, and hence it encapsulates pose, expression and shape information. We have used pretrained model[1] to calculate the ground truth of heatmap and have applied mean square error loss between predicted heatmaps and ground truth. We have done the joint training of keypoint predictor architecture along with the generator architecture and fed the output of keypoint predictor architecture in the multi-modal normalization network to learn the affine parameters and have optimized it with mean square error loss with the output of pretrained model.

---

[1] https://github.com/raymon-tian/hourglass-facekeypoints-detection.

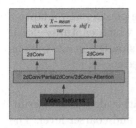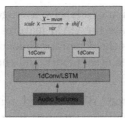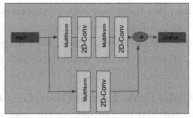

Fig. 3. Left: video features in MultiNorm architecture, Centre: audio features in Multi-Norm architecture Right: MultiNorm residual block. This consists of two MultiNorm block and two 2D convolution layers.

Optical Flow Predictor. The predictor is based on Encoder-Decoder architecture to predict the optical flow of the next frame. We are giving the previous frames and current audio mel-spectrogram as an input to the model with KL loss and reconstruction loss. The pretrained model is then used in the generator to calculate the affine parameters. The input of the optical flow is previous 5 frames along with 256 audio mel spectrogram-features and is jointly trained along with generator architecture and is optimized with mean square loss with the actual optical loss.

Multi-Modal Normalization. - The Eq. 5 and 6 formulate the setting of multi-modal normalization. In this experiment we have used the instance normalized feature map($\hat{x}_{IN}$). The multi modal data distribution($p(x)$) is learned through mixture of multivariate gaussian distribution corresponding to multi-modal features such as pitch, energy and audio mel-spectrogram features(AMF) from audio domain & static image and optical flow(OF)/facial keypoints heatmap(KH) features from video domain in the normalization to compute the different affine parameters (Eq. (7)). Multi-modal normalization gives the flexibility to learn the affine parameters by using various architectures namely 2D convolution, Partial Convolution [21] and attention model [33] for video related features and 1D convolution and LSTM layer for audio features as shown in Fig. 3. The parameter ρ follows the multinomial distribution (Eq. (8)) implemented by using softmax function.

$$y = \rho_1 \cdot (\Sigma_{Image}^{\frac{1}{2}} \cdot \hat{x}_{IN} + \mu_{Image}) + \rho_2 \cdot (\Sigma_{OF/KH}^{\frac{1}{2}} \cdot \hat{x}_{IN} + \mu_{OF/KH})$$
$$+ \rho_3 \cdot (\Sigma_{AMF}^{\frac{1}{2}} \cdot \hat{x}_{IN} + \mu_{AMF}) + \rho_4 \cdot (\Sigma_{pitch}^{\frac{1}{2}} \cdot \hat{x}_{IN} + \mu_{pitch}) + \rho_5 \cdot (\Sigma_{energy}^{\frac{1}{2}} \cdot \hat{x}_{IN} + \mu_{energy})$$
$$(7)$$

$$\sum_{j=1}^{5} \rho_j = 1 \tag{8}$$

Class Activation Map Based Layer. This layer is employed on third last layer of generator to capture the global and local features of face. In Class Activation

Map, we have done the concatenation of adaptive average pooling and adaptive max pooling of feature map to create the CAM features which captures global and local attention map, that helps the generator to focus on the image regions that are more discriminative such as eyes, mouth and cheeks.

Multi Scale Frame Discriminator. We have used multi-scale frame discriminator [34] to distinguish the fake and real image at finer and coarser level. The class activation map based layer is also used to distinguish the real or fake image by visualizing local and global attention maps.We have applied the adversarial loss(Eq. (9))on the information from the CAM output, n_{D_t} at different scale of the discriminator so that it will help the generator and discriminator to focus on local and global features and helps in generating more realistic image.

$$L_{\text{cam}} = E_{y \sim P_t}[\log(n_{D_t}(y))] + E_{x \sim P_s}[\log(D(1 - n_{D_t}(G(x))))] \tag{9}$$

5   Multi Modal Normalization in Video Emotion Detection

We have applied the proposed method in video emotion detection. We have leveraged 3D CNN based architecture to detect the number of classes. Figure 4 shows the architecture of video emotion detection. In our setting , we have used K = 2 and the data distribution is learnt from different multivariate Gaussian distribution corresponding to multi modal features namely mel -spectrogram and optical flow and $\widehat{x}_{IN}$ is the instance normalized feature map.

$$y_i = \rho_1 \cdot (\Sigma_1^{\frac{1}{2}} \cdot \widehat{x}_{IN} + \mu_1) + \rho_2 \cdot (\Sigma_2^{\frac{1}{2}} \cdot \widehat{x}_{IN} + \mu_2), \sum_{j=1}^{2} \rho_j = 1 \tag{10}$$

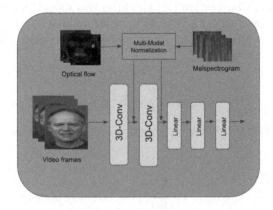

Fig. 4. Emotion detection architecture.

6   Experiments

6.1   Implementation Details

Audio to Video Generation

Datasets and PreProcessing Steps. We have used the GRID [10], LOM-BARD GRID [23], Crema-D [4] and VoxCeleb2 [8] datasets for the experiments and evaluation of different metrics. Videos are processed at 25fps and frames are cropped [3] and resized into 256X256 size and audio features are processed at 16khz. The ground truth of optical flow is calculated using farneback optical flow algorithm [12]. To extract the 15 channel keypoint heatmaps, we have used the pretrained hourglass face keypoint detection[2] with input and output size having $(15, 96, 96)$. We extract the pitch(F_0) using PyWorldVocoder[3] from the raw waveform with the frame size of 1024 and hop size of 256 sampled at 16khz to obtain the pitch of each frame and compute the L2-norm of the amplitude of each STFT frame as the energy.

Metrics. To quantify the quality of the final generated video, we have used the following metrics. PSNR(Peak Signal to Noise Ratio), SSIM(Structural Similarity Index), CPBD(Cumulative Probability Blur Detection) and ACD(Average Content Distance). PSNR, SSIM, and CPBD measure the quality of the generated image in terms of the presence of noise, perceptual degradation, and blurriness respectively. ACD [30] is used for the identification of the speaker from the generated frames by using OpenPose [5]. Along with image quality metrics, we have also calculated WER(Word Error Rate) using pretrained Lip-Net architecture [1], Blinks/sec using [28] and LMD(Landmark Distance) [6] to evaluate our performance of speech recognition, eye-blink reconstruction and lip reconstruction respectively.

Training and Inference. Our model is implemented in Pytorch and takes approximately 7 d to run on 4 Nvidia V100 GPUs for training. We have used the Adam optimizer [18] with learning rate(η) = 0.002 and β_1= 0.0 and β_2 = 0.90 for the generator and discriminators.

Video Emotion Detection

Datasets and PreProcessing Steps. We have used Crema-D [4] with 7436 videos having six classes namely Anger, Disgust, Fear, Happy, Neutral, Sadness. Videos are processed at 25fps and videos are resized as (t-dim, channels, x-dim, y-dim) = $(28, 3, 256, 342)$ with 28 frames of video to capture the temporal consistencies. The audio features are processed at 16khz to generate the 256 dimensional mel-spetrogram. The ground truth of optical flow is calculated used farneback optical flow algorithm [12].

---

[2] https://github.com/raymon-tian/hourglass-facekeypoints-detection.

[3] https://github.com/JeremyCCHsu/Python-Wrapper-for-World-Vocoder.

Training and Inference. We have trained the model on NVIDIA $V100$ machine with 4 GPU, the approximated time of training is around $18 - 20$ h. We ran the model for 25 epochs with a batch size of 4. We have used Adam optimizer and learning rate of $1e - 4$ with the cross entropy loss function. The test-train split ratio is 0.25.

Table 1. Comparison of the proposed method(MultiNorm-keypoint and MultiNorm-optical) with other previous works for GRID dataset

Method	SSIM↑	PSNR↑	CPBD↑	WER↓	ACD-C↓	ACD-E↓	Blinks/sec	LMD↓
OneShotA2V [19]	0.881	28.571	0.262	27.5	0.005	0.09	0.15	0.91
RSDGAN [32]	0.818	27.100	0.268	**23.1**	-	1.47×10^{-4}	0.39	–
Speech2Vid [9]	0.720	22.662	0.255	58.2	0.007	1.48×10^{-4}	-	–
ATVGnet [7]	0.83	**32.15**	–	–	–	–	–	1.29
X2face [35]	0.80	29.39	–	–	–	–	–	1.48
CascadedGAN [11]	0.81	27.1	0.26	23.1	-	1.47×10^{-4}	0.45	–
MultiNorm-optical	**0.908**	29.78	**0.272**	23.7	**0.005**	$\mathbf{1.41 \times 10^{-4}}$	0.45	**0.77**
MultiNorm-keypoint	0.887	29.01	0.269	25.2	0.006	1.41×10^{-4}	0.48	0.80

Table 2. Comparison of proposed method for CREMA-D dataset

Method	SSIM↑	PSNR↑	CPBD↑	WER↓	ACD-C↓	ACD-E↓	Blinks/sec	LMD↓
OneShotA2V [19]	0.773	24.057	0.184	NA	0.006	0.96	-	0.632
RSDGAN [32]	0.700	23.565	0.216	NA	–	1.40×10^{-4}	-	-
Speech2Vid [9]	0.700	22.190	0.217	NA	0.008	1.73×10^{-4}	–	–
MultiNorm-optical	0.826	27.723	0.224	NA	0.004	1.62×10^{-4}	–	0.592
MultiNorm-keypoint	**0.841**	**28.01**	**0.228**	NA	**0.003**	$\mathbf{1.38 \times 10^{-4}}$	–	**0.51**

Table 3. Comparison of proposed method for Grid Lombard dataset

Method	SSIM↑	PSNR↑	CPBD↑	WER↓	ACD-C↓	ACD-E↓	Blinks/sec	LMD↓
OneShotA2V [19]	0.922	28.978	0.453	NA	0.002	0.064	0.1	0.61
Speech2Vid [9]	0.782	26.784	0.406	NA	0.004	0.069	–	0.581
MultiNorm-optical	0.895	26.94	0.43	NA	0.001	0.048	0.21	0.588
MultiNorm-keypoint	**0.931**	**29.62**	**0.492**	NA	**0.001**	**0.046**	**0.31**	**0.563**

Table 4. Comparison of proposed method for voxceleb2 dataset

Method	SSIM↑	PSNR↑	CPBD↑	WER↓	ACD-C↓	ACD-E↓	Blinks/sec	LMD↓
OneShotA2V [19]	0.698	20.921	0.103	NA	0.011	0.096	0.05	0.72
MultiNorm-optical	0.714	21.94	0.118	NA	0.008	0.067	0.21	0.65
MultiNorm-keypoint	**0.732**	**22.41**	**0.126**	NA	**0.004**	**0.058**	**0.28**	**0.47**

6.2   Implementation Results

Quantitative Results

Audio to Video Generation. We have compared the proposed architecture with other prior works. Previous works such as Realistic Speech-Driven Facial Animation with GANs (RSDGAN) [32] and X2face [35] have used GAN based approach to produce quality videos. CascadedGAN [11] has used the L-GAN and T-GAN for motion(landmark) and texture generation in meta learning seeting(MAML) to generate the videos on an unseen person image. ATVGnet [7] uses an audio transformation network (AT-net) for audio to landmark generation and a visual generation network for facial generation. [37] has used Asymmetric Mutual Information Estimator (AMIE) to capture the mutual information to learn the cross-modal coherence for generating temporally expressive videos.

OneShotA2V [19] has used deep speech features into the generator architecture with spatially adaptive normalization layers in it along with lip frame discriminator, temporal discriminator and synchronization discriminator to generate realistic videos. The proposed method uses MultiNorm to capture the mutual relation between audio and video to generate expressive video. Tables 1, 2, 3 and 4 compare the proposed method with its competitors and shows better SSIM, PSNR, CPBD, Word error rate(WER), blinks/sec and LMD on GRID [10], Crema-D [4], Grid-lombard [23] and Voxceleb2 [8] datasets, suggesting highly expressive and realistic video synthesis. The average human blink rate of 0.28 blinks/second, especially when considering that the blink rate increases to 0.4 blinks/second during conversation. The proposed method produces blinks of 0.45 blinks/second which is close to natural human behaviour. The generated videos are present at following site:[4]

Emotion Detection. Table 5 presents the comparative results for 3D CNN based emotion detection with and without the proposed method. 3DCNN(base) has the base architecture of 2 layers of 3D convolution and 3 linear layers. 3DCNN(Base)+Optical Flow add 1 layer of 3D convolution and 2 linear layer to extract the affine parameters that goes into the normalization framework. The proposed model has used optical flow and mel-spectrogram to generate affine parameters in the normalization framework.

Table 5. Adding multi modal normalization increases the accuracy

Method	Accuracy
3DCNN(Base)	56.28
3DCNN(Base)+Optical Flow	67.81
Proposed model	**71.12**

---

[4] https://sites.google.com/view/iconip2021.

Qualitative Results: Audio to Video Generation

Expressive Aspect. Figure 5 displays the lip synchronized frames of speaker speaking the word *'bin'* and *'please'* as well as the blinking of the eyes and better image reconstruction and lip synchronization.

Fig. 5. Top: The speaker speaking the word 'bin', Middle: The speaker speaking the word 'please', Bottom: The speaker blinking his eyes

6.3 Ablation Study

Network Analysis in Multi Modal Normalization. We have done the ablation study on three architectures namely 2D convolution, Partial 2D convolution [21] and 2D convolution+Efficient channel Attention(ECA) [33] for extracting video features and two architectures namely 1D convolution and LSTM for audio features as shown in Fig. 3 to study its effect on multi-modal normalization with optical flow predictor in the proposed method. Table 6 shows that

Table 6. Ablation study of different networks of multi-modal normalization on Grid dataset

Method	SSIM↑	PSNR↑	CPBD↑	Blinks/sec	WER↓
2DConv+1dConv	0.875	28.65	0.261	0.35	25.6
Partial2DConv+1dConv	0.803	28.12	0.256	0.15	29.4
2DConv+ECA+1dConv	0.880	29.11	0.263	0.42	23.9
2DConv+LSTM	0.896	29.25	0.086	0.260	24.1
Partial2DConv+LSTM	0.823	28.12	0.258	0.12	28.3
2DConv+ECA+LSTM	**0.908**	**29.78**	**0.272**	**0.45**	**23.7**

2DConv+ECA+LSTM has improved the reconstruction metrics such as SSIM, PSNR and CPBD as well as word error rate and blinks/sec as compared to other networks.

Influence of Different K on MultiNorm. We studied the incremental effect of K on MultiNorm of the proposed model with optical flow predictor(OFP) and 2DConv+ECA(video features) and LSTM(audio features) combination in Multi-Norm on GRID dataset. Table 7 shows the impact of addition of mel-spectrogram features, pitch, predicted optical flow in the multi modal normalization. Base model consists generator and discriminator architecture with static image in the normalization(K=1).

Table 7. Incremental study of multi modal normalization on grid dataset

Method	SSIM↑	PSNR↑	CPBD↑	blinks/sec	WER↓
Base Model(BM)(K=1)	0.776	27.99	0.213	0.02	57.9
BM + OFP(K=2)	0.841	28.19	0.224	0.12	38.9
BM + OFP+mel(K=3)	0.878	28.43	0.244	0.38	27.4
BM + OFP+mel+pitch(K=4)	0.881	28.57	0.264	0.41	24.1
BM+OFP+mel+pitch+energy(K=5)	**0.908**	**29.78**	**0.272**	**0.45**	**23.7**

7 Conclusion

In this paper, we have proposed a novel multi-modal normalization for modeling the multi modal complex problems through the mixture of multivariate Gaussian distribution. We have seen how affine transformation in multi-modal normalization captures the mutual relationship across the domains. We have shown its effective in generative architecture (speech driven video generation) as well as in inference problem(video activity detection). Experimental evaluation of various datasets demonstrates superior performance compared to other related works on multiple quantitative and qualitative metrics.

References

1. Assael, Y.M., Shillingford, B., Whiteson, S., de Freitas, N.: Lipnet: end-to-end sentence-level lipreading. In: GPU Technology Conference (2017)
2. Ba, J., Kiros, J., Hinton, G.: Layer normalization (2016)
3. Bulat, A., Tzimiropoulos, G.: How far are we from solving the 2d & 3d face alignment problem? (and a dataset of 230,000 3d facial landmarks). In: International Conference on Computer Vision (2017)
4. Cao, H., Cooper, D., Keutmann, M., Gur, R., Nenkova, A., Verma, R.: Crema-d: Crowd-sourced emotional multimodal actors dataset. IEEE Trans. affective Comput. **5**, 377–390 (2014). https://doi.org/10.1109/TAFFC.2014.2336244

5. Cao, Z., Hidalgo, G., Simon, T., Wei, S.E., Sheikh, Y.: OpenPose: realtime multi-person 2D pose estimation using Part Affinity Fields. In: arXiv preprint arXiv:1812.08008 (2018)
6. Chen, L., Li, Z., Maddox, R.K., Duan, Z., Xu, C.: Lip movements generation at a glance (2018)
7. Chen, L., Maddox, R., Duan, Z., Xu, C.: Hierarchical cross-modal talking face generation with dynamic pixel-wise loss (2019)
8. Chung, J.S., Nagrani, A., Zisserman, A.: Voxceleb2: Deep speaker recognition. In: INTERSPEECH (2018)
9. Chung, J.S., Jamaludin, A., Zisserman, A.: You said that? In: British Machine Vision Conference (2017)
10. Cooke, M., Barker, J., Cunningham, S., Shao, X.: Grid AV speech corpus sample (2013)
11. Das, D., Biswas, S., Sinha, S., Bhowmick, B.: Speech-driven facial animation using cascaded GANs for learning of motion and texture. In: Vedaldi, A., Bischof, H., Brox, T., Frahm, J.-M. (eds.) ECCV 2020. LNCS, vol. 12375, pp. 408–424. Springer, Cham (2020). https://doi.org/10.1007/978-3-030-58577-8_25
12. Farnebäck, G.: Two-frame motion estimation based on polynomial expansion. In: Bigun, J., Gustavsson, T. (eds.) SCIA 2003. LNCS, vol. 2749, pp. 363–370. Springer, Heidelberg (2003). https://doi.org/10.1007/3-540-45103-X_50
13. Goodfellow, I., Bengio, Y., Courville, A.: Deep Learning. MIT Press, New York (2016). http://www.deeplearningbook.org
14. Huang, X., Belongie, S.: Arbitrary style transfer in real-time with adaptive instance normalization. In: Proceedings of the IEEE International Conference on Computer Vision, pp.1510–1519 (2017). https://doi.org/10.1109/ICCV.2017.167
15. Ioffe, S., Szegedy, C.: Batch normalization: Accelerating deep network training by reducing internal covariate shift (2015)
16. Karras, T., Laine, S., Aila, T.: A style-based generator architecture for generative adversarial networks (2019)
17. Kim, J., Kim, M., Kang, H.W., Lee, K.: U-gat-it: Unsupervised generative attentional networks with adaptive layer-instance normalization for image-to-image translation (2019)
18. Kingma, D., Ba, J.: Adam: A method for stochastic optimization. International Conference on Learning Representations (2014)
19. Kumar, N., Goel, S., Narang, A., Hasan, M.: Robust one shot audio to video generation. In: Proceedings of the IEEE/CVF Conference on Computer Vision and Pattern Recognition (CVPR) Workshops (2020)
20. Liu, D., Vu, M.T., Chatterjee, S., Rasmussen, L.: Neural network based explicit mixture models and expectation-maximization based learning. In: 2020 International Joint Conference on Neural Networks (IJCNN) pp. 1–10 (2020)
21. Liu, G., Shih, K.J., Wang, T.C., Reda, F.A., Sapra, K., Yu, Z., Tao, A., Catanzaro, B.: Partial convolution based padding. arXiv preprint arXiv:1811.11718 (2018)
22. Mallya, A., Wang, T.C., Sapra, K., Liu, M.Y.: World-consistent video-to-video synthesis (2020)
23. Alghamdi, N., Maddock, S., Marxer, R., Barker, J., Brown, G.J..: A corpus of audio-visual lombard speech with frontal and profile view. The Journal of the Acoustical Society of America 143, EL523–EL529 (2018). https://doi.org/10.1121/1.5042758
24. Nam, H., Kim, H.E.: Batch-instance normalization for adaptively style-invariant neural networks (05 2018)

25. Park, T., Liu, M.Y., Wang, T.C., Zhu, J.Y.: Semantic image synthesis with spatially-adaptive normalization. In: Proceedings of the IEEE Conference on Computer Vision and Pattern Recognition (2019)
26. Pfülb, B., Gepperth, A.: Overcoming catastrophic forgetting with gaussian mixture replay. ArXiv abs/2104.09220 (2021)
27. Richardson, E., Weiss, Y.: On GANs and GMMs. In: NeurIPS (2018)
28. Soukupova, T., Cech, J.: Real-time eye blink detection using facial landmarks (2016)
29. Storey, G., Bouridane, A., Jiang, R., Li, C.-T.: Atypical facial landmark localisation with stacked hourglass networks: a study on 3D facial modelling for medical diagnosis. In: Jiang, R., Li, C.-T., Crookes, D., Meng, W., Rosenberger, C. (eds.) Deep Biometrics. USL, pp. 37–49. Springer, Cham (2020). https://doi.org/10.1007/978-3-030-32583-1_3
30. Tulyakov, S., Liu, M.Y., Yang, X., Kautz, J.: MoCoGAN: Decomposing motion and content for video generation. In: IEEE Conference on Computer Vision and Pattern Recognition (CVPR), pp. 1526–1535 (2018)
31. Ulyanov, D., Vedaldi, A., Lempitsky, V.: Instance normalization: The missing ingredient for fast stylization (2016)
32. Vougioukas, K., Petridi, S., Pantic, M.: End-to-end speech-driven facial animation with temporal GANs. J. Foo **14**(1), 234–778 (2004)
33. Wang, Q., Wu, B., Zhu, P., Li, P., Zuo, W., Hu, Q.: ECA-Net: Efficient channel attention for deep convolutional neural networks (2019)
34. Wang, T.C., Liu, M.Y., Zhu, J.Y., Tao, A., Kautz, J., Catanzaro, B.: High-resolution image synthesis and semantic manipulation with conditional GANs. In: Proceedings of the IEEE Conference on Computer Vision and Pattern Recognition (2018)
35. Wiles, O., Koepke, A., Zisserman, A.: X2face: A network for controlling face generation by using images, audio, and pose codes. In: European Conference on Computer Vision (2018)
36. Zhou, B., Khosla, A., A., L., Oliva, A., Torralba, A.: Learning deep features for discriminative localization. In: CVPR (2016)
37. Zhu, H., Huang, H., Li, Y., Zheng, A., He, R.: Arbitrary talking face generation via attentional audio-visual coherence learning. arXiv: Computer Vision and Pattern Recognition (2020)

A Focally Discriminative Loss for Unsupervised Domain Adaptation

Dongting Sun[1], Mengzhu Wang[1], Xurui Ma[1], Tianming Zhang[2],
Nan Yin[1,2(✉)], Wei Yu[1,2], and Zhigang Luo[1]

[1] College of Computer, National University of Defense Technology, Hunan, China
[2] College of Computer, Harbin Engineering University, Harbin, China

Abstract. The maximum mean discrepancy (MMD) as a representative distribution metric between source domain and target domain has been widely applied in unsupervised domain adaptation (UDA), where both domains follow different distributions, and the labels from source domain are merely available. However, MMD and its class-wise variants possibly ignore the intra-class compactness, thus canceling out discriminability of feature representation. In this paper, we endeavor to improve the discriminative ability of MMD from two aspects: 1) we re-design the weights for MMD in order to align the distribution of relatively hard classes across domains; 2) we explore a focally contrastive loss to trade-off the positive sample pairs and negative ones for better discrimination. The intergration of both losses makes the intra-class features close as well as push away the inter-class features far from each other. Moreover, the improved loss is simple yet effective. Our model shows state-of-the-art compared to the most domain adaptation methods.

Keywords: Unsupervised domain adaptation · Maximum mean discrepancy · Class-wise MMD · Focally contrastive loss

1 Introduction

The success of computer vision applications mostly depends on whether there is enough labeled data available to train machine learning models. However, in practice, it is often labor-exhausting and time-consuming to collect amounts of labeled data. Unsupervised domain adaptation (UDA) [14,21,25–27] is a promising strategy for learning accurate classifiers in a target domain by leveraging the knowledge from a different but related, well-labeled source domain. Nevertheless, domain discrepancy is still a long-standing challenge for better classification performance in UDA.

To weaken the negative effect of domain discrepancy, a series of methods [3,20,29] have been proposed to reduce the gap between cross-domain distributions. Of them, most arts achieve this goal by learning domain-invariant features or classifiers through distribution alignment such as maximum mean discrepancy (MMD) [7] or adversarial learning [1,5]. However, this does not guarantee the

© Springer Nature Switzerland AG 2021
T. Mantoro et al. (Eds.): ICONIP 2021, LNCS 13108, pp. 54–64, 2021.
https://doi.org/10.1007/978-3-030-92185-9_5

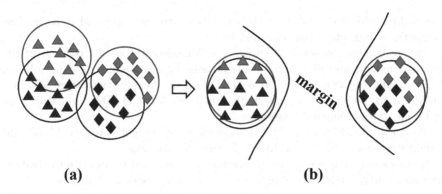

Fig. 1. (a) Traditional class-wise MMD has the risk of reducing intra-class discriminability in two domains, thereby inducing high overlap between samples of different classes; (b) The proposed method maximizes the margin between different classes to improve feature discriminability.

feasible discriminant ability of the features learned in the target domain or the classifier, because the distribution alignment is global and the fine-grained class information will be ignored. Towards this end, the class-wise feature alignment can be considered by simultaneously matching local joint distributions in both feature and category levels. For this purpose, a representative MMD variant, i.e., class-wise MMD [11], has been widely applied to quatify the conditional distribution difference across domains. Most recently, the work of [22] revisits class-wise MMD in theory, and find that minimizing it is equivalent to maximizing the source and target intra-class compactness so that feature discriminability could degrade. As shown in Fig. 1, traditional class-wise MMD may lead to the overlapping of different classes.

Following this stream of [22], we improve the discriminability of MMD from two aspects. Firstly, we assign the pseudo-label for the target domain. Then we calculate the weights for class-wise MMD from the two domains based on the pseudo-label, which can align the distribution of relatively hard class across domains. It considers the fine-grained information for each class. As can be shown in Fig. 1 (a), weight class-wise MMD ignores the discriminability between the two domains. Consider the adverse of the weight class-wise MMD, We suggest using all samples to construct a more informative structure and use it to learn more discriminant features, as shown in Fig. 1 (b). Specifically, given a selected sample, we sort all the other samples by distance, resulting in a sorted list. Ideally, all positive samples should be sorted prior the negative samples(we suppose the same class sample as the selected sample to be positive samples, vice-versa). To get this goal, we utilize focally contrastive loss to organize the samples of each selected sample. Given a selected sample, the optimisation of focally contrastive loss is to sort all positive samples before the negative samples and forcing a margin between the two domain samples. Meanwhile, we attempt to learn a hypersphere for each sample, which force the distance of a positive pair smaller than a threshold, it can preserve the similarity structure and lead to intra-class compactness and inter-class separability.

The evaluations on several benchmarks demonstrate the effectiveness of the proposed model, and the major contributions can be summarized as follows: (1) By taking the weight class-wise MMD and the focally contrastive loss into consideration, we propose a novel method named "A Focally Discriminative Loss for Unsupervised Domain Adaptation".
(2) We theoretically analyze the adverse effects of weight category MMD and explain the reasons for the decline in feature discernment.
(3) we propose a simple effective discriminability method to improve the feature discriminability, further enabling class-wise feature discriminativeness.

2 Method

2.1 Motivation

Inspired by [22], MMD equals to maximize intra-class distance between the source domain and target domain, and the relationship between intra- and inter-class distances is one down, the other up. To handle this problem, we propose an discriminability strategy to combat the adverse impacts of MMD. We show that the weight class-wise MMD ignore the feature discriminability and may result in degrade domain adaptation performance. Specifically, we propose to make a margin to improve discriminability for each class in our method, which can be shown in Fig. 2 (b), we force the distance of the positive samples to be less than the threshold. Under this setting, our method can preserve the similarity structure within each class while improving the discriminativeness of MMD.

2.2 Weight Class-Wise MMD

In order to calculate the distance between two domains, we adopted a non-parametric method commonly used in domain adaptation, the maximum mean difference (MMD) [11]. It can calculate the distance between samples in the source domain and the target domain. The formula is as follows:

$$
d_{\mathcal{H}}^1(p, q) = \left\| \frac{1}{n_s} \sum_{x_i^s \in D_s} A^T x_i^s - \frac{1}{n_t} \sum_{x_j^t \in D_t} A^T x_j^t \right\|_{\mathcal{H}}^2 \tag{1}
$$
$$
= \mathrm{tr}\left(A^T X_{st} M_0 X_{st}^T A \right)
$$

where $A \in R^{m \times k}$ is the transformation matrix, $x_i^s \in \mathcal{D}_s$, $x_j^t \in \mathcal{D}_t$ and X_{st} is composed of source domain samples X_s and X_t. But when solving the distance between conditional distributions, we cannot figure out the label of the target domain. Thus, we adopt the pseudo labels based on the basic network architecture ResNet50 [8]. and it can predict the target domain by training a classifier on the source data. Therefore, the MMD of the weight category can be as follows:

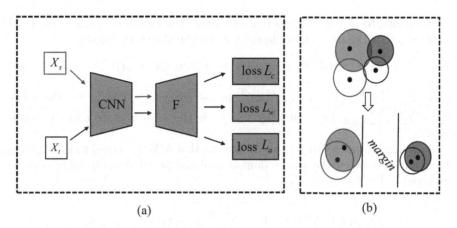

(a)   (b)

Fig. 2. An illustration of our method for UDA model. (a) The architecture of the proposed model consists of a classifier loss, a discriminability strategy loss, a weight class-wise MMD loss. (b) The advantage of the discriminability strategy, which make different classes have a margin.

$$d_{\mathcal{H}}^2(p,q) = \sum_{c=1}^{C} \left\| \frac{1}{n_s^{(c)}} \sum_{x_i^s \in D_s} A^T x_i^s - \frac{1}{n_t^{(c)}} \sum_{x_j^t \in D_t} A^T x_j^t \right\|_{\mathcal{H}}^2 \qquad (2)$$

$$= \mathrm{tr}\left(A^T X_{st} \sum_{c=1}^{C} M_c X_{st}^T A \right)$$

But it ignores the weight in each class. Eq. (2) make each class have the same weight. Based on Eq. (2), we change it as follows:

$$d_{\mathcal{H}}^3(p,q) = \sum_{c=1}^{C} \left\| \sum_{x_i^s \in D_s} w_i^{sc} A^T x_i^{sc} - \sum_{x_j^t \in D_t} w_j^{tc} A^T x_j^{tc} \right\|_{\mathcal{H}}^2 \qquad (3)$$

$$= \sum_{c=1}^{C} \mathrm{tr}\left(A^T X_{st}^c M_c \left(X_{st}^c \right)^T A \right)$$

where w_i^{sc} and w_j^{tc} are the weight of each class from the source domain and the target domain, and $\sum_{i=1}^{n_s} w_i^{sc} = 1, \sum_{i=1}^{n_t} w_i^{tc} = 1$ denotes the sum of the weight in source domain and target domain. M_c is computed as follows:

$$M_c = \begin{cases} w_i^{sc} \left(w_i^{sc} \right)^T, & (x_i, x_j \in X_s^c) \\ w_j^{tc} \left(w_j^{tc} \right)^T, & (x_i, x_j \in X_t^c) \\ -w_i^{sc} \left(w_j^{tc} \right)^T, & \begin{cases} x_i \in X_s^c, x_j \in X_t^c \\ x_j \in X_s^c, x_i \in X_t^c \end{cases} \\ 0, & (\text{otherwise}) \end{cases} \qquad (4)$$

According to [31], it deduces that the inter-class distance equals to the total variance minus the intra-class distance, as can be shown in follows:

$$\text{tr}\left(A^T S_b A\right) = \text{tr}\left(A^T S_v A\right) - \text{tr}\left(A^T S_w A\right) \tag{5}$$

where $S_v = \sum_{i=1}^{n} (x_i - m)(x_i - m)^T$ is the covariance matrix for the total, $S_w = \sum_{i=1}^{C} \sum_{x_j \in X^i} (x_j - m^i)(x_j - m^i)^T$ is the scatter matrix in the intra-class.

At the same time, Wang $et\ al.$ [22] means that MMD is equal to maximizing the distance between the source domain and the target domain, which can be displayed in Eq. (3).

$$
\begin{aligned}
\sum_{c=1}^{C} \text{tr}\left(A^T X M_c X^T A\right) &= \sum_{c=1}^{C} \frac{n_s^c + n_t^c}{n_s^c n_t^c} \text{tr}\left(A^T (S_{st})_b^c A\right) \\
&= \sum_{c=1}^{C} \frac{n_s^c + n_t^c}{n_s^c n_t^c} \text{tr}\left(A^T (S_{st})_v^c A\right) \\
&\quad - \sum_{c=1}^{c} \frac{n_s^c + n_t^c}{n_s^c n_t^c} \text{tr}\left(A^T (S_{st})_w^c A\right)
\end{aligned}
\tag{6}
$$

where $(S_{st})_b^c = \sum_{i \in \{s,t\}} n_i^c (m_i^c - m_{st}^c)(m_i^c - m_{st}^c)^T$, $(S_{st})_v^c = \sum_{i=1}^{n_{st}^c} (x_i - m_{st}^c)(x_i - m_{st}^c)^T$ and $(S_{st})_w^c = \sum_{i \in \{s,t\}} \sum_{j=1}^{n_i^c} (x_j - m_i^c)(x_j - m_i^c)^T$.

We change Eq. (6) as follows:

$$\text{tr}\left(A^T X M_c X^T A\right) = w_s^c (w_t^c)^T \text{tr}\left(A^T (m_s^c - m_t^c)(m_s^c - m_t^c)^T A\right) \tag{7}$$

then we can get the Eq. (6) as follows:

$$
\begin{aligned}
\text{tr}\left(A^T X M_c X^T A\right) &= \frac{n_s^c + n_t^c}{n_s^c n_t^c} w_s^c (w_t^c)^T \text{tr}\left(A^T (S_{st})_v^c A\right) \\
&\quad - \frac{n_s^c + n_t^c}{n_s^c n_t^c} w_s^c (w_t^c)^T \text{tr}\left(A^T (S_{st})_w^c A\right)
\end{aligned}
\tag{8}
$$

2.3 Discriminability Strategy

In order to handle the adverse of weight class-wise MMD, we intend to separate the different samples. To get this goal, we propose a discriminability strategy to handle the problem.

Given a selected image x_i, our goal is to move samples of different categories away from the boundary β and bring the same samples closer. At the same time m is the distance between the two borders. The mathematical expression is as follows:

$$
\begin{aligned}
\mathcal{L}_s &= (1 - y_{ij}) \cdot \max\left(0, \beta - d_{ij}\right) \\
&\quad + y_{ij} \cdot \max\left(0, d_{ij} - (\beta - m)\right)
\end{aligned}
\tag{9}
$$

where $y_{ij} = 1$ if $y_i = y_j$ and if $y_i \neq y_j$, $y_{ij} = 0$. $d_{ij} = \|f(x_i) - f(x_j)\|_2$ denotes the Euclidean distance between the two image features.

For the selected sample x_i, because different samples have different distances to the selection. In order to make full use of these samples, we use the distance between the sample and the selected sample to weight the sample. Then the weighting strategy is formally expressed as follows:

$$w_{ij}^+ = \exp(\beta - d_{ij}), x_j \in N_{c,i}$$
$$w_{ij}^- = \exp(d_{ij} - (\beta - m)), x_j \in P_{c,i} \tag{10}$$

For each selected sample x_i^c, we propose to make the selected sample closer to its positive sample set $P_{c,i}$ instead of a non-negative sample set $N_{c,i}$. At the same time, we force all counterexamples to be farther than the boundary β. Therefore, we pull all samples from the same class into the hypersphere. The diameter of each hypersphere is $\beta - m$. In order to collect all positive samples and set a boundary to learn a hypersphere-like surface, we minimize the loss of positive samples $\mathcal{L}_P$ as follows:

$$\mathcal{L}_P(x_i^c; f) = \sum_{x_j^k \in P_{c,f}} \frac{w_{ij}^+}{\sum_{x_j^k \in P_{c,j}} w_{ij}^+} \mathcal{L}_s(x_i^c, x_j^c; f) \tag{11}$$

Similarity, to push the negative samples set $N_{c,i}$ beyond the boundary β, we minimize the negative samples loss $\mathcal{L}_N$ as follows:

$$\mathcal{L}_N(x_i^c; f) = \sum_{x_j^k \in N_{c,i}} \frac{w_{ij}^-}{\sum_{x_j^k \in N_{c,i}} w_{ij}^-} \mathcal{L}_s(x_i^c, x_j^k; f) \tag{12}$$

Then discriminability strategy loss $\mathcal{L}_d$ can be expressed as follows:

$$\mathcal{L}_d(x_i^c; f) = \mathcal{L}_P(x_i^c; f) + \mathcal{L}_N(x_i^c; f) \tag{13}$$

where $\mathcal{L}_d$ equals to the sum pf positive samples loss and negative samples loss. In the sorted lost of x_i^c, we treat the features of other examples as constant. Therefore, we only update $f(x_i^c)$ in each iteration.

2.4   Total Network

Our aim is to reduce the domain gaps and improve the discriminability of the weight class-wise MMD. So we try to reduce the discrepancy between the two domains. We use the proposed method in the following Eq. 14:

$$\min_f \frac{1}{n_s} \sum_{i=1}^{n_s} J(f(x_i^s), y_i^s) + \lambda_1 d_{\mathcal{H}}^3(p, q) + \lambda_2 \mathcal{L}_d \tag{14}$$

Table 1. Recognition accuracy (%) on Office-Home dataset. (ResNet50)

OfficeHome	A→C	A→P	A→R	C→A	C→P	C→R	P→A	P→C	P→R	R→A	R→C	R →P	Avg.
ResNet50	34.9	50.0	58.0	37.4	41.9	46.2	38.5	31.2	60.4	53.9	41.2	59.9	46.1
DAN	43.6	57.0	67.9	45.8	56.5	60.4	44.0	43.6	67.7	63.1	51.5	74.3	56.3
DANN	45.6	59.3	70.1	47.0	58.5	60.9	46.1	43.7	68.5	63.2	51.8	76.8	57.6
JAN	45.9	61.2	68.9	50.4	59.7	61.0	45.8	43.4	70.3	63.9	52.4	76.8	58.3
CDAN	50.7	70.6	76.0	57.6	70.0	70.0	57.4	50.9	77.3	70.9	56.7	81.6	65.8
SAFN	52.0	71.7	76.3	**64.2**	69.9	71.9	63.7	51.4	77.1	70.9	57.1	81.5	67.3
TADA	53.1	72.3	77.2	59.1	71.2	72.1	59.7	53.1	78.4	72.4	60.0	82.9	67.6
SymNet	47.7	**72.9**	**78.5**	**64.2**	**71.3**	**74.2**	**64.2**	48.8	**79.5**	**74.5**	52.6	**82.7**	67.6
Ours	**56.0**	71.9	75.6	62.8	68.9	68.6	63.0	**56.8**	79.1	73.7	**61.7**	81.1	**68.4**

3   Experimental Evaluation

3.1   Result and Discussion

Results on Office-31. Office-31 is the most commonly used data set in domain use. The results of the Office-31 data set can be seen in Table 2. It can be seen that our method is effective. At the same time, it is worth noting that our method significantly improves the classification accuracy on difficult migration tasks. For example, the source domain of D→A is obviously different from the target domain, and the size of the source domain of W→A is even smaller than the target domain. And achieve comparable performance on simple transfer tasks.

Results on ImageCLEF-DA and Office-Home. All images in the Office-31 dataset are from Office scenes. Imageclef-da data sets are much more diverse than Office-31. In this article, we also test our method on imageclef-da and Office-Home datasets and prove the robustness of our method. In Imageclef-da and Office-Home, we also use ResNet-50 as our backbone network. The results can be seen in Table 1 and Table 3. From the table we can conclude that our model is superior to many of the most advanced methods. On some difficult data sets, our method generally improves accuracy.

3.2   Model Analysis

Ablation Study. Table 4 presents the results of ablation studies under different model variations and some strategies. The baseline for ResNet50 represents only the source classifier and no MMD. ResNet+class-wise MMD denotes class-wise MMD be taken into consideration, and the performance is increased from 76.1% to 84.3%. ResNet50+weight class-wise MMD denotes weight class-wise under consideration. From Table 4, it can be found that our model benefits from both weight class-wise MMD and discriminability strategy, it validates the effectiveness of our approach.

Table 2. Recognition accuracies (%) on the Office31 dataset.(ResNet50)

Office-31	A→W	D→W	W→D	A→D	D→A	W→A	Avg.
ResNet50	68.4	96.7	99.3	68.9	62.5	60.7	76.1
TCA	72.7	96.7	99.6	74.1	61.7	60.9	77.6
GFK	72.8	95.0	98.2	74.5	63.4	61.0	77.5
DDC	75.6	96.0	98.2	76.5	62.2	61.5	78.3
DAN	80.5	97.1	99.6	78.6	63.6	62.8	80.4
RTN	84.5	96.8	99.4	77.5	66.2	64.8	81.6
DANN	82.0	96.9	99.1	79.7	68.2	67.4	82.2
ADDA	86.2	96.2	98.4	77.8	69.5	68.9	82.9
JAN	85.4	97.4	99.8	84.7	68.6	70.0	84.3
MADA	90.0	97.4	99.6	87.8	70.3	66.4	85.2
SAFN	88.8	98.4	99.8	87.7	69.8	69.7	85.7
iCAN	92.5	98.8	**100.0**	90.1	72.1	69.9	87.2
CDAN	94.1	98.6	**100.0**	92.9	71.0	69.3	87.7
TADA	94.3	98.7	99.8	91.6	72.9	73.0	88.4
SymNet	90.8	98.8	**100.0**	**93.9**	74.6	72.5	88.4
Ours	**95.8**	**98.9**	100.0	93.6	**76.0**	**73.1**	**89.6**

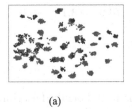
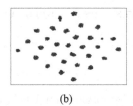
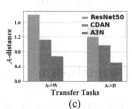

(a) (b) (c)

Fig. 3. Visualize the learned representations using ResNet50's t-SNE and the method we used on task A→W. We measure the $\mathcal{A}$-distance to quantity distribution between the two domains.

Feature Visualization. In Fig. 3 (a) and Fig. 3 (b), we visualize the network activation of ResNet50 learning task A→W and our method of using T-SNE embedding [4]. The red dots are the source domain and the blue dots are the target domain. It can be seen from the Fig. 3 (a) that the source and target domains are not well aligned and are difficult to distinguish. However, Fig. 3 (b) shows that the same class in source domain and target domain is very close and well aligned.

Quantitative Distribution Discrepancy. Figure 3 (c) shows that the $\mathcal{A}$-distance of ResNet50, CDAN and our method. $\mathcal{A}$-distance is a distribution measure that is often used in domain adaptation. $\mathcal{A}_{dis} = 2(1-2\epsilon)\epsilon$ smaller $\mathcal{A}$-distance indicates a better alignment of the source and target domains.

Table 3. Recognition accuracies (%) on ImageCLEF-DA.(ResNet50)

ImageCLEF-DA	I→P	P→I	I→C	C→I	C→P	P→C	Avg.
ResNet50	74.8	83.9	91.5	78.0	65.5	91.2	80.7
DAN	74.5	82.2	92.8	86.3	69.2	89.8	82.5
DANN	75.0	86.0	96.2	87.0	74.3	91.5	85.0
JAN	76.8	88.0	94.7	89.5	74.2	91.7	85.8
MADA	75.0	87.9	96.0	88.8	75.2	92.2	85.8
iCAN	79.5	89.7	94.7	89.9	78.5	92.0	87.4
CDAN	77.7	90.7	**97.7**	91.3	74.2	94.3	87.7
SAFN	78.0	91.7	96.2	91.1	77.0	94.7	88.1
Ours	**81.8**	**93.8**	97.3	**94.0**	81.2	**96.8**	**90.8**

Table 4. The ablation study on the Office-31 dataset.

Office-31	A→W	D→W	W→D	A→D	D→A	W→A	Avg.
ResNet50	68.4	96.7	99.3	68.9	62.5	60.7	76.1
ResNet50+class-wise MMD	85.4	97.4	99.8	84.7	68.6	70.0	84.3
ResNet50+weight class-wise MMD	93.2	98.4	100.0	91.3	74.5	72.6	88.3
Ours	**95.8**	**98.9**	**100.0**	**93.6**	**76.0**	**73.1**	**89.6**

4   Conclusion

In this paper, we propose a new method for UDA, called "A Focally Discriminative Loss for Unsupervised Domain Adaptation". Specifically, we propose a sample discriminant building module, a weight category MMD module, and a pseudo-label assignment module. We use discriminant loss and weight class loss based on ResNet50. It is different from the previous adaptive methods of antagonistic domain which are difficult to train and slow to converge. The effectiveness of the proposed method is verified by comprehensive experiments on domain adaptive data sets. In future work, we aim to construct more efficient discriminant strategies on weight categories MMD and extend our model to other in-depth unsupervised domain adaptation studies, such as individual re-identification, single-target tracking, and multi-target tracking.

References

1. Ajakan, H., Germain, P., Larochelle, H., Laviolette, F., Marchand, M.: Domain-adversarial neural networks. arXiv preprint arXiv:1412.4446 (2014)
2. Cui, S., Wang, S., Zhuo, J., Li, L., Huang, Q., Tian, Q.: Towards discriminability and diversity: Batch nuclear-norm maximization under label insufficient situations. In: Proceedings of the IEEE/CVF Conference on Computer Vision and Pattern Recognition, pp. 3941–3950 (2020)
3. Deng, W., Zheng, L., Sun, Y., Jiao, J.: Rethinking triplet loss for domain adaptation. IEEE Trans. Circuits Syst. Video Technol. **31**(1), 29–37 (2020)
4. Donahue, J., et al.: A deep convolutional activation feature for generic visual recognition. UC Berkeley & ICSI, Berkeley, CA, USA
5. Ganin, Y., et al.: Domain-adversarial training of neural networks. J. Mach. Learn. Res. **17**(1), 2030–2096 (2016)
6. Gong, B., Shi, Y., Sha, F., Grauman, K.: Geodesic flow kernel for unsupervised domain adaptation. In: 2012 IEEE Conference on Computer Vision and Pattern Recognition, pp. 2066–2073. IEEE (2012)
7. Gretton, A., Borgwardt, K.M., Rasch, M.J., Schölkopf, B., Smola, A.: A kernel two-sample test. J. Mach. Learn. Res. **13**(1), 723–773 (2012)
8. He, K., Zhang, X., Ren, S., Sun, J.: Deep residual learning for image recognition. In: Proceedings of the IEEE Conference on Computer Vision and Pattern Recognition, pp. 770–778 (2016)
9. Long, M., Cao, Y., Wang, J., Jordan, M.: Learning transferable features with deep adaptation networks. In: International Conference on Machine Learning, pp. 97–105. PMLR (2015)
10. Long, M., Cao, Z., Wang, J., Jordan, M.I.: Conditional adversarial domain adaptation. In: Advances in Neural Information Processing Systems, pp. 1640–1650 (2018)
11. Long, M., Wang, J., Ding, G., Sun, J., Yu, P.S.: Transfer feature learning with joint distribution adaptation. In: Proceedings of the IEEE International Conference on Computer Vision, pp. 2200–2207 (2013)
12. Long, M., Zhu, H., Wang, J., Jordan, M.I.: Unsupervised domain adaptation with residual transfer networks. In: Advances in Neural Information Processing Systems, pp. 136–144 (2016)
13. Pan, S.J., Tsang, I.W., Kwok, J.T., Yang, Q.: Domain adaptation via transfer component analysis. IEEE Trans. Neural Netw. **22**(2), 199–210 (2010)
14. Pan, S.J., Yang, Q.: A survey on transfer learning. IEEE Trans. knowl. Data Eng. **22**(10), 1345–1359 (2009)
15. Pei, Z., Cao, Z., Long, M., Wang, J.: Multi-adversarial domain adaptation. arXiv preprint arXiv:1809.02176 (2018)
16. Saenko, K., Kulis, B., Fritz, M., Darrell, T.: Adapting visual category models to new domains. In: Daniilidis, K., Maragos, P., Paragios, N. (eds.) ECCV 2010. LNCS, vol. 6314, pp. 213–226. Springer, Heidelberg (2010). https://doi.org/10.1007/978-3-642-15561-1_16
17. Tzeng, E., Hoffman, J., Saenko, K., Darrell, T.: Adversarial discriminative domain adaptation. In: Proceedings of the IEEE Conference on Computer Vision and Pattern Recognition, pp. 7167–7176 (2017)
18. Tzeng, E., Hoffman, J., Zhang, N., Saenko, K., Darrell, T.: Deep domain confusion: maximizing for domain invariance. arXiv preprint arXiv:1412.3474 (2014)

19. Venkateswara, H., Eusebio, J., Chakraborty, S., Panchanathan, S.: Deep hashing network for unsupervised domain adaptation. In: Proceedings of the IEEE Conference on Computer Vision and Pattern Recognition, pp. 5018–5027 (2017)

20. Wang, H., Yang, W., Wang, J., Wang, R., Lan, L., Geng, M.: Pairwise similarity regularization for adversarial domain adaptation. In: Proceedings of the 28th ACM International Conference on Multimedia, pp. 2409–2418 (2020)

21. Wang, M., Zhang, X., Lan, L., Wang, W., Tan, H., Luo, Z.: Improving unsupervised domain adaptation by reducing bi-level feature redundancy. arXiv preprint arXiv:2012.15732 (2020)

22. Wang, W., Li, H., Ding, Z., Wang, Z.: Rethink maximum mean discrepancy for domain adaptation. arXiv preprint arXiv:2007.00689 (2020)

23. Wang, X., Li, L., Ye, W., Long, M., Wang, J.: Transferable attention for domain adaptation. In: Proceedings of the AAAI Conference on Artificial Intelligence, vol. 33, pp. 5345–5352 (2019)

24. Xu, R., Li, G., Yang, J., Lin, L.: Larger norm more transferable: an adaptive feature norm approach for unsupervised domain adaptation. In: Proceedings of the IEEE International Conference on Computer Vision, pp. 1426–1435 (2019)

25. Yang, X., Dong, J., Cao, Y., Wang, X., Wang, M., Chua, T.S.: Tree-augmented cross-modal encoding for complex-query video retrieval. In: Proceedings of the 43rd International ACM SIGIR Conference on Research and Development in Information Retrieval, pp. 1339–1348 (2020)

26. Yang, X., Feng, F., Ji, W., Wang, M., Chua, T.S.: Deconfounded video moment retrieval with causal intervention. In: SIGIR (2021)

27. Yang, X., He, X., Wang, X., Ma, Y., Feng, F., Wang, M., Chua, T.S.: Interpretable fashion matching with rich attributes. In: Proceedings of the 42nd International ACM SIGIR Conference on Research and Development in Information Retrieval, pp. 775–784 (2019)

28. Zhang, W., Ouyang, W., Li, W., Xu, D.: Collaborative and adversarial network for unsupervised domain adaptation. In: Proceedings of the IEEE Conference on Computer Vision and Pattern Recognition, pp. 3801–3809 (2018)

29. Zhang, W., Zhang, X., Liao, Q., Yang, W., Lan, L., Luo, Z.: Robust normalized squares maximization for unsupervised domain adaptation. In: Proceedings of the 29th ACM International Conference on Information & Knowledge Management, pp. 2317–2320 (2020)

30. Zhang, Y., Tang, H., Jia, K., Tan, M.: Domain-symmetric networks for adversarial domain adaptation. In: Proceedings of the IEEE Conference on Computer Vision and Pattern Recognition, pp. 5031–5040 (2019)

31. Zheng, S., Ding, C., Nie, F., Huang, H.: Harmonic mean linear discriminant analysis. IEEE Trans. Knowl. Data Eng. **31**(8), 1520–1531 (2018)

32. Zhu, Y., Zhuang, F., Wang, J., Ke, G., Chen, J., Bian, J., Xiong, H., He, Q.: Deep subdomain adaptation network for image classification. IEEE Trans. Neural Netw. Learn. Syst. **32**(4), 1713–1722 (2020)

Automatic Drum Transcription with Label Augmentation Using Convolutional Neural Networks

Tianyu Xu[1,2], Pei Dai[1,2], Baoyin He[1,2], Mei Zhang[3], and Jinhui Zhu[1,2(✉)]

[1] School of Software Engineering, South China University of Technology,
Guangzhou, China
sexutianyu@mail.scut.edu.cn, csjhzhu@scut.edu.cn
[2] Key Laboratory of Big Data and Intelligent Robot (South China University
of Technology) Ministry of Education, Guangzhou, China
[3] School of Automation Science and Control Engineering,
South China University of Technology, Guangzhou, China
zhangmei@scut.edu.cn

Abstract. Automatic drum transcription (ADT) aims to generate drum notes from drum music recordings. It is a significant subtask of music information retrieval. The successful transcription of drum instruments is a key step in the analysis of drum music. Existing systems use the target drum instruments as a separate training objective, which faces the problems of over-fitting and limited performance improvement. To solve the above limitations, this paper presents a label augmentation approach with the use of convolutional neural network via joint learning and self-distillation. Joint learning is introduced to gather drum and music information with the adoption of joint label and joint classifier. Self-distillation is proposed to use to enhance the expressive ability of the target model. We evaluate the designed system on well-known datasets and make a careful comparison with other ADT systems. The ablation studies are also conducted and a detailed analysis of the experimental results is provided. The experimental results show that the proposed system can achieve more competitive performance and outperform the state-of-the-art approach.

Keywords: Automatic drum transcription · Convolutional neural networks · Joint learning · Label augmentation · Self-distillation

1 Introduction

Automatic music transcription (AMT) [2] system aims at generating symbolic representation from audio signals. AMT is an important high-level task in music information retrieval [3]. The transcription of drum instruments, called automatic drum transcription (ADT) is a sub-task of automatic music transcription. The goal of ADT is to generate drum events from audio signals [21]. While this is

T. Mantoro et al. (Eds.): ICONIP 2021, LNCS 13108, pp. 65–76, 2021.
https://doi.org/10.1007/978-3-030-92185-9_6

the first step for full analysis and fetches more music information of drum track. A robust ADT system plays an important role in automated drum education and entertainment systems.

Recently works mainly transcribe on three most commonly used drum instruments: kick drum (KD), snare drum (SD) and hi-hat (HH). These three instruments are the most common percussion instruments in drum music. The researchers use the performance of these instruments to evaluate the performance of their ADT systems.

Existing ADT systems can be categorized into two ways: *i. Segmentation and Classification based system, ii. Activation based system* [21]. The first category segments the drum onset from the audio then classifies segmentation respectively. This type of system requires a high quality on the segmentation part. Classification quality depends on segmentation quality. A loss detection or wrong location of segmentation would lead to misclassification. The second category generates the possibility called activation values on different drum instruments frame by frame then extracts the local maximum values from the activation values as the drum onset points. The *activation based system* takes care of the simultaneous events, which can achieve better performance. The second way is the prevalent way in recent years.

There are two main types of activation based systems: *Non-Negative Matrix Factorization (NMF) based system* and *Neural Network (NN) based system*. The NMF-based system decomposes the mix spectrogram into spectrum basis functions and corresponding activation values. This kind of method can achieve favourable performance under the condition of a small amount of annotated data, but its performance gains are limited. The main representations are *NMFD* [12], *SANMF* [5], *PFNMF, AM1* and *AM2* [22]. The NN-based system uses a trained neural network to generate activation values from the mix spectrogram. This system has good robustness when there is a large of amount of high quality training data, which possesses great potential for future development. Recent studies on NN-based systems can be mainly divided into recurrent neural network (RNN) and convolutional neural network (CNN). The RNN-based system treats the spectrogram as a series of subsequences and the typical examples are *RNN* [15,17], *tanhB* [15], lstmpB [16] and GRUts [18]. While the CNN-based method treats the context of spectrogram as an image to generate activation values frame by frame, as *CNN* [9,20], *Multi-task CNN* [19] and *Data-Aug CNN* [10]. The training objective is a multi-label of 0 or 1 tagging for each drum instrument at the frame level.

In recent years, many types of research are NN-based works. Most of these efforts achieve better transcription performance by changing the network structure. The majority of the new systems select the mix spectrogram as input, the annotation drum occurrences in the annotation file as training labels. It proves that spectrogram is a useful representation for ADT, and 0 or 1 label is beneficial for transcribing drum instruments. Using the transcriptional target as the training objective allows the neural network to focus on learning the transcription of drum instruments, but there is a risk of overfitting. Also, limited information

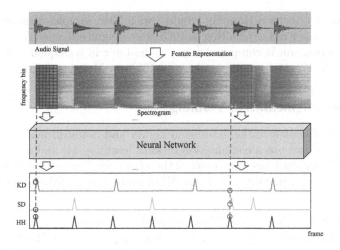

Fig. 1. System overview of the drum transcription architecture.

can be learned in such neural networks, which leads to the instability of transcriptional performance across scenarios. Based on the above reasons, a training method that incorporates more musical information is needed to be presented.

The main contributions of this paper are as follows: (i) A joint learning approach combining joint labels and joint classifier is proposed. Through the shared encoder, the neural network can learn more music information beyond the drum transcriptional target. (ii) We propose a self-distillation method to aggregate the joint label into the target model we needed. The performance can be improved greatly under the dual learning of target labels and aggregated labels. (iii) The label augmentation strategy is proposed with the use of joint learning and self-distillation, which is the first study in drum transcription. The proposed system works well across different scenarios and outperforms the state-of-the-art systems.

2   Drum Transcription Architecture

The overall drum transcription architecture is depicted as Fig. 1. The whole process is divided into three steps: Feature representation (FR), activation function (AF) and event segmentation (ES) [21]. First, the learnable representation named spectrogram is extracted from an audio signal, then the context of the spectrogram is input to neural network frame by frame to generate the possibility of three drum instruments. Finally, a local maximum peak picking algorithm is applied to extract drum events.

2.1   Feature Representation

First, the input audio with a 44.1kHz sampling rate is converted into a magnitude spectrogram using short-time Fourier transform (STFT). The STFT is calculated with *librosa* [13] using a Hanning window with 2048 samples window size and 512 samples hop length, resulting a 1025 × n magnitude spectrogram. Then, the Mel-spectrogram is computed using a Mel-filter in a frequency range of 20 20000 Hz with 128 Mel bands, resulting in 128 × n Mel-spectrogram.

2.2 Label Augmentation Training

In this work, to enhance the expression ability and detail processing ability of neural network, we use a combination of joint learning [11] and self-distillation [8] for label augmentation. Like most of the works, we consider the transcription of three drum instruments: kick drum (KD), snare drum (SD) and hi-hat (HH).

Joint Learning. The goal of joint learning is to construct a learning objective with more information and details. This idea was proposed by Hankook Lee [11] to improve the performance of image classification. In this paper, we construct a joint classifier, the working mechanism is shown in Fig. 2. The traditional training method is above the dotted line and it uses only drum instruments that need to be transcribed as the training target of the neural network. The proposed joint learning method combines the loudness information of sound and uses the aggregated labels as training objectives. As shown in the Fig. 2, the loudness labels contain four categories: origin audio (OG), the audio eliminating the part sound of KD and SD (EKS), the audio eliminating the part sound of KD and HH (EKH), the audio eliminating the part sound of SD and HH (ESH). Combining the four labels with the three transcriptional labels forms twelve more detailed labels, called joint labels. The main objective of the joint classifier is to be able to distinguish these joint labels.

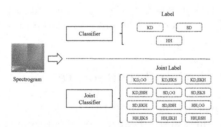

Fig. 2. The schematic diagram of joint learning and traditional learning. The traditional way of learning is above while the joint learning is under the dotted line.

Self-distillation. The system we need ultimately is the drum transcription model, so a way is needed to transfer the knowledge of the joint model to the objective model. Here we propose a self-distillation method, which is different from traditional knowledge distillation. The self-distillation model uses a shared encoder to connect the two classifiers, as shown in Fig. 3. One is a classifier that has learned complex information, the other one is our target classifier. By distilling into the target classifier we expect to get a model that is easy to transcribe drum records, and to some extent, the performance of the target model is improved. Through the shared encoder, we hope that the learning process of the complex model can influence the representation extraction of the encoder, thus beneficial to the target model.

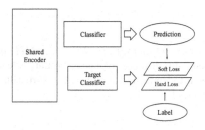

Fig. 3. The schematic diagram of self-distillation.

Label Augmentation. Figure 4 depicts the proposed label augmentation system. Let $x \in \{x_1,...,x_M\}$ be the input, $y \in \{1,...,N\}$ be its labels where $N = 3$ is number of drum instruments and $M = 4$ indicates the loudness labels mentioned in Sect. 2.2, $\tilde{y} \in \{1,...,N^*M\}$ be the joint labels of drum instruments with loudness, $\mathcal{L}_{CE}$ be the cross-entropy loss function, $z = f(x;w)$ be an embedding feature of x where f is an encoder neural network with the parameter w, $h(;v)$ be the joint classifier with the parameter v, and $g(;u)$ be the target classifier with the parameter u where $g(z;u) = 1/(1 + \exp(-(u^\top z)))$. Following section we use $i \in \{1,...,N\}$ as different drum instruments and $j \in \{1,...,M\}$ as different loudness.

The joint classifier can be described as multi-label classification with N^*M labels, where $h_{ij}(z;v) = 1/(1 + \exp(-(v_{ij}^\top z)))$. The training objectives of joint classifier can be written as:

$$\mathcal{L}_{LA}(x, \tilde{y}; w, v) = \frac{1}{M} \sum_{j=1}^{M} \mathcal{L}_{CE}(v^\top z_j, \tilde{y}) \tag{1}$$

where $\mathcal{L}_{CE}$ is the sigmoid cross-entropy loss function, then $\mathcal{L}_{LA}$ means the label augmentation loss function. Here we expect the neural network to be able to recognize not only the drum instruments but also the tune information.

To generate the soft label that matches the target classifier, we need to aggregate the joint label calculated by the joint classifier. Therefore, we make the aggregated label P as follow:

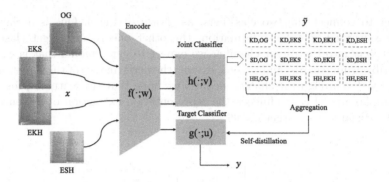

Fig. 4. An overview of label augmentation. We input Mel-spectrogram in different loudness of the same music simultaneously, then the network learns from joint labels with drum and loudness information. Finally, the self-distillation makes the target classifier learn from joint labels.

$$P(x; w, v) = 1/(1 + \exp(-(\frac{1}{M} \sum_{j=1}^{M} v^{\top} z_j))) \tag{2}$$

To compute the aggregated label, we first average the pre-sigmoid value, then compute the sigmoid activation function. The values after activation are served as the aggregated label generated by the joint classifier. At this point, we have completed the joint classification process.

Although the above steps can complete the drum transcription, it requires collecting M loudness data for the same music in the inference stage, which means M times data and computation. To accelerate the inference and make it easier, it needs a target classifier to finish that, as the $g(; u)$ shown in Fig. 4. The training objective of the whole model is as follow:

$$\mathcal{L}(x, y, \tilde{y}; w, u, v) = \mathcal{L}_{LA}(x, \tilde{y}; w, v)$$
$$+ \sum_{i=1}^{N} \lambda_i (\mathcal{L}_{CE}(g(z; u_i), y_i) \tag{3}$$
$$+ \mathcal{L}_{CE}(g(z; u_i), P(x; w, v_i)))$$

where λ is a hyperparameter that indicates different weights are applied to different drum instruments. Due to the SD contains the largest within-class variability, following HH then KD. The weights of KD, SD, HH are set to 0.5, 2.0, 1.0. When computing $\mathcal{L}(x, y, \tilde{y}; w, u, v)$, we consider $P(x; w, v_i)$ as a constant, where it is treated as the soft label and y is treated as the hard label.

2.3   Network Architecture

The network we use is a deep convolutional neural network as Fig. 5. Convolutional neural network has been successfully applied in many scenarios. In this case, a convolutional neural network is used in the Mel-spectrogram just as it works on images. For each frame, we consider the spectral context with 18 frames (about 209 ms) as the input feature. The encoder contains three groups of two layers 2×2 convolution and 2×2 max-pooling first, following is two layers 3×3 convolution and 3×3 max-pooling, and each convolution with nonlinear activation function Relu. The embedding features are flattened to vector. Then the fully connected layers are both used in joint classifier and target classifier, with Relu activation function in hidden layers and SigmoidSigmoid activation function in output layers. For the label y, we consider it as 1 whenever an annotation is present and 0 otherwise. To reduce the ambiguity, the vicinities of annotation are set to 0.5. For training, an Adam optimizer with 0.001 learning rate is used.

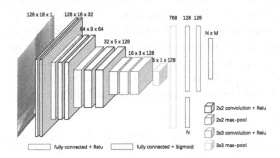

Fig. 5. Network architecture.

2.4   Event Segmentation

After generating the activation function frame by frame through the neural network, the last step is to segment the drum onsets through a peak picking method introduced in [1]. A point n in the activation function $A(n)$ is considered as a peak if it satisfies the three conditions:

$$
1. A(n) = \max(A(n-m), ..., A(n+m))
$$
$$
2. A(n) >= \text{mean}(A(n-l), ..., A(n+l)) + \delta \tag{4}
$$
$$
3. n - n_{lp} > w
$$

where δ is a variable threshold. A peak must be the maximum value within max window from $n - m$ to $n + m$, and exceed the mean value plus δ within the window from $n - l$ to $n + l$. Additionally, in order to prevent adjacent points from being redundant detection, a peak must have a distance w to last detected peak n_{lp}. Values for the parameters are set to $m = l = w = 3$.

3   Experimental Setup

To evaluate the performance of the proposed system, different models are individually trained on single data set and test on other data sets, known as Eval

Cross [21] strategy. As common, we consider the Precision, Recall and F-measure of three drum instruments kick drum (KD), snare drum (SD) and hi-hat (HH).

3.1 Tasks and Data Sets

The task of drum transcription and the description of the corresponding data set is shown in Table 1. The drum transcription can be roughly divided into three main tasks: DTD, DTP and DTM [21]. DTD represents detecting the three objective drum instruments from KD/SD/HH only drum tracks, which is suitable for IDMT-SMT-Drums (SMT) [6]. DTP indicates the transcription of KD/SD/HH from drum tracks with more instruments such as tom-toms, the data set is corresponding to ENST-Drums minus-one [7]. DTM is the most difficult task, it represents the transcription from drum tracks with accompaniment, ENST-Drums accompanied [7] fits this task. The individual training and testing of the dataset can constitute a total of six scenarios. But due to the lack of corresponding training data in SMT, we conduct experiments of four scenarios. Here we use abbreviations for each scenario, for example, DTD w DTP indicates that evaluation on DTD with system trained on DTP.

Table 1. Tasks and data sets of our experiment. The last column shows a round description of the data set. OT means other instruments and ACC means accompaniment.

Task	Data set	Audio description
DTD	IDMT-SMT-Drums [6]	KD + SD + HH
DTP	ENST-Drums minus-one [7]	KD + SD + HH + OT
DTM	ENST-Drums accompanied [7]	KD + SD + HH + OT + ACC

3.2 Implementation Detail

For training and evaluation, in each experiment, the data set is split into 70% training set and 15% validation set, the entire other data set serves as the test set. We mainly use F-measure along with Precision and Recall using mir_eval [14] with the tolerance window 50 ms. The training is aborted as soon as the resulting loss on the validation set has not decreased for 7 epochs.

3.3 Result and Discussion

Table 2 shows the result of the proposed system with the comparison of neural network based systems in recent years. It shows that the proposed approach achieves the best overall performance of 0.836, which outperforms the state-of-the-art systems.

Table 2. Precision, Recall and F-measure results for experiments in different scenarios. The second line from the bottom is ADT-CNN without using label augmentation, which means using the encoder and target classifier only and the last line is our approach. The best F-measure is in red and the second is in blue.

	DTD w DTP			DTD w DTM			DTP w DTM			DTM w DTP			Average		
	P	R	F	P	R	F	P	R	F	P	R	F	P	R	F
RNN [15,17] (baseline)	0.754	0.947	0.809	0.761	0.909	0.801	0.799	0.901	0.831	0.606	0.799	0.660	0.730	0.889	0.775
tanhB [15]	0.815	0.941	0.847	0.736	0.955	0.799	0.790	0.870	0.804	0.755	0.743	0.727	0.774	0.877	0.794(0.019↑)
ReLUts [17]	0.758	0.944	0.805	0.810	0.978	0.862	0.784	0.895	0.810	0.658	0.814	0.697	0.753	0.908	0.794(0.019↑)
lstmpB [16]	0.815	0.918	0.831	0.838	0.878	0.838	0.845	0.839	0.813	0.783	0.740	0.739	0.820	0.844	0.806(0.031↑)
GRUts [18]	0.821	0.941	0.851	0.798	0.943	0.830	0.810	0.932	0.850	0.634	0.796	0.677	0.766	0.903	0.802(0.027↑)
DrummerNet [4]	0.881	0.897	0.869	-	-	-	-	-	-	-	-	-	-	-	-
ADT-CNN	0.843	0.869	0.829	0.872	0.850	0.836	0.902	0.838	0.844	0.728	0.785	0.735	0.836	0.836	0.811(0.036↑)
ADT-CNN-LA	0.873	0.896	0.863	0.867	0.892	0.858	0.872	0.882	0.860	0.802	0.760	0.764	0.854	0.858	0.836(0.061↑)

Firstly, compared with the baseline, other newly proposed systems achieve better average F-measure, but limited. The system with the best overall performance is lstmpB, which improves 0.031, while other systems obtain single-digit increases also. As for DrummerNet, it achieves the best performance in the first scenario, but its usage scenario is quite limited and its training data set is in-house. This proves that the overall performance improvement for ADT can be very challenging. Compared with other systems, we can know that the proposed label augmentation method has achieved a significant performance improvement of 0.061. Compared with the 0.036 promotion of ADT-CNN, it is also a competitive performance boost. For each scenario, the proposed approach achieves the highest or second-highest F-measure, which provides a more stable transcriptional capacity than other systems across scenarios. This robust transcriptional ability is significant for practical applications. For many systems, most of them have a high recall, with low precision. It means that many non-events are also detected, resulting in a high false positive. It can be known that the models tend to recognize the input as an onset in ambiguous cases, which limits the generalization ability. Compare with them, our approach performs relatively stable in precision and recall, resulting in higher F-measure, which demonstrates the stability within a scenario.

It can be shown from the experimental results that testing on DTM will yield lower performance, this may be due to the interference of the accompaniment in such dataset. The proposed system yields better improvement while testing on DTP and DTM. One reason is that the DTD dataset is simpler. The performance on this dataset has reached relative saturation. On the other hand, it also shows that the ADT system still has a large room for improvement on the complex dataset. The enhancement effect of the proposed system is in line with the demand for real scene transcription.

Figure 6 depicts the detailed performance on different drum instruments of each system. The average transcriptional performance is determined by the transcriptional performance of the three drum instruments. From the figure, we can know that detecting KD is the easiest, following HH then SD. It is reasonable due to the large within-class variability of SD. Compared with other systems, the proposed method mainly improves the performance of KD and SD. The box size of KD is reduced, which implies a reduction of the with-class diversity in transcribing KD. The tendency for SD is the box moving up, which means that the performance of transcription is improving. In the experiment, we found that training in fast music tends to generate low SD precision while testing with slow music, and the recall is low in the opposite case. This may be due to rapid changes in frequency over time. To sum up, the proposed system has a robust transcription performance in the above three drum instruments. Meanwhile, it achieves the most stable and reliable transcriptional capacity.

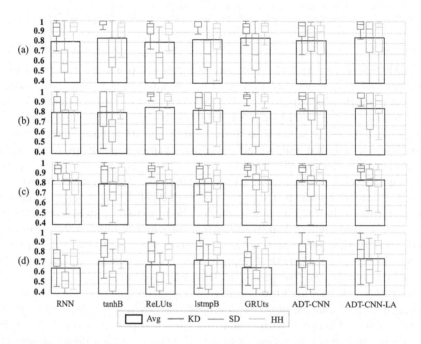

Fig. 6. Experimental detail for each drum instruments.(a) Evaluation on DTD with system trained on DTP. (b) Evaluation on DTD with system trained on DTM. (c) Evaluation on DTP with system trained on DTM. (d) Evaluation on DTM with system trained on DTP. The black bar means the average F-measure while the different color boxes represent the F-measure range across different drum instruments.

4 Conclusion

In this work, an automatic drum transcription approach with label augmentation via joint learning and self-distillation is introduced and compared with other neural network based ADT systems. The evaluation result shows that the new method can outperform the state-of-the-art system. Furthermore, a comparison with the system without label augmentation is carried out and the result shows the benefit of our approach. We can conclude that label augmentation for the neural network is beneficial for performance boost in ADT and provides better stability across scenarios.

Due to the extra M-1 inputs required, more data are needed in the training stage. In the future, a self-supervised approach with label augmentation can be studied to reduce the need for training data and more joint labels can be employed to improve performance. Detecting the events directly on the waveform instead of the spectrogram is also challenging and expectant work that can be conducted in the future. More advanced self-distillation methods are also an interesting direction that can be developed in the next step.

Acknowledgments. This work is partly supported by the Science and Technology Planning Project of Guangdong Province, China (Grant no.2018B010108002)

References

1. Bck, S., Widmer, G.: Maximum filter vibrato suppression for onset detection. In: 16th International Conference on Digital Audio Effects (DAFx) (2013)
2. Benetos, E., Dixon, S., Duan, Z., Ewert, S.: Automatic music transcription: an overview. IEEE Signal Process. Mag. **36**(1), 20–30 (2019)
3. Benetos, E., Dixon, S., Giannoulis, D., Kirchhoff, H., Klapuri, A.: Automatic music transcription: challenges and future directions. J. Intell. Inf. Syst. **41**(3), 407–434 (2013)
4. Choi, K., Cho, K.: Deep unsupervised drum transcription. In: Proceedings of the 20th International Society for Music Information Retrieval Conference, ISMIR, pp. 183–191 (2019)
5. Dittmar, C., Gärtner, D.: Real-time transcription and separation of drum recordings based on NMF decomposition. In: Proceedings of the International Conference on Digital Audio Effects (DAFx), pp. 187–194 (2014)
6. Dittmar, C., Gärtner, D.: Real-time transcription and separation of drum recordings based on NMF decomposition. In: Proceedings of the 17th International Conference on Digital Audio Effects, pp. 187–194 (2014)
7. Gillet, O., Richard, G.: Enst-drums: an extensive audio-visual database for drum signals processing. In: ISMIR 2006, 7th International Conference on Music Information Retrieval (2006)
8. Hinton, G.E., Vinyals, O., Dean, J.: Distilling the knowledge in a neural network. CoRR (2015)
9. Jacques, C., Roebel, A.: Automatic drum transcription with convolutional neural networks, pp. 80–86 (2018)
10. Jacques, C., Roebel, A.: Data augmentation for drum transcription with convolutional neural networks. In: 2019 27th European Signal Processing Conference (EUSIPCO) (2019)
11. Lee, H., Hwang, S.J., Shin, J.: Self-supervised label augmentation via input transformations. In: Proceedings of the 37th International Conference on Machine Learning, ICML 2020, vol. 119, pp. 5714–5724. PMLR (2020)

12. Lindsay-Smith, H., McDonald, S., Sandler, M.: Drumkit transcription via convolutive NMF. In: Proceedings of the International Conference on Digital Audio Effects (DAFx) (2012)
13. McFee, B., et al.: Librosa/librosa: 0.8.0 (2020)
14. Raffel, C., Mcfee, B., Humphrey, E.J., Salamon, J., Ellis, D.P.W.: mir_eval: a transparent implementation of common MIR metrics. In: Proceedings - 15th International Society for Music Information Retrieval Conference (ISMIR 2014) (2014)
15. Southall, C., Stables, R., Hockman, J.: Automatic drum transcription using bidirectional recurrent neural networks. In: Proceedings of the International Society for Music Information Retrieval Conference (ISMIR), pp. 591–597 (2016)
16. Southall, C., Stables, R., Hockman, J.: Automatic drum transcription for polyphonic recordings using soft attention mechanisms and convolutional neural networks, pp. 606–612 (2017)
17. Vogl, R., Dorfer, M., Knees, P.: Recurrent neural networks for drum transcription. In: Proceedings of the International Society for Music Information Retrieval Conference (ISMIR), pp. 730–736 (2016)
18. Vogl, R., Dorfer, M., Knees, P.: Drum transcription from polyphonic music with recurrent neural networks. In: Proceedings of the IEEE International Conference on Acoustics, Speech, and Signal Processing (ICASSP), pp. 201–205 (2017)
19. Vogl, R., Dorfer, M., Widmer, G., Knees, P.: Drum transcription via joint beat and drum modeling using convolutional recurrent neural networks. In: 18th International Society for Music Information Retrieval Conference (2017)
20. Vogl, R., Widmer, G., Knees, P.: Towards multi-instrument drum transcription. CoRR (2018)
21. Wu, C., et al.: A review of automatic drum transcription. IEEE/ACM Trans. Audio Speech Lang. Process. **26**(9), 1457–1483 (2018)
22. Wu, C.W., Lerch, A.: Drum transcription using partially fixed non-negative matrix factorization. In: The 23rd European Signal Processing Conference (EUSIPCO 2015) (2015)

Adaptive Curriculum Learning for Semi-supervised Segmentation of 3D CT-Scans

Obed Tettey Nartey[1,4]($\boxtimes$) (iD), Guowu Yang[1],
Dorothy Araba Yakoba Agyapong[3], JinZhao Wu[2], Asare K. Sarpong[5],
and Lady Nadia Frempong[3]

[1] Big Data Research Center, School of Computer Science and Engineering,
University of Electronic Science and Technology of China, Chengdu 611731, China
ashong.nartey@std.uestc.edu.cn, guowu@uestc.edu.cn

[2] The School of Computer Science and Electronic Information, Guangxi University,
Nanning 530004, China

[3] Biomedical Engineering Program, Kwame Nkrumah University of Science
and Technology, Kumasi, Ghana
gxmdwjzh@aliyun.com

[4] School of Information and Software Engineering, University of Electronic Science
and Technology of China, Chengdu, China

[5] School of Medicine, Department of Radiology and Imaging Sciences, Emory
University, 1364 Clitton Rd., Atlanta, GA 30322, USA

Abstract. Semi-supervised learning algorithms make use of both labelled training data and unlabelled data. However, the visual domain gap between these sets poses a challenge which prevents deep learning models from obtaining the results they have achieved most especially in the field of medical imaging. Recently, self-training with deep learning has become a powerful approach to leverage labelled training and unlabelled data. However, a challenge of generating noisy pseudo-labels and placing over-confident labelling belief on incorrect classes leads to deviation from the solution. To solve this challenge, the study investigates a curriculum-styled approach for deep semi-supervised segmentation which relaxes and treats pseudo-labels as continuous hidden variables by developing an adaptive pseudo-label generation strategy to jointly optimized the pseudo-label generation and selection process. A regularization scheme is further proposed to smoothen the probability outputs and sharpen the less represented pseudo-label regions. The proposed method was evaluated on three publicly available Computer Tomography (CT) scan benchmarks and extensive experiments on all modules have demonstrated the efficacy of the proposed method.

Keywords: Medical image segmentation · Curriculum learning · Self-training · Model regularization

© Springer Nature Switzerland AG 2021
T. Mantoro et al. (Eds.): ICONIP 2021, LNCS 13108, pp. 77–90, 2021.
https://doi.org/10.1007/978-3-030-92185-9_7

1   Introduction

Deep learning models, especially convolutional neural networks (CNNs), have attained state-of-the-art performances in a wide range of visual recognition tasks including medical imaging. The generalization capabilities of these networks rely on large and well-annotated datasets, which, in the case of segmentation, involves pixel-level annotations. The process of obtaining expert annotations in medical imaging is a laborious, capital intensive, time consuming and requires the clinical expertise of radiologists. Techniques such as fine-tuning have been used to train models for small data situations utilizing pre-trained models on large dataset like the ImageNet [8] which has been crucial for medical imaging and analysis. However, the performance deteriorates when there is divergence between the well-annotated training data and the testing data caused by factors such as variance in illumination, image background and objects' viewpoint [7,18]. The problem is further compounded with semantic gaps due to non-universal standard for training data annotation [7]. This work focuses on semi-supervised learning, where there is a small amount of fully-annotated data and an abundant amount of unlabelled data. Recently, many works have been implemented in this field [1,2,12,20–22] boosted by deep learning. Self-training is a wrapper method which involves iteratively generating a set of pseudo-labels corresponding to highly prediction scores on unlabelled test data, and then re-training the network models based on the selected pseudo-labels with their corresponding unlabelled data samples [10,14,27]. This learning strategy augments the labelled training set with the pseudo-labelled data together with their generated pseudo-labels. Although it has an advantage of leveraging unlabelled data, One major drawback of this learning method is reinforcing wrongly generated pseudo-labels through the network during retraining. Many works have been done [10,14,27] to overcome this issue, using methods such as co-training [12], fuzzy means [1,22]

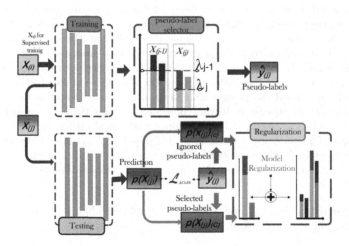

Fig. 1. Framework of the proposed model regularization self-training technique.

and adversarial methods to leverage unlabelled data and mitigate reinforcing mistakenly labelled sample. These methods have addressed the problem as an entropy minimization that sought to optimize the generation and selection of pseudo-labels to attenuate the misleading effect raised by incorrect or ambiguous supervision.

Generating pseudo-labels is characterise by a redundancy where the pixels with highly confident scores kept and pixels with low confident scores are ignored. Yang et al. [25] had proposed a curriculum learning strategy for domain adaptation for Semantic Segmentation of Urban Scenes to match the global label distributions of labelled training source data and the unlabelled target data by minimizing the KL-divergence existing between the two distributions. Their method learns from easy-to-hard informative features to adapt the labelled-source data to the unlabelled target data.

Yang et al.'s work inspires this research, an adaptive curriculum strategy for deep semi-supervised segmentation for CT-scans via self-training. It employs an alternative way of achieving the same goal of label regularization to mitigate the misleading effect of reinforcing mistakenly generated pseudo-labels and predict pixel-level information in the unlabelled test data. At the apex, the proposed method is an entropy minimization scheme that jointly learns a model and generates pseudo-labels by encouraging the smoothness of output probabilities and avoid over-confident predictions in the network training. The main contributions of the proposed work are as follows;

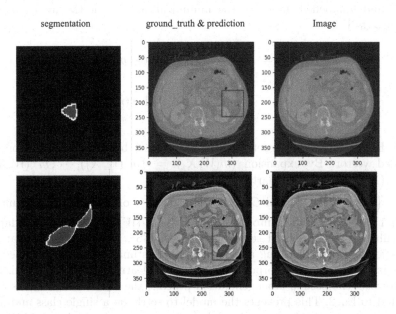

Fig. 2. Visual Representation of the selected regions of confidence with a comparison to the ground truth location in the image (Color figure online)

– A new self-training framework: adaptive curriculum semi-supervised segmentation(ACuSS) is proposed to jointly learn a model and invariant feature representation by optimizing the pseudo-label generation process and solving the entropy minimization problem from easy-to-hard tasks. The framework is designed to perform network learning and pseudo-label estimation under a unified loss minimization tasks in a joint manner.
– An adaptive curriculum learning pseudo-label generator and selector to incorporate more informative features for network training is developed.The pseudo-labels are treated as hidden variables that are learned via maximum likelihood estimation (MLE) to improve the quality of pseudo-labels.
– A cross entropy loss regularized by an output smoothness term is proposed to regularize the model. Specifically, an L_2 model regularizer is introduced to smoothen the network softmax output probabilities to training of the network.
– Comprehensive evaluations on CT-Scan segmentation tasks demonstrate that, the proposed framework's high and competitive results significantly outperforms other state-of-the-art techniques on three publicly available CT-scan benchmarks.

2   Adaptive Curriculum Semi-supervised Segmentation (ACuSS)

Given a semi-supervised learning task which utilizes both labelled training data X_i, Y_i and unlabelled data X_j, the minimization loss for the model m with parameters W is

$$
\min_W L_c(W) = -\frac{1}{X_i} \sum_{i \in S} \sum_{p \in \Omega} Y_{i,p} \bullet \log \left(m(W, X_i) \right)_p
$$
$$
-\frac{1}{X_j} \sum_{j \in U} \sum_{p \in \Omega} \hat{Y}_{j,p} \bullet \log \left(m(W, X_j) \right)_p
\tag{1}
$$

From Eq. 1, $\hat{Y}_j$ is the "pseudo-label" generated by the model as one-hot encoded vector. By expansion $(m(W, X_j) = \left(m^1(W, X_j), \cdots, m^C(W, X_j) \right)$, $Y_{i,p} = \left(Y_{i,p}^1, \cdots, Y_{i,p}^C \right)$, similarly $\hat{Y}_{j,p} = \left(\hat{Y}_{j,p}^1, \cdots, \hat{Y}_{j,p}^C \right)$, and C is the total number of classes. The formulation in Eq. 1 can be simplified into the form: $\min_w L_Y(W) + L_{\hat{Y}}(W)$. The adaptive curriculum semi-supervised segmentation (ACuSS), is a framework that performs joint network learning and pseudo-label estimation under a unified loss minimization problem. Incorporating curriculum learning into the minimization loss to relax the pseudo-label generation process, the general loss Eq. 1 is reformulated to Eq. 2. This prevents the model to settle on a single class and avoid propagating overconfident mistaken labels during network retraining, which hurt performance and behaviour.

$$\min_{W,\hat{Y}_j} L(W, \hat{Y}) = -\frac{1}{X_i} \sum_{i \in S} \sum_{p \in \Omega} Y_{i,p} \bullet \log\left(m(W, X_i)\right)_p$$

$$-\frac{1}{X_j} \sum_{j \in U} \sum_{p \in \Omega} \hat{Y}_{j,p} \bullet \log\left(\frac{(m(W, X_j))}{\lambda_c}\right)_p \qquad (2)$$

$$s.t. \quad \hat{Y}_{j,p} \in \Delta^{c-1}, \forall\{j\}.$$

From Eq. 2, λ_c controls the selection of pseudo-labelled samples, so that if a prediction is confident with $(m(W, X_j)) > \lambda_{c^*}$, then the sample is selected and labelled as class $C^* = argmax_c \left\{\frac{(m(W, X_j))}{\lambda_c}\right\}$ whereas less confident labels with $(m(W, X_j)) \leq \lambda_{c^*}$ are ignored. λ_c is a crucial parameter that is used to control the proportion of pseudo-labels to be learned and selected. It is a class-balancing strategy that is introduced in [11]. Particularly, for each class c, λ_c determines the single portion of the parameter indicating how many samples are to be selected from the unlabelled data. The confidence for a sample to be selected is set as the maximum value of its output probabilities from the softmax loss function. This is used to determine λ_c by selecting the most confident amount of class c predictions in the entire test data. The entire process to determine λ_c is given in Algorithm 1. It should be noted that, in this work, the only parameter to determine all of λ_c is the parameter p as given in [11]. For the feasible set, a probability simplex $\delta^{c-1} \in \Delta^{c-1}$, a one-hot encoded vector is used in the proposed self-training method to allow the generation of soft pseudo-labels to set the basis for the model regularized self-training.

Algorithm 1: Algorithm for determining λ_c

---

input : DNN $f(w)$, unlabeled Samples X_j, selected pseudo-labels p
output: λ_c
for $j \leftarrow 1$ **to** J **do**
    $P_{X_j} = f(w, X_j)$;
    $LP_{X_j} = argmax(P, axis = 0)$;
    $MP_{X_j} = \max(P, axis = 0)$;
    $M = \left[M, from - matrix - to - vector(MP_{X_j})\right]$ **for** $c \leftarrow 1$ **to** C **do**
        $MP_c, X_j = MP_{X_j}(LP_{X_j} == c)$;
        $M_c = [M_c, Matrix - to - vector(MP_c, X_j)]$
    end
end
for $c \leftarrow 1$ **to** C **do**
    $M_c = sort(M_c, order = descending)$;
    $len_{c,th} = length(M_c) \times p$;
    $\lambda_c = -\log(M_c[len_{c,th}])$
end
$return(\lambda_c)$

---

a. **Pseudo-label generation**: Initialize W and minimize the loss with respect to $\hat{Y}_j$. This involves fixing W and solving Eq. 3

$$\min_{\hat{Y}_j} -\frac{1}{X_j} \sum_{j \in U} \sum_{p \in \Omega} \hat{Y}_{j,p} \bullet \log\left(\frac{m(W, X_j))}{\lambda_c}\right)_p \tag{3}$$

$$s.t. \quad \hat{Y}_{j,p} \in \Delta^{c-1} \bigcup\{0\}, \forall\{j\}.$$

With Eq. 3 being optimized, the pseudo-labels are normalized class-wise and the ones with highest confidence are selected to retrain the network in step b).

b. **Updating the network**: In this step, $\hat{Y}_j$ is fixed and the objective function is optimized. Gradient descent is used to solve Eq. 4.

$$\min_W -\frac{1}{X_i} \sum_{i \in S} \sum_{p \in \Omega} Y_{i,p}^c \bullet \log m(W, X_i))_p$$
$$-\frac{1}{X_j} \sum_{j \in U} \sum_{p \in \Omega} \hat{Y}_{j,p}^c \bullet \log m(W, X_j))_p \tag{4}$$

Executing steps **a)** and **b)** once is considered to be a single iteration of the **self-training process**. Solving step **a)** requires an optimizer as defined in Eq. 5 for a random $\hat{\mathbf{Y}}_\mathbf{j} = \left(\hat{\mathbf{y}}_\mathbf{j}^{(1)}, \ldots, \hat{\mathbf{y}}_\mathbf{j}^{(C)}\right)$

$$\hat{Y}_j^{(c*)} = \begin{cases} 1, & \text{if } c = \underset{c}{argmax}\left\{\frac{m(W,X_j)}{\exp(\lambda_c)}\right\}, \\ & \text{and } m\left(W; X_\mathbf{j}\right) > \lambda_c \\ 0, & \text{otherwise.} \end{cases} \tag{5}$$

The general form of the regularizer can be defined as $MR_c\left(\mathbf{W}, \hat{\mathbf{Y}}_\mathbf{j}\right) = \phi mr_c\left(\mathbf{W}, \hat{\mathbf{Y}}_\mathbf{j}\right)$. Adding the model regularizer to the network retraining of step b) of the self-training procedure, the optimization problem in step **b)** is redefined as follows:

$$\min_{W,\hat{Y}} L_Y(W) + L_{\hat{Y}}(W, \hat{Y}) + MR_c\left(\mathbf{W}, \hat{\mathbf{Y}}_\mathbf{j}\right)$$
$$= L_Y(W) + L_{\hat{Y}}(W, \hat{Y}) + (\phi_r mr_c\left(m\left(\mathbf{W}, \mathbf{X_j}\right)\right)_r \tag{6}$$
$$+ \phi_s mr_c\left(m\left(\mathbf{W}, \mathbf{X_j}\right)\right)_s)$$

where $\phi \geq 0$ is a regularization weight coefficient. The regularizer based on the L_2 norm is added as a negation to prevent the model from ignoring a large amount of the generated pseudo-labels in step a).

$\left(\phi_r mr_c\left(m\left(\mathbf{W}, \mathbf{X_j}\right)\right)_r + \phi_s mr_c\left(m\left(\mathbf{W}, \mathbf{X_j}\right)\right)_s\right)$ are regularization of the rejected (r) and selected (s) predictions stemming from the confidence parameter λ. Because the method is a curriculum strategy, the ACuSS uses two thresholds $\{\lambda_j, \lambda_{j+1}\}$ defined in Eq. 8 and Eq. 9 respectively, which are adaptive and

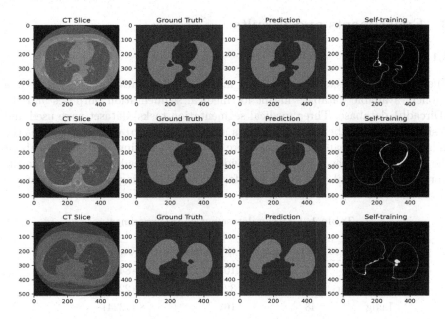

Fig. 3. Output of Sample axial slices of Lungs data; View of segmentation from the lungs data with the predicted mask compared to the ground truth

combine the local and global informative features of each instance. The adaptive threshold $\{\lambda_j\}$ stands for the threshold of the current instance X_j and $\{\lambda_{j+1}\}$ is used to progressively update $\{\lambda_j\}$ after each iteration.

$$\lambda_j^{(c)} = \beta\lambda_{(j-1)}^{(c)} + (1-\beta)\Psi\left(X_j, \lambda_{(j-1)}^{(c)}\right),$$
$$where, \Psi\left(X_j, \lambda_{(j-1)}^{(c)}\right) = \mathbb{P}_{X_j}^{(c)}\left[\phi_{j-1}^{(c)}{}^{\kappa}|\mathbb{P}_{X_j}^{(c)}|\right] \tag{7}$$

$$\lambda_j = -\frac{1}{X_j}\sum_{x_j \in X_j}\Pi_{x_j}\sum_{c=1}\frac{1}{C}\log\left(\frac{m(W, X_j)}{\lambda_c}\right)_p \tag{8}$$

The exponentially weighted average in Eq. 7 smoothens the adaptive threshold, incorporating past information and preventing the interference of noisy information. $\Psi\left(X_j, \lambda_{(j-1)}^{(c)}\right)$ represents the threshold for obtaining the instance X_j. The momentum factor β is used to maintain historical information threshold informative features, which smoothens, $\lambda_j^{(c)}$ the more it increases.

$$\lambda_{j+1} = -\frac{1}{X_j}\sum_{x_j \in X_j}\Pi_{x_j}^C\sum_{c=1}m(W, X_j)\log\left(\frac{m(W, X_j)}{\lambda_c}\right)_p \tag{9}$$

3   Experiments and Results

3.1   Datasets

The training datasets are from the medical segmentation decathlon [17], a publicly available database meant for medical segmentation purposes and the 3Dircadb1 database. Three medical segmentation tasks' dataset were used and designated as the labelled training data and the 3Dircadb1 dataset is used as the unlabelled data because the labels were dropped at testing time. Details about the data setup for the experiment provided in Table 1 Readers are to refer to the work by Amber et al. [17] for more information concerning the training data.

Table 1. CT-scan datasets used for experiments

Organ	#Training set	#Validation set	#Unlabelled set
Spleen	31	10	5 (844)
Colon cancer	94	32	1 (225)
Lung	48	16	8 (1324)

3.2   Implementation Details

U-NET [13] was adopted as the baseline network and the parameters were randomly initialization with an end-to-end training approach. For the supervised training phase the ratio of fully annotated data used to train and validate the model is 75 : 25. To remove irrelevant details and information, the image intensity values of all CT scans were clipped to the range $[-400, 450]$ HU and the images resized to 256×256. Specifically the ACuSS training process is made up of three phases.

a. Initial supervised learning phase. This is the warm-up phase where the model is trained with the labelled training sets. This is to train the initial segmentation model as the first pseudo-label generator as depicted in Fig. 1 in the first row.
b. The second phase involves pseudo-label generation. The pre-trained model from phase 1 is used to predict and get the predicted labels of the unlabelled. The pseudo-label selector as shown in Fig. 1 is used to select the most confident pseudo-labels generated.
c. The self-training phase, where the network is updated according to Eq. 6 by fine-tuning the pre-trained model. The pre-trained model is trained using both the labelled training set and the selected pseudo-labels together with their corresponding CT-Scans from the unlabelled set. And for this work the number of self-training iterations are 20, each having 100 epochs.

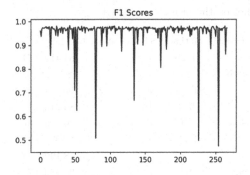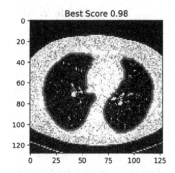

Fig. 4. ACuSS performance on a 3D Lung sample. The red contour is the ground truth and the green is the ACuSS segmentation contour. Best view in colours (Color figure online)

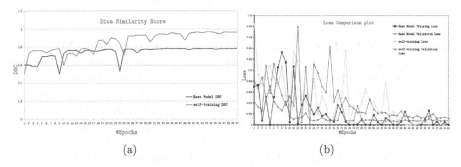

Fig. 5. (a) Comparison of DSC between base model and self-training model. (b) Comparison of loss curves between base model and self-training model. Best viewed in colours (Color figure online)

The Dice similarity loss function was utilized, with a batch size of 1. The Adam optimizer is adopted with a 3×10^{-3} learning rate. The pseudo-label generation parameters λ, β were set to 20% and 90% respectively, meaning that for $\lambda = 20\%$ and $\beta = 90\%$ as given in Fig. 2, the algorithm takes 20% of images from the unlabelled set and 90% of the pixels (red bounding box in Fig. 2) from the highly-confident predicted image regions. For the regularization weights λ_j and λ_{j+1}, are set to 5.0 and 10% respectively.

To evaluate the model, the Dice Score (DSC), is used to assess the different aspects of the performance of each task and region of interest. Other evaluation metrics such as Jaccard similarity coefficient (JSC), volumetric overlap error (VOE), relative volume difference (RVD), and average symmetric surface distance (ASD) are used to evaluate the performance of the segmentation models. Specifically the mean and standard deviation of the evaluation metrics are calculated and presented on the 3Dircadb1 dataset designated as the unlabelled data. For the DSC and JSC metrics, the greater the value, the better the model's performance whereas for VOE, ASD, and RVD metrics, the smaller the value the better the model's performance.

Table 2. Ablation studies on the proposed method

	λ	β	DSC(%)
Spleen	20	0	87±0.46
	20	50	89.97±0.17
	20	90	**95.8±1.93**
Colon cancer	20	0	89.2 ±1.96
	20	50	91.5±0.05
	20	90	**92.7±0.91**
Lungs	20	0	94.83±2.71
	20	50	**98.57±1.85**
	20	90	96.76±2.93

(a) Analysis of λ and β

Task	Method	ST	λ	$(\lambda + \beta)$	DSC%
Spleen	base	-	-	-	84.41±0.71
	ST	✓	✓	-	87.0±0.17
	ST+MR	✓	✓	✓	95.8 ±1.93
Colon	base	-	-	-	85.21±3.11
	ST	✓	✓	-	89.2±1.96
	ST+MR	✓	✓	✓	92.7 ±0.91
Lungs	base	-	-	-	88.13±7.11
	ST	✓	✓	-	94.83±2.71
	ST+MR	✓	✓	✓	98.57±1.85

(b) Impact of Schemes

3.3 Results and Discussion

Impact of Model Regularization: The model regularization (**MR**) term encourages smoothing the output, lower the over-confident mistakes on false positives and increase the probability scores of other classes. In Table 2a, the sensitivity analysis on the hyper parameters λ and β are provided. It can be observed that the more the amount of pixels (β) predicted in selected images λ with high prediction confidence, the greater the Dice Score for the tasks. Providing the model with diverse information to learn from, it is able to obtain high DSC values as provided in Fig. 4. From Fig. 2, the proposed region of confidence in the red bounding box contains the pixels for the spleen and other organs in the image presenting the model with diverse content to learn from. For instance, in Fig. 3, the predicted segmented masks by the self-training method (fourth column of Fig. 3) is similar to the ground truth. Figure 5b, illustrates comparison of the loss curves for the base model and the regularized self-training model. From the accuracy plot in Fig. 5, it could be seen how the proposed ACuSS model's curve kept ascending even when the base-model's curve flattened. The impact is further provided in Table 2b.

Segmentation Results Comparison with Other Works: The performance and ablation studies of the proposed ACuSS in Sect. 2 are reported in Table 2a, Table 2b and Table 3 respectively. By a one-by-one approach, all modules were tested and their performance studied on the test set (3Dicardb1 dataset). From Table 2a, λ and β control the amount of pseudo-labels to be selected with regards to images and pixels, saw the model obtain a higher performance of almost 6.0% as compared to using the self-training with only λ on the spleen dataset and there was also a performance gain of 1.2% for the colon segmentation task likewise the lungs dataset.

Additionally, Table 2b, provides the ablation analysis of the methods with their corresponding modules. Compared with other evaluation metrics in Table 3, the regularized self-training performs decently so far as ASD and RVD metrics

Table 3. Model performance over other evaluation metrics.

Tasks	DSC	JSC	VOE	ASD	RVD
Spleen	95.8 ± 1.93	92.11 ± 1.32	0.73 ± 2.62	5.77 ± 0.94	18.05 ± 3.82
Colon	92.7 ± 0.91	88.74 ± 0.08	1.26 ± 0.03	4.81 ± 3.07	11.23 ± 3.48
Lungs	98.57± 1.85	97.29 ± 0.70	0.10 ± 0.03	1.31 ± 1.81	1.93 ± 2.66

are concerned and this can be due to the existence of anatomical variations in the unlabelled test set.

The results of the method compared with state-of-the-art methods on the various CT-scans are presented in Table 4a, Table 4b and Table 4c. This study investigates a semi-automatic based deep learning algorithm to segment organs in CT scans utilizing an adaptive curriculum-styled strategy and regularization to prevent the model from making confident mistakes of incorrectly labelling

Table 4. Performance comparison with related works. (a) Comparison of ACuSS to other state-of-the-art lungs segmentation works. (b) Comparison of state-of-the-art spleen segmentation to the proposed work. (c) Comparison of ACuSS to other state-of-the-art colon segmentation works.

Method	DSC(%)	RVD(%)	VOE(%)	JSC(%)	ASD(mm)
3D-UNet [28]	92.6±0.016	—	20.91	—	0.94±0.193
MV-SIR [23]	92.6±0.04	0.01± 0.02	3.57±1.66	3.58± 6.58	0.072±0.033
Tziritas G [19]	86.70±0.047	−0.01	0.07	2.303	2.157±0.503
Shahzad R et al. [16]	88.5±0.028	—	10.53	89.46	1.553±0.376
Zeng G et al. [24]	90.5±0.028	—	10.53	89.46	1.050±0.360
ACuSS	98.57±1.85	1.93 ±2.66	0.10 ±0.03	97.29 ±0.70	1.31 ±1.81

(a)

Method	DSC	VOE	ASD	RVD	JSC
Hu et al. [6]	94.2	-	1.3	-	-
Larson et al. [9]	93	-	-	-	-
Roth et al. [15]	91	-	-	-	-
Zhou et al. [26]	92	-	-	-	-
Gibson et al. [4]	95	-	-	-	-
ACuSS	95.8 ±1.93	0.73 ±2.62	5.77 ±0.94	18.05 ±3.82	92.11 ±1.32

(b)

Method	DSC	VOE	ASD	RVD	JSC
2.5-D DL [5] w/o Prior	0.82±0.06	-	1.3	-	-
2.5-D DL [5] with Prior	0.88±0.02	-	-	-	-
CFFNN [3]	94	-	-	-	-
ACuSS	92.7 ±0.91	1.26 ±0.03	4.81 ±3.07	11.23 ±3.48	88.74 ±0.08

(c)

unlabelled data samples. Table 4b shows the mean DSC for the spleen and it could be seen that, the proposed work yields higher DSC. Although, the primary objective of the study is to minimise the visual gap between medical training data and unlabelled test samples, an addition of higher DSC has been obtained. To further demonstrate the generalizing capability and robustness of the proposed method, the results on the colon cancer and lungs are compared with other related studies and provided in Table 4c. In Fig. 4, the model achieved DCS of 98% emphasizing the robustness of the proposed method. For the colon segmentation, the model achieves less when compared with the existing state-of-the-art methods which could be due to the test set being so different from the training set although a higher performance was obtained in the lungs segmentation task.

4   Conclusion

To conclude, a novel adaptive curriculum learning for semi-supervised segmentation of CT-scans framework was proposed. It relaxes the pseudo-label generation and selection from a hard labelling scheme to a soft labelling scheme using simplex probability. Additionally a regularization term to smoothen the probability outputs was proposed to help self-training models generalize well on unseen test data and reduce the diverging visual gap between labelled training sets and unlabelled data. Compared to other learning methods, ACuSS is general framework with no special structural dependency which can easily be implemented to other semi-supervised semantic segmentation problems and still obtain a significant improvement in performance. Extensive experiments demonstrates that the proposed method outperforms several other methods. As part of future work, more target properties and regularization schemes for semi-supervised tasks shall be investigated.

Acknowledgements. This work was partially supported by the National Natural Science Foundation of China under Grant No. 61772006, Sub Project of Independent Scientific Research Project under Grant No. ZZKY-ZX-03-02-04, and the Special Fund for Bagui Scholars of Guangxi.

References

1. Anter, A.M., Hassanien, A.E., ElSoud, M.A.A., Tolba, M.F.: Neutrosophic sets and fuzzy c-means clustering for improving CT liver image segmentation. In: Körner, P., Abraham, A., Snášel, V. (eds.) Proceedings of the Fifth International Conference on Innovations in Bio-Inspired Computing and Applications IBICA 2014. AISC, vol. 303, pp. 193–203. Springer, Cham (2014). https://doi.org/10.1007/978-3-319-08156-4_20
2. Chung, M., Lee, J., Lee, J., Shin, Y.G.: Liver segmentation in abdominal CT images via auto-context neural network and self-supervised contour attention. ArXiv arXiv:2002.05895 (February 2020)

3. Devi, K.G., Radhakrishnan, R.: Automatic segmentation of colon in 3d CT images and removal of opacified fluid using cascade feed forward neural network. Comput. Math. Methods Med. **2015**, 670739–670739 (2015)
4. Gibson, E., et al.: Automatic multi-organ segmentation on abdominal CT with dense v-networks. IEEE Trans. Med. Imaging **37**(8), 1822–1834 (2018). https://doi.org/10.1109/TMI.2018.2806309
5. Gonzalez, Y., et al.: Semi-automatic sigmoid colon segmentation in CT for radiation therapy treatment planning via an iterative 2.5-d deep learning approach. Med. Image Anal. **68**, 101896 (2021)
6. Hu, P., Wu, F., Peng, J., Bao, Y., Chen, F., Kong, D.: Automatic abdominal multi-organ segmentation using deep convolutional neural network and time-implicit level sets. Int. J. Comput. Assist. Radiol. Surg. **12**(3), 399–411 (2016). https://doi.org/10.1007/s11548-016-1501-5
7. Kalluri, T., Varma, G., Chandraker, M., Jawahar, C.: Universal semi-supervised semantic segmentation. In: Proceedings of the IEEE/CVF International Conference on Computer Vision (ICCV) (October 2019)
8. Krizhevsky, A., Sutskever, I., Hinton, G.E.: Imagenet classification with deep convolutional neural networks. In: Advances in Neural Information Processing Systems, pp. 1097–1105 (2012)
9. Larsson, M., Zhang, Y., Kahl, F.: Robust abdominal organ segmentation using regional convolutional neural networks. In: Sharma, P., Bianchi, F.M. (eds.) SCIA 2017. LNCS, vol. 10270, pp. 41–52. Springer, Cham (2017). https://doi.org/10.1007/978-3-319-59129-2_4
10. Nartey, O., Yang, G., Wu, J., Asare, S.: Semi-supervised learning for fine-grained classification with self-training. IEEE Access 1 (December 2019).https://doi.org/10.1109/ACCESS.2019.2962258
11. Nartey, O., Yang, G., Wu, J., Asare, S., Frempong, L.N.: Robust semi-supervised traffic sign recognition via self-training and weakly-supervised learning. Sensors **20**(9), 2684 (2020). https://doi.org/10.3390/s20092684
12. Peng, J., Estrada, G., Pedersoli, M., Desrosiers, C.: Deep co-training for semi-supervised image segmentation. Pattern Recogn. 107–269 (2020).https://doi.org/10.1016/j.patcog.2020.107269
13. Ronneberger, O., Fischer, P., Brox, T.: U-Net: convolutional networks for biomedical image segmentation. In: Navab, N., Hornegger, J., Wells, W.M., Frangi, A.F. (eds.) MICCAI 2015. LNCS, vol. 9351, pp. 234–241. Springer, Cham (2015). https://doi.org/10.1007/978-3-319-24574-4_28
14. Rosenberg, C., Hebert, M., Schneiderman, H.: Semi-supervised self-training of object detection models. In: 2005 Seventh IEEE Workshops on Applications of Computer Vision (WACV/MOTION 2005), vol. 1, pp. 29–36 (2005)
15. Roth, H., et al.: Hierarchical 3d fully convolutional networks for multi-organ segmentation. CoRR (April 2017)
16. Shahzad, R., Gao, S., Tao, Q., Dzyubachyk, O., van der Geest, R.: Automated cardiovascular segmentation in patients with congenital heart disease from 3D CMR scans: combining multi-atlases and level-sets. In: Zuluaga, M.A., Bhatia, K., Kainz, B., Moghari, M.H., Pace, D.F. (eds.) RAMBO/HVSMR -2016. LNCS, vol. 10129, pp. 147–155. Springer, Cham (2017). https://doi.org/10.1007/978-3-319-52280-7_15
17. Simpson, A.L., et al.: A large annotated medical image dataset for the development and evaluation of segmentation algorithms. CoRRabs arXiv:1902.09063 (2019)

18. Tsai, Y.H., Hung, W.C., Schulter, S., Sohn, K., Yang, M.H., Chandraker, M.: Learning to adapt structured output space for semantic segmentation. In: Proceedings of the IEEE Conference on Computer Vision and Pattern Recognition (CVPR) (June 2018)

19. Tziritas, G.: Fully-automatic segmentation of cardiac images using 3-D MRF model optimization and substructures tracking. In: Zuluaga, M.A., Bhatia, K., Kainz, B., Moghari, M.H., Pace, D.F. (eds.) RAMBO/HVSMR -2016. LNCS, vol. 10129, pp. 129–136. Springer, Cham (2017). https://doi.org/10.1007/978-3-319-52280-7_13

20. Vorontsov, E., Tang, A., Pal, C., Kadoury, S.: Liver lesion segmentation informed by joint liver segmentation. In: 2018 IEEE 15th International Symposium on Biomedical Imaging (ISBI 2018), pp. 1332–1335 (2018)

21. Wang, X., Yang, J., Ai, D., Zheng, Y., Tang, S., Wang, Y.: Adaptive mesh expansion model (AMEM) for liver segmentation from CT image. PLoS ONE **10**(3), e0118064 (2015). https://doi.org/10.1371/journal.pone.0118064

22. Wu, W., Wu, S., Zhou, Z., Zhang, R., Zhang, Y.: 3d liver tumor segmentation in CT-images using improved fuzzy c-means and graph cuts. Biomed. Res. Int. **2017**, 5207685 (2017). https://doi.org/10.1155/2017/5207685

23. Xianling, D., et al.: Multi-view secondary input collaborative deep learning for lung nodule 3d segmentation. Cancer Imaging **20**, 53 (2020). https://doi.org/10.1186/s40644-020-00331-0

24. Zeng, G., Zheng, G.: Holistic decomposition convolution for effective semantic segmentation of medical volume images. Med. Image Anal. **57**, 149–164 (2019)

25. Zhang, Y., David, P., Gong, B.: Curriculum domain adaptation for semantic segmentation of urban scenes. In: The IEEE International Conference on Computer Vision (ICCV), vol. 2–5, p. 6 (October 2017)

26. Zhou, X., Ito, T., Takayama, R., Wang, S., Hara, T., Fujita, H.: Three-dimensional CT-image segmentation by combining 2d fully convolutional network with 3d majority voting. In: MICCAI Workshop Large-Scale Annotation Biomed. Data Expert Label Synth, pp. 111–120 (October 2016). https://doi.org/10.1007/978-3-319-46976-812

27. Zou, Y., Yu, Z., Vijaya Kumar, B., Wang, J.: Unsupervised domain adaptation for semantic segmentation via class-balanced self-training. In: The European Conference on Computer Vision (ECCV) (September 2018)

28. Çiçek, Ö., Abdulkadir, A., Lienkamp, S.S., Brox, T., Ronneberger, O.: 3D U-Net: learning dense volumetric segmentation from sparse annotation. In: Ourselin, S., Joskowicz, L., Sabuncu, M.R., Unal, G., Wells, W. (eds.) MICCAI 2016. LNCS, vol. 9901, pp. 424–432. Springer, Cham (2016). https://doi.org/10.1007/978-3-319-46723-8_49

Genetic Algorithm and Distinctiveness Pruning in the Shallow Networks for VehicleX

Linwei Zhang[✉] and Yeu-Shin Fu

Research School of Computer Science, Australian National University,
Canberra, Australia
{u6758419,guyver.fu}@anu.edu.au

Abstract. Building well-performing neural network models often require a large amount of real data. However, synthetic datasets are favoured as the collection of real data often brings up privacy and data security issues. This paper aims to build a shallow neural network model for the pre-trained synthetic feature dataset, VehicleX. Using genetic algorithm to reduce the dimensional complexity by randomly selecting a subset of features from before training. Furthermore, distinctiveness pruning techniques are used to reduce the network structure and attempt to find the optimal hidden neuron size. Both techniques improve the model performance in terms of test accuracy. The baseline model achieves 36.07% classification accuracy. Integrating genetic algorithm and the distinctiveness pruning achieve approximately 37.26% and 36.12% respectively while combining both methods achieve 37.31%.

Keywords: Synthetic dataset · Genetic algorithm · Distinctiveness pruning

1 Introduction and Related Work

With the advancement of machine learning (ML) and the improvement of computing power available in the modern days, many real-world applications are built based on the ML tools and approach [11]. The neural network (NN), being a computationally expensive approach of ML, can be applied to work in numerous fields. NN is commonly applied to vehicle detection in recent years [9]. Additionally, autonomous driving is an increasingly popular field of study [10]. Both involve the role of vehicles. As part of the deep learning branch in ML, the NN approach requires datasets to train its model before practical deployment. More often than not, NN needs a large dataset to achieve the high accuracy of the model. However, there are many difficulties in obtaining a large dataset, such as the restriction of data collection due to privacy issues, data security concerns, the cost of labor, and the potential error in the processing of labelling the data [16]. The synthetic dataset is favored for the highly practicality and flexibility in generating the synthetic data [10,16]. Although synthetic datasets might not

© Springer Nature Switzerland AG 2021
T. Mantoro et al. (Eds.): ICONIP 2021, LNCS 13108, pp. 91–102, 2021.
https://doi.org/10.1007/978-3-030-92185-9_8

accurately mimic the actual object, synthetic datasets might potentially play an important role in ML. The paper works on building a shallow NN model that trains on the synthetic VehicleX dataset and maps its feature to the corresponding vehicle type. The paper would explore adjusting the hyperparameters to build a baseline prediction model. Furthermore, the paper aims to improve the baseline model with input feature selection based on genetic algorithm, distinctiveness pruning, and fine-tuning on the hidden layer to increase the prediction accuracy of the model on classifying the vehicle instances. With the increase in test accuracy, the model might be able to learn from the synthetic dataset while put into practical usage in the future. Given the convenience of generating the synthetic data [10,16], such models reduce their reliance on the large-scale real dataset for the success of the NN-based application [6,7,10]. The model is tuned and evaluated on the validation and test sets respectively to provide a reliable result that boosts the confidence of the model.

Yao et al. [16] worked on narrowing the content domain gap between synthetic and real data using a large and synthetic vehicle-related dataset, VehicleX. They proposed the attribute descent approach to minimize the discrepancy between the vehicle and the real data calculated using Fréchet Inception Distance. The modified dataset is extended by classifying the dataset according to its vehicle ID. The performance result is evaluated using the mean average precision. The performance has a significant improvement contributed jointly by the synthetic and real data, with the use of the attribute descent approach [16].

2   Methods

We improve the model's performance using the synthetic features of VehicleX that are extracted from Resnet which are pre-trained on ImageNet. The methods involve input coding analysis, input feature selection with a genetic algorithm, and distinctiveness pruning and fine-tuning on the hidden layer techniques. The model is tuned and evaluated using the validation and test set respectively.

2.1   Input Coding Techniques

Data Analysis and Decision Making. The dataset contains a total of 75516 vehicle instances which is divided into training, testing, and validation datasets, with 45438, 15142, and 14936 images for training, testing, and validation respectively. Each vehicle instance has 2048 feature data with a detailed labelled on other attributes such as the vehicle type, colour, and orientation. The dataset has 1362 unique vehicle IDs and 11 unique vehicle types and colours. With the choice between 1362 and 11 classes, classification on 11 vehicle type classes is preferred in the situation of limited computational resource support. Given a large number of data instances, an 11-class classification requires fewer hidden and output neurons, and a smaller number of hidden layers which allows realistic and efficient training in resource-limited situations. By visualizing the dataset, the train, validation, and test sets are fairly split. Figure 1 shows the approximate ratio between the vehicle types to the total number of vehicle instances of the three datasets. Given the large dataset and the fair split of data, the experiments can use the 3 sets directly such that the model is built using the train set, tuned using the validation set, and evaluated using the test set. The value derived from the validation or the test set is the average value taken from running the model, under the same setting, three times to counter the influence of the model's random weight initialization on its performance. The use of the validation set for modelling prevents over-fitting.

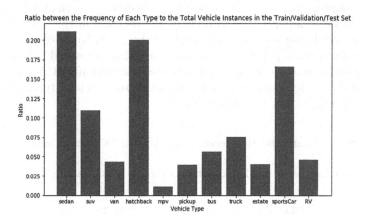

Fig. 1. Mapping of vehicle instances to 11 types on train, test and validation set

As each vehicle instance has 2048 low-dimensional feature data extracted from Resnet that is pre-trained on ImageNet, these features can represent the actual image for the multi-class classification task sophistically. However, 2048 features are a large number of features which is difficult to learn for shallow models [13]. Principal Component Analysis (PCA), being one of the popular dimensionality reduction techniques, can be applied to the feature data [13]. As PCA calculates the covariance matrix for dimensionality reduction, the input data has to be normalized for PCA to be properly performed. Hence, Z-score normalization is applied to the data points to ensure that the data are gaussian distributed for PCA to function well.

Neuron Network Layer Size. The number of neurons in the input layer depends on the number of features of a vehicle instance. Since we apply PCA on the input feature data, the number of input features is determined on a trial and error basis. When applying PCA, about 250 selected components in the train set could explain approximately 90% variance of the train set. As such, 250 draws the baseline for the PCA component selection. As shown in Table 1, with the increase in component size, the accuracy increases simultaneously. This is a result of having more features retained by a larger PCA component size. A higher number of features remains suggests a higher dimensional of the feature data in a vehicle instance. The shallow NN would require a longer time to learn the complex feature [13]. Table 1 shows a clear trade-off between the accuracy and the training time where the training time increases by approximately 2 s for every 50 additional components in the validation set. The rate of increase in the accuracy decreases when the component size reaches 400. The mean accuracy difference from 300 to 350, 350 to 400 and 400 to 450 is 0.37%, 0.2% and 0.17% respectively. With a trade-off between the training time and the accuracy, the model uses the PCA on 400 components as it still increases 0.2% for the additional 50 components from PCA on 350 as compared to less than 0.2% increase for PCA on 450 components. Additionally, the decrease in the rate of increase in the mean validation accuracy suggests that the model has tried its maximum abilities to learn the input data. The model performance would not be heavily dependent on the number of input features after adapting 400 PCA components. Given that the model classifies vehicle instances into 11 vehicle types, the NN is built with 400 input and 11 output neurons.

Table 1. Mean validation accuracy for different PCA component size.

Component size	Mean validation accuracy (in %)	Mean training time (in sec)
250	34.14	17.09
300	34.65	21.33
350	35.02	24.52
400	**35.22**	**26.51**
450	35.39	29.32

Due to the black-box nature of the NN [1], it is difficult to determine the hidden layer size and the number of training epochs that allow fast converge while avoiding over-fitting of the model. As the model would be implemented with distinctiveness pruning that will be discussed in Sect. 2.3, we can simply fix the hidden neuron with a large number. Since the model is a simple NN, the hidden neurons should be set between 400 and 11. With the limited computational resource, the hidden neurons are heuristically set as 100. By analyzing Fig. 2, the L2 norm weight regularization is implemented at a rate empirically set as 0.1 to reduce the over-fitting of the model. The model converges at around 50 epochs using the 100 hidden neurons.

Activation and Loss Functions and Optimiser. To perform a multi-class classification, the SoftMax activation function is typically used [14] on the output layer as it calculates the categorical probabilities distribution with the total sum to 1. Hyperbolic Tangent (Tanh) and Rectified Linear Unit (ReLU) activation functions are considered when deciding the activation functions for the hidden

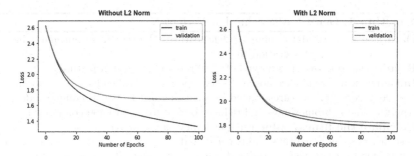

Fig. 2. Loss against iterations using 100 hidden neurons.

layer. However, Tanh is preferred as it has zero-centered [14] property which would be discussed in Sect. 2.3, and that the difference in the result accuracy between them is approximately 1%. The cross-entropy loss is used to calculate the error between the predicted and the true value. It provides a higher penalty and a steeper gradient for larger errors and reduces the impact of the vanishing gradient problem. Adaptive Moment Estimation (Adam) optimizer is chosen as it computes the adaptive learning rate for each parameter based on the lower-order moments and is less sensitive to the parameters [8]. Less sensitivity to the parameters reduces the amount of time required to tune all hyper-parameters into their optimal combination for the model to be well-learned.

2.2 Feature Selection Based on Genetic Algorithm

Although the number of features of each vehicle instance is reduced to 400 by PCA, the feature might not be the best component set as it is affected by the overall distribution of the data. Genetic algorithms (GA), being an approach of the evolutionary algorithms, are especially useful in the feature selection [3] as it discards the poorly performed subsets and uses the better-performed datasets to generate the new datasets in the effort of obtaining N features that further improve the model performance. It works better than brute force to find the best combination of N features as the search space is very large. For instance, trying to choose 1000 features from the 2048 feature data leads to a total of 3.247×10^{614} combinations. With smaller features set, the computation complexity for PCA to form the covariance matrix and extract the highest 400 components would also be lower [13]. In general, the GA approach is performed as shown in Algorithm 1 [3].

From Table 2, the higher the population (K) and the number of iterations, the more time is required. The K and the iteration values are chosen with the trade-off between the GA's effectiveness and the computational cost. As compared to the baseline model, GA leads to an increase in the validation accuracy within a reasonable time, with the K and the iteration values set as 2 and 5 respectively. From Table 3, N is selected as 1700 by experiment. Therefore, the model trains on the dataset using K as 2, the number of iterations as 5, and N as 1700.

Algorithm 1: Genetic Algorithm For VechicleX

Data: Dataset with data points of 2048 features; A pre-defined optimal value;
Pop size, K=2; Iteration=5

Result: Dataset with data points of 1700 randomly selected features

1 **Initialization** Size of chromosome is data points with 1700 features randomly
selected from 2048 features, sorted in ascending order of their column index.

2 Apply PCA to chromosomes and reduce the feature size to 400 components.

3 **Calculate the fitness value** using accuracy as the metric.

4 best fitness = highest fitness value calculated

5 **while** *iteration not reached and best fitness < pre-defined optimal value* **do**

6   Splits chromosomes into two groups. Each pair of parents is formed by
taking one chromosome from each group.

7   Perform single point **crossover** forming 2 new chromosomes [3].

8   **Mutation** rate is 50%. The chromosome would be replaced by a randomly
generated chromosome [3].

9   Apply PCA to the new chromosome.

10   Calculate the fitness value and store the best fitness value.

11   **Select** and retain only the top-K best accuracy chromosomes as the new
population, discard the rest.

Table 2. Relationship between GA parameters, the time taken and the best validation accuracy.

K	Number of iterations	Time taken (in sec)	Best validation accuracy (in %)
2	2	337.56	35.46
5	2	870.41	35.43
2	**5**	**810.82**	**35.53**
5	5	1574.49	35.54
2	10	1467.87	35.41

Table 3. Relationship between N features, best validation acuracy and time taken.

N	Best validation accuracy (in %)	Time taken (in sec)
1400	34.87	311.67
1500	35.01	321.71
1600	35.30	330.86
1700	**35.46**	**337.56**
1800	35.37	349.12

The random split in line 6 of Algorithm 1 avoids the redundant generation of the same chromosome while maintaining the randomness nature of GA. Another K new chromosomes are formed via crossover in this way.

Mutation with a rate of 0.5. A feature column in the chromosome is randomly picked and each feature value is modified by multiplying a number that is randomly generated between -1 and 1. A random scaling between -1 and 1 is unlikely to produce outliers. Mutating only on a feature column would largely maintain the features' characteristics.

Line 11 of Algorithm 1 indirectly uphold elitism as it has the chance of retaining a portion of the better-performed parent dataset. By experiment, the training epochs are set as 70 because the model equipped with GA converges at around 70 training epochs.

2.3 Distinctiveness Pruning

Distinctiveness pruning is a technique that removes the hidden neuron according to the angle computed between output activation vectors of hidden neurons on the data. As discussed in Sect. 2.1, the black-box nature of the NN [1] causes difficulty in determining the hidden neuron size. Finding the best-fit hidden neuron size experimentally might be time-consuming. Hence, Algorithm 2 is proposed to reduce the hidden neuron size while maintaining the model performance [4].

Algorithm 2: Modified Pruning Algorithm

Data: Dataset; The model's parameters
Result: A trained model with the same or smaller network structure
1 remove = True
2 Train the model with 50 epochs
3 **while** *remove* **do**
4   remove = False
5   Compute θs between each output activation vector with the rest of the output activation vectors use the cosine similarity equation.
6   **for** θ *in* θs **do**
7   **if** $\theta < 30$ **then**
8   remove = True
9   Remove one of the two neurons as they have similar functionality.
10   Adjust the remained neuron with the addition of the removed neuron's weight.
11   **else if** $\theta > 155$ **then**
12   remove = True
13   Remove both neurons as complementary pair.
14   Train the model with 1 epoch.

No normalization is required as the Tanh activation function is symmetrical and centered at the origin which allows the angle of any two neurons to be between 0 and $180°$. Algorithm 2 is modified such that if two neurons are similar to each other while one of them is complementary with another neuron, lines 9 and 10 would be performed and none of the neurons is to be removed in line 11. This is because the comparison between neurons are done in line 5 and the changes made in line 9 and 10 might change the neurons' functionality, whereby the previously observed complementary pair of neurons might no longer in complement of each other.

By experiment, the minimum and the maximum angles that the baseline model reaches upon completion of training are approximately 30° and 150° respectively. Empirically, the model prunes too much and does not perform well when the minimum threshold is set as 35°, while 150° is much smaller than the suggested angle at 165° for a complementary pair of neurons [4,5]. The model is unlikely to have complementary pairs of neurons. Hence, the minimum and maximum thresholds are chosen as 30° and 155° respectively to counteract the uncommon cases and to analyze the effect of pruning on the model performance.

Although Gedeon's research [4,5] states that no further training is required for the network, the accuracy decreases slightly without retraining in the experiment. Fine-tuning is introduced in Algorithm 2's line 14 to recover the lost accuracy by retraining the model with the remaining neurons and an extra training epoch for the model to converge [2]. Distinctiveness pruning and fine-tuning work in tandem to reduce the adverse effect of pruning. While the time for pruning might affect the overall performance, the functionality of the hidden neurons is best-trained when the model converges. Therefore, pruning would occur upon completion of training.

2.4 Hybrid Approach

The hybrid approach is the combination of feature selection based on GA with distinctiveness pruning. The hyperparameters pre-defined for GA in Sect. 2.2 are retained. Distinctiveness pruning serves as an extension to the model integrated with GA to analyze the possible improvement in the model performance. Pruning works in the model training phase in GA. Upon completion of training, Algorithm 2 applies the distinctiveness pruning.

By experiment, the initial minimum angle generated by output activation vectors of the model's hidden neurons with GA integrated is usually at approximately 20° while the maximum angle varies a lot with the minimum upper bound of 150°. The large-angle variance could be a result of GA's random feature selection. However, 165° is the maximum angle that allowed to be perceived as the non-complementary pair of vectors [4,5]. Thus, the minimum and maximum angles in Algorithm 2 are set as 20° and 165° to counteract the possible unusual cases. By training on the validation set, the average validation accuracy is approximately 36.40% at around 1890 s.

3   Results and Discussion

The paper successfully built a shallow NN that classifies the vehicle instances into their vehicle type. All techniques proposed have improved the test accuracy from the baseline model as shown in Table 4.

Table 4. Mean test accuracy of the model using 400 PCA components.

Model	Mean test accuracy (in %)
Baseline	36.07
Baseline + GA	37.26
Baseline + Prune	36.12
Baseline + GA + Prune (Hybrid)	37.31

3.1   Feature Selection Based on Genetic Algorithm

The feature selection based on GA is widely used as a wrapper feature selection technique [3]. In the paper, GA is used to select multiple subsets of 1700 features from the 2048 features and find the optimal subset evaluated on the model performance in classifying the vehicle instances according to their vehicle type. As shown in Table 4, the model integrated with GA has its mean test accuracy increases by 1.19% from the baseline model. The boost in the test accuracy is reasonable as GA tends to find the optimal set of features by generating the new datasets using the previously better-performed datasets in the effort of obtaining 1700 features that further improve the model performance. As the iteration goes, GA selects the more relevant features and reduces the less significant feature for modelling. The test accuracy varies approximately from 37.02% to 37.39% inclusively. The 0.37% difference might be a result of the randomness in both the weight initialization of the model and the randomness in the GA. Nonetheless, the increase in the test accuracy suggests the usefulness of GA in selecting a good subset of features before applying PCA which boosts the model's performance.

3.2   Distinctiveness Pruning

The distinctiveness pruning should improve the prediction of the model [17]. As shown in Table 4, the model integrated with distinctiveness pruning technique has its mean test accuracy increases by 0.05% from the baseline model.

As discussed in Sect. 2.3, the model performs the best using the minimum threshold at 30° and performs badly when the threshold is set as 35°. When the minimum angle sets at 35°, the model removes some useful hidden neurons that fall between 30° and 35°. Although the mean pruning ratio for the angle at 30° is approximately 3%, the pruning rate is affected by the random initialization of the hidden neuron. In Table 5, the model achieves an initial test accuracy that ranges approximately between 35.7% and 36.4%, which is a 0.7% accuracy difference. Such a small accuracy difference suggests the impact of the random weight initialization of the NN on its model and the technique's performance. Sets of weight that better fit the feature data are better trained and reached a higher accuracy. With reference to Table 5's last row, the model has a relatively good accuracy which suggests its good initialization of the weight. Since the weights fit well with the feature data, the output activation vectors often have unique functionality. As none of them are perceived as similar or complementary, pruning might not occur. However, the model's performance should not rely on random weight initialization. In fact, a lower test accuracy suggests overlapping or complementary hidden neuron functionality as a result of unpleasant weight initialization. Pruning occurs to remove the redundant neurons and retrain to retain or improve the model performance even if the weight initialized does not fit well with the feature data.

Table 5. Relationship between angles and accuracy.

Initial accuracy (in %)	Minimum angle (in degrees)	Maximum angle (in degrees)	Prune (Neuron remains)	Final accuracy (in %)
35.87	29.5	148.3	Yes (97)	36.28
35.73	29.8	152.2	Yes (96)	36.11
36.36	31.9	144.5	No	36.36

With the elimination of more hidden neurons, the model would have a smaller size and a lower cost of training [17]. The distinctiveness pruning speeds up the time for the model to retrain with a smaller number of hidden neurons trained on an extra epoch. The increase in the training efficiency is reasonable as less time is required for a smaller model to retrain in each and every epoch. Retrain on 1 epoch does not take too much computational resource while the model is optimized by the technique. The distinctiveness of pruning and fine-tuning work in tandem, by eliminating the unnecessary hidden neurons and retraining for a finer model, contribute to the improvement of the model performance.

3.3 Hybrid Approach

The hybrid approach of the feature selection based on GA and distinctiveness prune has the highest test among all models. It outperformed the baseline model by the mean test accuracy of 1.24%.

The test accuracy of the hybrid approach ranges between 37.25% to 37.47%. As feature subsets are randomly generated before applying to PCA, there are occasions when the generated feature subsets do not well-represent the original input dataset. The model would not learn well on a less relevant feature set and results in low classification accuracy. As discussed in Sect. 3.2, the model's performance might be influenced by the random weight initialization. A lower model performance occurs when the weight initialized does not fit well with the feature data. Distinctiveness pruning acts to recover the model performance that is negatively influenced by the random weight initialization. When pruning occurs on the seemingly poorly performed feature set, the accuracy is improved by pruning as it removes hidden neurons that share similar or complementary functionality and retrains the model. This keeps the feature subset robust against the potential negative impact of the random weight initialization. The highest accuracy feature subset is then selected as it best explains the vehicle instances. As such, the mean test accuracy is stable at above 37.2% while applying GA alone still has the tendency of obtaining accuracy lower than 37.2%, under the same parameter setting. GA and distinctiveness pruning techniques work in tandem to provide a better feature subset for the model to learn and perform better.

4   Conclusion and Future Work

The objective of the paper is to build a NN model that classifies the examples in the synthetic VehicleX dataset according to its vehicle type. A baseline prediction model is built using one hidden layer NN using Tanh and SoftMax activation functions with Adam optimizer for faster converge. Due to the large dataset size, the model is tuned using the validation set before being applied to the test set. The use of the average accuracy when deriving the model performance provides a reliable result for tuning and evaluating the model performance respectively. In addition, feature selection based on GA and distinctiveness pruning followed by the fine-tuning technique is proposed in an effort of improving the overall performance of the shallow NN model. The distinctiveness pruning compresses the model and improves the training efficiency. Among all approaches, the model integrated with the hybrid approach works the best with an average test accuracy reaches 37.31%. When we want to improve the model's performance with sufficient computational resources, the hybrid approach is recommended for its effectiveness in boosting the model performance. If the experiment environment has limited resources, the distinctiveness pruning is favored.

With sufficient computational resources, the population size and the number of iterations can be larger to explore the more possible combinations of 1700 features. GA can also be extended to find the optimal set of thresholds for the distinctiveness pruning to boost the model performance with the optimal solution on a hybrid approach. This work can also extend to multi-task learning by including additional attributes, such as the vehicle orientation and light intensity [15]. The stratified and repeated k-fold cross-validation might be applied to split the dataset with a similar proportion of the vehicle instance according to their vehicle type with repeated k-fold decreases the variance of the prediction error [12].

With reference to Table 4, a simple NN model's performance can be improved with proper algorithms applied in its classification of the synthetic dataset. As the accuracy of predicting the actual object using the model trained on the synthetic dataset increases [6,10], the synthetic dataset would play an important role in the advancement of the ML and the utilization of NN would be in a greater scale in the near future.

References

1. Buhrmester, V., Münch, D., Arens, M.: Analysis of explainers of black box deep neural networks for computer vision: A survey. arXiv preprint arXiv:1911.12116 (2019)
2. Chandakkar, P.S., Li, Y., Ding, P.L.K., Li, B.: Strategies for re-training a pruned neural network in an edge computing paradigm. In: 2017 IEEE International Conference on Edge Computing (EDGE), pp. 244–247. IEEE (2017)
3. El-Maaty, A.M.A., Wassal, A.G.: Hybrid GA-PCA feature selection approach for inertial human activity recognition. In: 2018 IEEE Symposium Series on Computational Intelligence (SSCI), pp. 1027–1032. IEEE (2018)
4. Gedeon, T.D.: Indicators of hidden neuron functionality: the weight matrix versus neuron behaviour. In: Proceedings 1995 Second New Zealand International Two-Stream Conference on Artificial Neural Networks and Expert Systems, pp. 26–29. IEEE (1995)
5. Gedeon, T.D., Harris, D.: Network reduction techniques. In: Proceedings International Conference on Neural Networks Methodologies and Applications, vol. 1, pp. 119–126 (1991)
6. Kakoulli, E., Soteriou, V., Theocharides, T.: Intelligent hotspot prediction for network-on-chip-based multicore systems. IEEE Trans. Comput. Aided Des. Integr. Circuits Syst. 31(3), 418–431 (2012)
7. Kim, E., Gopinath, D., Pasareanu, C., Seshia, S.A.: A programmatic and semantic approach to explaining and debugging neural network based object detectors. In: Proceedings of the IEEE/CVF Conference on Computer Vision and Pattern Recognition, pp. 11128–11137 (2020)
8. Kingma, D.P., Ba, J.: Adam: A method for stochastic optimization. arXiv preprint arXiv:1412.6980 (2014)
9. Li, B., Zhang, T., Xia, T.: Vehicle detection from 3d lidar using fully convolutional network. arXiv preprint arXiv:1608.07916 (2016)
10. Nowruzi, F.E., Kapoor, P., Kolhatkar, D., Hassanat, F.A., Laganiere, R., Rebut, J.: How much real data do we actually need: Analyzing object detection performance using synthetic and real data. arXiv preprint arXiv:1907.07061 (2019)
11. Ou, G., Murphey, Y.L.: Multi-class pattern classification using neural networks. Pattern Recogn. 40(1), 4–18 (2007)
12. Rodriguez, J.D., Perez, A., Lozano, J.A.: Sensitivity analysis of k-fold cross validation in prediction error estimation. IEEE Trans. Pattern Anal. Mach. Intell. 32(3), 569–575 (2009)
13. Sehgal, S., Singh, H., Agarwal, M., Bhasker, V., et al.: Data analysis using principal component analysis. In: 2014 International Conference on Medical Imaging, M-Health and Emerging Communication Systems (MedCom), pp. 45–48. IEEE (2014)
14. Sharma, S., Sharma, S.: Activation functions in neural networks. Towards Data-Science 6(12), 310–316 (2017)
15. Tang, Z., et al.: Pamtri: pose-aware multi-task learning for vehicle re-identification using highly randomized synthetic data. In: Proceedings of the IEEE/CVF International Conference on Computer Vision, pp. 211–220 (2019)
16. Yao, Y., Zheng, L., Yang, X., Naphade, M., Gedeon, T.: Simulating content consistent vehicle datasets with attribute descent. In: Vedaldi, A., Bischof, H., Brox, T., Frahm, J.-M. (eds.) ECCV 2020. LNCS, vol. 12351, pp. 775–791. Springer, Cham (2020). https://doi.org/10.1007/978-3-030-58539-6_46
17. Zhu, M., Gupta, S.: To prune, or not to prune: exploring the efficacy of pruning for model compression. arXiv preprint arXiv:1710.01878 (2017)

Stack Multiple Shallow Autoencoders into a Strong One: A New Reconstruction-Based Method to Detect Anomaly

Hanqi Wang[1], Xing Hu[2], Liang Song[1], Guanhua Zhang[1], Yang Liu[1], Jing Liu[1], and Linhua Jiang[1(✉)]

[1] Academy for Engineering & Technology, Fudan University, Shanghai 200433, China
jianglinhua@fudan.edu.cn
[2] University of Shanghai for Science and Technology, Shanghai 200093, China

Abstract. Anomaly detection methods based on deep learning typically utilize reconstruction as a proxy task. These methods train a deep model to reconstruct the input from high-level features extracted from the samples. The underlying assumption of these methods is that a deep model trained on normal data would produce higher reconstruction error for abnormal input. But this underlying assumption is not always valid. Because the neural networks have a strong capacity to generalize, the deep model can also reconstruct the unseen abnormal input well sometimes, leading to a not prominent reconstruction error for abnormal input. Hence the decision-making process cannot distinguish the abnormal samples well. In this paper, we stack multiple shallow autoencoders (StackedAE) to enlarge the difference between reconstructions of normal and abnormal inputs. Our architecture feeds the output reconstruction of prior AE into the next one as input. For abnormal input, the iterative reconstruction process would gradually enlarge the reconstruction error. Our goal is to propose a general architecture that can be applied to different data types, e.g., video and image. For video data, we further introduce a weighted loss to emphasize the importance of the center frame and its near neighbors because it is unfair to treat all frames in a 3D convolution frame cuboid equally. To understand the effectiveness of our proposed method, we test on video datasets UCSD-Ped2, CUHK Avenue, and the image dataset MNIST. The results of the experiments demonstrate the effectiveness of our idea.

Keywords: Image processing and computer vision · Anomaly detection · Stack architecture · Autoencoder · Weighted reconstruction loss

The research was partly supported by the National Natural Science Foundation of China (No. 61775139), Shanghai Science and Technology Innovation Action Plan (No. 20JC1416503), and Shanghai Key Research Laboratory of NSAI.

T. Mantoro et al. (Eds.): ICONIP 2021, LNCS 13108, pp. 103–115, 2021.
https://doi.org/10.1007/978-3-030-92185-9_9

1 Introduction

Anomaly detection is an attractive problem in many areas, such as video surveillance [11]. The goal of anomaly detection is to identify the occurrence of anomaly events. In real life, it is generally an unsupervised binary classification problem, for the reason that labeled large anomaly dataset is costly or inaccessible [23,25,26]. So, the anomaly detection models are supposed to learn normal distribution on a large normal dataset and then identify anomalies that are not consistent with the normal distribution learned by the model in the test.

Reconstruction-based [2,8] methods are commonly used to detect anomaly. These methods first learn high-level representations from the input, reconstructing the input from the learned representation. Finally, they use the reconstruction error as an indicator to find the abnormal samples. These methods are based on the assumption that a model trained on normal data is not able to reconstruct abnormal input well [26] because it has never seen them before. Subsequently, the bad reconstruction quality leads to a higher reconstruction error of the abnormal input, indicating its abnormality. In the past years, the weak representation learning ability limits the performance of the reconstruction-based method. With the rising of deep learning, the deep neural network allows the model to extract better representations [15,26], which greatly improves the performance of the reconstruction-based methods. Now, reconstruction-based deep models have proven their effectiveness and achieved very competitive results in many anomaly detection tasks.

However, the underlying assumption mentioned above is critical to the performance of these methods. However, it is not always true in practice [4,15]. The deep learning models are usually able to work on unseen samples due to their strong generalization capacity, especially when the unseen samples have many compositions common to the training samples. So, a reconstruction-based deep model is likely to reconstruct an abnormal input well because the abnormal sample and normal samples usually share many common compositions, e.g., walking pedestrians in the UCSD-Ped2 dataset. In this case, the reconstruction error of the abnormal input would be close to or even lower than the normal ones. As a result, the reconstruction error fails to function as an indicator.

In this paper, we adopt iterative reconstruction to enlarge the error of abnormal reconstructions. In the beginning, the first autoencoder (AE) gives a reconstruction of input. Then the second AE continues to reconstruct the original input from its reconstruction given by the prior AE. For an abnormal case, the reconstruction given by the first AE somewhat loses more critical information than the normal cases, and the second AE trained on reconstructions of normal inputs cannot handle reconstructions of abnormal inputs. As a result of these two points, the reconstruction of an anomaly given by the second AE naturally incurs more information loss than normal ones in these processes, leading to a more prominent reconstruction error. Consequently, the abnormal samples have a greater chance of being detected by the decision-making process. Through this, the problem of undesirable generalization capacity is relieved by the proposed architecture. Furthermore, our goal is to offer a general anomaly detection architecture applicable to different types of data. So, we try to prove the effectiveness of our idea on both video and images. For better performance on video data, we design 3D-StackedAE which introduces 3D convolution to capture both spatial and temporal information in video data. As for images, we build 2D-StackedAE for them, which generally use 2D convolution. Each sample in the 3D convolution-based model is a temporal cuboid stacked by multiple frames. For a cuboid, the frames of interest locate close to the center one because that is the one whose abnormality we want to measure. Therefore, we incorporate such centering prior via a simple weighting kernel during the training process. It assigns larger weights to the frames close to the center of a cuboid, and the experiment will show its boost in the result. In summary, there are three contributions as follows:

- We design StackedAE architecture to enlarge difference of the reconstruction error between abnormal and normal samples.
- For 3D convolution, we design a weighted loss to focus more on the center frame and those ones nearby it in a cuboid, which are more relevant to the abnormality of the measured center frame.
- We outperform the current state of the art result on image dataset MNIST and video anomaly detection dataset UCSD-Ped2. We also reach a competitive result on CUHK Avenue dataset.

Our code is available online: https://github.com/hanqiw/StackedAE

2 Related Work

The common unsupervised anomaly detection methods include one-class classification, clustering, distance-based measure, reconstruction, and prediction. One-class classification methods try to seek for a hyperplane to separate the anomalies from normal samples. For example, in [3,18], authors train a deep one-class network. Unfortunately, these one-class classification methods usually work ineffectively when the distribution of normal data is complex [14]. There are also some methods based on clustering [25], like k-means or gaussian mixture model. However, the performance of the clustering-based model is heavily dependent on the clustering result, which would be biased if a few anomaly samples exist in training data [14]. And distance-based methods are easy to implement. They show their effectiveness in some areas like trajectory anomaly analysis [9,22], but the inborn weaknesses of the distance-based anomaly measures [14] restrict their better performance in other areas.

The reconstruction-based method assumes that the anomaly samples cannot be reconstructed well if the model only learns from normal data. Thus the anomaly can be identified by reconstruction error. Some useful techniques were applied in the past years, such as sparse coding [11], autoencoder [2,8] or PCA-based methods [7]. In the early days, the performance of reconstruction-based methods is limited by the weak representation learning ability, but the rising of neural networks made a breakthrough. Recently, reconstruction-based methods combined with neural networks have achieved fruitful outcomes. Ravanbakhsh [17] design a generative adversarial network for video anomaly detection. Hasan [5] propose a convolutional AE (ConvAE) structure to detect anomaly in video sequence. Zhao et al. [24] refine Hasan's work by introducing 3D convolution into ConvAE structure. Based on previous works, Gong et al. [4], and Park et al. [15] try to solve the problem of too strong generalization capacity. They add an external memory module into the middle of AE to damage the reconstruction of abnormal samples. In [10], the author argues that this problem can also be alleviated by using prediction instead of reconstruction. The work of [10] assumes that anomalous frames in video sequences are unpredictable, but the normal frames are predictable, so this work use input to predict the next frames and measures abnormality on prediction error. Although it achieves a remarkable performance and reconstruction-based AE can be easily transferred to make a prediction, we will not discuss it in this paper. Because the prediction-based method can only be applied to time series such as video, it cannot work when the samples are not temporally correlated, such as image anomaly detection.

3 Our Method

This section begins with introducing our stacked architecture and strategy to weight frames in a cuboid, and then we present our objective loss and the decision-making process. Notably, to capture the rich temporal information in video data, we build 3D-StackedAE to deal with video. Considering the basic mechanism of 2D-StackedAE is well covered by the 3D-StackedAE, we will take 3D-StackedAE as an example to demonstrate our work in this section. The 2D-StackedAE designed for image anomaly detection will be demonstrated in the next section.

3.1 Stacked Architecture

As can be seen in Fig. 1, we build our model by stacking two autoencoders end-to-end. The whole model is trained on the normal dataset. As shown in Eq. 1 and Eq. 2, the first AE $f_{AE1}(\cdot)$ reconstructs the input from the input sample itself, outputting intermediate reconstruction, and the second AE $f_{AE2}(\cdot)$ reconstructs input from reconstruction generated by the prior AE, outputting final reconstruction.

$$\hat{x}_{inter} = f_{AE1}(x; \theta_1), \tag{1}$$

$$\hat{x}_{final} = f_{AE2}(\hat{x}_{inter}; \theta_2), \tag{2}$$

where x denotes input, $\hat{x}_{inter}$ and $\hat{x}_{final}$ denote intermediate reconstruction of x and final reconstruction of x respectively, θ_1 denotes the parameters of the first AE, θ_2 denotes the parameters of the second AE.

The first AE is trained on normal samples, which means abnormal events are unseen for it. Thus, the generated intermediate reconstruction of the abnormal sample gets worse quality than normal ones. Although in practice, it is often not

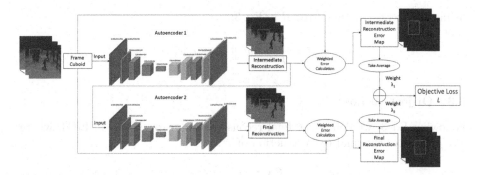

Fig. 1. An illustration for our proposed 3D-StackedAE architecture. The blue box is 3D convolution module, and the orange box is 3D deconvolution module. The anomaly events are highlighted by a red bounding box in cropped frames. (Color figure online)

bad enough to be reflected in a corresponding rise in the reconstruction error, it hinders the second AE from generating final reconstruction of the abnormal sample. Because the intermediate reconstruction of an abnormal sample loses much more information necessary for the next reconstruction compared with normal cases. Furthermore, the second AE is trained on the intermediate reconstruction of normal samples, and it inherently cannot handle abnormal ones well. As a Hence the distance between reconstruction errors of normal input and abnormal input is enlarged by the second AE.

In the training process, we apply intermediate supervision by adding the intermediate reconstruction error into the loss function. Through this loss function, the first AE is requested to give reconstruction as good as it can. Otherwise, we cannot guarantee the intermediate output, and the stacked architecture would degrade toward a single deep AE.

3.2   Weighted Reconstruction Error

When building 3D-StackedAE, we design an uneven weight kernel to adjust weights for multiple frames in a cuboid. Although we realize that evenly weighting every frame in a cuboid makes the abnormality or regularity of the center frame prominent if the label of the center frame is consistent with the majority in cuboid, it prevents the model from correctly identifying those abnormal cases when the frames are abnormal most and the center frame is normal, or otherwise. The abnormality score of the center frame would be obscured under this circumstance. Thus, when dealing with video data, we design a weighted reconstruction loss to pay more attention to the center frame and its near neighbor, but it still brings the other frames in cuboid far away from the center one into consideration.

Given a dataset $\{x^t\}_{t=1}^{T}$ with T samples, x^t denotes the sample, $\hat{x}^t$ denotes the reconstruction of x^t, and w^t denotes the tth element of weight kernel. We define the reconstruction error R in Eq. 3:

$$R = \frac{1}{T} \sum_t w^t \left\| \hat{x}^t - x^t \right\|_2^2. \tag{3}$$

3.3 Objective Loss

Following Eq. 3, we define the final reconstruction error R_{final} in Eq. 4, where $\hat{x}^t_{final}$ denotes the final reconstruction of x^t.

$$R_{final} = \frac{1}{T} \sum_t w^t \left\| \hat{x}^t_{final} - x^t \right\|_2^2. \tag{4}$$

In the same way, we define the intermediate reconstruction error R_{inter} in Eq. 5.

$$R_{inter} = \frac{1}{T} \sum_t w^t \left\| \hat{x}^t_{inter} - x^t \right\|_2^2, \tag{5}$$

where $\hat{x}^t_{inter}$ denotes the intermediate reconstruction of x^t.

Finally, we use the weighted sum of intermediate reconstruction and final reconstruction as loss function to guide the training process, as shown in Eq. 6.

$$L = \lambda_1 R_{inter} + \lambda_2 R_{final}, \tag{6}$$

where λ_1 denotes the weight for the intermediate reconstruction, λ_2 denotes the weight for the final reconstruction, and L denotes the objective loss function.

3.4 Regularity Score

In test phase, we first calculate the reconstruction error as we mention above. And then we score the regularity by Eq. 7:

$$s = 1 - \frac{e - \min(e)}{\max(e) - \min(e)}, \tag{7}$$

where e denotes the reconstruction error, and s denotes the regularity score. In decision-making process, the model will use the regularity score s as the basis for judgment.

4 Experiment

In this section, we first evaluate our proposed architecture on both video anomaly detection datasets UCSD-Ped2, CUHK Avenue, and image dataset MINIST. In the next, we make a comprehensive comparison with state-of-the-art methods to show the effectiveness of the proposed method. To further measure the improvement of our model, we implement ablation studies and make discussion at the end.

4.1   Experiment on MNIST Dataset

Dataset. MNIST is a hand-written digit image dataset. It contains 10 classes, from digit 0 to digit 9. The MNIST dataset contains 60,000 training images and 10,000 testing images. We alternately use one class in the training dataset as the normal dataset to train the model, and all the 10 classes in the test dataset to test the performance, then we take the average of 10 results to compare with other methods.

Table 1. Experiment results on MINIST. The AUC performances of all the methods we compare with are listed below, and the best performance is marked in **bold**.

Method	AUC
OC-SVM [19]	0.9499
KDE [16]	0.8116
VAE [1]	0.9632
PixCNN [13]	0.6141
DSEBM [23]	0.9554
Gong-MemAE [4]	0.9751
2D-StackedAE	**0.9810**

Implementation Details. The experiments are implemented using Pytorch. We read in images from MINIST dataset as 28×28 gray-scale images. In our experiment, we set training batch size as 64, λ_1 as 1, λ_2 as 0.3. The optimizer is Adam with earning rate 0.0002, the scheduler is MultiStepLR. The training and testing are run on an NVIDIA 2080Ti.

We use conv(k,s,c) and deconv(k,s,c) to represent convolution and deconvolution, where k is kernel size, s is stride, and c is channel numbers. We build 2D autoencoder which 2D-StackedAE is stacked with as following: Conv(3, 1, 1), Conv(3, 2, 8), Conv(3, 2, 16), Conv(3, 3, 32), Dconv(3, 3, 32), Dconv(3, 2, 16), Dconv(3, 2, 1), Conv(3, 1, 1).

Result Comparison with Existing Methods. We use the AUC (Area Under Curve), area under ROC (Receiver Operation Characteristic) curve as the measurement for performance evaluation. As we can see in Table 1, our proposed StackedAE outperforms all the other methods, which proves the effectiveness of our idea.

Visualizing How StackedAE Helps Distinguish Abnormal Inputs from Normal Inputs. We demonstrate how our proposed architecture works by visualizing the changing process from input to final reconstruction. As we can see in Fig. 2, the intermediate reconstruction quality of the normal sample and abnormal sample seem roughly equivalent by naked eyes. However, the second AE remarkably worsens the final reconstruction quality of the abnormal sample. We trained our model on normal samples "5" and test it on all kinds of hand-written digits. In the end, we can see a clear reconstruction for normal sample "5", but the reconstructions for abnormal samples "0", "1", "7" are significantly damaged. Thus, the normal and abnormal samples can be easily separated by our model.

4.2   Experiment on UCSD-ped2 and CUHK Avenue Datase

Dataset. The UCSD-Ped2 and CUHK Avenue are well-known video anomaly datasets. UCSD-Ped2 has 16 short video clips in the training dataset and 12 clips in the testing dataset. It includes 12 abnormal events, like riding a bike on the sidewalk. CUHK Avenue has 16 short video clips in the training dataset and 21 clips in the testing dataset. It contains 47 kinds of abnormal behaviors such as throwing and running.

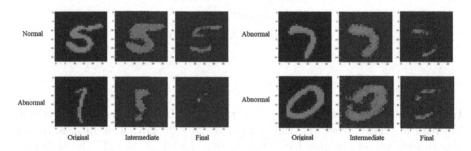

Fig. 2. The (normalized) visualization of intermediate reconstruction and final reconstruction result. The normal sample is "5", the abnormal samples are "1", "7", "0".

Implementation Details. The experiments are implemented using Pytorch. The 3D-StackedAE network structure for video data can be seen in Fig. 1. The frames are read in as 256×256 gray-scale images, and we set batch size as 10. In training phase, we adopt Adam as optimizer with learning rate 0.0002, and we set scheduler as MultiStepLR. The λ_1 is 1, λ_2 is 0.3. The weight kernel $w = [0.3, 0.3, 0.5, 0.5, 0.7, 0.7, 1, 1, 2, 1, 0.7, 0.7, 0.5, 0.5, 0.3, 0.3]$. All the programs are run on 4 Nvidia 2080Ti.

Visualization. We output a few examples to make a comparison among intermediate reconstruction, final reconstruction, and ground truth in Fig. 3(a). The anomaly area is highlighted in a red bounding box, we can see that the bike and the person on it are still recognizable in the intermediate reconstruction, and it

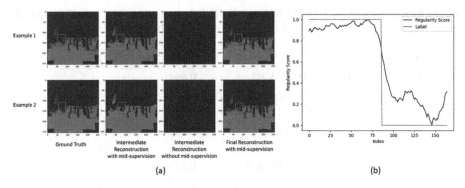

Fig. 3. Left: The (normalized) visualization of intermediate reconstruction and final reconstruction. The intermediate reconstruction with intermediate supervision and intermediate reconstruction without intermediate supervision are given, respectively. The given two examples are two consecutive frames. The anomaly is marked by a red bounding box. Right: The regularity score curve over 175 samples. (Color figure online)

almost disappears in the final reconstruction. But the people who walk on the sidewalk normally are unchanged.

And we plot the regularity score curve over 175 test samples in Fig. 3(b). First, we can see that our architecture successfully lowers the regularity score on abnormal frames, which makes them different from normal samples. Second, the regularity decreases significantly since the beginning of the anomaly event. It means that our model can sensitively capture the occurrence of an anomaly.

Result Comparison. We adopt AUC as our measurement to make the comparison. As can be seen in Table 2 below, our proposed StackedAE outperforms all the other reconstruction-based methods on UCSD-Ped2. Although it fails to defeat the state-of-the-art in [9] on Avenue dataset, the performance of StackedAE is still competitive. Notably, we are not pursuing the highest score in this paper. Our focus is on pointing out a general technique instead.

Although the performance of prediction-based methods exceeds the best result of reconstruction-based methods on the video dataset, our result is still comparable to the best result of the non-reconstruction methods. And the prediction-based method is limited to specific applications, and it cannot be generalized to other applications in many cases.

4.3 Ablation Studies

Study of the Impact of Stacked Architecture and Intermediate Supervision. We build a deep AE which has the same number of layers as the proposed stacked architecture. And we also disassemble our stacked architecture to get a single AE. To study if our technique can be generalized to other AE, we introduce a simple AE reported in [4] and stack two simple AEs in the same manner.

Table 2. Experiment on UCSD-Ped2 and CUHK Avenue. The AUC of all compared methods are listed below, the best results are marked in **bold**.

Method		UCSD-Ped2	CUHK avenue
Non-Reconstruction	MPPCA [7]	0.693	-
	MPPCA+SFA [12]	0.613	-
	MDT [12]	0.829	-
	AMDN [21]	0.908	-
	Unmasking [20]	0.822	0.806
	MT-FRCN [6]	0.922	-
	Frame-Pred [11]	0.954	0.849
	ParkMem-Pred [15]	**0.970**	**0.885**
Reconstrcution	ConvAE [5]	0.850	0.800
	ConvAE-3D [24]	0.912	0.771
	TSC [11]	0.910	0.806
	StackedRNN [11]	0.922	0.817
	Abnormal GAN [17]	0.935	-
	ParkMem-Recon [15]	0.902	0.828
	GongMem [4]	0.941	**0.833**
	3D-StackedAE	**0.950**	0.826

Table 3. Ablation studies of stacked architecture based on UCSD-Ped2 dataset.

Method	AUC
Simple AE [4]	0.917
Deep AE	0.923
Single AE	0.920
Stacked simple AEs	0.944
StackedAE w/o intermediate supervision	0.921
3D-StackedAE	**0.950**

To further discuss the importance of intermediate supervision, we remove it and train a stacked architecture model without intermediate supervision. In the end, we try different weighted loss strategies using the same StackedAE architecture.

Table 3 shows the AUC result of ablation studies on the UCSD-Ped2 dataset. Notably, our proposed StackedAE outperforms all the other models, it proves that the techniques we used play vital roles in the model.

We can see that the StackedAE works much better than a single AE. The AUC improves from 0.920 to 0.950, which means the stack makes an improvement on the performance. And the deep AE works better than single AE as well. We think it shows that the deeper structure brought improvement on anomaly detection performance to certain extents. However, the StackedAE outperforms deep AE, and this result demonstrates that the improvement of our design is not entirely brought by a deeper structure. Compared with StackedAE with intermediate supervision, the performance of StackedAE without intermediate supervision clearly degrades. It proves that intermediate supervision functions a lot. We can see the variation of intermediate reconstruction if we do not apply intermediate supervision in Fig. 3(a). It is shown that the intermediate reconstruction of the first AE would be meaningless if we do not apply intermediate supervision. Finally, we can look at the comparison between simple AE in [4] and stacked simple AEs, the AUC increase from 0.917 to 0.944. This comparison represents the fact that our idea can be generalized to other models.

Study of the Impact of Different Weighted Loss. To study what the suitably weighted loss is, we try different options. We train two models using gaussian distribution weighted loss and uniform distribution weighted loss, respectively. We can see in Table 4, the gaussian weighted loss get the worst result. As we mentioned above, the neighbor frames help identify the regularity of the center frame if the label of the center frame is consistent with the majority in a cuboid. In practice, we think this case is common in video anomaly detection scenarios. The gaussian weighted loss put too much focus on the frames around the center, but it ignores the frames far away from the center, so it suffers great degradation in AUC result, from 0.950 to 0.848. And the uniform distribution weighted loss works not well, too. The regularity of the center frame would be wrongly measured if it is opposed to the majority in the cuboid. While we think this situation does not frequently happen in video anomaly detection, it still hampers further improvement.

Table 4. Ablation studies of different weight kernel based on UCSD-Ped2 dataset.

Method	AUC
Gaussian weighted StackedAE	0.848
Uniform weighted StackedAE	0.948
Our weight kernel	**0.950**

5 Conclusion

We propose to stack multiple shallow autoencoders end-to-end based on the assumption that the reconstruction error of the abnormal samples accumulates during iterative reconstruction. Therefore the abnormal and normal samples are more separable in the eyes of our model. By carefully adjusting the weights of multiple frames in cuboid for 3D-StackedAE, our model can focus on the center frame and its near neighbor while not ignoring other frames. Extensive experiments have been done on three datasets (including one image dataset and two video datasets) to test the effectiveness of our method, where very good results are achieved for both image and video data. From this, we can conclude that our proposed new method is effective to distinguish the abnormal samples well and might be generalized to other models while needs more researches to test.

References

1. An, J., Cho, S.: Variational autoencoder based anomaly detection using reconstruction probability. Spec. Lect. IE **2**(1), 1–18 (2015)
2. Bengio, Y., Lamblin, P., Popovici, D., Larochelle, H., et al.: Greedy layer-wise training of deep networks. Adv. Neural. Inf. Process. Syst. **19**, 153 (2007)
3. Chalapathy, R., Menon, A.K., Chawla, S.: Anomaly detection using one-class neural networks. arXiv preprint arXiv:1802.06360 (2018)
4. Gong, D., et al.: Memorizing normality to detect anomaly: memory-augmented deep autoencoder for unsupervised anomaly detection. In: ICCV, pp. 1705–1714 (2019)
5. Hasan, M., Choi, J., Neumann, J., Roy-Chowdhury, A.K., Davis, L.S.: Learning temporal regularity in video sequences. In: CVPR, pp. 733–742 (2016)
6. Hinami, R., Mei, T., Satoh, S.: Joint detection and recounting of abnormal events by learning deep generic knowledge. In: ICCV (2017)
7. Kim, J., Grauman, K.: Observe locally, infer globally: a space-time MRF for detecting abnormal activities with incremental updates. In: CVPR, pp. 2921–2928. IEEE (2009)
8. Kingma, D.P., Welling, M.: Auto-encoding variational bayes. arXiv preprint arXiv:1312.6114 (2013)
9. Li, C., Xu, Q., Peng, C., Guo, Y.: Anomaly detection based on the global-local anomaly score for trajectory data. In: Gedeon, T., Wong, K.W., Lee, M. (eds.) ICONIP 2019. CCIS, vol. 1143, pp. 275–285. Springer, Cham (2019). https://doi.org/10.1007/978-3-030-36802-9_30
10. Liu, W., Luo, W., Lian, D., Gao, S.: Future frame prediction for anomaly detection-a new baseline. In: CVPR, pp. 6536–6545 (2018)
11. Luo, W., Liu, W., Gao, S.: A revisit of sparse coding based anomaly detection in stacked RNN framework. In: ICCV, pp. 341–349 (2017)
12. Mahadevan, V., Li, W., Bhalodia, V., Vasconcelos, N.: Anomaly detection in crowded scenes. In: 2010 IEEE Computer Society Conference on Computer Vision and Pattern Recognition, pp. 1975–1981. IEEE (2010)
13. van den Oord, A., Kalchbrenner, N., Vinyals, O., Espeholt, L., Graves, A., Kavukcuoglu, K.: Conditional image generation with pixelcnn decoders. In: NeurIPS, pp. 4797–4805 (2016)
14. Pang, G., Shen, C., Cao, L., Hengel, A.V.D.: Deep learning for anomaly detection: a review. ACM Comput. Surv. (CSUR) **54**(2), 1–38 (2021)
15. Park, H., Noh, J., Ham, B.: Learning memory-guided normality for anomaly detection. In: ICCV, pp. 14372–14381 (2020)
16. Parzen, E.: On estimation of a probability density function and mode. Ann. Math. Stat. **33**(3), 1065–1076 (1962)
17. Ravanbakhsh, M., Nabi, M., Sangineto, E., Marcenaro, L., Regazzoni, C., Sebe, N.: Abnormal event detection in videos using generative adversarial nets. In: ICIP, pp. 1577–1581. IEEE (2017)

18. Ruff, L., et al.: Deep one-class classification. In: ICML, pp. 4393–4402. PMLR (2018)
19. Schölkopf, B., Williamson, R.C., Smola, A.J., Shawe-Taylor, J., Platt, J.C., et al.: Support vector method for novelty detection. In: NeurIPS (1999)
20. Tudor Ionescu, R., Smeureanu, S., Alexe, B., Popescu, M.: Unmasking the abnormal events in video. In: ICCV, pp. 2895–2903 (2017)
21. Xu, D., Ricci, E., Yan, Y., Song, J., Sebe, N.: Learning deep representations of appearance and motion for anomalous event detection. arXiv preprint arXiv:1510.01553 (2015)
22. Yan, B., Han, G.: Effective feature extraction via stacked sparse autoencoder to improve intrusion detection system. IEEE Access **6**, 41238–41248 (2018)
23. Zhai, S., Cheng, Y., Lu, W., Zhang, Z.: Deep structured energy based models for anomaly detection. In: ICML, pp. 1100–1109. PMLR (2016)
24. Zhao, Y., Deng, B., Shen, C., Liu, Y., Lu, H., Hua, X.S.: Spatio-temporal autoencoder for video anomaly detection. In: ACM MM, pp. 1933–1941 (2017)
25. Zimek, A., Schubert, E., Kriegel, H.P.: A survey on unsupervised outlier detection in high-dimensional numerical data. Stat. Anal. Data Min. ASA Data Sci. J. **5**(5), 363–387 (2012)
26. Zong, B., et al.: Deep autoencoding gaussian mixture model for 3 unsupervised anomaly detection. In: ICLR (2018)

Learning Discriminative Representation with Attention and Diversity for Large-Scale Face Recognition

Zhong Zheng, Manli Zhang, and Guixia Kang[(✉)]

Key Laboratory of Universal Wireless Communications, Ministry of Education,
Beijing University of Posts and Telecommunications, Beijing, China
{zhongzheng,gxkang}@bupt.edu.cn

Abstract. Recent improvements of face recognition mainly come from investigations on metric or loss functions rather than the structure of CNNs and interpretation of face representation. By analyzing the application of face representation (e.g. age estimation), we find some principles (utilizing low-level features and preserving enough localization information) of pixel-level prediction approaches would help to enhance the discriminative ability of face representation. Therefore, we propose a new plug-and-play attention module to integrate low-level features and propose two types of diversity regularizers to maintain localization information while reduce redundant correlation. Moreover, our diversity regularizers can achieve decorrelation without eigenvalue decomposition or the approximation process. Visualization results illustrate that models with our attention module and diversity regularizers capture more critical localization information. And competitive performance on large-scale face recognition benchmark verifies the effectiveness of our approaches.

Keywords: Face recognition · Attention · Diversity

1 Introduction

The performance of face recognition has been rapidly improved due to the prodigious progress of deep learning [25] and the investigation on metrics for face recognition. Specifically, many works [7,17,23,27] delve into metrics and loss functions used in the training phase, aiming at enlarging the distance among inter-class samples and lessening the distance among intra-class samples. Even though the breakthrough on metric function, the structure of CNNs and the interpretation of face representation remain less explored.

Other than face recognition, face representation has been widely used in applications such as age estimation [8], facial landmarks [2] and so on. It implies that fine-grained and localization information are crucial in face representation. Meanwhile,

This research was supported by Fundamental Research Funds for the Central Universities (2020XD-A06-1), the State Key Program of the National Natural Science Foundation of China (82030037), the National Science and Technology Major Project of China (No. 2017ZX03001022).

T. Mantoro et al. (Eds.): ICONIP 2021, LNCS 13108, pp. 116–128, 2021.
https://doi.org/10.1007/978-3-030-92185-9_10

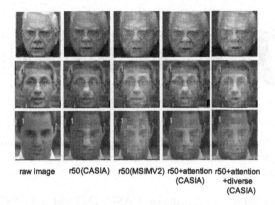

raw image   r50(CASIA)   r50(MS1MV2) r50+attention r50+attention
                                        (CASIA)   +diverse
                                                     (CASIA)

Fig. 1. Response visualization. Darker red region refers to the higher effec of this region on face representation. Higher response mainly focus on sense organs such as eyes, nose and lips. Models with different performance (trained on CASIA and MS1MV2) have different response. Our proposed attention module can emphasize the key regions and our diversity regularizers can generate comprehensive response. (Color figure online)

localization information is also important in pixel-level prediction tasks such as segmentation [15] and large instance identification [16]. Therefore, we assume that the mechanisms used in pixel-level prediction will help improving face representation. There are two mechanisms we tend to apply to face recognition [16]:

1. Integrating information of low-level features to the representation of samples.
2. Preserving localization information as much as possible.

We employ these mechanisms based on the study following: we visualize the disparity of embedding by occluding the regions of raw images to explore the effect of different regions at face images on the face representation (visualization details described in Sect. 4.4). The results shown in Fig. 1. We can see that, sense organs such as eyes, nose and lips have higher response in terms of other regions, which is align to the tendency of face information capturing of human. Moreover, we find that the response of regions are different from models with different performance: response of model trained with large dataset MS1MV2 [7] is different from model trained with CASIA [32]. These phenomena all suggest the importance of localization information (in the low-level) and the influence of localization information from different regions. The code and demo will be publicly available.

To sum up, our contributions are listed as follows:

1. Considering the localization information in the intermediate layer of CNNs, we propose a new attention module to incorporate localization information of low-level features into high-level representation. It can enhance the effect of key regions on face representation and reduce the variance caused by noise regions.
2. We propose two types of diversity regularizers to preserve localization information from different regions and reduce the redundancy in the representation within limited computational cost.

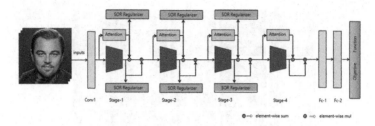

Fig. 2. The overall structure of our methods. We propose the attention module over stage, and employ diversity regularizers after attention module and trunk branch.

3. We conduct extensive experiments on CASIA [32] and MS1MV2 [7,9] datasets. The state-of-the-art performance of evaluation on IJB-C [19], MegaFace [13] shows the effectiveness of our methods.

2   Related Work

Attention mechanism has been widely used in computer vision tasks [11,31] since it can automatically adjust the focal of feature like the human visual system. To improve the representation, [11] introduces squeeze-and-excitation block to constrain the interdependencies between the channels. Different from [31], we employ 3D attention maps to integrate localization and global information. Specifically, our 3D attention module perform downsampling to obtain global information, and our 3D attention module is over stage which can efficiently extract low-level or localization information. Meanwhile, our attention mechanism tends to capture global and localization information of the images in the low-level features, inspired by pixel-level prediction tasks [15,16] and non-local mechanism [28]. Different from non-local method [28], we perform integration of information over stage rather than the same stage.

Many previous works have shown that the orthogonality of features and decorrelation of features can enhance the performance of CNNs [1,26,33]. [33] employs diversity regularizers to improve the interpretation ability of face representation. In [26], through SVD, the measurement of embeddings is equivalent to the measurement of the multiplication of features and US of fully connected layer. [1] investigate the norm preserving of features and soft orthogonality regularization to encourage the orthogonality of features and weights. However, these methods need much computation to perform eigenvalue decomposition and the approximation process. We propose two types of diversity regularizers inspired by [21] and [14] which can reduce the computational cost and enhance the diversity of representation the same time.

3   Approach

3.1   Overall Network Architecture

As we discussed above, localization information will help fine-grained tasks such as face recognition. Thus, to emphasize the representation of localization, we incorporate

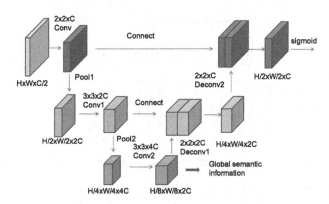

Fig. 3. Our attention module is divided into two parts. In the first part, the input feature maps are first downsampled by convolution and pooling operations to extract global and localization information. In the second part, stride 2 deconvolution is used for upsampling, and the localization information extracted in the first part is used to guide importance selection of attention module.

low-level features into attention of high level representation, as shown in Fig. 2. Our attention module is a plug and play module based on different stages of ResNet [10]. In every stage of ResNet, the input feature maps of the current stage contain more localization information than the output of that. Based on that, we take the output of last stage or the input of the current stage as the input of attention module. The output of attention module is the mask for trunk branch of the current stage.

To avoid degradation of feature values by dot production and to preserve good property of trunk branch, we employ attention residual learning in our architecture, i.e. the residual of trunk branch is the multiplication of the attention maps and trunk branch. Given $X \in \mathbb{R}^{C_i \times H_i \times W_i}$ as the input feature maps of stage i, where C_i is the number of channels and $H_i \times W_i$ is the size of feature maps. The output of trunk branch of this stage i is $\tilde{X}_i = F_i(X_i)$, $\tilde{X}_i \in \mathbb{R}^{C_{i+1} \times H_{i+1} \times W_{i+1}}$, the output of attention module of this stage is $M_i = A_i(X_i)$, $M_i \in \mathbb{R}^{C_{i+1} \times H_{i+1} \times W_{i+1}}$. Thus the output of stage i is

$$X_{i+1} = \tilde{X}_i + \tilde{X}_i * M_i \tag{1}$$

$*$ is elementwise multiply. The range of M_i is [0,1].

In addition, to enhance the diversity of localization information in our representation, we employ two types of response maps diversity loss in our attention module and the trunk branch. The diversity loss used after attention maps is to enforce the diversity of attention masks while that used after the trunk branch is to encourage the decorrelation of feature maps.

3.2 Attention Module

Previous attention modules focus on separate channel attention (the attention feature map is 1D, $M_c \in \mathbb{R}^{C \times 1 \times 1}$) and spatial attention (the attention feature map is 2D, $M_c \in \mathbb{R}^{1 \times H \times W}$), such as [11,31]. The noise from background or redundant information from feature maps will be relieved in this way. However, since localization and

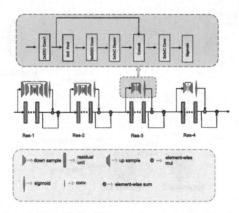

Fig. 4. Attention modules for four stages. The downsampling ratio and downsample modules are changed according to the size of feature maps.

global information are more important for fine-grained tasks such as face recognition, we employ 3D attention module, i.e. the attention feature map $M \in \mathbb{R}^{C \times H \times W}$ is 3D, to integrate and utilize localization and global information.

To better incorporate localization and global information in low-level features, we design our attention module inspired by encoder-and-decoder approaches such as FCN [18] and U-Net [22], which are widely used in segmentation tasks. The architectures of FCN and U-Net first perform convolution or downsampling to shrink the spatial size of feature maps and then employ deconvolution or upsampling to expand the spatial size of feature maps. There are skip connection between downsampling and upsampling feature maps. These architectures will be more robust against small input variance and the larger receptive field while retain the resolution. Our attention module also contains downsampling, upsampling and skip connection, as shown in Fig. 3. We employ concatenation in skip connections.

Constrained by the spatial size, we design different attention module architectures for different stages, as shown in Fig. 4. The ratio of downsampling will increase in bottom stages since it needs more downsampling to obtain global and localization information. In addition, to customize attention module to face recognition tasks, we first employ a stride 2 residual unit to obtain the same spatial size as the output, then employ 3×3 kernels in all convolution layer and use stride 2 deconvolution to perform upsampling. Figure 4 illustrates the attention modules for four stages.

3.3 Response Maps Diversity Loss

Even though incorporating low-level features into attention of high level representation can help to integrate localization and global information, the attention mechanism is still prone to focus on a more compact subspace which loses sight of low feature correlations [5]. To handle this, we employ response maps diversity loss to make feature space more comprehensive [5] and enhance the interpretability of face representation [33].

In [33] the author use spatial activation diversity loss (SAD) to decorrelate the response maps of different channels:

$$L_{SAD}^{response} = \sum_{i \neq j} \left\| \frac{< x_i, x_j >}{\|x_i\|_F \|x_j\|_F} \right\|^2 \tag{2}$$

x_i is the response map of filter i in the layer. If x_i is a unit-norm vector of $\mathbb{R}^N$ where $N = H \times W$ ($H \times W$ is the spatial size of response maps), the SAD loss is equivalent to soft orthogonality regularization [1,5]:

$$L_{SO} = \left\| XX^T - I \right\|_F^2 \tag{3}$$

where $X \in \mathbb{R}^{C \times N}$ is response maps of the layer, C is the number of channels or filters. We can see that, when $C > N$, the rank of gram matrix of X is equal to or less than N, which cannot reach identity. And in most of layers in CNNs, C is larger than N. To solve this problem, [1] proposes the spectral norm-based regularizer and [5] regularizes the condition number of gram matrix. However, the regularizers of [1,5] all need to solve eigenvalue decomposition (EVD) or the approximation process.

To avoid the expensive computation of EVD or the approximation process, we propose two types of response maps diversity loss which can enhance the diversity of response maps and save computational cost.

We employ a new formulation proposed by [21], this formulation is different from the often-used method that considers response maps at a layer of one sample:
Considered over a dataset X with m examples, a neuron is a vector in $\mathbb{R}^m$. A layer is the subspace of $\mathbb{R}^m$ spanned by its neurons' vector.

Follow this formulation, we assume that the neurons' vector can be the orthonormal set of the subspace. The orthogonality of orthonormal set implies the decorrelation between the response maps of different neurons. It can reduce the redundancy in the response maps and enhance the diversity of response maps from different filters. Thus, we propose the subspace orthogonality regularizer (SOR):

$$L_{SOR} = \left\| FF^T - I \right\|_F^2 \tag{4}$$

where $F \in \mathbb{R}^{C \times M}$ is response maps of a layer, and $M = B \times H \times W$, B is the number of samples in a batch. In SOR, since $M > C$, the identity of gram matrix of F is able to be approached. The previous methods such as SAD and SO take the number of samples into the computation of the expectation of the loss. In our method, we consider the number of samples as the formulation of the subspace. We compute the gradient of SOR via auto-differentiation. Our SOR can achieve response maps diversity while using limit computational cost.

4 Experiments

In this section, we first introduce the implementation details and evaluation metrics of our experiments. Then we perform ablation study of our proposed methods to illustrate the effect of every method. After that, we compare our approach with existing attention and diversity module and state-of-the-art face recognition. Finally, we visualize and analyze our networks to further validate the effectiveness of our approach.

4.1   Implementation Details and Evaluation

In our experiments, we employ ResNet-50 and ResNet-100 [10] as our backbone and conduct experiments on MXNet [6]. We follow the benchmark of insightface [7]: we employ arcface as classification loss whose margin is 0.5 and scale is 64 and perform evaluation using the final 512 embedding. We train our networks on CASIA [32] and MS1MV2 [7,9]. There are 10K identities and 0.5M images in CASIA [32] dataset and 85K identities and 5.8M images in MS1MV2 [7]. We evaluate our models on LFW [12], CFP-FP [24], AgeDB-30 [20], CPLFW [34], CALFW [35]. And we further test our models on large-scale face recognition dataset MegaFace [13] and IJB-C [19] for validating the effectiveness of our approach.

4.2   Ablation Study

We perform ablation study on MS1MV2 Dataset [7] and ResNet-100 [10]. In the baseline, we set the batchsize of every GPU to 48, and we use 8 NVIDIA GTX 1080Ti GPUs. We employ arcface as our baseline. We add proposed methods on baseline and compared with existing attention module SE block [4] and diversity method SVD-Net [26]. The results are shown in Table 1.

Table 1. Ablation Study of our proposed methods on MS1MV2. Verification results(%) on different evaluation datasets demonstrate the effectiveness of our proposed methods.

Method	CPLFW	CALFW	MegaFace(ID/Ver)	IJB-B	IJB-C
baseline	92.08	95.54	98.35/98.48	94.2	95.6
+SE[4]	92.54	95.70	98.38/98.51	94.2	95.7
+SVD-Net[31]	92.41	95.56	98.38/98.50	94.1	95.6
+our attention	92.96	96.08	98.43/98.54	94.5	95.8
+SOR	92.78	95.72	98.40/98.51	94.3	95.7
+our attention+SOR	**93.20**	**96.18**	**98.48/98.60**	**94.7**	**96.0**

There are six experiments shown in Table 1: 1) baseline experiment conducted according to description above; 2) add SE block on ResNet-50 based on baseline; 3) add SVD layer on the top of ResNet-50; 4) add our proposed attention module on ResNet-50; 5) add SOR on the output of every stage; 6) add our attention module SOR on the output of every stage. We can see that, the performance of our proposed methods (experiment 4, 5) are superior to SE block and SVD-Net (experiment 2, 3). The combinations of attention module and diversity regularizers (experiment6) have better performance than single method which validates our ideas.

In Table 2, we use the same train set MS1MV2 and verification results on different evaluation datasets demonstrate the effectiveness of our proposed methods.

Table 2. Use the same train set MS1MV2. Verification results(%) on different evaluation datasets demonstrate the effectiveness of our proposed methods.

Method	CPLFW	CALFW	MegaFace(ID/Ver)	IJB-B/IJB-C	FLOPS(G)
R50	91.78	94.37	98.30/98.40	93.3/95.0	12.6
R50+Our Method	92.41	96.07	98.40/98.52	94.3/95.6	14.6
R100	92.08	95.54	98.35/98.48	94.2/95.6	24.4
R100+Our Method	**93.20**	**96.18**	**98.48/98.60**	**94.7/96.0**	**28.3**

4.3   Compared with State-of-the-Art Face Recognition Methods

To compare with existing face recognition tasks, we trained our models on MS1MV2 Dataset [7], and employ ResNet-100 [10] as our backbone. We conduct our experiments on 8 GPUs, the batchsize of each GPU is 42. We still employ arcface loss as our classification loss and follow the parameters settings in [7].

Table 3. Results of MegaFace challenge 1 on FaceScrub dataset.

Method	ID (%)	Ver (%)
Softmax [17]	54.85	65.92
Contrastive Loss [25]	65.21	78.86
Triplet [23]	64.79	78.32
CenterLoss [29]	85.48	77.48
SphereFace [17]	90.30	81.40
FaceNet [23]	70.49	86.47
CosFace [27]	82.72	96.65
MS1MV2, R100, ArcFace, R [7]	98.35	98.48
Our Method, R	**98.48**	**98.60**

The MegaFace dataset [13] includes 1M images of 690K different individuals in the gallery set and 100K photos of 530 individuals from FaceScrub in the probe set. On MegaFace, there are two testing scenarios (identification Id(%) and verification Ver(%)) under two protocols (large or small training set). R refers to iBUG clean version of MegaFace. The training set is defined as large if it contains more than 0.5M images. For the fair comparison, we train our method on MS1MV2 under large protocol. The results of evaluation on MegaFace are shown in Table 3. Our method achieves the best single-model identification and verification performance, not only surpassing the strong baseline arcface but also outperforming other existing methods.

The IJB-B dataset [30] contains 1, 845 subjects with 21.8 K static images and 55 K frames from 7, 011 videos. In total, there are 12, 115 templates with 10, 270 genuine matches and 8M impostor matches. The IJB-C dataset [19] is a further extension of

Table 4. 1:1 verification TAR(@FAR = 1e−4) on the IJB-B and IJB-C dataset.

Method	IJB-B	IJB-C
ResNet50 [3]	0.784	0.825
SENet50 [3]	0.800	0.840
ResNet50+SENet50 [3]	0.800	0.841
MS1MV2, R100, ArcFace [7]	0.942	0.956
Our Method	**0.947**	**0.960**

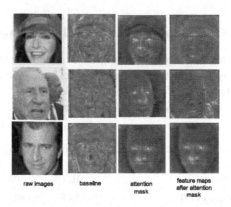

raw images   baseline   attention   feature maps
                                         mask   after attention
                                                             mask

Fig. 5. Visualization of the output of first stage. The first column is the raw images, the second column is the feature maps obtained by the baseline, the third column is the attention mask obtained from our attention model, and the fourth column is the feature maps obtained by fusion of attention masks and feature maps of trunk branch.

IJB-B, having 3, 531 subjects with 31.3 K still images and 117.5 K frames from 11, 779 videos. In total, there are 23, 124 templates with 19, 557 genuine matches and 15, 639 K impostor matches. In Table 4, we compare the TAR (@FAR = 1e−4) of our model with the previous state-of-the-art models. Our method achieves better verification results compared with the results in previous papers.

4.4   Visualization

We take the feature maps $(B \times C \times H \times W)$ from the output of the first stage of the model, averaging it in the dimension of C, and obtaining a response visualization image, Fig. 5 shows the visualization results. We can observe that our attention mechanism enables the model to obtain global and localization information in the shallow structure and uses these information to standardize the response and make the response clearer. The response of images has no highlight in different regions in the baseline, while attention masks of our attention module emphasize foreground and important organs of human. By using our attention module, noise and redundant information can be suppressed, and critical regions will have higher response.

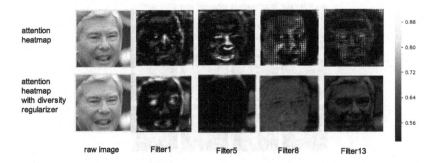

Fig. 6. Visualization of attention heatmaps. The first line shows attention heatmaps obtained by the model which employs our attention mechanism. The second line shows attention heatmaps obtained by the model which employs both our attention mechanism and the diversity mechanism. Even through critical localization information can be captured by our attention mechanism, diversity regularizers still help to preserve more localization information and reduce the correlation between filters.

We draw the attention heat map of the attention module at first stage, shown in Fig. 6. Since our attention mechanism is a unique three-dimensional structure, we randomly selected some filters for visualization. We can see that there are differences between the filters in our attention heatmaps, and captured information shown in our attention heatmaps is more complex and richer than the captured information of attention mechanisms in other related work. Moreover, the correlation between the filters in the attention heatmaps of the model which employ our diversity mechanism is decreasing. Our diverse mechanism enables our model to obtain more diverse attention information.

We employ a $N \times N$ mask to block a part of the original input picture, and obtain the 512-D embedding of this picture. We calculate the distance between the embeddings before and after the occlusion and visualize it on Fig. 7. Large distance between response refers to the effect of this region for face representation. The red area represents the area that the model pay special attention to, and the blue area represents the area that the model has no special attention. We can see that, the regional response of baseline only cover few regions of organs while our attention mechanism makes the response of the model more concentrated on the areas which are important for the recognition task (sense organs). The use of the diversity regularizers makes our model's attention area more comprehensive.

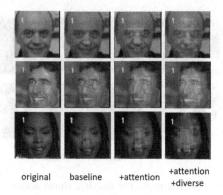

original   baseline   +attention   +attention +diverse

Fig. 7. Regional response visualization. The red region corresponds to the effect of this region to representation. Our attention module can emphasize the key regions and our diversity regularizers encourage the comprehensive of localization information. (Color figure online)

5   Conclusions

In this paper, we propose a new attention structure and two diversity regularizers for improving the discriminative ability of face representation. Our attention module can incorporate low-level information into features and emphasize global and localization information. Our diversity regularizers can decorrelate filters and encourage the comprehensive of localization information. Through extensive experiments on large-scale face recognition benchmark and visualization of features maps and representation, we validate that the effectiveness and efficiency of our proposed methods. In the future, we will extend our approaches to enhance the interpretation of CNNs and the robustness of face representation under occlusions.

References

1. Bansal, N., Chen, X., Wang, Z.: Can we gain more from orthogonality regularizations in training deep networks? In: Advances in Neural Information Processing Systems, pp. 4261–4271 (2018)
2. Burgos-Artizzu, X.P., Perona, P., Dollár, P.: Robust face landmark estimation under occlusion. In: Proceedings of the IEEE International Conference on Computer Vision, pp. 1513–1520 (2013)
3. Cao, Q., Shen, L., Xie, W., Parkhi, O.M., Zisserman, A.: Vggface2: a dataset for recognising faces across pose and age. In: 2018 13th IEEE International Conference on Automatic Face & Gesture Recognition (FG 2018), pp. 67–74. IEEE (2018)
4. Cao, Y., Xu, J., Lin, S., Wei, F., Hu, H.: Gcnet: non-local networks meet squeeze-excitation networks and beyond. In: Proceedings of the IEEE International Conference on Computer Vision Workshops (2019)
5. Chen, T., et al.: Abd-net: attentive but diverse person re-identification. In: Proceedings of the IEEE International Conference on Computer Vision, pp. 8351–8361 (2019)
6. Chen, T., et al.: Mxnet: a flexible and efficient machine learning library for heterogeneous distributed systems. arXiv preprint arXiv:1512.01274 (2015)

7. Deng, J., Guo, J., Xue, N., Zafeiriou, S.: Arcface: additive angular margin loss for deep face recognition. In: Proceedings of the IEEE Conference on Computer Vision and Pattern Recognition, pp. 4690–4699 (2019)

8. Fu, Y., Guo, G., Huang, T.S.: Age synthesis and estimation via faces: a survey. IEEE Trans. Pattern Anal. Mach. Intell. **32**(11), 1955–1976 (2010)

9. Guo, Y., Zhang, L., Hu, Y., He, X., Gao, J.: MS-Celeb-1M: a dataset and benchmark for large-scale face recognition. In: Leibe, B., Matas, J., Sebe, N., Welling, M. (eds.) ECCV 2016. LNCS, vol. 9907, pp. 87–102. Springer, Cham (2016). https://doi.org/10.1007/978-3-319-46487-9_6

10. He, K., Zhang, X., Ren, S., Sun, J.: Deep residual learning for image recognition. In: Proceedings of the IEEE Conference on Computer Vision and Pattern Recognition, pp. 770–778 (2016)

11. Hu, J., Shen, L., Sun, G.: Squeeze-and-excitation networks. In: Proceedings of the IEEE Conference on Computer Vision and Pattern Recognition, pp. 7132–7141 (2018)

12. Huang, G.B., Mattar, M., Berg, T., Learned-Miller, E.: Labeled faces in the wild: a database for studying face recognition in unconstrained environments (2008)

13. Kemelmacher-Shlizerman, I., Seitz, S.M., Miller, D., Brossard, E.: The megaface benchmark: 1 million faces for recognition at scale. In: Proceedings of the IEEE CConference on Computer Vision and Pattern Recognition, pp. 4873–4882 (2016)

14. Kornblith, S., Norouzi, M., Lee, H., Hinton, G.: Similarity of neural network representations revisited. arXiv preprint arXiv:1905.00414 (2019)

15. Li, H., Xiong, P., An, J., Wang, L.: Pyramid attention network for semantic segmentation. arXiv preprint arXiv:1805.10180 (2018)

16. Liu, S., Qi, L., Qin, H., Shi, J., Jia, J.: Path aggregation network for instance segmentation. In: Proceedings of the IEEE Conference on Computer Vision and Pattern Recognition, pp. 8759–8768 (2018)

17. Liu, W., Wen, Y., Yu, Z., Li, M., Raj, B., Song, L.: Sphereface: deep hypersphere embedding for face recognition. In: Proceedings of the IEEE Conference on Computer Vision and Pattern Recognition, pp. 212–220 (2017)

18. Long, J., Shelhamer, E., Darrell, T.: Fully convolutional networks for semantic segmentation. In: Proceedings of the IEEE Conference on Computer Vision and Pattern Recognition, pp. 3431–3440 (2015)

19. Maze, B., et al.: Iarpa janus benchmark-c: face dataset and protocol. In: 2018 International Conference on Biometrics (ICB), pp. 158–165. IEEE (2018)

20. Moschoglou, S., Papaioannou, A., Sagonas, C., Deng, J., Kotsia, I., Zafeiriou, S.: Agedb: the first manually collected, in-the-wild age database. In: Proceedings of the IEEE Conference on Computer Vision and Pattern Recognition Workshops, pp. 51–59 (2017)

21. Raghu, M., Gilmer, J., Yosinski, J., Sohl-Dickstein, J.: Svcca: singular vector canonical correlation analysis for deep learning dynamics and interpretability. In: Advances in Neural Information Processing Systems, pp. 6076–6085 (2017)

22. Ronneberger, O., Fischer, P., Brox, T.: U-Net: convolutional networks for biomedical image segmentation. In: Navab, N., Hornegger, J., Wells, W.M., Frangi, A.F. (eds.) MICCAI 2015. LNCS, vol. 9351, pp. 234–241. Springer, Cham (2015). https://doi.org/10.1007/978-3-319-24574-4_28

23. Schroff, F., Kalenichenko, D., Philbin, J.: Facenet: a unified embedding for face recognition and clustering. In: Proceedings of the IEEE Conference on Computer Vision and Pattern Recognition, pp. 815–823 (2015)

24. Sengupta, S., Chen, J.C., Castillo, C., Patel, V.M., Chellappa, R., Jacobs, D.W.: Frontal to profile face verification in the wild. In: 2016 IEEE Winter Conference on Applications of Computer Vision (WACV), pp. 1–9. IEEE (2016)

25. Sun, Y., Wang, X., Tang, X.: Deep learning face representation from predicting 10,000 classes. In: Proceedings of the IEEE Conference on Computer Vision and Pattern Recognition, pp. 1891–1898 (2014)
26. Sun, Y., Zheng, L., Deng, W., Wang, S.: Svdnet for pedestrian retrieval. In: Proceedings of the IEEE International Conference on Computer Vision, pp. 3800–3808 (2017)
27. Wang, H., et al.: Cosface: large margin cosine loss for deep face recognition. In: Proceedings of the IEEE Conference on Computer Vision and Pattern Recognition, pp. 5265–5274 (2018)
28. Wang, X., Girshick, R., Gupta, A., He, K.: Non-local neural networks. In: Proceedings of the IEEE Conference on Computer Vision and Pattern Recognition, pp. 7794–7803 (2018)
29. Wen, Y., Zhang, K., Li, Z., Qiao, Yu.: A discriminative feature learning approach for deep face recognition. In: Leibe, B., Matas, J., Sebe, N., Welling, M. (eds.) ECCV 2016. LNCS, vol. 9911, pp. 499–515. Springer, Cham (2016). https://doi.org/10.1007/978-3-319-46478-7_31
30. Whitelam, C., et al.: Iarpa janus benchmark-b face dataset. In: Proceedings of the IEEE Conference on Computer Vision and Pattern Recognition Workshops, pp. 90–98 (2017)
31. Woo, S., Park, J., Lee, J.Y., So Kweon, I.: Cbam: convolutional block attention module. In: Proceedings of the European Conference on Computer Vision (ECCV), pp. 3–19 (2018)
32. Yi, D., Lei, Z., Liao, S., Li, S.Z.: Learning face representation from scratch. arXiv preprint arXiv:1411.7923 (2014)
33. Yin, B., Tran, L., Li, H., Shen, X., Liu, X.: Towards interpretable face recognition. arXiv preprint arXiv:1805.00611 (2018)
34. Zheng, T., Deng, W.: Cross-pose lfw: a database for studying cross-pose face recognition in unconstrained environments. Beijing University of Posts and Telecommunications, Technical report 5 (2018)
35. Zheng, T., Deng, W., Hu, J.: Cross-age lfw: a database for studying cross-age face recognition in unconstrained environments. arXiv preprint arXiv:1708.08197 (2017)

Multi-task Perceptual Occlusion Face Detection with Semantic Attention Network

Lian Shen[1,2], Jia-Xiang Lin[1,2(✉)], and Chang-Ying Wang[1]

[1] College of Computer and Information Sciences,
Fujian Agriculture and Forestry University, Fuzhou 350028, China
{shenlian,linjx,wangchangying}@fafu.edu.cn
[2] Key Laboratory of Smart Agriculture and Forestry,
Fujian Agriculture and Forestry University, Fujian Province University,
Fuzhou 350028, China

Abstract. Face detection has been well studied for many years and one of the remaining challenges is to detect complex occluded faces in a real-world environment. Hence, this paper introduces a Multi-task Perceptual Occlusion Face Detection framework with a semantic attention network (MTOFD), which can detect face under complex occlusion conditions, especially, it takes the discrimination of occlusion type as a learning task, and use occlusion semantic information to improve face detection. In addition, an adaptive semantic attention network is employed to solve the conflict problem caused by multi-task in feature fusion, in which the potential semantic information of the occlusion task is learned adaptively, and the most important semantic information is selected and aggregated automatically to the task of occlusion face detection. Finally, MTOFD is tested and compared with some typical algorithms, such as FAN and AOFD, and it is found that our algorithm achieves state-of-the-art performance on dataset MAFA.

Keywords: Occlusion face detection · Multi-task learning · Semantic attention

1 Introduction

Face detection is a fundamental and essential task in various face applications. From the pioneering work of Viola-Jones [14] face detectors to recent state-of-the-art CNN-based methods [3,12,19], the research on face detection has achieved remarkable success and is widely used in automation, medical, security fields. However, face detection in occlusion scenarios is still a challenging problem. More importantly, face occlusion caused by masks, glasses, etc. is widespread in our lives. The traditional face detection algorithm will be greatly affected under the occlusion condition.

To solve the occlusion problem, in recent, many methods [15,18,22] mainly focus on unoccluded face regions. Extract highly discriminative features from

© Springer Nature Switzerland AG 2021
T. Mantoro et al. (Eds.): ICONIP 2021, LNCS 13108, pp. 129–140, 2021.
https://doi.org/10.1007/978-3-030-92185-9_11

it for detection. However, occluded faces have significant intra-class variation, leading to a poor generalization of features based on non-occluded face region extraction. Meanwhile, the lack of a high-quality occlusion face database also restricts the research of this type of algorithm. Inspired by the human multi-modal vision mechanism, the occlusion face detection problem can be regarded as a multi-task problem [13,16]. Only by understanding the semantic information of both the occluded object and the face can we effectively solve the occlusion problem in face detection. Obviously, face occlusion objects are endless in reality, and it is impossible to learn complete occlusion type. However, high-order semantic information can abstract representative relation between occluded object and face. So, the semantic information can be integrated into the face detection task to solve the occlusion problem.

Based on the above problem and reflections, in this paper, we focus on the semantic relationship between occlusion and face and use a neural network to learn more robust mixed features of occlusion. To be specific, we propose a method MTOFD to practice our idea. MTOFD takes the occlusion classification task as an auxiliary task to learn occlusion semantic information. Occlusion information is shared in each detection header through a multi-task framework. However, the lower the correlation between tasks, the lower the fusion feature discriminative ability. In order to effectively solve the problem of feature fusion in multi-task, a semantic attention network (SAN) is designed in MTOFD. SAN can adaptively select effective features of occlusion into object task.

Main contributions of our work are listed as follows:

1. An end-to-end framework (MTOFD) is proposed for face detection under complex occlusion conditions. Our method provides a new idea to solve complex face occlusion detection problem by using occlusion semantic information.
2. Feature fusion problem caused by multi-task weight sharing to be solved through the semantic attention network (SAN). The ablation experiment confirmed the effectiveness of SAN, which could easily be generalized to other multi-tasking learning frameworks.
3. According to experiments on MAFA and WIDER FACE, Our approach not only achieves state-of-the-art performance under complex occlusion conditions but also can perform well in other special conditions. Such as illumination, rotation, blur, etc.

2   Related Work

2.1   Occlusion Face Detection

DPM [5] is an earlier face detection algorithm to deal with the occlusion problem. It perceives the occlusion area information through a deformable model, which improves the detection accuracy but reduces the real-time performance. In recent years, face detection algorithms based on CNN have been continuously proposed, and face detection performance has been greatly improved, and people have also turned their attention to more complex occlusion face detection problems. FAN [15] algorithm proposed based on the idea of enhancing the non-occluded face region, by introducing the attention mechanism, greatly reduces the false positives (FP) increased due to occlusion. [21] proposed a context-based multi-scale regional convolutional neural network CMS-RCNN that refers to the human visual mechanism, that is, when the face and body are detected at the same time and meet a specific spatial relationship, the person is judged The face exists. It enriches the semantic features of the face area and suppresses the influence of occlusion information on normal face features. Based on the idea of integrated detection. Although the above methods have achieved good results under the condition of partially occluding the human face, they perform poorly in the case of complex occlusion and large-area occlusion. AOFD [2] adds semantic segmentation branch to enhance the semantic information of occlusion face detection, and proposes a mask generation module, using this module, let the network learn discriminative features.

2.2 Multi-task Learning

Caruana [1] first conducted a detailed analysis of multi-task learning (MTL). Since then, there have been several ways to use MTL to solve different problems in computer vision. One of the earlier approaches for jointly addressing the tasks of face detection, pose estimation, and landmark localization was proposed in [23]. This method take global feature mixtures to capture the topological changes due to viewpoint variations. A joint cascade-based approach was proposed in [17] for simultaneously detecting faces and landmark points. This method improves the detection performance by adding face alignment steps into the cascade structure. Eigen and Fergus [4] proposed a multi-scale CNN to simultaneously predict depth, surface normals and semantic labels of pictures. They use three different scales of CNN networks, where the output of the smaller network is fed as input to the different scales of the larger network. UberNet [8] uses a similar concept to train medium-low, medium and high-level vision tasks at the same time. It combines CNN multi-task training with three image pyramids of different proportions in all middle layers.

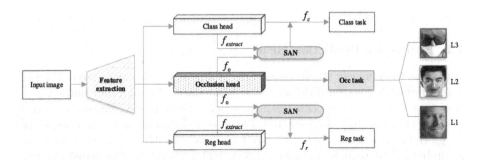

Fig. 1. Overall framework of multi-task occlusion perception network (MTOFD).

3 Method

3.1 Overall Framework

Figure 1 shows the framework of the MTOFD face detector. MTOFD is based on the RetinaNet [9] structure, which is mainly composed of two modules: feature extraction and multi-task detection head. It has three prediction heads: class head (face vs background), Occ head (Occlusion classification), reg head (bounding box regression). Each predictive head obtains shared features from the feature extraction module, feature extraction module adopts feature parymid network to extract features based on RetNet50.

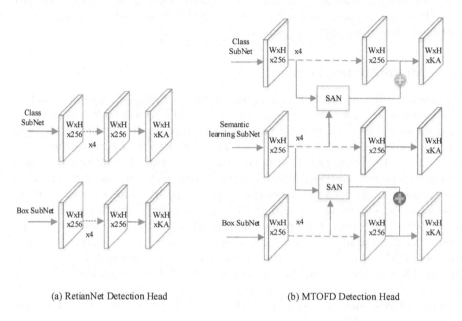

(a) RetianNet Detection Head   (b) MTOFD Detection Head

Fig. 2. Network architecture of Multi-task prediction head

3.2 Multi-task Prediction Head

Multi-task learning can share the features of different tasks, and shared features have different impact on different tasks. Retinaface [3] design four mutual promotion tasks to extract specific features, which have more discrimination relative to single task learning. However, multi-task learning framework requires specific design to balance the impact of feature fusion and RHSOD [11] found that there is gap between classification and regression tasks. Therefore, the shared features based on the backbone network are not necessarily a promotion of the task. Based on the above research, we improve the multi-task detection heads of RetinaNet (see Fig. 2a). Specifically, we add an auxiliary task to solve specific problems (occlusion face detection). The auxiliary detection head extract occlusion information from the shared feature, and occlusion information through SAN to be selected. Finally by the residual structure connect corresponding task prediction head network and occlusion information tranform to object task (see Fig. 2b).

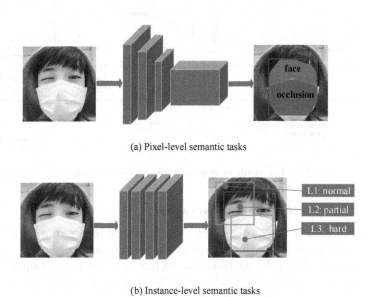

(a) Pixel-level semantic tasks

(b) Instance-level semantic tasks

Fig. 3. Different orders of semantic recognition task. (a) Recognition and segmentation of pixels (occlusion vs face) in the face box. (b) Determine the occlusion category of the face box. All instances of face box are anchor box of $IoU > 0.4$

3.3 Semantic Learning Task

Different previous occlusion face detection algorithms focus on non-occlusion area features, MTOFD purpose is extracting effective occlusion information to improve the accuracy of complex occlusion face detection. But how to get effective occlusion information is a problem? AOFD [20] extracts semantic information through pixel-level semantic segmentation task to improve detection accuracy (see Fig. 3a). High-order semantic information can be regarded as some combination of low-order visual information, and semantic information can handle abstract occlusion detection issues. Inspired by it, we learn semantics information through an instance semantic classification task. Specifically, we identify each face anchor box instance of a semantic label: L1, L2, L3 (see Fig. 3b), representing: no occlusion, partial occlusion, difficult occlusion. Then, the face data of the given lables were learned through the instance classification task, and the occlusion semantic information was obtained. The specific loss function is introduced in Sect. 3.5.

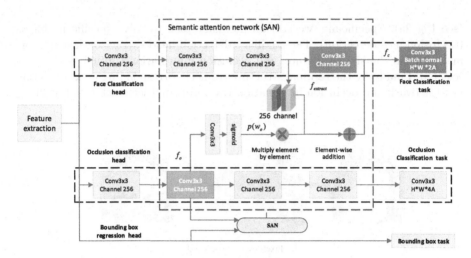

Fig. 4. Details of Semantic attention network

3.4 SAN Semantic Attention Network

The main function of SAN is to extract the semantic information of auxiliary tasks and selectively put it into the features of the main task. Specific network structure see Fig. 4. In the black dotted box is: face classification head and occlusion classification head. Set current training dataset is $D\{x^+, x^-\}$, where x^+ denote the occluded face and x^- denote the non-occluded face. The training sample $x \in D$ is sent to the three prediction heads after the feature extraction module. Face classification head input is $f_{extract} \in \mathbb{R}^{c \times m \times n}$, occlusion classification head input is the $f_o \in \mathbb{R}^{c \times m \times n}$. The probability map $p(w_o) \in \mathbb{R}^{256 \times m \times n}$ of the feature response of the occlusion classification task is obtained by convolution of f_o. The calculation formula of the probability map is as follows:

$$p(w_o) = softmax(conv(f_o)) \tag{1}$$

$p(w_o)$ and $f_{extract}$ are multiplied by each channel to get the feature fusion map $\mathbf{S}$. Finally, add $\mathbf{S}$ and $conv(f_{extract})$ to get f_c. So, the feature map f_c contains the corresponding occlusion semantic information:

$$f_c = f_{extract} * p(w_o) + conv(f_{extract}) \tag{2}$$

The above is the workflow of SAN. When $x \in x^-$, $\mathbf{S}$ containing occlusion semantic information. When $x \in x^+$, $\mathbf{S}$ containing face semantic information. f_c adaptively obtains features that are beneficial to face classification from $\mathbf{S}$.

3.5 Network Loss Function

Anchor parameters are consistent with RetinaNet. For any training anchor i, we minimise the following multi-task loss:

$$loss = L_{cls} + L_{reg} + L_{occ} \tag{3}$$

Where L_{cls} is an face binary classification loss, L_{box} is a regression loss of face box positions, L_{occ} is a occlusion three classification loss (no occlusion, partial occlusion, complex occlusion).

There is a problem of unbalanced positive and negative samples in face classification task. Hence, we use a focal loss [9] For face classification:

$$L_{cls} = \frac{-1}{N_{pos}} * [\alpha \sum_{i \in Pos} (1 - p_i)^\gamma \log p_i] \\ + (1 - \alpha) \sum_{i \in Neg} p_i^\gamma log(1 - p_i) \tag{4}$$

We apply a smooth version of L_1 loss for bounding box regression task:

$$L_{box} = \frac{1}{N_{pos}} \sum_{i \in Pos} smooth_{L_1}(t_i - t_i^*) \tag{5}$$

For occlusion classification loss we apply a cross entropy loss:

$$L_{occ} = \frac{1}{N_{pos}} \sum_{i \in Pos} -[o_i * \log o_i^* \\ + (1 - o_i) * \log (1 - o_i^*)] \tag{6}$$

Face classification loss L_{cls}, where p_i is a predicted probability of anchor i being a face and p_i is 0 or 1 as the corresponding negative anchors or positive anchors. α and γ are balancing parameter between the classification loss of positive and negative anchors. Parameters α and γ are set to 0.2 and 1.8. For loss L_{box}, $t_i = \{t_w, t_h, t_x, t_y\}_i$ represent coordinates of a predicted bounding box, and $t_i^* = \{t_w^*, t_h^*, t_x^*, t_y^*\}_i$ is that of the ground-truth box associated with a positive anchor i. For occlusion classification loss L_{occ}, k represents the number of occlusion categories and $k = 3$. o_i is a vector representing the 3 parameterized type of a predicted occlusion level and o_i^* is that of the ground-truth occlusion type associated with a positive anchor i.

4   Experiment

In this section, we conduct experiments on MAFA and WIDE RFACE. Firstly, we compare the experimental results of our MTOFD with state-of-the-art methods on MAFA dataset. Next, we aim to further validate the method proposed thorough ablation experiments. In addition, we compared face detection on challenge dataset WIDE RFACE. Finally, the experiments on MAFA and WIDERFACE show that the proposed method has good performance and can be generalized easily.

4.1 Data Set

WIDER FACE: The WIDER FACE dataset consists of 393,703 annotated face bounding boxes in 32,203 images, these images have variations in facial expression, lighting, occlusion, pose and scale condition. The dataset was divided into verification (10%), training (40%) and testing(50%) sets, and three difficulty levels were defined: Hard, Medium and Easy.

MAFA: The MAFA dataset consists of 30,811 images and 35806 images mask faces collected from the Internet. This is a benchmark occlusion face detection covers 60 cases of occlusion face. In addition, in the MAFA data set, human faces are divided into three categories: All, Mask, and Ignored. Specifically, the ignored faces are those with severe blur or distortion and those with dimensions less than 32 pixels.

4.2 Result on MAFA

We compare the typical occlusion face detection algorithm HPM [7], MTCNN [17], LLE-CNNs [6], FAN [15], Retinaface [3], AOFD [2] on the MAFA dataset. MTOFD was implemented based on the RetinaNet and used (Average precision) AP as the evaluation criteria. As can be seen from Table 1, MTOFD achieves the best accuracy on all three subsets of MAFA, respectively 82.1%, 92.8%, 92.3%. Compared with the advanced AOFD algorithm, MTOFD improves 0.8%, 9.3%, and 0.4% on the all, mask, ignore subsets respectively. The result proved that the proposed algorithm performs well on complex occlusion data sets. Figure 5 show some result on MAFA, these images are very challenging, such as images $\{(d), (h), (j), (m), (n)\}$, but our detector handles them well.

4.3 Result on WIDER FACE

WIDER FACE is a challenging face data set, which not only have occlusion face image, but also includes other challenge factors for face data, such as illumination, rotation, blur, etc. In order to verify the robustness of the proposed algorithm, we compared the typical face detection algorithms SSH [10], MTCNN [17], FAN [15], RetinaFace [3], Pyramid [12]. In Table 3, MTOFD achieves a good performance of 89.3% AP on the Hard subset. The result proved that our algorithm can not only deal with the special condition of face detection, but also has a good robustness performance for face detection.

4.4 Ablation Study on MAFA

In this section, we verified the various modules of the proposed algorithm. Specifically, from two aspects to analyze, one is the effect of comparing the occlusion semantic learning task on the occlusion detector, and is the verification of the adaptive semantic learning module.

Table 1. Comparison with different occlusion face detection algorithms on MAFA

Method	Backbone	MAFA (all)	MAFA (mask)	MAFA (ignore)
HPM	–	–	–	50.9%
MTCNN	–	–	–	60.3%
LLE-CNNs	–	–	–	76.7%
FAN	ResNet-101	–	76.6%	88.3%
RetinaFace	ResNet-50	75.6%	84.1%	85.2%
AOFD	Vgg16	81.3%	83.5%	91.9%
our(MTOFD)	ResNet-50	81.5%	**92.3%**	91.7%
out(MTOFD)	ResNet-101	**82.1%**	**92.8%**	**92.3%**

Instance Semantic Learning vs Pixel Semantics Learning: in Table 2, $Task_i$ represents the instance level semantic learning, $Task_P$ represents pixel semantics (see Fig. 3). As can be seen from the ablation experiment, instance-level semantics learning can extract more effective occlusion information. At the MAFA data set, $Task_i$ and $Task_p$ can improve the accuracy of the face occlusion detection. However, $Task_i$ relative to $Task_p$ increased by 1.7%, 0.7%, 0.6% respectively, which explains that instance semantic information relative to pixel semantics is more abundant. The result of the ablation experiment also further explained that the pixel level information is difficult to represent the high order high occlusion semantic and verify the limitations of low-order visual information deal with occlusion face detection on open scenes.

Semantic Attention Network: in Table 2, $Baseline+Task_P+SAN$ increased by 1.3%, 1.6% and 1.8% respectively relative to $Baseline + Task_p$ on MAFA. $Baseline+Task_i+SAN$ increased by 2.8%, 1.8% and 2.6% respectively relative to $Baseline + Task_i$ on MAFA. The results show that the semantic attention network can effectively solve the feature fusion problem and fuse the learned occlusion semantic information into the corresponding detection task. Compared with $Task_i+SAN$ and $Task_p+SAN$, $Task_i+SAN$ have greater improvement. It is proved $Task_i$ learned more abundant occlusion information relative to $Task_i$.

Table 2. Effectiveness of various designs for MTOFD on MAFA

Method	$Task_i$	$Task_p$	SAN	MAFA (all)	MAFA (mask)	MAFA (ignore)
RetinaNet (Baseline)	–	–	–	75.1%	85.8%	86.1%
RetinaNet+$Task_p$	–	✓	–	76.6%	89.6%	88.6%
RetinaNet+$Task_i$	✓	–	–	78.3%	90.3%	89.2%
RetinaNet+$Task_p$+SAN	–	✓	✓	77.9%	91.2%	90.4%
RetinaNet+$Task_i$+SAN	✓	–	✓	81.1%	92.1%	91.8%

Table 3. Comparison with different face detection algorithms on WIDERFACE

Method	Backbone	Easy	Medium	Hard
MTCNN	Vgg16	84.2%	82.3%	60.1%
SSH	Vgg16	93.2%	92.6%	84.5%
FAN	ResNet-50	95.3%	94.2%	88.8%
RetinaFace	ResNet-50	95.5%	**95.7%**	89.1%
Pyramid	Vgg16	**96.1%**	95.2%	88.1%
our(MTOFD)	ResNet-50	94.6%	94.1%	88.9%
out(MTOFD)	ResNet-101	95.5%	95.2%	**89.3%**

Fig. 5. Sample of result on MAFA.

5   Conclusion

This paper has proposed a face detection model named MTOFD to Improved occlusion face detection performance. Our model can provide rich occlusion semantic information to guide the detection task. By embedding the proposed semantic attention network into the Multi-task learning framework, the detection performance has been further improved on the MAFA benchmark. The ablation study demonstrates that instance-level semantic learning is more effective than pixel-level semantic learning. Our model also achieves superior performances than several state-of-the-art algorithms on the MAFA and good performance on WIDER FACE that have the most non-occlusion face data.

Acknowledgments. The work was supported in part by the Natural Science Foundation of Fujian Province, China (2018J01644, 2019J05048); the industrial project of Third Institute of Oceanography, Ministry of Natural Resources (KH200047A) and the industrial project of State Grid Fujian Electric Power Company (KH200129A).

References

1. Caruana, R.A.: Multitask learning: a knowledge-based source of inductive bias. Mach. Learn. Proc. **10**(1), 41–48 (1993)
2. Chen, Y., Song, L., Hu, Y., He, R.: Adversarial occlusion-aware face detection. In: 2018 IEEE 9th International Conference on Biometrics Theory, Applications and Systems (BTAS), pp. 1–9 (2018)
3. Deng, J., Guo, J., Ververas, E., Kotsia, I., Zafeiriou, S.: Retinaface: single-shot multi-level face localisation in the wild. In: Proceedings of the IEEE/CVF Conference on Computer Vision and Pattern Recognition (CVPR), pp. 5203–5212, June 2020
4. Eigen, D., Fergus, R.: Predicting depth, surface normals and semantic labels with a common multi-scale convolutional architecture. In: Proceedings of the IEEE International Conference on Computer Vision (ICCV), pp. 2650–2658, December 2015
5. Felzenszwalb, P.F., Girshick, R.B., McAllester, D., Ramanan, D.: Object detection with discriminatively trained part-based models. IEEE Trans. Pattern Anal. Mach. Intell. **32**(9), 1627–1645 (2010)
6. Ge, S., Li, J., Ye, Q., Luo, Z.: Detecting masked faces in the wild with lle-cnns. In: Proceedings of the IEEE Conference on Computer Vision and Pattern Recognition, pp. 426–434 (July 2017)
7. Ghiasi, G., Fowlkes, C.C.: Occlusion coherence: Detecting and localizing occluded faces. CoRR abs/1506.08347 (2015). http://arxiv.org/abs/1506.08347
8. Kokkinos, I.: Ubernet: training a universal convolutional neural network for low-, mid-, and high-level vision using diverse datasets and limited memory. In: Proceedings of the IEEE Conference on Computer Vision and Pattern Recognition (CVPR), pp. 5454–5463, July 2017
9. Lin, T.Y., Goyal, P., Girshick, R., He, K., Dollar, P.: Focal loss for dense object detection. IEEE Trans. Pattern Anal. Mach. Intell. **99**, 2999–3007 (2017)
10. Najibi, M., Samangouei, P., Chellappa, R., Davis, L.S.: Ssh: single stage headless face detector. In: Proceedings of the IEEE International Conference on Computer Vision (ICCV), pp. 4875–4884, October 2017
11. Song, G., Liu, Y., Wang, X.: Revisiting the sibling head in object detector. In: Proceedings of the IEEE/CVF Conference on Computer Vision and Pattern Recognition (CVPR), pp. 11563–11572, June 2020
12. Tang, X., Du, D.K., He, Z., Liu, J.: Pyramidbox: a context-assisted single shot face detector. In: ECCV, pp. 797–813 (2018)
13. Vandenhende, S., Georgoulis, S., Proesmans, M., Dai, D., Gool, L.: Revisiting multi-task learning in the deep learning era. ArXiv abs/2004.13379 (2020)
14. Viola, P., Jones, M.J.: Robust real-time face detection. Int. J. Comput. Vision **57**(2), 137–154 (2004)

15. Wang, J., Yuan, Y., Yu, G.: Face attention network: an effective face detector for the occluded faces. CoRR abs/1711.07246 (2017), http://arxiv.org/abs/1711.07246

16. Zhang, C., Yang, Z., He, X., Deng, L.: Multimodal intelligence: representation learning, information fusion, and applications. IEEE J. Sel. Top. Signal Process. **14**(3), 478–493 (2020)

17. Zhang, K., Zhang, Z., Li, Z., Qiao, Y.: Joint face detection and alignment using multitask cascaded convolutional networks. IEEE Signal Process. Lett. **23**(10), 1499–1503 (2016)

18. Zhang, S., Chi, C., Lei, Z., Li, S.Z.: Refineface: refinement neural network for high performance face detection. CoRR abs/1909.04376 (2019), http://arxiv.org/abs/1909.04376

19. Zhang, S., Wen, L., Shi, H., Lei, Z., Lyu, S., Li, S.Z.: Single-shot scale-aware network for real-time face detection. Int. J. Comput. Vision **127**(6), 537–559 (2019)

20. Zhang, Y., Ding, M., Bai, Y., Ghanem, B.: Detecting small faces in the wild based on generative adversarial network and contextual information. Pattern Recogn. **94**, 74–86 (2019)

21. Zhu, C., Zheng, Y., Luu, K., Savvides, M.: CMS-RCNN: contextual multi-scale region-based CNN for unconstrained face detection. In: Bhanu, B., Kumar, A. (eds.) Deep Learning for Biometrics. ACVPR, pp. 57–79. Springer, Cham (2017). https://doi.org/10.1007/978-3-319-61657-5_3

22. Zhu, M., Shi, D., Zheng, M., Sadiq, M.: Robust facial landmark detection via occlusion-adaptive deep networks. In: Proceedings of the IEEE Conference on Computer Vision and Pattern Recognition, 18–22 June, pp. 2675–2683

23. Zhu, X., Ramanan, D.: Face detection, pose estimation, and landmark localization in the wild. In: 2012 IEEE Conference on Computer Vision and Pattern Recognition, pp. 2879–2886 (2012)

RAIDU-Net: Image Inpainting via Residual Attention Fusion and Gated Information Distillation

Penghao He, Ying Yu$^{(\boxtimes)}$, Chaoyue Xu, and Hao Yang

School of Information Science and Engineering, Yunnan University,
Kunming 650500, China
`penghaohe@mail.ynu.edu.cn`

Abstract. Most existing deep learning image inpainting methods cannot extract the effective information from broken images and reduce interference of masked partial information in broken images. To address this issue, we propose an image inpainting model based on a residual attention fusion block and a gated information distillation block. Our model adopts U-net as the backbone of the generator to realize the encoding and decoding operations of the image. The encoder and decoder employ the residual attention fusion block, which can enhance the utilization of practical information in the broken image and reduce the interference of redundant information. Moreover, we embed the gated information distillation block in the skip-connection of the encoder and decoder, which can further extract useful low-level features from the generator. Experiments on public databases show that our RAIDU-Net architecture achieves promising results and outperforms the existing state-of-the-art methods.

Keywords: Image inpainting · Deep learning · Residual attention · Information distillation

1 Introduction

Image inpainting aims to restore semantically correct and visually realistic images by filling pixels in masked regions. In recent years, image inpainting has become a challenging and essential task in computer vision. It has been applied in many practical scenarios, such as damaged artifact restoration, image editing, image object removal, etc.

In general, image inpainting methods can be divided into traditional approaches and deep learning-based approaches. Traditional image inpainting methods can be mainly divided into diffusion-based methods [2] and patch-based methods [1]. Diffusion-based methods extract background information features by differential equations to fill the pixels in the missing region. Patch-based

---

Supported by organization x.

T. Mantoro et al. (Eds.): ICONIP 2021, LNCS 13108, pp. 141–151, 2021.
https://doi.org/10.1007/978-3-030-92185-9_12

methods copy the best-matched background block image information to the corresponding missing regions in turn. These methods have a better ability to restore larger missing image areas and backgrounds. However, they cannot accurately repair the texture and detail features of the image for complex missing regions.

With the rapid development of convolutional neural networks, deep learning-based methods have been widely used for image inpainting. These methods usually rely on U-net [11] (encoder-decoder) structures on the generator to enable it to capture high-level semantic features and restore details of image textures and edges more accurately. By using the U-net structure, Iizuka et al. [4] proposed global and local discriminators to maintain the consistency of the whole image. Yan et al. [12] proposed a particular shift-connection layer, connecting the encoder and decoder parts of the U-net to better integrate patch-based methods with deep learning methods. Ma et al. [9] introduced a two-stage U-net network from coarse to fine to obtain more accurate results. Although these methods achieve better results in large-scale and complex damaged images, they still suffer from blurred image edges and inadequate detail generation. The main reasons are (1) the convolutional operations capture the valid information in the image background and the damaged information in the masked region. As a result, the results of image restoration are adversely affected by the damaged information. (2) Correlation information between distant image regions cannot be accurately captured by simple skip-connection, which is crucial for image inpainting. To illustrate, in facial restoration, where the facial features are almost entirely invisible, image recovery requires capturing information between correlated distant image features to restore the results perfectly.

In this paper, we propose a Residual Attention Fusion Block (RAFB) and a Gated Information Distillation Block (GIDB) based on the U-net framework. The RAFB replaces all convolutional layers in the U-net layer, composed of residual connections and Channel Space Attention Fusion Blocks (CSAFB). The RAFB can extract multi-scale information and enhance the effective information extracted from residual blocks and suppress redundant information through CSAFB. In addition, we embed the GIDB into the skip-connection of the generator network. This block can efficiently utilize the useful low-level features in the encoder network and accurately capture the information related to the image features at long distances. Experiments on the Celeb-HQ and Pairs datasets show that our method outperforms the state-of-the-art approaches. Specifically, the main contributions of this paper are as follows:

- We propose a new generator backbone of residual attention fusion block, ensuring the maximum utilization of practical information and reducing broken information in the masked region.
- We design a gated information distillation block embedded in the generator, which further distills and filters the image information features. In this way, the valid information in the image is obtained by using this block and the broken information in the masked region of the image is discarded.

- Experiments show that the proposed method can generate reasonable content and delicate details for damaged images. We also provide ablation studies in this paper in order to verify the effectiveness of each proposed block.

2 Related Work

Deep learning-based image inpainting trains adversarial neural networks through a large amount of data to obtain the mapping from incomplete images to the complete ones. Pathak et al. [10] introduced a contextual encoder that first uses an encoder-decoder structure to extract richer semantic features of images. Based on the context encoder, Liu et al. [6] proposed partial convolution to replace all convolutions of the U-net structure to reduce the effect of incorrect mask information on image restoration results. Later, Yu et al. [15] proposed gated convolution to address the issue of irregular masks affecting image inpainting.

The attention mechanism can efficiently allocate the available computational resources according to the importance of image features. According to the attention mechanism dimension, attention can be divided into spatial attention (SA) and channel attention (CA). A number of studies [7,8,14] have shown that spatial attention mechanisms have been applied to image inpainting tasks. Yu et al. [14] designed a contextual attention layer that uses SA to find the most reasonable contextual content for the missing image regions. Liu et al. [8] proposed a new coherent semantic attention layer to improve the spatial coherence of the missing regions of damaged images by SA. Liu et al. [7] proposed a multi-scale attention layer in the encoder to capture multi-scale and effective features in damaged images better.

3 Proposed Approach

In this section, we first introduce the overall network architecture of the proposed RAIDU-Net. Then we describe the proposed two effective blocks, i.e., the Residual Attention Fusion Block (RAFB) and the Gated Information Distillation Block (GIDB), respectively.

3.1 Network Architecture

We propose a novel end-to-end image inpainting network, based mainly on the Gated Information Distillation Block (GIDB) and the Residual Attention Fusion Block (RAFB). This proposed network is referred to as RAIDU-Net. Figure 1 depicts the overall network architecture of our model, which consists of two subnetworks, i.e., generator network G and discriminator network D. The discriminator network is a pix2pix which contains six convolutional layers. To increase the apparent texture of the generated images and satisfy their semantic content visually, we combine multiple loss functions as the joint training loss of my model, which is defined as

$$L = \lambda_{l_1} L_{l_1} + \lambda_{cont} L_{content} + \lambda_{sty} L_{sty} + \lambda_{tv} L_{tv} + \lambda_{adv} L_{adv}. \tag{1}$$

where λ_{l_1}, λ_{cont}, λ_{sty}, λ_{tv}, λ_{adv} are the corresponding tradeoff parameters.

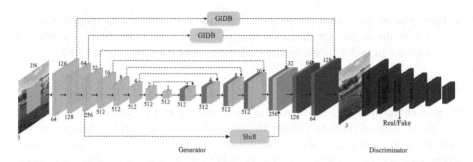

Fig. 1. The architecture of RAIDU-Net Network. We add gated information distillation block (GIDB) at outmost two skip-connection layers, and adopt the Shift-connection layer of Shift-Net.

3.2 Residual Attention Fusion Block

Compared with traditional convolution, residual convolution contains two receptive fields of different sizes that can extract rich multi-scale image information. This information includes not only sufficient background details, but also partially redundant information. For the image inpainting task, the masked pixels in the damaged image are useless, and extracting this pixel information may cause the restoration result to be ghosted or even fail. Additionally, each image to be restored contains more or less noisy information. Therefore, as illustrated in Fig. 2 and Fig. 3, we propose the RAFB, which not only alleviates the interference of redundant information and enhances the importance of effective features but also can better capture feature-related information across long distances between image regions.

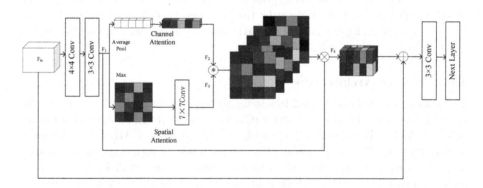

Fig. 2. The basic block of RAIDencoder

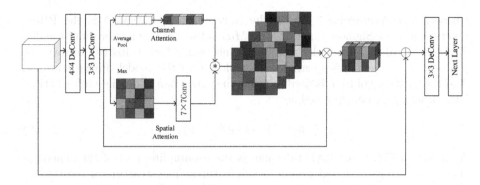

Fig. 3. The basic block of RAIDdecoder

In this paper, the encoder and decoder are called RAFB-encoder and RAFB-decoder, respectively. RAFB-encoder is the down-sampling part of the generator, which is used to extract the detailed features of the damaged image, as shown in Fig. 2. The image is first downsampled by a 4×4 convolution in the down-sampling process with a step size of 2. After that, the multi-scale information in the damaged image can be extracted by increasing the receptive field of the RAFB module through a 3×3 convolution. This process can be expressed as

$$F_1 = Conv_{3 \times 3}(Conv_{4 \times 4}(F_{in})) \tag{2}$$

where F_{in} is the input layer, F_1 is the output result, and $Conv$ is the convolution operation. To reduce the interference of redundant information in F_1 and to better utilize the effective information, F_1 is input to the attention fusion module in this paper. In the attention fusion module, we adopt a side-by-side connection approach. To begin with, F_1 is input to the channel attention module and spatial attention module to obtain the weights of channels and the weights of spatial pixel values, respectively. The AveragePool operation is used in the channel attention branch to integrate the information within the channel. And different weights are assigned to each channel, which is helpful to reduce the interference of redundant information. Because high-frequency details, which are primarily reflected in the form of extreme values, are essential for image inpainting. So in the spatial attention branch, we only employ the Max operation first, instead of the average operation, and then perform 7×7 convolution to generate the spatial attention map, which can enhance the essential detail features in the image. Finally, the channel attention map and spatial attention map of two side-by-side branches are subjected to the Hadamard multiplication operation and the matrix multiplication operation with F_1. This process can be formulated as

$$F_2 = \sigma(\delta(F_{fc}(F_{gp}(F_1)))) \tag{3}$$

$$F_3 = \sigma(F_{fc}(Max(F_1))) \tag{4}$$

$$F_4 = F_1 \times (F_2 \odot F_3) \tag{5}$$

where F_{gp} is AveragePool, F_{fc} is the convolution operation, δ is the PRelu function, σ is the Sigmoid function, and Max is the maximum operation. Finally, F_4 is added to F_{in} using the residual connection, which helps to improve the problem of gradient disappearance or explosion of the model during training. And the 3×3 convolution is used to refine the extraction of detailed information. This process can be expressed as

$$F_{out} = Conv_{3 \times 3}(F_4 + F_{in}) \tag{6}$$

As shown in Fig. 3, the RAFB-decoder is the upsampling part of the generator that is used to recover texture and detail features. The structure of the RAFB-decoder and the RAFB-encoder are roughly similar. The only difference in the upsampling section is that we use dilated convolution instead of the standard convolution in the downsampling section. This expands the receptive field of the model and facilitates the model to capture a wider range of inter-feature correlation information when repairing the missing image content.

3.3   Gated Information Distillation Block

The Gated Information Distillation Block (GIDB) is proposed in this paper to extract low-level features in the encoder more effectively, as shown in Fig. 4(a). First, the low-level features in the encoder are divided into two parts by the channel split operation to obtain two branches: the distillation branch and the fine branch. Inspired by Ref [15], we use a gated convolution in the distillation branch in this paper, as shown in Fig. 4(b). Gated convolution is achieved by adding a sigmoid function as a soft gating function behind the convolution layer, which weights the output of the current convolution layer before the input to the next convolution layer. The distillation branch extracts the most effective feature information in the encoder and reduces the interference of redundant information by using the gated convolution and channel split operation. The refinement branch refines the coarse information by 3×3 convolution. After three distillation and refinement extractions, the output results are subjected to contact operation, and cross-channel information interaction is realized through 1×1 convolution, which is helps to capture long-distance information. This process can be expressed as

$$distilled\_c_1, refined\_c_1 = GConv(F_{in}), Conv(F_{in}) \tag{7}$$

$$distilled\_c_2, refined\_c_2 = GConv(distilled\_c_1), Conv(distilled\_c_1) \tag{8}$$

$$distilled\_c_3, refined\_c_3 = GConv(distilled\_c_2), Conv(distilled\_c_2) \tag{9}$$

$$distilled\_c_4 = Conv(distilled\_c_3) \tag{10}$$

$$out = Conv(Cat(distilled\_c_1, distilled\_c_2, distilled\_c_3, distilled\_c_4)) \tag{11}$$

where $GConv$ is the gated convolution operation, $Conv$ is the convolution operation, Cat is the concatenation operation along the channel dimension.

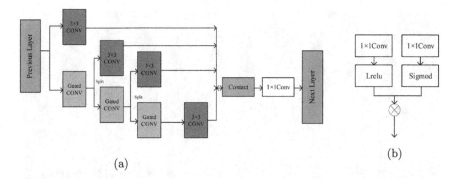

Fig. 4. (a) The structure of our proposed Gated Information Distillation Block (GIDB) (b) The structure of Gated Convolution.

4 Experiments

The experiments use the publicly available datasets CelebA-HQ [5] and Pairs [3]. CelebA-HQ is a face dataset containing 30,000 face images with 1024×1024 resolution, 28000 of which are used as the training set and 2,000 as the test set. Pairs is a street view dataset containing 15,000 images, 14900 of which are used as the training set and 100 as the test set. The image size of all datasets is adjusted to 256×256 pixels and the center mask size is 128×128 pixels. The Adam algorithm optimizes our model during the training process. The initial learning rate is set to 0.0002 and the BatchSize is set to 1. The weights of each part of the loss function are rec = 1.0, cont = 1.0, sty = 12.0, tv = 0.01, and adv = 0.2. Our model is implemented on PyTorch 1.3.1 framework CUDA10.2 and runs on NVIDIA GeForce RTX 2080Ti GPU.

4.1 Qualitative Comparison

We compare our model RAIDU-Net to three state-of-the-art models in experiments: shift-net [12], PEN [16], and RES2-Unet [13]. The experimental inpainting results in Fig. 5. It can be seen that in the second and third rows of images, this model perfectly restores the sunglasses, while the other three models have distorted images and missing sunglasses. In the first and fourth rows of images, this model restores the eyes of the human face better and symmetrically compared with the other three models. In the 5–8 rows, we can notice that the model can restore the edge and texture details of the windows and walls. Overall, the model does well in restoring the facial features of the face, and the missing glasses are fully recovered. This demonstrates that the model can extract and utilize the effective features well while also maintaining the correlation between long-distance feature regions. For Pairs datasets, our model can recover the texture details of doors and windows better. Therefore, our model are significantly better than other comparable models in both semantic structure and texture details.

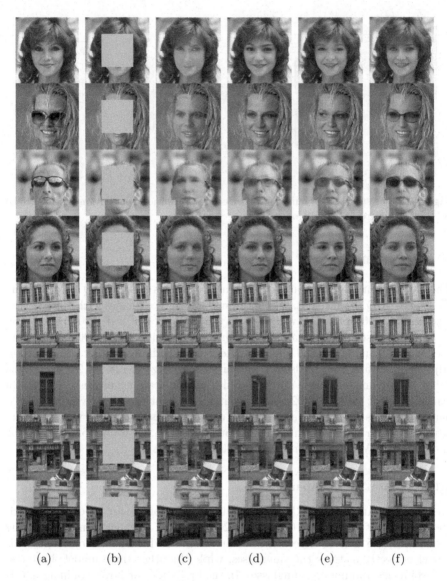

(a)   (b)   (c)   (d)   (e)   (f)

Fig. 5. Qualitative comparisons on CelebA-HQ (1–4 rows) and Paris Street-view (5–8 rows), from left to right: (a) Ground Truth (b) input, (c) PEN, (d) Shift-Net, (e) RES2-Unet and (f) ours.

4.2   Quantitative Comparisons

To evaluate the image inpainting model, we use the peak signal-to-noise ratio (PSNR), Multi-Scale-Structural Similarity Index (MS-SSIM), L_1Loss, L_2Loss, and Fr'echet Inception Distance (FID). PSNR and MS-SSIM are higher indexes that show better model restoration, whereas L_1loss, L_2loss and FID are higher indexes that show the worse model restoration. Table. 1 shows the quantitative evaluation results for the CelebA-HQ and Pairs test datasets. It can be seen that our model is optimal in all quantitative evaluation metrics on both datasets compared to the other three models.

Table 1. Center mask numerical comparison on CelebA-HQ and Paris street-view.

Method		Shift-NET	PEN	RES2-Unet	Our method
CelebA-HQ	PSNR	26.50	25.36	26.80	**26.91**
	MS-SSIM	0.911	0.8882	0.9162	**0.9183**
	L1	0.0208	0.0230	0.0188	**0.0179**
	L2	0.0033	0.0034	0.0028	**0.0025**
	FID	7.677	7.844	7.157	**6.811**
Paris street-view	PSNR	25.14	24.34	25.59	**26.30**
	MS-SSIM	0.8245	0.8095	0.8294	**0.8624**
	L1	0.0241	0.0260	0.0225	**0.0190**
	L2	0.0043	0.0051	0.0025	**0.0024**
	FID	51.67	58.85	43.95	**41.14**

4.3 Ablation Study

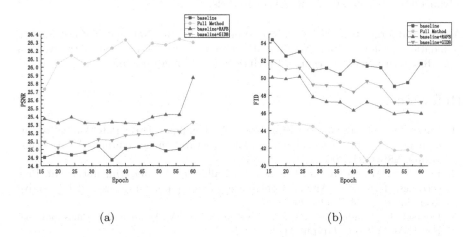

(a)   (b)

Fig. 6. (a): Compare the test PSNR values of the inpainting networks on the Paris dataset. (b): Compare the test FID values of the inpainting networks on the Paris dataset.

We conducted ablation experiments on the Pairs dataset to verify the utility of the blocks proposed in this paper. Shiftnet is taken as the base model, and the proposed blocks RAFB and GIDB are added to the shiftnet. The test image size is 256×256 pixels, and the mask size is 128×128 pixels, and all ablation experiments are trained with the same settings. Each ablation experiments is trained separately for 50 epochs. The qualitative evaluation results of the ablation experiments are shown in Fig. 6. Fig. 6(a) compares PSNR results for each control model on the Pairs datasets. It can be seen that RAFB and GIDB has a significant effect on the improvement of the PSNR index. Fig. 6(b) compares the FID metrics results of each control model on the Pairs dataset, where the RAFB and GIDB affect the decrease of the FID metrics. This indicates that the RAFB and GIDB can significantly improve the image inpainting quality at both the pixel and perceptual evaluation.

5   Conclusion

This paper proposed an image inpainting model based on residual attention fusion and gated information distillation. It enables the generator to extract effective features and reduce redundant information by using the residual attention fusion block, thereby enhancing the generator's ability to capture long-range information in the broken image. The gated information distillation block is embedded in the skip-connection, which gradually utilizes low-level features by use of distillation extraction and fine feature process, and facilitates the generation of texture details of the restored image. Experiments show that our model can obtain realistic details and reasonable semantic structures in various image inpainting tasks, and that the qualitative and quantitative evaluations reach advanced levels compared with other models.

Acknowledgements. This work was supported by the National Natural Science Foundation of China (Grant No. 62166048, Grant No. 61263048) and by the Applied Basic Research Project of Yunnan Province (Grant No. 2018FB102).

References

1. Barnes, C., Shechtman, E., Finkelstein, A., Goldman, D.B.: PatchMatch: a randomized correspondence algorithm for structural image editing. ACM Trans. Graph. **28**(3), 24 (2009)
2. Bertalmio, M., Sapiro, G., Caselles, V., Ballester, C.: Image inpainting. In: Proceedings of the 27th Annual Conference on Computer Graphics and Interactive Techniques, pp. 417–424 (2000)
3. Doersch, C., Singh, S., Gupta, A., Sivic, J., Efros, A.: What makes Paris look like Paris? ACM Trans. Graph. **31**(4) (2012)
4. Iizuka, S., Simo-Serra, E., Ishikawa, H.: Globally and locally consistent image completion. ACM Trans. Graph. (ToG) **36**(4), 1–14 (2017)
5. Karras, T., Aila, T., Laine, S., Lehtinen, J.: Progressive growing of GANs for improved quality, stability, and variation. arXiv preprint arXiv:1710.10196 (2017)
6. Liu, G., Reda, F.A., Shih, K.J., Wang, T.C., Tao, A., Catanzaro, B.: Image inpainting for irregular holes using partial convolutions. In: Proceedings of the European Conference on Computer Vision (ECCV), pp. 85–100 (2018)

7. Liu, H., Jiang, B., Song, Y., Huang, W., Yang, C.: Rethinking image inpainting via a mutual encoder-decoder with feature equalizations. arXiv preprint arXiv:2007.06929 (2020)
8. Liu, H., Jiang, B., Xiao, Y., Yang, C.: Coherent semantic attention for image inpainting. In: Proceedings of the IEEE/CVF International Conference on Computer Vision, pp. 4170–4179 (2019)
9. Ma, Y., Liu, X., Bai, S., Wang, L., He, D., Liu, A.: Coarse-to-fine image inpainting via region-wise convolutions and non-local correlation. In: IJCAI, pp. 3123–3129 (2019)
10. Pathak, D., Krahenbuhl, P., Donahue, J., Darrell, T., Efros, A.A.: Context encoders: feature learning by inpainting. In: Proceedings of the IEEE Conference on Computer Vision and Pattern Recognition, pp. 2536–2544 (2016)
11. Ronneberger, O., Fischer, P., Brox, T.: U-Net: convolutional networks for biomedical image segmentation. In: Navab, N., Hornegger, J., Wells, W.M., Frangi, A.F. (eds.) MICCAI 2015. LNCS, vol. 9351, pp. 234–241. Springer, Cham (2015). https://doi.org/10.1007/978-3-319-24574-4_28
12. Yan, Z., Li, X., Li, M., Zuo, W., Shan, S.: Shift-Net: image inpainting via deep feature rearrangement. In: Proceedings of the European Conference on Computer Vision (ECCV), pp. 1–17 (2018)
13. Yang, H., Yu, Y.: Res2U-Net: image inpainting via multi-scale backbone and channel attention. In: Yang, H., Pasupa, K., Leung, A.C.-S., Kwok, J.T., Chan, J.H., King, I. (eds.) ICONIP 2020. LNCS, vol. 12532, pp. 498–508. Springer, Cham (2020). https://doi.org/10.1007/978-3-030-63830-6_42
14. Yu, J., Lin, Z., Yang, J., Shen, X., Lu, X., Huang, T.S.: Generative image inpainting with contextual attention. In: Proceedings of the IEEE Conference on Computer Vision and Pattern Recognition, pp. 5505–5514 (2018)
15. Yu, J., Lin, Z., Yang, J., Shen, X., Lu, X., Huang, T.S.: Free-form image inpainting with gated convolution. In: Proceedings of the IEEE/CVF International Conference on Computer Vision, pp. 4471–4480 (2019)
16. Zeng, Y., Fu, J., Chao, H., Guo, B.: Learning pyramid-context encoder network for high-quality image inpainting. In: Proceedings of the IEEE/CVF Conference on Computer Vision and Pattern Recognition, pp. 1486–1494 (2019)

Sentence Rewriting with Few-Shot Learning for Document-Level Event Coreference Resolution

Xinyu Chen, Sheng Xu, Peifeng Li[(✉)], and Qiaoming Zhu

School of Computer Science and Technology, Soochow University, Suzhou, China
{xychennlp,sxu}@stu.suda.edu.cn, {pfli,qmzhu}@suda.edu.cn

Abstract. The existing event coreference resolution models is hard to identify the coreferent relation between non-verb-triggered event mention and verb-triggered event mention, due to their different expressions. Motivated by the recent successful application of the sentence rewriting models on information extraction and the fact that event triggers and arguments are beneficial for event coreference resolution, we employ the sentence rewriting mechanism to boost event coreference resolution. First, we rewrite the sentences containing non-verbs-triggered event mentions and convert them to verb-triggered by the fine-tuning pre-training model and few-shot learning. Then, we utilize semantic roles labeling to extract the event arguments from the original sentences with verb-triggered event mention and the rewritten sentences. Finally, we feed the event sentences, the triggers, and the arguments to BERT with a multi-head attention mechanism to resolve those coreferent events. Experimental results on both the KBP 2016 and KBP 2017 datasets show that our proposed model outperforms the state-of-the-art baseline.

Keywords: Event coreference resolution · Sentence rewriting · Few-shot learning

1 Introduction

Document-level event coreference resolution aims to discover the event mentions in a document that refer to the same event ontology in reality and gather them into the coreferent chains. Event mentions are located in the sentence that express them (event sentence) and referred by triggers, which are the main words that can most clearly express the occurrence of events. Arguments, participants of events, usually appear in event sentences rely on triggers, such as agents (ARG0), patients (ARG1), time (ARGM-LOC), and place (ARGM-TMP). Those coreferent event mentions generally have similar triggers and arguments in event sentences. Taking the coreferent event mention pair E1 and E2 and their event sentences S1 and S2 as example.

*S1: With {**Palestinian cameramen**}$_{E1\text{-}ARG1}$, {**shot**}$_{E1\text{-}trigger}$ and killed by {**an Israeli soldier**}$_{E1\text{-}ARG0}$ on the west bank in mid-April.*

© Springer Nature Switzerland AG 2021
T. Mantoro et al. (Eds.): ICONIP 2021, LNCS 13108, pp. 152–164, 2021.
https://doi.org/10.1007/978-3-030-92185-9_13

S2: They believe {he}$_{E2\text{-}ARG1}$ was {shot}$_{E2\text{-}trigger}$ by {an armored Israeli personnel carrier}$_{E2\text{-}ARG0}$.

The event mentions E1 and E2 are both triggered by the word "shot", and the agents of E1 and E2 are "an Israeli soldier" and "an armored Israeli personnel carrier", respectively, which have similar semantics. Additionally, the patient "he" of E2 refers to the patient "Palestinian cameramen" of E1. Hence, if we extract arguments through semantic role labeling (SRL) [1] center on their trigger "shot" can accurately find their similar arguments, in this way, employing a pairwise-similarity-based event coreference resolution model can easily and correctly regard them as coreferent.

However, in most cases, arguments information in an event sentence is hard to extract exactly, and this phenomenon often occurs in situations like the following example, which shows the drawback of pairwise-similarity-based model on event coreference resolution.

S3: And second, there must be legal buffs who can tell me if the United States could legally stop him on the {journey}$_{E3\text{-}trigger}$.

S4: {He}$_{E4\text{-}ARG0}$ could {go}$_{E4\text{-}trigger}$ from {Murmansk}$_{E4\text{-}ARG1}$ which is ice free year round but he would have to pass through Scandinavian waters.

S3 contains an event mention E3 triggered by "journey" and S4 contains an event mention E4 triggered by "go" and in fact they refer to the same event. Although "journey" and "go" have similar semantics, their parts-of-speech are different (i.e. "journey" is a non-verbal (nominal) trigger, while "go" is a verbal trigger), which will lead to the fact that the SRL tool cannot extract the arguments of this nominal trigger correctly due to the poor performance of SRL on nominal predicates. If we use such extracting results to judge whether the event mentions E3 and E4 are coreferent, the mismatch of argument information will lead to the model mistakenly regarding them as non-coreferent. Hence, the gap between the event mentions triggered by non-verbs (most of them are nouns or adjectives) and those triggered by verbs is a critical issue that hinders the event coreference resolution.

Motivated by the recent successful application of the sentence rewriting models on information extraction [2], we propose an sentence rewriting mechanism to rewrite event sentences containing non-verbal triggers to address the above issue. We first transform those non-verbal triggers into verb by the pre-training Transformer model, which is fine-tuned on a labeled dataset with few-shot learning, to obtain rewritten sentences. Then, we utilize the SRL tool to extract the arguments from the rewritten sentences and finally introduce them into our event coreference resolution model, which consists of a BERT encoder and a multi-head attention mechanism. Experimental results on both the KBP 2016 and KBP 2017 datasets show that our proposed model outperforms the state-of-the-art baseline [13]. The contributions of this paper are summarized as follows:

- We propose a sentence rewriting mechanism which can accurately rewrite the event sentences containing non-verbal triggers to extract arguments effectively. This mechanism can eliminate differences between the event mentions triggered by non-verbs and those triggered by verbs.

- We successfully combined the pre-training model with few-shot learning to improve event coreference resolution. In our knowledge, we are the first to apply the idea of few-shot learning to event coreference resolution.
- Our model outperforms the state-of-the-art baseline both on the KBP 2016 and 2017 datasets.

2   Related Work

Most previous studies on event coreference resolution were performed on two popular corpora ACE [4] and KBP [5]. Early research usually directly use the annotation features of events [6–8]. Recent studies pay more attention to the coreference resolution on unlabeled raw texts. Specifically, vectorization methods [9] and markov logic network [10] are machine learning methods for event coreference resolution. Compared with them, Huang et al. [11] incorporated knowledge of argument compatibility from a large number of the unlabeled corpus. Fang et al. [12] expanded a semantically sparse dataset using data augmentation, and improved the quality of the expanded dataset by reinforcement learning. Lu et al. [13] investigate span-based models for event coreference resolution. These neural network methods outperform those machine learning methods.

The work [13] is similar to ours, which helps resolve coreferent events through argument extraction. However, it ignores the gap between the event mentions triggered by non-verbs and those triggered by verbs. In contrast, we propose a sentence rewriting mechanism with few-shot learning to rewrite sentences containing non-verbal triggers, which improves the performance of argument extraction and the improvement further provides help for event coreference resolution.

3   Event Coreference Resolution on Sentence Rewriting

This paper proposes an event coreference resolution model on sentence rewriting (CR-RW), which takes the event sentence pair as input and calculates the semantic similarity between them to predict whether the event mentions they contained are coreferent. The model architecture is shown in Fig. 1. First, we check whether the triggers of two event mentions are both verb or not. Then, the event sentences containing non-verbal trigger will be rewritten to the sentences containing semantically similar verbal trigger in the document. Finally, we encode event sentences by BERT and utilize the verbal predicate-center SRL to extract event arguments and send them with their event sentence and trigger into an attention network to capture their similarity for coreference prediction.

It is worth mentioning that our sentence rewriting mechanism aims to select the most suitable sentence in the document instead of generating a sentence by seq2seq model, which avoids the introduction of more irrelevant information.

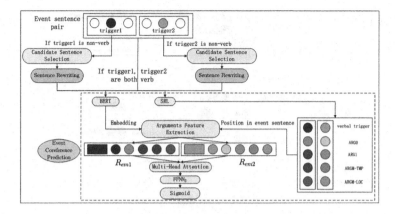

Fig. 1. Architecture of CR-RW model.

3.1 Candidate Sentence Selection

In input event sentence pair, we must rewrite those sentences containing non-verbal triggers as verb-triggered expressions to extract event arguments more accurately. Those sentences which mention coreferent events usually have more overlapping information than others, on the basis of this fact and inspired by Bao et al. [2]. For those sentences to be rewritten, we select a set of sentences with verbal and semantically similar trigger within document as their candidate rewritten sentence by comparing ROUGE [3] score. Different from Bao et al. [2], we use the average precision, recall, and F1-measure of ROUGE-1/2/L as scoring function respectively to match the sentences in the document, after which we choose three sentences with the maximum average precision, recall, and F1-measure score, respectively, of ROUGE-1/2/L as candidate rewritten sentences.

3.2 Sentence Rewriting

Argument information is the critical cue for event coreference resolution. However, the poor performance of nominal predicate-center SRL on extracting arguments harms the downstream task event coreference resolution. Hence, we propose a sentence rewriting mechanism to convert those non-verb-triggered event mentions expressions to verb-triggered form to address the above issue.

Different from other sentence rewriting methods, we introduce few-shot learning to sentence rewriting. Specially, we first select three candidate sentences for each of those non-verb-triggered event mentions to form the support set, mentioned in the above subsection. Then, we use a model shown in Fig. 2 to select the most suitable rewritten sentence from the support set.

Input Representation. Let the original sentence, which is to be rewritten, and three candidate sentences represent as $Sent_o$, $Sent_{cd1}$, $Sent_{cd2}$, and $Sent_{cd3}$ respectively, and we add a special $[CLS]$ token to the beginning of each sentence representation, and a $[SEP]$ token to the end, to join them together as X_{inp},

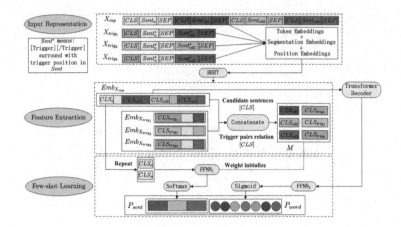

Fig. 2. Architecture of sentence rewriting model.

which is inspired by [14]. Different from [14], we splice each one of $Sent^*_{cd1}$, $Sent^*_{cd2}$, and $Sent^*_{cd3}$ behind "$[CLS]Sent^*_o[SEP]$" respectively (Here $Sent^*$ is $Sent$ surrounding with special symbol [Trigger][/Trigger] in trigger position) for extracting relation between non-verbal trigger and verbal trigger, denoted as X_{trig_1}, X_{trig_2}, and X_{trig_3}. The process above can be formalized as follows.

$$X_{inp} = Concatenate([CLS], Sent_o, [SEP], [CLS], Sent_{cd1}, [SEP],$$
$$[CLS], Sent_{cd2}, [SEP], [CLS], Sent_{cd3}, [SEP]), \tag{1}$$

$$X_{trig_i} = Concatenate([CLS], Sent^*_o, [SEP], Sent^*_{cdi}), i \in \{1, 2, 3\}. \tag{2}$$

The token, segmentation, and position embedding of X_{inp} and X_{trig_i} are the input of our sentence rewriting model.

Feature Extraction. We use BERT to encode the token, segmentation, and position embedding of X_{inp}, and X_{trig_i} as $Emb_{X_{inp}}$, and $Emb_{X_{trig_i}}$, respectively. For $Emb_{X_{inp}}$, we take four positions of the $[CLS]$ embedding as the representation of $Sent_o$, and $Sent_{cdi}$ and donate them as CLS_o, and CLS_{cdi} respectively. For $Emb_{X_{trig_i}}$, its $[CLS]$ embedding is the trigger relation expression between $Sent_o$ and $Sent_{cdi}$, denoted as CLS_{trig_i} as follows.

$$CLS_o = Emb^{(0)}_{X_{inp}}, CLS_{cdi} = Emb^{(p_i)}_{X_{inp}}, i \in \{1, 2, 3\}, \tag{3}$$

$$CLS_{trig_i} = Emb^{(0)}_{X_{trig_i}}, i \in \{1, 2, 3\}, \tag{4}$$

where p_i denote the $[CLS]$ token position of $Sent_{cdi}$.

Finally, we concatenate the representation $[CLS]$ of three candidate sentences and the trigger pairs relation to form a feature matrix as below.

$$M = \begin{bmatrix} CLS_{cd1} \ CLS_{trig_1} \\ CLS_{cd2} \ CLS_{trig_2} \\ CLS_{cd3} \ CLS_{trig_3} \end{bmatrix}. \tag{5}$$

Here, the matrix M will play the role of the support set feature in the following few-shot learning stage, which provides help for selecting the rewritten sentence from three candidate sentences.

Few-Shot Learning. Currently, few-shot learning is used in a few NLP tasks [15], the main idea of which is to use task-specific data to fine-tune the pre-training language model. In this paper, we introduce it into the event coreference resolution task. Different from [15], we utilize the support set feature to help rewrite sentences for event coreference resolution. Specifically, we feed the feature of the original sentence CLS_o with repeating twice to a feedforward neural network $FFNN_1$ layer, it is worth mentioning that we initialize the parameter matrix W and bias b of $FFNN_1$ layer with the support set feature matrix M and 0, respectively. We can follow obtain the sentence-level scores through the Softmax layer, and the candidate sentence with highest score is most likely to be selected as the rewritten sentence as follows.

$$P_{sent} = Softmax(W \cdot [CLS_o, CLS_o] + b), initialize : W = M, b = 0. \quad (6)$$

Meanwhile, the feature $Emb_{X_{inp}}$ is sent to Transformer Decoder layer, $FFNN_2$ layer in turn and finally the Sigmoid layer is used to obtain the word-level scores of each sentence, denoted as P_{word}, which act as auxiliary score for selecting the most suitable sentence. As shown follows:

$$P_{word} = Sigmoid(W_{FN} \cdot TransformerDecoder(Emb_{X_{inp}}) + b_{FN}), \quad (7)$$

where W_{FN} and b_{FN} are the weight matrix and bias of $FFNN_2$ layer, respectively.

Loss Functions. We train our sentence rewriting model by minimizing the linear combination of the sentence-level and word-level loss, denoted as L_{sent} and L_{word} respectively. We use the KL-divergence for word-level loss function and the cross entropy for sentence-level as below.

$$L_{word} = D_{KL}(P_{word}||Y_{word}), \quad (8)$$

$$L_{sent} = -\sum_{i=1}^{N}[Y_{sent}^{(i)} \cdot log(P_{sent}^{(i)})], \quad (9)$$

$$FinalLoss = a \cdot L_{word} + b \cdot L_{sent}, \quad (10)$$

where a is set to 1, b is set to 2 and N is the number of the candidate sentences, Y_{word} and Y_{sent} are gold rewritten sentence label of word-level and sentence-level, respectively, for supervised training.

Due to the fact that the ACE corpus annotated the gold event sentences but KBP corpus does not, we train our sentence rewriting model on the ACE corpus and apply it to the KBP corpus. When applying model on KBP corpus after minimizing the $FinalLoss$ by fine-tuning on the ACE corpus, candidate sentence with maximum score in P_{sent} is selected to the rewritten sentence.

3.3 Event Coreference Prediction

In this stage, we compare the similarity between event mention pair by their event sentences, event trigger and arguments extracted from SRL to predict whether they are coreferent. Here, all event sentences with non-verbal triggers have been rewritten, i.e., all the non-verbal triggers have been converted into semantically similar verbs. Specifically, we first send two event sentences to the BERT layer and SRL layer, respectively. The BERT layer outputs the embeddings of two event sentences and the SRL layer extracts the event arguments on their verbal triggers. In this paper, we choose the following items as event arguments including the agent (ARG0), recipient (ARG1), time (ARGM-TMP) and location (ARGM-LOC). We extract all argument feature vectors from the embedding by the arguments position in the event sentence. Then we concatenate the features of the sentence, verbal trigger, agent, recipient, time, and location for two event mentions and represent them as R_{em1} and R_{em2} respectively. Then we feed R_{em1} and R_{em2} into the multi-head attention layer, whose output is an similarity expression of two event mentions. Finally, transforming similarity expression through the FFNN$_3$ layer and Sigmoid layer, we obtain the score of coreference to judge whether two events are coreferent as follows.

$$Score = Sigmoid(W_{out} \cdot Attention(R_{em1}, R_{em2}) + b_{out}), \qquad (11)$$

where W_{out} is the weight matrix and b_{out} is the bias of FFNN$_3$ layer.

Since the positive and negative samples in the dataset are unbalanced, we choose focal loss [16] as the loss function.

$$FocalLoss = -\alpha(1 - p)^\gamma log(p). \qquad (12)$$

For handling the pairwise event coreference predictions, we perform best-first clustering [8] on the pairwise results to build the event coreference chain.

4   Experimentation

4.1   Experimental Settings

Sentence Rewriting Settings. We use the annotated event sentences in the ACE 2005 corpus to build the dataset for training the sentence rewriting model. Specifically, we take the event sentences with non-verbal triggers on each coreference chain as the original sentences, and select three candidate sentences with verbal triggers in the coreference chain for them by the scoring function mentioned in Subsect. 3.1. If there are less than three in the chain, we will match other sentences in the document to complete. We regard the three candidate sentences as the support set for few-shot learning. For labeling the gold rewritten sentence, we first reproduce the existing pairwise-similarity-based event coreference resolution model [12], and then apply it to calculate the coreference score between origin sentence and three candidate sentences, and finally we label the candidate sentence with the best score as gold rewritten sentence for supervised training. Apart from that, the training epochs is set to 5 and the learning rate is set to 10^{-5}, and we use Adam optimizer to update the parameters.

Event Coreference Resolution Settings. Following previous work, we use the KBP 2015 dataset as the training set and the official complete test set of KBP 2016 and KBP 2017 as the test set and choose 10% of the training set as the development set. Note that we did not use any annotated information in the official test set. Following the previous work [10–13], we use MUC [17], B^3 [18], BLANC [19] and $CEAF_e$ [20] to evaluate the performance of our event coreference resolution model CR-RW and also report the average scores (AVG) of the above four metrics, which can more objectively measure the performance of the model. We set the training epochs of CR-RW as 10 rounds, set the learning rate as 10^{-5}, and use Adam optimizer to update the parameters, the heads number of attention mechanism is set to 3. PyTorch is choosed as our deep learning framework for CR-RW.

Upstream Event Extraction. We first extract event mentions with their event sentences, triggers from raw texts and then feed them to CR-RW for coreference prediction. Our event extractor is a BERT-based classifier, which is similar to Yang et al. [21], and we reproduce it on KBP 2016 and 2017 datasets, respectively. We compare our event extractor with the state-of-the-art baseline Lu et al. [13] and Table 1 shows the F1-scores of event extraction on the KBP 2016, KBP 2017 datasets, which obviously shows that our event extractor achieves comparable performance. It is fair to compare the performance of the event coreference resolution based on such similar results.

Table 1. F1-scores of upstream event extraction.

	KBP 2016			KBP 2017		
System	P	R	F	P	R	F
Lu	**63.9**	59.4	**61.6**	68.6	**60.8**	**64.4**
Ours	60.5	**61.9**	60.7	**70.2**	58.9	63.3

4.2 Experimental Results

To verify the performance of CR-RW, we select the state-of-the-art model Lu et al. [13] as the baseline. The performance comparison between our model and the baseline is shown in Table 2. The results show that our CR-RW outperforms the SOTA baseline significantly and this indicates that the sentence rewriting mechanism can unify the trigger parts-of-speech of two event mentions, which is helpful for extracting event arguments, and the improvement of event arguments extraction is beneficial for event coreference resolution.

Table 2. Performance of event coreference resolution on KBP 2016 and 2017.

System	KBP 2016					KBP 2017				
	MUC	B³	BLANC	CEAF$_e$	AVG	MUC	B³	BLANC	CEAF$_e$	AVG
Lu	38.9	52.6	35.0	51.9	44.6	42.8	53.7	**36.4**	51.5	46.1
CR-RW	**40.5**	**54.1**	**35.2**	**52.9**	**45.7**	**44.2**	**55.0**	36.0	**54.2**	**47.4**

From an overall perspective, our CR-RW improve the AVG by 1.1 and 1.3 on KBP 2016 and KBP 2017 datasets, respectively, compared with the best baseline Lu. From the perspective of four different evaluation metrics, our CR-RW improve almost all of the score on KBP dataset. In particular, the improvement of MUC(+1.6) on KBP 2016 and CEAF$_e$(+2.7) on KBP 2017.

5   Analysis

To further analyze our CR-RW model, we conduct the comparative experiments under two different settings: ablation study on different event mention pairs and analysis on few-shot learning.

5.1   Ablation Study

To justify the effectiveness of the sentence rewriting mechanism, we performed ablation experiments on KBP 2016 and KBP 2017 datasets. Specifically, we compare the event coreference resolution performance of extracted arguments w/o (without) and w/(with) the sentence rewriting mechanism. In addition, since event mention pairs can be divided into three categories according to the parts-of-speech of the triggers, i.e., two non-verbal triggers (NN, 21.45/22.50% in KBP 2016/2017), one non-verbal trigger and one verbal trigger (NV, 47.30/47.11% in KBP 2016/2017), and two verbal trigger (VV, 31.25/30.39% in KBP 2016/2017). The experimental results are shown in Table 3. As we expect, the sentence rewriting mechanism on NN, NV, and all pairs can improve the performance of event coreference resolution.

Compared with the Baseline which removes the argument information from CR-RW(i.e., only take origin event sentence pair as the input of BERT), adding the arguments to the VV set (+VV) can improve the performance of event coreference resolution. This indicates that argument information is the effective evidence to detect coreferent events and the extracted arguments by verbal predicate-center SRL are accurate.

If we add the arguments in the NN set (+NN), the NV set (+NV) or both of them (All), which is extracted from the original sentences (w/o rewriting), to the model +VV, respectively, the performance will drop rapidly. If we do not rewrite the non-verb-triggered sentences and directly perform SRL to extract arguments, most of the extracted arguments are pseudo ones and will harm our model. In particular, when adding the NV set causes the greatest damage to the

Table 3. An ablation study for CR-RW.

KBP 2016

System	MUC	B³	BLANC	CEAFₑ	AVG
Baseline (w/o arguments)	38.0	52.3	33.9	50.6	43.7
+VV	38.8	52.9	34.3	51.4	44.4
+NN(w/o|w/rewriting)	37.5|38.5	51.9|52.7	33.6|34.2	50.0|51.0	43.3|44.1
+NV(w/o|w/rewriting)	36.0|39.4	52.4|53.2	33.3|34.5	50.6|51.7	43.1|44.7
+All(w/o|w/rewriting)	37.0|**40.5**	51.5|**54.1**	33.2|**35.2**	49.4|**52.9**	42.8|**45.7**

KBP 2017

System	MUC	B³	BLANC	CEAFₑ	AVG
Baseline(w/o arguments)	42.2	53.2	34.9	50.9	45.3
+VV	42.8	53.9	35.1	52.2	46.0
+NN(w/o|w/rewriting)	41.2|42.6	52.4|53.7	34.4|35.0	49.5|52.0	44.4|45.8
+NV(w/o|w/rewriting)	41.4|43.2	52.6|54.2	34.5|35.3	49.8|52.8	44.6|46.4
+All(w/o|w/rewriting)	41.0|**44.2**	52.1|**55.0**	34.2|**36.0**	49.2|**54.2**	44.1|**47.3**

performance because it accounting for the largest proportion. The reason is due to the poor performance of the nominal predicate-center SRL.

Most importantly, compared with the results without rewriting sentences in the NN set (+NN), the NV set (+NV) and both of them (All), the results with rewriting sentences (w/rewriting) can improve the performance significantly. This indicates that the rewriting sentences mechanism is effective for SRL to extract correct arguments and further help the event coreference resolution.

5.2   Analysis on Few-Shot Learning

To verify the effectiveness of few-shot learning in CR-RW, we designed another rewriting strategy, i.e. regarding event mentions in the most similar sentence as coreferent event. Following the work of [2], we treat the sentence achieves the best average recall score of ROUGE-1/2/L as the most similar sentence. The experimental result is shown in Table 4, where Sim refers to the CR-RW model using the above strategy, not few-shot learning. Obviously, our rewriting strategy of few-shot learning outperforms Sim, which proves the effectiveness of our few-shot learning rewriting strategy. Further study about cases is discussed in Subsect. 5.3.

Table 4. Comparison on different rewriting strategies.

Strategy	KBP 2016					KBP 2017				
	MUC	B³	BLANC	CEAF_e	AVG	MUC	B³	BLANC	CEAF_e	AVG
Sim	39.1	53.1	34.4	51.5	44.5	43.0	54.1	35.2	52.5	46.2
Few-shot learning	**40.5**	**54.1**	**35.2**	**52.9**	**45.7**	**44.2**	**55.0**	**36.0**	**54.2**	**47.4**

5.3 Case Study

In this subsection, we give the examples to analyze the effectiveness of our model CR-RW. Take the sentences S3 and S4 in Sect. 1, we need judge whether two event mentions triggered by "journey" and "go", respectively, are coreferent. Their SRL results SRL_{S3} and SRL_{S4} are shown follows.

SRL_{S3}: *And second, there must be legal buffs who can tell me if* {*the United States*} $_{ARG0}$ *could legally* {*stop*} $_{verb}$ {*him*}$_{ARG1}$ *on the journey.*

SRL_{S4}: {*He*}$_{ARG0}$ *could* {*go*}$_{verb}$ *from* {*Murmansk*}$_{ARG1}$ *which is ice free year round but he would have to pass through Scandinavian waters.*

If we directly utilize SRL to extract event arguments as above, the mismatching of ARG0, predicate, and ARG1 will mislead our model to identify them as non-coreferent. If we rewrite S3 with our rewriting strategy and then find three candidate sentences C1, C2, and C3 with their SRL results as follows, where C1 is selected as the rewritten sentence of S3 by our rewriting model.

SRL_{C1}: *So it is entirely conceivable* {*he*}$_{ARG0}$ *could just* {*sail*}$_{verb}$ *out of* {*Murmansk*}$_{ARG1}$ *and ...*

SRL_{C2}: *The Russians even gave the United States some advice when* {*we*}$_{ARG0}$ *first* {*went*}$_{verb}$ *in.*

SRL_{C3}: {*I*}$_{ARG0}$ *have never* {*sailed*}$_{verb}$ *on the ocean, merely freshwater lakes, and I'm not certain how practical this journey would be.*

We can find that the ARG0 "he" and ARG1 "Murmansk" extracted from the rewritten sentence C1 can match the corresponding arguments of S4. Therefore, our model can easily and correctly detect them as coreferent.

Finally, we analyze the other two candidate sentences C2 and C3. C2 has the verb "went" which is similar to "journey". However, the verb "went" here does not trigger any event mentions but it was mistakenly recognized as an event trigger by event extractor upstream, thus it is not suitable to regard C2 as the rewritten sentence of S3. C3 achieves the best average recall score of ROUGE-1/2/L with S3 in three candidate sentences. However, the event mention triggered by "sail" in C3 is different from that "journey" triggered, therefore it is not suitable to directly treat the most similar sentence as rewritten sentence and regard event mentions it contained as coreferent without few-shot learning, which also indicates the drawback of rewriting strategy Sim.

6   Conclusion

This paper proposed a sentence rewriting mechanism with few-shot learning to improve the performance of event coreference resolution. We first transform those non-verbal triggers into verb by the pre-training model Transformer, which is fine-tuned on a labeled dataset with few-shot learning. Finally, we utilize the SRL tool to extract the arguments from the rewritten sentences and then introduce them into our event coreference resolution model, which consists of a BERT and a multi-head attention mechanism. Experimental results on both the KBP 2016 and KBP 2017 datasets show that our proposed model CR-RW outperforms the state-of-the-art baseline. In the future, we will focus on how to improve the performance of argument extraction.

Acknowledgments. The authors would like to thank the two anonymous reviewers for their comments on this paper. This research was supported by the National Natural Science Foundation of China (Nos. 61772354, 61836007 and 61773276.), and the Priority Academic Program Development of Jiangsu Higher Education Institutions (PAPD).

References

1. Gardner, M., Grus, J., Neumann, M.: AllenNLP: a deep semantic natural language processing platform. In: Proceedings of the ACL Workshop for Natural Language Processing Open Source Software, pp. 1–6 (2017)
2. Bao, G., Zhang, Y.: Contextualized rewriting for text summarization. In: Proceedings of the Thirty-Fifth AAAI Conference on Artificial Intelligence (2021)
3. Lin, C.-Y.: ROUGE: a package for automatic evaluation of summaries. In: Proceedings of the Workshop on Text Summarization Branches Out, pp. 74–81 (2004)
4. Walker, C., Strassel, S., Medero, J., Maeda, K.: ACE2005 multilingual training corpus. Progress of Theoretical Physics Supplement (2006)
5. Mitamura, T., Liu, Z., Hovy, E.: Overview of TAC KBP 2015 event nugget track. In: Proceedings of the TAC (2015)
6. Ahn, D.: The stages of event extraction. In: Proceedings of ACL 2006, pp. 1–8 (2006)
7. Bejan, C.A., Harabagiu, S.M.: Unsupervised event coreference resolution with rich linguistic features. In: Proceedings of ACL 2010, pp. 1412–1422 (2010)
8. Ng, V., Claire, C.: Identifying anaphoric and nonanaphoric noun phrases to improve coreference resolution. In: Proceedings of ACL 2002, pp. 1–7 (2002)
9. Peng, H., Song, Y., Dan, R.: Event detection and coreference with minimal supervision. In: Proceedings of EMNLP 2016, pp. 192–402 (2016)
10. Lu, J., Ng, V.: Joint learning for event coreference resolution. In: Proceedings of ACL 2017, pp. 90–101 (2017)
11. Huang, Y.J., Lu, J., Kurohashi, S., Ng, V.: Improving event coreference resolution by learning argument compatibility from unlabeled data. In: Proceedings of ACL 2019, pp. 785–795 (2019)
12. Fang, J., Li, P.: Data augmentation with reinforcement learning for document-level event coreference resolution. In: Zhu, X., Zhang, M., Hong, Yu., He, R. (eds.) NLPCC 2020. LNCS (LNAI), vol. 12430, pp. 751–763. Springer, Cham (2020). https://doi.org/10.1007/978-3-030-60450-9_59
13. Lu, J., Ng, V.: Span-based event coreference resolution. In: Proceedings of the Thirty-Fifth AAAI Conference on Artificial Intelligence (2021)

14. Liu, Y., Lapata, M.: Text summarization with pretrained encoders. In: Proceedings of the 2019 Conference on Empirical Methods in Natural Language Processing and the Ninth International Joint Conference on Natural Language Processing (EMNLP-IJCNLP) (2019)
15. Chen, Z., Eavani, H., Chen, W., Liu, Y., Wang, Y.: Few-Shot NLG with pre-trained language model. In: Proceedings of ACL 2020, pp. 183–190 (2020)
16. Lin, T., Goyal, P., Girshick, R., He, K., Dollár, P.: Focal loss for dense object detection. IEEE Trans. Patt. Anal. Mach. Intell. **42**, 2999–3007 (2017)
17. Vilain, M.B., Burger, J.D., Aberdeen, J.S., Connolly, D., Hirschman, L.: A model-theoretic coreference scoring scheme. In: Proceeding of the 6th MUC (1995)
18. Bagga, A., Baldwin, B.: Algorithms for scoring coreference chains. In: Proceedings of LREC 1998, pp. 563–566 (1998)
19. Recasens, M., Hovy, E.: BLANC: Implementing the rand index for coreference evaluation. Nat. Lang. Eng. **17**, 485–510 (2017)
20. Luo, X.: On coreference resolution performance metrics. In: Proceedings of EMNLP 2005, pp. 25–32 (2005)
21. Yang, S., Feng, D., Qiao, L., Kan, Z., Li, D.: Exploring pre-trained language models for event extraction and generation. In: Proceedings of ACL 2019, pp. 5284–5294 (2019)

A Novel Metric Learning Framework for Semi-supervised Domain Adaptation

Rakesh Kumar Sanodiya[1]([✉]), Chinmay Sharma[2], Sai Satwik[1],
Aravind Challa[1], Sathwik Rao[1], and Leehter Yao[3]

[1] Indian Institute of Information Technology Sri City, Sri City, India
rakesh.s@iiits.in
[2] Birla Institute of Technology Mesra, Ranchi, India
[3] National Taipei University of Technology, Taipei 10608, Taiwan
ltyao@ntut.edu.tw

Abstract. In many real-life problems, test and training data belong to different distributions. As a result, a classifier trained on a particular data distribution may produce unsatisfactory results on a different test set. Domain Adaptation is a branch of machine learning that has established methods to classify different but related target data by leveraging information from a labeled source distribution. In this paper, we propose a novel Adaptive Metric Learning Framework (AMLF) to encode metric learning in a domain adaptation framework. In AMLF, feature matching and instance re-weighting are performed jointly to reduce the distribution difference between the domains, while Mahalanobis distance is learned simultaneously to reduce the distance between samples of the same class and to maximize the distance between samples of different classes. Specifically, we incorporate the Maximum Mean Discrepancy (MMD) criterion for feature matching and to construct new domain-invariant feature representations for both distribution differences and irrelevant instances. To validate the effectiveness of our approach we performed experiments on all tasks of the PIE face real-world dataset and compared the results with several state-of-the-art domain adaptation methods. Comprehensive experimental results verify that AMLF can significantly outperform all the considered domain adaptation methods.

Keywords: Domain adaptation · Transfer learning · Semi-supervised learning · Classification

1 Introduction

According to theoretical assumptions, many traditional machine learning techniques depend on the primitive idea that the test and training data are drawn from the same distribution and belong to the same feature space [1,2]. But, in real-life scenarios, this idea does not always holds true. We may have training data from one distribution (source domain) and the test data from another distribution (target domain). For instance, a classifier trained for recognizing objects

© Springer Nature Switzerland AG 2021
T. Mantoro et al. (Eds.): ICONIP 2021, LNCS 13108, pp. 165–176, 2021.
https://doi.org/10.1007/978-3-030-92185-9_14

in an image using a particular data distribution might not perform well while classifying same objects in some other data distribution because of the differences between the distributions. Varying factors like pose, resolution, angle, illumination and location of subject within the image, result in discrepancy between datasets. This problem of distribution divergence can be addressed using transfer learning. The present work in this field can be categorized broadly into three groups: (a) *feature matching* which aims to align distributions by minimizing conditional and marginal distribution divergence between the target and the source domain, or use the geometrical structure of the subspace to perform subspace learning; (b) *instance re − weighting* which assigns weights to the source domain samples to better match the data distribution of the target domain; and (c) *metric learning* which addresses the learning problem by formulating a positive semi-definite distance matrix and use that to reduce discrepancy between the domains. Thus, for initiating the learning process we use patterns that have been already learned for a different but related task.

Domain Adaptation [3,4] can be used to address the issues arising due to the discrepancy between domains. An effective classifier is formulated by leveraging information from the abundant labeled data present in the source domain and using that to classify target domain data with no or very few labels, aiming to reduce the difference between the target and the source domain data. In general, Domain Adaptation methods can be categorized into the following types: (i) Unsupervised Domain Adaptation and (ii) Semi-supervised Domain Adaptation. In Unsupervised Domain Adaptation the source domain contains sufficient labeled data, while the target domain data is completely unlabeled. Whereas, in Semi-supervised Domain Adaptation methods, the target domain distribution contains a few labeled samples alongside the abundantly labeled source domain data. In this paper, we focus on Semi-supervised Domain Adaptation.

Many previous studies reduce divergence between distributions by learning new representations of the data in both the domains or trying to match the target data distribution by re-weighting source domain data. Also, most existing domain adaptation methods measure dissimilarity between target and source data points using Euclidean distance. But, the shortcomings corresponding to the Euclidean distance metric may hinder transfer of knowledge across the domains. As a result, for classification problems, objective functions formulated by using Euclidean distance produce inadequate results: neither the between-class distances are maximized nor the within-class distances are minimized. Therefore, the distribution divergence between the domains has to be reduced and at the same time, for the target domain, a suitable distance metric must be learned.

Therefore, we develop an adaptive metric learning framework, where the source domain Mahalanobis distance metric is learned, both domain feature matching and instance re-weighting are performed simultaneously in the parallel framework. Mahalanobis distance is used for maximizing the distance between different classes and minimizing the distance within the same class. Whereas instance-re-weighting and feature matching reduce the divergence between the distribution of source and target domains.

In summary, the proposed work aims to contribute in the following ways:

1. Introduce a new metric learning framework AMLF which learns the Mahalanobis distance to preserve discriminative information of the source domain and considers both feature matching and instance re-weighting to reduce the distribution difference between both domains.
2. Use the real-world dataset: the PIE face dataset to create source-target task pairs and evaluate the performance of our approach for the classification problem. Furthermore, provide comprehensive analysis and comparison to validate the efficacy of AMLF over other existing transfer learning approaches.

2   Related Work

In recent years, researchers have proposed many novel Domain Adaptation (DA) approaches with the purpose of producing training models that can perform well and significantly increase accuracy across different types of data. This section briefly describes the existing methods that conceptually contributed to our approach.

Zhang et al. proposed Joint Geometrical and Statistical Alignment (JGSA) [5] method, which attempts to reduce the differences between domains both statistically and geometrically by exploiting both the shared and domain-specific features. It projects data onto a lower-dimensional subspace to reduce geometrical and distribution shifts simultaneously. Chen et al. [6] proposed a Manifold Embedded Distribution Alignment (MEDA) approach that obtains feature representations by transforming the subspaces. Furthermore, it assigns relative weights to the conditional and marginal distributions instead of equal weights while doing distribution alignment for better knowledge transfer across domains.

In Robust Transfer Metric Learning (RTML) Ding et al. [7] proposed an approach that reduces conditional and marginal distribution differences between the target and source domain as well as the classes using a low-rank metric. It also includes a marginalized denoising scheme for achieving a better cross-domain metric. Cross-Domain Metric Learning (CDML) [8] approach finds a shared Mahalanobis distance across domains to transfer knowledge. In order to learn the Mahalanobis distance in CDML, source domain data is aligned with its labeled information, the structure of the target domain data is preserved and the distribution divergence between the target and source domain is minimized. Transfer Joint Matching (TJM) [9] re-weights source domain instances and minimizes Maximum Mean Discrepancy (MMD) between the domains for feature matching. It reduces the discrepancy between domains by generating domain-invariant feature representations, which are produced by combining Principal Component Analysis (PCA) with the re-weighted source domain instances and the minimization of MMD.

Metric Transfer Learning Framework (MTLF) [10] proposed by Xu et al. encodes transfer learning with metric learning. It proposes to learn instance weights and utilize them for reducing the gap between different distributions, and simultaneously, in a parallel framework, learn a Mahalanobis distance for minimizing the intra-class distance and maximising the inter-class distances in the target domain. We follow a similar approach in our method.

3   Adaptive Metric Learning Framework for Semi-supervised Domain Adaptation

In this section, a novel adaptive metric learning framework (AMLF) for semi-supervised domain adaptation is proposed, which addresses the limitation of existing approaches by including all the objective functions required to construct domain invariant feature space.

3.1   Formulation of the Model

In this paper, we focus on learning a Mahalanobis distance metric for the target domain. As the Mahalanobis distance metric is learned by a pair of must-link (pair of samples that belong to the same class) and cannot-link (pair of samples that belong to different classes) constraints, we need labeled information to generate such constraints. However, the labeled information for the target domain is not provided. Therefore, for learning the metric, we use the pairwise constraints of the source domain as we have its labeled information. However, pairwise constraints of the source domain cannot be directly utilized because of distribution gap between the source domain and the target domains. Therefore, to minimize the distribution gap we consider MMD and instance re-weighting strategies. In our proposed framework, learning a Mahalanobis distance metric and minimizing the distribution difference between the source and the target domain are performed simultaneously. The description of each component of our framework is as follows:

Instance Re-Weighting for Source Domain. We define the source domain data as $D_S = \{(x_1, y_1), ..., (x_{N_S}, y_{N_S})\}$ and the target domain data as $D_T = \{(x_{N_S+1}, y_{N_S+1}), ..., (x_{N_S+N_T}, y_{N_S+N_T})\}$. Since, we have different distributions for the target and source domains, i.e., $P_S(x) \neq P_T(x)$, so for the target domain, a Mahalanobis distance metric cannot be directly learned using labeled data from the source domain. Similar to the idea proposed in many instance-based transfer learning methods, our approach uses pairwise must-links and cannot link constraints of the source domain labeled data after re-weighting them to learn the metric for the target domain data.

We learn instance weights η for the source data distribution.

A regularisation term $\psi(\eta)$ is applied on the instance weights which is defined as follows:

$$\psi(\eta) = \|\eta - \eta_o\|^2 \tag{1}$$

where $\eta_0(x)$ are the initial weights of the source domain instances under Euclidean distance. It basically depicts the relative similarity of an instance x_i to both the domains. As proposed by Xu et al. in [10], $\eta_0(x_i)$ can be derived by the summation of products of Gaussian kernel functions $\{\phi_j\}$'s and their respective non-negative parameters $\{\alpha_j\}$'s as: $\eta_0(x_i) = \sum_{j=1}^{b} \alpha_j \phi_j(x_i)$. Following similar approach as [10], we can find the weights $\eta_0(x)$ by reducing the KL-divergence between the target distribution and the re-weighted source distribution using the given convex optimization problem:

$$\max_{\alpha} \sum_{x_i \in D_T} log \sum_{j=1}^{b} \alpha_j \phi_j(x_i) \tag{2}$$

s.t.

$$\sum_{x_i \in D_S} \sum_{j=1}^{b} \alpha_j \phi_j(x_i) = N_S, \quad and \quad \alpha > 0 \tag{3}$$

Mahalanobis Distance Metric. For classification problems Mahalanobis distance metric provides better between-class separability and within-class compactness.

According to [10], even with a unified scale, different distributions can be easily distinguished using a learned Mahalanobis distance. Hence, for the target domain, we learn a Mahalanobis distance metric. We define G as a positive semi-definite matrix. So for a pair of data samples x_i and x_j, the Mahalanobis distance metric is defined as:

$$d_{ij} = \sqrt{(x_i - x_j)^T G (x_i - x_j)} \tag{4}$$

The matrix G can be decomposed as $H^T H$, where $H \in R^{d \times d}$ (since G is positive semi-definite).

To control the generalisation error of the metric in terms of H, we form a regularization term $r(H)$ which is defined as:

$$r(H) = tr(H^T H) \tag{5}$$

Minimization of Distribution Divergence Using MMD. Even with the re-weighted source domain instances, the discrepancy between data distributions of both the domains is significant. Therefore, we need to further minimize the gap in feature distributions using a distance measure.

Marginal distribution divergence can be minimized using the Maximum Mean Discrepancy (MMD) criterion. MMD estimates the distance between the distributions of target and source domain in a d-dimensional space using the given formula:

$$\left\| \frac{1}{N_S} \sum_{i=1}^{N_S} H^T x_i - \frac{1}{N_T} \sum_{j=N_S+1}^{N_S+N_T} H^T x_j \right\|^2 = tr(H^T D M_o D^T H) \tag{6}$$

where M_o is the MMD matrix and can be computed as discussed in paper [11], and D is the input data.

Our Approach for Classification Problems. Our approach can be used to solve classification problems by using the distance metric H and instance weights $\widehat{\eta}$ in the K-Nearest Neighbour (KNN) classifier.

The loss function can be defined as:

$$L(f, H, \widehat{\eta}, D_S, D_T) = L_{in}(H, \widehat{\eta}) - L_{out}(H, \widehat{\eta}) \tag{7}$$

where $L_{in}(H, \widehat{\eta})$ is the sum of differences between weighted instances within a class and $L_{out}(H, \widehat{\eta})$ is the sum of differences between weighted instances between classes and are defined as:

$$\begin{cases} L_{in}(H, \widehat{\eta}) = \sum_{y_i = y_j} \widehat{\eta}(x_i)\widehat{\eta}(x_j) \|H(x_i - x_j)\|^2 \\ L_{out}(H, \widehat{\eta}) = \sum_{y_i \neq y_j} \widehat{\eta}(x_i)\widehat{\eta}(x_j) \|H(x_i - x_j)\|^2 \end{cases}$$

Our adaptive objective function for classification problem is obtained by combining Eqs. (1), (5), (6) and (7) as follows:

$$\min_{H,\widehat{\eta}} \; tr(H^T H) + \lambda_1 \|\widehat{\eta} - \widehat{\eta}_0\|^2 + \lambda_2(H^T D M_o D^T H)$$
$$+\beta \sum_{i,j} \widehat{\eta}(x_i)\widehat{\eta}(x_j) \|H(x_i - x_j)\|^2 \delta_{ij} \tag{8}$$

where,

$$\delta_{ij} = \begin{cases} 1, & \text{if } y_i = y_j \\ -1, & \text{if } y_i \neq y_j \end{cases}$$

3.2   Optimization of Adaptive Objective Function

In our approach, we first use an unified optimization problem using the loss function $L(f,\ H,\ \widehat{\eta},\ D_S,\ D_T)$ and then form an unconstrained optimization problem as follows:

$$\min_{H,\widehat{\eta}} \mathcal{J} = r(H) + \lambda_1 \|\widehat{\eta} - \widehat{\eta}_0\|^2 + \lambda_2(H^T D M_o D^T H) + \beta L(f, H, \widehat{\eta}, D_S, D_T)$$
$$+\rho \left((\widehat{\eta}^T e - N_S)^2 + \sum_{i=1}^{N_S}(max(0, -\widehat{\eta}(x_i)))^2 \right) \tag{9}$$

where ρ is considered as the non-negative penalty coefficient and $e \in R^{(N_S + N_T^l) \times 1}$,

and

$$e_i = \begin{cases} 1, & \text{if } i \leq N_S \\ 0, & \text{if } N_S < i \leq N_S + N_T^l \end{cases}$$

Since, a non-parametric form is being utilized for the classification model, we do not need to explicitly optimize our model with respect to f. Hence, to learn H and $\hat{\eta}$ alternatively and iteratively, we use an alternating optimization algorithm. To be precise, the matrix H_t is first fixed at the t-th iteration and the value of $\hat{\eta}_t$ is updated using gradient descent as follows:

$$\hat{\eta}_{t+1} = \hat{\eta}_t - \gamma \left.\frac{\partial J}{\partial \hat{\eta}}\right|_{\hat{\eta}_t} \tag{10}$$

where γ depicts the step-size and $\gamma > 0$.
The derivative of objective J with respect to $\hat{\eta}$ can be written as

$$\frac{\partial J}{\partial \hat{\eta}} = 2\lambda_1(\hat{\eta} - \hat{\eta}_o) + \beta\zeta + \rho\left[2(\hat{\eta}^T e - N_S)e + \hat{\eta}^2\xi\right] \tag{11}$$

where ξ is a vector and $\xi_i = \text{sign}(\max(0, -\hat{\eta}(x_i)))$, and

$$\zeta_i = \sum_{i,j} \hat{\eta}(x_j)\|H(x_i - x_j)\|^2 \delta_{i,j} \tag{12}$$

Now, as the value of $\hat{\eta}_{t+1}$ is updated, we now fix $\hat{\eta}_{t+1}$ alternatingly and H_t is updated as follows:

$$H_{t+1} = H_t - \gamma_w \left.\frac{\partial J}{\partial H}\right|_{H_t} \tag{13}$$

where $\gamma_w > 0$ is the step-size.
To solve classification problems, the derivative of the objective J with respect to H is written as follows:

$$\frac{\partial J}{\partial H} = 2\beta \sum_{i,j} \hat{\eta}(x_i)\hat{\eta}(x_j)Hv_{ij}v_{ij}^T\delta_{ij} + 2H + \lambda_2(DM_oD^TH) \tag{14}$$

here, $v_{ij} = (x_i - x_j)$.
The values of $\hat{\eta}$ and H are alternatingly and iteratively updated until the difference of values from the objective function J is less than a specified threshold value ε.

4   Experiments

A. Features and Benchmarks
The PIE ("Pose, Illumination, Expression") Face data-set [11] comprises of 41,368 images taken under 21 different lighting conditions of 68 people. The data set has been divided into 5 groups based on the pose of the person in the image, that are PIE (C29), PIE (C27), PIE (C09), PIE (C07) and PIE (C05) where C29 represents right pose, C27 for frontal pose, C09 for downward pose, C07 for upward pose and C05 indicates left pose. The images were produced under varying illuminations and poses, and thus have significant differences in their distribution. One of the poses is randomly set as the target domain and some other pose is chosen as the source domain. Using this methodology we

Table 1. Accuracy(%) on PIE face dataset

Tasks	Unsupervised domain adaptation approaches					Semi-supervised domain adaptation approaches					
	TJM (2014)	CDML (2014)	JGSA (2017)	RTML (2017)	MEDA (2018)	MTLF (2017) (3 labels)	MTLF (2017) (4 labels)	MTLF (2017) (5 labels)	AMLF (our approach) (3 labels)	AMLF (our approach) (4 labels)	AMLF (our approach) (5 labels)
PIE 1 5 → 7	29.52	53.22	68.07	60.12	49.60	56.05	66.11	69.12	54.14	67.03	**73.48**
PIE 2 5 → 9	33.76	53.12	67.52	55.21	48.40	48.04	61.34	70.89	56.37	66.18	**75.49**
PIE 3 5 → 27	59.20	80.12	82.87	**85.19**	77.23	66.33	76.21	77.11	80.20	84.20	84.98
PIE 4 5 → 29	26.96	48.23	46.50	52.98	39.82	26.96	41.48	52.14	42.10	56.37	**64.64**
PIE 5 7 → 5	39.40	52.39	25.21	58.13	58.49	68.67	73.11	80.91	67.71	**79.32**	76.86
PIE 6 7 → 9	37.74	54.23	54.77	63.92	55.27	58.52	71.57	79.17	60.23	71.63	**82.54**
PIE 7 7 → 27	49.80	68.36	58.96	76.16	81.25	80.74	82.37	82.40	79.69	**86.18**	86.12
PIE 8 7 → 29	17.09	37.34	35.41	40.38	44.05	37.99	54.84	67.34	39.09	56.86	**66.42**
PIE 9 9 → 5	37.39	43.54	22.81	53.12	56.24	75.21	79.83	81.45	75.90	78.48	**81.51**
PIE 10 9 → 7	35.29	54.87	44.19	58.67	57.82	62.74	71.03	76.06	62.12	72.25	**76.80**
PIE 11 9 → 27	44.03	62.76	56.86	69.81	76.23	83.87	84.47	85.67	83.09	**86.54**	84.53
PIE 12 9 → 29	17.03	38.21	41.36	42.13	53.06	43.00	62.01	68.63	43.50	62.32	**71.08**
PIE 13 27 → 5	59.51	75.12	72.14	81.12	85.95	73.17	77.61	86.31	73.86	82.41	**88.21**
PIE 14 27 → 7	60.58	80.53	88.27	8.92	78.20	76.98	82.07	88.09	77.59	83.86	**89.20**
PIE 15 27 → 9	64.88	83.72	86.09	89.51	80.20	89.40	89.77	93.69	89.34	90.69	**94.36**
PIE 16 27 → 29	25.06	52.78	74.32	56.26	67.70	57.23	62.93	69.18	59.19	70.22	**74.14**
PIE 17 29 → 5	32.86	27.34	17.52	29.11	57.71	65.13	65.55	75.21	70.02	68.34	**75.93**
PIE 18 29 → 7	22.89	30.82	41.06	33.28	49.66	54.27	66.91	74.40	55.37	68.69	**75.26**
PIE 19 29 → 9	22.24	36.34	49.20	39.85	62.13	54.66	67.46	76.35	57.97	71.20	**77.70**
PIE 20 29 → 27	30.72	40.61	34.75	47.13	72.15	74.11	75.37	**78.01**	74.62	76.12	76.96
AVERAGE	37.29	53.69	53.39	58.80	62.56	62.65	70.60	76.60	65.10	73.95	**78.81**

made 20 combinations of source-target task pairs. For example, 27→09 depicts that PIE (C27) is being used as the source domain and PIE (C09) is the target domain.

We incorporated 3, 4 and 5 labeled samples from the target domain respectively at a time in our training data, which has 49 labeled instances per class from the source domain, for conducting experiments with our semi-supervised approach (Fig. 1).

B. Comparison with other methods

We compared our proposed approach with other existing methods such as: Joint Geometrical and Statistical Alignment (JGSA), Transfer Joint Matching (TJM), Manifold Embedded Distribution Alignment (MEDA), Cross-Domain Metric Learning (CDML), Robust Transfer Metric Learning (RTML) and Metric Transfer Learning Framework (MTLF)

C. Parameter Sensitivity Test

Experiments were conducted using different parameter values to demonstrate the effectiveness of our proposed approach on all tasks of the PIE Face dataset. A variation in the accuracy is observed due to the values of these parameters. With the optimal values of these parameters, the model is able to produce the best possible results for all tasks of the dataset. Therefore, we evaluate the parameter sensitivity to establish the optimal trade-off parameter values.

We do preprocessing on the PIE Face dataset, which has 1024 dimensions, by transforming the data to a lower-dimensional space from a higher-dimensional space to reduce the time taken for computation. Principle Component Analysis (PCA) is used in our approach for this purpose. The effect of different dimensions d on the accuracy is studied to arrive at an ideal value. For all PIE Face dataset tasks, we varied the value of d from 50 to 150. Finally, keeping the computational time in consideration, the optimal value of d was set as 100.

To study the impact of number of nearest neighbours K on the accuracy, we varied its value from 1 to 10 while other parameters were fixed at a constant value. Highest average accuracy for all tasks of the dataset was achieved with K set as 1. The effect of number of instance pairs C on the overall accuracy was observed by varying its value from 0 to 1500. The accuracy increases with the value of C, but considering the computational time cost we did not conduct experiments with higher values of C and fixed it as 1500.

The optimal values of λ_1, which balances impact of instance re-weighting for source domain, and λ_2, which controls the impact of MMD loss, were determined by varying their values between 10^{-3} and 10^3 while other parameters were fixed. Through this process we obtained best results on all tasks of the dataset with λ_1 and λ_2 were set as 10^2 and 10^{-2} respectively. Similarly, the optimal values of learning rates γ and γ_W were obtained by varying the value of one of them at a time between 10^{-7} and 10^{-1}, other parameters were kept constant. Our algorithm produced the maximum accuracy on the PIE Face dataset with γ set as 10^{-3} and γ_W set as 10^{-4}.

Similar analysis for β and ρ provided us with their optimal values as $\beta = 1$ and $\rho = 1$.

5 Results and Discussion

Table 1 presents the results obtained by our proposed approach AMLF and other existing approaches on all 20 tasks of the PIE Face dataset along with the average accuracy for all methods. All parameters for AMLF were set to their optimal values obtained from the parameter sensitivity test. A summary of the obtained results is presented in this section.

Our proposed semi-supervised method, AMLF, when trained on training data consisting of 5 labeled samples from the target domain in addition to the source domain data, outperforms other methods by achieving an average accuracy of 78.81%. It is followed by MTLF, trained on the same training data, with an average accuracy of 76.60%. We observe a trend of significant increase in accuracy, for both AMLF and MTLF, when trained on training data consisting more

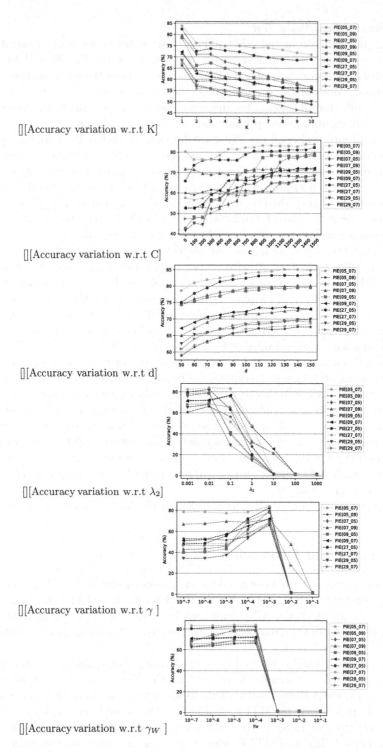

[][Accuracy variation w.r.t K]

[][Accuracy variation w.r.t C]

[][Accuracy variation w.r.t d]

[][Accuracy variation w.r.t λ_2]

[][Accuracy variation w.r.t γ]

[][Accuracy variation w.r.t γ_W]

Fig. 1. Variation in the accuracy of our model for the PIE Face dataset based on the values of parameters and learning rates

labeled target domain samples. However, our method produces better overall results than MTLF in each such case because of better knowledge transfer facilitated by better feature matching.

Also, AMLF produced best results for almost all PIE Face dataset source-target task pairs, except PIE 3 ($5 \rightarrow 7$), for which RTML achieves a slightly higher accuracy, and PIE 20 ($29 \rightarrow 27$) for which MTLF performs better.

6   Conclusion and Future Work

In this paper, we present a novel Adaptive Metric Learning Framework (AMLF) for Semi-supervised Domain Adaptation. By defining adaptive objective based on Mahalanobis distance, instance re-weighting and maximum mean discrepancy, AMLF makes it possible to more efficiently preserve and use intrinsic geometric information between instances from different domains with similar or dissimilar labels. Therefore, AMLF can minimize the distance between the similar class samples and maximize the distance between dissimilar class samples for the target domains. In this way, AMLF improves the performance of the target domain classifier. Experiments on a real-world dataset were conducted and detailed analysis of the results verify the effectiveness of AMLF over other existing domain adaptation approaches.

Our future work would focus on leveraging knowledge from multiple source domains, as in the real-world applications the training data may be collected from different source domains. Also, we would produce pseudo labels for the unlabeled target domain data to minimize conditional distribution divergence between the domains.

References

1. Pan, S.J., Yang, Q.: A survey on transfer learning. IEEE Trans. Knowl. Data Eng. **22**(10), 1345–1359 (2010)
2. Sanodiya, Rakesh Kumar, Sharma, Chinmay, Mathew, Jimson: Unified framework for visual domain adaptation using globality-locality preserving projections. In: Gedeon, Tom, Wong, Kok Wai, Lee, Minho (eds.) ICONIP 2019. LNCS, vol. 11953, pp. 340–351. Springer, Cham (2019). https://doi.org/10.1007/978-3-030-36708-4_28
3. Sanodiya, R.K., Mathew, J.: A novel unsupervised globality-locality preserving projections in transfer learning. Image Vis. Comput. **90**, 103802 (2019)
4. Sanodiya, R.K., Mathew, J.: A framework for semi-supervised metric transfer learning on manifolds. Knowl.-Based Syst. **176**, 1–14 (2019)
5. Zhang, J., Li, W., Ogunbona, P.: Joint geometrical and statistical alignment for visual domain adaptation. In: Proceedings of the IEEE Conference on Computer Vision and Pattern Recognition, pp. 1859–1867 (2017)
6. Wang, J., Feng, W., Chen, Y., Yu, H., Huang, M., Yu, P.S.: Visual domain adaptation with manifold embedded distribution alignment. In: Proceedings of the 26th ACM International Conference on Multimedia, pp. 402–410 (2018)
7. Ding, Z., Fu, Y.: Robust transfer metric learning for image classification. IEEE Trans. Image Process. **26**(2), 660–670 (2017)

8. Wang, H., Wang, W., Zhang, C., Xu, F.: Cross-domain metric learning based on information theory. In: Proceedings of the AAAI Conference on Artificial Intelligence, vol. 28, no. 1 (2014)
9. Long, M., Wang, J., Ding, G., Sun, J., Yu, P.S.: Transfer joint matching for unsupervised domain adaptation. In: Proceedings of the IEEE Conference on Computer Vision and Pattern Recognition, pp. 1410–1417 (2014)
10. Xu, Y., et al.: A unified framework for metric transfer learning. IEEE Trans. Knowl. Data Eng. 29(6), 1158–1171 (2017)
11. Long, M., Wang, J., Ding, G., Sun, J., Yu, P.S.: Transfer feature learning with joint distribution adaptation. In: Proceedings of the IEEE International Conference on Computer Vision, pp. 2200–2207 (2013)

Generating Adversarial Examples by Distributed Upsampling

Shunkai Zhou[1], Yueling Zhang[1,3]($\boxtimes$), Guitao Cao[1], and Jiangtao Wang[1,2]($\boxtimes$)

[1] School of Software Engineering, East China Normal University,
Shanghai 200062, China
51194501087@stu.ecnu.edu.cn, {ylzhang,gtcao,jtwang}@sei.ecnu.edu.cn
[2] National Trusted Embedded Software Engineering Technology Research Center,
Shanghai 200062, China
[3] Computer Science and Information System, Singapore Management University,
Singapore 188065, Singapore

Abstract. The development of neural networks provides state-of-the-art results for many tasks. However, much research has shown that deep neural networks are vulnerable to adversarial attacks that fool deep models by adding small perturbations into original images. The defense of the neural networks also benefits from a better understanding of the attacks. In consequence, generating adversarial examples with a high attack success rate is worth researching. Inspired by single image super-resolution, this paper treats adversarial attacks as an image generation task and designs a new model based on generative adversarial networks (GANs). The latent feature maps have been divided into low-level and high-level in this research. We exploit low-level features with noises to add perturbations during the upsampling process. To further generate perturbed images, we reconsider and make use of checkerboard artifacts caused by deconvolution. We illustrate the performance of our method using experiments conducted on MNIST and CIFAR-10. The experiment results prove that adversarial examples generated by our method achieve a higher attack success rate and better transferability.

Keywords: Adversarial example · Generative adversarial networks · Deep neural network · Deconvolution

1 Introduction

Deep neural networks have been widely researched and applied in a large number of tasks including segmentation, object recognition, face recognition. Among these tasks, the security-critical ones have higher requirements on the robustness of deep models. Unfortunately, although deep neural networks are powerful, recent work has demonstrated that they are vulnerable to adversarial attacks, which can mislead the deep models by adding small perturbations to the original images [3,18]. This security hazard affects the actual application of neural networks seriously. For example, traffic accidents are prone to occur for autopilot when using adversarial examples to attack deep networks.

© Springer Nature Switzerland AG 2021
T. Mantoro et al. (Eds.): ICONIP 2021, LNCS 13108, pp. 177–189, 2021.
https://doi.org/10.1007/978-3-030-92185-9_15

There are various attack methods in the literature to generate adversarial examples. Goodfellow et al. [3] first proposed Fast Gradient Sign Method(FGSM) based on gradient. Projected gradient descent [10] was developed to achieve higher attack success rates. Other optimization-based algorithms [1,21] rely on optimization schemes of simple pixel space metrics to encourage the generation of more realistic images. These attack methods normally use proper distance metrics L_0, L_2 and L_∞ to restrict the magnitudes of perturbations and make the presented adversarial examples visually natural.

Different from the methods above, AdvGAN [20] used a GAN [2] to generate adversarial perturbations. Once the generation network is trained, this model-based method could generate adversarial examples without revisiting the target model. GAN has demonstrated compelling capabilities in the field of image generation. Direct generating perturbations separates perturbations from images and limits the representation ability of GAN. It fails to exploit latent features of original images.

To take original images as prior knowledge to generate perturbed images directly and use different latent features better, we propose Super-Resolution Inspired GAN (SRIGAN). Inspired by single image super-resolution, we treat the adversarial attacks as a special image generation problem. Normal image generation problems always try to generate high-quality images. Unlike them, the main objective of adversarial image generation is to fool the deep neural networks. SRIGAN generates final adversarial examples by disturbing upsampling process. It tries to add extra noises to the low-level feature maps and exploits the checkerboard effect of deconvolution.

Our contributions can be summarized as follows:

- In this work, GAN with an encoder-decoder based generator has been adopted to generate adversarial examples rather than adversarial perturbations. We add noises into low-level features that mixed context information to produce perturbations in the upsampling process. Finally, we design a new model to produce adversarial examples.
- We reconsider checkerboard artifacts caused by deconvolution and show that it has a positive effect on the generation of adversarial examples. A Higher attack success rate could be achieved by exploiting the property of checkerboard artifacts since perfect images are not required.
- We compare SRIGAN with AdvGAN on MNIST and CIFAR-10. Our experiments show that SRIGAN can achieve a higher attack success rate and better transferability than AdvGAN in different experimental settings.

2   Related Work

2.1   Adversarial Examples

The concept of adversarial examples was introduced by Szegedy et al. [18]. Adversarial examples which can mislead networks are formed by applying small but intentional perturbations to examples from the dataset. Adversarial attacks are broadly classified into white-box and black-box attacks. Adversarial attacks also can be divided into target attacks and untarget attacks.

A number of attack strategies are proposed. Fast Gradient Sign Method takes advantage of a first-order approximation of the loss function [3]. Formally, the perturbations could be expressed as $\eta = \epsilon \ \text{sign}(\nabla_x J(\theta, x, y))$, where x is the input image, J is the cost used to train the target network and ϵ represents perturbation magnitude. FGSM can be interpreted as a simple one-step scheme for maximizing the inner part of the saddle point formulation. A more powerful adversary is the multi-step variant–BIM [7] and PGD [10]. These iterative methods execute FGSM many times with smaller ϵ. The goal of optimization-based methods is to minimize the distance between the perturbed images and the original images on the premise that the attack can be achieved [1]. However, the optimization process is slow.

In order to accelerate the generation process, AdvGAN [20] was developed to generate adversarial perturbations by using GAN. AdvGAN++ [11] was proposed to use latent features to generate adversarial examples.

2.2   Generative Adversarial Networks (GANs)

GAN was proposed by Goodfellow et al. and has been used in various data generations successfully [2]. A typical GAN consists of a generator and a discriminator. The generative adversarial networks train generators and discriminators simultaneously via an adversarial process. Finally, the model distribution could be improved by the generator during the conflict with the discriminator.

DCGAN [16] first applies fully convolutional neural networks to GANs. In order to improve user control of GANs, Mehdi Mirza et al. [13] proposed CGAN, which takes extra information as a prior condition to produce the special generation we need. Pix2Pix [5] is an image-to-image conditional GAN that takes images as inputs. CycleGAN [24] further enhances the quality of generated images with unpaired data. The mapping from an original image to a perturbed image has been learned from a similar image-to-image architecture.

2.3   Checkerboard Artifacts

We can often see checkerboard pattern of artifacts when we look at images generated by neural networks closely [14]. This is because deconvolution can easily put more of the metaphorical paint in some places than others, especially when the kernel size is not divisible by the stride. In order to avoid the phenomenon, resize-convolution [14] and PixelShuffle [17] have been proposed to replace deconvolution modules.

3   Proposed Approach

In fact, adversarial examples are perturbed images. If GAN can produce a perfect image without noises, we can generate perturbed images by interfering with its generation process. Then, adversarial examples could be achieved with the bound of the target model.

3.1 Problem Definition

Suppose that M is a deep neural network trained on a dataset from a certain distribution and can achieve high accuracy. Let (x, y) be one instance from the same distribution, where x is an original input and y is the corresponding true class label. The overall system aims to generate x' to mislead the deep model when x' is similar to x. Mathematically:

$$x' = G(f_1(x)|(f_2(x) + z)) \tag{1}$$

such that

$$M(x') \neq y \tag{2}$$

$$\|x' - x\|_p \leq \epsilon \tag{3}$$

where Eq. (3) ensures that the distance between x and x' is within allowable perturbation range.

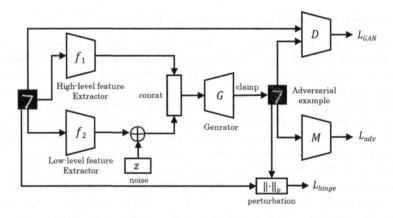

Fig. 1. Overview of SRIGAN

3.2 Proposed Framework

The network we proposed consists of five parts(see Fig. 1): two feature extractors f_1, f_2, a generator G, a discriminator D, and a target model M. Humans are not affected by the adversarial perturbations because they can still extract high-level information from the original images such as semantic information. On the other hand, deep neural networks are affected by perturbated images due to the misleading semantic information extracted from the perturbated images. High-level semantic information is the abstraction of low-level information. Therefore, more attention should be paid to perturb low-level feature maps to generate adversarial examples efficiently. As shown in Fig. 3, less amount and more concentrate perturbations are generated by SRIGAN. Most of the perturbations are next to the digits in the images. It is because we mainly mutate the edge features to disturb the semantic information. Firstly, we use two feature extractors to separate the high-level and low-level information apart. f_1 is a deep neural network to extract high-level feature maps for the later upsampling process to construct the original image. The depth and the capacity of f_1 are greater than f_2. Extractor f_2 is designed to extract low-level detailed representations, such as edge and texture. f_2 also takes advantage of Pyramid Pooling Module [23] to extract global context information for later feature fusion. In this way, we are able to separate high-level and low-level features apart. Secondly, the mixed feature maps are used by the generator G for upsampling. The aim of D is to make these generated images less different from the original images. Finally, the adversarial images are passed to the target model. To successfully cheat the target model, we need to minimize the possibility that the adversarial example will be classified as the correct label. Besides, SRIGAN should bound the magnitude of perturbations. The final loss function that satisfies all the above demands is demonstrated in Eq. 4.

$$\mathcal{L} = \mathcal{L}_{adv} + \alpha\mathcal{L}_{GAN} + \beta\mathcal{L}_{hinge} \tag{4}$$

Here α and β are hyper-parameters to control the weight of each objective. We can adjust them to balance three losses to improve training results.

Adversarial loss is designed to produce more realistic images. G aims to minimize the loss to generate images close to true data, and D tries to maximize the loss to distinguish fake images from true images. The optimal parameters would be obtained by solving the min-max problem.

$$\mathcal{L}_{GAN} = \min_{G} \max_{D} \mathbb{E}_x[log(D(x))] + \mathbb{E}_x[log(1 - D(G(f_1(x)|(f_2(x) + z)))] \tag{5}$$

$\mathcal{L}_{adv}$ is designed for a higher attack success rate. It can be written as:

$$\mathcal{L}_{adv} = \mathbb{E}_x \max[M_t(G(f_1(x)|(f_2(x) + z))) - M_{\hat{t}}(G(f_1(x)|(f_2(x) + z))), 0] \tag{6}$$

where t is the true label of x while $\hat{t}$ is the label with the highest classification probability among other incorrect labels.

The loss for limiting the magnitude of perturbations within a user-specified bound is:

$$\mathcal{L}_{hinge} = \mathbb{E}_x \|G(f_1(x)|(f_2(x) + z)) - x\|_2 \tag{7}$$

Here we choose l_2 loss between the adversary and input image as distance metric to make sure that the difference between them is minimal.

3.3   Reconsider Deconvolution

The upsampling module is an indispensable part of image-to-image translation tasks. When we used neural networks to generate images, we often reconstructed them with low resolution and high-level descriptions. Unfortunately, deconvolution can easily cause "uneven overlap" called checkerboard artifacts [14].

Unlike other image generation tasks such as single image super-resolution, the perturbed images we need in adversarial attacks do not have to be high-quality. We find that deconvolution is more suitable for our network. Through deconvolution, the noises added to the detail information can further disturb feature maps in the upsampling process. In addition, due to checkerboard artifacts, each pixel has a different degree of perturbation during the upsampling process. For example, four interference values are used in the convolution operation of a pixel point, while only one of its adjacent pixel points is the value of disturbance in the convolution process (see Fig. 2). Moreover, decoders typically consist of a multi-layer deconvolution network. Consequently, they often compound and create artifacts on a variety of scales when building a larger image iteratively in the upsampling process. As a result, the degree of interference of each pixel of the final generated image is different or even greatly different, which greatly challenges the classification ability of the model. The experiment proves that in our method, the success rate of attack achieved by the upsampling module using deconvolution is higher than that of Resize-Convolution and PixelShuffle. All of the three upsampling modules have exactly same representation power.

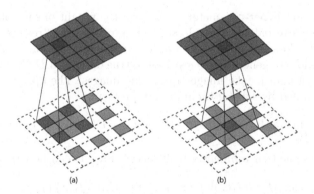

Fig. 2. Upsampling with deconvolution. In deconvolution above, size of input is 3, kenerl size is 3, stride is 2 and padding is 1. The Dotted squares are zeros. (a) Using four real pixels to calculate. (b) Only one real pixel is involved in calculation.

4   Experiments

In this section, we evaluate the performance of SRIGAN we proposed on white-box and black-box settings and compare the attack success rate of different upsampling modules. To demonstrate the advanced performance of SRIGAN, we compare it with AdvGAN on the benchmarks of MNIST [8] and CIFAR-10 [6]. We always use the training dataset to train our generative model and the validation dataset for evaluation for each dataset. For a fair comparison, unless otherwise stated, all adversarial examples for different attack methods under an L_∞ bound of 0.3 on MNIST and 0.03 on CIFAR-10.

4.1 Experimental Setup

Target Model Architectures. For MNIST and CIFAR-10, we both select two different models to perform experiments. We adopt model A from [19] and model B from [1] for MNIST. We use ResNet-32 [4] and Wide-Resnet-34-10 [22] for CIFAR-10. These models are pretrained on the original training dataset. Models A and B get 99% accuracy on test dataset while ResNet-32 and Wide-Resnet achieve 92.4% and 95.1% accuracy respectively.

Implementation Details. We adopt an encoder and decoder based architecture of discriminator and generator similar to those of [5,24]. Adam optimizer is selected to optimize generator and discriminator with learning rate 0.001. We exploit the last convolution layer of the target model as the output of f_1 to simplify the training process of our model. f_2 is a simple decoder module combined with Pyramid Pooling Module [23]. We sample the noise vector from a normal distribution and make use of the least squares objective proposed by LSGAN [12] to stabilize the training procedure of GAN. The batch size we select to train the model is 128.

4.2 Attack in White-Box Setting

We consider the adversary could get the parameters of the model as well as architecture. AdvGAN uses semi-whitebox attacks in their experiments. The advantage of semi-whitebox setting is that we don't have to access the original target model any more after the generator is trained. However, in essence, both white-box and semi-whitebox can only attack when the target model's structure and parameters are obtained.

Attack Without Defenses. We evaluate SRIGAN on different models without defense. We first apply SRIGAN to perform target attacks against Model A and Model B on MNIST dataset. Table 1 shows that SRIGAN could attack successfully even if using different target classes. Next is untarget attacks. In traditional experiments, the perturbation bound of MNIST is 0.3. We perform experiments with different bounds on MNIST considering 0.3 is too large to distinguish. From Table 2, we can see SRIGAN achieves higher attack success rates within 0.2 perturbation bound than AdvGAN with 0.3 perturbation bound. What's more, as shown in Table 3, we find SRIGAN achieves higher attack rates when against most models both on CIFAR-10 and MNIST than AdvGAN. The original images and corresponding adversarial examples generated by SRIGAN on CIFAR-10 shown in Fig. 4 are selected randomly.

Table 1. Target attack success rate of SRIGAN with different models on MNIST

Target class	0	1	2	3	4	5	6	7	8	9
Model A	99.1%	99.1%	99.8%	99.8%	99.9%	99.9%	99.0%	99.3%	99.8%	99.9%
Model B	99.1%	99.2%	99.8%	99.8%	99.6%	99.9%	98.2%	99.7%	99.9%	99.9%

Table 2. Attack success rate with different bound on MNIST when using SRIGAN

	Perturbation bound	0.1	0.2	0.3
Target model	Model A	24.8%	98.4%	99.6%
	Model B	36.8%	98.7%	99.8%

Table 3. Attack success rate of SRIGAN and AdvGAN when using white-box setting on MNIST and CIFAR-10

Dataset	Target model	AdvGAN	SRIGAN
Mnist	Model A	97.9%	**99.6%**
	Model B	98.3%	**99.8%**
CIFAR-10	Resnet-32	94.7%	**95.2%**
	Wide-Resnet	**99.3%**	97.1%

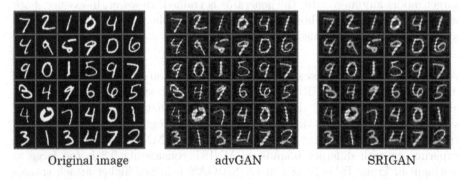

Original image   advGAN   SRIGAN

Fig. 3. Adversarial examples generated by AdvGAN and SRIGAN for Model B on MNIST when perturbation bound is 0.2

Fig. 4. Adversarial examples generated by SRIGAN for Resnet-32 on CIFAR-10. Left are original images and right are adversarial examples

Attack Under Defenses. To defend against more and more attack strategies, a number of defenses have been proposed [9,10,15]. Among them, adversarial training is widely accepted as the most effective way. Goodfellow et al. proposed to improve the robustness of DNNs by using adversarial images along with clean images to train models [3].

Actually, most of the current defense strategies are not robust when attacking against them. We set the first threat model analyzed in [1] as an evaluation standard. We suppose the adversary would directly attack the original learning model because of less attention to defenses. In such a case, if adversarial examples can still achieve a high attack success rate, it means the attack strategy is powerful because of its robustness. Under the setting mentioned above, we adopt an attack strategy to generate adversarial examples based on target models without any defense firstly and evaluate their attack performance against the models trained with defense strategies later. Given the fact that SRIGAN could generate different adversarial instances each time because of the randomness of sampling noises, it could produce resilient adversarial examples successfully under defense.

Here we apply three adversarial training methods for experiments on MNIST and CIFAR-10 (see Tabel 4). The first method for adversarial training is FGSM [3]. The second defense strategy is Ensemble Adversarial Training [19] extended from the first method. This strategy uses multiple models to generate adversarial examples for training. Moreover, the last method is iterative training, which employs PGD [10] as a universal first-order adversary.

Table 4. Attack success rate of FGSM, AdvGAN, AdvGAN++ and SRIGAN when target model is under defense

Dataset	Target model	Defense	FGSM	AdvGAN	AdvGAN++	SRIGAN
MNIST	Model A	Adv	4.3%	8.0%	14.9%	**16.7%**
		Ens	1.6%	6.3%	8.0%	**13.1%**
		Pgd	4.4%	**5.6%**	4.1%	5.2%
	Model B	Adv	2.7%	18.7%	10.9%	**20.9%**
		Ens	1.6%	13.5%	9.0%	**14.6%**
		Pgd	1.6%	12.6%	9.5%	**13.9%**
CIFAR-10	Resnet-32	Adv	13.1%	16.0%	21.3%	**24.2%**
		Ens	10.0%	14.3%	20.2%	**23.4%**
		Pgd	22.8%	29.4%	32.5%	**35.3%**
	Wide-Resnet	Adv	5.0%	14.2%	19.3%	**21.1%**
		Ens	4.7%	13.9%	22.9%	**26.0%**
		Pgd	14.9%	20.7%	32.4%	**35.6%**

4.3   Transferability in Black-Box Setting

The property that the adversarial example generated from one model can fool another different model is called transferability. Considering we have no prior knowledge of target model when using black-box setting, we set transferability as the evaluation standard. A powerful adversarial attack strategy should have the cross-model generalization ability. Table 5 shows that the adversarial examples produced by SRIGAN are significantly transferable to other models. To further prove the better transferability of SRIGAN, we select Net A, Net B and Net C from [19] as target models and test them against resnet-20 [4] on MNIST respectively. From Table 6, we find that the attack success rate gradually increases with combining multiple models into one target model. This is because the generator enhanced generalization ability of attack to attack multiple models at the same time.

Table 5. Transferability of adversarial examples generated by AdvGAN, AdvGAN++ and SRIGAN

Dataset	Target model	Attack model	AdvGAN	AdvGAN++	SRIGAN
MNIST	Model A	Model B	24.1%	**32.6%**	29.7%
	Model B	Model A	85.3%	90.2%	**94.4%**
CIFAR-10	Resnet-32	Wide-Resnet	52.0%	48.2%	**68.1%**
	Wide-Resnet	Resnet-32	34.6%	35.2%	**39.0%**

Table 6. Transferability of SRIGAN generated with different target models on MNIST

Attack model	Target model	AdvGAN	SRIGAN
resnet 20	Net A	53.8%	**70.9%**
	Net B	67.1%	**75.4%**
	Net C	66.9%	**81.7%**
	A+B	71.6%	**87.6%**
	A+C	77.4%	**83.1%**
	B+C	81.9%	**87.9%**
	A+B+C	86.3%	**90.0%**

4.4 Comparison of Upsampling Methods

Here, we select deconvolution, resize-convolution [14] and PixelShuffle [17] as upsampling methods of SRIGAN for comparison. Except for the generator, other settings are the same. Figure 5 shows the attack success rate of adversarial examples generated by deconvolution is the highest.

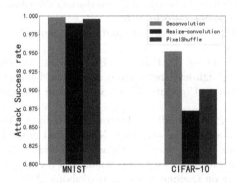

Fig. 5. Adversarial examples generated with different upsampling modules. For MNIST, target model is Model B. For CIFAR-10, target model is Resnet-32

5   Conclusion

In this paper, we transform adversarial attacks into image generations. We propose a new architecture SRIGAN to exploit low-level feature maps with noises to perform attack. Therefore, a higher attack success rate would be achieved by disturbing the generation process of the image-to-image network. In addition, we find that checkerboard artifact sometimes is an advantage for image generation. Experiments performed on MNIST and CIFAR-10 in different experimental settings validate the effectiveness of our approach.

Acknowledgements. This work was supported by National Natural Science Foundation of China (Grant No. 61871186) and the National Trusted Embedded Software Engineering Technology Research Center (East China Normal University).

References

1. Carlini, N., Wagner, D.: Towards evaluating the robustness of neural networks. In: 2017 IEEE Symposium on Security and Privacy (SP), pp. 39–57. IEEE (2017)
2. Goodfellow, I., et al.: Generative adversarial nets. In: Advances in Neural Information Processing Systems, pp. 2672–2680 (2014)
3. Goodfellow, I.J., Shlens, J., Szegedy, C.: Explaining and harnessing adversarial examples. arXiv preprint arXiv:1412.6572 (2014)
4. He, K., Zhang, X., Ren, S., Sun, J.: Deep residual learning for image recognition. In: Proceedings of the IEEE Conference on Computer Vision and Pattern Recognition, pp. 770–778 (2016)
5. Isola, P., Zhu, J.Y., Zhou, T., Efros, A.A.: Image-to-image translation with conditional adversarial networks. In: Proceedings of the IEEE Conference on Computer Vision and Pattern Recognition, pp. 1125–1134 (2017)
6. Krizhevsky, A., Hinton, G., et al.: Learning multiple layers of features from tiny images (2009)
7. Kurakin, A., Goodfellow, I., Bengio, S.: Adversarial examples in the physical world. arXiv preprint arXiv:1607.02533 (2016)
8. LeCun, Y.: The MNIST database of handwritten digits. http://yann.lecun.com/exdb/mnist/ (1998)
9. Liao, F., Liang, M., Dong, Y., Pang, T., Hu, X., Zhu, J.: Defense against adversarial attacks using high-level representation guided denoiser. In: Proceedings of the IEEE Conference on Computer Vision and Pattern Recognition. pp. 1778–1787 (2018)
10. Madry, A., Makelov, A., Schmidt, L., Tsipras, D., Vladu, A.: Towards deep learning models resistant to adversarial attacks. arXiv preprint arXiv:1706.06083 (2017)
11. Mangla, P., Jandial, S., Varshney, S., Balasubramanian, V.N.: AdvGAN++: harnessing latent layers for adversary generation. In: Proceedings of the IEEE International Conference on Computer Vision Workshops (2019)
12. Mao, X., Li, Q., Xie, H., Lau, R.Y., Wang, Z., Smolley, S.P.: Least squares generative adversarial networks. In: Proceedings of the IEEE International Conference on Computer Vision, pp. 2794–2802 (2017)
13. Mirza, M., Osindero, S.: Conditional generative adversarial nets. arXiv preprint arXiv:1411.1784 (2014)
14. Odena, A., Dumoulin, V., Olah, C.: Deconvolution and checkerboard artifacts. Distill **1**(10), e3 (2016)
15. Papernot, N., McDaniel, P., Wu, X., Jha, S., Swami, A.: Distillation as a defense to adversarial perturbations against deep neural networks. In: 2016 IEEE Symposium on Security and Privacy (SP), pp. 582–597. IEEE (2016)
16. Radford, A., Metz, L., Chintala, S.: Unsupervised representation learning with deep convolutional generative adversarial networks. arXiv preprint arXiv:1511.06434 (2015)
17. Shi, W., et al.: Real-time single image and video super-resolution using an efficient sub-pixel convolutional neural network. In: Proceedings of the IEEE Conference on Computer Vision and Pattern Recognition, pp. 1874–1883 (2016)

18. Szegedy, C., et al.: Intriguing properties of neural networks. arXiv preprint arXiv:1312.6199 (2013)
19. Tramèr, F., Kurakin, A., Papernot, N., Goodfellow, I., Boneh, D., McDaniel, P.: Ensemble adversarial training: attacks and defenses. arXiv preprint arXiv:1705.07204 (2017)
20. Xiao, C., Li, B., Zhu, J.Y., He, W., Liu, M., Song, D.: Generating adversarial examples with adversarial networks. arXiv preprint arXiv:1801.02610 (2018)
21. Xiao, C., Zhu, J.Y., Li, B., He, W., Liu, M., Song, D.: Spatially transformed adversarial examples. arXiv preprint arXiv:1801.02612 (2018)
22. Zagoruyko, S., Komodakis, N.: Wide residual networks. arXiv preprint arXiv:1605.07146 (2016)
23. Zhao, H., Shi, J., Qi, X., Wang, X., Jia, J.: Pyramid scene parsing network. In: Proceedings of the IEEE Conference on Computer Vision and Pattern Recognition, pp. 2881–2890 (2017)
24. Zhu, J.Y., Park, T., Isola, P., Efros, A.A.: Unpaired image-to-image translation using cycle-consistent adversarial networks. In: Proceedings of the IEEE International Conference on Computer Vision, pp. 2223–2232 (2017)

CPSAM: Channel and Position Squeeze Attention Module

Yuchen Gong[1], Zhihao Gu[1(✉)], Zhenghao Zhang[2], and Lizhuang Ma[1,2(✉)]

[1] Department of Computer Science and Engineering, Shanghai Jiao Tong University,
Shanghai, China
{gyc1997930,ellery-holmes}@sjtu.edu.cn
[2] Department of Computer Science and Software Engineering,
East China Normal University, Shanghai, China
ma-lz@cs.sjtu.edu.cn

Abstract. In deep neural networks, how to model the remote dependency on time or space has always been a problem for scholars. By aggregatingpioneering method of capturing remote dependencies. However, the NL network faces many problems; 1) For different query positions in the image, the long-range dependency modeled by the NL network is quite similar so that it's a wates of computation cost to build pixel-level pairwise relations. 2) The NL network only focuses on capturing spatial-wise lo a ng-range dependencies and neglects channel-wise attention. Therefore, in response to thesquery-specific global context of each query location, Non-Local (NL) networks propose e problems, we propose the Channel and Position Squeeze Attention Module (CPSAM). Specifically, for a feature map of the middle layer, our module infers attention maps along channel and spatial dimensions in parallel. The Channel Squeeze Attention Module selectively joins the feature of different position by a query-independent feature map. Meanwhile, the Position Squeeze Attention Module uses both avg and max pooling to compress the spatial dimension and Integrate the correlation characteristics between all channel maps. Finally, the outputs of two attention modules are combine together through the conv layer to further enhance feature representation. We have achieved higher accuracy and fewer parameters on the cifar100 and ImageNet1k compared to the NL network. The code will be publicly available soon.

Keywords: Attention mechanism · Image classification · Non-local network

Supported by National Key Research and Development Program of China (Grant 2019YFC1521104), National Natural Science Foundation of China (Grant 61972157), Shanghai Municipal Science and Technology Major Project (Grant 2021SHZDZX0102), Shanghai Science and Technology Commission (Grant 21511101200), Art major project of National Social Science Fund (Grant I8ZD22)
Y. Gong, Z. Gu—Equal Contribution.

T. Mantoro et al. (Eds.): ICONIP 2021, LNCS 13108, pp. 190–202, 2021.
https://doi.org/10.1007/978-3-030-92185-9_16

1   Introduction

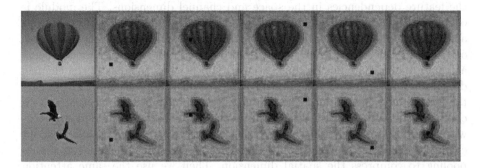

Fig. 1. For different red query positions (after the second column), the heatmaps are almost same. The pictures are based on NL block on ImageNet1K classification. (Color figure online)

As the Non-Local (NL) network [25] shows a powerful ability to improve performance in computer vision task such as image classification, object detection, semantic segmentation and video classification. More and more works want to apply the ability of capturing long-range dependencies of NL networks to more downstream scenarios. However, NL networks still face two problems.

Firstly, the NL network pays more attention to the pair-wise relationship between spatial dimension, while ignoring the attention enhancement in the channel dimension. The most popular channel-wise attention is Squeeze-and-Excitation (SE) attention [17]. It computes channel attention through global pooling. CBAM [26] adds position-wise attention by using large-size convolution kernel. However, these kind of position-wise attention can only capture local relations but failed to modeling long-range dependencies.

Second, the NL network needs to calculate the pair-wise relationship between a query position and all other position in each query position, which incurs a large computation cost. [4,28] use mathematical method to reduce computation cost and optimize the non-local block. Cao et al. [2] observer that, for each query position, the attention maps are almost same. So they simplify the NL network by computing the affinity matrix via a convolution layer. However, these approaches either result in a less accurate outcome or still missing the channel-wise attention.

In this paper, we propose Channel and Position Squeeze Attention Module (CPSAM) to both consider channel-wise attention and reduce the computation cost of NL network. We provide a self-attention mechanism to separately capture the feature dependencies in the space and channel dimensions. The module is composed of two parallel attention modules, Channel Squeeze Attention Module and Position Squeeze Attention Module. First we simplify the NL network base on the observation of [2], which can be shown in Fig. 1 and create channel squeeze attention module. In this module, we use a 1 × 1 conv to squeeze the channel dimension and then calculate a query-independent attention map to aggregate the feature at each position. The module has a significant smaller computation cost than NL Network while still maintain no decrease accuracy. Besides spatial-wise attention, we further introduce position squeeze attention module to adopt channel-wise attention. We use both global average pooling and global maximum pooling to compress the spatial dimension. Then we use the same attention mechanism to Grasp the channel dependency between any two channel-wise feature map. Finally, the output of these two attention modules are multiplied with a learnable parameter to adaptively enhance the feature representations.

Contributions. Our main contribution includes three aspects.

1) We raise a simple and valid attention module, Position and Channel Squeeze Attention Module, which can be inserted to any CNNs.
2) We simplify NL network and propose channel squeeze attention module to learn long-range spatial-wise attention. We also design a position squeeze attention module to learn channel-wise attention.
3) We achieve higher accuracy and fewer parameters on the cifar100 and ImageNet1K compared to the NL network. And not inferior to the start-of-the-art methods

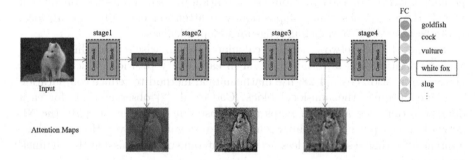

Fig. 2. General picture of the proposed network. We insert our CPSAM at the end of stage 1, 2 and 3. The feature maps of CPSAM at different stages gradually focus on high-level semantics when stage is rising.

2   Related Work

Network Architecture. Since CNNs have recently had a great a achievement in planty of computer vision tasks, many innovations in academia are now focused on improving the network structure to obtain higher accuracy. As the network structure becomes more and more complex, in order to improve the performance of deep networks, a very important structural adjustment direction is to improve the basic blocks. ResNeXt [27] and Xception [6] increase cardinality by splitting the convolution layers. Deformable Conv [7,30] apply offsets to improve geometric modeling ability. Recently Ding et al. [10] uses model reparameterization to make RepVGG faster and more accurate.

Attention Mechanism. As we all know, the attention mechanism [23,24] plays a very critical role in computer vision tasks. For example, in face recognition, it is found that if you pay more attention to eyes or other features with significant identity information, the recognition accuracy will be higher.

In image classification [1,16,17,26] and segmentation [11,14,19], Hu et al. [17] propose a compact 'Squeeze-and-Excitation' to exploit the inter-channel relationships. Woo et al. [26] introduce the CBAM module on the basis of SENet, which increases the attention of the spatial dimension through the large-size convolution kernel. Recently, Hou et al. [15] capture direction-aware and position-sensitive information by compressing the feature along the x-axis and y-axis.

Self-attention mechanism [2,4,12,19,21] has also performed well on many tasks recently. Wang et al. [25] introduce Non-local Network to capture the long-range dependency through modeling pixel-level pairwise relations. Cao et al. [2] however, find that the attention map calculated from different query position is almost same, so he simplifies the NLNet and combine it with SENet. Other examples included CCNet [19], SCNet [21], A^2Net [4] and GSoPNet [12] exploit self-attention mechanisms to capture different types of spatial information.

Different from the previous work, we have considered the spatial-wise attention and channel-wise attention at the same time on the basis of simplifying the NL network. The experimental results from cifar100 and ImageNet1K verify the effectiveness of our proposed method.

3   Method

We first introduce the overall framework of our method in Sect. 3.1, then we elaborate on how to design the channel squeeze attention module in Sect. 3.2 and position squeeze attention module in Sect. 3.3. Finally we discuss how to combine two attention modules in Sect. 3.4.

3.1   Overview

As shown in Fig. 3, our CPSAM is composed of two different modules, position squeeze attention module and channel squeeze attention module. For a given intermediate feature map, our module will learn channel-wise and spatial-wise attention and adaptively fused them with original feature map to further enhance the feature representation. Channel squeeze attention module is a simplified NL network, it uses 1×1 conv to compress the channel dimension so that we can gather channel independent spatial-wise attention. Position squeeze attention module uses maximum and average pooling to squeeze the spatial dimension so that we can have a spatial independent feature map. The ablation study in Sect. 4.2 shows that two attention modules perform best when operated in parallel. Also, sharing unary function ϕ makes the model more compact.

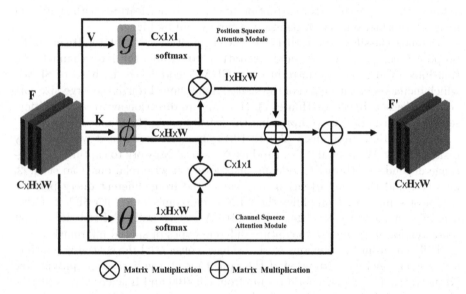

Fig. 3. Detailed architecture of Channel and Position Squeeze Attention Module. CPSAM is composed of position squeeze attention module and channel squeeze attention module. The Channel squeeze attention module is to gather spatial-wise information. The position squeeze attention module uses global pooling to compress spatial dimension to get channel-wise dependencies.

3.2   Channel Squeeze Attention Module

In this section, we first revisit the NL [25] blocks and then we present the channel squeeze attention module.

Non-Local Block. The core idea of the NL network is that the response value of a certain location is weighted by the responses of all locations, and the weight is computed according to the similarity between the current location and other locations. The more comparable the location, the greater the contribution to the current location. Following [2], the general equation of non-local block can be expressed as

$$y_i = \sum_{j=1}^{H*W} \frac{f(x_i, x_j)}{C(x)} g(x_j) \tag{1}$$

where $x \in R^{C*H*W}$ denotes the input feature map, i is the query position while j indexes all the possible positions in x. The unary function g is used to compute the representation of the input at the position j. $f(x_i, x_j)$ indexes the output of the relationship between x_i and x_j. $C(x)$ is a normalization factor.Taking the most frequently used Embedded Gaussian as an example:

$$\frac{f(x_i, x_j)}{C(x)} = \frac{exp(f(x_i, x_j))}{\sum_m exp(f(x_i, x_m))} \tag{2}$$

Channel Squeeze Attention Module. However, Cao et al. [2] discover that the attention maps calculated for every query position are almost same, this not only greatly increase the computation expense but also adds some useless blocks. Based on this discovery, we drop our channel squeeze attention module. Our main idea is to calculate a common long-range global attention map for every query position. Since we only need to consider the position relationship, we use a convolution layer to squeeze the channel to reduce computation cost and avoid channel information.

As illustrated in Fig. 3, for a given input $x \in R^{C \times H \times W}$, we will first transfer it through a 1×1 conv block to produce a new feature $x_c \in R^{1 \times H \times W}$ and use softmax to generate the output attention map $Q \in R^{1 \times H \times W}$. Meanwhile, we feed the feature map x through a shared network and reshape it to $K \in R^{C \times N}$. The shared network is composed of multi-layer preceptron (MLP) with one hidden layer. Then we calculate the global attention through a matrix multiplication between K and Q and perform an element-wise sum operation with the original input feature x. In short the equation of our adaptive channel squeeze is

$$x'_j = x_j + \alpha \sum \frac{exp(\theta(x_j))}{\sum exp(\theta(x_m))} \phi(x_j) \tag{3}$$

where θ is a 1 conv block and ϕ is a MLP with one hidden layer. The reduction ratio of the hidden layer is set to 4. α is a learnable parameter which allow us to adaptively learning the weights. Since we set α to zero at the beginning, we can insert the module to any pre-trained model.

3.3 Position Squeeze Attention Module

Besides spatial attention, channel wise attention is also very important to improve feature representation. We can use each kernel in the conv layer as a feature extractor [29] and focus on what is meaningful for a given picture. Therefore, we hope that by obtaining the attention of the position-independent channel dimension, the network can adaptively emphasize the interdependence channel feature maps to improve feature representation.

Our main idea is to obtain position-independent inter-channel relationship, so we use global pooling to squeeze the spatial dimension. This will not only reduce the computation cost but also to avoid spatial information. Average pooling is usually used to accumulate spatial information, but max-pooling can also retain important feature map information, especially rotation invariance. So we use both max-pooling and average-pooling to squeeze the spatial dimension.

As illustrated in Fig. 3, for a given input $x \in R^{C \times H \times W}$, we will first use average pooling and max pooling and apply softamx to generate two different channel attention maps $V_a \in R^{C \times 1 \times 1}$ and $V_m \in R^{C \times 1 \times 1}$ which donate average-pooled attention map and max-pooled attention map. Meanwhile, we feed the input feature x through the shared network mentioned above to generate feature map $K \in R^{C \times N}$. Then we separately calculate the result through a matrix multiplication between K and V_a, V_m

$$x'_j = x_j + \beta(\sum \frac{exp(g_m(x_j))}{\sum exp(g_m(x_m))}\phi(x_j) + \sum \frac{exp(g_a(x_j))}{\sum exp(g_a(x_m))}\phi(x_j)) \quad (4)$$

where g_m is global maximum pooling and g_a is global average pooling. β is a learnable parameter which allows us to adaptively learn the weights. Since we set β to zero at the beginning, we can insert the module to any pre-trained model.

3.4 Combination of Two Attention Modules

After obtaining the channel attention and location attention of a given feature map, we hope to find the best way to merge two attention modules. Since we can place two modules in parallel or serially, we test channel-first order, position-first order and parallel arrangement. We find that parallel arrangement is better than sequential arrangement, detail experiment will be discussed in Sect. 4.2.

4   Experiments

We excute experiments on basic computer vision tasks such as image classification on cifar100 [22] and ImageNet1K [8] to verify the performance of our proposed method. We compare not only effectiveness but also GFLOPs and the amount of parameters to evaluate the efficiency. The results demonstrate that we achieve less GFLOPs and parameters with higher accuracy compared with the baseline Non-local network. Moreover, we are not inferior to the latest state of art method.

4.1   Image Classification

We use ResNet as backbone and all networks are trained on two 1080tis using Pytorch. To achieve the best result, we add our CPSAM at the end of each stage except stage 4, as shown in Fig. 2. The training setup is almost the same for cifar100 and ImageNet.

Cifar100. The CIFAR-100 datasets [22] is a subset of the 80 million tiny images. Dataset contains 60000 32 × 32 color images in total. And We use 80% of the data for training and the remaining data for testing.

For cifar100 experiments, since the original 7 * 7 conv block has a bad performance, we modify the first conv with kernel size 3, stride 1 and padding 1 and delete the first max pooling layer according to [9]. The initial learning rate is set to 0.1 and decrease by 0.2 times after 60,120,160 epoches.

ImageNet1K. The ImageNet1K dataset [8] has 1000 categories. It contains a total of 1.7 million data, of which 70% is used as training data, and the rest is used as test data.

For ImageNet experiments, to speed up the training process, we first train a resnet backbone for 100 epoches on two 1080tis with batchsize 64. The initial learning rate is set to 0.1 and decrease by 0.1 times after 30,60,90 epoches. Second, we insert the CPSAM block into the pretrained model and continue training for other 40 epoch with a 0.01 original learning rate.

4.2   Ablation Studies on CIFAR-100

We first do some ablation study on cifar100 using ResNet as backbone.

Ablation Study for Attention Modules. Our CPSAM module contains channel squeeze attention module (CSAM) and position squeeze attention module (PSAM). To validate the performance of our block, we eliminate each branch separately to verify the effectiveness of utilizing both channel and spatial attention branches. In this experiment,we add only one CPSAM block at the end of stage3 and the result is demonstrated below.

As shown in Table 1, both adaptive channel squeeze module and adaptive position squeeze module can improve the performance with a slight increase of GFLOPs and Params (0.32% ↑ by CSAM and 0.23% ↑ by PSAM). When we combine them together, the improvement is even more significant (0.58% ↑ by resnet50 and 0.35% ↑ by resnet101). This illustrates that both channel and spatial attention play a key role in concluding the final output.

Table 1. Ablation study for attention modules on cifar-100 dataset.

Backbone	CSAM	PSAM	GFLOPs	Params	Top-1	Top-5
ResNet-50			1.31	23.71M	78.23	94.38
	✓		1.33	24.23M	78.54	94.47
		✓	1.34	24.73M	78.46	94.51
	✓	✓	1.34	24.75M	**78.61**	**94.65**
ResNet-100			2.52	42.70M	80.16	94.89
	✓		2.54	43.24M	80.42	94.94
		✓	2.55	43.65M	80.29	94.92
	✓	✓	2.55	43.70M	**80.51**	**94.97**

Ablation Study for Combining Method. After showing that both spatial attention and channel attention improves the performance, we do ablation study to find the best combination of spatial and channel attention modules. We arrange them in sequential and parallel way and add them in both stage1, stage2 and stage3. The result is show in Table 2. The table shows that generating an attention map in parallel gets a more accurate attention map then sequentially arranges the attention modules (0.7% ↑ by channel first , 0.19% ↑ by spatial first and 1.01% ↑ by parallel). Also, parallel addition of the attention module can achieve the purpose of reducing the amount of calculation by sharing the parameters of the MLP part.

Table 2. Ablation Study for combining method on cifar-100 dataset

	Top-1	Top-5
ResNet50	78.23	94.38
ResNet50 + channel + spatial	78.93	94.72
ResNet50 + spatial + channel	78.42	94.53
ResNet50 + channel + spatial (Parallel)	**79.24**	**94.8**

Ablation Study for Different Stages. Table 3 explores different number of CPSAM modules inserted to different locations of resnet50. First, we separately insert the module to the end of different stage and then, we add it to multi stages.

Table 3. Ablation Study for adding CPSAM on different stages on cifar-100 dataset

Stage1	Stage2	Stage3	Stage4	GFLOPs (Δ)	Params (Δ)	Top-1 (%)	Top-5 (%)
				–	–	78.23	94.38
✓				0.02	0.06	78.98	94.73
	✓			0.02	0.25	78.63	94.69
		✓		0.03	1.04	78.61	94.65
			✓	0.03	3.19	78.08	94.25
✓	✓	✓		0.1	2.35	**79.24**	textbf94.8
✓	✓	✓	✓	0.13	5.57	78.72	94.7

In Table 3, we can see that if we just add one CPSAM module, inserting it to the end of stage1 is the best choice (0.75% ↑). However, CPSAM will have negative effect if it is placed in the end of last stage (0.15% ↓). This may be because the size of the feature map of the last stage is only 7 * 7. For each position of the feature map, the whole image is already felt, so the attention module cannot be promoted very effectively.

4.3 Comparing with NL-Block and State-of-the-Art on ImageNet1k

In this section, we compete our CPSAM with baseline NL network and some state-of-the-art methods on Imagenet1k.

Comparing with NL-Block. Non-local network has been proved to remarkably improve capability of multiple tasks so we compare it with our CPSAM to show the superiority of ours. We insert NL block to the end of stage1,2, and 3. The result is shown in Table 4, we can see that our module achieves higher accuracy (0.7% ↑ in res18 and 1.3% ↑ in res50) with less parameters than NL network.

Table 4. Compared with NL network on Imagenet1k

Backbone	Module	GFLOPs (Δ)	Params (Δ)	Top-1 (%)
ResNet-18	Baseline	–	–	70.2
	+ NL	0.02	0.16	70.5
	+ CPSAM	0.02	0.09	**70.9**
ResNet-50	Baseline	–	–	76.3
	+ NL	0.32	2.15	76.9
	+ CPSAM	0.31	1.38	**77.6**

Comparing with State-of-the-Art. To verify generalization performance of CPSAM, we conduct experiment on many famous deep neural network and some latest attention module. As shown in Table 5, CPSAM achieves the same performance even if the model is very deep.

Table 5. Compared with State-of-the-Art methods on Imagenet1k

Module	GFLOPs	Params	Top-1
SE-ResNet-50 [17]	4.2	28.1	76.9
GE-ResNet-50 [16]	–	31.1	76.8
SRM-ResNet-50 [20]	–	25.6	77.1
A^2-Net [4]	–	–	77.0
DenseNet-201 [18]	4.4	20.0	77.4
Oct-ResNet-50 [3]	2.4	25.6	77.3
ResNet-101 [13]	7.9	44.6	77.4
ResNet-152 [13]	11.6	60.2	78.3
SE-ResNet-152 [17]	11.7	67.2	78.4
ResNet-50 + NL [25]	4.1	27.66	76.9
ResNet-50 + GC [2]	3.9	28.08	77.7
ResNet-50 + BAT [5]	4.2	26.72	77.8
ResNet-50 + CPSAM	4.1	26.94	77.6
ResNet-101 + BAT [5]	6.6	43.54	78.5
ResNet-101 + CPSAM	6.4	45.26	**78.6**

5   Conclusion

We design CPSAM, a simple and effective attention module which both consider channel and spatial wise attention. The core operation is to use 1x1 conv or global pooling to squeeze the unnecessary dimension to reduce computation cost and avoid interference information. It can be easily optimized by connecting to most existing networks. A large number of image classification experiments have verified the superiority of our method in terms of accuracy and efficiency.

References

1. Bello, I., Zoph, B., Vaswani, A., Shlens, J., Le, Q.V.: Attention augmented convolutional networks. In: Proceedings of the IEEE/CVF International Conference on Computer Vision, pp. 3286–3295 (2019)
2. Cao, Y., Xu, J., Lin, S., Wei, F., Hu, H.: GCNet: non-local networks meet squeeze-excitation networks and beyond. In: Proceedings of the IEEE/CVF International Conference on Computer Vision Workshops (2019)

3. Chen, Y., et al.: Drop an octave: Reducing spatial redundancy in convolutional neural networks with octave convolution. In: Proceedings of the IEEE/CVF International Conference on Computer Vision, pp. 3435–3444 (2019)
4. Chen, Y., Kalantidis, Y., Li, J., Yan, S., Feng, J.: $a^\wedge$2-nets: double attention networks. arXiv preprint arXiv:1810.11579 (2018)
5. Chi, L., Yuan, Z., Mu, Y., Wang, C.: Non-local neural networks with grouped bilinear attentional transforms. In: Proceedings of the IEEE/CVF Conference on Computer Vision and Pattern Recognition, pp. 11804–11813 (2020)
6. Chollet, F.: Xception: Deep learning with depthwise separable convolutions. In: Proceedings of the IEEE Conference on Computer Vision and Pattern Recognition, pp. 1251–1258 (2017)
7. Dai, J., et al.: Deformable convolutional networks. In: Proceedings of the IEEE International Conference on Computer Vision, pp. 764–773 (2017)
8. Deng, J., Dong, W., Socher, R., Li, L.J., Li, K., Fei-Fei, L.: ImageNet: a large-scale hierarchical image database. In: 2009 IEEE Conference on Computer Vision and Pattern Recognition, pp. 248–255. IEEE (2009)
9. DeVries, T., Taylor, G.W.: Improved regularization of convolutional neural networks with cutout. arXiv preprint arXiv:1708.04552 (2017)
10. Ding, X., Zhang, X., Ma, N., Han, J., Ding, G., Sun, J.: RepVGG: making VGG-style convnets great again. arXiv preprint arXiv:2101.03697 (2021)
11. Fu, J., et al.: Dual attention network for scene segmentation. In: Proceedings of the IEEE/CVF Conference on Computer Vision and Pattern Recognition, pp. 3146–3154 (2019)
12. Gao, Z., Xie, J., Wang, Q., Li, P.: Global second-order pooling convolutional networks. In: Proceedings of the IEEE/CVF Conference on Computer Vision and Pattern Recognition, pp. 3024–3033 (2019)
13. He, K., Zhang, X., Ren, S., Sun, J.: Deep residual learning for image recognition. In: Proceedings of the IEEE Conference on Computer Vision and Pattern Recognition, pp. 770–778 (2016)
14. Hou, Q., Zhang, L., Cheng, M.M., Feng, J.: Strip pooling: rethinking spatial pooling for scene parsing. In: Proceedings of the IEEE/CVF Conference on Computer Vision and Pattern Recognition, pp. 4003–4012 (2020)
15. Hou, Q., Zhou, D., Feng, J.: Coordinate attention for efficient mobile network design. arXiv preprint arXiv:2103.02907 (2021)
16. Hu, J., Shen, L., Albanie, S., Sun, G., Vedaldi, A.: Gather-Excite: exploiting feature context in convolutional neural networks. arXiv preprint arXiv:1810.12348 (2018)
17. Hu, J., Shen, L., Sun, G.: Squeeze-and-excitation networks. In: Proceedings of the IEEE Conference on Computer Vision and Pattern Recognition, pp. 7132–7141 (2018)
18. Huang, G., Liu, Z., Van Der Maaten, L., Weinberger, K.Q.: Densely connected convolutional networks. In: Proceedings of the IEEE Conference on Computer Vision and Pattern Recognition, pp. 4700–4708 (2017)
19. Huang, Z., Wang, X., Huang, L., Huang, C., Wei, Y., Liu, W.: CCNet: criss-cross attention for semantic segmentation. In: Proceedings of the IEEE/CVF International Conference on Computer Vision, pp. 603–612 (2019)
20. Lee, H., Kim, H.E., Nam, H.: SRM: a style-based recalibration module for convolutional neural networks. In: Proceedings of the IEEE/CVF International Conference on Computer Vision, pp. 1854–1862 (2019)
21. Liu, J.J., Hou, Q., Cheng, M.M., Wang, C., Feng, J.: Improving convolutional networks with self-calibrated convolutions. In: Proceedings of the IEEE/CVF Conference on Computer Vision and Pattern Recognition, pp. 10096–10105 (2020)

22. Torralba, A., Fergus, R., Freeman, W.T.: 80 million tiny images: a large data set for nonparametric object and scene recognition. IEEE Trans. Patt. Anal. Mach. Intell. **30**(11), 1958–1970 (2008)
23. Tsotsos, J.K.: A Computational Perspective on Visual Attention. MIT Press, Cambridge (2011)
24. Tsotsos, J.K., et al.: Analyzing vision at the complexity level. Behav. Brain Sci. **13**(3), 423–469 (1990)
25. Wang, X., Girshick, R., Gupta, A., He, K.: Non-local neural networks. In: Proceedings of the IEEE Conference on Computer Vision and Pattern Recognition, pp. 7794–7803 (2018)
26. Woo, S., Park, J., Lee, J.Y., Kweon, I.S.: CBAM: convolutional block attention module. In: Proceedings of the European Conference on Computer Vision (ECCV), pp. 3–19 (2018)
27. Xie, S., Girshick, R., Dollár, P., Tu, Z., He, K.: Aggregated residual transformations for deep neural networks. In: Proceedings of the IEEE Conference on Computer Vision and Pattern Recognition, pp. 1492–1500 (2017)
28. Yue, K., Sun, M., Yuan, Y., Zhou, F., Ding, E., Xu, F.: Compact generalized non-local network. arXiv preprint arXiv:1810.13125 (2018)
29. Zeiler, M.D., Fergus, R.: Visualizing and understanding convolutional networks. In: Fleet, D., Pajdla, T., Schiele, B., Tuytelaars, T. (eds.) ECCV 2014. LNCS, vol. 8689, pp. 818–833. Springer, Cham (2014). https://doi.org/10.1007/978-3-319-10590-1_53
30. Zhu, X., Hu, H., Lin, S., Dai, J.: Deformable ConvNets v2: more deformable, better results. In: Proceedings of the IEEE/CVF Conference on Computer Vision and Pattern Recognition, pp. 9308–9316 (2019)

A Multi-Channel Graph Attention Network for Chinese NER

Yichun Zhao🆔, Kui Meng[✉]🆔, and Gongshen Liu🆔

School of Electronic Information and Electrical Engineering,
Shanghai Jiao Tong University, Shanghai, China
{zhaoyichun,mengkui,lgshen}@sjtu.edu.cn

Abstract. Incorporating lexicons into Chinese NER via lattice inputs is proven to be effective because it can exploit both boundary and semantic information of words to achieve better results. Previous works have tried various structures such as RNN, CNN, GNN and Transformer to accommodate lattice inputs with great success. In this paper, we propose a Multi-Channel Graph Attention Network (MCGAT), which consists of three word-modified graph attention networks. MCGAT considers relative position relations between characters and words, and combines statistical information of word frequency and pointwise mutual information to further improve the performance of the model. Experiments on four datasets show that our model achieves comprehensive improvements on the original model and outperforms the state-of-the-art model on three datasets using the same lexicon. MCGAT is competitive in terms of efficiency and has a faster inference speed than the transformer-based model. We also demonstrate that MCGAT can significantly improve the performance of pre-trained models like BERT as a downstream network.

Keywords: Chinese named entity recognition · Graph attention network · Lexicon · Statistical information

1 Introduction

Named Entity Recognition (NER) refers to the identification of entities with specific meanings in the text, such as persons, locations, organizations and times. Downstream tasks including information retrieval, relation extraction and question answering require NER as an essential preprocessing step. For languages like English which has natural separators, NER is generally described as a word-level sequence tagging problem. Compared with English, NER is a more challenging work for languages like Chinese of which words are not naturally separated.

There are two main processing methods for Chinese NER tasks. One intuitive way is the pipeline model from Chinese Word Segmentation (CWS) to NER, which means using a CWS tools to segment the source sentence first and then tag the sentence word by word. However, the propagation of word segmentation errors is unavoidable and the out-of-vocabulary (OOV) problem is also difficult to

© Springer Nature Switzerland AG 2021
T. Mantoro et al. (Eds.): ICONIP 2021, LNCS 13108, pp. 203–214, 2021.
https://doi.org/10.1007/978-3-030-92185-9_17

be dealt with through the method. The other way is to directly regard Chinese NER as a character-level sequence tagging problem. Character-level Chinese NER does not make good use of both boundary information and semantics information of words in the source sentence. Since that the boundaries of named entities often coincide with the boundaries of words and towards transliterated words such as "可乐", it's difficult for characters "可" and "乐" to describe the combination meaning of cola.

To overcome the above drawbacks, in 2018 Zhang and Yang proposed Lattice LSTM [15], which uses an external lexicon to match all the words in the sentence, converting the sentence into a directed acyclic graph(DAG), and leverages a variant of LSTM to automatically integrate word information into characters. In 2019, Sui proposed CGN [12] which uses graph attention network(GAT) [13] to adapt to lattice inputs. But the model does not distinguish between different edges so that it may confuse the flow of information between characters and words. In this paper, we divide the edges into 10 categories, using trainable embeddings to represent the relations between characters and words. We also integrate statistical information in the dataset through word frequency count and unsupervised new words extraction, which are proven to be beneficial in our model.

The contributions of our work can be summarized as follows:

- We propose a Multi-Channel Graph Attention Network for Chinese NER with three word-modified GATs to integrate word information based on relational and statistical level.
- Experiments on four datasets show that our model achieves comprehensive improvements on the original model and outperforms the state-of-the-art model on three datasets using the same lexicon with faster inference speed.
- Our model can also further improve the performance of pre-trained models like BERT as a downstream network.
- Our code and data will be released at https://github.com/cdxzyc/MCGAT.

2   Related Work

Incorporating lexicons into Chinese NER has received continued research attention recent years. Lattice LSTM [15] firstly uses an external lexicon to integrate word information into characters. Based on Lattice LSTM [15], WC-LSTM [8] designs four strategies to fuse word information. LR-CNN [3] uses CNN as the basic unit and enhances with a rethinking mechanism. CGN [12] and LGN [4] represent lattice inputs as graph structures and feed them into GNNs. FLAT [7] and PLT [10] use transformer to adapt to the lattice input by using special relative position encoding methods. Simple-Lexicon [9] weights the matched word embeddings at different positions by word frequency and directly concatenates them with the character embeddings.

However, Chinese NER models with lexicon usually ignore statistical information within the corpus and models based on GNNs rarely consider the relations between character and word nodes. Based on the original model CGN [12], our model therefore incorporate both the relations of the nodes and the statistical information of the dataset into GATs.

3   Model

A Multi-Channel Graph Attention Network includes a text encoder layer, a graph layer, a scalar fusion layer and a tag decoder layer. The overall architecture of MCGAT is shown as Fig. 1.

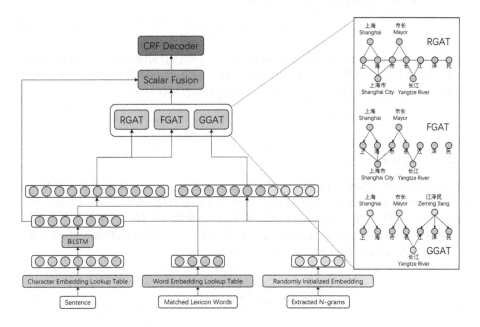

Fig. 1. Overall architecture of MCGAT

3.1   Text Encoder Layer

The input to MCGAT is a source sentence and all the words in the external lexicon that match a consecutive subsequence of the sentence. We denote the sentence of length n as a sequence $c = \{c_1, c_2, \cdots, c_n\}$ where c_i is the i-th character, and denote the matched words list of length m as a sequence $w = \{w_1, w_2, \cdots, w_m\}$, where w_i is the ith word matched from the beginning.

Each character c_i is represented by $x_i = e^c(c_i)$ where e^c denotes a character embedding lookup table. To extract sequence features of the sentence, we firstly use a BiLSTM to encode character embeddings of the sentence and obtain the output hidden state h_i as follows:

$$h_i = \overrightarrow{LSTM}(x_i, \overrightarrow{h_{i-1}}) \oplus \overleftarrow{LSTM}(x_i, \overleftarrow{h_{i-1}}) \tag{1}$$

$H^C = \{h_1, h_2, \cdots, h_n\}$ denotes hidden states list of the source sentence.

In recent years, pre-trained language models such as BERT [2] have shown powerful performance in NLP domain and have obtained competitive results on NER tasks. In order to explore the effectiveness of MCGAT as a downstream network of pretrained models, we also try to use BERT [2] as an encoder to directly obtain hidden states of characters:

$$h_i = BERT(c_i) \tag{2}$$

Each matched word w_i is represented by $l_i = e^w(w_i)$ where e^w denotes a word embedding lookup table. We concatenate H^C with embeddings of the matched words list to obtain H^L as the final input to the graph layer:

$$H^L = [h_1, h_2, \cdots, h_n, l_1, l_2, \cdots, l_m] \tag{3}$$

3.2   Graph Layer

The graph layer consists of three subgraphs, each using a word-modified GAT to capture different channel of word information. In an L-layer vanilla GAT [13], the input of the lth layer are a sequence of N node features $H_l = \{h_1^l, h_2^l, \cdots, h_N^l\}$ where $h_i^l \in \mathbb{R}^{D^l}$ and an adjacency matrix $A \in \mathbb{R}^{N \times N}$. The GAT operation with K independent attention heads can be written as:

$$h_i^{l+1} = \overset{K}{\underset{k=1}{\|}} \sigma(\sum_{j \in N_i} \alpha_{ij}^k W^k h_j^l) \tag{4}$$

where $\|$ is the concatenation operation, σ denotes a nonlinear activation function as which we use ELU in this paper, N_i is the set of neighboring nodes of node i in the graph, α_{ij}^k denotes the attention coefficient, $W^k \in \mathbb{R}^{D^l \times D^{l+1}}$ are trainable parameters to transform the dimensionality of node features each layer.

Based on the vanilla GAT [13], we construct different graphs and design specific attention mechanism for each GAT in the graph layer. The graphs constructed by three GATs for the same example sentence are also shown in Fig. 1. In this subsection, we will introduce three GATs in order.

Relation Graph Attention Network. As shown in Fig. 1, we construct an undirected graph by making connections between each matched word and characters it contains. Consecutive characters in the sentence are also connected.

For vanilla GAT [13] in CGN [12], the status of all nodes are the same, which means all edges are equivalent and the attention mechanism only considers the similarity between neighboring nodes. Since the roles of character nodes and word nodes in the graph are different, we summarize ten relative position relations as shown in Fig. 2.

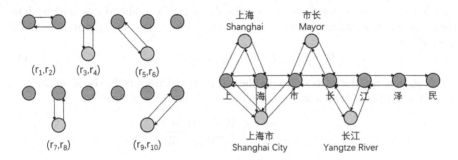

Fig. 2. Ten relative position relations: r_1 and r_2 are relations between consecutive characters, $r_3 - r_{10}$ are relations between the matched word and characters it contains at the position of S,B,M,E respectively

For ten relative position relations, we trained different embeddings for them and design the following equation to calculate the attention coefficients:

$$\alpha_{ij} = \frac{exp(\sigma(a_1 W_1 h_i + a_2 W_2 h_j + a_3 e_{ij}))}{\sum_{v \in N_i} exp(\sigma(a_1 W_1 h_i + a_2 W_2 h_v + a_3 e_{iv}))} \tag{5}$$

where $h_i, h_j \in \mathbb{R}^{D^l}$ are features of node i and node j, $e_{ij} \in \mathbb{R}^{D_R}$ denotes embedding of the relation between node i and node j, σ denotes a nonlinear activation function as which we use $LeakyReLU$, $W_1, W_2 \in \mathbb{R}^{D^l \times D^{l+1}}$, $a_1, a_2 \in \mathbb{R}^{D^{l+1} \times 1}$ and $a_3 \in \mathbb{R}^{D_R \times 1}$ are all trainable parameters.

Frequency Graph Attention Network. Inspired by Simple-Lexicon [9] which weights the matched words at different positions by word frequency, we construct a directed graph as shown in Fig. 1 by only connecting each matched word and characters it contains. The weight of each matched word is set as its count in the dataset.

The attention coefficients of FGAT are calculated as follows:

$$\alpha_{ij} = \frac{exp(count_j)}{\sum_{v \in N_i} exp(count_v)} \tag{6}$$

where $count_j$ and $count_v$ denote counts of word j and word v in the dataset, N_i is the set of all neighboring word nodes of the character node i in the graph. The important thing is when we count the matched words in the dataset, we use a prior knowledge that only the longer word is counted when a shorter word are fully contained by a longer word. Because we believe that in this case the longer word will reflect the correct boundary information with higher probability.

Gram Graph Attention Network. As shown in Fig. 1, the graph of Gram Graph Attention Network(GGAT) is similar to FGAT. In this part, we use PMI [1] to extract a n-grams list from the dataset instead of using external lexicon as before. PMI [1] calculates the Mutual Information (MI) and Left and Right Entropy between consecutive characters and then adds these three values as scores of possible n-grams. We then obtain the n-grams list based on a score threshold. For these n-grams, we randomly initialize embeddings and update them during the training process. Therefore, we change the input to GGAT from Eq. 3 into:

$$H^G = [h_1, h_2, \cdots, h_n, g_1, g_2, \cdots, g_m] \quad (7)$$

where g_i denotes the embedding of the ith matched n-gram in the sentence.

The attention coefficients of GGAT are calculated as follows:

$$\alpha_{ij} = \frac{exp(\sigma(W_1 h_i \cdot W_2 h_j))}{\sum_{v \in N_i} exp(\sigma(W_1 h_i \cdot W_2 h_v))} \quad (8)$$

where $h_i, h_j \in \mathbb{R}^{D^l}$ are node features of node i and node j, N_i is the set of neighboring n-gram nodes of the character node i in the graph, $W_1, W_2 \in \mathbb{R}^{D^l \times D^{l+1}}$ are trainable parameters.

Since PMI [1] focus on word boundary information instead of semantic information, in GGAT we do not directly incorporate the embeddings of n-grams into character nodes like Eq. 4. We train four different position embeddings $\{e_B, e_M, e_E, e_S\}$ which corresponds to four types of character position in n-grams. Finally we design the following equation to calculate the GGAT output:

$$h_i^{l+1} = \overset{K}{\underset{k=1}{\|}} \sigma(\sum_{j \in N_i} \alpha_{ij}^k e_j) \quad (9)$$

where $e_j \in \mathbb{R}^{D^{l+1}}$ denotes the position embedding of the n-gram node j relative to the character node i.

Two sets of input nodes are used for three graphs, and the initial node features for GATs are H^L and H^G as shown in Eq. 3 and 7. Adjacency matrices of three graphs are $A^R, A^F, A^G \in \mathbb{R}^{(n+m) \times (n+m)}$. The outputs of three graphs are denoted as N_1, N_2 and N_3:

$$N_1 = RGAT(H^L, A^R) \quad (10)$$

$$N_2 = FGAT(H^L, A^F) \quad (11)$$

$$N_3 = GGAT(H^G, A^G) \quad (12)$$

where $N_k \in \mathbb{R}^{D_{out} \times (n+m)}, k \in \{1, 2, 3\}$. Finally only the first n columns of N_k are kept because we just use the representations of character nodes for the tag decoder layer:

$$F_k = N_k[:, 0:n] \in \mathbb{R}^{D_{out} \times n}, k \in \{1, 2, 3\} \quad (13)$$

3.3 Scalar Fusion Layer

Since the GATs are from different channels, we use a scalar fusion layer to integrate original character features from the encoder and word features from graphs. The scalar fusion weighs how much information the network obtains from character features or word features. All features are weighted as a whole in order to preserve the internal associations. The fused representation of F_k and H^C is obtained as follows:

$$O = w_1 F_1 + w_2 F_2 + w_3 F_3 + w_4 W_{out} H^C \tag{14}$$

where w_1, w_2, w_3, $w_4 \in \mathbb{R}^1$ and $W_{out} \in \mathbb{R}^{D_{in} \times D_{out}}$ are trainable parameters.

3.4 Tag Decoder Layer

To model the dependencies between consecutive tags, we use a standard CRF [5] layer on the top of the scalar fusion layer's output $O = \{o_1, o_2, \cdots, o_n\}$ as the final tag decoder. The probability of a certain sequence of tags $y = \{t_1, t_2, \cdots, t_n\}$ is defined as

$$P(y|s) = \frac{exp(\sum_i (W_{CRF}^{t_i} o_i + b_{CRF}^{(t_{i-1}, t_i)}))}{\sum_{y'} exp(\sum_i (W_{CRF}^{t_i'} o_i + b_{CRF}^{(t_{i-1}', t_i')})))} \tag{15}$$

Here $\sum_{y'}$ is the traversal of all possible tag sequences, $W_{CRF}^{t_i}$ is a parameter only related to t_i, and $b_{CRF}^{(t_{i-1}, t_i)}$ is a parameter describing the dependency between t_{i-1} and t_i. For decoding, we use the first-order Viterbi algorithm to find the best sequence of tags.

4 Experiments

4.1 Experimental Settings

Datasets. We test MCGAT on four Chinese NER datasets, including OntoNotes 4.0 [14], MSRA [6], Weibo [11] and Resume [15]. Statistics of four datasets are shown in Table 1. We use the same data split strategy as CGN [12] in detail.

Embeddings. We use the same character embeddings as Lattice LSTM [15] and the same lexicon embeddings as CGN [12]. When we use pre-trained language models as the text encoder, we choose BERT$_{base}$ [2] as the baseline.

Hyper-parameters. Table 2 shows the values of hyper-parameters for our models. We set the dimensionality of LSTM hidden states to 300 and set the number of LSTM layers to 1. The dimensionality of GAT hidden states is set to 30 and the number of GAT heads is set to 5. The layer number of RGAT is set to 2 while that of FGAT and GGAT are both set to 1. We use SGD as the optimizer with the momentum set to 0.9. The initial learning rate is set to 0.001, and the decay rate is set to 0.01. Since the size of each dataset is different, we use the batch size from 10 to 30 for different datasets.

Table 1. Statistics of datasets

Datasets	Type	Train	Dev	Test
OntoNotes	Sentence	15.7k	4.3k	4.3k
	Char	491.9k	200.5k	208.1k
MSRA	Sentence	46.4k	–	4.4k
	Char	2169.9k	–	172.6k
Weibo	Sentence	1.4k	0.27	0.27k
	Char	73.8k	14.5k	14.8k
Resume	Sentence	3.8k	0.46k	0.48k
	Char	124.1k	13.9k	15.1k

Table 2. Hyper-parameter values

Parameter	Value	Parameter	Value
Char emb size	50	Lexicon emb size	300
Gram emb size	300	LSTM hidden size	300
LSTM layer	1	GAT hidden size	30
GAT layer	[1,2]	Learning rate lr	0.001
lr decay	0.01	Optimizer	SGD
Momentum	0.9	Batch size	[10,20,30]

4.2   Results and Analysis

We compare the proposed model with several competing models on four datasets and the overall results of F1 scores are shown in Table 3. Since the unsupervised PMI [1] algorithm generally works better on bigger corpus, to compare the experimental results more comprehensively, we denote the model using only RGAT and FGAT as MCGAT-V1, and the model adding GGAT as MCGAT-V2.

On four datasets of OntoNotes, MSRA, Resume and Weibo, MCGAT-V1 and MCGAT-V2 together achieve great performance of obtaining 75.77, 93.95, 95.18 and 64.28 F1 scores respectively. It can be seen that MCGAT performs significantly better than the original model CGN [12] and gets absolute F1 score improvements of 0.98%, 0.48%, 1.06% and 1.19% on four datasets respectively. At the same time, the joint result of MCGAT-V1 and MCGAT-V2 outperform the state-of-the-art model FLAT [7] on three datasets of OntoNotes, Resume and Weibo when using the same Lexicon as CGN [12]. The above results show the effectiveness of our model and suggest that MCGAT is able to better leverage word information in Chinese NER tasks.

Comparing the results of MCGAT-V1 and MCGAT-V2 on different datasets, we find that the F1 score of MCGAT-V2 drops significantly relative to MCGAT-V1 on small datasets like Weibo, while improves on large datasets like MSRA. It is consistent with the inherent characteristic of unsupervised algorithms, where PMI [1] works better on larger corpus. Like the example sentence in Fig. 1,

Table 3. Experimental results (%) on four datasets. All models' results are from original papers except that the superscript * means the result is obtained by running the public source code and BiLSTM results are from Lattice LSTM [15]. YJ denotes the lexicon used in Lattice LSTM [15] and LS denotes the lexicon used in CGN [12]

BiLSTM	Lexicon	OntoNotes	MSRA	Resume	Weibo
	–	71.81	91.87	94.41	56.75
Lattice LSTM [15]	YJ	73.88	93.18	94.46	58.79
LR-CNN [3]	YJ	74.45	93.71	95.11	59.92
LGN [4]	YJ	74.85	93.63	95.41	60.15
Simple-Lexicon [9]	YJ	75.64	93.66	95.53	61.42
PLT [10]	YJ	74.60	93.26	95.40	59.92
FLAT [7]	YJ	76.45	94.12	95.45	60.32
CGN [12]	LS	74.79	93.47	94.12*	63.09
FLAT [7]	LS	75.70	**94.35**	94.93	63.42
MCGAT-V1	LS	**75.77**	93.72	95.02	**64.28**
MCGAT-V2	LS	75.42	93.95	**95.18**	63.32

PMI [1] is able to find proper nouns or people names with obvious boundaries which are not found in the lexicon. We believe that if there is a large unlabeled same-domain corpus, then MCGAT-V2 will achieve even greater improvements.

4.3   Combing Pre-trained Model

We change the context encoder of MCGAT-V1 from BiLSTM to BERT [2] and compared its performance with the common BERT+CRF tagger on four datasets. The results of F1 scores are shown in Table 4.

Table 4. Experimental results (%) on four datasets with BERT [2]

	OntoNotes	MSRA	Resume	Weibo
BERT [2]	79.01	94.20	95.79	67.77
BERT+MCGAT-V1	79.78	94.51	95.95	68.99

It can be seen that adding MCGAT-V1 as a downstream network performs better than the common BERT [2] on all datasets, which means MCGAT works for powerful pre-trained language models like BERT [2] as well.

4.4   Ablation Study

To explore the contribution of each component in MCGAT-V1, we conduct an ablation experiment on the Weibo dataset. The results are shown in Table 5.

Firstly, we remove the relation embeddings in RGAT which leads the RGAT degenerating to a vanilla GAT [13] and the F1 score drops to 61.19%. It suggests

Table 5. Ablation study (%) on Weibo dataset

Model	P	R	F1
MCGAT-V1	70.65	58.96	64.28
- Relation	69.21	54.83	61.19
- RGAT	68.16	50.97	58.32
- FGAT	70.05	57.47	63.14
- Scalar Fusion	70.22	54.11	61.12

that relative position relations play a great role in the use of word information. Then we remove the RGAT and FGAT respectively and the F1 score drops to 58.32% and 63.14% which suggests both GATs are essential components. To verify the effectiveness of the fusion strategy, we replaced the scalar fusion with a simple addition of features and found that the F1 score drops to 61.12%.

4.5   Efficiency Study

As an important upstream task of NLP, NER requires high speed of training and testing. To study the computation efficiency of our model, we conduct an experiment of relative speed on the Resume dataset. In the original code of CGN [12], the graphs are constructed at each batch. It is a time-consuming step which can be loaded directly during the data pre-processing stage. To compare the speed of different models more accurately, we only measure the time spent on forward and backward propagation for each model. Figure 3 shows the relative train and test speed of different lexicon-based models.

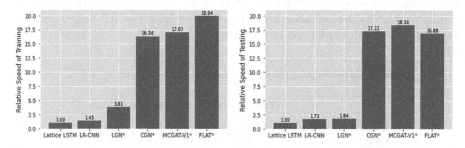

Fig. 3. Relative speed of different models compared with Lattice LSTM [15]. Only the time spent on forward and backward propagation are measured. Lattice LSTM [15] and LR-CNN [3] are non-batch parallel with batch size set to 1, while batch size of other models with '*' are set to 16

It can be seen that MCGAT-V1 is very competitive in terms of efficiency. Compared with CGN [12], MCGAT-V1 has a slight improvement in both training and testing time. It probably because MCGAT-V1 only have two graphs and FGAT is single-layer with lower computational complexity. The easier scalar fusion may also be a reason. Unlike previous studies, we find that the inference speed of CGN [12] and MCGAT-V1 are both faster than FLAT [7] which is a transform-based model.

4.6   Sentence Length Analysis

To analyze the influence of different sentence lengths, we conduct an experiment to compare the F1 score and inference speed of Lattice LSTM [15], CGN [12] and MCGAT-V1 against sentence length on the OntoNotes dataset, the results are shown in Fig. 4.

F1 scores of all three models drop as the sentence length increases, which may be a result of the rapidly increasing number of matched words in the sentence. Compared with Lattice LSTM [15] and CGN [12], MCGAT-V1 is more stable and shows greater robustness to longer sentence length.

For sentence inference, MCGAT-V1 has a faster speed for shorter sentences than CGN [12], and shorter sentences are much more common than longer ones, which may be the reason why MCGAT-V1 performs better in the above efficiency study.

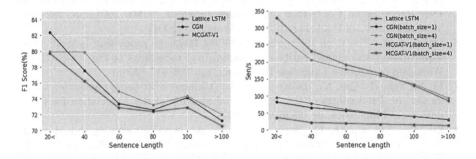

Fig. 4. F1 score and inference speed against sentence length

5   Conclusion

In this paper, we propose a lexicon-based Chinese NER model named Multi-Channel Graph Attention Network (MCGAT). MCGAT uses three word-modified GATs to integrate word information based on relational and statistical level. We evaluate MCGAT on four Chinese NER datasets, the results show that the proposed model achieves comprehensive improvements on the original model and outperforms the state-of-the-art model on three datasets using the same lexicon. MCGAT is competitive in terms of efficiency and has a faster inference speed than the transformer-based model. It can also significantly improve the performance of pre-trained models like BERT as a downstream network.

Acknowledgements. This research work has been funded by the National Natural Science Foundation of China (No. 61772337).

References

1. Bouma, G.: Normalized (pointwise) mutual information in collocation extraction. In: Proceedings of GSCL, pp. 31–40 (2009)
2. Devlin, J., Chang, M.W., Lee, K., Toutanova, K.: Bert: pre-training of deep bidirectional transformers for language understanding. In: Proceedings of the 2019 Conference of the North American Chapter of the Association for Computational Linguistics: Human Language Technologies, Volume 1 (Long and Short Papers), pp. 4171–4186 (2019)
3. Gui, T., Ma, R., Zhang, Q., Zhao, L., Jiang, Y.G., Huang, X.: CNN-based Chinese NER with lexicon rethinking. In: IJCAI, pp. 4982–4988 (2019)
4. Gui, T., et al.: A lexicon-based graph neural network for Chinese NER. In: Proceedings of the 2019 Conference on Empirical Methods in Natural Language Processing and the 9th International Joint Conference on Natural Language Processing (EMNLP-IJCNLP), pp. 1039–1049 (2019)
5. Lafferty, J.D., McCallum, A., Pereira, F.C.N.: Conditional random fields: probabilistic models for segmenting and labeling sequence data. In: Proceedings of the 18th International Conference on Machine Learning, ICML 2001, pp. 282–289. Morgan Kaufmann Publishers Inc., San Francisco (2001)
6. Levow, G.A.: The third international Chinese language processing bakeoff: word segmentation and named entity recognition. In: Proceedings of the 5th SIGHAN Workshop on Chinese Language Processing, pp. 108–117 (2006)
7. Li, X., Yan, H., Qiu, X., Huang, X.J.: Flat: Chinese NER using flat-lattice transformer. In: Proceedings of the 58th Annual Meeting of the Association for Computational Linguistics, pp. 6836–6842 (2020)
8. Liu, W., Xu, T., Xu, Q., Song, J., Zu, Y.: An encoding strategy based word-character LSTM for Chinese NER. In: Proceedings of the 2019 Conference of the North American Chapter of the Association for Computational Linguistics: Human Language Technologies, Volume 1 (Long and Short Papers), pp. 2379–2389 (2019)
9. Ma, R., Peng, M., Zhang, Q., Wei, Z., Huang, X.J.: Simplify the usage of lexicon in Chinese NER. In: Proceedings of the 58th Annual Meeting of the Association for Computational Linguistics, pp. 5951–5960 (2020)
10. Mengge, X., Yu, B., Liu, T., Zhang, Y., Meng, E., Wang, B.: Porous lattice transformer encoder for Chinese NER. In: Proceedings of the 28th International Conference on Computational Linguistics, pp. 3831–3841 (2020)
11. Peng, N., Dredze, M.: Named entity recognition for Chinese social media with jointly trained embeddings. In: Proceedings of the 2015 Conference on Empirical Methods in Natural Language Processing, pp. 548–554 (2015)
12. Sui, D., Chen, Y., Liu, K., Zhao, J., Liu, S.: Leverage lexical knowledge for Chinese named entity recognition via collaborative graph network. In: Proceedings of the 2019 Conference on Empirical Methods in Natural Language Processing and the 9th International Joint Conference on Natural Language Processing (EMNLP-IJCNLP), pp. 3821–3831 (2019)
13. Velickovic, P., Cucurull, G., Casanova, A., Romero, A., Lio, P., Bengio, Y.: Graph attention networks. Stat **1050**, 4 (2018)
14. Weischedel, R., et al.: Ontonotes release 4.0. LDC2011T03, Philadelphia, Penn.: Linguistic Data Consortium (2011)
15. Zhang, Y., Yang, J.: Chinese NER using lattice LSTM. In: Proceedings of the 56th Annual Meeting of the Association for Computational Linguistics (Volume 1: Long Papers), pp. 1554–1564 (2018)

GSNESR: A Global Social Network Embedding Approach for Social Recommendation

Bing-biao Xiao and Jian-wei Liu[✉]

Department of Automation, College of Information Science and Engineering,
China University of Petroleum, Beijing, Beijing 102249, China
liujw@cup.edu.cn

Abstract. Recommender systems are extensively utilized on the internet for helping customers to pick on the items which strike his fancy. Along with the fast progress of online social networks, how to use the additional social information for recommendation has been intensively investigated. In this article, we devise a graph embedding technology to incorporate the customers' social network side information into conventional matrix factorization model. More specifically, first we introduce the graph embedding approach Node2Vec to obtain the customer social latent factor. Then we utilize the matrix factorization technique to find the customer scoring latent factor. Finally we think of recommendation problem as a successive task of social network embedding and integrate customer social latent factor and customer scoring latent factor into our recommendation model. We select the dominant scoring predict task as the evaluation scenario. The effectiveness for our proposed social recommendation (GSNESR) model is validated on three benchmark real world datasets. Experimental results indicate that our proposed GSNESR outperform other state-of-the-art methods.

Keywords: Social recommender systems · Matrix factorization · Network embedding · Scoring prediction

1 Introduction

In recent two decades, recommender systems are playing an increasingly vital role in various online services. It helps customers pick on items capturing their fancy from a large number of ones. In pace with the rapid development of online social networks, gathering ubiquitous social information to help improve recommendation effect has gained a great amount of interest of researchers and practitioners. For example, we always enquire our classmates and good friends for recommendations of nice goods or movies in our daily life. That is one of form of social recommendations. There are a plenty of various social recommendation approaches have been created in the past two decades. In the light of the ways of gathering social information, all these recommendation approaches can

© Springer Nature Switzerland AG 2021
T. Mantoro et al. (Eds.): ICONIP 2021, LNCS 13108, pp. 215–226, 2021.
https://doi.org/10.1007/978-3-030-92185-9_18

be roughly divided into three main categories: conventional social recommender system, social recommender system via graph embedding technology and social recommender system by means of graph neural network (GNN) [1, 3–10].

Conventional social recommender system regards social information as a regularization term [5] or a surrogate of social matrix factorization term [11] added in their recommender model's loss function and control the generation process of customers' scoring latent vector.

A drawback for most introducing regularization term into loss function of recommender systems is that it only considers directly-linked customers as a customer's friends in social network. And the internet friendships are only used to derive the scoring similarity. These recommendation systems do not make full use of social relationship information and only consider the direct dependency between internet friendships. The recommending results are dominant by scoring information, the indirectly-linked internet friendships are ignored in modeling of whole recommending process, resulting in the insufficiency information abstraction and the low-dimensional problem formulation.

To circumvent the drawback for previous recommendation approaches, we create a social recommendation algorithm named global social network embedding approach for social recommendation (GSNESR). The novel idea of GSNESR is to establish the global social network embedding vector, we dub customer social latent factor. The customer social latent factor abstracting by social network embedding algorithm collects the social context information about customers. Due to the introduction of the customer social latent factor matrix, our GSNESR can dig out the customers' social information from a viewpoint of social network. Specifically, to relieve the information insufficiency problem mentioned above, Our GSNESR model establishes two types of customer latent factor, i.e., customer scoring latent factor and customer social latent factor. Customer scoring latent factor is devoted to abstract the semantic information in customer scorings. Customer social latent factor is prescribed to understand the semantic information in customer social network. Furthermore, in order to overcome the imperfection that previous models only make use of the directly-linked structure of social network, we draw social latent factor matrix built from all customers into our scoring predict process. The score prediction process is governed by the customer social latent factor matrix. The global and local weight vector are generated, i.e., the unbiased preference weight vector for all customer and personalized trust preference weight matrix, we utilize two ways to fusion the customer social latent factor matrix. In addition, GSNESR also adds a social regulation term to pick up the local social semantic information of customers. Finally, comparison experiments are carried out to confirm the proposed GSNESR. In a nutshell, our contributions can be sum up as follows:

1. We introduce the graph embedding Node2Vec algorithm to generate new customer latent factor, it introduce the semantic information in social network. Then our GSNESR method utilizes two kinds of latent factor to acquire sufficient information for building recommendation system.

2. We consider the impact of every other customer in the social network on target customer, construct unbiased preference weight matrix for all customers and personalized trust preference weight matrix.
3. The proposed GSNESR approach introduces both the global and local social semantic information to relieve the information insufficiency problem existing in superlarge, extreme spare customer-item scoring matrix.

2 Preliminary for Matrix Factorization Recommendation Framework

2.1 Low Rank Matrix Factorization Model

A conventional approach for recommender systems is to factorize the customer-item scoring matrix, and leverages the factorized customer-specific and item-specific matrices to fill missing data in the customer-item scoring matrix.

In this paper, we consider a $m \times n$ customer-item scoring matrix R describing m customers' scorings on n goods. The low-rank matrix factorization approach attempts to secure the l-rank factors such that the product of the l-rank factors should be equal to $\mathbf{R}$:

$$\mathbf{R} \approx \mathbf{C}^T \mathbf{D} \tag{1}$$

where $\mathbf{C} \in \mathbb{R}^{l \times m}$ and $\mathbf{D} \in \mathbb{R}^{l \times n}$ with $l < min(m, n)$. In real life datasets, the quantity of goods is huge, and very few customers are willing to score the quality of goods and the service quality of merchants, resulting in the matrix $\mathbf{R}$ usually has the property of the extreme data sparseness.

The objective function of matrix factorization methods is as follow:

$$\min_{\mathbf{C,D}} \frac{1}{2} \sum_{i=1}^{m} \sum_{j=1}^{n} I_{ij} (R_{ij} - \mathbf{c}_i^T \mathbf{d}_j)^2 \tag{2}$$

where I_{ij} is the indicator matrix that is equal to 1 if customer $\mathbf{c}_i$ rated item $\mathbf{d}_j$ and equal to 0 otherwise. Usually, two regularization terms are added into loss function so as to avoid overfitting:

$$\min_{\mathbf{C,D}} \frac{1}{2} \sum_{i=1}^{m} \sum_{j=1}^{n} I_{ij} (R_{ij} - \mathbf{c}_i^T \mathbf{d}_j)^2 + \frac{\lambda_1}{2} ||\mathbf{C}||_F^2 + \frac{\lambda_2}{2} ||\mathbf{D}||_F^2 \tag{3}$$

where $\lambda_1, \lambda_2 > 0$.

2.2 The Social Recommendation Model

In this subsection, we briefly present the individual-based social regularization paradigm which is devised by Hao Ma [7]. Because it can figure out close-related social relation and generate personalized recommendation results.

The individual-based social regularization paradigm assumes that for directly-connected customers, their latent vector would have some similar characteristic, and the latent vector similarity among customers is determined by the scoring similarity of the two customers. Therefore the individual-based social regularization term can be expressed as:

$$\frac{\beta}{2} \sum_{i=1}^{m} \sum_{j \in \mathcal{F}+(i)} \mathrm{Sim}(i,j) \|\mathbf{c}_i - \mathbf{c}_j\|_F^2 \tag{4}$$

where $\mathcal{F}^+(i)$ is the set of customers that is directly connected with customer i in the social network. $Sim(i,j)$ is the rating similarity function, the dominant similarity functions are Vector Space Similarity (VSS) and Pearson Correlation Coefficient (PCC).

Utilizing above definitions, the individual-based social regularization paradigm can be expressed as follow:

$$\min_{\mathbf{C},\mathbf{D}} \frac{1}{2} \sum_{i=1}^{m} \sum_{j=1}^{n} I_{ij} \left(R_{ij} - \mathbf{c}_i^T \mathbf{d}_j\right)^2$$
$$+ \frac{\lambda_1}{2} \|\mathbf{C}\|_F^2 + \frac{\lambda_2}{2} \|\mathbf{D}\|_F^2 + \frac{\beta}{2} \sum_{i=1}^{m} \sum_{l \in \mathcal{F}+(i)} \mathrm{Sim}(i,l) \|\mathbf{c}_i - \mathbf{c}_l\|_F^2 \tag{5}$$

3   The Proposed Embedding Framework

Unfortunately, the preference shifted problem has been encumbered with the individual-based social regularization paradigm mentioned above. More specifically, for optimizing the loss function in Eq. 5, we need to acquire the latent factor of customer i, therefore we need to solve the social regularization term:

$$\sum_{j \in \mathcal{F}+(i)} \mathrm{sim}(i,j) \|\mathbf{c}_i - \mathbf{c}_j\|_F^2 \tag{6}$$

If similarity scoring between customer i and j is high, in Eq. 6 the values of $\mathbf{c}_i$ and $\mathbf{c}_j$ will be alike. Meanwhile, in the process of training the latent factor of customer k, we also need to optimize the social regularization term:

$$\sum_{k \in \mathcal{F}+(j)} \mathrm{Sim}(j,k) \|\mathbf{c}_j - \mathbf{c}_k\|_F^2 \tag{7}$$

It asks for that $\mathbf{c}_j$ and $\mathbf{c}_k$ will be alike. This implies that $\mathbf{c}_j$ will approximately equal to $\mathbf{c}_k$, so the solution step given by Eq. 6 will be nonsensical pursuit. This phenomenon is denoted as preference shifted problem. The cause leading to this phenomenon is that we usually get stuck in local minima in optimizing the social regularization term.

3.1   Global Social Latent Factor

To circumvent the preference shift problem, we build a global social latent factor to distil the global social information existing in social network.

Node2Vec attempts to learn the vertex vector representation from network of vertexes [2]. It is a natural extension from the vector representation learning technology in natural language processing, such as Skip-gram. Node2Vec extends Skip-gram to vertex vector learning on graph. In this paper we intend to use Node2Vec to acquire the customer social feature on the social network.

The Node2Vec approach aims to optimize the following objective function, which maximizes the log-probability of observing a network neighborhood $N_s(u)$ for a node u conditioned on its feature representation, which is given by mapping function f:

$$\max_f \sum_{u \in \mathbf{G}} \log \Pr\left(N_S(u) \mid f(u)\right) \mid \tag{8}$$

where $\mathbf{G}$ represents the set of vertexes in the social network, i.e., the set of customers. We attach conditional independence assumption. We suppose that the likelihood of a neighborhood of target customer is independent of any other neighborhood of target customers given the hidden representation of the target customer, applied this hypothesis yields:

$$\Pr\left(N_S(u) \mid f(u)\right) = \prod_{u_i \in N_S(u)} \Pr\left(u_i \mid f(u)\right) \tag{9}$$

We formulate the conditional likelihood of every scored neighborhood customer pair as a softmax function parameterized by an inner product of their feature mapping functions:

$$\Pr\left(u_i \mid f(u)\right) = \frac{\exp\left(f\left(u_i\right) \cdot f(u)\right)}{\sum_{v \in \mathbf{G}} \exp(f(v) \cdot f(u))} \tag{10}$$

Equipping with the conditional independence assumption the loss function can be recombined as

$$\max_f \sum_{u \in G} \left[-\log Z_u + \sum_{u_i \in N_S(u)} f\left(u_i\right) \cdot f(u) \right] \tag{11}$$

$$W = f^* = \arg\max_f \sum_{u \in G} \left[-\log Z_u + \sum_{u_i \in N_s(u)} f\left(u_i\right) f(u) \right] \tag{12}$$

where $Z_u = \sum_{v \in \mathbf{G}} \exp(f(u) \cdot f(v))$

Finally, by means of above feature learning process, we can derive the feature mapping function f. That is what we call the customer social latent matrix. We denote it as $\mathbf{W}$. $\mathbf{W}$ condenses the directly and indirectly dependencies among customers from the entire social network.

3.2 Unbiased Weight Vector for All Customers

From above discussion we have derived the social latent matrix $\mathbf{W}$. To combine this social latent matrix $\mathbf{W}$ and the scoring latent factor matrix obtaining from scoring matrix factorization, we incorporate an unbiased weight vector $\omega \in \mathbb{R}^{1 \times d}$. The unbiased weight vector defines the contributions of all other customers in the social network to target customer.

With the help of this unbiased weight vector, our loss function can be rewritten as:

$$
\min_{\mathbf{C},\mathbf{D}} \frac{1}{2} \sum_{i=1}^{m} \sum_{j=1}^{n} I_{ij} \odot \left(R_{ij} - \mathbf{c}_i^T \mathbf{d}_j - \omega^T \mathbf{w}_j \right)^2
$$

$$
+ \frac{\lambda_1}{2} \|\mathbf{C}\|_F^2 + \frac{\lambda_2}{2} \|\mathbf{D}\|_F^2 + \frac{\lambda_3}{2} \|\mathbf{W}\|_F^2 \tag{13}
$$

$$
+ \frac{\beta}{2} \sum_{i=1}^{m} \sum_{l \in \mathcal{F}+(i)} \mathrm{Sim}(i, l) \|\mathbf{c}_i - \mathbf{c}_l\|_F^2
$$

$$
\text{s.t. } \omega \geq 0
$$

where $\otimes$ denotes the Hadamard product between matrices, and $\| \cdot \|_F$ is the Frobenius norm. I is an indicator matrix with $I_{i,j} = 1$ if customer i has scored goods j, and otherwise $I_{i,j} = 0$.

3.3 Personalized Trust Preference Weight Matrix

The results from Subsect. 3.2 are based on the assumption: all customers have the unbiased weight parameters, i.e., all customers' social characteristic on the same dimension has the same contribution on target customer's scoring. However, in terms of practical application scenarios, each customer has own personal trust preference. The trust preferences of different customers are dissimilar for the same person in the social network. Unbiased weights don't consider personalized characteristics. To fulfill personalized recommendation, we need to distribute each customer a different personalized preference weight vector.

We draw the social latent matrix into our scoring matrix factorization model to establish the social recommendation model. After introducing the personalized preference weight matrix, the loss function can be rearranged as follow:

$$
\min_{\mathbf{C},\mathbf{D}} \frac{1}{2} \sum_{i=1}^{m} \sum_{j=1}^{n} I_{ij} \left(R_{ij} - \mathbf{c}_i^T \mathbf{d}_j - \mathbf{x}_i^T \mathbf{w}_j \right)^2
$$

$$
+ \frac{\lambda_1}{2} \|\mathbf{C}\|_F^2 + \frac{\lambda_2}{2} \|\mathbf{D}\|_F^2 + \frac{\lambda_3}{2} \|\mathbf{W}\|_F^2 \tag{14}
$$

$$
+ \frac{\beta}{2} \sum_{i=1}^{m} \sum_{l \in \mathcal{F}+(i)} \mathrm{Sim}(i, l) \|\mathbf{c}_i - \mathbf{c}_l\|_F^2
$$

where $\mathbf{x}_i \in \mathbf{R}^{d \times 1}$ is the personalized weight matrix. d is the dimension for the pretrained social latent vector. We implement gradient descent to find local optimization for hidden matrices $\mathbf{C}$ and $\mathbf{D}$ in the loss function Eq. (14).

In a nutshell, for our loss function, the local social information is abstracted by the aid of the social regularization term, and the global social representation is grasped by means of the pretrained social hidden factor matrix and personalized trust preference weight matrix. This effectiveness of introducing these measures is confirmed by the experimental results. Another additional merit for incorporating the personalized trust preference weight matrix is that the disproportionate problem between customer latent matrix $\mathbf{C}$ and social latent matrix $\mathbf{X}$ is indirectly eliminated.

4   Experimental Results

4.1   Datasets

We evaluate our models on three datasets: Douban, Epinions and Ciao. All of them are collected from real world social network.

The first dataset we choose is Douban. Douban is a Chinese social network site. It is also the largest online database of Chinese books, movies and music and one of the largest online communities in China. It provides customer scoring, review and recommendation services for movies, books and music. Douban members can assign 5-scale integer scorings (from 1 to 5) to movies, books and music. It also provides both following mechanism based social network services and real world friendship based social network services. Douban not only has rating information but also social information. Hence, Douban is an ideal source for our research on social recommendation.

The second dataset we employ for evaluation is Epinions. Epinions is a consumer review site. Visitors read reviews about various items to help them decide on a purchase. They can also write reviews that may earn them reward on Epinions. Every member of Epinions maintains a "trust" list which presents a network of trust relationships between customers, and a "block(distrust)" list which presents a network of distrust relationships. This network is called the "Web of trust". All the trust relationships interact and form the Web of Trust which is then combined with review ratings to determine which reviews are shown to the user.

The third dataset we choose is Ciao. It contains movie scoring records of customers and trust relations between them. The scoring ranging from 1 to 5. This dataset contains all the information Epinions has except the time points when the trust relations are established. The statistics of these three datasets are shown in Table 1.

Table 1. Statistics of three datasets.

Dataset	Epinions	Douban	Ciao
Users	40,163	2,848	7,375
Items	139,738	39,586	105,114
Ratings	664,824	894,887	284,086
Density	0.0118%	0.0794%	0.0365%

4.2   Metrics

The performance of our model is evaluated by the error between the true rating and the predicted rating. The two performance metrics used for this evaluation are Mean Absolute Error (MAE) and Root Mean Square Error (RMSE). The objective of the model is to minimize these error values. It is also worth noting that even small reductions in these metrics can significantly improve Top-k recommendations.

The metric MAE is defined as:

$$MAE = \frac{1}{T} \sum_{i,j} \left| R_{ij} - \hat{R}_{ij} \right| \tag{15}$$

The metric RMSE is denoted as:

$$RMSE = \sqrt{\frac{1}{T} \sum_{i,j} \left(R_{ij} - \hat{R}_{ij} \right)^2} \tag{16}$$

4.3   Experimental Comparisons

In this section, in order to verify the effectiveness of our proposed recommendation approach, we compare the recommendation results of the following methods:

UserMean: UserMean method uses the mean value of every user to predict the missing scores.

ItemMean: ItemMean utilizes the mean value of every item to predict the missing scores.

SoMF: SoMF introduces social trust mechanism into matrix factorization process to constraint the target user's latent factor being close to the representations of users he or she trusts.

CUNE-MF: CUNE-MF is another network embedding technology based on social recommendation method. It produces Top-K semantic friends for every user and uses these Top-K semantic users for basic MF and BPR approach to solve the recommendation problem.

PMF: PMF is proposed by Salakhutdinov and Minhin [11]. It only uses customer-item matrix for recommendations.

SoReg: SoReg is proposed by Ma et al. It uses the rating similarity information between directly connected users in the social network to provide rating prediction.

The experimental results are shown in Table 2, 3, 4 and 5, respectively.

We call our method proposed in Subsect. 3.2 GSNESR-1, and the method proposed in Subsect. 3.3 GSNESR-2. From the experimental results in Table 1 and 2, firstly we can observe that our method outperforms other approaches in both 80% or 60% settings on three datasets even only use unbiased preference weight vector for all customers. With the personalized preference weight matrix designed by GSNESR-2, we can further improve the performance of model. We also can notice that, whichever method we use, the model performance on 60% dataset setting is always worse than 80% dataset setting.

Table 2. Performance comparisons of MAE (dimensionality of social latent factor = 10)

Datasets	Algorithms				
	UserMean	ItemMean	CUNE_MF	GSNESR-1	GSNESR-2
Epinions (80%)	0.9098	0.9298	0.933	0.8876	0.8679
Improvement	4.61%	6.66%	6.98%		
Epinions (60%)	0.9167	0.9289	0.9741	0.8973	0.8846
Improvement	3.50%	4.77%	9.19%		
Douban (80%)	0.6791	0.6866	0.5994	0.5964	0.5948
Improvement	12.41%	13.37%	0.77%		
Douban (60%)	0.6882	0.6984	0.6055	0.6019	0.5977
Improvement	13.15%	14.42%	1.29%		
Ciao (80%)	0.7687	0.7703	0.9137	0.8785	0.7402
Improvement	3.71%	3.91%	18.99%		
Ciao (60%)	0.7761	0.7809	0.8428	0.9188	0.7605
Improvement	2.01%	2.61%	9.77%		

Table 3. Performance comparisons of MAE (dimensionality of social latent factor = 10)

Datasets	Algorithms				
	SoMF	PMF	SoReg	GSNESR-1	GSNESR-2
Epinions (80%)	0.8912	0.9171	0.9084	0.8876	0.8679
Improvement	2.61%	5.36%	4.46%		
Epinions (60%)	0.9002	0.9274	0.907	0.8973	0.8846
Improvement	1.73%	4.62%	2.47%		
Douban (80%)	0.6113	0.61	0.5961	0.5964	0.5948
Improvement	2.70%	2.49%	0.22%		
Douban (60%)	0.6272	0.624	0.6032	0.6019	0.5977
Improvement	4.70%	4.21%	0.91%		
Ciao (80%)	0.7844	0.7896	0.7603	0.8785	0.7402
Improvement	5.63%	6.26%	2.64%		
Ciao (60%)	0.8008	0.8082	0.9121	0.9188	0.7605
Improvement	5.03%	5.90%	16.62%		

Table 4. Performance comparisons of RMSE (dimensionality of social latent factor = 10)

Datasets	Algorithms				
	UserMean	Item-Mean	CUNE_MF	GSNESR-1	GSNESR-2
Epinions-80%	1.2023	1.1944	1.2235	1.1289	1.1122
Improvement	7.49%	6.88%	9.10%		
Epinions-60%	1.2156	1.2012	1.2716	1.1395	1.1226
Improvement	7.65%	6.54%	11.72%		
Douban-80%	0.8741	0.8064	0.7632	0.7682	0.7565
Improvement	13.45%	6.19%	0.88%		
Douban-60%	0.873	0.8095	0.7707	0.7634	0.76
Improvement	12.94%	6.11%	1.39%		
Ciao-80%	1.026	1.0216	1.1964	1.0086	0.9776
Improvement	4.717%	4.307%	18.288%		
Ciao-60%	1.0399	1.0395	1.1198	1.0469	1.0048
Improvement	3.38%	3.34%	10.27%		

Table 5. Performance comparisons of RMSE (dimensionality of social latent factor = 10)

Datasets	Algorithms				
	SoMF	PMF	SoReg	GSNESR-1	GSNESR-2
Epinions-80%	1.1462	1.2382	1.1996	1.1289	1.1122
Improvement	2.97%	10.18%	7.29%		
Epinions-60%	1.1709	1.2472	1.1972	1.1395	1.1226
Improvement	4.13%	9.99%	6.23%		
Douban-80%	0.7599	0.783	0.7604	0.7682	0.7565
Improvement	0.45%	3.38%	0.51%		
Douban-60%	0.765	0.8006	0.7692	0.7634	0.76
Improvement	0.65%	5.071%	1.196%		
Ciao-80%	1.0433	1.0888	1.0165	1.0086	0.9776
Improvement	6.297%	10.21%	3.83%		
Ciao-60%	1.0661	1.0998	1.1984	1.0469	1.0048
Improvement	5.75%	8.64%	16.15%		

4.4   Result and Analysis

The experiment result is consistent with the data sparsity problem in social recommendation problem. Sparse data will decrease the efficiency of model. In terms of MAE metric, we found that the simplest User-Mean and Item-Mean exhibit relatively poor performance in two settings on all datasets. In all three datasets, the method User-Mean outperforms Item-Mean. This is can be explained by Table 1, on three datasets the average number of ratings per user is larger compared with the average number of ratings per item. So the User-Mean has better performance. Besides we find that those methods based on collaborative filtering e.g. PMF, outperform the User-Mean and Item-Mean method, since the collaborative filtering based methods utilize the rating similarity of users to recommend. The network embedding based method CUNE_MF provides an improvement by a large margin on Douban dataset. But on other two dataset it performs worse than all other models. It is because the Top-K semantic friends generated by the CUNE_MF model become inaccurate when the social information dataset becomes very large, and the number of friends defined by CUNE_MF is still K which leads to the limited social context semantic information. While in our method GSNESR-2, we use the social latent factor matrix which stores all vectors for users to recommend instead of only Top-K semantic friends. We achieve best results on all three datasets compared with other baselines.

5   Conclusions and Future Work

In this paper, we investigate the pros and cons of current dominant social recommendation algorithms. We argue that despite existing social recommendation methods have obtained the promising results, up to the present day, it still difficult to take advantage of the additional social features of customers in the context of large extremely sparse data. Inspired by this, we incorporate a global social latent factor matrix, personalized trust preference weight matrix and social regularization term to design a new social recommendation algorithm called GSNESR. Specifically, the global social latent factor matrix module can effectively learn the customers' social latent representation by performing random walk on the social network to obtain training node sequence corpus and extend skip-gram skill on the training corpus to obtain node embedding vector, which can provide auxiliary social information for the MF recommendation framework. Furthermore, we also apply a social regularization term to extract close-connected customers' social features, which mean, our model takes advantage of both global and local social features of customers. Finally, we combine the classical scoring matrix factorization method, the global social latent factor and local social latent factor together through a trust preference matrix and a social regularization parameter.

There are some avenues for future study. For example, in this work we have shown the effectiveness of our GSNESR method in scoring prediction scenarios and we can further consider that GSNESR is performed outside of scoring prediction scenarios, such as Top-K recommendation domain. Besides, we also would attempt to compare our GSNESR algorithm with other network embedding method with the same experiment setting. Furthermore, note that heterogeneous graph neural network has captured many researcher' attention and achieved tremendous success recent years. Inspired by this, we also consider applying heterogeneous graph neural network to generate node embedding for our framework.

Acknowledgements. This work was supported by the Science Foundation of China University of Petroleum, Beijing (No. 2462020YXZZ023).

References

1. Chan, A.W., Yeh, C.J., Krumboltz, J.D.: Mentoring ethnic minority counseling and clinical psychology students: a multicultural, ecological, and relational model. J. Couns. Psychol. **62**(4), 592 (2015)
2. Grover, A., Leskovec, J.: node2vec: scalable feature learning for networks. In: Proceedings of the 22nd ACM SIGKDD International Conference on Knowledge Discovery and Data Mining, San Francisco, CA, USA, pp. 855–864 (2016)
3. Fan, W., et al.: Graph neural networks for social recommendation. In: The World Wide Web Conference, San Francisco, CA, USA, pp. 417–426. ACM (2019)
4. Guo, G., Zhang, J., Yorke-Smith, N.: TrustSVD: collaborative filtering with both the explicit and implicit influence of user trust and of item ratings. In: Proceedings of the AAAI Conference on Artificial Intelligence, vol. 29, no. 1. AAAI Press (2015)
5. Jamali, M., Ester, M.: A matrix factorization technique with trust propagation for recommendation in social networks. In: Proceedings of the 4th ACM Conference on Recommender Systems, pp. 135–142 (2010)
6. Lu, Y., et al.: Social influence attentive neural network for friend-enhanced recommendation. In: Dong, Y., Mladenić, D., Saunders, C. (eds.) ECML PKDD 2020. LNCS (LNAI), vol. 12460, pp. 3–18. Springer, Cham (2021). https://doi.org/10.1007/978-3-030-67667-4_1
7. Ma, H., Zhou, D., Liu, C., Lyu, M.R., King, I.: Recommender systems with social regularization. In: Proceedings of the 4th ACM International Conference on Web Search and Data Mining, pp. 287–296 (2011)
8. Wu, L., Sun, P., Hong, R., Fu, Y., Wang, X., Wang, M.: SocialGCN: an efficient graph convolutional network based model for social recommendation. arXiv preprint arXiv:1811.02815 (2018)
9. Yu, J., Yin, H., Li, J., Gao, M., Huang, Z., Cui, L.: Enhance social recommendation with adversarial graph convolutional networks. IEEE Trans. Knowl. Data Eng. (2020). (TKDM 2020)
10. Yu, J., Gao, M., Yin, H., Li, J., Gao, C., Wang, Q.: Generating reliable friends via adversarial training to improve social recommendation. In: 2019 IEEE International Conference on Data Mining (ICDM), pp. 768–777 (2019)
11. Ma, H., Yang, H., Lyu, M.R., King, I.: SoRec: social recommendation using probabilistic matrix factorization. In: Proceedings of the 17th ACM Conference on Information and Knowledge Management, CIKM 2008, pp. 931–940. Napa Valley, California, USA (2008)

Classification Models for Medical Data with Interpretative Rules

Xinyue Xu[1]($\boxtimes$), Xiang Ding[2], Zhenyue Qin[1], and Yang Liu[1,3]

[1] Australian National University, Canberra, Australia
{u6708515,zhenyue.qin,yang.liu3}@anu.edu.au
[2] Nanjing University, Nanjing, China
xding@smail.nju.edu.cn
[3] Data61-CSIRO, Canberra, Australia

Abstract. The raging of COVID-19 has been going on for a long time. Thus, it is essential to find a more accurate classification model for recognizing positive cases. In this paper, we use a variety of classification models to recognize the positive cases of SARS. We conduct evaluation with two types of SARS datasets, numerical and categorical types. For the sake of more clear interpretability, we also generate explanatory rules for the models. Our prediction models and rule generation models both get effective results on these two kinds of datasets. All explanatory rules achieve an accuracy of more than 70%, which indicates that the classification model can have strong inherent explanatory ability. We also make a brief analysis of the characteristics of different rule generation models. We hope to provide new possibilities for the interpretability of the classification models.

1 Introduction

The outbreak of SARS has become a global problem. Both the first-generation coronavirus in 2002 and the second-generation coronavirus in 2019 have caused huge casualties and property losses of human beings. SARS often begins with flu-like symptoms, but it is more than just a common cold [1]. Therefore, models that can distinguish between SARS and other diseases are especially important in these difficult times.

In recent years, machine learning has been widely used in the medical field. The same can be said for the prediction of SARS. By observing the dataset, classification models can be generated to determine whether the inputs are positive SARS cases. In this paper, some common classification models are applied for classification and prediction of SARS. We also analyze their performance on two different SARS datasets. However, the model of machine learning is arguably a black box model that lacks interpretability. High-stakes decisions made by black-box machine learning models in the medical field can cause problems. On the other hand, it is non-trivial to interpret the working mechanisms of machine learning models [2]. Ideally, the design of the model needs to be inherently interpretable [3].

T. Mantoro et al. (Eds.): ICONIP 2021, LNCS 13108, pp. 227–239, 2021.
https://doi.org/10.1007/978-3-030-92185-9_19

Based on the classification models survey of COVID-19 in Brazil [4], we introduce two different types of SARS datasets in this paper. One is the original dataset used for the Brazilian positive and negative cases classification (SARS-CoV-2), the other is the SARS-CoV-1 dataset [5]. We use these two datasets to verify the effectiveness of various classification models for recognizing the SARS cases. This paper evaluates the model's performance from two perspectives. We not only conduct assessment from the perspective of classification accuracy, but also gauge the model's capability from its interpretability.

The accuracy of the different models on these two datasets are evaluated. This paper also explores the inherent interpretability of the models. The prediction model and the explanation model are analyzed and compared respectively. The following are the main contributions of this paper:

1. We evaluate the performance of several classification models from the perspective of both accuracy and interpretability.
2. The impact of different types of datasets on rules generation performance is investigated. Thus, we provide an insight on how to select the applicable method based on data.
3. The resulting rules can provide health professionals with thresholds for pathological studies as well as intuitive testing criteria for potential patients.

This paper organizes as follows: The composition of the two datasets is very special, which will be discussed in more detail in Sect. 2. Section 3 discusses all the algorithms used in this paper. In Sect. 4, the experimental results are presented. In Sect. 5, the experimental results are discussed and the future work is proposed. Section 6 summarizes the full paper.

2   Dataset

In this paper, although both datasets are about SARS, their types of data are different. SARS-CoV-1 dataset is a fuzzy dataset with continuous data, while SARS-CoV-2 dataset is a Boolean dataset with categorical data. We intend to comprehensively assess the model's performance via observing their behaviors on processing these two very different data types.

2.1   SARS-CoV-1

In traditional Boolean logic, the truth value can only be the integer value of 0 or 1. However, in practice, the truth value of many events may vary between completely true and completely false. In order to make up for the deficiency of classical logic, Zadeh put forward the theory of fuzzy sets in 1965 [6]. Fuzzy logic is closer to human reasoning than the classical logic and is more proper to generate approximate information and uncertain decisions. It is thus a tool for dealing with many inherently imprecise problems [7].

SARS-CoV-1 dataset [5] is a fuzzy dataset that contains four diseases sets and 23 kinds of fuzzy values in each set. There are four diseases in total, which are hypertension, normal, pneumonia and SARS. There are a thousand observations for each disease. Raw data have no label values. Their file names represent the disease categories to which they belong. The 23 fuzzy values can be divided into four general symptoms, which are fever, blood pressure, nausea and whether there is abnormal pain. Fever data is monitored and collected four times a day at 8 a.m., 12 p.m., 4 p.m., and 8 p.m. Fever is divided into three fuzzy sets: slight, moderate, and high fever. Blood pressure is measured by systolic or diastolic blood pressure, which is divided into three grades: low, normal or high. Nausea, like fever, can be slight, moderate or high. Abnormal pain is divided into yes and no. This dataset will be trained using a variety of methods and the resulting model will be used to predict the disease that a patient has. It thus can be treated as a disease classification problem.

2.2   SARS-CoV-2

The dataset [4] collected and preprocessed COVID-19 data from more than 50,000 Brazilians. It contains patient information for both reverse transcriptase polymerase chain reaction (RT-PCR) and rapid tests (antibody and antigen). It excludes other factors that interferes with this, such as asymptomatic patients, undefined gender and so on. This paper selects the balanced dataset formed by the combination of the two detection methods. Patients who are labeled positive are considered to be given priority for testing. The result is labeled as 0 for being positive and 1 for being negative. This paper also classifies the positive and negative cases, and attempts to generate interpretable rules to reflect the inherent explanatory nature of the model.

The dataset contains ten features. Two of them are demographic characteristics: sex and whether or not they are health professional. Eight symptoms: fever, sore throat, dyspnea, olfactory disorders, cough, coryza, taste disorders and headache. All features have only two unique values, 0 and 1. 0 is symptomatic and 1 is asymptomatic. Hence, this dataset can be thought of as a completely categorical dataset. The combination of two detection types balanced dataset is used to in the hope of maximizing unbiased estimation. In view of the original data, the number of negatives is much higher than the number of positives. This will cause great disturbance to classification prediction and rule generation models.

3   Methods

3.1   Classification Models

Decision Tree is a flowchart-like structure based on strategic choice. It represents a mapping relationship between object attributes and object values. The branching path of a decision tree is the rule that interprets the decision tree model. The generating path is intuitive. The unique characteristic of decision trees compared with other machine learning models is the clarity of their information representation. The generated decision tree directly forms the hierarchical structure. This structure can be regarded as a set of if-else rules which can be easily understood even by non-experts.

Multilayer Perceptron is a very traditional classification method. In this paper, the common three-layer fully connected neural network is used to train the training set. Logistic activation function is used in the output layer, because it outputs more stable values. It performs well in digital representation, and avoids loss of precision. Cross-entropy loss is used to calculate the model loss, which is a very common loss function for classification problems. The weights of the trained neural network will be used to calculate the causal index, which will be explained in more detail in the rules generation section.

XGBoost is widely used to build highly effective models on many machine learning tasks [8]. It is an enhanced version of the plain gradient boosting decision tree (GBDT) algorithm. It introduces second-order Taylor expansion to loss function and add a regularization term to avoid overfitting. XGBoost also presents an advanced sparsity-aware algorithm for sparse data. Moreover, it speed up parallel learning to build a scalable tree boosting system by designing compressed data format and cache-aware access to speed up parallel learning.

CatBoost brings improved accuracy and reduced overfitting even with default parameters compared to other gradient boosting algorithms [9]. It inherently supports non-numerical data so no data-preprocessing is needed when using categorical data. In this paper, we use default parameters of CatBoost and it outperforms other classification algorithms in accuracy and precision. CatBoost is optimized for faster training on GPU and CPU. The GPU implementation of CatBoost runs faster than other GBDT models such as XGBoost and LightGBM.

Ensemble Learning is a strategic generation and combination of multiple models (such as classifiers) to improve the performance of the model [10]. Here we use the bagging method to combine decision tree and random forest to give predictions as well as generate rules. Although decision tree is simple and intuitive, it has the hidden trouble of over-fitting [11]. Learning an optimal decision tree is considered an NP-complete problem [12]. In practice, decision tree is built based on the heuristic greedy algorithm, which cannot guarantee the establishment of the global optimal decision tree. The library Skope-rules [13] introduces the random forest, which can alleviate this problem. Integrating the advantages of decision tree and random tree is the method adopted in this library.

Fuzzy Clustering is a data clustering technique wherein each data point belongs to a cluster to some degree that is specified by a membership level [14]. The membership level shows how strongly the data points are associated with a particular cluster. Fuzzy clustering is the process of calculating these member levels and deciding the data points belong to which clusters according to member levels. One of the most widely used fuzzy clustering algorithms is the Fuzzy C-means clustering algorithm [15].

3.2   Rules Generation

Decision Tree Rules. Decision tree is inherently a set of rules leading to different nodes. On each layer, nodes are split according to the segmentation point of the rule. The leaf node in the tree represents the result of a decision based on a series of judgments on the path from the root node to that leaf node. Rules of the leaf nodes representing the same cluster are then disjuncted to form the rule of this cluster.

Causal Index and Characteristic Input for MLP. Input patterns are classified according to their impact on each particular output [16]. We define these inputs as the characteristic input of the particular class. In view of the large number of characteristic inputs, the arithmetic mean of characteristic inputs is calculated and their average is defined as characteristic ON pattern, predictions with respect to other classes are considered characteristic OFF pattern, and vice versa. Each class has a characteristic ON pattern and the number of characteristic OFF patterns equals to the number of the remaining classes. The characteristic inputs are calculated in the training set of the network and does not involve any data from the test set.

The causal index depends on the differentiability of the network activation function. Based on the rate of change of the influence of the input value x_i on the output value y_k, we can determine which inputs have a greater influence on the classification. According to the weight of each neuron in the neural network, a weight matrix with the size of $features * classes$ is obtained. It can be clearly seen that the impact of each feature input on the output of each class. The calculation method [17] is to multiply and sum the weights of each layer to get the final causal index, as shown in Eq. 1:

$$\frac{dy_k}{dx_i} = \frac{dy_k}{dU_{k2}} \cdot \frac{dU_{k2}}{dh_j} \cdot \frac{dh_j}{dU_{j1}} \cdot \frac{dU_{j1}}{dx_i} \quad = f'(U_{k2}) \cdot f'(U_{j1}) \cdot \sum_j w_{jk} \cdot W_{ij}. \quad (1)$$

Since the derivatives $f'(U_{k2}) \cdot f'(U_{j1})$ will get constants, we can ignore them.

Using the fully connected neural network completed by training, the weight of each layer is extracted and the causal index matrix can be obtained by calculating according to the above formula. The causal index of each class is analyzed. The three causal indexes with the largest absolute value are extracted to generate rules. If the causal index is positive, the feature is positively correlated with the class, while a negative number is negatively correlated with the class. The boundary value is determined by the characteristic ON input. If the correlation is positive, the feature must be greater than the boundary value, while the negative correlation is less than the boundary value. The rules generated for each feature are concatenated using AND. Each class generates a discriminant rule to interpret the predicted results.

Ensemble Learning Rule. The approach of ensemble learning is a trade-off between the interpretability of decision trees and the modeling capability of random forests [13]. It introduces random forests to solve the over-fitting phenomenon caused by the over-complexity of decision trees in large datasets. Rules are extracted from a set of trees. A weighted combination of these rules is constructed by solving the L1 regularization optimization problem on the weights [18]. Extracting rules from a set of trees enables us to generate such a set of trees using existing fast algorithms, such as bagged decision trees or gradient enhancement. Rules that are too similar or duplicate are then removed based on the supported similarity thresholds. In this paper, Skope-Rules model is used to train the training set. The structure [19] of Skope-Rules consists bagging estimator, performance filtering and semantic deduplication. The Bagging estimator trains multiple decision tree classifiers. Within a tree, each node in the Bagging estimator can be thought of as a rule. Performance filtering uses out-of-bag accuracy and recall thresholds to select the best rule. Semantic deduplication applies similarity filtering and comparison operators are used to combine variable names [19]. In the classification task, Skope-rules are useful for describing each cluster. Each cluster can be postprocessed and approximated with a set of interpretable rules.

Fuzzy Classification System Rules. In addition to using neural networks and decision trees to automatically extract rules, fuzzy rules can also be extracted directly from data. For example, in [20], the input space is divided into fuzzy regions. The fuzzy rules are generated by determining which numerical data is contained in the fuzzy regions. While the above approach is very simple, it also requires pre-partitioning the input space. In [21], fuzzy rules are extracted for classification problems with different fuzzy regions. This method solves the problem of the fuzzy system mentioned above. The input area of each class is represented by a set of hyper-boxes that allow overlaps between hyper-boxes of the same class, but not between different classes. When multiple classes overlap, the algorithm will dynamically expand, divide, and shrink the hyper-box. The system which this paper used [22] introduces two types of hyper-boxes: activation hyper-boxes – which define the existence regions for classes, and inhibition hyper-boxes – which inhibit the existence of data within the activation hyper-boxes. The hyper-boxes are defined recursively. Firstly, calculating the minimum and maximum values of data for each class. If there is a overlapping region between class i and j, it should be defined as an inhibition box. If in the inhibition hyper-box exists classes i and j, the system will define additional activation hyper-boxes for these classes. Similarly, if there is an overlap between these activated hyper-boxes, we further define the overlap region as inhibition hyper-boxes. In this way, the overlap of the activation hyper-boxes is resolved recursively. The Fig. 1 is the schematic diagram of the fuzzy classification system. It is effective to use a set of hyper-boxes to represent the data presence domain of a class to deal with classification problems with many input variables. The overlap between different classes is solved by the method of recursive definition. The boundary of the hyper-boxes are the rules we aim to generate.

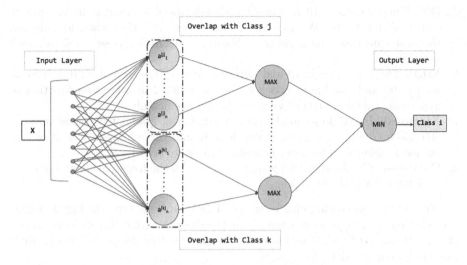

Fig. 1. Fuzzy classification system architecture

4   Experimental Analysis

4.1   Prediction

We adopt 5 classification models (Decision tree, MLP, XGBoost, CatBoost and Fuzzy c-means) on the numerical dataset of SARS-CoV-1. Each model achieves a perfect result of 100% accuracy on testing data, but the corresponding rules for each model are different. Different rules generated by these models are presented in Sect. 4.2.

On the boolean dataset, we adopt 4 classification models to predict the occurrence of SARS-CoV-2 (COVID-19). Since this is a binary classification, the prediction results are measured by accuracy, precision and recall as presented in Table 1. These 4 models show similar results in prediction. Among them, CatBoost model presents the highest accuracy (89.46%) and precision (87.01%), while DT presents the highest recall (94.21%).

Table 1. Results of the classification models using balanced dataset

Models	Accuracy (%)	Precision (%)	Recall (%)
DT	88.66	85.60	**94.21**
MLP	89.14	86.93	93.29
XGBoost	89.14	86.73	93.60
CatBoost	**89.46**	**87.01**	93.90

1. **DT.** Since there are 10 features for classification, we assign the maximum depth of tree to 10. We use Gini index as the criterion when looking for the best split, and the number of features to consider is set to 7 to avoid overfitting.
2. **MLP.** The input size is 10 and output size is 1. The size of hidden layer is set to 100 and maximum epoch rounds is 1200. The activation function is hyperbolic tangent. Other hyperparameters are default.
3. **XGBoost.** In XGBoost model, we set number of gradient boosting trees to 500. Each tree has a maximum depth of 80 and the minimum weight of 3 is needed to form a child node. Learning rate is set to 0.1.
4. **CatBoost.** CatBoost model is quite simple. All the hyperparameters are set to default without fine-tuning.

The receiver operating characteristic (ROC) curve is given in Fig. 2. These 4 prediction models perform similarly as the discrimination threshold is varied. Among them, MLP and CatBoost show slightly better diagnostic ability with area under curve (AUC) being 0.94.

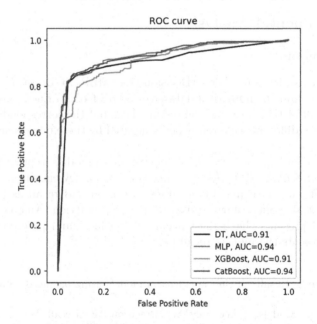

Fig. 2. Models' ROC curves and AUC score using balanced SARS-CoV-2 dataset

4.2   Rules Generation Results

Although we have strong interpretable rules on the SARS-CoV-1 dataset, our rule boundaries on the purely categorical SARS-CoV-2 dataset is trivial. SARS-CoV-2 is a binary classification dataset. For all the features in the dataset, it only have 0/1 as unique values. Therefore, all the rules generation methods give the similar rules and the same threshold which is 0.5. The result rules for SARS-CoV-2 is in Table 2. For positive cases, the precision for MLP rules is about 74.71% and the description degree (precision) for ensemble rules is about 84.79%. Because the decision tree is too large (over 100 leaf nodes), it will not be shown here. Fuzzy rules cannot be used in categorical data (especially 0/1) because it is hard to fuzz the data.

Table 2. SARS-CoV-2 dataset rules

MLP Rules for cluster 0:	Symptom-Coryza: > 0.5, Symptom-Fever: < 0.5, Symptom-Cough: > 0.5
MLP Rules for cluster 1:	Symptom-Coryza: < 0.5, Symptom-Fever: > 0.5, Symptom-Cough: < 0.5
Ensemble Rules for 0:	Symptom-Fever <= 0.5 and Symptom-Taste Disorders > 0.5 and Gender > 0.5
Ensemble Rules for 1:	Symptom-Fever > 0.5 and Symptom-Taste Disorders > 0.5 and Symptom-Olfactory Disorders > 0.5

Rules are generated on the SARS-CoV-1 dataset using 4 methods.

Causal Index Rules. Input size is 23 and output size is 4. The number of hidden neurons is set as 8, since the dataset has 8 classes. The batch size is set as 10 and learning rate is 0.01. The epoch number is set as 500, but it can achieve 100% accuracy and 0 loss after 50 epochs. According to the causal index calculation method, three features with the highest correlation were found for each class, and the positive and negative signs were used to judge whether they were greater than or less than the characteristic boundary value. The result is the following set of rules. The number before the colon represents the selected feature column.

For HighBP: 17: > 0.41692959, 14: > 0.41889875, 12: < 0.46736246
For normal: 14: < 0.41889875, 17: < 0.41692959, 12: > 0.46736246
For pneumonia: 14: < 0.41889875, 12: > 0.46736246, 17: < 0.41692959
For SARS: 14: > 0.41889875, 17: > 0.41692959, 16: > 0.30410143

Verified by the test set, the accuracy is 100%. It can be proved that this set of rules can explain the neural network model predictions for SARS-CoV-1 dataset.

Decision Tree Rules. The precision of the decision tree is also 100%, and the depth of the tree is 4. The explanatory rules it generates are much more straightforward.

The rules for decision tree:

IF [22] ≥ 0.25,   THEN   Class 4.
IF [17] ≥ 0.4,   THEN   Class 1.

IF [6] $\geq$ 0.499,   THEN   Class 2.
IF [6] $\leq$ 0.499,   THEN   Class 3.

It can be seen that the decision tree only uses three features to perfectly classify the data, which verifies that the SARS-CoV-1 dataset model trained by the decision tree is also very effective.

Ensemble Learning Rules. Compared with the above two methods of generating rules, the rules generated by ensemble learning are more detailed in Table 3. It also provides a variety of possibilities for classification rules. For each category, four rules with an accuracy of 100% are generated in this paper and any of them can be used to accurately classify data. The reason for doing this is to test whether the model can generate a variety of rules. This tends to do well in the face of less ideal datasets. Because the symptoms of SARS are too obvious, only three rules are generated, and the repeated rules will be deleted automatically according to the model design. It can be observed that there are two evaluation parameters after the rule, the meaning of which are respectively precision and recall for the current cluster. The third performance item is the number of extractions in the tree built during rule fitting. It can be seen that the integration model generates as many rules as possible while maintaining precision, which can be described in detail for the entire dataset.

Table 3. The rules for ensemble learning model.

Rules for SARS-COV-1 HBP
('fever_temp_12_s > 0.9368999898433685 and fever_temp_16_s > 0.9986999928951263', (1.0, 1.0, 2))
('blood_pre_dia_s <= 0.43650001287460327 and fever_temp_16_s > 0.9986999928951263', (1.0, 1.0, 6))
('fever_temp_8_s > 0.9999000132083893 and fever_temp_12_m <= 0.07154999673366547', (1.0, 1.0, 2))
('fever_temp_20_m <= 0.0010499999625608325 and blood_pre_dia_m > 0.13900000229477882', (1.0, 1.0, 2))
Rules for SARS-COV-1 NOR
('blood_pre_sys_h <= 0.25 and fever_temp_12_h <= 0.4000000059604645', (1.0, 1.0, 2))
('blood_pre_sys_h <= 0.25 and fever_temp_20_s > 0.5000000074505806', (1.0, 1.0, 2))
('blood_pre_dia_h <= 0.4000000059604645 and fever_temp_20_s > 0.5000000074505806', (1.0, 1.0, 2))
('fever_temp_8_m <= 0.4000000134110451 and blood_pre_sys_s > 0.4000000059604645', (1.0, 1.0, 2))
Rules for SARS-COV-1 PNE
('blood_pre_sys_m <= 0.0006500000017695129 and blood_pre_dia_s > 0.9850499927997589', (1.0, 1.0, 2))
('blood_pre_sys_m <= 0.0006500000017695129 and fever_temp_16_h > 0.4000000059604645', (1.0, 1.0, 8))
('blood_pre_dia_m <= 0.003650000086054206 and fever_temp_12_s <= 0.524349994957447', (1.0, 1.0, 2))
('fever_temp_20_m > 0.366100013256073 and blood_pre_dia_s > 0.9990499913692474', (1.0, 1.0, 2))
Rules for SARS-COV-1 SARS
('nausea_m > 0.25', (1.0, 1.0, 12))
('nausea_h > 0.25', (1.0, 1.0, 12))
('no_pain <= 0.5', (1.0, 1.0, 14))

Fuzzy System Rules. According to the hyperbox boundaries generated for each dimension we get the rules for each class, which is Table 4. These ranges of maximum and minimum values can be used to accurately locate each type of disease.

Table 4. The rules for fuzzy system

Column	1-min	1-max	2-min	2-max	3-min	3-max	4-min	4-max
fever_temp_8_s	1.0	1.0	0.8	1.0	0.0	0.2	0.0	0.2
fever_temp_8_m	0.0	0.0	0.0	0.2	0.6	1.0	0.4	1.0
fever_temp_8_h	0.0	0.0	0.0	0.0	0.8	1.0	0.5	1.0
fever_temp_12_s	1.0	1.0	0.8	1.0	0.0	0.2	0.0	0.1
fever_temp_12_m	0.0	0.0	0.0	0.2	0.6	1.0	0.5	1.0
fever_temp_12_h	0.0	0.0	0.0	0.0	0.8	1.0	0.5	1.0
fever_temp_16_s	1.0	1.0	0.8	1.0	0.0	0.2	0.0	0.0
fever_temp_16_m	0.0	0.0	0.0	0.196	0.6	1.0	0.5	1.0
fever_temp_16_h	0.0	0.0	0.0	0.0	0.8	1.0	0.5	1.0
fever_temp_20_s	1.0	1.0	0.8	1.0	0.0	0.2	0.0	0.0
fever_temp_20_m	0.0	0.0	0.0	0.2	0.6	1.0	0.5	1.0
fever_temp_20_h	0.0	0.0	0.0	0.0	0.8	1.0	0.5	1.0
blood_pre_sys_s	0.0	0.0	0.8	1.0	1.0	1.0	0.0	0.0
blood_pre_sys_m	0.2	0.5	0.0	0.2	0.0	0.0	0.5	1.0
blood_pre_sys_h	0.8	1.0	0.0	0.0	0.0	0.0	0.5	1.0
blood_pre_dia_s	0.0	0.0	0.8	1.0	1.0	1.0	0.0	0.2
blood_pre_dia_m	0.2	0.5	0.0	0.2	0.0	0.0	0.5	1.0
blood_pre_dia_h	0.8	1.0	0.0	0.0	0.0	0.0	0.5	1.0
nausea_s	1.0	1.0	1.0	1.0	1.0	1.0	0.0	0.2
nausea_m	0.0	0.0	0.0	0.0	0.0	0.0	0.5	1.0
nausea_h	0.0	0.0	0.0	0.0	0.0	0.0	0.5	1.0
no_pain	1.0	1.0	1.0	1.0	1.0	1.0	0.0	0.0
have_pain	0.0	0.0	0.0	0.0	0.0	0.0	0.5	1.0

5   Discussion and Future Work

All 5 classification models present a perfect 100% accuracy on SARS-CoV-1 dataset. Discussion about characteristics of rules generated by these models are in the next paragraph. Prediction results on SARS-CoV-2 dataset is presented in Table 1. These models show similar patterns in prediction results. In the context of clinical-related tasks such as SARS detection, high recall is important to control the spread of disease. Meanwhile, we want our model to have high precision to be effective, so there is a trade-off among accuracy, precision and recall [23]. DT model has the highest recall score, but the lowest accuracy and precision. In general, CatBoost model has the best performance. It achieves the highest accuracy and precision among them, and second highest recall over MLP and XGBoost.

Since the rules generated by SARS-CoV-2 dataset are not obvious enough (almost one rules for one cluster), this paper will discuss the application scenario of rules generation methods using the results of SARS-CoV-1. Decision trees are very intuitive models that generate rules through node splitting. MLP rules can reflect the weight of different features and highlight their importance in the network. Using arithmetical mean value as a threshold, has a strong ability to generalize numerical data. Ensemble rules using bagging estimators, enhance the ability of decision tree and random forest. The fuzzy box rules give the boundary of each cluster, it can be seen in some way as a representation of different categories in different dimensions. The structure of the network is determined automatically by obtaining fuzzy rules and overlapping between classes. The fuzzy parameters can be modified to improve the generalization ability of the model. The model is relatively easy to implement and the principle is easy to understand. Recursive definition can be a good solution to the overlap problem. The difficulty of feature generation of categorical type is much higher than that of numerical type. Their thresholds tend to be fixed because there is no continuous variable to get an accurate threshold. Since the dataset is medical data, some symptoms may not be typical of SARS, which leads to the possibility that the same characteristic rules may exist between different categories. In future work, we will explore a more effective rule-generation model for purely categorical data.

6   Conclusion

With the application of machine learning in more and more fields, the interpretability of machine learning model itself has gradually become the focus of attention. In order for models to have higher confidence in high-risk areas such as medical decision-making, we need not only to develop more accurate models to accomplish the task, but also to have inherent interpretability of the models themselves. To this end, we use a variety of classification models to predict SARS positive cases and generates interpretable rules while ensuring accuracy. It provides a new insight on the interpretability of classification models on different types of datasets. Effective rule generation can benefit both health workers and people who need pre-testing. The research for interpretability of machine learning models needs to continue, we believe there will be more models with inherent interpretability in the future.

References

1. Pedersen, S.F., Ho, Y.C.: SARS-CoV-2: a storm is raging. J. Clin. Invest. **130**(5), 2202–2205 (2020)
2. Doshi-Velez, F., Kim, B.: Towards a rigorous science of interpretable machine learning. arXiv: 1702.08608 (2017)
3. Rudin, C.: Stop explaining black box machine learning models for high stakes decisions and use interpretable models instead. Nat. Mach. Intell. **1**, 206–215 (2019)

4. Viana dos Santos Santana, Í., et al.: Classification Models for COVID-19 test prioritization in Brazil: machine learning approach. J. Med. Internet Res. **23**, e27293 (2021)
5. Mendis, B.S., Gedeon, T.D., Koczy, L.T.: Investigation of aggregation in fuzzy signatures. In: 3rd International Conference on Computational Intelligence, Robotics and Autonomous Systems (2005)
6. Zadeh, L.A.: Fuzzy sets. In: Fuzzy Sets, Fuzzy Logic, and Fuzzy Systems: Selected Papers by Lotfi A Zadeh, pp. 394–432 (1996)
7. Novák, V.: Fuzzy natural logic: towards mathematical logic of human reasoning. In: Seising, R., Trillas, E., Kacprzyk, J. (eds.) Towards the Future of Fuzzy Logic. SFSC, vol. 325, pp. 137–165. Springer, Cham (2015). https://doi.org/10.1007/978-3-319-18750-1_8
8. Chen, T., Guestrin, C.: XGBoost: a scalable tree boosting system. In: Proceedings of the 22nd ACM SIGKDD International Conference on Knowledge Discovery and Data Mining, pp. 785–794 (2016)
9. Dorogush, A.V., Ershov, V., Gulin, A. CatBoost: gradient boosting with categorical features support. arXiv:1810.11363 (2018)
10. Dietterich, T.G.: Ensemble learning. In: The Handbook of Brain Theory and Neural Networks, vol. 2, pp. 110–125 (2002)
11. Schaffer, C.: Overfitting avoidance as bias. Mach. Learn. **10**(2), 153–178 (1993)
12. Laurent, H., Rivest, R.L.: Constructing optimal binary decision trees is NP-complete. Inf. Process. Lett. **5**(1), 15–17 (1976)
13. Gautier, R., Jaffre, G., Ndiaye, B.: scikit-learn-contrib/skope-rules (2020). https://github.com/scikit-learn-contrib/skope-rules
14. Bora, D.J., Gupta, D., Kumar, A.: A comparative study between fuzzy clustering algorithm and hard clustering algorithm. arXiv preprint arXiv:1404.6059 (2014)
15. Bezdek, J.C., Ehrlich, R., Full, W.: FCM: the fuzzy c-means clustering algorithm. Comput. Geosci. **10**(2–3), 191–203 (1984)
16. Gedeon, T.D., Turner, H.S.: Explaining student grades predicted by a neural network. In: International Joint Conference on Neural Networks (1993)
17. Harry, S.T., Tamás, D.G.: Extracting Meaning from Neural Networks (2020). http://users.cecs.anu.edu.au/~Tom.Gedeon/pdfs/Extracting%20Meaning%20from%20Neural%20Networks.pdf
18. Friedman, J.H., Popescu, B.E.: Predictive learning via rule ensembles. Ann. Appl. Stat. **2**(3), 916–954 (2008)
19. Gautier, R., Jaffre, G., Ndiaye, B.: Interpretability with diversified-by-design rules; Skope-rules, a python package (2020). http://2018.ds3-datascience-polytechnique.fr/wp-content/uploads/2018/06/DS3-309.pdf
20. Simpson, P.K.: Fuzzy min-max neural networks-part 1: classification. IEEE Trans. Neural Netw. **3**(5), 776–786 (1992)
21. Fisher, R.A.: The use of multiple measurements in taxonomic problems. Ann. Eugen. **7**(2), 179–188 (1936)
22. Abe, S., Lan, M.S.: A method for fuzzy rules extraction directly from numerical data and its application to pattern classification. IEEE Trans. Fuzzy Syst. **3**(1), 18–28 (1995)
23. Lekhtman, A.: Data Science in Medicine - Precision & Recall or Specificity & Sensitivity? (2019). https://towardsdatascience.com/should-i-look-at-precision-recall-or-specificity-sensitivity-3946158aace1

Contrastive Goal Grouping for Policy Generalization in Goal-Conditioned Reinforcement Learning

Qiming Zou$^{(\boxtimes)}$ and Einoshin Suzuki$^{(\boxtimes)}$

Kyushu University, Fukuoka, Japan
zou.qiming.847@s.kyushu-u.ac.jp, suzuki@inf.kyushu-u.ac.jp

Abstract. We propose Contrastive Goal Grouping (COGOAL), a self-supervised goal embedding algorithm for learning a well-structured latent goal space to simplify goal-conditioned reinforcement learning. Compared to conventional reconstruction-based methods such as variational autoencoder, our approach can benefit from previously learnt goals and achieve better generalizability. More specifically, we theoretically prove a sufficient condition for determining whether goals share similar optimal policies, and propose COGOAL that groups goals satisfying the condition in the latent space via contrastive learning. The learnt goal embeddings enable a fully-trained policy for a goal to reach new goals which are adjacent in the latent space. We conduct experiments on visual navigation and visual object search tasks. COGOAL significantly outperforms the baseline methods in terms of sample efficiency in the visual object search task, in which a previously learnt policy is adaptively transferred to reach new goals with fine-tuning.

Keywords: Goal-conditioned reinforcement learning · Goal embedding · Self-supervised learning

1 Introduction

A goal-oriented task, which requires an agent to reach a certain goal state, is ubiquitous in real-world applications such as object manipulation [18], visual navigation [27], and object search [17]. Goal-conditioned Reinforcement Learning (RL) provides an elegant framework for agents to learn diverse goal-reaching behaviors [22]. However, current goal-conditioned RL approaches often learn each goal independently, even when these goals can be reached by similar policies. In this work, we propose Contrastive Goal Grouping (COGOAL) that groups similar goals in the latent space with contrastive learning such that a previously learnt policy can be adaptively transferred to reach new goals with fine-tuning.

A general method to embed goals optimizes latent variables to reconstruct the goals, and then uses the latent variables for goal embeddings [18,21]. This kind of method is effective when each goal contains rich task-relevant information,

T. Mantoro et al. (Eds.): ICONIP 2021, LNCS 13108, pp. 240–253, 2021.
https://doi.org/10.1007/978-3-030-92185-9_20

e.g., a goal image in a visual navigation task [27]. However, a goal is usually given in a form which is not informative to the agent. For example, when the goals are given in texts in an object search task [17] without prior knowledge of the environment, it would be impossible to exactly infer the spatial relationship between a bottle of "crystal hot sauce" and "a coca cola glass bottle".

ARC [9] and PSM [1] propose to obtain state embeddings such that the embeddings of two states are closely located if the long-term optimal behaviours starting from these states [1] or reaching to these states [9] are similar. This notion of proximity is applicable to the goal embedding, i.e., we can group the embeddings of goals which share similar optimal goal-reaching policies. However, fully-trained goal-reaching policies are required to determine the similarity.

In this work, we propose a sufficient condition to determine whether goals share similar optimal policies without using fully-trained goal-reaching policies. Based on this condition, we construct pairs of similar goals and group their embeddings with contrastive learning. As shown in Fig. 1, with our COGOAL, the current goal (blue star) is adjacent to a previously learnt goal (blue cross) in the latent space. Since the policy is conditioned on the similar goal embeddings, the initial trajectory for the current goal (black line in Fig. 1) is generated with the help of the previously learnt goal.

The primary contribution of our work is a self-supervised goal embedding algorithm called COGOAL. COGOAL enables previously learnt goal-reaching policies to be adaptively transferred to reach new goals. In our experimental evaluation, we demonstrate that COGOAL outperforms baseline methods in a visual navigation task and a visual object search task in terms of policy generalizability.

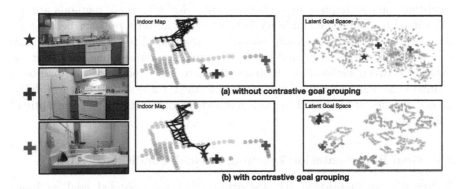

Fig. 1. Example of a visual navigation task. Left: current goal image (blue star) and previously learnt goal images (blue and red crosses). Right: the top plots (a) and the bottom plots (b) visualize agents' trajectories for reaching the current goal and goal embeddings without and with COGOAL, respectively. In the left plots in (a) and (b), the indoor map visualizes the trajectories with black lines and possible goals in real-world rooms with circles (the color represents the type of the room, e.g., the brown circles are in a kitchen). In the right plots in (a) and (b), we visualize the latent goal space with t-SNE [14], and each circle represents the embedding of a goal in the indoor map (Color figure online).

2   Related Works

2.1   Goal-Conditioned RL

Goal-conditioned RL learns from reward signals to predict action sequences which lead the agent towards the desired goals [11]. However, the reward signal depends on the goal and is sparse, i.e., it is typically received only when the goal is reached, which increases the requirement for the number of samples. To improve the sample efficiency, Hindsight Experience Replay (HER) [3] replaces the goal with a state which is included in the collected trajectories, even though the actual goal is not reached. The assumption is that each state can become an expected state for a goal in other episodes, which is general in goal-conditioned RL tasks and thus adopted in our method. Curriculum learning [4] is utilized to generate sub-goals which are close to the starting state and are possibly useful in guiding the agent to reach the actual goal in the long term [8,26]. A high-level policy or procedure is required for proposing valuable sub-goals or starting states for each goal, which is inefficient when the agent is required to reach diverse goals.

2.2   Self-supervised Learning for State Representation Learning in RL

Recent works propose to learn a compact state representation for RL with self-supervised learning. CURL [13] demonstrates how InfoNCE loss (which is a contrastive learning loss proposed in [19]) can be used to learn an efficient state representation. ATC [23] and LTW [6] learn state representations which encourage consistency across temporally adjacent states. However, these methods are not guaranteed to define the similarity between the long-term optimal behaviours, and thus the optimal behaviours can not be generalized to be effective on similar but new states. ARC [9] and PSM [1] obtain state embeddings such that the distance between embeddings matches the divergence between the long-term optimal behaviours reaching [9] or starting from [1] these states. However, fully-trained policies are required in both methods to obtain the long-term optimal behaviours.

2.3   Goal Representation Learning in Goal-Conditioned RL

A typical goal-conditioned RL algorithm takes the specified goal as input and updates the goal embedding with reward signals [3,22]. However, positive rewards are usually sparse in a goal-oriented task, and thus even a large number of trajectories might not provide meaningful learning signals. An auto-encoding model trains goal embeddings with an auxiliary goal reconstruction task in which the embeddings and a decoder network are optimized jointly to reconstruct the goals [18,21]. The goal embedding distribution is modeled at a point-wise level, i.e., such a method maximizes the log-likelihood of each goal. Point-wise goal embedding is general (with no assumptions for downstream tasks), however, since such a method embeds each goal independently, it fails to utilize the sequential property of Markov Decision Process (MDP), which is critical in decision making [1].

3   Background

3.1   Reinforcement Learning

RL is typically formalized with a Markov Decision Process (MDP) [24]. An MDP is defined as a tuple $<\mathcal{S}, \mathcal{A}, P, R, \gamma>$, where $\mathcal{S}$ is the state space, $\mathcal{A}$ is the action space, and $P\colon \mathcal{S} \times \mathcal{S} \times \mathcal{A} \to [0,1]$ denotes the state transition function. The agent is initialized to initial state s_0. At time step t, the agent at state $s_t \in \mathcal{S}$ executes action $a_t \in \mathcal{A}$, and then the state transits to $s_{t+1} \in \mathcal{S}$ with probability $P(s_{t+1}|s_t, a_t)$. Reward r_t is drawn from reward distribution $R\colon \mathcal{S} \times \mathcal{A} \to \mathbb{R}$. Interaction record $\tau = \{(s_t, a_t, r_t)\}_{t=0,1,\cdots,T^\tau-1}$ is called a trajectory, where T^τ is the length of τ.

The objective is to find optimal policy π^* that maximizes its expected discounted sum of rewards, or values, in one episode. Q-learning [25] is a commonly used method to achieve this objective when the action space is discrete. In Q-learning, the value of a state is defined as a conditional expectation[1] of the discounted sum of rewards given π and s_0 over τ, i.e.,

$$V^\pi(s) = \mathbb{E}_\tau \left[\sum_{t=0}^{\infty} \gamma^t r_t | \pi, s_0 = s \right], \tag{1}$$

where γ is a discount factor $(0 < \gamma < 1)$. Similarly, action-value (or Q-value) function $Q^\pi\colon \mathcal{S} \times \mathcal{A} \to \mathbb{R}$ is

$$Q^\pi(s,a) = \mathbb{E}_\tau \left[\sum_{t=0}^{\infty} \gamma^t r_t | \pi, s_0 = s, a_0 = a \right]. \tag{2}$$

The optimal Q-value function is defined as

$$Q^*(s,a) = \arg\max_{\pi \in \Pi} Q^\pi(s,a), \ \forall s \in \mathcal{S}, \ a \in \mathcal{A}, \tag{3}$$

where Π is a set of policies over $\mathcal{S}$ and $\mathcal{A}$. In Q-learning, given $Q^*(s,a)$, π^* is obtained by choosing actions greedily [25]:

$$\pi^*(s) = \arg\max_{a \in \mathcal{A}} Q^*(s,a), \forall s \in \mathcal{S}. \tag{4}$$

Deep Q Network (DQN) [16] directly estimates $Q^*(s,a)$ with a deep neural network, and obtains the optimal policy with Eq. (4).

---

[1] The conditional expectation is denoted as $\mathbb{E}_Z[X|Y]$ which represents the expected value of X given Y over Z.

3.2   Goal-Conditioned Reinforcement Learning

We consider a general problem of reaching goal g drawn from goal space $\mathcal{G}$. For each g, there is state region $\mathcal{S}^g \subseteq \mathcal{S}$ where $s \in \mathcal{S}^g$ fulfills the constraints for reaching g. We define the state at which g is reached as s^g. The reward function is an indicator function defined as

$$\mathbb{1}_{\mathcal{S}_g}(s) := \begin{cases} 1 & \text{if } s \in \mathcal{S}_g, \\ 0 & \text{if } s \notin \mathcal{S}_g. \end{cases} \tag{5}$$

We define T_g^τ as the number of time steps taken for reaching g in trajectory τ, i.e., the first hit time of goal g in trajectory τ. An episode is terminated when the given goal is reached, i.e., $T^\tau = T_g^\tau$. A universal value function [22] approximates different goal-conditioned value functions or action-value functions with a single model, e.g., deep neural network,

$$\hat{V}^\pi(s, g) \;=\; \mathbb{E}_\tau \left[\gamma^{T_g^\tau} | \pi, s_0 = s \right], \; s_{T_g^\tau} \in \mathcal{S}^g, \tag{6}$$

$$\hat{Q}^\pi(s, a, g) \;=\; \mathbb{E}_\tau \left[\gamma^{T_g^\tau} | \pi, s_0 = s, a_0 = a \right], \; s_{T_g^\tau} \in \mathcal{S}^g. \tag{7}$$

As shown in Fig. 2, the state and the goal are encoded by two separate neural networks $\phi \colon \mathcal{S} \to \mathbb{R}^n$ and $\psi \colon \mathcal{G} \to \mathbb{R}^n$, where n is the dimension of the state embedding and the goal embedding, respectively. The state embedding and the goal embedding are concatenated into vector $\phi(s) \oplus \psi(g)$, where $\oplus$ denotes the concatenation operator. Fully-connected network $h \colon \mathbb{R}^{2n} \to \mathbb{R}^{|\mathcal{A}|}$ maps $\phi(s) \oplus \psi(g)$ to vector $\boldsymbol{a}$, where each element is $\hat{Q}^\pi(s, a, g)$ for action a given state s and goal g.

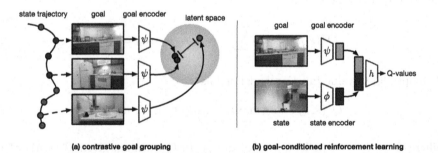

(a) contrastive goal grouping   **(b) goal-conditioned reinforcement learning**

Fig. 2. Goal-conditioned RL with COGOAL. (a) Goals are randomly sampled from a state trajectory because each state can possibly lead to one or multiple goals; the goal embeddings (encoded by network ψ) are learnt by bringing the embeddings of similar goals together, while pushing the other goal embeddings apart. (b) The goal and state images are encoded by network ψ and network ϕ, respectively, and then concatenated into a vector. Fully-connected network h takes the vector as input and outputs the Q-values.

4   Contrastive Goal Grouping

Our COGOAL aims to group similar goals in the latent space such that previously learnt policy can be generalized to new goals. In Sect. 4.1, we propose a sufficient condition for determining the goal similarity. In Sect. 4.2, we define a constrained optimization problem for the goal embedding.

4.1   Sufficient Condition for Optimal Policy Similarity

In this section, we first define transition distance, and then show an assumption that the MDP is reversible. Based on this assumption, we derive a theorem that goals which can be reached in a small number of time steps share similar optimal goal-reaching policies.

We introduce the definition of the transition distance.

Definition 1 (Transition distance). *Let* $s,\ s' \in \mathcal{S}$, *then the transition distance from s to s' is defined as the absolute value of the first hit time difference in a trajectory τ:*

$$d^\tau(s, s') = |T_s^\tau - T_{s'}^\tau|, \tag{8}$$

where T_s^τ and $T_{s'}^\tau$ are the first hit time of state s and s' in trajectory τ, respectively.

We set the following assumption.

Assumption 1. *Trajectory τ sampled from the MDP is reversible. An agent can generate reversed trajectory τ^{-1} by moving from final state s' of τ to initial state s of τ. The numbers of the steps in τ and τ^{-1} are equal, i.e.,*

$$d^\tau(s, s') = d^{\tau^{-1}}(s', s), \ s, s' \in \tau. \tag{9}$$

This assumption holds true in a wide range of robotic applications such as robot locomotion, pick-place, and assembly tasks. For example, a wheeled robot moves toward a position s' starting from a position s, and then the robot can reverse the trajectory to reach the position s starting from the position s'.

Based on Assumption 1, we derive a bound.

Theorem 1. *If there exists τ which satisfies*

$$d^\tau(s^g, s^{g^+}) \le k, \ g, g^+ \in \mathcal{G}, \ s^g, s^{g^+} \in \tau, \tag{10}$$

then

$$\left| Q^*(s, a, g) - Q^*(s, a, g^+) \right| < 1 - \gamma^k, \ \forall s \in \mathcal{S}, \forall a \in \mathcal{A}. \tag{11}$$

Proof. Since trajectory τ is reversible according to Assumption 1, we can assume that $T_{g^+}^\tau \geq T_g^\tau$ without loss of generality, and we have

$$
\begin{aligned}
&\left| Q^*(s,a,g) - Q^*(s,a,g^+) \right| \\
&\overset{\text{Eq.}(7)}{=} \mathbb{E}_\tau \left[\left| \gamma^{T_g^\tau} - \gamma^{T_{g^+}^\tau} \right| \middle| \pi, s_0 = s, a_0 = a \right] \\
&= \mathbb{E}_\tau \left[\gamma^{T_g^\tau} \left(1 - \gamma^{T_{g^+}^\tau - T_g^\tau} \right) \middle| \pi, s_0 = s, a_0 = a \right] \\
&\overset{\text{Eq.}(8)}{=} \mathbb{E}_\tau \left[\gamma^{T_g^\tau} \left(1 - \gamma^{d^\tau(s^{g^+}, s^g)} \right) \middle| \pi, s_0 = s, a_0 = a \right] \\
&< 1 - \gamma^k.
\end{aligned}
$$

Theorem 1 shows that if goals g and g^+ can be reached sequentially within k steps, then the optimal Q-value difference of g and g^+ is bounded by $1 - \gamma^k$.

4.2 Contrastive Goal Grouping

Our method aims to group similar goals in the latent space. To achieve this aim, we need two steps: i) sampling pairs of goals which share similar optimal goal-reaching policies; ii) bringing the embeddings of the similar goals together.

Since the optimal policy can be obtained with the optimal Q function (as shown in Eq. (4)), the difference $|Q^*(s,a,g) - Q^*(s,a,g^+)|$ reflects the dissimilarity of the corresponding optimal policies. Therefore, according to Theorem 1, a pair of similar goals g and g^+ should satisfy the sufficient constraint given in Eq. (10). This constraint can be satisfied by sampling goals which are reached at states s^g and s^{g^+}, and both states are reached within k time steps in a trajectory. Collecting diverse trajectories for different goals is challenging. Therefore, we make the same assumption as [3] that each potential goal can be input by user in some episodes and can be reached at one state in the state space at least. Due to the instability of RL [16,24], instead of updating the policies with RL over the potential goals as [3], we relabel the trajectories with potential goals to create dense supervision signals for goal grouping (as shown in Fig. 2(a)).

A straightforward idea to group the goal embeddings is to minimize the mean squared error (MSE) between the embeddings, i.e., $\psi^* = \arg\min_\psi \|\psi(g) - \psi(g^+)\|_2^2$. However, minimizing the MSE causes an undesired degenerated solution, i.e., all goal embeddings collapse to a single point. Contrastive learning [10] repulses negative pairs apart while attracting positive pairs with each other. A negative pair includes samples which have different semantic labels, and repulsing the embeddings of these samples apart precludes the undesired degenerated solution. Nevertheless, at the early stage of RL, the agent generates nearly random trajectories, and thus similar goals might be reached in different trajectories. Therefore, randomly sampling goals from different trajectories to construct the negative pairs might sample goals which are actually similar. The recently proposed Whitening-MSE (WMSE) [7] is utilized to avoid the need of constructing the negative pairs by constraining the goal embeddings to lie in a spherical distribution. Since WMSE makes no assumption for the neural network architecture or the way of constructing the positive pairs, it is straightforward to adopt WMSE in our problem.

The final optimization problem we aim to solve is

$$\min_{\psi} \quad \left\| \psi(g) - \psi(g^+) \right\|_2^2 \tag{12}$$

$$\text{s.t.} \; cov\left(\psi(g), \psi(g^+)\right) = cov\left(\psi(g^+), \psi(g)\right) = I, \tag{13}$$

$$d^\tau(s^g, s^{g^+}) \leq k, \; \exists \tau \; g, g^+ \in \mathcal{G}, \; s^g, s^{g^+} \in \tau. \tag{14}$$

Equation (13) constraints that the co-variance matrix between the goal embeddings is the identity matrix I. This constraint makes all the components of the goal embedding to be linearly independent to each other, hence avoiding the degenerated solution.

We further verify that grouping similar goals in the latent space would help a policy to be transferred to reach new goals. In standard RL, the order between estimated Q values at two states is preserved, i.e., if $Q^\pi(s, a_1) > Q^\pi(s, a_2)$ then $Q^\pi(s', a_1) > Q^\pi(s', a_2)$, if the Euclidean distance between state embeddings $\phi(s)$ and $\phi(s')$ is sufficiently small, which is stated as follows.

Definition 2 (Lipschitz continuous function [20]). *Given two metric spaces* (X, d_X) *and* (Y, d_Y), *where* d_X *denotes the metric on set* X *and* d_Y *is the metric on set* Y, *a function* $f : X \to Y$ *is Lipschitz continuous if there exists a real constant* $K \geqslant 0$ *such that, for all* x_1 *and* x_2 *in* X, $d_Y(f(x_1), f(x_2)) \leq K d_X(x_1, x_2)$.

Theorem 2 ([5]). *Let* Δ *be the estimated Q-value difference for actions* a_1 *and* a_2 *at state* s, *i.e.,* $\Delta = |Q^\pi(s, a_1) - Q^\pi(s, a_2)|$, *and* ϵ *be the Euclidean distance between state embeddings* $\phi(s)$ *and* $\phi(s')$. *Assume we have* $Q^\pi(s, a_1) > Q^\pi(s, a_2)$. *If the Q-value function is* K-*Lipschitz continuous and* $\epsilon < \Delta/(2K)$, *we have* $Q^\pi(s', a_1) > Q^\pi(s', a_2)$.

See *Proposition III.2* in [5] for the proof. In goal-conditioned RL, the Q-value function takes a state and a goal as input, and we have

$$\epsilon = \left\| \phi(s) \oplus \psi(g) - \phi(s) \oplus \psi(g^+) \right\|_2 = \left\| \psi(g) - \psi(g^+) \right\|_2. \tag{15}$$

Therefore, according to Theorem 2, at state s, the order of Q values of different actions for goal g would be preserved for goal g^+ if ϵ is sufficiently small. Since the policy is obtained by greedily choosing actions with the maximum Q-value (Eq. (4)), when ϵ is sufficiently small, the learnt policy for goal g can be transferred to reach new goal g^+.

The detailed algorithm is shown in Algorithm 1. The agent 1) collects state trajectories with random actions; 2) constructs positive pairs with the goals which are reached within k steps in the collected state trajectories; 3) updates the goal encoder by minimizing the loss function (in Eq. (12)) constrained by Eq. (13); and 4) updates the state encoder and the policy with an off-the-shelf RL algorithm.

Algorithm 1: COGOAL

Require: Goal encoder ψ, state encoder ϕ, policy π, environment $\mathcal{M}$,
time-window size k, number H of the pre-training steps, number N
of the RL episodes, maximum number T of the episode time steps.

```
/* Pre-train the goal encoder via COGOAL                          */
```
for *iterations* $= 0, 1, \cdots, H - 1$ **do**
> Collect trajectory τ by interacting with $\mathcal{M}$ using random actions.
> Uniformly sample goals reached within k steps in τ without replacement.
> Update ψ to minimize the WMSE loss between the embeddings of the
> sampled goals, i.e., Eq. (12)–(14).

end
```
/* Goal-conditioned RL                                            */
```
for *episodes* $= 0, 1, \cdots, N - 1$ **do**
> Uniformly sample goal g.
> Reset the agent to initial state s_0.
> Roll-out π in $\mathcal{M}$, producing transitions $\{(s_t, a_t, r_t, s_{t+1}, \psi(g))\}_{t=0,1,\cdots,T-1}$.
> Update π and ϕ with the collected transitions using an off-the-shelf RL
> algorithm.

end

---

5   Experiments

Benchmark Environments. The active vision dataset [2] simulates robotic vision tasks in everyday indoor environments using real imagery. This simulated environment contains an automobile wheeled robot which can take six kinds of actions: forward, backward, left, right, clockwise rotation, and counter-clockwise rotation. We evaluate COGOAL on two tasks based on the active vision dataset.

Visual Navigation: At each time step, the robot receives an RGB image as observation. In addition to the observation, a goal image is given to the robot. The robot is required to move to a specific position and an orientation to capture the same observation as the goal image within 1000 time steps.

Visual Object Search: The observation space and the action space of the robot is the same as the previous visual navigation task. The robot is required to search for several target objects within 100 time steps. The goals are given to the robot with the names of the target objects, e.g., "coca cola glass bottle".

In both tasks, the agent receives a positive reward (+1) only when the goal image or the selected object is found, otherwise the reward is a small negative value (-0.001 in the visual navigation task and -0.01 in the visual object search task). One episode is ended when the goal is found or the maximum time step is reached.

Baseline Methods. We compare COGOAL to the following algorithms.

DQN [16]: This is a standard value-based algorithm for discrete action tasks. The goal embedding is updated along with the policy update.

DQN+VAE [18,21]: This is the same as the DQN algorithm, but with auxiliary goal reconstruction tasks. Variational Autoencoder (VAE) [12] is a commonly used method to embed goals in goal-conditioned RL.

CURL [13]: This is an RL algorithm combined with contrastive learning which learns state embeddings by bringing augmented views of the same state together while pushing augmented views of different states apart in the latent space.

CURL (GOAL): We adopt CURL to embed goals, i.e., the method brings augmented views of the same goal together while pushing augmented views of different goals apart in the latent space.

Note that, for a fair comparison, both COGOAL and other baseline methods utilize DQN [16] to update policy and its state encoder with reward signals.

Settings. In the visual navigation task, the agent must reach one of the 15 goals located in different rooms, and then the fully-trained policies are fine-tuned on 100 new goals in the same environment. Similarly, in the visual object search task, the agent is required to search for one of the 15 target objects which are scattered in different rooms, and then the fully-trained policies are fine-tuned on new goal-oriented tasks which include 30 new target objects in the same environment. In both tasks, a goal is randomly sampled at the beginning of each episode, and each series of experiments is repeated 10 times with different random seeds.

5.1   Results and Analysis

Learnt Goal Embedding Visualization. As we explained in Sect. 4.2, if two goals can be reached by similar policies, similar goal embeddings will lead to better policy generalization. Based on this assumption, we evaluate the quality of embeddings by checking whether the goals located in the same room are grouped in the same cluster, since such goals tend to share similar optimal policies.

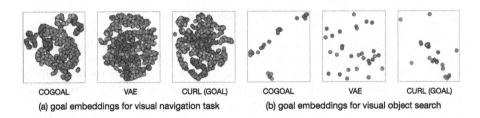

COGOAL   VAE   CURL (GOAL)   COGOAL   VAE   CURL (GOAL)

(a) goal embeddings for visual navigation task   (b) goal embeddings for visual object search

Fig. 3. Embedding visualization. We visualize the learnt goal embeddings using t-SNE, where a circle corresponds to a potential goal and the color indicates the room which the goal image or the target object is located in. (Color figure online)

In Fig. 3, the learnt goal embeddings are projected to a 2-dimensional space with t-SNE [14]. An embedded goal is visualized by a circle with color representing the room which the corresponding goal is located in. We can see that, in the latent space, COGOAL separates goals located in different rooms for both tasks. In the visual navigation task (as shown in Fig. 3(a)), although COGOAL achieves the best performance in grouping the embedded goals compared to the baselines, VAE and CURL (GOAL) are also capable of grouping a part of the goals in the latent space. The main reason is that adjacent positions in a room tend to have similar visual observations. Thus, embedding goals independently (as DQN+VAE and CURL (GOAL)) can still roughly reflect the adjacency relationship between goals. On the contrary, in the visual object search task, the goal is given in text, i.e., the name of the target object. As shown in Fig. 3(b), for VAE and CURL (GOAL), the embeddings of objects located in different rooms are mixed, implying that VAE and CURL (GOAL) fail to reflect the spatial relationship between goal objects. Since the policy takes the goal embedding as input, the policy would be misled when the spatially adjacent goals have extremely different embeddings or the goals located far apart in the environment share similar embeddings.

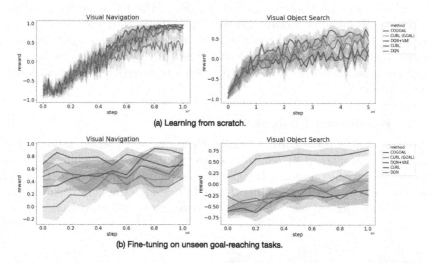

Fig. 4. Learning curves on robotic vision tasks. (a) The policies are learnt to reach goals from scratch. (b) Fully-trained policies are fine-tuned on unseen new goals. The x-axis represents the number of the time steps. The y-axis represents the accumulated reward in an episode. The solid lines and shaded regions represent the mean and the standard deviation over 10 random seeds, respectively.

Learning Performance on Robotic Vision Tasks. In Fig. 4(a), we compare the sample efficiency of all methods when learning from scratch, i.e., the state encoder, the goal encoder, and the policy are randomly initialized. In the visual navigation task, we observe that COGOAL achieves comparable performance, but the gap is relatively small. However, in the visual object search task, COGOAL significantly outperforms other baseline methods in terms of sample efficiency. This difference can be explained in part by the goal embedding spaces visualized in Fig. 3. The embedded goals of the baseline methods are mixed in the visual object search task but are more structured in the visual navigation task according to their room types.

Our experiments on both tasks in Fig. 4(b) show that COGOAL significantly outperforms baseline methods in terms of the generalizability. The fully-trained policies from the previous series of the experiments are transferred and fine-tuned on new goal-oriented tasks. As shown in Fig. 4(b), the initial performance of our method is better than other baselines, indicating that the policies are generalized to reach the new goals. Besides, our method converges to near-optimal policies with less samples, which demonstrates that previously learnt goal-reaching policies accelerate the learning of the new goal-oriented tasks.

Overall, due to the structured goal embedding space, COGOAL outperforms the baseline methods by a large margin in terms of the generalizability and achieves comparable sample efficiency when learning from scratch.

6   Conclusion

We proposed a self-supervised goal embedding method, Contrastive Goal Grouping (COGOAL), which achieves high generalizability in the goal-conditioned RL. The algorithm learns a well-structured goal embedding space where goals reached by similar optimal policies are grouped into a cluster. We demonstrated the performance of our method in real-world tasks, i.e., visual navigation and visual object search, and showed that COGOAL is capable of improving the sample efficiency in learning new goal-oriented tasks with the help of previously learnt goal-oriented tasks.

COGOAL pre-trains the goal embeddings with randomly collected trajectories and then fixes the goal embeddings during RL, which is not guaranteed to cover the entire goal space. A future direction is to update the goal embeddings along with the RL process and avoid the catastrophic forgetting [15] of the previously trained goal-reaching policies caused by the continuous updates of goal embeddings.

Acknowledgments. This work was supported by China Scholarship Council (Grant No. 202008050300).

References

1. Agarwal, R., Machado, M.C., Castro, P.S., Bellemare, M.G.: Contrastive behavioral similarity embeddings for generalization in reinforcement learning. In: Proceedings of the ICLR (2021)
2. Ammirato, P., Poirson, P., Park, E., Kosecka, J., Berg, A.C.: A dataset for developing and benchmarking active vision. In: Proceedings of the ICRA, pp. 1378–1385 (2017)
3. Andrychowicz, M., et al.: Hindsight experience replay. In: Proceedings of the NeurIPS, pp. 5048–5058 (2017)

4. Bengio, Y., Louradour, J., Collobert, R., Weston, J.: Curriculum learning. In: Proceedings of the ICML, pp. 41–48 (2009)
5. Bodnar, C., Hausman, K., Dulac-Arnold, G., Jonschkowski, R.: A metric space perspective on self-supervised policy adaptation. IEEE Robot. Autom. Lett. **6**(3), 4329–4336 (2021)
6. Ermolov, A., Sebe, N.: Latent world models for intrinsically motivated exploration. In: Proceedings of the NeurIPS (2020)
7. Ermolov, A., Siarohin, A., Sangineto, E., Sebe, N.: Whitening for self-supervised representation learning. In: Proceedings of the ICML (2021)
8. Florensa, C., Held, D., Wulfmeier, M., Zhang, M., Abbeel, P.: Reverse curriculum generation for reinforcement learning. In: Proceedings of the CoRL, pp. 482–495 (2017)
9. Ghosh, D., Gupta, A., Levine, S.: Learning actionable representations with goal conditioned policies. In: Proceedings of the ICLR (2019)
10. Hadsell, R., Chopra, S., LeCun, Y.: Dimensionality reduction by learning an invariant mapping. In: Proceedings of the CVPR, pp. 1735–1742 (2006)
11. Kaelbling, L.P.: Learning to achieve goals. In: Proceedings of the IJCAI, pp. 1094–1099 (1993)
12. Kingma, D.P., Welling, M.: Auto-encoding variational Bayes. In: Proceedings of the ICLR (2014)
13. Laskin, M., Srinivas, A., Abbeel, P.: CURL: contrastive unsupervised representations for reinforcement learning. In: Proceedings of the ICML, pp. 5639–5650 (2020)
14. Van der Maaten, L., Hinton, G.: Visualizing data using t-SNE. J. Mach. Learn. Res. **9**(11), 2579–2605 (2008)
15. McCloskey, M., Cohen, N.J.: Catastrophic interference in connectionist networks: the sequential learning problem. Psychol. Learn. Motiv. **24**, 109–165 (1989)
16. Mnih, V., et al.: Human-level control through deep reinforcement learning. Nature **518**(7540), 529–533 (2015)
17. Mousavian, A., Toshev, A., Fiser, M., Kosecká, J., Wahid, A., Davidson, J.: Visual representations for semantic target driven navigation. In: Proceedings of the ICRA, pp. 8846–8852 (2019)
18. Nair, A., Pong, V., Dalal, M., Bahl, S., Lin, S., Levine, S.: Visual reinforcement learning with imagined goals. In: Proceedinngs of the NeurIPS, pp. 9209–9220 (2018)
19. van den Oord, A., Li, Y., Vinyals, O.: Representation learning with contrastive predictive coding. arXiv arXiv:1807.03748 (2018)
20. O'Searcoid, M.: Metric Spaces. Springer, London (2006). https://doi.org/10.1007/1-84628-244-6
21. Pong, V., Dalal, M., Lin, S., Nair, A., Bahl, S., Levine, S.: Skew-Fit: state-covering self-supervised reinforcement learning. In: Proceedings of the ICML, pp. 7783–7792 (2020)
22. Schaul, T., Horgan, D., Gregor, K., Silver, D.: Universal value function approximators. In: Proceedings of the ICML, pp. 1312–1320 (2015)
23. Stooke, A., Lee, K., Abbeel, P., Laskin, M.: Decoupling representation learning from reinforcement learning. arXiv arXiv:2009.08319 (2020)
24. Sutton, R.S., Barto, A.G.: Reinforcement Learning: An Introduction. MIT Press (2018)
25. Watkins, C.J.C.H.: Learning from delayed rewards. Ph.D. thesis, King's College, Cambridge (1989)

26. Zhang, T., Guo, S., Tan, T., Hu, X., Chen, F.: Generating adjacency-constrained subgoals in hierarchical reinforcement learning. In: Proceedings of the NeurIPS (2020)
27. Zhu, Y., et al.: Target-driven visual navigation in indoor scenes using deep reinforcement learning. In: Proceedings of the ICRA, pp. 3357–3364 (2017)

Global Fusion Capsule Network with Pairwise-Relation Attention Graph Routing

Xinyi Li, Song Wu, and Guoqiang Xiao[✉]

College of Computer and Information Science,
Southwest University, Chongqing, China
gqxiao@swu.edu.cn

Abstract. Comparing with traditional convolutional neural networks, the recent capsule networks based on routing-by-agreement have shown their robustness and interpretability. However, the routing mechanisms lead to huge computations and parameters, significantly reducing the effectiveness and efficiency of capsule networks towards complex and large-scale data. Moreover, the capsule network and its variants only explore the PrimaryCaps layer for various reasoning tasks while ignoring the specific information learned by the low-level convolutional layers. In this paper, we propose a Graph Routing based on Multi-head Pairwise-relation Attention (GraMPA) that could thoroughly exploit both the semantic and location similarity of capsules in the form of multi-head graphs, to improve the robustness of the capsule network with much fewer parameters. Moreover, a Global Fusion Capsule Network (GFCN) architecture based on the Multi-block Attention (MBA) module and Multi-block Feature Fusion (MBFF) module is designed, aiming to fully explore detailed low-level signals and global information to enhance the quality of final representations especially for complicated images. Exhaustive experiments show that our method can achieve better classification performance and robustness on CIFAR10 and SVHN datasets with fewer parameters and computations. The source code of GraMPA is available at https://github.com/SWU-CS-MediaLab/GraMPA.

Keywords: Capsule networks · Graph routing algorithm · Object classification

1 Introduction

Recently, convolutional neural networks (CNNs) have become the mainstream in various computer vision applications [2, 4, 22, 24, 26–28, 31]. However, the weakness of CNNs has been of great concern, such as its inability to preserve part-whole spatial relationships of object features, making it difficult to recognize objects with changed perspectives [14, 16]. In addition, CNNs have been regarded as a black-box model with little interpretability [8, 14]. The introduction of the

T. Mantoro et al. (Eds.): ICONIP 2021, LNCS 13108, pp. 254–265, 2021.
https://doi.org/10.1007/978-3-030-92185-9_21

capsule networks has improved these deficiencies of CNNs. It performs well in detecting viewpoint changes and resisting attacks due to its ability to represent multidimensional semantic information of features. However, the core process of routing algorithm in capsule network is iterative clustering [10,14], which results in huge parameters and high computations, significantly downgrading its performance on complex and large-scale datasets. Moreover, it is difficult to adaptively determine the number of routing iterations, because too many iterations and too few iterations may lead to over-fitting and under-fitting, respectively.

In view of these problems, many studies are looking for more effective routing methods to reduce the complexity of routing iterations as well as the number of parameters [7,8,10,17,29]. The idea of graph is also included in capsule networks to optimize its amount of parameters and reasoning performance [7,19,25]. However, they fail to simultaneously explore deep into the spatial and semantic similarities of the low-level capsules during the routing process. To solve this problem, illuminated by some remarkable works of exploring similarity between features [3,18,21,30], a novel routing mechanism named Graph Routing Based on Multi-head Pairwise-Relation Attention (GraMPA) is proposed in this paper, where all the low-level capsules can be represented by multi-head graphs, and a learning model is established to explore the spatial and semantic correlations of capsules from the adjacency matrix for better clustering weights.

Furthermore, the semantic information from intermediate layers of the capsule network is ignored for the final reasoning task in previous work. Through the multiple blocks of the backbone, abstract feature representations with high-level semantics are generated for the input data. Specifically, in the high-level layers, some subtle features may be learned more coarsely and less obviously, while the global representations may not be effectively learned in the low-level layers [20]. Therefore, for various types of input data, the significance of each block may be different for the different reasoning tasks [2,4,26,27]. Consequently, inspired by some representative attention-based multi-scale feature fusion methods [20,23], a Global Fusion Capsule Network (GFCN) architecture is proposed to fuse multiple semantic features by using the Multi-Block Attention (MBA) and the Multi-Block Feature Fusion (MBFF) module, which could further enhance and optimize the final representations.

Our main contributions are summarized as follows:

(1) A novel routing mechanism named Graph Routing Based on Multi-head Pairwise-Relation Attention (GraMPA) is proposed, which uses the pairwise similarity between capsule vectors in the form of multi-head graphs to learn attention maps for clustering similar low-level capsules, with the advantage of reducing the number of parameters and avoiding under-fitting or over-fitting caused by traditional routing iterations mechanism.

(2) The Multi-Block Attention (MBA) mechanism is designed to learn the importance of each blocks' semantic information and enlarge the importance of relatively significant blocks. Moreover, the Multi-Block Feature Fusion (MBFF) module is designed to combine the generated semantic information from the intermediate layers of the backbone network according to the weights of the

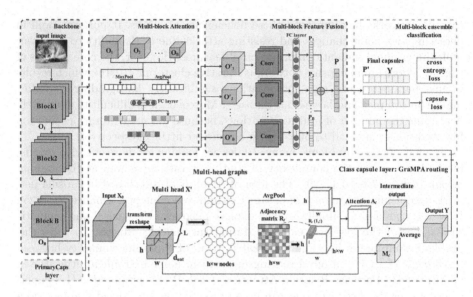

Fig. 1. Overview of our proposed Global Fusion Capsule Network with Pairwise-Relation Attention Graph Routing. It is consists of five components. The backbone network extracts the intermediate feature representations for input data. The class capsule layer clusters the low-level capsules into high-level capsules by using the GraMPA routing. The Multi-block Attention and Multi-block Feature Fusion is utilized to integrate the multi-layer semantic information into the final capsules to robust the classification performance.

multi-block attention matrix. The two modules robust the capability of reasoning of final capsules by complementing the global information.

(3) The proposed Multi-head Pairwise-Relation Attention (GraMPA) shows better classification performance than state-of-the-art baselines on CIFAR10 and SVHN datasets with fewer parameters and computations. Meanwhile, the experiment results also demonstrated the stability and robustness of the GraMPA. The GraMPA could effectively resist the attacks from adversarial examples, and it is also insensitive to the predefined parameters. The disentangled reconstruction and multi-block visualization further prove the effectiveness of our method.

2   The Proposed GFCN with GraMPA

The overview of our proposed Global Fusion Capsule Network (GFCN) with Pairwise-Relation Attention Graph Routing is illustrated in Fig. 1.

2.1   Multi-block Attention for Capsule

To adaptively combine signals of multiple semantic levels, we propose multi-block attention (MBA) module, which is applicable for the capsule network.

We denote the output of each block of backbone as $O_s \in \mathbb{R}^{c_s \times w_s \times w_s}$, where $s \in \{1, 2, ..., B\}$, c_s and w_s are the channel dimension, height and width of the sth feature map, respectively, B is the number of blocks. Some studies [5,23] show that the combination of max pooling and global average pooling can better capture multi-layer semantic information. Thus, we perform max pooling and global pooling on the O_s to get two descriptors $D_s^{max} \in \mathbb{R}^{c_s}$ and $D_s^{avg} \in \mathbb{R}^{c_s}$ to compress the information of the feature map, namely $D_s^{max} = MaxPool(O_s)$ and $D_s^{avg} = AvgPool(O_s)$.

Subsequently, all D_s^{max} and D_s^{avg} generated by various blocks are concatenated to obtain the global information $D^{max} \in \mathbb{R}^c$ and $D^{avg} \in \mathbb{R}^c$, c is the sum of all c_s. In short, the global descriptors are defined as $D^{max} = [D_1^{max}, D_2^{max}, ..., D_B^{max}]$ and $D^{avg} = [D_1^{avg}, D_2^{avg}, ..., D_B^{avg}]$.

Then, two fully connected layers are applied on D^{max} and D^{avg}, and the outputs of two fully connected layers are then merged by element-wise summation, followed by a sigmoid function to obtain the attention mask $a \in \mathbb{R}^c$. The process is expressed as:

$$a = Sigmoid(W_1(W_0(D^{max})) + W_1(W_0(D^{avg}))) \tag{1}$$

where the number of neurons in the two fully connected layers is set to $c \times \alpha$ and c respectively, α denotes the dimension reduction ratio which is set to $1/4$ in our experiments.

Finally, we divide the attention mask a into B parts, and each part is denoted as $a_s \in \mathbb{R}^{c_s}$, which has the same channel dimension as the corresponding O_s. Then, we can update the output of each block by multiplying attention mask a_s and O_s by channel to get O_s', which can enlarge the signal of important blocks and weaken the signal of relatively unimportant blocks.

2.2 Multi-block Feature Fusion

A Multi-block Feature Fusion (MBFF) module is utilized to fuse multiple features extracted from different blocks to assist the classification task. After being updated by multi-block attention, the output of the sth block is denoted as $O_s' \in \mathbb{R}^{d \times w_s \times w_s}$. The sth output feature map O_s' is utilized to learn the class information and obtain the prediction results $P_s \in \mathbb{R}^N$, where $s \in \{1, 2, ..., B\}$ and N is the number of object classes. The P_s is computed as:

$$P_s = Sigmoid(f_s(O_s')) \tag{2}$$

where $f_s(\cdot)$ represent a embedding function implemented by three convolutional layers with 64 1×1 kernels, 64 1×1 kernels and 16 $w_s \times w_s$ kernels respectively, followed by a FC layer with N nodes. A batch normalization (BN) and a ReLU activation are subsequent to each convolutional layer.

Then, a softmax function is applied to the sum of the predicted scores P_s of all blocks to get final prediction of MBFF module P, namely $P = Softmax(\sum_{s=1}^B P_s)$. The obtained $P \in \mathbb{R}^N$ combines the classification results predicted by all blocks and contains the multi-level category information. Subsequently, P is fused with the output of the capsule network to assist the classification decision, as described in Sect. 2.4.

2.3   Graph Routing Based on Multi-head Pairwise-Relation Attention

For the proposed graph routing based on multi-head pairwise-relation attention (GraMPA), the primary capsule layer extracts the capsules from the output of the backbone O_B, which is further used as the input of the class capsule layer. The low-level capsules can be clustered into high-level capsules by using the proposed routing algorithm GraMPA, which is stated as X_0.

The input $X_0 \in \mathbb{R}^{M \times d_{in}}$ for the routing algorithm is first transformed into a feature space $X \in \mathbb{R}^{M \times d_{out}}$ by the transformation matrix $W_t \in \mathbb{R}^{M \times d_{in} \times d_{out}}$, where M is the total number of low-level capsules, d_{in} and d_{out} refer to the dimension of low-level capsules and high-level capsules, respectively. Then, the X is reshaped into $X' \in \mathbb{R}^{L \times d_{out} \times h \times w}$, where L, d, h and w denote the number of heads, the channels, the height and width of the feature map, respectively. In other words, as shown in Fig. 1, X' contains L capsule tensors, namely L groups of d-dimensional capsule vectors, and the number of capsules in each group is $h \times w$. Following the GraCapsNet [7], we transform each group of capsules into a graph, where each node is a capsule vector of length d_{out}. In short, X' is divided into L capsule tensors, and each capsule tensor $X'_l \in \mathbb{R}^{1 \times d_{out} \times h \times w}$, where $l \in \{1, ..., L\}$, is constructed into one graph that includes $h \times w$ nodes. Note that we also refer to the nodes as features, feature nodes and feature vectors.

Subsequently, we measure the pair-wise similarity between the nodes in each graph to define the adjacency matrices, i.e., the edge information of each graph. The ith node of the lth graph is denoted as F_l^i, where $i \in \{1, ..., K\}$ and K is the number of nodes in each graph, i.e., $K = h \times w$. And $r_l^{i,j}$ is used to represent the affinity between the ith node and the jth node of the lth graph. The $r_l^{i,j}$ is calculated by:

$$r_l^{i,j} = f_s(F_l^i, F_l^j) = \gamma(F_l^i)^T \cdot \delta(F_l^j) \tag{3}$$

where the embedding functions $\gamma(\cdot)$ and $\delta(\cdot)$ are both implemented as a 1×1 convolutional layer with $d_{out} \times \beta$ kernels, where β is a dimension reduction ratio we set to $1/4$, subsequent to a BN and a ReLU activation. Similarly, the similarity between the jth node and the ith node of the lth graph is computed as $r_l^{j,i} = f_s(F_l^j, F_l^i)$. Then the adjacency matrix of the lth graph is defined as $R_l \in \mathbb{R}^{K \times K}$, and $R_l(i,j) = r_l^{i,j}$.

Then, the adjacency matrix R_l is used to learn attention for the clustering-like effect. For the ith node, the similarity relationships between the ith node and all other nodes are stacked into a vector representing the global structure information:

$$\mathbf{r}_l^i = [r_l^{i,1}, r_l^{i,2}, ..., r_l^{i,K}] = R_l(i,:) \tag{4}$$

where $R_l(i,:)$ is the ith row of R_l. The vector $\mathbf{r}_l^i$ can assist us to observe the correlation between nodes globally, and stacking the affinity $r_l^{i,j}$ in order can also indirectly introduce the position and spatial relationship between two nodes in the later learning.

Moreover, in addition to the global knowledge, we also include the original information of the node features itself:

$$Z_l^i = [AvgPool_c(F_l^i), \theta(\mathbf{r}_l^i)] \tag{5}$$

where $AvgPool_c(\cdot)$ represent the global average pooling along the channel dimension, $\theta(\cdot)$ is a 1×1 convolutional layer followed by a BN and a ReLU activation.

Then, a small learnable model is constructed to infer the attention values $A_l \in \mathbb{R}^{K \times N}$ (N is the number of object classes) from $Z_l = [Z_l^1, Z_l^2, ..., Z_l^K]$ that contains both global structure knowledge and original feature information:

$$A_l = Sigmoid(W_a(f_a(Z_l))) \tag{6}$$

where $f_a(\cdot)$ shrinks the channel dimension of Z_l to 1 and transform Z_l into the shape $(K \times 1)$, which is implemented by two 1×1 convolutional layers followed by a BN and a ReLU activation, and $W_a \in \mathbb{R}^{1 \times N}$ are trainable parameters. The 1×1 convolutional layer above including $\gamma(\cdot)$, $\delta(\cdot)$, $\theta(\cdot)$ and $f_a(\cdot)$ are shared among all heads.

The gained attention A_l can also be interpreted as clustering weights. Specifically, for N high-level capsules, each high-level capsule has K weights corresponding to K low-level capsules, which represents the contribution of each low-level capsule to the high-level capsule. Therefore, the clustering effect can be achieved by weighted summation of K low-level capsules with A_l as the weight. The intermediate output M_l of the lth graph is calculated by:

$$M_l = (A_l)^T X_l' \tag{7}$$

The final prediction capsules $Y \in \mathbb{R}^{N \times d_{out}}$ are stated as:

$$Y = squash(\frac{1}{L} \sum_{l=1}^{L} M_l) \tag{8}$$

where $squash(\cdot)$ follows CapsNet's setting [14].

2.4   Multi-block Ensemble Classification

The output prediction of MBFF module $P \in \mathbb{R}^{N \times 1}$ is utilized to enhance the classification performance of the class capsule layer's output Y. We feed the P containing multi-level class information into the cross-entropy loss function, enabling MBA and MBFF module to generate more compact and discriminative feature representation, and thus gaining more reliable classification result P.

To combine P and Y, firstly, P is masked by retaining only the maximum value (i.e. the score of the class predicted by P) and setting other values to 0:

$$p_x' = \begin{cases} p_x & p_x = max(P) \\ 0 & otherwise \end{cases} \tag{9}$$

where $x \in \{1, ..., N\}$ and p_x represents the xth value in P. The obtained $P' \in \mathbb{R}^{N \times 1}$ and $Y \in \mathbb{R}^{N \times d_{out}}$ are concatenated to generate $Y' \in \mathbb{R}^{N \times (d_{out}+1)}$ which is fed into the marginal loss function the same as CapsNet [14]:

$$L_k = T_k max(0, m^+ - \|v_k\|)^2 + \lambda(1 - T_k)max(0, \|v_k\| - m^-)^2 \tag{10}$$

where T_k is set to 1 if the kth class is the true class and 0 otherwise, m^+, m^- and λ are the hyper-parameters set to 0.9, 0.1 and 0.5 respectively.

The loss function is to enlarge the length of the true class's capsule vector, while suppressing the length of other capsules. And the capsule vector with the largest length is predicted as the true class. Intuitively, because the classification prediction scores of the MBFF module is taken as the $(d_{out} + 1)^{th}$ dimension of the capsules, the highest prediction score of MBFF can effectively increase the length of the capsule, thus enhancing the probability of being predicted as the true class.

3 Experiments

3.1 Implementation

The proposed GFCN with GraMPA is implemented by Pytorch. The parameters of the class capsule layer are set as follows: capsule tenor, heads and capsule dimensions are 4, 4 and 16, respectively, and each graph in GraMPA routing has 64 nodes. Following the setup of Self-Routing [8] and GraCapsNets [7], we use ResNet20 [9] as the backbone, divided into three blocks with six convolution layers per block. We experiment with the classification performance of our method on two benchmarks CIFAR10 [11] and SVHN [13], using random crop, horizontal flip and normalization as data preprocessing. Moreover, the MINIST [12] dataset is utilized to test the reconstruction performance.

3.2 Classification Performance

Comparison with Various CapsNet Architecture. We test the classification performance of our architecture by comparing it with some state-of-the-art routing algorithms. Resnet20 [9] is used as the backbone in all routing methods, whose last two layers, average pooling and fully connected layer, are replaced by a primary capsule layer (PrimaryCaps) and class capsule layer (ClassCaps). Additionally, we compare the capsule network with two CNN variants: (1) original ResNet20 denoted as Res+Avg+FC; (2) replace the final average pooling and FC layer of ResNet20 with a convolutional layer and an FC layer, stated as Res+Conv+FC. We also evaluate the classification performance of only using GraMPA routing and only using global fusion capsule network architecture to test the effect of the MBFF module and MBA module, and the results are shown in Table 1.

As shown in the Table 1, when only adopting GraMPA routing without MBA and MBFF module, our algorithm achieves the highest classification performance

Table 1. Comparison of parameter counts (M), FLOPs (M), and error rates (%) between various state-of-the-art routing algorithms and CNNs models on the benchmark datasets. DR, EM, SR and GCN denote dynamic routing [14], EM routing [10], self-routing [8] and GraCapsNets [7] respectively. Res+GraMPA represents that the Graph Routing Based on Multi-head Pairwise-relation Attention is built on the ResNet-20 backbone, while Res+GraMPA+MBFF means that MBFF module is also adopted without MBA module. Res+GraMPA+MBA+MBFF represents that we use the proposed overall model.

Method	# Param.(M)	# FLOPs(M)	CIFAR10	SVHN
Res+Avg+FC	0.27	41.3	7.94	3.55
Res+Conv+FC	0.89	61.0	10.01	3.98
Res+DR(NIPS 2017) [14]	5.81	73.5	8.46	3.49
Res+EM(ICLR 2018) [10]	0.91	76.6	10.25	3.85
Res+SR(NIPS 2019) [8]	0.93	62.2	8.17	3.34
Res+GCN(AAAI 2021) [7]	0.28	59.6	7.99	2.98
Res+GraMPA	0.37	**51.6**	7.53	2.70
Res+GraMPA+MBFF	0.61	53.3	7.49	2.63
Res+GraMPA+MBA+MBFF	0.64	54.5	**7.25**	**2.57**

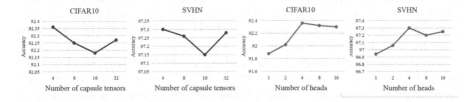

Fig. 2. Ablation study. The left two charts show the accuracy of CIFAR10 and SVHN datasets when adjusting the number of capsule tensors. The right two charts demonstrate the accuracy when modifying the number of heads in GraMPA routing.

among all routing methods with relatively low parameters and the lowest FLOPs. Compared with CapsNet [14] proposed by Sabour et al., our routing algorithm significantly reduces the number of parameters and improves the accuracy. Moreover, with the help of multi-layer semantic fusion of the MBFF module, the accuracy is further enhanced despite the slight increase of the number of parameters. And the MBA module can amplify the signals of more important blocks, so that the accuracy is further improved.

Ablation Study. In this experiment, we adjust the number of capsule tensors extracted by the primary capsule layer, and also modify the number of heads in multi-head pairwise-relation attention to evaluate the sensitivity of our model to parameters. The experimental results in the Fig. 2 show that the best classification performance can be achieved when the number of capsule tenors and heads

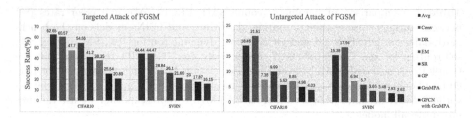

Fig. 3. Success rates (%) of untargeted and targeted FGSM attacks against different routing algorithms and CNN models.

are both set to 4. And in general, the accuracy increases with the number of heads, which shows the effectiveness of the multi-head mechanism. Overall, our routing algorithm is not sensitive to parameters since it maintains high accuracy under different parameter settings.

3.3 Robustness to Adversarial Examples

Adversarial examples refer to the input sample formed by deliberately adding slight interference to the dataset, which results in the model training a wrong output with high confidence. The capsule network has been proved to be more robust to the adversarial examples [14], so we use the targeted and untargeted white-box Fast Gradient Sign Method (FGSM) [6] to generate the adversarial examples to evaluate the robustness of our model. The lower success rate of the attack signifies the stronger robustness of the model. The purpose of targeted attacks is to mislead the model to classify the input into the specified class, while the untargeted attacks aim to misguide the model to obtain wrong and unspecified classification results. In FGSM, the disturbance coefficient is expressed as ϵ, which is set to 0.1 in the experiment.

As shown in Fig. 3, both targeted attack and untargeted attack achieve the lowest success rate in GraMPA routing algorithm among all routing methods and CNN variants, which proves the strongest robustness of GraMPA routing. The reason is that multi-head pairwise-relation attention mine the correlation between capsules more effectively to achieve better clustering results. Furthermore, the global information fusion mechanism uses multi-layer semantic information to assist classification decisions, thus further resisting attacks.

3.4 Visualization

We use the Grad-CAM method [15] to visualize the features learned by each block in the CIFAR10 dataset, as shown in the Fig. 4. For each class, our model can learn interpretable and obvious feature information, which shows the effectiveness of our model. Additionally, it can be observed that the features learned by low-level blocks are more specific and detailed, such as edge and blob, while the features learned by high-level blocks are more abstract and global. For example, the high-level semantic information uses frog legs as the main feature to

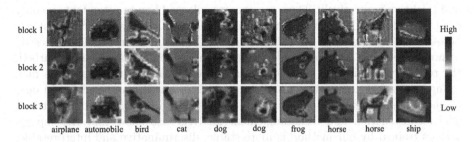

Fig. 4. The visualization of features learned by all blocks generated by Grad-CAM method on CIFAR10 dataset. It is shown that the features from the low-level to the high-level gradually change from specific and detailed to abstract and global.

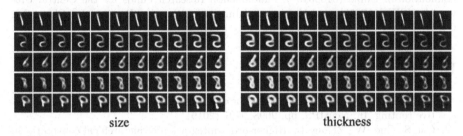

Fig. 5. Reconstruction of images in MINIST dataset generated by perturbing one dimension of final capsule vectors. The left and right images show the reconstruction results of perturbing the 5th and 15th dimensions of the output capsules respectively, corresponding to the change in size and thickness.

judge its category, but pays little attention to its edge and other body details. Therefore, it is likely to mislead the model's classification decisions if the frog legs are included in the non-frog images to generate adversarial examples to attack the model. As a result, the fusion of multi-layer semantics can use detailed low-level features to supplement the abstract and high-layer information, so that the model pays attention to both the global features and the details of the object, and it can effectively enhance the robustness of the model.

3.5 Reconstruction and disentangled Representations

We reconstruct the input image of MINIST dataset by using the capsule vectors trained by the model as the input of the decoder, which is implemented as two fully connected layers the same as CapsNets [14]. When the value of a certain dimension of these capsule vectors is disturbed, a certain characteristic of the reconstruction results can be changed. In other words, our algorithm can realize the correspondence between the disturbed dimension of the capsule vector and the changed reconstruction feature. As shown in the Fig. 5, when we tweak the 5th and 15th dimensions of the output capsule by intervals of 0.1 in the range $[-0.5, 0.5]$, the reconstruction results of various classes can be adjusted in size and thickness respectively. Thus, to generate enlarged handwritten font, we only need to modify the 5th dimension of the capsule, and bold handwritten font can be obtained by tweaking the 5th dimension. On the contrary, some studies [8] demonstrated that in CapsNets [14], if one dimension of a capsule is disturbed, the reconstructed images of different categories of handwritten numerals may change in size, style, rotation and so on, which means it cannot achieve a one-to-one correspondence between capsule dimensions and reconstruction features. The reconstruction and disentanglement analysis also show that the capsule vectors trained by our method is more stable, discriminative and interpretable, where various dimensions represent different features of the image.

Acknowledgements. This work was supported by the National Natural Science Foundation of China (61806168), Fundamental Research Funds for the Central Universities (SWU117059), and Venture and Innovation Support Program for Chongqing Overseas Returnees (CX2018075).

References

1. Ahmed, K., Torresani, L.: Star-Caps: capsule networks with straight-through attentive routing. In: NeurIPS, pp. 9098–9107 (2019)
2. Cai, S., Zuo, W., Zhang, L.: Higher-order integration of hierarchical convolutional activations for fine-grained visual categorization. In: ICCV, pp. 511–520. IEEE Computer Society (2017)
3. Cao, Y., Xu, J., Lin, S., Wei, F., Hu, H.: GCNet: non-local networks meet squeeze-excitation networks and beyond. In: ICCV Workshops, pp. 1971–1980. IEEE (2019)
4. Chang, X., Hospedales, T.M., Xiang, T.: Multi-level factorisation net for person re-identification. In: CVPR, pp. 2109–2118. IEEE Computer Society (2018)
5. Durand, T., Mordan, T., Thome, N., Cord, M.: WILDCAT: weakly supervised learning of deep convnets for image classification, pointwise localization and segmentation. In: CVPR, pp. 5957–5966. IEEE Computer Society (2017)
6. Goodfellow, I.J., Shlens, J., Szegedy, C.: Explaining and harnessing adversarial examples. In: ICLR (Poster) (2015)
7. Gu, J., Tresp, V.: Interpretable graph capsule networks for object recognition. CoRR abs/2012.01674 (2020)
8. Hahn, T., Pyeon, M., Kim, G.: Self-routing capsule networks. In: NeurIPS, pp. 7656–7665 (2019)
9. He, K., Zhang, X., Ren, S., Sun, J.: Deep residual learning for image recognition. In: CVPR, pp. 770–778. IEEE Computer Society (2016)
10. Hinton, G.E., Sabour, S., Frosst, N.: Matrix capsules with EM routing. In: ICLR (Poster). OpenReview.net (2018)
11. Krizhevsky, A., Hinton, G., et al.: Learning multiple layers of features from tiny images (2009)
12. LeCun, Y., Bottou, L., Bengio, Y., Haffner, P.: Gradient-based learning applied to document recognition. Proc. IEEE **86**(11), 2278–2324 (1998)
13. Netzer, Y., Wang, T., Coates, A., Bissacco, A., Wu, B., Ng, A.Y.: Reading digits in natural images with unsupervised feature learning (2011)
14. Sabour, S., Frosst, N., Hinton, G.E.: Dynamic routing between capsules. In: NIPS, pp. 3856–3866 (2017)

15. Selvaraju, R.R., Cogswell, M., Das, A., Vedantam, R., Parikh, D., Batra, D.: Grad-CAM: visual explanations from deep networks via gradient-based localization. In: ICCV, pp. 618–626. IEEE Computer Society (2017)
16. Szegedy, C., et al.: Intriguing properties of neural networks. In: ICLR (Poster) (2014)
17. Tsai, Y.H., Srivastava, N., Goh, H., Salakhutdinov, R.: Capsules with inverted dot-product attention routing. In: ICLR. OpenReview.net (2020)
18. Vaswani, A., et al.: Attention is all you need. In: NIPS, pp. 5998–6008 (2017)
19. Verma, S., Zhang, Z.: Graph capsule convolutional neural networks. CoRR abs/1805.08090 (2018)
20. Wang, H., Wang, S., Qin, Z., Zhang, Y., Li, R., Xia, Y.: Triple attention learning for classification of 14 thoracic diseases using chest radiography. Med. Image Anal. **67**, 101846 (2021)
21. Wang, X., Girshick, R.B., Gupta, A., He, K.: Non-local neural networks. In: CVPR, pp. 7794–7803. IEEE Computer Society (2018)
22. Wang, X., Zou, X., Bakker, E.M., Wu, S.: Self-constraining and attention-based hashing network for bit-scalable cross-modal retrieval. Neurocomputing **400**, 255–271 (2020). https://doi.org/10.1016/j.neucom.2020.03.019. https://www.sciencedirect.com/science/article/pii/S0925231220303544
23. Woo, S., Park, J., Lee, J.-Y., Kweon, I.S.: CBAM: convolutional block attention module. In: Ferrari, V., Hebert, M., Sminchisescu, C., Weiss, Y. (eds.) ECCV 2018. LNCS, vol. 11211, pp. 3–19. Springer, Cham (2018). https://doi.org/10.1007/978-3-030-01234-2_1
24. Wu, S., Oerlemans, A., Bakker, E.M., Lew, M.S.: Deep binary codes for large scale image retrieval. Neurocomputing **257**, 5–15 (2017). Machine Learning and Signal Processing for Big Multimedia Analysis. https://doi.org/10.1016/j.neucom.2016.12.070. https://www.sciencedirect.com/science/article/pii/S0925231217301455
25. Xinyi, Z., Chen, L.: Capsule graph neural network. In: ICLR (Poster), OpenReview.net (2019)
26. Yang, S., Ramanan, D.: Multi-scale recognition with DAG-CNNs. In: ICCV, pp. 1215–1223. IEEE Computer Society (2015)
27. Yu, W., Yang, K., Yao, H., Sun, X., Xu, P.: Exploiting the complementary strengths of multi-layer CNN features for image retrieval. Neurocomputing **237**, 235–241 (2017)
28. Yu, Z., Dou, Z., Bakker, E.M., Wu, S.: Self-supervised asymmetric deep hashing with margin-scalable constraint for image retrieval (2021)
29. Zhang, L., Edraki, M., Qi, G.: CapProNet: deep feature learning via orthogonal projections onto capsule subspaces. In: NeurIPS, pp. 5819–5828 (2018)
30. Zhang, Z., Lan, C., Zeng, W., Jin, X., Chen, Z.: Relation-aware global attention for person re-identification. In: CVPR, pp. 3183–3192. IEEE (2020)
31. Zou, X., Wang, X., Bakker, E.M., Wu, S.: Multi-label semantics preserving based deep cross-modal hashing. Sig. Process. Image Commun. **93**, 116131 (2021). https://doi.org/10.1016/j.image.2020.116131. https://www.sciencedirect.com/science/article/pii/S0923596520302344

MA-GAN: A Method Based on Generative Adversarial Network for Calligraphy Morphing

Jiaxin Zhao, Yang Zhang, Xiaohu Ma$^{(\boxtimes)}$, Dongdong Yang, Yao Shen, and Hualiang Jiang

School of Computer Science and Technology, Soochow University, Suzhou 215006, China
xhma@suda.edu.cn

Abstract. Some applications can be based on the image transfer method in the development of contemporary related fields, which can make the incomplete Chinese font complete. These methods have the ground truth and train on paired calligraphy image from different fonts. We propose a new method called mode averaging generative adversarial network (MA-GAN) in order to generate some variations of a single Chinese character. As the data set of calligraphy font creation through samples of a given single character is usually small, it is not suitable to use conventional generative models. Therefore, we designed a special pyramid generative adversarial network, using the weighted mean loss in the loss function to average the image and the adversarial loss to correct the topology. The pyramid structure allows the generator to control the rendering of the topology at low resolution layers, while supplementing and changing details at high resolution layers without destroying the image topology resulting from the diversity requirements. We compared MA-GAN with other generation models and proved its good performance in the task of creating calligraphy fonts based on the small data sets.

Keywords: Machine learning · Generative adversarial network · Calligraphy morphing · Mode averaging

1 Introduction

Chinese calligraphy is a very unique visual art from ancient China, and it is also a form of Chinese cultural expression loved by many people around the world. People prefer to use personalized fonts when designing posters and logos than the standard fonts used. However, most of the time it is difficult for us to find a type designers and describe in detail the fonts we need. With the development of generative adversarial network [1], there have been many researches on the

---

Supported by Project Funded by the Priority Academic Program Development of Jiangsu Higher Education Institutions.

T. Mantoro et al. (Eds.): ICONIP 2021, LNCS 13108, pp. 266–278, 2021.
https://doi.org/10.1007/978-3-030-92185-9_22

generation of Chinese calligraphy fonts such as CalliGAN [2], SCFont [3], the method proposed by Zhang [4], the method proposed by Lyu [5] and others [2,6–8]. These methods are based on transfer model, when training at least two paired font sets must be provided, when generating only specific original font images can be used to transfer to corresponding trained fonts. The approaches adopted in the above mentioned studies prefer to generating missing characters in a font by a trained correspondence from a standard font to target font (e.g. SimSun).

In this paper, we propose a generative model, rather than a transfer model to generate some variants of a single character by given a set of character images extracted from different fonts. Due to the data set of a single character is usually very small, it may only include dozens of samples. Although it is similar to the normal generation against network training, we add restrictions to prevent it from over-fitting like a normal one-stage generative model such as DCGAN [9], LSGAN [10] even a multi-stage network like LAPGAN [11]. Figure 1 describes the entire training and generation process.

The main contributions of this work are as follows.

(1) We use the mean square error loss of the images generated by the generator and the images in the data set to obtain an average mode. The adversarial loss of the generative adversarial network is used to maintain the topological structure of the image output by the generator.
(2) We propose the Mode Averaging Generative Adversarial Network (MA-GAN), which has a generator (tower generator) with a pyramid structure and a corresponding training method for the model.
(3) We conduct related experiments to demonstrate the feasibility of the generator structure and the supporting training method applied to the average task of the model.

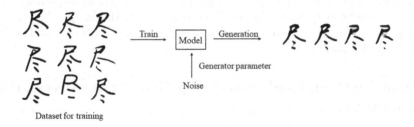

Fig. 1. The entire process of our method

2   Related Work

2.1   Laplacian Pyramid of Adversarial Networks

Due to images are complex and high dimensional, it is hard to build a model well, despite extensive efforts. Considering the multi-scale structure of natural images, Emily et al. proposed a pyramid generative model to synthesize high-resolution image, the entire model has a series of sub-models, and each sub-model uses the Laplacian pyramid [11] to capture image structures of different scales, and finally generates a target image together.

To train this model, we have to prepare a series of low-resolution down-sampled images $I_1...I_k$ for the origin images I_0 in the training set. Then we build $k+1$ generators to process images from low-resolution to high-resolution $G_k...G_0$ and paired discriminators $D_k...D_0$. Each generator G_i except the first level G_k accepts a random vector $\mathbf{z}_i$ and a low-pass image from the previous level and output a synthetic high-pass feature image $\hat{I}_i$, and the first generator output a complete image not a high-pass feature image. In detail, the generation of the complete image $\tilde{I}_i$ synthesized each layer follows the process described in Eq. 1.

$$\tilde{I}_i = \tilde{I}_{i+1} + G_i(\mathbf{z_i}, \tilde{I}_{i+1}) \quad i \neq k$$
$$\tilde{I}_i = G_i(\mathbf{z_i}) \quad i = k \tag{1}$$

2.2 Markov Discriminator (Patch GAN)

Markov discriminator [12] is a kind of discriminant model. There are many types of CNN-based classification models. Many networks introduce a fully connected layer at the end, and then output the results of the discrimination. The Markov composed of convolutional layers, the final output is an $n \times n$ matrix, the mean value of the output feature matrix is taken as the actual result. In fact, each value in the output matrix represents a receptive field in the original image, corresponding to a patch of the original image, so a GAN with this kind of discriminator is also named Patch GAN.

Markov discriminator has some advantages especially in style transfer and high-resolution image generation. Markov discriminator in the generative adversarial network can make the generated high-resolution images more clearer, while it is also beneficial to maintain the texture details of the image. At present, Markov discriminators are often used in GAN networks such as Pix2Pix [12] and CycleGAN [13].

3 Our Method: Mode Averaging Generative Adversarial Network

We propose a mode averaging algorithm to solve the problem that over-fitting and failure of conventional generative adversarial networks when training by a data set with small samples. We use a pyramid generator model, but adjust the proportion of low-pass images of each layer. At the same time, in order to adapt to the multi-level generator structure, the Markov discriminator we use does not have a feature map size transformation.

3.1 The Algorithm of Mode Averaging

The main idea of the algorithm is optimizing the generator through the mean square loss of the image generated by the generator and the image in the data set to obtain an average pattern, and using the adversarial loss of the network to maintain the topology of the image output by the generator. Since the output of the generator is limited by the above loss, there should be only one output "mode" (topological structure) or some relatively similar output "modes" (topological structure), the output distribution p_G of the generator should be part of the training set distribution p_{data}, which means $p_G \subseteq p_{data}$.

Therefore, we can use the generator's output distribution p_G to define an average pattern generator that maintains local diversity (generating the training target of the adversarial network): Given a set of input patterns $X = \{x_1, x_2, ..., x_n\}$ and the weight set corresponding to each input $A = \{a_1, a_2, ..., a_n\}$, when the training is completed, the output distribution p_G of the generator G with the sample $G(\mathbf{z})$ therein, and the sample t in the distribution $p_{data} - p_G$, at this time p_G should satisfy Eq. 2.

$$p_G : \forall G(\mathbf{z}) \in p_G \subseteq p_{data}, t \in p_{data} - p_G$$
$$\sum_{x_i \in X} a_i \|G(\mathbf{z}) - x_i\|_2^2 < \sum_{x_i \in X} a_i \|t - x_i\|_2^2 \tag{2}$$

Among them, the first line of Eq. 2 expresses the training goal of generative adversarial network, and the output of the generator will be regarded as a real image by the discriminator, which maintains the topology structure. The second line of Eq. 2 expresses the goal of mode averaging, which leads the output images (modes) closest to the weighted average of the input mode under the premise of maintaining the topology.

3.2 Network Structure

Overview. Our model is built with full convolution, it can adapt to images of various resolutions as input and output. When introducing the model's related structure, 64×64 images are used as the input and output of the model.

Due to it is a pyramid generator, naturally the resolution of the image (and feature map) processed at each level is different. For the model in this article, we set the down-sampling factor to 0.8, the minimum processing resolution is set to 25×25.

Taking training and outputting 64×64 images as an example, it can be calculated that the generator contains a total of 5 levels, processing 25×25, 32×32, 40×40, 51×51, and 64×64 images respectively. Then the overall generator can be constructed according to the model shown in Fig. 2.

Following the model structure in Fig. 2, when we train the first layer, the model activates the parameters of G_1 and D_1 and optimizes it. When we train the second layer, the model activates the parameters of G_1, G_2, and G_2, but fixing the parameters of G_1, only optimizes the parameters of G_2 and G_2, and so on, when training the layer i, the model will activate the parameters of G_1 G_i and D_i, but only optimize the parameters of G_i and D_1. After training, only the generator parameters of all levels need to be saved.

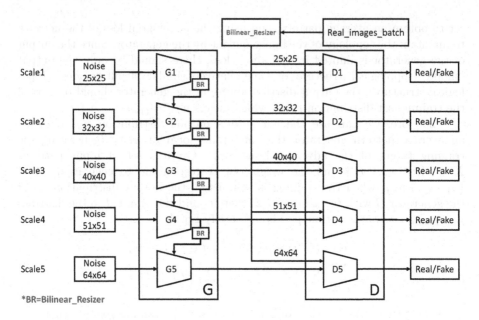

Fig. 2. The overall structure diagram of the model with 64×64 image

Generator of Each Layer. The independent generator of each level only needs to input the random feature map and the image generated by the previous level to obtain the image output of this level. In this regard, each level generator of the model in this paper has the same convolutional layer parameters except for the different size of the feature map processed and the different factor of the residual connection.

A unified convolution module is designed [15] since the generators and discriminators of each layer do not have the scale transformation of the feature map, as shown in Fig. 3(a).

The generator structure of each layer in this paper is shown in Fig. 3(b). Among them, conv-block is the module described in Fig. 3(a), and conv-2d only performs convolution operations. If φ represents the process from the first conv-block to the final $tanh$ in the current generator, $\hat{x}_i$ represents the image generated by the layer generator, ε represents the noise amplification rate, $\mathbf{z}_i$ represents the random noise feature of the layer, and br() represents the billinear interpolation, assuming that the generator has a total of i layers, the data flow of each level of the generator can be expressed by Eq. 3.

$$\hat{x}_i = \begin{cases} \varphi_i(\mathbf{z}_i) & i = 1 \\ \varphi_i[\varepsilon \cdot \mathbf{z}_i + \text{br}(\hat{x}_{i-1})] + \text{br}(\hat{x}_{i-1}) & i > 1 \,\&\, i < n \\ \varphi_i[\varepsilon \cdot \mathbf{z}_i + \text{br}(\hat{x}_{i-1})] + 0.25 \cdot \text{br}(\hat{x}_{i-1}) & i = n \end{cases} \tag{3}$$

The initial layer only needs to generate images based on the random noise map. In the last layer of the generator, we reduce the reference of the final generation layer to the previous layer, so that the output of the final generation layer can pay more attention to the transformation of the random noise map to describe the varieties in image details more, and as the result, the diversity of results can be increased.

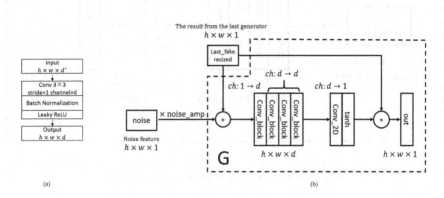

Fig. 3. (a) A unified convolution module (b) The generator structure of each layer

Discriminator of Each Layer. We use Markov discriminator in this paper, so the final output of discriminator should be the mean value of the output feature map (single channel). The discriminators of each layer in this article still use the convolution module in Fig. 3(a), and the discriminator structure of each layer is shown in Fig. 4.

Here the output of the discriminator is a feature map of scale $h \times h \times 1$. The value applied to the objective function is the mean value of this feature map. We put the process of averaging outside the discriminator, when calculating the loss function, the actual output value R of this feature map F is calculated as Eq. 4.

$$R = \frac{1}{h \cdot w} \sum_{i=1}^{h} \sum_{j=1}^{w} F(i,j) \tag{4}$$

3.3 Loss Function

In this paper, we use Eq. 5 to describe the generate loss. In the equation, $\hat{X}_i$ represents a batch images generated by G_i, and X_i represents a batch images with the same scale with $\hat{X}_i$ from real data set.

$$\text{Loss}G_i = -D_i(\hat{X}_i) + \alpha \cdot \text{rec}(\hat{X}_i, X_i) \tag{5}$$

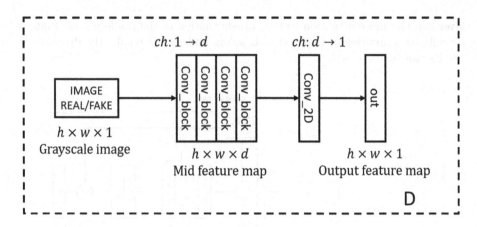

Fig. 4. The generator structure of each layer

Among them, rec() represents the consistency loss function. The single input is a batch of images generated by the generator and a batch of images in the training set. The training set images will be disrupted during the training process, the final optimization result is to generate the fake image with the smallest sum of consistency loss of all images in the same training set. The loss is finally the mean value of the sum of rec() loss of each image in a batch. For a single image $\hat{x} \in \hat{X}_i$ and $x \in X_i$, the loss consists of two parts, one part is the mean square loss between the generated image and the real image (Eq. 6), the other part is the edge structural loss of the generated image and the real image (SSIM, Eq. 7). The purpose of using edge features and calculating structural loss is to reduce the ghosting of strokes.

In the Eq. 7, $\text{Sobel}_{dx}(X)$ represents the image after the action of the sobel operator in the x direction; $\text{Sobel}_{dy}(X)$ represents the image after the action of the sobel operator in the y direction. Since the range of SSIM structure similarity value is $[0, 1]$, 1 means that the two image structures are completely the same, and 0 means that the two image structures are completely inconsistent. Therefore, we define the rec() loss of a single image according to Eq. 8. α in Eq. 5 is a pre-set hyper-parameter, which is set to 10 in this model. The value of a in Eq. 5 is set 5 for the last layer of the model. The hyper-parameter β in Eq. 8 is used to control the strength of the edge structure loss, which can be manually specified according to the number of the specific Chinese characters to be trained. The recommended value range is $[0, 1]$, or the empirical formula $\beta = 1 - \text{sigmoid}[0.2(N - 10)]$ can be used to determine, where N is the number of the samples in the training set.

$$\text{MSE}(\hat{x}, x) = \|\hat{x} - x\|_2^2 \tag{6}$$

$$\underset{EDGE}{\text{SSIM}}(\hat{x}, x) = \frac{\text{SSIM}[\text{Sobel}_{dx}(\hat{x}), \text{Sobel}_{dx}(x)] + \text{SSIM}[\text{Sobel}_{dy}(\hat{x}), \text{Sobel}_{dy}(x)]}{2} \tag{7}$$

$$\text{rec}(\hat{x}, x) = \text{MSE}(\hat{x}, x) + \beta \cdot [1 - \underset{EDGE}{\text{SSIM}}(\hat{x}, x)] \qquad (8)$$

If $\hat{X}_i$ represents a batch of fake images generated at the level i, and X_i represents the a batch of real images zoomed to the level i, the loss of the discriminator D_i can be expressed by Eq. 9, and the $\text{GP}(X)$ in the equation means that this model uses the method named gradient penalty [14] to optimize the network.

$$\text{Loss}D_i = D_i(\hat{X}_i) - D_i(X_i) + \text{GP}(\hat{X}_i, X_i) \qquad (9)$$

4   Experiment

4.1   Character Synthesis

In this section, we select multiple different Chinese characters for generation experiment. For each Chinese character, 16 different font samples are selected as input to obtain an average mode. After the model training has completed, use the models of each Chinese character to generate multiple times, observe the consistency of the topological structure of each generated picture and the diversity shown in the local details. We aim to generate the phrase "wildfire burns endlessly, spring breeze blows again" as the goal, and conducts generation experiments. The results are shown in Fig. 5.

Fig. 5. The result of character synthesis experiment (Color figure online)

The red box in the above figure shows the local diversity of the image. The basic consistency of the stroke skeleton can also be observed from the figure. We calculated the average SSIM of each Chinese character in Fig. 5 and made statistics. The results are shown in Table 1. After maintaining the consistency of the basic structure, a completely consistent image is not generated but some changes are still retained, and the original goal of the task is better achieved.

Table 1. SSIM mean value of the images generated by each Chinese character

Character	YE	HUO	SHAO	BU	JING	CHUN	FENG	CHUI	YOU	SHENG
SSIM mean value	0.79	0.82	0.77	0.88	0.84	0.75	0.73	0.74	0.8	0.86

4.2   Character Synthesis with Specific Weight

When using non-equivalent weights, we select a Chinese character and set the weight of one character in its training set to 1, 2, 3, and 4 respectively, and keep the weight of other (characters)modes as 1, and observe the average results of the modes under the specific weight mode. The experimental results are shown in Fig. 6.

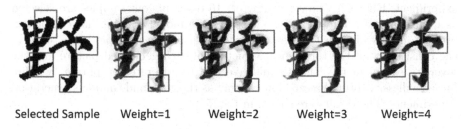

Selected Sample   Weight=1   Weight=2   Weight=3   Weight=4

Fig. 6. The result of character synthesis with different weight (Color figure online)

It can be clearly observed from the experimental results that increasing the weight of a certain sample can make the result generated by the final generator closer to that sample, while the generated result still maintains local diversity. In this experiment, we calculated the generated results under each weight and the SSIM value of the selected sample. The results are shown in Table 2.

Table 2. SSIM value between the selected sample and synthesized result with different weight

Weight	1(default)	2	3	4
SSIM value	0.57	0.65	0.71	0.81

4.3 The Effect of Noise Amplification Rate on Generation

It can be seen from the experiment on noise amplification rate in Fig. 7 under that the lower the noise amplification rate, the clearer image generated by the sample, but it will reduce the diversity of the generation. When the noise rate increases, the diversity of the output image increases, however blurring and artifacts of strokes are more likely to occur. Therefore, the setting of noise amplification rate is usually selected according to the different Chinese characters that need to be created, and generally set in the range of $[0.05, 0.15]$. The average SSIM structure similarity results are shown in Table 3.

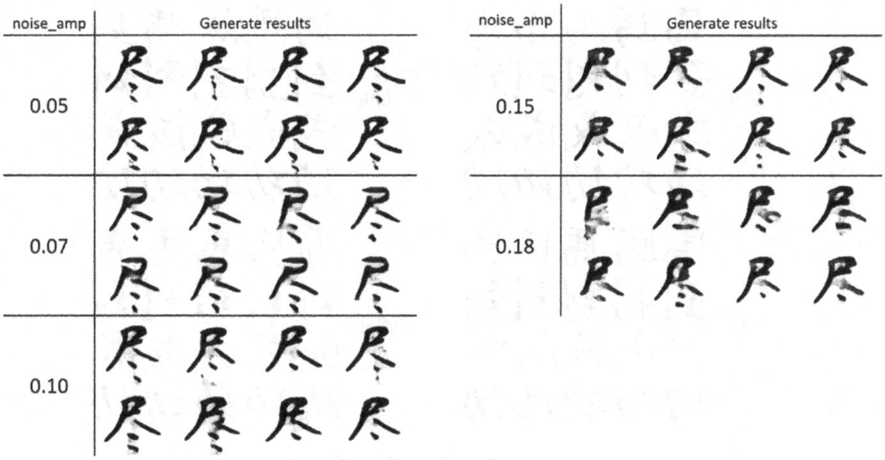

Fig. 7. The result with different noise amplification rate

Table 3. SSIM mean value between the synthesized result with different noise amp

Noise amp	0.05	0.07	0.10	0.15	0.18
SSIM mean value	0.6032	0.5646	0.5124	0.4735	0.4382

4.4 Compare with Other Generative Model

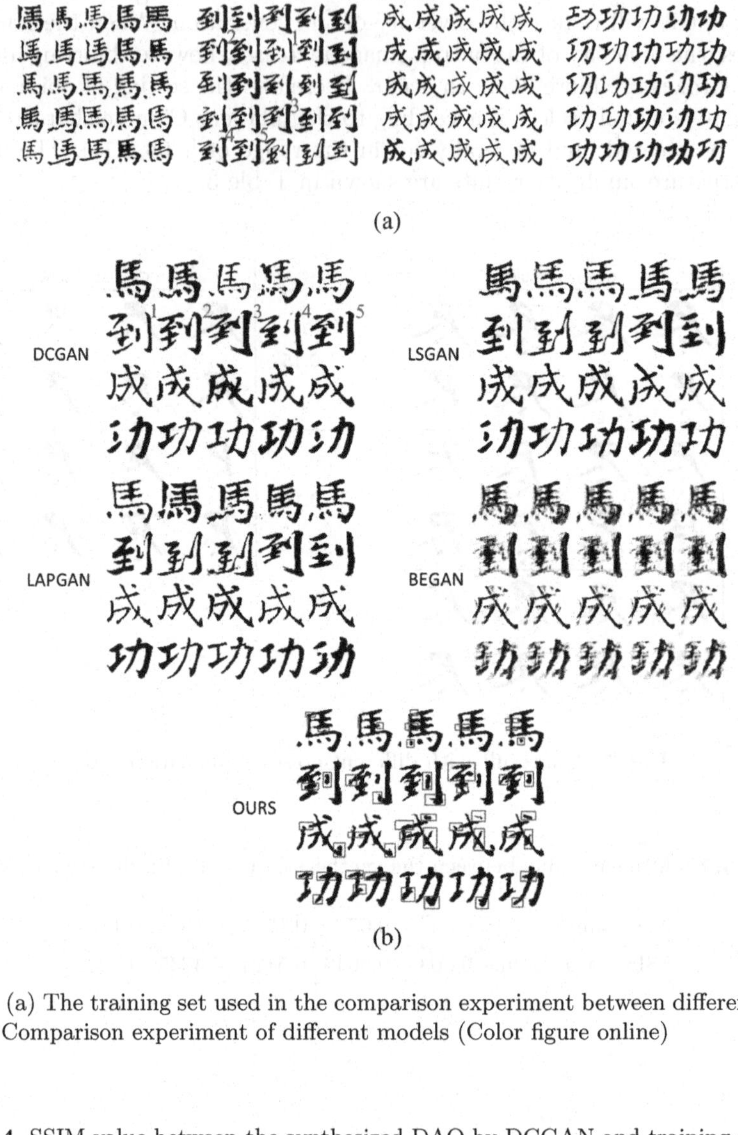

(a)

(b)

Fig. 8. (a) The training set used in the comparison experiment between different models (b) Comparison experiment of different models (Color figure online)

Table 4. SSIM value between the synthesized DAO by DCGAN and training samples

No	1	2	3	4	5
SSIM value	0.9994	0.9992	0.9974	0.9829	0.9988

We select four Chinese characters "MA DAO CHENG GONG" and set the weight of each input mode to 1 concurrently. When the model training has completed, we generate each word 5 times and observe the experimental results. Figure 8(a) is the training set used in this experiment.

Figure 8(b) is image generation result of this experiment. When comparing the generated results with the training set used, it can be found that except for our model and BEGAN in this chapter, the results generated by the rest of the models are close to a specific sample, and even over-fitting, this means the generated results almost "copy" the training sample in the training set, take the word "DAO" generated by DCGAN as an example. Pay attention to the numbers 1 to 5 in Fig. 8(a) and Fig. 8(b). The SSIM value between the generated sample and the most similar sample in the training set are shown in Table 4. Although BEGAN can perform a certain image averaging under this number of samples, it is far inferior to the model in this paper in terms of image generation results, and its generation results cannot show diversity locally. It can also be seen from the experimental results that the generated results of the model proposed in this article are not too close to a specific sample, and the results generated by using different random variables as model inputs are basically consistent in structure, which can be observed at the stroke level. To the difference, which also shows that the goal of this model is to generate an average style, and to retain a certain amount of this style variety.

5   Conclusion

In order to accomplish the task of synthesizing calligraphy characters, we proposed a concept of the mode average, and designed a pyramid generator model and corresponding objective function to complete this target. The pyramid model places the up-sampling process outside the generator, and modifies the conventional feature map up-sampling to the up-sampling of the image generated by each layer of the generator. So that it can still keep the generated image relatively clear, and to a certain extent solve the problem that the generated image is relatively fuzzy when the mean square error loss is used to train the generative model.

The model we proposed can be trained with a small data set to obtain the average mode, while the conventional generative adversarial network will have the phenomenon of "copying" the sample over-fitting. We selected some data sets suitable for the task of synthesizing calligraphy for comparative experiments. We explained the functions and some hyper-parameters in the model by experiments. At the same time, we also verified the mode average function of the model in this paper. Finally we compare the difference between the model designed in this paper and the general generative adversarial network in terms of image generation with small data set.

References

1. Goodfellow, I.J., et al.: Generative adversarial networks. arXiv preprint arXiv: 1406.2661 (10 June 2014)
2. Gao, Y., Wu, J.: CalliGAN: unpaired mutli-chirography Chinese calligraphy image translation. In: Jawahar, C.V., Li, H., Mori, G., Schindler, K. (eds.) ACCV 2018. LNCS, vol. 11362, pp. 334–348. Springer, Cham (2019). https://doi.org/10.1007/978-3-030-20890-5_22
3. Jiang, Y., Lian, Z., Tang, Y., Xiao, J.: Scfont: structure-guided Chinese font generation via deep stacked networks. In: Proceedings of the AAAI Conference on Artificial Intelligence 2019, Honolulu, pp 4015–4022. AAAI (2019)
4. Zhang, G., Huang, W., Chen, R., et al.: Calligraphy fonts generation based on generative adversarial networks. ICIC Express Lett. Part B Appl. Int. J. Res. Surv. **10**(3), 203–209 (2019)
5. Lyu, P., Bai, X., Yao, C., Zhu, Z., Huang, T., Liu, W.: Auto-encoder guided GAN for Chinese calligraphy synthesis. In: 2017 14th IAPR International Conference on Document Analysis and Recognition (ICDAR), Kyoto, pp 1095–1100. IEEE (2017)
6. Zong, A., Zhu, Y.: Strokebank: automating personalized Chinese handwriting generation. In: Twenty-Sixth IAAI Conference 2014, Quebec, pp 3024–3029. AAAI (2014)
7. Li, M., Wang, J., Yang, Y., Huang, W., Du, W.: Improving GAN-based calligraphy character generation using graph matching. In: 2019 IEEE 19th International Conference on Software Quality, Reliability and Security Companion (QRS-C), Sofia, pp 291–295. IEEE (2019)
8. Sun, D., Ren, T., Li, C., Su, H., Zhu, J.: Learning to write stylized Chinese characters by reading a handful of examples. In: 2018 International Joint Conference on Artificial Intelligence, Stockholm, pp 920–927. IJCAI (2018)
9. Radford, A., Metz, L., Chintala, S.: Unsupervised representation learning with deep convolutional generative adversarial networks. arXiv preprint arXiv: 1511.06434. Nov 19 (2015)
10. Mao, X., Li, Q., Xie, H., Lau, R.Y., Wang, Z., Paul Smolley, S.: Least squares generative adversarial networks. In: Proceedings of the IEEE International Conference on Computer Vision, pp. 2794–2802 (2017)
11. Denton, E., Chintala, S., Szlam, A., Fergus, R.: Deep generative image models using a laplacian pyramid of adversarial networks. In: Advances in Neural Information Processing Systems 2015, Cambridge, pp. 1486–1494. MIT Press (2015)
12. Isola, P., Zhu, J.Y., Zhou, T., Efros, A.A.: Image-to-image translation with conditional adversarial networks. In: Proceedings of the IEEE Conference on Computer Vision and Pattern Recognition 2017, Honolulu, pp 1125–1134. IEEE (2017)
13. Zhu, J.Y., Park, T., Isola, P., Efros, A.A.: Unpaired image-to-image translation using cycle-consistent adversarial networks. In: Proceedings of the IEEE International Conference on Computer Vision 2017, Venice, pp. 2223–2232. IEEE (2017)
14. Gulrajani, I., Ahmed, F., Arjovsky, M., Dumoulin, V., Courville, A.: Improved training of wasserstein gans. In: Conference on Neural Information Processing Systems 2017, Long Beach, pp. 5768–5778. MIT Press (2017)
15. He, K., Zhang, X., Ren, S., Sun, J.: Deep residual learning for image recognition. In: Proceedings of the IEEE Conference on Computer Vision and Pattern Recognition 2016, Las Vegas, pp 770–778. IEEE (2016)

One-Stage Open Set Object Detection with Prototype Learning

Yongyu Xiong[1,2], Peipei Yang[1,2(✉)], and Cheng-Lin Liu[1,2,3]

[1] National Laboratory of Pattern Recognition, Institute of Automation,
Chinese Academy of Sciences, Beijing, People's Republic of China
xiongyongyu2019@ia.ac.cn, {ppyang,liucl}@nlpr.ia.ac.cn
[2] School of Artificial Intelligence, University of Chinese Academy of Sciences, Beijing,
People's Republic of China
[3] CAS Center for Excellence of Brain Science and Intelligence Technology, Beijing,
People's Republic of China

Abstract. Convolutional Neural Network (CNN) based object detection has achieved remarkable progress. However, most existing methods work on closed set assumption and can detect only objects of known classes. In real-world scenes, an image may contain unknown-class foreground objects that are unseen in training set but of potential interest, and open set object detection aims at detecting them as foreground, rather than rejecting them as background. A few methods have been proposed for this task, but they suffer from either low speed or unsatisfactory ability of unknown identification. In this paper, we propose a one-stage open set object detection method based on prototype learning. Benefiting from the compact distributions of known classes yielded by prototype learning, our method shows superior performance on identifying objects of both known and unknown classes from images in the open set scenario. It also inherits all advantages of YOLO v3 such as the high inference speed and the ability of multi-scale detection. To evaluate the performance of our method, we conduct experiments with both closed & open set settings, and especially assess the performance of unknown identification using recall and precision of the unknown class. The experimental results show that our method identifies unknown objects better while keeping the accuracy on known classes.

Keywords: Object detection · Open set recognition · Prototype learning

1 Introduction

As a fundamental problem of computer vision, object detection aims at detecting objects of a certain class of interest in images, and has been studied for many years. Recently, the development of convolutional neural network (CNN) leads to remarkable breakthroughs in object detection, and CNN based methods have become the mainstream. They can be roughly grouped into two-stage detection

© Springer Nature Switzerland AG 2021
T. Mantoro et al. (Eds.): ICONIP 2021, LNCS 13108, pp. 279–291, 2021.
https://doi.org/10.1007/978-3-030-92185-9_23

and one-stage detection [19], where the former [5,15] usually gives better accuracy while the latter [11,14] excels in high detection speed. Thanks to all those efforts, object detection has been successfully applied in many areas including auto-driving, robot vision, smart retail supermarkets, etc. [19].

Standard object detection approaches can detect only objects of known classes, assuming that all the foreground classes to be detected should exist in the training set. However, this assumption can be violated by a common scenario in real-world applications: a test image may contain novel foreground objects of unknown classes that never appear in training images but are also of interest. This is the problem of *open set object detection* (OSD). Unfortunately, standard detection algorithms cannot deal with this case, which either neglect novel objects as background, or even identify them as a known class by mistake. To solve this problem, we focus on OSD that is able to detect both objects of known classes and objects of interest beloning to some unknown class.

Open set object detection is closely related to *open set recognition* (OSR) [3], which aims at understanding the real world with incomplete knowledge. Although both of them aim at identifying unknown classes, the OSR methods cannot be directly exploited for OSD due to two important reasons. First, for OSR, any sample outside of known classes is regarded as unknown, while for OSD, there is an extra background class in addition to the known classes that is also available during training but of no interest. Thus, a candidate object rejected by all classes of interest may be either background or unknown-class object. Second, even if we explicitly treat the background as an additional class of interest, it is still difficult for identifying a novel object, because the background has diverse appearance and is easily confused with novel objects which are not trained.

Due to the novelty and difficulty of this topic, only very few works focus on OSD. Dhamija et al. [1] first proposes the concept of OSD and defines several evaluation metrics. A series of experiments are conducted to evaluate the performance of several popular object detection methods under the open set condition using such metrics. Joseph et al. [6] proposes to use the energy model to conduct unknown category discovery and example replay for continual incremental learning. Miller et al. [12] proposes to use dropout sampling to extract label uncertainty for increasing the performance of object detection under open set condition. These works proposed the concept of OSD and some effective ideas to solve this problem, but there are still some important issues to be solved. First, existing methods are mostly designed based on two-stage object detection framework. They first generate a large number of potential foreground boxes and then adopt common open set recognition model to identify unknown objects. This strategy suffers from the drawback of most two-stage detection methods that the speed cannot satisfy the requirement of real-time applications. Second, previous OSD methods focus on avoiding classifying an unknown object as a known class by mistake. In practical open set scenarios, it is also valuable to distinguish novel objects that may be of interest from background.

To overcome the above problems, we propose a novel high-speed open set detection method that is able to correctly detect known objects while identifying unknown-class objects of interest. We first build a base model of open set detection using YOLO v3 [14], which is a representative one-stage detector with obvious higher speed than two-stage methods. By exploiting the objectness score to determine the foreground regions and setting class confidence thresholds for sigmoid outputs, we can identify the novel objects from these foreground regions when the confidence score of each class is below a preset threshold.

To improve the compactness of distribution of known classes in feature space and avoid confusion of unknown objects with known classes, we introduce the prototype classifier into YOLO v3 [14] to explicitly model the feature distribution of objects. The obtained model is trained to make features of each class concentrate in a separated and compact region in the feature space, which makes it simpler to distinguish unknown classes from known classes. Our method benefits from all advantages of YOLO v3 and gives a favorable performance on unknown object detection.

To evaluate our proposed method, we conduct experiments on both closed and open set settings. For the open set test, we divide the PASCAL VOC [2] into two parts according to classes. One part is used as known classes while the other as unknown. While existing works assess the performance of open set detection by counting how many unknown objects are misclassified as known, we adopt a more comprehensive evaluation criterion including recall and precision of the unknown objects. The experimental results show that our method can effectively identify the foreground objects of unknown classes while keeping satisfactory performances on known classes.

The main contributions of this work can be summarized as follows.

- We propose an open set object detection method within the one-stage detection framework. It has obviously higher speed than existing two-stage methods.
- We propose to construct open set object detection by integrating the prototype classifier into the framework. Exploiting the ideal feature distribution generated by the prototype classifier, it performs well on detecting both known and unknown objects.
- We evaluate our proposed method by both closed and open set experiments. We also assess the capability of the algorithm to discover unknown objects from background, which has not been considered before.

2   Related Work

In this section, we will give a brief review on the development of CNN-based object detection methods, followed by recent advances on open set object detection. Then, a short introduction to prototype learning and its application to open set recognition are present.

Object Detection and OSD. Object detection is a fundamental task in computer vision and has been studied for many years. Due to the great progress of deep neural networks in computer vision and pattern recognition, CNN-based object detection has been the mainstream method. They can be generally divided into two-stage and one-stage methods [19]. The former first generates a large number of potential foreground boxes, from which the objects are obtained by classification and bounding box regression. The latter outputs class labels and locations of the objects directly in one shot.

Faster R-CNN [15] is the most representative work on two-stage object detection. It firstly proposes to use the anchor as an object prior to generate class-agnostic foreground regions and then obtain the bounding boxes of the object by classification and regression. After Faster R-CNN [15], there are many works to improve the performance of two-stage object detection, including feature fusion and enhancement [7], etc. Nowadays, the state-of-the-art results are still held by two-stage methods.

With the advent of SSD [11], one-stage object detectors have attracted much attention because of their high computational efficiency. SSD [11] directly uses anchors with different aspect ratios and areas on multi-layer feature maps to generate category predictions and bounding box locations. Due to the real-time efficiency, many efforts have been devoted to future improve the performance of one-stage object detectors, including loss function to solve sample imbalance problem [8], and new architectures for object detection [13].

All the above methods work under the closed set assumption. Recently, a few methods are proposed to solve the open set object detection problem. Dhamija et al. [1] firstly propose the concept of open set object detection and design evaluation metrics to evaluate the performance of several popular object detectors under open set conditions. Joseph et al. [6] proposes open world object detection, which first identifies both known and unknown objects and then dynamically updates knowledge by continually incremental learning. Because it is based on two-stage object detection framework, the speed cannot satisfy the requirement of real-time applications. Miller et al. [12] proposes to use dropout sampling as an approximation to Bayesian inference over the parameters of deep neural network to extract label uncertainty, which increases the performance of object detection under open set conditions. Although SSD [11] is adopted in this method, it is still computationally expensive since it requires multiple inference passes per image.

Prototype Learning. Prototype learning is a classical and representative method in pattern recognition which uses prototype to represent the main characteristics of classes. The earliest prototype learning method is k-nearest-neighbor (K-NN). In order to reduce the heavy burden of storage space and computation requirement of K-NN, an online method called learning vector quantization (LVQ) is proposed. The LVQ has been studied in many works and there are a lot of variants [4, 10]. After the arrival of the deep learning era, prototype learning can be incorporated into the deep neural network and trained in an end-to-end manner. It has played an important role in few-shot learning [16], robust representing learning [17], open set recognition [18], etc.

3   Method

3.1   Open Set Object Detection

Before introducing our method, we first formalize the problem of Open Set Object Detection (OSD). In a common object detection setting, all classes in the label space **Y** can be broadly categorized into three types [12]: **1) Known** classes in K are labeled in the training dataset and the detector is trained to detect them. **2) Known Unknown** classes in U_K exist in the training dataset but are not labeled. The detector is trained to ignore these objects which typically appear in the background. **3) Unknown Unknown** classes in U_U are not present in the training dataset. The detector has never seen objects of those classes during training and therefore has not learned to identify them. It is a challenge to identify the unknown unknowns under open set conditions.

Traditional detection only considers the problem of detecting known objects in K and neglecting objects in U_K, without paying attention to U_U. Existing works on OSD [1,12] consider the existence of U_U and make efforts on preventing misclassifying the objects in U_U as known classes. Different from the aforementioned methods, we consider OSD as a more challenging task, which is able to not only prevent misclassifying U_U as K but also distinguish them from U_K. Thus, we adopt both precision and recall of unknown classes as well as known classes to evaluate the OSD algorithm.

3.2   Open Set Object Detection Using YOLO V3

YOLOv3 [14] has proven an efficient object detection algorithm in a lot of applications. Thus, we first adapt it to the OSD setting by a simple modification. The architecture of YOLOv3 consists of three components including feature extraction network (Darknet53), across scale multi-layer feature fusion, and detection head. Darknet53 first produces powerful feature representations and then combines intermediate layer feature maps to produce multi-scale feature maps. The detection head predicts class condition probability, objectness score, and location offset for each predefined anchor on every pixel of the feature map.

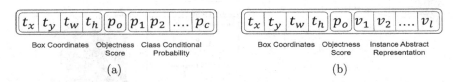

(a)   (b)

Fig. 1. Comparison of the outputs of detection heads. (a) Original YOLO v3, where c is the number of classes. (b) Our method based on prototype classifier, where l is the dimension of abstract representations.

The detailed output of the prediction head is illustrated in Fig. 1(a). YOLO v3 [14] regards the objectness score p_0 as the confidence of foreground and uses

C sigmoid functions to act as one-vs-all classifiers giving the class conditional probabilities $\{p_i\}_{i=1}^{C}$. During detection, only the candidate boxes with $p_0 > \tilde{p}_0$ are considered and the probability that the object belongs to class i is calculated by $p_0 \times p_i$. In this process, since all classifiers give a low response, the unknown objects are usually simply neglected as background and cannot be detected.

We can easily adapt YOLO v3 [14] to an open set object detector by exploiting the output of the detection head in a different way. Considering that in the training phase, the objectness score regards objects of all foreground classes as positive and background as negative, it is expected to give a higher response to similar foreground objects in the test phase. Therefore, if a candidate box is given low responses by all classifiers but a high objectness score, there is a large probability that the box contains an object of unknown class. Based on the analysis, we can identify an object with a high objectness score but lower responses from all classifiers as an unknown class. Although such a strategy provides a straightforward way to identify objects of unknown classes with YOLO v3, an inherent defect of the method affects its accuracy on unknown classes. The C sigmoid classifiers define C linear classifiers in an implicit feature space determined by the last 1×1 convolution, which cannot produce a compact distribution for each class. Therefore, features of unknown objects are liable to overlap with known classes, which makes it difficult to obtain satisfactory performance of unknown identification. In order to solve this problem, we resort to generative model and exploit prototype classifier to obtain a compact feature representation for each class.

3.3   Prototype Based Open Set Object Detection

In this section, we will present our prototype based open set object detection, which is illustrated in Fig. 2, after a brief introduction on prototype learning. By

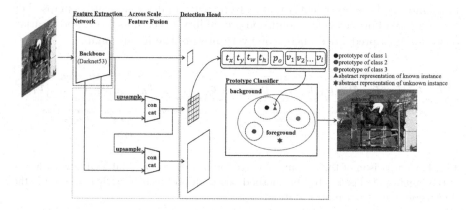

Fig. 2. The overall framework of our proposed open set object detection. Person class is known and horse class is unknown. For clarity, only the prototype classifier on mid level of feature maps is shown.

introducing prototype classifier, which has proven effective on open set recognition [18], into the framework of YOLO v3, the detector obtains the ideal capability of discovering unknown objects. Meanwhile, all advantages inherently built-in one-stage detector are inherited.

Different from sigmoid classifiers which give the class conditional probabilities directly, prototype learning maintains several prototypes in the feature space for each class and uses prototype matching for classification [17]. The prototypes are feature vectors representing the main characteristics of each class, which are denoted as m_{ij} where $i \in \{1, 2, ..., C\}$ represents the class index and $j \in \{1, 2, ..., n\}$ represents the prototype index in each class. Here we assume each class having the equal number of n prototypes and this assumption can be easily relaxed in the real application.

To obtain the class of an input X, we first calculate its abstract representation $f(X)$ by the feature extractor. Then, $f(X)$ is compared with all prototypes in each class and the nearest prototype according to the distance metric defined in the space is found. This sample is assigned by the class of the nearest prototype using Eq. 2.

$$d(f(X), m_{ij}) = \|f(X) - m_{ij}\|^2 \tag{1}$$

$$y^* = \arg\min_i \{\forall_j \ d(f(X), m_{ij})\} \tag{2}$$

Recently, some works [17, 18] attempt to integrate the prototype learning with convolutional neural network, and obtain remarkable progress in robust classification and open set recognition. Following these works, we adopt a prototype classifier to replace the sigmoid classifiers in the detection head of YOLO v3. The output of detection head with respect to each predicted box x is composed of three components: location offset, objectness score, and abstract representation of this box $f(x) \in \mathbb{R}^l$, as illustrated in Fig. 1(b). During training and inference, we use this feature vector to compare with prototypes of each class. Considering that YOLO v3 uses multi-scale feature maps to detect objects of different scales and the feature distributions are different in three levels, we set a prototype classifier on each level of feature maps.

In the inference stage, the model accepts an image as input and generates feature maps. The detector generates predicted boxes based on the predefined anchors per pixel in every level of feature maps. We use the objectness score as measurement of foreground and select the candidate boxes with $p_0 > \tilde{p}_0$ as potential foreground regions. Then, the distances between $f(x)$ and all prototypes are calculated according to Eq. 1. Finally, we assign the class label of each candidate box according to Eq. 4, where T_{dis} is the predefined distance threshold.

$$(i^*, j^*) = \arg\min_{i,j} d(f(x), m_{ij}) \tag{3}$$

$$y^* = \begin{cases} i, & \text{if } d(f(x), m_{i^*j^*}) \leq T_{dis}; \\ \text{unknown}, & \text{if } d(f(x), m_{i^*j^*}) > T_{dis}. \end{cases} \tag{4}$$

3.4   Model Training

The proposed model contains two parts of parameters to learn. One is the parameters of the CNN extractor θ, and the other is the prototypes in each class $M = \{m_{ij} | i = 1, ..., C; j = 1, ..., n\}$. During the training stage, the two parts of parameters are trained jointly in an end-to-end manner.

In our proposed prototype based open set object detection, the distance Eq. 1 is used as a measurement of similarity between instances and prototypes. Thus, we use distance to define the probability $p(m_{ij} | x)$ that an instance x belongs to a prototype m_{ij}.

$$p(m_{ij} | x) = \frac{e^{-d(f(x), m_{ij})}}{\sum_{p=1}^{C} \sum_{q=1}^{n} e^{-d(f(x), m_{pq})}} \tag{5}$$

According to Eq. 5 we can easily infer the posterior probability of each input instance $p(y = i | x) = \sum_{j=1}^{n} p(m_{ij} | x)$. Then, we can use cross entropy (CE) loss (called as distance-based CE (DCE) in this case [17]) to optimize the model. Although training with DCE loss can make the model classify the data correctly, its robustness is insufficient under open set conditions since it cannot make instances of the same class gather compact enough. To improve the robustness, we add a regularization named prototype loss (PL) [17]

$$pl((x, y); \theta, M) = \| f(x) - m_{yj} \|^2, \tag{6}$$

where m_{yj} is the nearest prototype to $f(x)$ belonging to ground truth class y. Thus, the classification loss of our proposed model can be defined as

$$L_{cls}((x, y); \theta, M) = -\log p(y | x) + \lambda pl((x, y); \theta, M). \tag{7}$$

Using the classification loss defined in Eq. 7 and keeping other terms from the original YOLO v3, we obtain the total loss of our proposed open set object detection as:

$$L((x, y); \theta, M) = L_{cls}((x, y); \theta, M) + \lambda_{obj} L_{obj}((x, y); \theta) + \lambda_{box} L_{box}((x, y); \theta), \tag{8}$$

where $\lambda_{obj}, \lambda_{box}$ are hyper parameters. In the training stage, we first randomly initialize the prototypes in each class of different feature levels and then optimize parameters of network and class prototypes jointly in an end-to-end manner according to Eq. 8.

4   Experiments and Results

4.1   Experimental Setting

In order to comprehensively evaluate the performance of the proposed method, we conduct both closed set object detection that does not include unknown classes and open set object detection with unknown classes. We introduce the experimental settings for both of them in the following.

Closed Set Experiments. For closed set detection experiments, we evaluate the proposed approach on PASCAL VOC [2] benchmark following standard training and test protocols. We use the training and validation sets of PASCAL VOC 2007 and 2012 for training and PASCAL VOC 2007 test data for test. The performance is measured by average precision (AP) 0.5 [2].

Open Set Experiments. Open set object detection is a relatively new task. Existing datasets [2,9] are unsuitable for evaluation of OSD algorithms since they do not explicitly label the objects of unknown classes. Previous methods [1,6] adopt twenty classes from PASCAL VOC [2] as training set and choose sixty classes from Microsoft COCO [9] that are different from training classes as the test set. Since the scenes and styles of these two datasets are obviously different, this setting cannot accurately evaluate the performance of the detector under open set conditions. Therefore, we have to propose a protocol to adapt the existing dataset to open set conditions.

In order to build a dataset containing labels of unknown unknowns, we divide PASCAL VOC [2] into two parts. The first part $\{\mathcal{D}_K^{train}, \mathcal{D}_K^{test}\}$ **only** contains training and test images belonging to the first N_1 classes that act as known classes, while the second part $\{\mathcal{D}_U^{train}, \mathcal{D}_U^{test}\}$ contains the training and test images of the remaining N_2 classes that act as unknown unknowns. We use $\mathcal{D}_K^{train}$ to train the model, which ensures that the model does not see any unknowns during training. Then, $\mathcal{D}_U^{train}$ is used as evaluation set to select suitable threshold for identifying unknowns. Finally, we conduct a closed set test using $\mathcal{D}_K^{test}$ and an open set test using $\mathcal{D}_K^{test} \cup \mathcal{D}_U^{test}$ to evaluate the OSD algorithms.

Previous works [1,6] focus on preventing misclassification of a unknown objects as known class, they use the Absolute Open Set Error (A-OSE) [12] or Wilderness Impact (WI) [1] as measurement. In this paper, we consider an OSD algorithm should not only reduce the misclassification of unknowns but also distinguish unknowns from the background. Therefore, we test the ability of algorithms to discover the candidate unknown objects from background by regarding unknown objects as a special class and calculating their recall and precision.

4.2   Implementation Details

We use PyTorch for implementation[1], adopt 4 GPUs for training with a batch size of 64 (16 images per GPU) using SGD, and optimize for 300 epochs in total. First three epochs are used for warmup and the initial learning rate is set to 0.01. Then, onecycle learning rate scheduler is used and the final learning rate is 0.002. We use a weight decay of 0.0005 and a momentum of 0.937. Input images are resized to 640×640, and we also perform random horizontal image flipping, mosaic, color space transformation, and random scale for data augmentation.

---

[1] https://github.com/ultralytics/yolov3.

4.3 Main Results

Results on Closed Set Detection. At first, in order to evaluate the performance of our proposed method on closed set object detection, we conduct common object detection experiments on PASCAL VOC [2] datasets. For the original YOLO v3 with a sigmoid classifier, we follow the standard setting. For our proposed method, we use one prototype with the 128-dimensional feature for each class on each level of features, keeping other settings unchanged. From Table 1, we can see that our method has comparable performance as the original YOLO v3. Assessing with precision and recall, our proposed method can surpass by 1% than original YOLO v3.

Table 1. Detection performance under closed set setting

Model	Precision	Recall	mAP@0.5
Sigmoid classifier (original yolo v3)	0.777	0.822	**0.821**
Prototype classifier (ours)	**0.781**	**0.831**	0.818

Results on Open Set Detection. As presented in Sect. 3.2, YOLO v3 can be used as an open-set classifier by modifying the classification strategy. A candidate box with a high objectness score but low responses on all sigmoid classifiers is identified as unknown. In contrast, our proposed method uses the objectness score to select the foreground candidate boxes and exploit the prototype classifier to identify the unknowns.

According to the open set setting, we choose the first ten classes from PASCAL VOC as known and the remaining classes as unknown. For both methods, we train the model using the set of known classes and test on the union of known and unknown classes. During test, we adopt the same threshold of objectness score for both methods and use $\mathcal{D}_U^{train}$ as the evaluation set to determine the suitable threshold. Table 2 shows that our proposed method performs better than YOLO v3 in discovering unknown classes, which verifies the advantage of compact feature representations on unknown discovery. We also use t-SNE to visualize the feature distribution of test samples of unknown class and known classes to further verify the effectiveness of our proposed method. From Fig. 3, we can see the features of each known class gather together, while the features of unknown class are distributed far away from the center.

Table 2. Detection performance under open set setting

Model	Category	Precision	Recall
Sigmoid classifier (original yolo v3)	Known Classes	0.572	**0.785**
Prototype classifier (ours)	Known Classes	**0.595**	0.747
Sigmoid classifier (original yolo v3)	Unknown Classes	0.271	0.162
Prototype classifier (ours)	Unknown Classes	**0.322**	**0.210**

(a)   (b)   (c)

Fig. 3. Feature distribution of samples. Black points and colored points represent unknown class and known classes respectively. Subfigures (a-c) correspond to level 1–3 of the feature map. (Color figure online)

4.4   Configuration of Prototype on Different Scales

YOLO v3 [14] adopts multi-scale feature map fusion strategy similar to FPN [7] to improve the object detection performance. Each level of multi-scale feature maps is responsible for detecting objects of a scale, and we assume that feature maps of different scales should have different feature distributions. In order to verify this assumption, we design two different settings for prototypes: 1) shared setting: multi-scale features share the same class prototypes; 2) separated setting: each scale of feature maps owns a separate set of class prototypes. From Table 3, we can see that the separated setting performs much better than the shared setting, which indicates that each scale of the feature map indeed has a different distribution. In order to further investigate the distributions of abstract representations of samples, we adopt t-SNE to visualize them. From Fig. 4, the separated setting can obtain a more separable classification surface than the shared setting, and the samples within the same class gather more compact. Thus, it gives better performance.

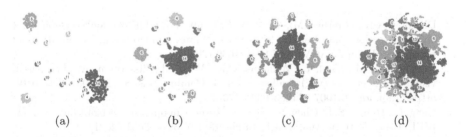

(a)   (b)   (c)   (d)

Fig. 4. Visualization of the feature distribution for different configurations of prototype in closed set test. Subfigures (a-c) correspond to the first to third level in the separated setting, while (d) corresponds to the shared setting.

Table 3. Detection performance with different prototypes setting

Model	Precision	Recall	mAP@0.5
Separated setting	**0.781**	0.831	**0.818**
Shared setting	0.764	**0.834**	0.809

5   Conclusion

In this paper, we focus on the problem of open set object detection which intends to detect unseen objects from images that may be of interest. To this aim, we propose a method by incorporating the idea of prototype learning into the framework of popular YOLO v3. It inherits the outstanding detection performance from YOLO v3 and obtains the ability of identifying foreground objects of unknown classes by exploiting the prototype classifiers. As a one-stage detection approach, it benefits from higher inference speed than existing OSD methods that are mostly two-stage. The experimental results show that our method is able to effectively identify unseen objects of unknown classes, while keeping the performance on known objects. By visualizing the distribution of feature representations, we see that samples of different classes are well separated in the feature space. This characteristic is in favor of unknown class identification, which explains the effectiveness of our method from one perspective. Furthermore, our method can be easily adapted to other one-stage detection methods and is expected to be effective too.

Acknowledgments. This work has been supported by the National Key Research and Development Program Grant No. 2018AAA0100400, the National Natural Science Foundation of China (NSFC) grant 61721004.

References

1. Dhamija, A.R., Gunther, M., Ventura, J., Boult, T.E.: The overlooked elephant of object detection: open set. In: 2020 IEEE Winter Conference on Applications of Computer Vision (WACV), pp. 1021–1030 (2020)
2. Everingham, M., Gool, L., Williams, C.K., Winn, J., Zisserman, A.: The pascal visual object classes (voc) challenge. Int. J. Comput. Vis. **88**(2), 303–338 (2010). https://doi.org/10.1007/s11263-009-0275-4
3. Geng, C., Huang, S.J., Chen, S.: Recent advances in open set recognition: a survey. IEEE Trans. Pattern Anal. Mach. Intell. **43**(10), 3614–3631 (2021)
4. Geva, S., Sitte, J.: Adaptive nearest neighbor pattern classification. IEEE Trans. Neural Netw. **2**(2), 318–322 (1991)
5. Girshick, R.: Fast r-cnn. In: 2015 IEEE International Conference on Computer Vision (ICCV), pp. 1440–1448 (2015)
6. Joseph, K.J., Khan, S., Khan, F.S., Balasubramanian, V.N.: Towards open world object detection. In: Proceedings of the IEEE/CVF Conference on Computer Vision and Pattern Recognition (CVPR) (2021)

7. Lin, T.Y., Dollar, P., Girshick, R., He, K., Hariharan, B., Belongie, S.: Feature pyramid networks for object detection. In: 2017 IEEE Conference on Computer Vision and Pattern Recognition (CVPR), pp. 936–944 (2017)
8. Lin, T.Y., Goyal, P., Girshick, R., He, K., Dollar, P.: Focal loss for dense object detection. IEEE Trans. Pattern Anal. Mach. Intell. **42**(2), 318–327 (2020)
9. Lin, T.-Y., et al.: Microsoft COCO: common objects in context. In: Fleet, D., Pajdla, T., Schiele, B., Tuytelaars, T. (eds.) ECCV 2014. LNCS, vol. 8693, pp. 740–755. Springer, Cham (2014). https://doi.org/10.1007/978-3-319-10602-1_48
10. Liu, C.L., Eim, I.J., Kim, J.: High accuracy handwritten Chinese character recognition by improved feature matching method. In: Proceedings of the Fourth International Conference on Document Analysis and Recognition, vol. 2, pp. 1033–1037 (1997)
11. Liu, W., et al.: SSD: single shot multibox detector. In: Leibe, B., Matas, J., Sebe, N., Welling, M. (eds.) ECCV 2016. LNCS, vol. 9905, pp. 21–37. Springer, Cham (2016). https://doi.org/10.1007/978-3-319-46448-0_2
12. Miller, D., Nicholson, L., Dayoub, F., Sunderhauf, N.: Dropout sampling for robust object detection in open-set conditions. In: 2018 IEEE International Conference on Robotics and Automation (ICRA), pp. 3243–3249 (2018)
13. Redmon, J., Divvala, S., Girshick, R., Farhadi, A.: You only look once: unified, real-time object detection. In: 2016 IEEE Conference on Computer Vision and Pattern Recognition (CVPR), pp. 779–788 (2016)
14. Redmon, J., Farhadi, A.: Yolov3: An incremental improvement, Computer Vision and Pattern Recognition arXiv preprint arXiv:1804.02767 (2018)
15. Ren, S., He, K., Girshick, R., Sun, J.: Faster r-cnn: towards real-time object detection with region proposal networks. IEEE Trans. Pattern Anal. Mach. Intell. **39**(6), 1137–1149 (2017)
16. Snell, J., Swersky, K., Zemel, R.S.: Prototypical networks for few-shot learning. In: Advances in Neural Information Processing Systems, vol. 30, pp. 4077–4087 (2017)
17. Yang, H.M., Zhang, X.Y., Yin, F., Liu, C.L.: Robust classification with convolutional prototype learning. In: 2018 IEEE/CVF Conference on Computer Vision and Pattern Recognition (CVPR), pp. 3474–3482 (2018)
18. Yang, H.M., Zhang, X.Y., Yin, F., Yang, Q., Liu, C.L.: Convolutional prototype network for open set recognition. IEEE Trans. Pattern Anal. Mach. Intell. 1 (2020, early access)
19. Zou, Z., Shi, Z., Guo, Y., Ye, J.: Object detection in 20 years: A survey. arXiv preprint arXiv:1905.05055 (2019)

Aesthetic-Aware Recommender System for Online Fashion Products

Bei Zhou, Basem Suleiman[(✉)], and Waheeb Yaqub

School of Computer Science, The University of Sydney, Sydney, Australia
{bei.zhou,basem.suleiman,waheeb.yaqub}@sydney.edu.au

Abstract. Recommender systems become widely used on the web, especially in e-commerce. In online fashion platforms, personalization becomes prevalent as user's decisions are driven not only by textual features but also the aesthetic features of products. We propose a novel machine learning approach that makes personalized fashion recommendations based on aesthetic and descriptive features of fashion products. We also introduce a new model that discovers the hidden correlations between fashion products that are frequently bought together as pairs. Our experiments on four real fashion datasets demonstrated that our aesthetic-aware recommender can achieve recommendations with accuracy up-to 89% when compared to related fashion recommenders.

Keywords: Recommender systems · Machine learning · E-commerce · Personalization

1 Introduction

The exponential growth of information generated by online platforms, especially social media and e-commerce has demanded intelligent ways to help users (or customers) to deal with information overload. One of the most popular and successful approaches is recommendation systems which have been applied in a variety of online platforms. In online shopping applications, for example, a recommender system suggests a few products or items deemed to be interesting for users based on their profile or shopping activity.

Recommender system algorithms fall into three major categories, content-based filtering (CBF), collaborative filtering (CF) and hybrid filtering that is the conjunction of the former two [11]. CBF utilizes metadata derived from both users and items to predict possible user preferences. CF, on the other hand, makes recommendations based on explicit or implicit user-item interactions. In the explicit approach, recommendations are made based on preferences information provided directly by users. The implicit interaction approach involves devising recommendations by predicting user preferences based on intelligent algorithms. The work in this paper focuses on the implicit CF as users often do not provide their preferences explicitly. Furthermore, it is challenging to maintain updated explicit user preferences from existing users who already provided initial

© Springer Nature Switzerland AG 2021
T. Mantoro et al. (Eds.): ICONIP 2021, LNCS 13108, pp. 292–304, 2021.
https://doi.org/10.1007/978-3-030-92185-9_24

preferences. With the ever-growing e-commerce applications, making implicit CF recommendations becomes much more prevalent and significant.

Large variety of recommendation algorithms [5, 7, 10] have demonstrated significant value in various e-commerce applications where recommendations are driven based on product related features. However, such algorithms would fail to make quality results for certain e-commerce products such as fashion products. Unlike many other products, users are interested not only in the description or technical details of the product but also its aesthetic aspects (e.g., color, design, and style) which are captured in product images. Aesthetic aspects are relative, a color or design of a shirt might be deemed beautiful by a user but not attractive for another user. This is due to the natural differences in human preferences where people might make different choices based on their style and/or personality. Beside aesthetic aspects, users are often interested in details that might not be possible to devise from product images such as material and size. This information is often provided in the product title or description and would be subject to user preferences, some users may prefer cotton material over silk material. Therefore, both product details and its aesthetic aspects are significant for designing personalized recommendations for fashion products.

In this paper, we propose a novel personalized recommendation system for fashion products. Our approach personalizes recommendations based on aesthetic features and descriptive features of fashion products obtained from real datasets (Amazon fashion and Tradesy.com). We developed a composite neural network architecture that captures aesthetic and descriptive features of real fashion products. Specifically, we utilized Brain-inspired Deep Network (BDN) model and Bag of Words (BoWs) model for extracting and learning aesthetic features and descriptive features respectively. Furthermore, we utilized Stacked Denoising Auto-Encoder (SDAE) to discover the hidden correlations between fashion products that are frequently bought together in pairs. By leveraging the item-pairs often purchased together by users, the system can learn the latent correlations between these item pairs based on the SDAE models. *These hidden relationships captured by the model represent and reflect the inner implicit associations between item-pairs generated by massive explicit user-item interactions.* Some of the frequently purchased pairs might be exception because they were purchased together solely to waive delivery fee from common manufacturers. Additionally, some users may prefer to buy some fashion products together merely based on their own interest that cannot be generalized. Thus, our model effectively lower the impact of noise on its performance while extracting the authentic correlations between frequently bought together items.

The current body of work in the area of fashion recommendation focused on features extracted either from product description or from product images [19, 22], or learning features from both[6, 12], however aesthetic features are rarely extracted from fashion images as a means to make more accurate recommendations. Unlike current studies, our models extract and learn aesthetic and descriptive features from product description and images in a combined neural network model which improves the quality of the recommendations made to users. However, these extracted features fail to capture the aesthetic perspectives of human beings, directly leading to the circumstances that the recommended products are product-oriented, rather human-oriented. In our approach, we employed BDN which is a type of aesthetic CNN to extract the features from the images that can capture the aesthetic aspects of fashion product. Therefore, our model achieves better results in contrast to common machine learning approaches or traditional CNN methods [24]. Our experiments have been conducted to work out which features that contribute the most to accurate recommendations are, image features, title features or the conjunction of these two. Although from our experiments, it is observed that title based features are not sufficient to make good recommendations but they are still useful when used as a complement to features extracted from images. The results from the experiments have shown that the combination of these two types of features can yield better recommendations than independently.

The contributions of this paper are threefold: *(1) a novel recommender model for personalized fashion products that extract and learn both aesthetic features and descriptive features of fashion products.* Our model structure combines advanced neural network models namely; Aesthetic CNN (BDN) and BoWs, *(2) aesthetic-aware fashion recommender algorithm that describe the logic of the proposed fashion recommender system based on the first contribution,* and *(3) experimental evaluation and analysis of the effectiveness of our proposed model and algorithm using four real fashion datasets.*

2   Related Work

Factoring image features into recommendation systems has exhibited some strengths and merits [7,19]. For example, Chen et al.[3] utilized textual and visual information to recommend tweets. Incorporating visual factors of fashion products is pivotal because users would not purchase a t-shirt from *Amazon* without having a glimpse of the actual image of it [8]. Although various visual features were exploited in recommendation tasks, they are conventional features and low-level non-aesthetic features [8]. Due to the advancement of computer vision techniques in recent years, image processing related tasks, such as image recognition and object detection, etc., have gained tremendous success. However, the assessment of the aesthetic aspects of images is still challenging because of the high subjectivity of aesthetics [22]. In this paper, we utilized high-level aesthetic features extracted by the aesthetic neural network to take the user's aesthetic preference into consideration when making recommendations.

The aesthetic networks were originally proposed for image aesthetic assessment. To reflect the subjective and complicated aesthetic perception, various deep neural networks were exploited to mimic the underlying mechanisms of human perception [14]. Among these, Brain-inspired Deep Network (BDN) model is the state-of-the-art deep learning model that is capable of extracting the features from images that can be used as an aesthetic assessment [24]. The Brain-Inspired Deep Network (BDN) [22] was designed and developed to extract the aesthetic features of images and then predicts a rating score that reflects the degree of aesthetics. The BDN was inspired by Chatterjee's visual neuroscience model [2] and differs from the normal neural network. BDN takes raw image features as input that reflect human aesthetic perceptions, such as hue, duotones, and saturation and can extract high-level aesthetic features, which are colours, styles, and structure. BDN has been successfully applied to building a visual-awareness recommendation system that is able to model and predict users' preferences [24].

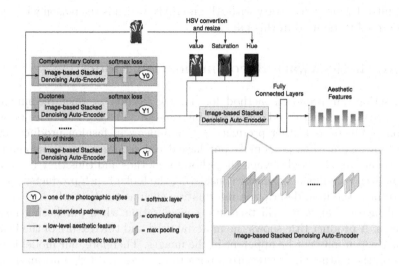

Fig. 1. Brain-inspired deep network architecture, adopted from [24]

Association rules can be applied to datasets with many records to uncover the items that are frequently bought together in one transaction [15]. These association rules are normally composed of a given set of items that connect with one specific item with a certain probability. Normally, association rules based methods require to be adapted before they could be applied to recommendation tasks because they are originally designed to be leveraged as exploratory tools which can help to discover previously unknown correlations [1]. For instance, in the context of book recommendations, recommending book titles rather than categories begets the situation where a large number of association rules would be applied. However, matching all combinations of the items can not guarantee the discovery of proper rules, therefore a suitable aggregator function must be applied to group the rules into a single one, which will be used for further processing [9]. In the recommendation systems, even though associations with low frequencies could also be taken into consideration, if no other rules with high confidence are discovered, this could make the mining process computationally expensive because it entails the extractions of as many rules as possible [26].

Auto-Encoders have been explored for their application on the recommendation system [16]. For instance, Wu et al.[23] used denoising auto-encoder to build a top-N recommendation system based on mere image features. One of the pitfalls of this classic Auto-Encoder (AE) is its inability to capture the hidden representations of complex features, such as features extracted from images or texts. Stacked Auto-Encoder (SAE) contains multiple hidden layers and is capable of extracting more accurate hidden representations that could be utilized to recover the original input [18]. Building upon the SAE, the Stacked Denoising Auto-Encoder (SDAT) [20] is an extension of SAE that is trained to reconstruct the original input from its corrupted version, and thereby SDAT is more robust and fault-tolerance in contrary with AE and SAE, which is the reason why SDAT was selected to be used in this paper.

3   Aesthetic-Aware Recommendation Model

We describe our proposed method for making fashion product recommendations based on aesthetic and contextual information. The key difference between fashion products and other products is that images in fashion products carry significant weight when users make purchase decisions. Users have personal preferences and dressing styles when they shop for clothes and thus they need more personalized recommendations by considering the look of such products. Equally important is the item description information such as the title which is another critical factor that users will take into consideration when purchasing fashion products. A product title shows the material used and the texture of the fashion product which may not be apparent in the images. Therefore, our model is built based on the features extracted from both the images and title information.

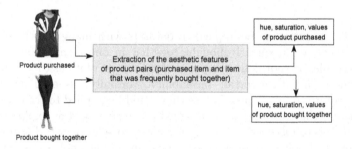

Fig. 2. Aesthetic features extractions of product pairs

3.1   Model Construction and Training

To factor in both images and titles, we construct two separate Stacked Denoising Auto-Encoder (SDAE) models; (a) one for reconstructing image features from purchased items to items that are frequently bought together, named Image-based Stacked Denoising Auto-Encoder (ISDAE), and the other for title features, name Title-based Stacked Denoising Auto-Encoder (TSDAE). Due to the distinctive characteristics of image and text data, different models have to be built to extract unique features and representations from them [5]. First, we construct a pre-trained Brain-inspired Deep Network to extract aesthetic features from the product image pairs (bought together). The architecture of this extraction process is adopted from [24] and illustrated in Fig. 1.

The input image is processed by parallel pipelines, each of which consists of independent convolutional networks beside the top three pathways that are used to extract the features such as hue, saturation and values. The intermediate features produced by the BDN model are used as the aesthetic features fitting into the later model because the final output of the BDN model should be geared towards specific applications. The image pairs consist of one purchased item and another item that was frequently bought together with it as shown in Fig. 2. These features represent the aesthetic aspects of these image pairs, incorporating hue, saturation, and value, etc. These extracted feature pairs are saved to be used in ISDAE model training, which are denoted as $(\mathcal{M}, \mathcal{M}')$ where (m, m') denotes each single pair in the pairs. The feature extraction process is run once for each item pair. Thus, it has one time initial overhead.

These extracted aesthetic image feature pairs were fitted in the ISDAE model where the first element of the pair, the image features of bought items, was used as input of the model and the other as output, as shown in Fig. 3. The ISDAE model consists of multiple layers including the input layer, the hidden layers, and the output layer. There are $\mathcal{I}$ nodes in both input and output layer, where the number $\mathcal{I}$ corresponds to the dimension of the features extracted from the BDN model. The model had seven hidden layers with the number of nodes of 256, 192, 128, 64, 128, 192, 256 respectively. The number of hidden layers has an impact on the model performance where more hidden layers contribute to the overfitting problem, which cannot be effectively mitigated by normal approaches like dropout or regularization. ISDAE model first reconstruct the 2D input to a vector whose dimension is identical to the input and is calculated as follows:

$$y^m = h(\boldsymbol{W}^\top \boldsymbol{m} + \boldsymbol{b}) \tag{1}$$

Algorithm 1: Learning algorithm for ISDAE

Initialize parameters with random values;
iter $\leftarrow$ 0;
while *iter* < *MaxIter* **do**
    for $(m, m') \in (\mathcal{M}, \mathcal{M}')$ **do**
        Compute y^m using Equation 2;
        Compute gradients $\frac{\partial \mathcal{L}(W, b)}{\partial W}$ and $\frac{\partial \mathcal{L}(W, b)}{\partial b}$ using Equation 6;
        Update parameters $\boldsymbol{W}$ and $\boldsymbol{b}$ using Equation 7;
    end
    iter $\leftarrow$ iter + 1
end

---

where $h(\cdot)$ is an element-wise mapping function, and W and b are the parameters of the model.

$$\mathcal{L}(W, b) = \frac{1}{N} \sum_{i=1}^{n} \ell(y^m, m') + \mathcal{R}(W, b) \tag{2}$$

where $\mathcal{R}$ is the regularization term (L2-Norm) to restrain the model complexity and suppress over-fitting.

$$\mathcal{R}(W, b) = \frac{\lambda}{2} \cdot (\|W\|_2^2 + \|b\|_2^2) \tag{3}$$

The parameters of ISDAE model are learnt by minimizing the average reconstruction error:

$$\underset{W, b}{\arg\min} \, \mathcal{L}(W, b) \tag{4}$$

We apply Stochastic Gradient Descend (SGD) to learn the parameters. Algorithm 1 shows the detailed procedure how the model is trained. The η in Eq. 7 is the learning rate set to 0.005 in our model's training process.

$$\begin{aligned}
\frac{\partial \mathcal{L}(W, b)}{\partial W} &= \frac{\partial \mathcal{L}(W, b)}{\partial y^m} \cdot \frac{\partial y^m}{\partial W} + \lambda W \\
\frac{\partial \mathcal{L}(W, b)}{\partial b} &= \frac{\partial \mathcal{L}(W, b)}{\partial y^m} \cdot \frac{\partial y^m}{\partial b} + \lambda b
\end{aligned} \tag{5}$$

$$\begin{aligned}
W &= W - \eta \cdot \frac{\partial \mathcal{L}(W, b)}{\partial W} \\
b &= b - \eta \cdot \frac{\partial \mathcal{L}(W, b)}{\partial b}
\end{aligned} \tag{6}$$

Besides visual and aesthetic features of items, prior research has shown that visual features alone do not substitute the textual features, instead, they act as complementary features to textual content [5]. Therefore, we constructed another model for extracting textual information for making fashion product recommendations. For extracting important features from product title, we used Bag of Words (BoWs) method as illustrated in the Fig. 4 because BoWs is a statistical framework that is popular for object categorization [25] and is relevant to our application. In this process, we remove the stop words (e.g., a, an, the, of, in) and the punctuation by creating a stop words list and excluding the words of titles that exist in that list. This will reduce the dimension of the input data for the TSDAE model. Hence, only words with high frequency are retained in the resulting vocabulary list (corpus).

Following the similar idea of image feature association model whose architecture is shown in the Fig. 3, the TSDAE model was built to uncover the hidden correlations between extracted title feature pairs. The hidden layers of the TSDAE model had 128, 96, 32, 96, 128 nodes in each layer respectively. The first elements of these features pairs are fitted into the model as input and the second elements of these pairs are regarded as labels the model needs to reconstruct.

Image Feature
of Item Bought

Image Feature
of Item Bought Together

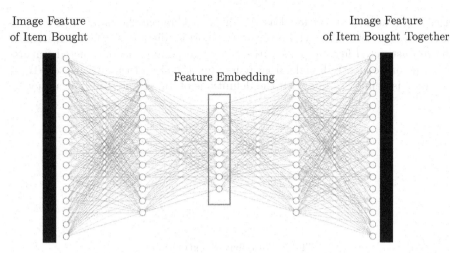

Fig. 3. Aesthetic feature pairs association

The feature embedding obtained contains the correlation information between the feature pair. The training of the TSDAE model is identical to those of the ISDAE model. The title features extracted from the item bought are used as the input of the model, and the features extracted from the items bought together are used as the label that the model requires to reconstruct by training.

3.2   Computing Recommendations

Before measuring the accuracy of these models, for all the items in the test set, the BDN model and the BoWs approach were applied to extract their image features and title features, which then are concatenated together, called **feature set**. After that both the ISDAE model and title TSDAE model are trained using image features pairs and title feature pairs respectively. The subsequent procedures were created to evaluate these trained models. For item pairs that were frequently bought together, the image features and title features of the first elements in every part are fitted into the well-trained ISDAE model and TSDAE model independently which then yield two outputs. These outputs are then concatenated, called **reconstructed features**.

Top-N recommendations have gained popularity from many shopping websites because this classic item-based approach showed excellent performance [21]. Nearest Neighbour-based method [4] is leveraged to evaluate the performance of the recommendations in which each of the reconstructed features is used to calculate the Euclidean Distance with all the features in the set.

Top-1 accuracy is defined as the division between the number of features in the feature set whose shortest distances calculated with all the reconstructed features. These features are the pairs with their corresponding reconstructed features in terms of frequently bought item pairs and the size of the feature size. Top-3 accuracy is defined as the division between the number of features in the feature set one of whose top three shortest distances calculated with all the reconstructed feature pairs with their corresponding reconstructed features in terms of frequently bought item pairs, and the size of the feature set. Top-5 accuracy is defined in line with the definitions of Top-1 and Top-3 accuracy.

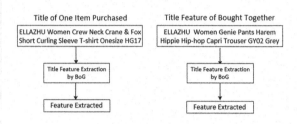

Fig. 4. Title feature extraction

4   Experiments and Results

We evaluated our context-aware visual recommendation method using four real-world datasets namely Amazon Fashion, Amazon Men, Amazon Woman, and Tradesy.com [24]. These datasets share a common characteristic that frequently-bought-together items are provided, and serve the purpose of our model construction and training. These datasets are pre-processed as follows. As our model is based on pairwise items, an item with the highest rating score is solely retained as the item with multiple frequently-bought-together items. Items with a higher rating are commonly better quality and popularity than the lower ones and thus form a pair with the original item. These item pairs are then stored for later feature extractions. The resulting datasets after pre-processing is shown in Table 1. Each dataset is associated with the number of users, items and item pairs. Each user bought at least 1 item from the item list. Each item has 0, 1 or multiple items bought with it. Each pre-processed dataset is then leveraged to train our proposed ISDAT and TSDAT models.

We implemented our context-aware visual recommendation models described in the Learning algorithm for ISDAE (Sect. 3.1) using Python 3.6 and Tensorflow deep learning framework 2.0. We used a NVIDIA GTX 950 GPU with 2 GB memory to train our model and a PC with 16 GB memory and Intel Core i5 processor to predict recommendations using our models. We evaluated the performance of our model using the four datasets described in Table 1. Specifically, we run three experiment sets to evaluate the accuracy of our context-aware model in terms of making fashion product recommendations. In the first experiment set, we evaluated the accuracy of our model in terms of making recommendations based on both image and title features (image+title) of fashion products compared to image-based recommendations (image only) and title-based recommendations

Table 1. Datasets summary (after pre-processing)

Dataset	No. of Users	No. of Items	No. of Item Pairs
Amazon Fashion	64,583	234,892	62,589
Amazon Woman	97,678	347,591	86,931
Amazon Men	34,244	110,636	28,746
Tradesy.com	33,864	326,393	83,627

Table 2. Results on individual and combined features

Feature Type	Top-1 Accuracy	Top-3 Accuracy	Top-5 Accuracy
Image Feature Only	51.6 %	69.1 %	76.8 %
Title Feature Only	42.5 %	46.9 %	51.4 %
Image+Title	60.7 %	77.9 %	86.4 %

(title only) using Amazon fashion dataset. Thus, we run 3 experiments in total and for each experiment and we computed the overall accuracy of all recommendations made by each of the approaches. The result is shown in the Table 2. Top-N recommendation is computed as described in Sect. 3.2. The accuracy is calculated by the ratio between the number of top-N recommendations and the overall number of recommendations. For example, the accuracy for the top-5 recommendation using both image and title features is 86.4% (54,076 (accurate recommendations)/62,589 (total number of recommendations).

As shown in Table 2, the increase in the accuracy of recommendation based on image features shows that image features carry more weight when pairing the frequently-bought fashion items, while the recommendations based on the title features did not score higher accuracy as users may put less weight on the title features alone when they want to purchase items. However, the results from combining both image and title features to make recommendations demonstrate a significant improvement. When complementing image features with title features, the accuracy score reaches the pinnacle. The observations derived in the results align with our hypothesis of the behaviour of buying fashion products where people first get influenced by the image of the product and item titles serve as an auxiliary contributing factor. Besides, the top-1 recommendation accuracy is lower than top-3 and top-5. More than one item recommended for them to choose from is beneficial to make comparison and sound decisions. This explains the reason why the more recommendations are offered, the higher the accuracy may be achieved. In the second experiment set, we evaluated the consistency of our model (image+title recommendations) in terms of making accurate recommendations across the four datasets (described in Table 1). For each dataset, we computed three types of recommendations for each fashion item in that dataset namely top-1, top-3 and top-5. Thus, we run 4 separate experiments in total. As shown in Table 3, top-5 recommendations scored the highest accuracy scores and top-1 recommendations were the least accurate across all the datasets. As shown in Table 1, the number of item pairs in the Amazon Woman and Tradesy.com datasets are much larger than those in the Amazon Fashion and Amazon Men datasets. However, the results in Table 3 demonstrate the dataset size influences the accuracy of the recommendations slightly. This indicates that our recommendation model can maintain high accuracy when the dataset size scale. The results also demonstrate that our recommendation model is capable of capturing the correlations between frequently-bought items and recommending proper items to users in accordance with their unique preferences.

Table 3. Results of aesthetic-aware recommender across four datasets

Dataset	Top-1 Accuracy	Top-3 Accuracy	Top-5 Accuracy
Amazon Fashion	60.7 %	77.9 %	86.4 %
Amazon Woman	60.3 %	76.6 %	85.8 %
Amazon Men	62.1 %	78.4 %	89.3 %
Tradesy.com	58.9 %	77.5 %	86.6 %

Table 4. The results on SIFT, Traditional CNN and Aesthetic CNN

Approach	Top-1 Accuracy	Top-3 Accuracy	Top-5 Accuracy
SIFT	44.2 %	52.9 %	56.7 %
VGG16 CNN	49.5 %	68.6 %	73.1 %
Aesthetic CNN	51.6 %	69.1 %	76.8 %

To compare our proposed aesthetic-aware model with relevant models, we considered machine learning and CNN algorithms that are widely used to extract valuable features from images such as VGG neural network [17]. We choose the Scale-Invariant Feature Transform (SIFT) [13] and VGG16 CNN as representative benchmarks algorithms for extracting features from image data. We implemented both algorithms as alternatives to our aesthetic-aware CNN and computed the recommendations accuracy on the Amazon fashion dataset as per the previous experiments. In this experiment, we extract the features from the product images alone and use them to make recommendations without using the title features. The goal is to find the best model to extract image features and thus could contribute to devising better fashion recommendations.

Table 4 summarizes the accuracy of the recommendations resulting from the benchmark algorithms and our model. As shown, SIFT scored the worst in all recommendation categories. In contrast, our aesthetic-aware model achieved the best accuracy compared to the other models and across the top-1, top-3 and top-5 recommendations. Therefore, as visual features are crucial in a user's decision when buying fashion products, our fashion recommender system would best benefit from the aesthetic feature extraction presented in our model.

5   Conclusion

In this paper, we propose a novel aesthetic-aware fashion recommender system that devises recommendations based on aesthetic and descriptive features extracted from real fashion products. We developed a novel model by utilizing Brain-inspired Deep Network (BDN) model and Bag of Words (BoWs) model for extracting and learning aesthetic and descriptive features respectively. We also utilized Stacked Denoising Auto-Encoder (SDAE) to discover the hidden correlations between fashion products that are frequently bought together. We evaluated our aesthetic-aware recommender using four real fashion datasets. Our results demonstrated the effectiveness of our aesthetic-aware recommender in terms of making more accurate recommendations when compared to descriptive features only and aesthetic features only across the four datasets. For future work, it could be beneficial to evaluate our proposed model on other fashion datasets and compare the results with similar studies. Furthermore, it would be worth improving the proposed model by investigating the use of transformer model, which could help to spot which regions in the product images intrigued the users to purchase them together.

References

1. Atkinson, J., Figueroa, A., Pérez, C.: A semantically-based lattice approach for assessing patterns in text mining tasks. Comp. y Sistemas **17**(4), 467–476 (2013)
2. Chatterjee, A.: Prospects for a cognitive neuroscience of visual aesthetics (2003)
3. Chen, T., He, X., Kan, M.Y.: Context-aware image tweet modelling and recommendation. In: Proceedings of the 24th ACM International Conference on Multimedia, pp. 1018–1027 (2016)
4. Deshpande, M., Karypis, G.: Item-based top-n recommendation algorithms. ACM Trans. Inf. Syst. (TOIS) **22**(1), 143–177 (2004)
5. Guan, Y., Wei, Q., Chen, G.: Deep learning based personalized recommendation with multi-view information integration. Decis. Supp. Sys. **118**, 58–69 (2019)
6. Han, X., Wu, Z., Jiang, Y.G., Davis, L.S.: Learning fashion compatibility with bidirectional lstms. In: ACM International Conference on Multimedia, pp. 1078–1086 (2017)
7. He, R., McAuley, J.: Ups and downs: modeling the visual evolution of fashion trends with one-class collaborative filtering. In: WWW, pp. 507–517 (2016)
8. He, R., McAuley, J.: Vbpr: visual bayesian personalized ranking from implicit feedback. In: 13th AAAI Conference on Artificial Intelligence (2016)
9. Ji, C.R., Deng, Z.H.: Mining frequent ordered patterns without candidate generation. In: Fuzzy System and Knowledge Discovery, vol. 1, pp. 402–406 (2007)
10. Kang, W.C., Fang, C., Wang, Z., McAuley, J.: Visually-aware fashion recommendation and design with generative image models. In: IEEE International Conference on Data Mining (ICDM), pp. 207–216. IEEE (2017)
11. Karimi, M., Jannach, D., Jugovac, M.: News recommender systems-survey and roads ahead. Inf. Process. Manage. **54**(6), 1203–1227 (2018)
12. Li, Y., Cao, L., Zhu, J., Luo, J.: Mining fashion outfit composition using an end-to-end deep learning approach on set data. IEEE Trans. Multimedia **19**(8), 1946–1955 (2017)

13. Lindeberg, T.: Scale invariant feature transform (2012)
14. Lu, X., Lin, Z., Jin, H., Yang, J., Wang, J.Z.: Rapid: rating pictorial aesthetics using deep learning. In: Proceedings of ACM International Conference on Multimedia, pp. 457–466 (2014)
15. Osadchiy, T., Poliakov, I., Olivier, P., Rowland, M., Foster, E.: Recommender system based on pairwise association rules. Expert Syst. App. **115**, 535–542 (2019)
16. Ouyang, Y., Liu, W., Rong, W., Xiong, Z.: Autoencoder-based collaborative filtering. In: Loo, C.K., Yap, K.S., Wong, K.W., Beng Jin, A.T., Huang, K. (eds.) ICONIP 2014. LNCS, vol. 8836, pp. 284–291. Springer, Cham (2014). https://doi.org/10.1007/978-3-319-12643-2_35
17. Simonyan, K., Zisserman, A.: Very deep convolutional networks for large-scale image recognition. arXiv preprint arXiv:1409.1556 (2014)
18. Suk, H.I., Lee, S.W., Shen, D., Initiative, A.D.N., et al.: Latent feature representation with stacked auto-encoder for ad/mci diagnosis. Brain Struct. Funct. **220**(2), 841–859 (2015)
19. Veit, A., Kovacs, B., Bell, S., McAuley, J., Bala, K., Belongie, S.: Learning visual clothing style with heterogeneous dyadic co-occurrences. In: IEEE International Conference on Computer Vision, pp. 4642–4650 (2015)
20. Vincent, P., Larochelle, H., Lajoie, I., Bengio, Y., Manzagol, P.A.: Stacked denoising autoencoders: learning useful representations in a deep network with a local denoising criterion. J. Mach. Learn. Res. **11**, 3371–3408 (2010)
21. Wang, Y., Zhao, X., Chen, Y., Zhang, W., Xiao, W., et al.: Local top-N recommendation via refined item-user bi-clustering. In: Bouguettaya, A. (ed.) WISE 2017. LNCS, vol. 10570, pp. 357–371. Springer, Cham (2017). https://doi.org/10.1007/978-3-319-68786-5_29
22. Wang, Z., Chang, S., Dolcos, F., Beck, D., Liu, D., Huang, T.S.: Brain-inspired deep networks for image aesthetics assessment. arXiv preprint arXiv:1601.04155 (2016)
23. Wu, Y., DuBois, C., Zheng, A.X., Ester, M.: Collaborative denoising auto-encoders for top-n recommender systems. In: ACM on Web Search and Data Mining, pp. 153–162 (2016)
24. Yu, W., He, X., Pei, J., Chen, X., Xiong, L., Liu, J., Qin, Z.: Visually-aware recommendation with aesthetic features. arXiv preprint arXiv:1905.02009 (2019)
25. Zhang, Y., Jin, R., Zhou, Z.H.: Understanding bag-of-words model: a statistical framework. Int. J. Mach. Learn. Cybern. **1**(1–4), 43–52 (2010)
26. Zheng, Z., Kohavi, R., Mason, L.: Real world performance of association rule algorithms. In: ACM on Knowledge Discovery and Data Mining, pp. 401–406 (2001)

DAFD: Domain Adaptation Framework for Fake News Detection

Yinqiu Huang[1], Min Gao[1(✉)], Jia Wang[1], and Kai Shu[2]

[1] School of Big Data and Software Engineering, Chongqing University,
Chongqing 401331, China
{yinqiu,gaomin}@cqu.edu.cn
[2] Illinois Institute of Technology, Chicago, IL 60616, USA

Abstract. Nowadays, social media has become the leading platform for news dissemination and consumption. Due to the convenience of social media platforms, fake news spread at an unprecedented speed, which has brought severe adverse effects to society. In recent years, the method based on deep learning has shown superior performance in fake news detection. However, the training of this kind of model needs a large amount of labeled data. When a new domain of fake news appears, it usually contains only a small amount of labeled data. We proposed a novel Domain Adaptation framework for Fake news Detection named DAFD. It adopts a dual strategy based on domain adaptation and adversarial training, aligns the data distribution of the source domain and target domain during the pre-training process, and generates adversarial examples in the embedding space during the fine-tuning process to increase the generalization and robustness of the model, which can effectively detect fake news in a new domain. Extensive experiments on real datasets show that the proposed DAFD achieves the best performance compared with the state-of-the-art methods for a new domain with a small amount of labeled data.

Keywords: Fake news · Domain adaptation · Adversarial training

1 Introduction

With the continuous development of the internet, all kinds of social media gradually enter people's lives. However, while these social media platforms bring convenience, they also become a breeding ground for fake news. The wide dissemination of fake news has brought severe adverse effects to society, such as weakening the public's trust in the government and journalism, and causing severe economic losses. Thus, timely detection of fake news has great significance.

However, the identification of fake news usually has many challenges because the data is large-scale and multi-modal. To solve this problem, researchers have proposed fake news detection models based on deep learning [8,13–15]. A large

© Springer Nature Switzerland AG 2021
T. Mantoro et al. (Eds.): ICONIP 2021, LNCS 13108, pp. 305–316, 2021.
https://doi.org/10.1007/978-3-030-92185-9_25

number of manual labeled data is the premise of training a deep learning network. However, when a new domain appears, only domain experts can make accurate manual annotations. Expensive and time-consuming manual labeling leads to a lack of labeled data in a new domain. Directly training the deep networks on a small amount of labeled data usually causes the models to overfit. In response to this problem, some researchers use pre-trained models to help detection. However, these pre-trained models do not have good detection performance because of the discrepancy between source data and target data.

To tackle these challenges, we propose a Domain-Adaptive Fake news Detection framework (DAFD) based on a dual strategy to improve detection performance. First, we perform domain adaptation operations in the pre-training process. The domain adaptation loss is introduced to align the data distributions of the two domains to learn a news representation with semantic information and domain alignment. We use a domain adaptation loss based on the maximum mean difference (MMD) [1] to learn a representation that optimizes fake news detection and domain invariance. Second, we use adversarial training to generate adversarial examples in the embedding space for additional training during the fine-tuning process to enhance the model robustness and generalization capabilities. Our main contributions are summarized as follows:

- The proposed DAFD can automatically align the data distribution of the source domain and the target domain during the pre-training process to ensure the model's detection performance in the target domain.
- We use adversarial training to make the model more robust and generalized in fine-tuning and further improve the performance of detection.
- We have conducted extensive experiments to show the performance of DAFD on the detection of fake news with a small amount of labeled data and analyzed why adversarial training and domain adaptation work.

2   Related Work

2.1   Fake News Detection

Recently, researchers have proposed many fake news detection technologies, which can be roughly divided into methods based on statistical learning and methods based on deep learning. Statistical learning methods usually extract features from news, and then train the model through these features. Tacchini et al. [17] tried to determine the authenticity of a news article based on the users who interacted with it or liked it. The deep learning model often has better performance because of its strong ability to automatic learning information representation. Ma et al. [8] proposed a new method to capture news features based on rumor life cycle time series. Shu et al. [15] developed an interpretable fake news detection framework using the mutual attention of news and corresponding comments.

However, these methods all require many labeled data and cannot detect with only a little labeled data. Therefore, We pay more attention to how to detect fake news in a new domain.

2.2 Transfer Learning

Our research is related to transfer learning, which can be roughly divided into four categories. Instance based Transfer Learning generates rules based on some weights and reuses data samples. Feature based Transfer Learning reduces the distance between different domains by feature transformation. For example, Zhang et al. [21] proposed to train different transformation matrices between different domains to achieve better results. Model based Transfer Learning finds the shared parameter information between different domains for information transfer. Relation Based Transfer Learning focuses on the relationship between samples in different domains, and researchers usually use Markov Logic Net to explore the similarity of relationships between different domains [3].

With the idea of feature based transfer learning, we propose a pretraining-finetuning framework to continuously align the data distribution during the pre-training process to improve the performance of fake news detection in a new domain.

3 Methodology

We will simply introduce the domain adaptation framework of fake news detection in this section and then introduce each component in detail.

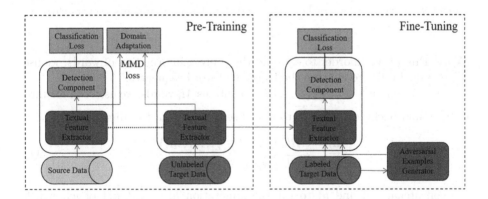

Fig. 1. The architecture of the DAFD, including pre-training and fine-tuning.

3.1 Overview

The question is set up as follows. We regard fake news detection as a binary text classification problem. That is, each news article can be real ($y = 0$) or fake ($y = 1$), and the target domain data used for fine-tuning contains only a small amount of labeled data. Our goal is to predict the labels of other news data in the target domain.

Figure 1 shows the structure of the DAFD. It can be divided into two parts: pre-training and fine-tuning. The former comprises three parts: textual feature extractor, domain adaptation, and detection part. The latter consists of adversarial examples generator, textual feature extractor, and detection part.

The pre-training model's input is source data and unlabeled target data, and then the textual feature extractor is used to model the news text from language features to hidden feature space. Meanwhile, we use domain adaptation to learn a classification representation that aligns the data distribution and can extract semantics. Finally, the fake news detection component performs fake news detection. The fine-tuning module takes labeled target data as input, extracts features through a pre-trained textual feature extractor, and then detects fake news through a fake news detection component that initializes parameters randomly, and then uses discriminative fine-tuning and gradual unfreezing [5] to finetune the model parameters. To enhance the robustness and generalization capabilities, we also generate adversarial examples for adversarial training. In this paper, the fake news detection component is an MLP network.

3.2 Textual Feature Extractor

Some researchers have found that the hierarchical attention neural network [19] has great advantages for learning document representations that emphasize important words or sentences. Inspired by this, we propose to learn document representation through a hierarchical structure. Concretely, we first learn sentence vectors through a word encoder and then learn documents vectors through a sentence encoder.

Word Encoder. In order to consider the context information of words, we use bidirectional GRU to capture the features of word sequences.

Given a sentence $s_i = \{w_1^i, \ldots, w_{M_i}^i\}$ contains M_i words, we use both forward GRU $\overrightarrow{f}$ and backward GRU $\overleftarrow{f}$ to model sentences from two directions:

$$
\begin{aligned}
\overrightarrow{\mathbf{h}_t^i} &= \overrightarrow{GRU}\left(\mathbf{w}_t^i\right), t \in \{1, \ldots, M_i\}, \\
\overleftarrow{\mathbf{h}_t^i} &= \overleftarrow{GRU}\left(\mathbf{w}_t^i\right), t \in \{M_i, \ldots, 1\}.
\end{aligned}
\tag{1}
$$

We concatenate $\overrightarrow{h_t^i}$ and $\overleftarrow{h_t^i}$ to get the annotation $h_t^i = \left[\overrightarrow{h_t^i}, \overleftarrow{h_t^i}\right]$ of w_t^i, which contains the contextual information centered around w_t^i. Not every word has the same influence on a sentence. Hence, we propose an attention mechanism to measure word importance, and the sentence vector v^i is

$$
\mathbf{v}^i = \sum_{t=1}^{M_i} \alpha_t^i \mathbf{h}_t^i, \quad \mathbf{u}_t^i = \tanh\left(\mathbf{W}_w \mathbf{h}_t^i + \mathbf{b}_w\right), \quad \alpha_t^i = \frac{\exp\left(\mathbf{u}_t^i \mathbf{u}_w^\top\right)}{\sum_{k=1}^{M_i} \exp\left(\mathbf{u}_k^i \mathbf{u}_w^\top\right)}, \tag{2}
$$

where α_t^i represents the importance of tth word for the sentence s_i, u_t^i is a hidden representation of h_t^i, and u_w is a parameter matrix representing a word-level context vector, which will be initialized randomly and updated with other parameters.

Sentence Encoder. Similar to word encoder, we learn the document representation by capturing the context information in the sentence-level. Given the sentence vectors v^i, we use a bidirectional GRU to encode the sentences:

$$\begin{aligned}
\overrightarrow{\mathbf{h}^i} &= \overrightarrow{GRU}\left(\mathbf{v}^i\right), i \in \{1, \ldots, N\}, \\
\overleftarrow{\mathbf{h}^i} &= \overleftarrow{GRU}\left(\mathbf{v}^i\right), i \in \{N, \ldots, 1\}.
\end{aligned} \tag{3}$$

We concatenate $\overrightarrow{h^i}$ and $\overleftarrow{h^i}$ to get the annotation $h^i = \left[\overrightarrow{h^i}, \overleftarrow{h^i}\right]$, which captures the context from neighbor sentences around sentence s_i. Similarly, an attention mechanism is used to measure sentence importance, and we calculate the document vector v by

$$\mathbf{v} = \sum_i \alpha^i \mathbf{h}^i, \quad \mathbf{u}_i = \tanh\left(\mathbf{W}_s \mathbf{h}^i + \mathbf{b}_s\right), \quad \alpha_i = \frac{\exp\left(\mathbf{u}_s \mathbf{u}_i^\top\right)}{\sum_i \exp\left(\mathbf{u}_s \mathbf{u}_i^\top\right)}, \tag{4}$$

where α^i represents the importance of ith sentence for the document, u_s is the weight parameter that represents the sentence-level context vector. It also will be initialized randomly and updated with other parameters.

3.3 Domain Adaptation Component

In the new domain scenario, the labeled data of the target domain is usually very scarce, so we cannot directly train the model with the target data. Besides, news in the new domain (target domain) usually has a different data distribution. If it is simply pre-trained, it will lead to the classifier overfits the source data distribution. Suppose we can get a representation that aligns the source data distribution with the target data distribution during the pre-training process, we can be better compatible with the target data.

We use MMD distance to measure the distance between two domains. We calculate the distance by a specific representation $\varphi(\cdot)$. We use this representation $\varphi(\cdot)$ to operate on source data, $x_s \in X_S$, and target data, $x_t \in X_T$. In our work, we use the Gaussian kernel function, then an empirical approximation to this distance is computed as:

$$\text{MMD}\left(X_S, X_T\right) = \left\| \frac{1}{|X_S|} \sum_{x_s \in X_S} \phi\left(x_s\right) - \frac{1}{|X_T|} \sum_{x_t \in X_T} \phi\left(x_t\right) \right\|, \tag{5}$$

As shown in Fig. 1, we not only need to minimize the distance between the source domain and the target domain (or align the data distribution of the two domains) but also need an effective classification representation that is conducive to detection. Such a classification representation will enable us to transfer fake news detection models across domains easily. We achieve this goal by minimizing the following loss during the pre-training process:

$$L = L_C\left(X_S, Y_S\right) + \lambda MMD^2\left(X_S, X_T\right), \tag{6}$$

where $L_C(X_S, Y_S)$ represents the classification loss of the labeled data X_S and the real label Y_S in the source domain, $MMD(X_S, X_T)$ represents the distance between the two domains. The hyperparameter λ represents the degree to which we want to align the data distribution.

Existing studies have shown that the meanings of features extracted from different layers of neural networks are different [20]. Specifically, some neural networks in the first few layers capture relatively general features, while some neural networks in the latter layers capture more specific features. Tzeng et al. [18] proved that domain adaptation in the previous layer of the classifier could achieve the best results, so we perform domain adaptation operations after feature extraction.

3.4   Adversarial Examples Generator

Since there is usually less labeled data in the target domain, direct fine-tuning will result in poor model generalization. Adversarial training is commonly used to build robust deep models. Existing research shows that adversarial training on the language model can increase generalization ability and robustness [2,11].

Pre-trained models such as Bert [4] and Albert [7] have been proved to impact downstream tasks positively. Our goal is to further the generalization ability of fake news detection in a new domain by enhancing the robustness of model embeddings. It is challenging to create actual adversarial examples for the language directly because the context determines the semantics of each word [22]. Some scholars have proposed that embedding-based confrontation is more effective than text-based confrontation [22]. Therefore, we create appropriate adversarial examples in the embedded space and then update the parameters on these adversarial examples to implement adversarial training.

Specifically, we use a gradient-based method to add norm-bounded adversarial perturbations to the embedding of the input sentence. We use $W = [w_1, w_2, \cdots, w_n]$ to represent the sequence of one-hot representations of the input words. V represents the embedding matrix, $y = f_\theta(X)$ represents the language model function, $X = VW$ is the word embeddings, and y is the output (class probabilities). θ represents all the parameters of the model. We get new prediction $\hat{y} = f_\theta(X + \delta))$ by adding adversarial perturbations δ to the embedding. To maintain the original semantics, we limit the norm of δ and assume that the detection results of the model will not change after perturbation.

For any δ within a norm ball, we expect to minimize the maximum risk as:

$$\min_\theta \mathbb{E}_{(Z,y)\sim\mathcal{D}} \left[\max_{\|\delta\|\le\epsilon} L\left(f_\theta(\boldsymbol{X} + \boldsymbol{\delta}), y\right) \right], \tag{7}$$

where D is the data distribution, L is the loss function. Madry et al. [10] demonstrated that SGD and PGD could reliably solve the saddle-point problem in neural networks.

With the idea of PGD, we perform the following steps in each iteration:

$$x_{t+1} = \Pi_{x+S}\left(x_t + \alpha g\left(x_t\right) / \|g\left(x_t\right)\|_2\right), \tag{8}$$

where α is the step size, $S = r\epsilon\mathbb{R}^d : \|r\|_2 \leq \varepsilon$ is the constraint space of disturbance, and $g(x)$ is the gradient of embedding. The loss function during the fine-tuning process is as follow:

$$L = L_C(X_T, Y_T) + L_C(X_{T_{adv}}, Y_T),$$ (9)

where $X_{T\,adv}$ is the adversarial examples generated by Eq. 8.

4 Experiments

Table 1. The statistics of datasets

	PolitiFact	GossipCop	Covid
True news	145	3,586	$5,600$
Fake news	270	2,230	$5,100$
Total	415	5,816	$10,700$

Our complete code can be obtained from the link[1]. In this section, we conducted a lot of experiments on three datasets to validate and analyze the performance of DAFD on fake news detection. Specifically, we aim to answer the following evaluation questions:

- EQ.1 Compared with other existing methods, can DAFD improve the fake news detection performance in a new domain?
- EQ.2 How effective are domain adaptation and adversarial training in improving DAFD detection performance?
- EQ.3 What is the impact of the amount of labeled data on the detection performance of DAFD?

4.1 Experiments Setup

In this experiment, We use the FakeNewsNet [16] dataset, which is a comprehensive dataset for fake news detection. It contains the data of two fact verification platforms: PolitiFact and GossipCop, including news content, labels, and social information. In addition, we also use a dataset Covid [12] for pretraining. Our model only uses the news text in the dataset, and some baseline methods use the news comment data additionally. The detailed statistics of these datasets are shown in Table 1. We use four metrics commonly used in classification tasks to measure the effect of fake news detection.

---

[1] https://github.com/964070525/DAFD-Domain-Adaptation-Framework-for-Fake-News-Detection.

4.2 Baselines

To validate the effectiveness of the DAFD, we choose both traditional machine learning algorithms and deep learning models as baseline methods. The representative advanced fake news detection baselines compared in this paper are as follows:

- Traditional machine learning: TF-IDF is a statistical method, which is often used for feature extraction. Based on TF-IDF features, we detect fake news with different traditional machine learning algorithms, including Naive Bayesian (NB), Decision Tree (DT) and Gradient Boosting Decision Tree (GBDT).
- Text-CNN [6]: It uses the convolution neural network to extract news features and capture different granularity feature information through different size filters.
- TCNN-URG [13]: It uses two components to extract different information features, uses a convolutional neural network to learn the representation of content, and uses a variational automatic encoder to learn the feature information of user comments.
- dEFEND [15]: It is an interpretable fake news detection framework using the mutual attention of news and corresponding comments.

4.3 Fake News Detection Performance (EQ.1)

To answer EQ.1, we compare our method with representative baselines introduced in Sect. 4.2 under the lack of labeled data in a new domain. To simulate the lack of labeled data, we artificially decrease the ratio of labeled data to 5% in GossipCop dataset and named it GossipCop_05. The experimental results of other ratios are shown in Sect. 4.6. Because the PolitiFact dataset is small, we keep 75% of the labeled data and named it PolitiFact_75. The amount of data used for pre-training should be greater than the target data amount. For PolitiFact_75 data, GossipCop data is used as the source domain for pre-training, and for GossipCop_05 data, Covid data is used for pre-training. The average performance is shown in Table 2.

Among the traditional machine learning methods, GBDT has the best effect because GBDT continuously fits the model residuals of the previous stage, making the model have a stronger generalization ability. In the deep learning methods, dEFEND and TCNN-URG use the comments of each news. Among them, dEFEND has the best effect. dEFEND uses two separate encoders and a co-attention layer to capture the article's representation to achieve better detection results. The effects of Text-CNN and TCNN-URG are relatively poor, which also show that traditional deep learning models cannot effectively detect with a small amount of labeled data.

We can observe that DAFD has achieved the best results in almost all indicators. All baselines are trained directly on the target data. When we use dEFEND to pretraining on GossipCop and then finetuning on PolitiFact_75, the F1 score of DAFD is still better than dEFEND (7.4%), which prove the DAFD can effectively detect fake news in a new domain.

Table 2. The performance comparison between DAFD and baselines

Datasets	Metric	NB	DT	GBDT	Text-CNN	TCNN-URG	dEFEND	DAFD
PolitiFact_75	Accuracy	0.760	0.712	0.808	0.653	0.712	0.904	**0.966**
	Precision	0.942	0.768	0.928	0.678	0.711	0.902	**0.965**
	Recall	0.756	0.791	0.810	0.863	0.941	0.956	**0.960**
	F1	0.838	0.779	0.864	0.760	0.810	0.928	**0.962**
GossipCop_05	Accuracy	0.627	0.630	0.681	0.630	0.591	0.689	**0.703**
	Precision	0.156	0.502	0.438	0.505	0.428	0.604	**0.614**
	Recall	0.556	0.521	**0.622**	0.604	0.196	0.548	0.610
	F1	0.244	0.511	0.514	0.516	0.269	0.575	**0.602**

4.4   Effects of Framework Components (EQ.2)

To answer EQ.2, we evaluate the dual strategy of domain adaptation and adversarial training. Specifically, we investigate the effects of these strategies by defining three variants of DAFD:

- DAFD\D: DAFD without domain adaptation. In the process of pre training, it removes the domain adaptation part, only uses the source domain data, uses cross entropy as the loss function.
- DAFD\A: DAFD without adversarial training. In the process of fine-tuning, it removes the adversarial training part and keeps the rest components.
- DAFD\All: DAFD without the pre-training and fine-tuning parts. It uses the feature extractor to extract the features and then directly performs classification detection, using cross entropy as the loss function.

The best performances are shown in Fig. 2, we make the observations:

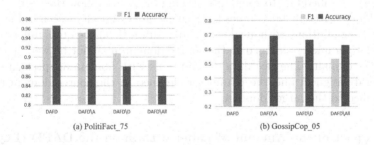

(a) PolitiFact_75   (b) GossipCop_05

Fig. 2. Impact analysis of framework components for fake news detection.

- Without the adversarial training part, the performance will be reduced, which indicates that the adversarial training helps to improve the generalization ability of the model.
- Without the domain adaptation process,the performance reduces 5.6% and 8.9% in terms of F1 and Accuracy metrics on PolitiFact, 8.8% and 5.4% on GossipCop.

- Without the pre-training and fine-tuning parts, the performance degradation is the biggest, which indicates the importance of DAFD framework in fake news detection in a new domain.

Through the component analysis of DAFD, we conclude that (1) both components of domain adaptation and adversarial training are conducive to improve the performance of DAFD; (2) it is necessary to use a pre-training and fine-tuning framework to detect fake news in a new domain because it can effectively use other knowledge to assist detection.

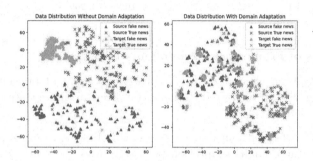

Fig. 3. The distribution of news embedding with domain adaptation and without domain adaptation

4.5   Analysis of Domain Adaptation Effectiveness

To observe the result of domain adaptation more intuitively, we use t-Distribution Stochastic Neighbor Embedding (t-SNE) [9] to decrease the dimension of news' embedding to 2 and draw them in Fig. 3. To show the results better, we randomly selected 200 samples for each category. As can be seen in Fig. 3, the two data distributions without domain adaptation (left) are quite different, and the model parameters learned in the source domain may not be well applied to the target domain. The two data distributions with domain adaptation (right) are roughly aligned, and the parameters learned in the source domain can be well applied to the target domain.

4.6   Impact of the Amount of Labeled Data on the DAFD (EQ.3)

This section answers EQ.3, we explore the impact of the amount of labeled data on the model performance. We decrease the dataset GossipCop again and name it GossipCop_10 according to the proportion of training samples of 10%, in the same way, we get GossipCop_15 and GossipCop_75. We also use Covid data for pre-training. The average performance is reported in Fig. 4.

As shown in Fig. 4, when the amount of labeled data is 5%, the F1 value increases by the framework is the largest (9.6%). With the increasing amount

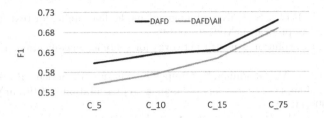

Fig. 4. The influence of labeled data amount on the performance

of labeled data, the F1 value benefits gradually decrease. Until the amount of labeled data reaches 75%, the F1 value benefit is the lowest (3.1%). This is because when there are enough labeled data in the new domain, the deep learning model can learn enough information from the data in the new domain, and the benefit of the pretraining model will decrease.

5   Conclusions and Future Work

In this paper, we investigate an important problem of fake news detection. Due to the suddenness of news in a new domain, it is challenging to obtain a large amount of labeled data, and it is difficult for models based on deep learning to cope with this situation. Therefore, we propose a new framework that can use knowledge in other domains to assist detection, align data distribution through domain adaptation technology, and enhance model robustness and generalization capabilities through adversarial training. Experiments on real-world datasets demonstrate the effectiveness of the proposed framework. In the future, we will capture the specificity of each domain in the process of domain adaptation to further improve the detection performance in each domain. Besides, we will introduce social information to assist detection.

Acknowledgments. Yinqiu Huang, Min Gao, and Jia Wang are supported by the Natural Science Foundation of Chongqing (cstc2020jcyj-msxmX0690). Kai Shu is supported by the John S. and James L. Knight Foundation through a grant to the Institute for Data, Democracy & Politics at The George Washington University.

References

1. Borgwardt, K.M., Gretton, A., Rasch, M.J., Kriegel, H.P., Schlkopf, B., Smola, J.A.: Integrating structured biological data by kernel maximum mean discrepancy. Bioinformatics **22**, e49–e57 (2006)
2. Cheng, Y., Jiang, L., Macherey, W.: Robust neural machine translation with doubly adversarial inputs (2019). arXiv preprint arXiv:1906.02443
3. Davis, J., Domingos, P.: Deep transfer via second-order markov logic. ACM (2009)
4. Devlin, J., Chang, M.W., Lee, K., Toutanova, K.: Bert: pre-training of deep bidirectional transformers for language understanding (2018). arXiv preprint arXiv:1810.04805

5. Howard, J., Ruder, S.: Universal language model fine-tuning for text classification (2018). arXiv preprint arXiv:1801.06146

6. Kim, Y.: Convolutional neural networks for sentence classification. Eprint Arxiv (2014)

7. Lan, Z., Chen, M., Goodman, S., Gimpel, K., Sharma, P., Soricut, R.: Albert: a lite bert for self-supervised learning of language representations (2019). arXiv preprint arXiv:1909.11942

8. Ma, J., Gao, W., Wei, Z., Lu, Y., Wong, K.F.: Detect rumors using time series of social context information on microblogging websites. In: Proceedings of the 24th ACM International on Conference on Information and Knowledge Management, pp. 1751–1754 (2015)

9. Van der Maaten, L., Hinton, G.: Visualizing data using t-sne. J. Mach. Learn. Res. **9**(11), 2579–2605 (2008)

10. Madry, A., Makelov, A., Schmidt, L., Tsipras, D., Vladu, A.: Towards deep learning models resistant to adversarial attacks (2017). arXiv preprint arXiv:1706.06083

11. Miyato, T., Dai, A.M., Goodfellow, I.: Adversarial training methods for semi-supervised text classification (2016). arXiv preprint arXiv:1605.07725

12. Patwa, P., et al.: Fighting an infodemic: Covid-19 fake news dataset (2020). arXiv preprint arXiv:2011.03327

13. Qian, F., Gong, C., Sharma, K., Liu, Y.: Neural user response generator: fake news detection with collective user intelligence. In: IJCAI, vol. 18, pp. 3834–3840 (2018)

14. Ruchansky, N., Seo, S., Liu, Y.: Csi: a hybrid deep model for fake news detection. In: Proceedings of the 2017 ACM on Conference on Information and Knowledge Management, pp. 797–806 (2017)

15. Shu, K., Cui, L., Wang, S., Lee, D., Liu, H.: Defend: explainable fake news detection. In: Proceedings of the 25th ACM SIGKDD International Conference on Knowledge Discovery & Data Mining, pp. 395–405 (2019)

16. Shu, K., Mahudeswaran, D., Wang, S., Lee, D., Liu, H.: Fakenewsnet: a data repository with news content, social context, and spatiotemporal information for studying fake news on social media. Big Data **8**(3), 171–188 (2020)

17. Tacchini, E., Ballarin, G., Della Vedova, M.L., Moret, S., de Alfaro, L.: Some like it hoax: automated fake news detection in social networks (2017). arXiv preprint arXiv:1704.07506

18. Tzeng, E., Hoffman, J., Zhang, N., Saenko, K., Darrell, T.: Deep domain confusion: Maximizing for domain invariance. Comput. Sci. (2014)

19. Yang, Z., Yang, D., Dyer, C., He, X., Smola, A., Hovy, E.: Hierarchical attention networks for document classification. In: Proceedings of the 2016 Conference of the North American Chapter of the Association for Computational Linguistics: Human Language Technologies, pp. 1480–1489 (2016)

20. Yosinski, J., Clune, J., Bengio, Y., Lipson, H.: How Transferable are Features in Deep Neural Networks? MIT Press, Cambridge (2014)

21. Zhang, J., Li, W., Ogunbona, P.: Joint geometrical and statistical alignment for visual domain adaptation. In: Proceedings of the IEEE Conference on Computer Vision and Pattern Recognition, pp. 1859–1867 (2017)

22. Zhu, C., Cheng, Y., Gan, Z., Sun, S., Goldstein, T., Liu, J.: Freelb: enhanced adversarial training for language understanding (2019)

Document Image Classification Method Based on Graph Convolutional Network

Yangyang Xiong[1], Zhongjian Dai[1], Yan Liu[2(✉)], and Xiaotian Ding[2]

[1] Beijing Institute of Technology, Beijing 100081, China
[2] Taikang Insurance Group, Beijing 100031, China
liuyan146@taikanglife.com

Abstract. Automatic and reliable document image classification is an essential part of high-level business intelligence. Previous studies mainly focus on applying Convolutional Neural Network (CNN)-based methods like GoogLeNet, VGG, ResNet, etc. These methods only rely on visual information of images but textual and layout features are ignored, thereby their performances in document image classification tasks are limited. Using multi-modal content can improve classification performances since most document images found in business systems carry explicit semantic and layout information. This paper presents an innovative method based on the Graph Convolutional Network (GCN) to learn multiple input image features, including visual, textual, and positional features. Compared with the CNN-based methods, the proposed approach can make full use of the multi-modal features of the document image to lead the model competitive with other state-of-the-art methods with much fewer parameters. In addition, the proposed model does not require large-scale pre-training. Experiments show that the proposed method achieves an accuracy of 93.45% on the popular RVL-CDIP document image dataset.

Keywords: Graph convolutional network · Document classification · Image processing

1 Introduction

Document digitization plays a critical role in the automatic retrieve and management of document information. Most of these documents are still processed manually, with billions of labor costs each year in industry. Thus, researches on automatic document image classification have great practical value. The document image classification task attempts to predict the type of a document image by analyzing the document's appearance, layout, and content representation. Traditional solutions to this challenge mainly include the image-based classification method and the text-based classification method. The former tries to extract patterns in the pixels of the image to match elements with a specific category, such as shapes or textures. The latter tries to understand the text printed in the document and associate it with its corresponding class.

ⓒ Springer Nature Switzerland AG 2021
T. Mantoro et al. (Eds.): ICONIP 2021, LNCS 13108, pp. 317–329, 2021.
https://doi.org/10.1007/978-3-030-92185-9_26

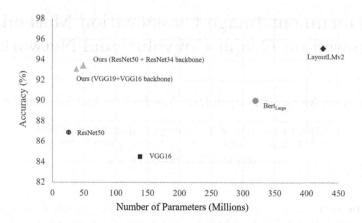

Fig. 1. Model size vs. classification accuracy on the RVL-CDIP dataset. The LayoutLMv2 [1] is currently the state-of-the-art method. However, this model has much more parameters (426M) and requires tens of millions of data for model pre-training to achieve the best accuracy.

However, in real-business applications, the same kind of document often contains different layouts. This intra-class difference makes visual-based classification difficult, and it is impossible to perform rigid feature detection and feature matching, like SURF [2], SIFT [3], and ORB [4]. In addition, different kinds of documents sometimes show high visual similarity, which increases the difficulty of classification. For example, some news articles contain tables and figures, make them look like scientific publications. Therefore, it is difficult for pure visual methods, including CNN, to classify document images with ideal classification accuracy.

If judging from the content of the text, these documents have a similar structure: the address and date usually appear at the top, and the signature usually appears at the bottom. Making full use of the information in the document images, including visual, positional, and textual features, can improve document classification accuracy. In recent years, researchers have started to use the graph concept, including the GCN [5], to do some graph node classification and link prediction tasks with the feature aggregation capabilities of GCN. Therefore, we propose a framework based on the GCN architecture, which can make full use of the multimodal characteristics of the document image. The model incorporates three types of input features: (1) Compact image feature representations for the slice of each text block and the whole document image; (2) Textual features from the text content of each text block; and (3) Positional features denoting the positions of texts within a document image. By doing so, the model can aggregate the visual features and textual features in the document image, and the accuracy of document image classification can be effectively improved.

To sum up, the contribution of this work lies in three folds:

(1) A one-step, end-to-end approach is developed to handle document image classification tasks by a single GCN-based model. The model possesses great scalability to take such a task across various document images with complex layouts.
(2) The model uses the concept of graphs to classify documents and innovatively proposes a method for constructing node features that combine visual, positional, and textual features, which can greatly improve the model performances with fewer parameter sizes, and the best accuracy-speed trade-off is achieved. As shown in Fig. 1.
(3) In practical applications, the model can be trained from scratch and does not require large-scale pre-training.

2   Related Work

Document image classification tasks were generally solved using semantic-based methods in the past. And Bag of Words (BOW)-based methods have shown great success in document image classification [6,7]. However, the primary mechanism of the BOW-based process is to calculate the frequency information of the corresponding word dictionary and ignore the unique layout position information between the document image components, which limits the ability to describe document images.

With the development of deep learning methods in various fields of computer vision, such as target recognition, scene analysis, and natural language processing, deep learning methods show better performance than traditional methods. Some scholars use deep CNN in the field of document image classification and achieve satisfying performances. For the first time, Le Kang et al. use CNN to classify document images [8]. Their results prove that the performance of the CNN is better than the traditional methods. Later, Afzal et al. propose to design a deeper neural network [9], pre-train the network on the ImageNet dataset [10], and then perform transfer learning on the document image dataset. They get better results on the same document image classification dataset with a 12.25% improvement of accuracy. Their experiments show that training a CNN requires many data, and the transfer learning techniques are practical and feasible. However, the CNN-based model can only handle visually different documents, and the performance is deficient on visually similar documents.

To classify document images from the content, some researchers combine the Optical Character Recognition (OCR) [11–14] with Natural Language Processing (NLP) [15]. These methods can deal with visually similar documents well, but do not make full use of the visual information of the document images. Moreover, the document images usually contain defects, including rotation, skew, distortion, scanning noise, etc. All of these bring significant challenges to the OCR system and directly affect subsequent NLP modules. Although enormous efforts have been paid, the OCR + NLP approaches are still short of satisfying performance for the above reasons.

Recently, some researchers notice that the classification of complex document images requires multi-modal feature fusion. For example, LayoutLMv2 [1] realizes to combine textual, visual, and positional information for the document classification task, achieving state-of-the-art performance. Still, it has many parameters (426M) to achieve the optimal result, and requires tens of millions of pre-training data.

3   Proposed Approach

We propose a document image classification framework, which constructs a graph representation for each document image, and the overall architecture is shown in Fig. 2. The first CNN sub-module (CNN1) is responsible for extracting the whole image's visual features. For each OCR text block, the second CNN sub-module (CNN2) is used for extracting local-aware visual features for the text image slice of the block. Textual features are extracted by a Tokenize-Embedding-GRU (Gated Recurrent Unit) pipeline from text contents. Positional features are extracted by a Fully Connected layer (FC1) from text block coordinates. The GCN sub-module is designed to fuse and update the above visual, textual, and positional features and extract graph representations for the document image. At last, the graph representations are passed to a Fully Connected layer (the classification layer, FC3 in Fig. 2) to get the specific category of the document image.

The input of the model includes four parts from the document image, which are: (1) the full image of the document; (2) the image slices of each text block; (3) the text contents of each text block; and (4) the coordinates of each text block. In practice, the text block information is generated by an off-the-shelf OCR system, from which we can get the text content and the coordinates of the four vertices for each text block. One text block from the OCR results is taken as one graph node. Based on this information, an innovative graph node feature construction method is proposed, which combines the full image feature and the feature of each text block.

3.1   Graph Node Feature Extraction

Node features of the graph are constructed from two parts. They are full input image features and text block features, where the text block features include text image features, text content features, and text position features.

The whole image features are extracted by a CNN sub-module (CNN1 in Fig. 2). In our experiments, we attempt to use different CNN backbones, including ResNet50 and VGG19. For these backbones, the final Fully Connected layer is removed, and the size of the Adaptive Average Pooling layer is changed to 7×7. The full document image is resized to a fixed size and then passed to this module to get a $7\times7 \times C$ feature map, where C is the image feature channel. Then, this feature map is split into 7×7 parts along the x-direction and y-direction, so 49 parts of features are obtained along the channel-direction ($1\times1\times C$). Finally, each part of the features is squeezed and taken as one node feature of the graph.

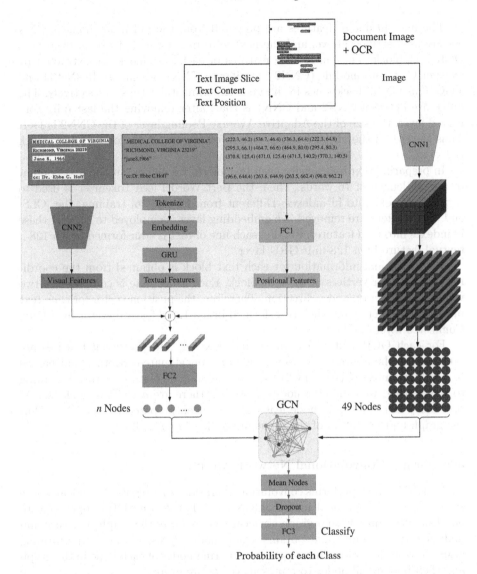

Fig. 2. Overall design of the proposed model. The model employs CNN1 and CNN2 as the backbone network for extracting the full-image visual features and the local-aware visual features. The embedding layer is responsible for converting text information into textual features. The FC1 converts the position vector into the positional feature. The GCN sub-module is designed to fuse and update node features and extract graph representations for the document image. FC1, FC2, and FC3 is Fully Connected layer.

From a computer vision point of view, this is similar to dividing the original image into 7×7 sections and then extracting a node feature by the CNN for each section.

The first 49 nodes' features are prepared from the full input image's CNN feature as described above, and the next is to prepare node features from each OCR text block. The image slice features of each text block are extracted by another CNN sub-module (CNN2). Similar to CNN1, we choose ResNet34 and VGG16 as CNN2 backbones in different experimental setups, respectively. The difference between CNN2 and CNN1 is that, after removing the last fully connected layer, the size of the Adaptive Average Pooling layer of the CNN2 is 1×1. Thus the size of the visual features generated by the CNN2 for each text block is $1 \times 1 \times C$.

In preparing text features for each text block, we pad or cut the text content to a fixed length of 16 words. Then, the Bert Word Piece Tokenizer is used to convert the text into id indexes. Different from BERT [16] training, the [CLS] and [SEP] tokens are removed. An embedding layer is employed to convert these id indexes into 64-d features. Finally, each line of text is transformed into a 128-d textual feature by a 128-unit GRU layer.

The positional information for each text block is obtained from the coordinates of the four vertices of the text block. Each coordinate is composed of two values in x-direction and y-direction. Therefore, the position vector for each text block is constructed and then transformed into a 128-d feature vector by a Fully Connected layer (FC1).

For each OCR text block, the visual, textual, and positional features are prepared by the above steps. Next, they are concatenated together and passed to a Fully Connected layer (FC2) to get the final node feature vector. According to this setting, we can get n nodes' feature if there are n OCR text blocks. As previously introduced, 49 node features have been prepared from CNN1. Thus the graph representation of the input image has 49+n nodes.

3.2   Graph Convolutional Network Module

Unlike CNN, which performs convolution operations in a regular Euclidean space such as a two-dimensional matrix, GCN extends the convolution operation to non-Euclidean data with a graph structure. GCN takes the graph structure and node features as input and obtains a new node representation by performing graph convolution operations on the neighboring nodes of each node in the graph and then pooling all nodes to represent the entire graph.

A multi-layer GCN is defined by the following layer-wise propagation rule [5]:

$$H^{(l+1)} = \sigma(\widetilde{D}^{-1/2} \widetilde{A} \widetilde{D}^{-1/2} H^{(l)} W^{(l)}) \tag{1}$$

Therefore, as long as the input feature X and the adjacency matrix A are known, the updated node feature can be calculated. In our model, the input feature X is the n+49 nodes' features. Since the graph in our model is a Fully Connected graph, every two nodes have a connection, so the adjacency matrix A is $N \times N$ full-one matrix. We build a GCN module with two graph convolutional layers, as shown in Fig. 3. Each layer of graph convolution is followed by a SiLU activation function. The graph is defined by the fully connected N nodes and initialized from the node features prepared by the above steps. States and features are propagated across the entire graph by the two graph convolutional layers. The final node states vector of the graph is the $N \times 512$ vector. Then, the final node states are averaged to a 1×512 vector, which is the graph representation of the input data. Finally, the 512-d enriched graph representation is then passed to a $512 \times k$ FC layer (FC3 in Fig. 2), where k is the number of the classes of document images.

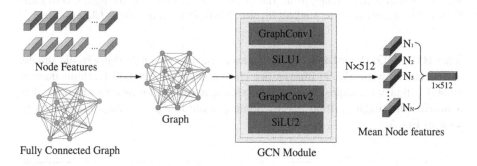

Fig. 3. Schematic depiction of multi-layer Graph Convolutional of aggregating node characteristics. The model's input includes a graph definition with a total of N nodes and the node features.

4   Experiments

4.1   Datasets Description

The model is applied to the document image classification task on the Medical Insurance Document Image (MIDI) dataset and the Ryerson Vision Lab Complex Document Information Processing (RVL-CDIP) dataset [17].

The MIDI Dataset. This dataset contains scanned and photo images collected from the real business system. It has a total of 160,000 images in 20 categories, and sample images are shown in Fig. 4. We split these images into 120,000 training images, 20,000 validation images, and 20,000 testing images. These images are collected from various provinces and cities in China. This dataset has the characteristics of significant intra-class differences and slight inter-class differences.

Fig. 4. OverallSample Images from the MIDI dataset. From left to right: Claim form, Personal information form, Medical invoice, Medical imaging report, Claim notice.

The RVL-CDIP Dataset. This dataset consists of 400,000 grayscale images in 16 classes, with 25,000 images per class. There are 320,000 training images, 40,000 validation images, and 40,000 testing images. The images are resized to a maximum length of 1000 pixels. Some sample images of this dataset can be seen in Fig. 5.

Fig. 5. Sample images from the RVL-CDIP dataset. From left to right: Letter, Form, Email, Handwritten, Advertisement, Scientific report, Scientific publication, Specification, File folder, News article, Budget, Invoice, Presentation, Questionnaire, Resume, Memo.

4.2   Model Training and Evaluation

For each experiment, a trainable end-to-end pipeline was built according to Fig. 2, and the output of the pipeline was the enriched feature of the original image. After the final classifier (FC3), the category to which the image belongs was predicted. To test the impact of different visual backbones (CNN1 and CNN2 in Fig. 2) on the model performances, we tested ResNet50 and VGG19 for CNN1 and tested ResNet34 and VGG16 for CNN2. In order to compare our model with the CNN-based visual model, we also tested the performance of the VGG16 and ResNet50 on the MIDI dataset. For the RVL-CDIP dataset, we followed the same model and hyper-parameter setups with the MIDI experiments.

The training epochs were set to 20 for all experiments with gradient accumulation technology to ensure stable convergence of the model. All models were trained on an NVIDIA Tesla V100 machine, using the Cross-Entropy Loss function and AdamW optimizer. The max learning rate was set to 8e-5, and the cosine learning rate scheduler was set. In addition, the learning rate warm-up steps were set to 50000 for the RVL-CDIP dataset, 10000 for the MIDI dataset, respectively. During training, all input data were shuffled at each epoch begin.

5 Results and Discussion

On the MIDI dataset, the classification accuracies of the proposed models and the CNN-based models are shown in Table 1. The results suggest that our models with different backbone setups significantly surpass CNN-based methods. The experiments reach the best classification accuracy of 99.10%, with 6.58% and 5.71% accuracy improvement than the CNN-based VGG16 and ResNet50. The outstanding performance means that the proposed models can be directly used in industrial applications since this dataset is the actual business dataset. The proposed models have much fewer parameters than the VGG16 because we removed the large-parameter FC layer.

Table 1. Classification accuracy of different models on the MIDI dataset.

Models	Accuracy(%)	Parameter size (millions)
VGG16	92.62	130
ResNet50	93.39	26
Ours (VGG 19 + VGG 16 backbone)	**99.04**	**38**
Ours (ResNet50 + ResNet34 backbone)	**99.10**	**49**

Table 2 shows the result of our model compared with VGG16, ResNet50, and other models, including text-only models and image-only models on the RVL-CDIP dataset. The table shows that the proposed model outperforms those text-only or image-only models as it leverages the multi-modal information within the documents. The proposed model uses the fewest parameters but shows the best classification accuracy.

It is worth noting that although the RVL-CDIP dataset is larger than the MIDI dataset. Due to the higher image resolution, higher OCR character recognition accuracy, and color images, the classification accuracy on the MIDI dataset is higher than on the RVL-CDIP dataset when using the same model setup and training setups. The OCR engine in our experiments is a general multi-language engine and not specially optimized for English data. Thus, the OCR character recognition accuracy is unsatisfactory due to the OCR engine optimization and the low pixel resolution of texts in several images.

Table 2. Comparison of accuracies on RVL-CDIP of best models from other papers.

	Models	Accuracy (%)	Parameter size (millions)
Text-only models	BERT-Base [16]	89.81	110
	UniLMv2-Base [18]	90.06	125
	BERT-Large [16]	89.92	340
	UniLMv2-Large [18]	90.20	355
Image-only models	VGG16	84.52	138
	ResNet50	86.83	26
	Document section-based models + AlexNet transfer learning [17]	89.80	–
	AlexNet + spatial pyramidal pooling + image resizing [19]	90.94	–
	Transfer Learning from AlexNet, VGG16,GoogLeNet and ResNet50 [20]	90.97	–
	Transfer Learning from VGG16 trained on Imagenet[21]	92.21	–
Proposed models	Ours (VGG19 + VGG16 backbone)	**93.06**	**38**
	Ours (ResNet50 + ResNet34 backbone)	**93.45**	**49**

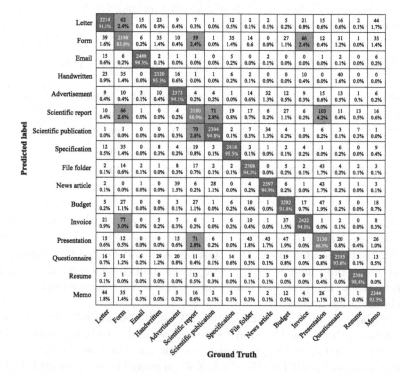

Fig. 6. Confusion matrix of the proposed model on the RVL-CDIP dataset.

Figure 6 reports the confusion matrix of the proposed model on the RVL-CDIP dataset. It shows that the proposed model performs very well on most categories of images. However, the classification accuracy for the three categories is less than 90%, which is form, scientific report, and presentation. This is because there are overlaps of definitions among the three categories. For example, some pages of scientific reports usually contain data forms, which make them be defined as the "form" category.

6 Conclusion

This paper presents a document image classification framework based on GCN. We propose a novel multi-modal graph node feature construction method to combine the visual, textual and positional features of each text block in the image and the visual feature of the full document image. All of these make the feature expression more abundant. By transmitting information to the GCN network, the meaningful features are enriched for classification. Experiments were carried out on the MIDI dataset and the RVL-CDIP dataset. The proposed model obtained classification accuracies of 99.10% and 93.45% on the two datasets, respectively, which are superior to CNN algorithms. Experimental data have shown that our model is effective and efficient. Moreover, our end-to-end pipeline does not require handcrafted features or largescale pre-training as other works.

In our experiments, the OCR engine we can obtain is not optimized for the English data. The lower gain on the RVL-CDIP dataset is directly affected by the high error rate of OCR recognition and the low image resolution of several images. Therefore, we will further find commercial OCR systems suitable for English text recognition to tackle this problem. We also consider adding more features to the GCN model, such as learning the relationship between text blocks, to make full use of the GCN capabilities and various information of the document image.

References

1. Xu, Y., Xu, Y., Lv, T., Cui, L., Wei, F., Wang, G., et al.: LayoutLMv2: multi-modal pre-training for visually-rich document understanding, pp. 1–16 (2020)
2. Bay, H., Ess, A., Tuytelaars, T., Van Gool, L.: Speeded-up robust features (SURF) original publication. Comput. Vis. Image Underst. **110**, 346–359 (2008)
3. Low, D.G.: Distinctive image features from scale-invariant keypoints. Int. J. Comput. Vis. **60**, 91–110 (2004)
4. Rublee, E., Rabaud, V., Konolige, K., Bradski, G.: ORB: an efficient alternative to SIFT or SURF. In: Proceedings of IEEE International Confernce Computing Vision, pp. 2564–2571 (2011)
5. Kipf, T.N., Welling, M.: Semi-supervised classification with graph convolutional networks. In: 5th International Conference Learning Representation ICLR 2017 - Conference Track Proceedings, pp. 1–14 (2017)

6. Barbu, E., Héroux, P., Adam, S., Trupin, É.: Using bags of symbols for automatic indexing of graphical document image databases. In: Liu, W., Lladós, J. (eds.) GREC 2005. LNCS, vol. 3926, pp. 195–205. Springer, Heidelberg (2006). https://doi.org/10.1007/11767978_18

7. Kumar, J., Prasad, R., Cao, H., Abd-Almageed, W., Doermann, D., Natarajan, P.: Shape codebook based handwritten and machine printed text zone extraction. In: ProcSPIE (2011). https://doi.org/10.1117/12.876725

8. Kang, L., Kumar, J., Ye, P., Li, Y., Doermann, D.: Convolutional neural networks for document image classification. In: 2014 22nd International Conference on Pattern Recognition, p. 3168–3172 (2014)

9. Afzal, M.Z., Capobianco, S., Malik, M.I., Marinai, S., Breuel, T.M., Dengel, A., et al.: Deepdocclassifier: document classification with deep convolutional neural network. In: Proceedings of International Conference Document Analysis and Recognition, ICDAR, pp. 1111–1115. IEEE (2015)

10. Krizhevsky, A., Sutskever, I., Hinton, G.E.: ImageNet classification with deep convolutional neural networks. Commun. ACM **60**, 84–90 (2017)

11. Zhou, X., Yao, C., Wen, H., Wang, Y., Zhou, S., He, W., et al.: EAST: an efficient and accurate scene text detector. In: Proceedings - 30th IEEE Conference oComputer Vision Pattern Recognition, CVPR 2017, 2017-January, pp. 2642–2651 (2017)

12. Shi, B., Bai, X., Yao, C.: An end-to-end trainable neural network for image-based sequence recognition and its application to scene text recognition. IEEE Trans. Pattern Anal. Mach. Intell. **39**, 2298–304 (2017)

13. Tian, Z., Huang, W., He, T., He, P., Qiao, Yu.: Detecting text in natural image with connectionist text proposal network. In: Leibe, B., Matas, J., Sebe, N., Welling, M. (eds.) ECCV 2016. LNCS, vol. 9912, pp. 56–72. Springer, Cham (2016). https://doi.org/10.1007/978-3-319-46484-8_4

14. Zhang, C., Liang, B., Huang, Z., En, M., Han, J., Ding, E., et al.: Look more than once: an accurate detector for text of arbitrary shapes. In: Proceedings of IEEE Computer Society Conference on Computer Vision and Pattern Recognition, 2019-June, pp. 10544–10553 (2019)

15. Noce, L., Gallo, I., Zamberletti, A., Calefati, A.: Embedded textual content for document image classification with convolutional neural networks. In: DocEng 2016 - Proceedings 2016 ACM Symposium Document Engineering, pp. 165–73 (2016)

16. Devlin, J., Chang, M.W., Lee, K., Toutanova, K.: BERT: pre-training of deep bidirectional transformers for language understanding. In: NAACL HLT 2019–2019 Conference North America Chapter Association Computer Linguistius Human Language Technology - Proceedings Conference, pp. 1:4171–1:4186 (2019)

17. Harley, A.W., Ufkes, A., Derpanis, K.G.: Evaluation of deep convolutional nets for document image classification and retrieval. In: Proceedings of International Conference on Document Analysis Recognition, ICDAR, 2015-November, pp. 991–995 (2015)

18. Bao, H., Dong, L., Wei, F., Wang, W., Yang, N., Liu, X., et al.: Unilmv2: pseudo-masked language models for unified language model pre-Training. In: 37th International Conference on Machine Learning, ICML 2020, Part F16814, pp. 619–629 (2020)

19. Tensmeyer, C., Martinez, T.: Analysis of convolutional neural networks for document image classification. In: Proceedings of International Conference on Document Analysis Recognition, ICDAR, vol. 1, pp. 388–393 (2017)

20. Afzal, M.Z., Kolsch, A., Ahmed, S., Liwicki, M.: Cutting the error by half: investigation of very deep cnn and advanced training strategies for document image classification. In: Proceedings of International Conference on Document Analysis Recognition, ICDAR, vol. 1, pp. 883–888 (2017)
21. Das, A., Roy, S., Bhattacharya, U., Parui, S.K.: Document image classification with intra-domain transfer learning and stacked generalization of deep convolutional neural networks. In: Proceedings - International Conference on Pattern Recognition, 2018-August, pp. 3180–3185 (2018)

Continual Learning of 3D Point Cloud Generators

Michał Sadowski[1]([✉]), Karol J. Piczak[1]([✉]), Przemysław Spurek[1],
and Tomasz Trzciński[1,2,3]

[1] Jagiellonian University, Kraków, Poland
m.sadowski@doctoral.uj.edu.pl, {karol.piczak,przemyslaw.spurek}@uj.edu.pl
[2] Warsaw University of Technology, Warszawa, Poland
tomasz.trzcinski@pw.edu.pl
[3] Tooploox, Wrocław, Poland

Abstract. Most continual learning evaluations to date have focused on fully supervised image classification problems. This work for the first time extends such an analysis to the domain of 3D point cloud generation, showing that 3D object generators are prone to catastrophic forgetting along the same vein as image classifiers. Classic mitigation techniques, such as regularization and replay, are only partially effective in alleviating this issue. We show that due to the specifics of generative tasks, it is possible to maintain most of the generative diversity with a simple technique of uniformly sampling from different columns of a progressive neural network. While such an approach performs well on a typical synthetic class-incremental setup, more realistic scenarios might hinder strong concept separation by shifting task boundaries and introducing class overlap between tasks. Therefore, we propose an autonomous branch construction (ABC) method. This learning adaptation relevant to parameter-isolation methods employs the reconstruction loss to map new training examples to proper branches of the model. Internal routing of training data allows for a more effective and robust continual learning and generation of separate concepts in overlapping task setups.

Keywords: Continual learning · 3D point clouds · Generative models · Reconstruction loss

1 Introduction

Catastrophic forgetting is a well-known phenomenon of incremental training. Numerous studies have been performed in supervised image classification problems [5,10,18] showing that models trained on a sequence of disjoint tasks lose their discriminative capability very rapidly. This decline is especially profound when the task identity is not accessible during evaluation (*class-incremental learning*).

As continual learning of generative models is still a scarcely researched subject, most works have concentrated on images. In this paper, we shift our focus to

T. Mantoro et al. (Eds.): ICONIP 2021, LNCS 13108, pp. 330–341, 2021.
https://doi.org/10.1007/978-3-030-92185-9_27

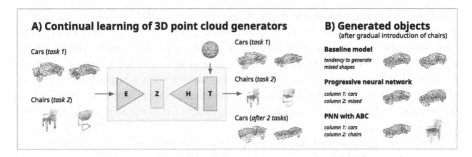

Fig. 1. *Panel A* summarizes continual learning of *HyperCloud*-based point cloud generators. A convolutional encoder (E) converts training examples into a latent representation (Z) that is used by a hypernetwork decoder (H) to create dedicated networks (T) transforming uniform spheres into the original shapes. After introducing a new class, the ability to generate instances of the old class is quickly forgotten. *Panel B* shows examples of objects generated in the *gradual introduction* task setup. Autonomous branch construction helps maintain concept separation between columns of a progressive neural network.

generative models of 3D point clouds trained incrementally on an object reconstruction task. This problem still requires a certain level of complexity from the model but is much easier to evaluate than its image counterparts relying on the FID [8] metric.

Our goal is to create a generator that will provide us with a diverse pool of objects based on all the classes in the training dataset. As the number of encountered classes grows, we expect that introducing new concepts will not require a complete re-training of the model. Unfortunately, in contrast to reconstruction tasks selected by Thai et al. [26], our experiments show that point cloud generators trained sequentially with new classes rapidly lose their ability to generate examples of previously seen objects, similarly to their image counterparts.

A typical approach to combat the loss of previous knowledge is to use various regularization techniques to maintain model parameters close to their values suitable for previous tasks. Another even more effective solution is to support the model with some form of memory. For this purpose, we can either use a memory buffer replaying exact samples from the past or an auxiliary generative model. Unfortunately, classic regularization and memory-based techniques prove to be only partially effective in alleviating catastrophic forgetting in point cloud generators.

Therefore, our first proposition is to apply a simple and effective mechanism that maintains most of the model's generative diversity by sampling generated objects uniformly from all the columns of a *progressive neural network* [21]. This approach can also be employed with other similar parameter isolation techniques, such as PackNet [17] or SupSup [27].

This sampling procedure performs well on a typical class-incremental setup where task boundaries are clearly defined, and there is no overlap between classes in different training exposures. Such an assumption might not hold in more realistic scenarios, where new concepts can appear gradually, and examples of previously seen classes can intermix with the current task. While more realistic task definitions are a still-developing area of research in continual learning, a couple of papers have proposed methods more suitable for such setups. However, they mainly concentrate on image tasks and classification problems. In our case, we can propose a different solution, tailored explicitly for generative tasks involving a reconstruction loss.

As our main contribution, we propose an adaptation to the learning procedure, dubbed ABC (*autonomous branch construction*), applicable to various parameter isolation techniques which create separate branches of the model. This method allows for effective learning of new tasks with unknown overlap with previously seen classes by dynamically matching training examples with the most corresponding model branch.

2 Related Work

2.1 Point Cloud Generation

Although a 3D point cloud has a seemingly simple structure, its generation with deep learning is challenging. Gadelha et al. [7] and Stypułkowski et al. [25] introduced methods based on variational and adversarial auto-encoders, which are trained by directly optimizing the chamfer (CD) or earth mover's distance (EMD) between two point sets, producing a fixed number of points for each shape. PointFlow, a probabilistic framework proposed by Yang et al. [29], uses a continuous normalizing flow for the distribution of shapes and points given a shape for point cloud generation. Spurek et al. [24] introduced the Hyper-Cloud model based on a hypernetwork, which returns weight definitions for a second network. This object-specific target network then transforms a uniform 3D ball into a given shape. The advantage of this approach lies in the possibility of generating a fully adjustable number of points for each shape, effectively parametrizing its surface. For this reason, we choose HyperCloud as our baseline model.

2.2 Continual Learning for Generative Models

Works in this area focus mainly on GAN and VAE models trained on MNIST, Fashion MNIST, SVHN, and CIFAR10. Several authors [16,22,28] apply EWC [13] and generative replay [23] showing good results from the latter, provided that the dataset is simple enough.

A different approach in VAE models is latent regularization. Achille et al. [2] introduce a VASE model for learning nonoverlapping disentangled representations with an auxiliary classifier inferring the most likely environment and a replay feedback loop. BooVAE [14] learns the approximation of the aggregated posterior as a prior for each task, while Keng et al. [11] present a similar approach with maximization of the mutual information between classes and latent variables during training.

2.3   Unsupervised Continual Learning

Several papers that have tackled continual learning with more diverse task setups and unsupervised learning show various resemblances to our approach. Lee et al. [15] describe a model expansion approach for task-free setups. However, their problem is multi-class and requires label knowledge. Khare et al. [12] introduce a similar method for routing training data, though focusing on classification tasks. Abati et al. [1] use task-specific gating modules inferring the task identity with an auxiliary task classifier. Continual unsupervised representation learning [20] also shares many similarities with our work, but it requires deep generative replay to mitigate forgetting in the shared representation. De Lange and Tuytelaars [4] use dynamic memory partitioning for replaying prototypes. The importance of task data distribution has been described by Hsu et al. [9].

3   Our Method

In this section, we first characterize continual learning scenarios considered in our work, motivating why a typical sequence of disjoint tasks might be overly simplifying. We follow with a description of progressive neural networks, showing how we can use them in continual learning of point cloud generators, and noting their potential weaknesses in more elaborate task setups. Finally, we introduce our method that can mitigate these issues.

3.1   Continual Learning Setting

There are many scenarios of continual learning experiments. Most of them are designed specifically for supervised classification problems. Only class-incremental learning (defined in [10]) is compatible with generative models without the need to impose additional constraints. It can be easily applied by training a model sequentially on separate classes chosen from the dataset. Although it perfectly demonstrates the phenomenon of catastrophic forgetting, such a setup may oversimplify the problem when applying parameter isolation techniques, which might easily use different model branches for generative purposes.

Moreover, it is unlikely that we will encounter a perfect separation between exposures to different classes in real-world problems. More realistic scenarios should assume the potential for classes to overlap and repeat. Examples of such more complex data splits for continual learning were introduced in [6,9]. We follow this approach, and besides standard disjoint class-incremental learning, we experiment on tasks with shifted boundaries, task repetition, and gradual class introduction according to the Dirichlet distribution.

3.2   Parameter Isolation Techniques

One of the most common continual learning parameter isolation methods is the use of progressive neural networks [21]. This approach works at the architectural level by incorporating agents that learn a series of tasks and transfer knowledge to improve convergence speed. For each new task, we instantiate a new column (a neural network) and freeze the weights of the rest of the model. In order to prevent the model from overgrowing, we can decrease layer sizes in consecutive columns. Knowledge transfer is enabled via lateral connections to the corresponding layers of previously learned columns. This aspect differentiates progressive neural networks from simply learning an ensemble of separate models, allowing for faster knowledge acquisition through forward transfer. A progressive neural network starts with only one column. When training of the first task completes, this column is frozen. Then, we add a second column. During training, both columns process the input. In consecutive tasks, we add new columns analogically. Progressive neural networks are the most effective when the task identity is known during the testing phase because it is possible to choose an appropriate column to process the data. We assume that such information is not available in our experiments. However, as our goal is not to classify point clouds but to generate a diverse pool of samples from the model in the testing phase, we propose to sample an equal number of 3D shapes from each column.

As long as each column learns a single class, this approach performs remarkably well. However, it deteriorates when we move to tasks in which classes can overlap. Similar to joint training, it shifts the distribution of the generated objects in the direction of a shape averaged between classes. To disentangle class information, we propose autonomous branch construction.

3.3   Autonomous Branch Construction

We introduce autonomous branch construction (ABC) as an extension for parameter isolation methods in generative models. The general idea of ABC can be described as routing of training examples based on a thresholded reconstruction loss. This allows for effective learning of new tasks with unknown overlap with previously seen classes.

More specifically, we split each training iteration into two steps: parameter selection and restricted model weights update.

During the parameter selection step, we pass the whole input batch through all the branches (columns in PNN nomenclature) of the model and calculate reconstruction loss for each one of them. Then, we assign each sample to a branch for which it has the lowest loss. This means that training examples are preferably assigned to branches that have already encountered a similar concept. If the loss value exceeds a predefined threshold, we assign them to a new model branch (column). Throughout each task, only one new branch can be created. If the new branch has not been selected with a required frequency (e.g. in task repetition where all concepts can be mapped to previous branches), it is discarded.

In the second phase, we perform backpropagation for standard model loss and update model weights. However, each reconstructed sample influences only the branch assigned to it in the parameter selection step. It means that each branch of the model specializes in specific class reconstruction and generation.

4   Evaluation Protocol

4.1   Point Cloud Generation

We evaluate our continual learning setups on a HyperCloud [24] neural network - a state-of-the-art model for generating 3D point clouds. Its architecture consists of three neural networks working in tandem, as depicted in Fig. 1a. The encoder part maps an input point cloud to a lower-dimensional latent space. The hypernetwork decoder maps values from the latent space to a vector of weights, constructing a separate target network for each object. This network models a function $T : \mathbb{R}^3 \longrightarrow \mathbb{R}^3$, which transforms points from the prior distribution (unit uniform ball) to the elements of the given object. Therefore, a target network fully defines a 3D object and can produce any number of points representing its surface.

The whole system can be trained with a reconstruction loss using a selected distance metric (e.g., chamfer or earth mover's distances). In addition to the reconstruction loss, an additional term ensures that the latent values are distributed according to the standard normal density.

In order to validate that the results are not specific to the application of a hypernetwork approach, we repeat the baseline experiment on an analogous model with the decoder mapping latent values directly to 3D shapes (with a predefined number of points).

4.2   Model Evaluation

As our main evaluation metric we choose Jensen-Shannon Divergence (JSD) which measures the distance between P_r and P_g, the marginal distributions of points in the S_r (set of reference point clouds) and S_g (set of generated point clouds), computed by discretizing the space into 28^3 voxels:

$$\text{JSD}(P_r, P_g) = \frac{1}{2}D_{KL}(P_r\|M) + \frac{1}{2}D_{KL}(P_g\|M) \qquad (1)$$

where $M = \frac{1}{2}(P_r + P_g)$ and $D_{KL}(\cdot\|\cdot)$ is the Kullback-Leibler divergence between two distributions. We expect the JSD for an ideal model to approach 0, although a model always generating the average shape can also achieve a perfect score, which is a drawback of this metric. Therefore, we also support our findings with additional metrics used for assessing the quality of point cloud generators [3,29]: coverage (COV), minimum matching distance (MMD) and 1-nearest neighbor accuracy (1-NNA). These metrics come in two variants employing chamfer (CD) or earth mover's distance (EMD) as a measure of similarity.

Besides parameter isolation methods, for the first time, we evaluate point cloud generators on representative continual learning methods from the other two general families: replay methods and regularization-based methods. As an example of regularization techniques, we employ Elastic Weight Consolidation (EWC) [13], which tries to safeguard values of parameters essential to good performance on prior tasks by assigning a penalty to deviations from their previous value. For comparison, we also employ an identity importance matrix resulting in L_2 regularization on the distance to old parameters.

In terms of memory-based approaches, we evaluate generative replay [23], which, apart from the main task solving network, uses a generator model to recreate samples of previously seen classes. HyperCloud is a generative model itself, so there is no need to make this distinction. During training, a copy of the main model can generate additional data samples, which are concatenated with each batch of the training data. The ratio r between sampled and real data depends on the desired importance of a new task compared to older tasks. In our experiments, we test three different values 20%, 40%, 60%. For comparison, we also perform experiments with exact replay. Instead of using an old copy of a generative model to generate data, we sample real data of previous tasks from different memory buffer sizes (128 or 512) with a 1:1 or 1:2 ratio of old to new examples.

5   Experiments

5.1   Implementation Details

We use HyperCloud architecture as described in [24] where the encoder is a PointNet-like [19] network consisting of 1D convolutional layers with 64, 128, 256, 512, 512 channels, and fully connected layers with 512 and 2048 neurons. The latent dimension is set to 2048. The decoder is a fully connected neural network with layer widths equal to 64, 128, 512, 1024, 2048 and a final layer mapping directly to the target network. The target network is a small multilayer perceptron with 32, 64, 128, and 64 neurons. All networks use ReLU activations.

We train the HyperCloud model with Adam optimizer, setting the learning rate to 0.0001, with standard betas of 0.9 and 0.999, and no weight decay. We utilize chamfer loss for reconstruction, setting the reconstruction loss coefficient to 0.05 and KLD loss coefficient to 0.5. We also define the time span of the progressive unit sphere radius normalization [24] as 100 epochs. The training batch size is set to 64 and 32 for computing validation metrics. We perform validation every 10 epochs for the first 50 epochs of a given task, then decay to a frequency of 25 (past 50 epochs) and 50 (past 100 epochs). Each epoch processes 5000 examples from the current task. The ABC threshold is set at 8.0.

Our main experiment evaluates models trained on point clouds of *cars*, *chairs*, *airplanes* and *tables* from the ShapeNet dataset. Each point cloud has 2048 points. 85% of the objects are selected for training purposes, 5% of instances in the whole dataset are designated for metric evaluation.

We train the first class for 1000 epochs, and subsequent tasks are limited to 500 epochs. For regularization methods, we also perform a hyperparameter sweep over independent encoder and hypernetwork regularization strengths.

5.2   Task Setups

In our baseline class-incremental learning setup, each task consists of samples from one class (1: *cars*, 2: *chairs*, 3: *airplanes* and 4: *tables*). In the task repetition sequence, we train models sequentially on three tasks (1: *cars*, 2: *chairs*, 3: *cars*). The overlapping sequence presents a shift in the task boundary where 30% examples of the last class are still seen in the new task (1: *100% cars*, 2: *30% cars, 70% chairs*). This is akin to baseline training with exact replay. The gradual introduction of new concepts presents new classes with a decreasing frequency derived from the expected value of Dirichlet distribution with $\alpha = 2$ (1: *100% cars*, 2: *69% cars, 31% chairs*). This would have a similar effect as an introduction of a very aggressive replay technique.

5.3   Results

We present detailed results for our evaluation protocols in Table 1. An evolution of the JSD metric in the main experiment can be seen for selected methods in Fig. 2.

Baseline training without any mitigation technique shows strong signs of catastrophic forgetting across all metrics. While regularization techniques can influence this process only marginally, memory-based approaches show a much more favorable behavior. A progressive neural network (PNN) effectively maps distinct concepts in a single-class per task scenario. This can be seen in the PNN generating examples only from the last column, which rapidly diverge from the mixed distribution. A uniform sampling technique has an excellent profile that almost matches the best available baseline of joint training on all classes for 2500 epochs.

A significant drawback of progressive neural networks is the cost of model expansion. In a standard setting, each new column has the same size as the original model. In order to reduce this impact, we also evaluate a *decaying PNN* where the first column is of standard size, but each consecutive column has a 50% reduced size compared to the previous column. For a sequence of four tasks, we can safely use such a reduction (from 4 model sizes to 1.875) without impacting the performance.

We also verify additional variants of the baseline model. Replacing the hyper-network with a standard VAE decoder has no favorable influence on the final metrics, similarly with unsupervised pre-training on 30 separate classes. When compared with a control experiment (additional 1000 epochs on the first task), it seems that slight improvements are only due to the elongated training procedure.

When we introduce task repetition, vanilla PNN still performs much better than the baseline model. However, it loses some of its edge due to a shifting balance in the generated classes. Introducing ABC helps the performance by preventing the model from creating a new column for a repeated concept.

Similar behavior can be seen with last class overlap. The baseline model mostly shifts to the new class with some decrease in forgetting due to the replay characteristic of this setup. Vanilla ABC will learn the new column in the same way, but it maintains a previous column that helps compensate for this setback. In contrast, PNN with ABC is capable of better concept isolation, creating two separate columns, one for each class.

Table 1. Evaluation of models trained sequentially with different continual learning techniques. Generative capability is assessed across four metrics (with either chamfer or earth mover's distances). Cells represent final metric values after training on all tasks and validating on an equal mix of all seen classes. Values for regularization and replay techniques show average results across hyperparameter sweeps.

Results for incremental learning with single-class tasks

Method	JSD	MMD		COV%		1-NNA%	
		CD	*EMD*	*CD*	*EMD*	*CD*	*EMD*
Baseline (HyperCloud)	35.3	14.8	16.7	23.8	22.7	80.3	80.5
Baseline (VAE)	41.4	12.5	17.4	23.4	17.2	88.5	91.6
Baseline with pre-training	33.7	15.6	16.6	28.1	26.6	82.2	82.4
Pre-training control	31.7	14.2	16.2	29.3	29.3	79.1	79.9
L_2 regularization	28.6	9.1	14.4	25.1	26.3	91.7	93.9
EWC regularization	28.8	8.9	13.9	28.0	27.0	87.9	88.6
Exact replay	18.1	6.0	**11.6**	41.8	39.8	75.5	77.1
Generative replay	19.2	10.5	15.1	46.5	**48.4**	**69.1**	**72.3**
PNN (*first column*)	27.2	17.4	15.0	21.5	19.9	76.8	81.8
PNN (*last column*)	38.4	15.9	19.7	28.9	27.0	89.3	96.5
PNN (*uniform sampling*)	9.5	**5.4**	12.2	51.2	**48.4**	**69.1**	83.8
Decaying PNN (*uniform*)	**8.0**	5.5	12.2	**52.7**	**48.4**	74.8	82.6
Joint training (best case)	8.0	5.0	10.6	54.7	53.5	63.5	65.0
Ideal metric behavior:	→ 0	→ 0		→ 100%		→ 50%	

Results for different task setups

Task repetition							
Baseline (HyperCloud)	15.3	10.0	11.9	39.5	39.5	71.3	69.7
PNN (*uniform sampling*)	7.7	3.5	9.3	**52.3**	**50.0**	64.6	70.3
PNN with ABC (*uniform*)	**3.8**	**3.3**	**8.8**	51.6	**50.0**	**63.1**	**67.4**
30% overlap of the last class							
Baseline (HyperCloud)	9.1	3.9	9.9	42.6	44.9	67.8	70.7
PNN (*uniform sampling*)	8.0	3.5	9.7	50.0	**46.9**	66.4	77.1
PNN with ABC (uniform)	**4.9**	**3.0**	**8.9**	50.8	46.1	**58.8**	70.3
Gradual introduction of new concepts							
Baseline (HyperCloud)	**6.2**	3.5	**9.0**	48.8	**52.7**	68.2	**69.7**
PNN (*uniform sampling*)	9.2	4.9	10.3	44.1	49.6	66.8	74.6
PNN with ABC (uniform)	6.6	**2.6**	9.1	**49.2**	39.8	**59.6**	77.5

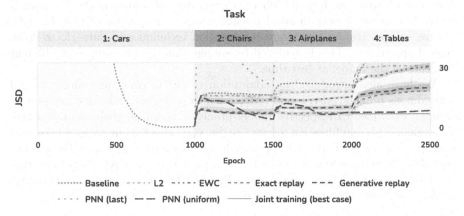

Fig. 2. JSD values (multiplied by 10^2) computed between point clouds generated by respective models and reference examples from the validation set (a balanced mix of all seen classes). Shaded color bands indicate maximal and minimal metric values for different settings of hyperparameters. Best viewed in color. (Color figure online)

The most challenging setup for PNN is the gradual introduction of new concepts. This task sequence creates a solid baseline, equivalent to replaying 69% of old examples with an unlimited memory buffer. On a positive note, memory-based techniques will not be able to improve it further. In this case, PNN can only compete with this baseline when using ABC. However, one additional fact in support of ABC can be inferred by looking at the generated objects in Fig. 1b. Both the HyperCloud baseline and vanilla PNN tend to create objects that resemble a shape mixed between classes (cars with chair legs instead of wheels), whereas concepts generated from the ABC-augmented model more closely resemble the original classes. The JSD metric does not easily capture this difference.

6   Conclusion

In this work, for the first time, we analyzed 3D point cloud generative models in a continual learning scenario. We extended our protocol to sequences of exposures where task boundaries are not clearly defined alongside a typical disjoint single-class incremental evaluation. We have shown that point cloud generators are prone to catastrophic forgetting, similarly to models trained continually in classification tasks. This phenomenon contrasts with the findings of Thai et al. [26], where continual reconstruction tasks did not exhibit such a behavior. These discrepancies might be introduced through the specifics of task definitions or the type of model used.

We also highlight that in continual learning of generative models, an overlooked potential lies in applying classic parameter isolation techniques, such as progressive neural networks. This family of methods is often discarded in supervised image classification due to the required knowledge of the task identity at test time. A different formulation of a generative problem, where the goal is to create diverse objects from all possible classes irrespective of task identity, creates an opportunity for parameter isolation techniques to outperform other typical approaches. This is achieved by simple uniform sampling from different branches of the model at test time.

We have also introduced an adaptation to the learning procedure, named autonomous branch construction (ABC), which performs a selective mapping of training examples to respective model branches based on their reconstruction loss value. This modification proved to be very effective in tasks where new concepts appear gradually and previously seen examples mix with the current task data. Such a technique could also be applied to other parameter isolation techniques, like models pruned through PackNet [17], to mitigate the impact on model size due to the expanding nature of progressive neural networks.

Acknowledgments. This research was funded by Foundation for Polish Science (grant no POIR.04.04.00-00-14DE/18-00 carried out within the Team-Net program co-financed by the European Union under the European Regional Development Fund), National Science Centre, Poland (grant no 2020/39/B/ST6/01511) and Warsaw University of Technology (POB Research Centre for Artificial Intelligence and Robotics within the Excellence Initiative Program - Research University). The author has applied a CC BY license to any Author Accepted Manuscript (AAM) version arising from this submission, in accordance with the grants' open access conditions.

References

1. Abati, D., et al.: Conditional channel gated networks for task-aware continual learning. In: IEEE/CVF Conference on Computer Vision and Pattern Recognition (CVPR) (2020)
2. Achille, A., et al.: Life-long disentangled representation learning with cross-domain latent homologies. In: Conference on Neural Information Processing Systems (NeurIPS) (2018)
3. Achlioptas, P., et al.: Learning representations and generative models for 3D point clouds. In: International Conference on Machine Learning (ICML) (2018)
4. De Lange, M., Tuytelaars, T.: Continual prototype evolution: learning online from non-stationary data streams. In: International Conference on Computer Vision (ICCV) (2021)
5. De Lange, M., et al.: A continual learning survey: defying forgetting in classification tasks. IEEE Trans. Pattern Anal. Mach. Intell. (2021)
6. Deja, K., et al.: Multiband VAE: latent space partitioning for knowledge consolidation in continual learning (2021). arXiv:2106.12196
7. Gadelha, M., Wang, R., Maji, S.: Multiresolution tree networks for 3D point cloud processing. In: European Conference on Computer Vision (ECCV) (2018)
8. Heusel, M., et al.: GANs trained by a two time-scale update rule converge to a local Nash equilibrium. In: International Conference on Neural Information Processing Systems (NeurIPS) (2017)

9. Hsu, T.M.H., Qi, H., Brown, M.: Measuring the effects of non-identical data distribution for federated visual classification (2019). arXiv:1909.06335
10. Hsu, Y.C., et al.: Re-evaluating continual learning scenarios: a categorization and case for strong baselines (2018). arXiv:1810.12488
11. Kang, W.Y., Han, C.H., Zhang, B.T.: Discriminative variational autoencoder for continual learning with generative replay (2019)
12. Khare, S., Cao, K., Rehg, J.: Unsupervised class-incremental learning through confusion (2021). arXiv:2104.04450
13. Kirkpatrick, J., et al.: Overcoming catastrophic forgetting in neural networks. Proc. Natl. Acad. Sci. (PNAS) **114**(13), 3521–3526 (2017)
14. Kuzina, A., Egorov, E., Burnaev, E.: BooVAE: a scalable framework for continual VAE learning under boosting approach (2019). arXiv:1908.11853
15. Lee, S., et al.: A neural Dirichlet process mixture model for task-free continual learning. In: International Conference on Learning Representations (ICLR) (2020)
16. Lesort, T., et al.: Generative models from the perspective of continual learning. In: International Joint Conference on Neural Networks (IJCNN) (2019)
17. Mallya, A., Lazebnik, S.: PackNet: adding multiple tasks to a single network by iterative pruning. In: IEEE Conference on Computer Vision and Pattern Recognition (CVPR) (2018)
18. Masana, M., et al.: Class-incremental learning: survey and performance evaluation (2020). arXiv:2010.15277
19. Qi, C.R., et al.: PointNet: Deep learning on point sets for 3D classification and segmentation. In: IEEE Conference on Computer Vision and Pattern Recognition (CVPR) (2017)
20. Rao, D., et al.: Continual unsupervised representation learning. In: Conference on Neural Information Processing Systems (NeurIPS) (2019)
21. Rusu, A.A., et al.: Progressive neural networks (2016). arXiv:1606.04671
22. Seff, A., et al.: Continual learning in generative adversarial nets (2017). arXiv:1705.08395
23. Shin, H., et al.: Continual learning with deep generative replay. In: Conference on Neural Information Processing Systems (NeurIPS) (2017)
24. Spurek, P., et al.: Hypernetwork approach to generating point clouds. In: International Conference on Machine Learning (ICML) (2020)
25. Stypułkowski, M., et al.: Conditional invertible flow for point cloud generation (2019). arXiv:1910.07344
26. Thai, A., et al.: Does continual learning = catastrophic forgetting? (2021). arXiv:2101.07295
27. Wortsman, M., et al.: Supermasks in superposition. In: Conference on Neural Information Processing Systems (NeurIPS) (2020)
28. Wu, C., et al.: Memory replay GANs: Learning to generate images from new categories without forgetting. In: Conference on Neural Information Processing Systems (NeurIPS) (2018)
29. Yang, G., et al.: PointFlow: 3D point cloud generation with continuous normalizing flows. In: IEEE/CVF International Conference on Computer Vision (ICCV) (2019)

Attention-Based 3D ResNet for Detection of Alzheimer's Disease Process

Mingjin Liu, Jialiang Tang, Wenxin Yu$^{(\boxtimes)}$, and Ning Jiang

School of Computer and Science, Southwest University of Science and Technology, Mianyang, China
yuwenxin@swust.edu.cn

Abstract. Alzheimer's disease is a classic form of dementia that is progressive and irreversible. It is crucial to use 3D magnetic resonance imaging (MRI) to diagnose Alzheimer's disease (AD) in the early treatment, which is beneficial for controlling the disease and allows the patient to receive a proper cure. Previous studies use a deeper model with more parameters and a tedious training process. These methods neglect the local changes in brain regions, and their performance is unsatisfactory. Therefore, in this paper, we propose a 3D Attention Residual Network (3D ARNet) to classify 3D brain MRI into 4-way classification. Specifically, our proposed 3D ARNet is a shallow network with only 10 layers, which is compact and converges fast. Moreover, we propose the attention mechanism to utilize the expressive information in brain MRI and apply the Instance-Batch Normalization (IBN) to highlight the global changes and local changes in MRI at the same time. We conduct extensive experiments on benchmark datasets (i.e., ADNI dataset). The experimental results demonstrate that our method is more efficient in diagnosing Alzheimer's disease.

Keywords: Alzheimer's disease · Deep learning · Image classification · Attention mechanism · Magnetic resonance imaging

1 Introduction

Alzheimer's disease (AD) is a common neurodegenerative brain disease among the elderly [12]. AD's pathological features include senile plaque precipitation outside brain cells, fibrous tangles inside nerve cells, synaptic dysfunction, and massive loss of neurons. Clinically, AD is manifested as memory impairment, aphasia, cognitive impairment, and behavioral changes. Unfortunately, Alzheimer's disease is difficult to detect in the early stages. Meanwhile, the clinical detection accuracy of AD is very low. It leads the patient to miss the best treatment time. It is estimated that by 2050, AD patients will rise to 3 million [23]. Mild Cognitive Impairment (MCI) is a transitional stage between normal cognitive decline and AD [11], which is the earliest clinically measurable stage of AD. About 15% of MCI patients will convert to AD every year [14]. Therefore,

© Springer Nature Switzerland AG 2021
T. Mantoro et al. (Eds.): ICONIP 2021, LNCS 13108, pp. 342–353, 2021.
https://doi.org/10.1007/978-3-030-92185-9_28

timely diagnosis of early MCI (EMCI) and late MCI (LMCI) play a crucial role in reducing the incidence of AD.

In recent years, with the development of artificial intelligence, many researchers have combined MRI with traditional machine learning to develop an AD automatic diagnosis system. Researchers used random forest (RF) [2], support vector machines (SVM) [15] and principal component analysis (PCA) [22] to diagnose AD. Subsequently, the use of deep learning methods for the diagnosis of AD increases. In terms of deep learning, there is a shift from the 2D convolutional neural networks (CNNs) to 3D CNNs to diagnose AD. In related studies in the past two years [16,19], the researchers used deeper models with more parameters, using dense connections to maximize information and gradient flow. However, there may be some obvious problems. Previous studies did not focus on changes in brain regions from the features of the disease. Moreover, through MRI analysis, we find that the changes in the brain are mainly concentrated in the expansion of the ventricle, the atrophy of the hippocampus, and the atrophy of brain tissue. And they use a complicated deep network model blindly, which is easily overfit and significantly drops the model's ability. Meanwhile, the training of the complicated model is time-consuming. For the network model, we can simplify the complexity of the network model.

Therefore, in this paper, we propose a 3D Attention Residual Network (3D ARNet) to improve the diagnosis of Alzheimer's disease process. The shallow 3D ARNet network with only 10 layers is more compact and converges faster. Specifically, we introduce the attention mechanism into 3D ARNet aims to capture the local brain changes, which is beneficial to extract more expressive features in MRI. At the same time, instead of normalizing the global or local features individual, we use Instance-Batch Normalization (IBN) to regularize global and local features together. IBN can reflect the global dependence and local relationship of features and effectively improve 3D ARNet's performance. In addition, it is crucial to reduce misdiagnosis of the diagnosis of brain diseases. Therefore, we use the method of probability fusion to improve the diagnosis effect, and the diagnosis result is determined by the three network models, rather than a single network. Hence, the main contributions of this paper are as follows:

1. We propose a shallow 3D residual network based on the attention mechanism. While improving the diagnostic performance of the Alzheimer's disease stages, the network incorporates disease characteristics and is easy to train.
2. We combine Instance Normalization (IN) and Batch Normalization (BN) as Instance-Batch Normalization (IBN), regularize global and local features simultaneously.
3. For reducing misdiagnosis, we use a probabilistic fusion method to integrate the three 3D network models, instead of a single network determining the diagnosis result.

The rest of this paper is organized as follows. Section 2 reviews the literature. Section 3 provides the details of our method. Experimental results and discussion are presented in Sect. 4. Finally, Sect. 5 concludes our work.

2 Related Works

In the initial study, various machine learning methods were applied to this field. Gerardin et al. [15] use a support vector machine to calculate and compare the brain atrophy volume of patients to classify stages of the disease. Liu et al. [9] use an unsupervised learning algorithm of locally linear embedding to convert multivariate MRI data of regional brain volume and cortical thickness into a local linear space with fewer dimensions, and then use brain features to train the classifier. Dimitriadis et al. [2] realize in a four-class classification of 61.9% accuracy by combining MRI-based features with a random forest-based ensemble strategy. However, machine learning methods all require complex feature engineering.

With the development of AD diagnosis problems, deep learning (DL) has emerged as one of the most promising tools to address AD diagnosis. Cheng et al. [1] propose a classification framework based on the combination of 2D CNN and recurrent neural networks (RNNs), which learns the features of 3D positron emission tomography (PET) images by decomposing the 3D images into a sequence of 2D slices. In their framework, 2D CNN is built to capture the intra-slice features while RNN is used to extract the inter-slice features for final classification. But, most 2D method studies may have information loss due to 3D MRI translate into the 2D slices.

Hence, in recent years, some 3D neural networks were proposed to solve the problem of insufficient information in 2D slice-level methods. Liu et al. [8] extract features from MRI and PET images using the deep learning architecture of stacked autoencoders, and achieve an accuracy of 47.42% on the 4-way classification task. Hosseini-Asl et al. [5] propose a deep 3D CNN for the prediction of AD, which can capture AD biomarkers and adapt to the common features of data sets in different domains. Korolev et al. [7] show using the residual and plain 3D CNN architectures to accelerant feature extraction on 2-way classification tasks without complex preprocessing or model stacking. Lu et al. [10] propose multi-scale and multi-modal deep neural network (MMDNN) to incorporate multiple scales of information from multiple regions in the gray matter of the brain MRI. Wang et al. [19] propose an ensemble of 3D densely connected convolutional networks (3D-DenseNets) for AD and MCI diagnosis. They introduce dense connections to maximize the flow of information, to prevent the gradient of the deep network from disappearing. Ruiz et al. [16] obtain tolerable performance in 4 classification by using 3D integration method. Their methods consist of three Densenet-121 classifiers with different growth rates and parameters and generate probability output through the softmax layer. Finally, probabilistic fusion is performed by the previous probability scores to obtain the final classification results. In the above two works, researchers may not have incorporated features of the disease and ignore the local changes in the brain, and blindly use deeper networks for feature extraction. In our work, we use attention mechanisms and instance-batch normalization in the shallow neural network to solve these problems.

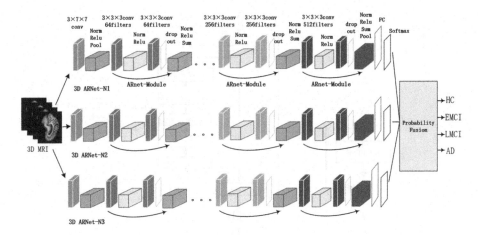

Fig. 1. The structure of probability fusion classification based on 3D ARNet-10

3   Proposed Method

In this section, we describe our method. The overall structure diagram is shown in Fig. 1. First, the input 3D MRI passes through three independent 3D ARNet10, which are with fewer parameters (10.15M) and easy to train. Then attention module is used to more extract expressive features. In these networks, IBN is used to regularize global and local features at the same time. Finally, the outputs of the three networks are fused based on probability to get the final classification results of healthy control (HC), EMCI, LMCI, and AD stages.

3.1   Attention Mechanism

As shown in Fig. 4, the brain changes in a few concentrated areas during the 4 stages from a HC to AD. For example, brain atrophy, ventricle enlargement, and hippocampus atrophy. However, it is difficult to find the difference between adjacent stages of the disease when doctors perform manual diagnosis by MRI. To capture the changes in the brain, we propose the attention mechanism to improve the efficiency and accuracy of classification. Specifically, we inspired by Squeeze-and-Excitation (SE) block [6] to build ARNet to capture the expressive local features. For ease of description, we define the input feature as $X \in R^{C \times H \times W \times L}$, and define the output feature as $Y \in R^{C \times H \times W \times L}$, C, H, W, L represent the number of channels, height, width, length of X and Y. And we assume that the number of channels and height and width unchanged after the process by the ResNet module and ARNet module. Figure 2 shows the structure of ResNet module and ARNet module. For the ResNet module, the input X is passed through residual block $\mathcal{F}$ and the residual connection. The output of the ResNet module is got as:

$$X^* = \mathcal{F}(X) + X. \tag{1}$$

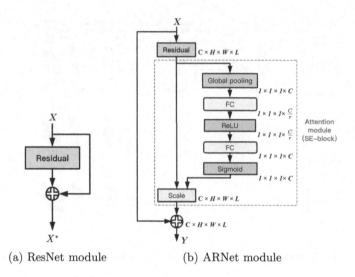

(a) ResNet module   (b) ARNet module

Fig. 2. Attention module embedded in ARNet module

For the ARNet module, we have added an SE branch to the ARNet module, X passes through the residual connection, residual block, and SE block respectively. As shown in Fig. 3, the SE branch can divide into squeeze stage and excitation stage. In the squeeze stage, the input $X \in R^{C \times H \times W \times L}$ is squeezed as the $Z \in R^{C \times 1 \times 1 \times 1}$ by the average pooling:

$$Z = \mathbf{F}_{avgpool}(X) = \frac{1}{H \times W \times L} \sum_{i=1}^{H} \sum_{j=1}^{W} \sum_{k=1}^{L} X(i, j, k). \tag{2}$$

There is a C channels in Z and each value in Z represents a channel. In the excitation stage, the squeezed feature Z is used to capture the interdependence of each channel. We first input Z into two $1 \times 1 \times 1$ convolutional layers and the sigmoid function to implement a gating mechanism as follow:

$$S = F_{ex}(Z, W) = \sigma(g(Z, W)) = \sigma(W_2 \operatorname{ReLU}(W_1 Z)), \tag{3}$$

$\mathbf{W}_1 \in R^{\frac{C}{r} \times C}$ and $\mathbf{W}_2 \in R^{C \times \frac{C}{r}}$ are the weights of two $1 \times 1 \times 1$ convolutional layers, r is the parameter to control the complexity of SE branch. The obtained S is the activation of each channel in X, it reflect the interrelationship of each channel. In the end, we excite the X by the activation S:

$$X_{se} = F_{scale}(X, S) = F \cdot S, \tag{4}$$

The final output of ARNet module is obtained by adding the outputs ResNet block, SE block, and X:

$$Y = \mathcal{F}(X) + X + X_{se}, \tag{5}$$

Y is very expressive and can explore changes in the brain, which is more helpful for AD detection.

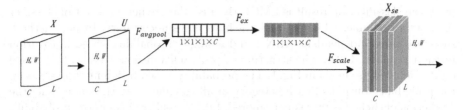

Fig. 3. The squeeze and excitation stage of the attention module

3.2 Instance-Batch Normalization

The processing of the changed local areas of the brain is very important for precise AD detection. In CNNs for traditional image classification, Batch Normalization (BN) is widely used for global normalization. The local feature normalization often is ignored. Instance Normalization (IN) is widely used in generative adversarial networks and style transfer tasks, which is more suitable for the scenes that focus on local areas. Therefore, we combine IN and BN as IBN [13], which can regularize global features and local features at the same time. Specifically, we define the input as $X \in R^{C \times H \times W}$, define the convolutional layer and Rectified Linear Unit (ReLU) as $\mathcal{C}$ and $\mathcal{R}$, define IN and BN as $\mathcal{IN}$ and $\mathcal{BN}$. To simplify the description, we assume that all operations do not change the shape of the feature. The X is first input to the convolutional layer as $X' = \mathcal{C}(X)$, $X' \in R^{C \times H \times W}$. Then, X' is divide into $X'_{in} \in R^{C/2 \times H \times W}$ (first C/2 channels) and $X'_{bn} \in R^{C/2 \times H \times W}$, Y_{in} and Y_{bn} are normalized by IN and BN respectively:

$$Y_{in} = \mathcal{IN}(X'_{in}), \quad Y_{bn} = \mathcal{BN}(X'_{bn}), \tag{6}$$

In the end, $Y_{in} \in R^{C/2 \times H \times W}$ and $Y_{bn} \in R^{C/2 \times H \times W}$ are connected together as $Y \in R^{C \times H \times W}$:

$$Y = \mathcal{R}([Y_{in}, Y_{bn}]), \tag{7}$$

Y is normalized from the global and local perspective, which is more beatific for AD detection. In Sect. 4.2, we show that applying IBN to our task consistently outperforms Batch BN, as well as Group Normalization (GN).

3.3 Probabilistic Fusion Method

In the traditional majority voting method, the prediction results of most classifiers are used as the final predictions. Each classifier is independent, and the error rate between different classifiers is not correlated, so the performance of the integrated model is better than that of a single classifier. However, majority voting is not suitable for multi-classification tasks. For some subjects that are difficult to categorize, the error rate is increased because of uncertainty between multiple categories. For example, there are three classifiers $\mathcal{N}_1$, $\mathcal{N}_2$, and $\mathcal{N}_3$ to classificate the input data as [HC, EMCI LMCI, AD], the output probabilities are $\mathcal{N}_1$: [0.7, 0.1, 0.2, 0.0], $\mathcal{N}_2$: [0.3, 0.1, 0.5, 0.1], $\mathcal{N}_3$: [0.4, 0.0, 0.5, 0.1]. If majority voting is used, the classification result is LMCI. Because the prediction result of classifier $\mathcal{N}_1$ is more reliable, while $\mathcal{N}_2$ and $\mathcal{N}_3$ have more uncertainty. In our method, there are three independent ARNet and a simple probability-based integration method is adopted [20]. The structure of probability fusion classification based on ResNet10 [4] is shown in Fig. 1. The probability-based fusion method sums up the probability output of each category of all classifiers. At the same time, the prediction of each classifier is not ignored. Each classifier outputs the probability of the test sample being classified into each disease stage. The probability of the class is assigned to the test sample by the classifier is:

$$U^n = (P_1^n, P_2^n, P_3^n, P_4^n), \tag{8}$$

Where P_i^n is the probability of the ith stage disease class of the test sample. And then U^n is normalized by:

$$U^n = \frac{U^n}{max(U^n)}, \tag{9}$$

Where $max(U^n)$ is the maximum element value of $(P_1^n, P_2^n, P_3^n, P_4^n)$. After calculating the output of all classifiers, the final disease class of the test sample is determined by the probability-based fusion method is as follows:

$$y = arg\ max \left(\prod_{n=1}^{m} P_1^n, \prod_{n=1}^{m} P_2^n, \prod_{n=1}^{m} p_3^n, \prod_{n=1}^{m} p_4^n \right). \tag{10}$$

4   Experiments and Results

4.1   Datasets

Magnetic resonance imaging (MRI) is used to analyze the anatomy of the brain due to its high spatial resolution and soft-tissue contrast ability. Over the past few decades, magnetic resonance imaging has made great strides in assessing brain injury and exploring brain anatomy [18]. MRI is known to be non-invasive, free of ionizing radiation, and generally carries lower health risks than other procedures such as computed tomography (CT) and positron emission tomography (PET) [17].

The Alzheimer's Disease Neuroimaging Initiative (ADNI) seeks to develop biomarkers of the disease and advances the understanding of AD pathophysiology, improves diagnostic methods for early detection of AD, and the clinical trial design. Additional goals are examining the rate of progress for both mild cognitive impairment and Alzheimer's disease, as well as building a large repository of clinical and imaging data. Three phases were successively launched: ADNI 1, ADNI GO/2, and ADNI 3. ADNI GO (2009) included 200 EMCI subjects. ADNI 2 (2011) included information on 150 healthy controls (HC), 100 subjects with early mild cognitive impairment(EMCI), 150 subjects with late mild cognitive impairment (LMCI), and 150 patients with AD.

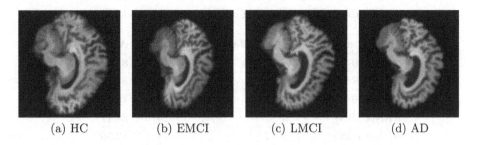

    (a) HC   (b) EMCI   (c) LMCI   (d) AD

Fig. 4. MRIs at different stages of Alzheimer's disease

In our study, we use 3T structural brain MRI scans from 600 images of the ADNI dataset[1]. Data is obtained from MRI scans downloaded by ADNI 2 and ADNI GO. The data is subjected to specific image preprocessing steps, such as mask, intensity normalization, redirection, and spatial normalization. Each MRI is 110 * 110 * 110 voxel data. Data leakage caused by the use of the same subjects in training and validation sets is shown to artificially improve model performance significantly [21]. Therefore, avoid any data leakage. We retain 80% of the MRI scans as the training set and 20% as the validation set. The training set consisted of 480 brain MRI scans and the validation set consisted of 120 brain MRI scans. In order to obtain the optimal network data set, the training set and the verification set are balanced. Figure 4 shows MRI data from patients at 4 stages of Alzheimer's disease. All experiments are conducted on the NVIDIA RTX 2080 GPU system.

4.2 Implementation Details

In our research, we use different variants of the network based on ResNet-10. Then select three network classifiers for probabilistic fusion. Classifier 1 is ResNet10-2p1d [3], classifier 2 is Pre-act-ResNet10, classifier 3 is ResNet10, classifier 4 is ARNet (BN), classifier 5 is ARNet (GN), classifier 6 is ARNet1 (IBN), and classifier 7 is ARNet2 (IBN). The results of the classifier model and performance are shown in Table 1. The first convolution of some classifiers is different. The performance of ARNet2 (IBN) is the best, and the accuracy on the verification set is 85%. The training time of ARNet2 (IBN) only needs 72m 30s. We repeat the experimental results of paper [16] and found that the integration results of three Densenet-121 were 81% accuracy, and the three models were 51.67%, 54.17%, and 66.67% respectively, and the training time of each model was more than 3h. The training time of ARnet2 (IBN) is about 1/3 of that of Densenet-121 [16,19].

In contrast to most standard architectures for image classification, we use smaller kernels in the first convolutional layer to prevent early spatial downsampling for ARNet2 (IBN). Previous ResNets [4] used a relatively large kernel size

---

[1] http://adni.loni.usc.edu/.

Table 1. Comparison of different network models

Network model	First convolution kernel	Layers	Accuracy
ResNet10-2p1d	1 * 7 * 7	10	81.67%
Pre-act-ResNet10	1 * 3 * 3	10	81.67%
ResNet10	7 * 7 * 7	10	79.17%
ARNet (BN)	7 * 7 * 7	10	83.17%
ARNet (GN)	7 * 7 * 7	10	82.50%
ARNet1 (IBN)	7 * 7 * 7	10	84.67%
ARNet2 (IBN)	3 * 3 * 3	10	85.00%

and stride in their first layer, which greatly reduced the spatial dimension of the input. In natural image classification tasks, speeding up the calculation is usually not harmful. However, for medical images, early downsampling results in a significant loss of performance. Therefore, in our experiment, we prefer to use the smaller 3 * 3 * 3 convolution kernel in the first layer.

In addition, the network model deals with the local changing areas of the brain and extracts the local features of the brain. But both global and local features cannot be ignored. Therefore, we combine Instance Normalization (IN) and Batch Normalization (BN) as Instance-Batch Normalization (IBN) to regularize both global and local features. In the experiment, we introduce Group Normalization (GN) as a reference, comparing BN, IBN, and GN. As shown in Table 1, the experimental results show that IBN is more suitable for 3D brain MRI. IBN has the best effect, and BN is superior to GN.

4.3   Comparison to Other Methods

In this section, we compare the performance of our study with the deep learning work of recent years about Alzheimer's disease detection. Table 2 shows the comparison between the proposed model and the previous methods on 4-way classification tasks. It can be seen that the proposed integrated model based on the attention mechanism performs best in 4-way classification tasks. In order to make a more accurate diagnosis of the disease, we use a probabilistic fusion method to integrate the classifiers 1, 2, and 6 three 3D network models, instead of a single network determining the diagnosis result. The experimental result of probability fusion for classifiers ResNet10-2p1d, Pre-act-ResNet10, and ARNet2 (IBN) is 84.17%. Our integrated model achieves better performance than the Densenet-121. It is worth noting that, as shown in Fig. 5, compared with the best single DenseNet-121 model, ARNet2 (IBN) of 85% accuracy is far better than the corresponding accuracy of 66.7%. In diagnosing the two stages of EMCI and LMCI, our model performs better and can diagnose s earlier. Our method uses shallower networks and a shorter training time for the network model. On the parameters of the network model, the parameter of ARNet2 (IBN) is 10.15M, while the parameter of DenseNet-121 is 11.2M. These demonstrate the advancement of our method in the diagnosis of Alzheimer's disease.

Table 2. Comparison with previous research works

Architecture	Datasets				Accuracy (%)
	CN	[s/E]MCI	[p/c/L]MCI	AD	
3D SAE [8]	52	56	43	51	47.42
Random Forest [2]	60	60	60	60	61.90
MSDNN [10]	360	409	217	238	75.44
3D DenseNet En. [16]	120	120	120	120	83.33
3D ARNet En. (Our work)	120	120	120	120	84.17

Legend: s = stable, p = progressive, c = converting, En. = Ensemble

However, from the analysis of Fig. 5, we still have some inadequacy in the diagnosis of EMCI, and there is a relatively large error rate. This also shows that the earliest detection of Alzheimer's disease is still a difficult point.

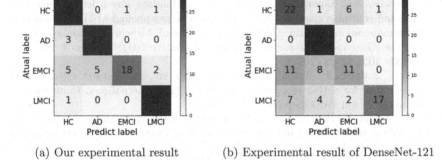

(a) Our experimental result   (b) Experimental result of DenseNet-121

Fig. 5. Comparison of experimental result with previous experimental result

5   Conclusion

In this study, we propose a 3D Attention Residual Network (3D ARNet) to improve the diagnosis of Alzheimer's disease stages. Because several brain regions of the patient have local changes. We combine the characteristics of the disease, add an attention mechanism into the network model to extract more expressive features in MRI. Therefore, our model can better diagnose the four stages of HC, EMCI, LMCI, and AD, especially the two critical stages of EMCI and LMCI. Our method can better help identify the various stages of the and enable patients to receive treatment early. In terms of feature extraction, since global features and local features can't be ignored, we use IBN to process global and local features at the same time. In addition, for accurate diagnosis of the disease, we use probabilistic fusion methods to integrate the three models, instead of letting a single network model determine the diagnosis result. Compared with the 3D Densenet-121 network, 3D ARNet10 has fewer parameters, resulting in becoming easier to train, and achieves better performance due to its shallower network. At the same time, these are proven on fully retained data and independent datasets. Through this work, we show the potential of the ARNet in neuroimaging. In the future, we will apply ARNet to the diagnosis of other brain diseases.

Acknowledgement. This work was supported in part by the Sichuan Science and Technology Program under Grant 2020YFS0307, Mianyang Science and Technology Program 2020YFZJ016, SWUST Doctoral Foundation under Grant 19zx7102. *Data used in the preparation of this article were obtained from the Alzheimer's Disease Neuroimaging Initiative (ADNI) database (http://adni.loni.usc.edu/). For up-to-date information, see http://adni-info.org/.

References

1. Cheng, D., Liu, M.: Combining convolutional and recurrent neural networks for Alzheimer's disease diagnosis using pet images. In: 2017 IEEE International Conference on Imaging Systems and Techniques (IST), pp. 1–5. IEEE (2017)
2. Dimitriadis, S.I., Liparas, D., Tsolaki, M.N.: Random forest feature selection, fusion and ensemble strategy: combining multiple morphological MRI measures to discriminate among healthy elderly, MCI, cMCI and Alzheimer's disease patients: from the Alzheimer's disease neuroimaging initiative (ADNI) data. J. Neurosci. Methods **302**, 14–23 (2017)
3. Hara, K., Kataoka, H., Satoh, Y.: Learning spatio-temporal features with 3D residual networks for action recognition. In: 2017 IEEE International Conference on Computer Vision Workshop (ICCVW) (2017)
4. He, K., Zhang, X., Ren, S., Sun, J.: Deep residual learning for image recognition. In: Proceedings of the IEEE Conference on Computer Vision and Pattern Recognition, pp. 770–778 (2016)
5. Hosseini-Asl, E., Keynton, R., El-Baz, A.: Alzheimer's disease diagnostics by adaptation of 3D convolutional network. In: 2016 IEEE International Conference on Image Processing (ICIP), pp. 126–130. IEEE (2016)
6. Hu, J., Shen, L., Sun, G.: Squeeze-and-excitation networks. In: Proceedings of the IEEE Conference on Computer Vision and Pattern Recognition, pp. 7132–7141 (2018)
7. Korolev, S., Safiullin, A., Belyaev, M., Dodonova, Y.: Residual and plain convolutional neural networks for 3D brain MRI classification. In: 2017 IEEE 14th International Symposium on Biomedical Imaging (ISBI 2017), pp. 835–838. IEEE (2017)
8. Liu, S., Liu, S., Cai, W., Pujol, S., Kikinis, R., Feng, D.: Early diagnosis of Alzheimer's disease with deep learning. In: 2014 IEEE 11th International Symposium on Biomedical Imaging (ISBI), pp. 1015–1018. IEEE (2014)
9. Liu, X., Tosun, D., Weiner, M.W., Schuff, N., Initiative, A.D.N., et al.: Locally linear embedding (LLE) for MRI based Alzheimer's disease classification. Neuroimage **83**, 148–157 (2013)
10. Lu, D., Popuri, K., Ding, G.W., Balachandar, R., Beg, M.F., Initiative, A.D.N.: Multimodal and multiscale deep neural networks for the early diagnosis of Alzheimer's disease using structural MR and FDG-pet images. Sci. Rep. **8**(1), 5697 (2018)

11. Markesbery, W.R.: Neuropathologic alterations in mild cognitive impairment: a review. J. Alzheimers Dis. **19**(1), 221–228 (2010)

12. Minati, L., Edginton, T., Grazia Bruzzone, M., Giaccone, G.: Reviews: current concepts in Alzheimer's disease: a multidisciplinary review. Am. J. Alzheimer's Dis. Other Dementias® **24**(2), 95–121 (2009)

13. Pan, X., Luo, P., Shi, J., Tang, X.: Two at once: enhancing learning and generalization capacities via IBN-net. In: Proceedings of the European Conference on Computer Vision (ECCV), pp. 464–479 (2018)

14. Petersen, R.C., et al.: Mild cognitive impairment: ten years later. Arch. Neurol. **66**(12), 1447–1455 (2009)

15. Plant, C., et al.: Automated detection of brain atrophy patterns based on MRI for the prediction of Alzheimer's disease. Neuroimage **50**(1), 162–174 (2010)

16. Ruiz, J., Mahmud, M., Modasshir, M., Shamim Kaiser, M., for the Alzheimer's Disease Neuroimaging Initiative: 3D DenseNet ensemble in 4-way classification of Alzheimer's disease. In: Mahmud, M., Vassanelli, S., Kaiser, M.S., Zhong, N. (eds.) BI 2020. LNCS (LNAI), vol. 12241, pp. 85–96. Springer, Cham (2020). https://doi.org/10.1007/978-3-030-59277-6_8

17. Rutegrd, M.K., Btsman, M., Axelsson, J., Brynolfsson, P., Riklund, K.: PET/MRI and PET/CT hybrid imaging of rectal cancer - description and initial observations from the RECTOPET (REctal cancer trial on PET/MRI/CT) study. Cancer Imaging **19**(1), 1–9 (2019)

18. Smith-Bindman, R., et al.: Use of diagnostic imaging studies and associated radiation exposure for patients enrolled in large integrated health care systems, 1996–2010. J. Am. Med. Assoc. **307**, 2400–2409 (2012)

19. Wang, H., et al.: Ensemble of 3D densely connected convolutional network for diagnosis of mild cognitive impairment and Alzheimer's disease. Neurocomputing **333**, 145–156 (2019)

20. Wen, G., Hou, Z., Li, H., Li, D., Jiang, L., Xun, E.: Ensemble of deep neural networks with probability-based fusion for facial expression recognition. Cogn. Comput. **9**(5), 597–610 (2017)

21. Yi, R.F., Guan, Z., Kumar, R., Wu, J.Y., Fiterau, M.: Alzheimer's disease brain MRI classification: Challenges and insights (2019)

22. Zhang, X., Li, W., Shen, W., Zhang, L., Pu, X., Chen, L.: Automatic identification of Alzheimer's disease and epilepsy based on MRI. In: 2019 IEEE 31st International Conference on Tools with Artificial Intelligence (ICTAI), pp. 614–620. IEEE (2019)

23. Zhu, X., Suk, H.-I., Zhu, Y., Thung, K.-H., Wu, G., Shen, D.: Multi-view classification for identification of Alzheimer's disease. In: Zhou, L., Wang, L., Wang, Q., Shi, Y. (eds.) MLMI 2015. LNCS, vol. 9352, pp. 255–262. Springer, Cham (2015). https://doi.org/10.1007/978-3-319-24888-2_31

Generation of a Large-Scale Line Image Dataset with Ground Truth Texts from Page-Level Autograph Documents

Ayumu Nagai[✉]

Gunma University, Kiryu, Japan
nagai@gunma-u.ac.jp

Abstract. Recently, Deep Learning techniques help to recognize Japanese historical cursive with high accuracy. However, most of the known cursive dataset have been gathered from printed documents which are written for the general public and easy to read. Our research aims to improve the recognition of autograph documents, which are more difficult to recognize than printed documents because they are often private and written in various writing styles. To create a useful autograph document dataset, this paper devises a technique to generate many line images accompanied by the corresponding ground truth (GT) texts, given an autograph document whose GT transcription is available only at the page-level.

Our method utilizes HRNet for line detection and CRNN for line recognition. HRNet is used to decompose the page image into lines that is mapped to GT text, which is decomposed separately from GT transcription, by similarity-based alignment solved by beam search. We introduce two ideas to the alignment: to allow out-of-order mapping of the lines not adjacent to each other and to allow many-to-many mapping. With these orthogonal two ideas, we obtained a dataset consisting of 43,271 reliable autograph line images mapped to GT texts. By training CRNN from scratch on this dataset together with printed dataset, recognition accuracy for autograph documents is improved.

Keywords: Handwriting recognition · Alignment · Deep Learning

1 Introduction

Our main research target is to recognize Japanese historical cursive of various documents. In recent years, CODH (Center for Open Data in the Humanities) has released a large amount of character-level image data on Japanese historical cursive. Competitions using these data were also held, and the accuracy rate reached around 95%. However, most of these data are easy to read because they are extracted from woodblock-printed documents for the general public. The characters in printed documents are written with readability in mind. Deep Learning may have brought cursive character recgonition of printed documents to the expert level, but not the autograph documents so far. Autograph documents are more difficult to read than printed documents, since they are private

© Springer Nature Switzerland AG 2021
T. Mantoro et al. (Eds.): ICONIP 2021, LNCS 13108, pp. 354–366, 2021.
https://doi.org/10.1007/978-3-030-92185-9_29

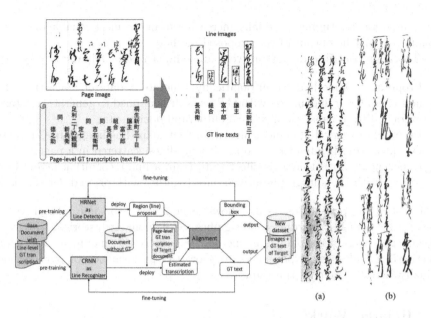

Fig. 1. (top left) Our goal is to extract line images from a page image and extract GT line texts from GT transcription, and to create a dataset consisted of line image annotated with its corresponding GT line text. **(bottom left)** Outline of our model how to train. **(right)** Documents with sub-lines. (a) has regular lines starting from the very top to the bottom of the page. However, (b) is a two-columned (or two-rowed) document consisting of sub-lines that are short lines.

documents which have various writing styles. In Fact, Masuda mentions that the (autograph) letters are one of the most difficult historical documents to read [9]. Nakano also pointed out that Japanese traditional poetries and letters are the most difficult to read even among experts, and that start learning from them is like a mountaineering beginner aspiring to climb the Himalayas, and therefore recommended that beginners should read from printed documents [12]. Now deep network can recognize 95% of the printed documents, meaning that the next target should be the autograph documents.

This paper proposes a method to automatically generate a dataset of line images accompanied by ground truth (GT) text, and apply it to Japanese historical autograph documents, as shown in Fig. 1(top left), useful to learn various writing styles of autograph documents. Figure 1(bottom left) is an overview to generate a dataset from autograph page images and the corresponding page-level GT transcription. HRNet [16] as a line detector and CRNN [14] as a line recognizer are pre-trained on existing base data separately, and then fine-tuned on the target document repeatedly and alternately. In each iteration, the line detector predicts bounding boxes in the page image, and the line recognizer predicts text strings on the cropped line images. The cropped line image is mapped to GT text extracted from page-level GT transcription by the proposed alignment algorithm, and a pair of line image and GT text is obtained. These pairs are reused for fine-tuning, gradually increasing from the most reliable ones, and finally become the output of the proposed method.

The main contributions of our work are summarized as follows:

- Proposal of a method to automatically generate a dataset consisting of a large number of line images with GT texts. It is applied to generate a dataset extracted from Japanese historical autograph documents.
- Proposal of a novel alignment algorithm that maps line images to GT texts. Our algorithm is intended to cope with sub-lines scattered in a page.
- Experiment shows that it is effective to train a line recognition model from scratch with the autograph-derived data together with printed-derived data.

The rest of the paper consists as follows: Sect. 2 refers to some related works. Section 3 explains the hardness and significance of Japanese historical autograph documents. Section 4 explains our method, especially the alignment algorithm. Section 5 includes our experiments to show the performance of our method. Finally, Sect. 6 concludes this paper.

2   Related Works

AnyOCR [5] is an OCR system which semi-automatically trains line recognizer repeatedly with the help of human expert. AnyOCR trains line recognizer, but not line detector.

Strecker [15] developed a system to semi-automatically generate a newspaper image with ground truth. The system layouts the collected text contents and photos, and outputs a PDF file, which is then actually printed, scanned, and aligned with the text data to create the ground truth.

There have been researches on recognizing Japanese historical cursive such as [3,8,11], but to our knowledge, there have been no research on automatically generating dataset for training a model for recognition so far.

3   Hardness and Significance of Autograph Documents

In this chapter, we will give five factors that make Japanese historical cursive in autograph documents more difficult than printed documents and the significance of transcription of autograph documents. Figure 2 shows the five factors of difficulty: First, various ways to break up characters; second, more broken, often even omitted; third, many similar characters as a result of broken up[1]; fourth, a strong habit of writing style; and fifth, concatination of characters.

---

[1] Hence, there is a limit to character-level recognition. It is often impossible to recognize without looking at the characters before and after; we are forced to recognize characters by line-level. In other words, Sayre's paradox often occurs in autograph documents. [10] also gives some other reasons. In fact, Aburai points out that focusing too much on the character alone is counterproductive [1], and advises to read the character by considering the context [2].

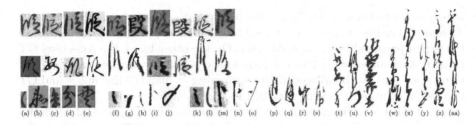

Fig. 2. Five factors that make autograph documents difficult to read. First, wide variety of breaking up characters: First and second lines are the characters of "段" in printed and autograph documents, respectively. Second, more broken: Lines (a) to (e) and (f) to (j) are one of the most broken examples of "候", "衛", "兵", "分", and "貫" in printed and autograph documents, respectively; often even omitted: (t) is "罷在候ニ付" but "候" is omitted. (u) is "伝左衛門", but "衛" is omitted. (v) is "佐野五郎兵衛", but "兵" is omitted. Third, many similar characters: (k) and (l) are both "之", similar to "候" in (f), "衛" in (g), and "兵" in (h). (m) is "歩" and (n) is "ト", both of which are similar to "分" in (i). (o) is "メ", similar to "貫" in (j). Fourth, strong habit of writing: (p) to (s) are examples of "通", "趣", "此", and "致" that are not in "Cursive Examples Dictionary" (くずし字用例辞典) [6]. Fifth, characters are often concatenated: (w) is "異国船", (x) is "可然旨今日", (y) is "可被下候以上", (z) is "高畠次郎左衛門組", and (aa) is "不罷越内自然".

The difference comes from the fact that a craftsman writes a fair copy of printed documents. An autograph document is a private document that is read only by the writer himself and those close to the writer, so it is the handwriting of the writer himself. A printed document, on the other hand, is for the general public with varying levels of education, so it is the handwriting by a craftsman who makes a fair copy of the text [12]. As a result, printed documents consist of handwritten characters that are easier to read than autograph documents.

In fact, in the study of [3], more than half of the documents in the dataset which we call CODH have F1-score over 0.87, while the F1-score of the autograph document called "Ryouri-kata Kokoroe-no-koto" (料理方心得之事) is only 0.5865. This fact indicates that the characters in the autograph documents are quite difficult for Deep Learning (as well as human) to recognize.

According to Nakano [12], the total number of Japanese historical documents is over one million, of which less than 1% have been transcribed, and there are far more autograph documents than printed documents. This means that improving the recognition rate of characters in autograph documents will surely contribute greatly to the transcription of Japanese historical documents.

4   Proposed Method

We developed a method to create a dataset consisting of line images annotated with corresponding GT text, given an autograph document whose GT transcription is available only at page-level. Figure 1(bottom left) shows the outline.

HRNet as a line detector and CRNN as a line recognizer are pre-trained with the *base data* reconstructed from CODH44, which is a public dataset consisting mostly of printed characters. In the following, we describe fine-tuning for a *target document*. The line images are obtained by cropping the page image according to the region proposed by the line detector. On the other hand, the page-level GT transcription corresponding to the page image is also decomposed into GT text using space and line break as delimiters. In order to map the line image and GT text, we maximized the similarity in the alignment framework. The similarity *sim* is defined as a linear sum of two types of similarity: positional similarity sim_{loc} and string similarity sim_{ed}. sim_{loc} is the similarity of the location on the page. sim_{ed} is the similarity between GT text *gt* and the predicted string *pt* by the line recognizer. *sim* is defined between each component of the series $\{i\}$ of GT texts and the series $\{j\}$ of line images, where both $\{i\}$ and $\{j\}$ are sorted in descending order of x-coordinates. $\{i\}$ and $\{j\}$ are mapped by alignment so that the product of *sim* is maximized. The amount of training data is gradually increased with each iteration of fine-tuning by filtering the line image and GT text pairs obtained by alignment to the reliable ones that there is a line image corresponding to every GT text on the page with similarity β or higher.

4.1   Alignment and Its Formulation

Some autograph documents are filled with short texts which we call sub-lines. They can be names, dates, official positions, or money amounts. It is often in two-column (or two-row) as shown in Fig. 1(b). Although alignment is an algorithm that aligns two series, a two-column document is troublesome, since one series is separated into two series in parallel which is not assumed by ordinary alignment. To cope with this problem, two ideas are introduced to increase flexibility. One is to allow out-of-order mapping between sub-lines that are not adjacent to each other. This solves the above-mentioned two-column problem. The other is to allow many-to-many mapping when sub-lines are vertically close to each other. In Japanese historical documents, spaces may be inserted even in the middle of a sentence or a name. Many-to-many mapping can reduce the negative effects of such spaces. In this section, we formulate a naive alignment, and then modify the formulation while introducing our ideas one by one.

Naive Formulation of Alignment as a Starting Point. The similarity $sim(i, j)$ mentioned at the beginning of this section can be formulated as follows:

$$sim(i, j) = \alpha \cdot sim_{loc}(i, j) + (1 - \alpha) \cdot sim_{ed}(i, j) \tag{1}$$

$$s.t. \quad sim_{loc}(i, j) = GF(rx_i, rx_j, \sigma_x) \cdot GF(ry_i, ry_j, \sigma_y) \cdot GF(rh_i, rh_j, rh_j/4) \tag{2}$$

$$GF(x, x', \sigma) = \exp\left(-\frac{1}{2\sigma^2}(x - x')^2\right) \tag{3}$$

$$sim_{ed}(i, j) = \frac{max\_sim}{leng}\left(leng - ed(gt, pt)\right) \tag{4}$$

$$max\_sim = \min\left(sim_0 + \frac{1 - sim_0}{step\_num}(leng - 1), 1\right) \tag{5}$$

$$leng = length(gt), \tag{6}$$

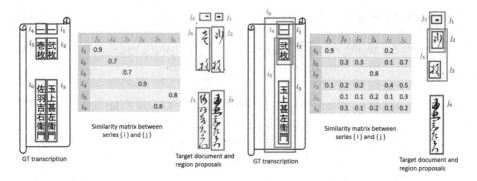

Fig. 3. **(left)** Allow out-of-order mapping between sub-lines that are not adjacent to each other, such as j_5 and j_6. **(right)** Many-to-many mapping is allowed when sub-lines are vertically close to each other. Mapping between j_6, which is a union of j_2 and j_3, and the elements in series $\{i\}$, say i_2, is allowed.

where rx, ry, rh are position and length of the sub-line or GT text in relative coordinates within the page normalized to $(0,0,1,1)$. max_sim is to discount the score for shorter string length. The naive formulation of alignment is as follows:

$$\underset{align}{\textbf{argmax}} \ L_1(align) \tag{7}$$

$$s.t. \ L_1(align) = \prod_{(i_a, j_b) \in align} sim(i_a, j_b) \tag{8}$$

$$(i_a, j_b) \in align \land (i_c, j_d) \in align \rightarrow i_a \neq i_c \land j_b \neq j_d \tag{9}$$

$$(i_a, j_b) \in align \land (i_c, j_d) \in align \land i_a \prec i_c \rightarrow j_b \prec j_d \tag{10}$$

$$(i_a, j_b) \in align \land (i_c, j_d) \in align \land j_b \prec j_d \rightarrow i_a \prec i_c, \tag{11}$$

where $i_a \prec i_c$ denotes that i_a appears before (i.e., right side of) i_c in series $\{i\}$.

Allow Out-of-Order Mapping of Non-adjacent Sub-lines. Alignment is a method of mapping between elements of two series. Usually the order of the elements within each series is fixed, and no mapping is done to swap the order. However, there is little point in defining the (total) order by x values between the sub-lines in upper and lower columns of a two-column document, for example. Sub-lines of a two-column document are ordered only within each column, so the ordinary alignment does not work well. Our idea is that sub-lines in the same column should be mapped in-order, but sub-lines across columns should be mapped out-of-order. For example in Fig. 3(left), j_1 and j_4, j_2 and j_6, and j_3 and j_5 are mapped in-order, but out-of-order for other combinations. Introducing this idea, the formulation of alignment becomes as follows:

$$\underset{align}{\text{argmax}} \ L_2(align) \tag{12}$$

$$s.t. \ L_2(align) = \prod_{(i_a,j_b)\in align} \text{sim}(i_a,j_b) \cdot \prod_{\substack{(i_a,j_b)\in align \\ (i_c,j_d)\in align \\ i_a \prec i_c}} \text{ord}(i_a,j_b,i_c,j_d) \tag{13}$$

$$(i_a,j_b) \in align \land (i_c,j_d) \in align \rightarrow i_a \neq i_c \land j_b \neq j_d \tag{14}$$

$$\text{ord}(i_a,j_b,i_c,j_d) = \begin{cases} 0, & \text{if } i_a \not\prec i_c \\ 0, & \text{if bbox of } j_b \text{ is adjacent to the left of } j_d \\ 1, & \text{otherwise} \end{cases} \tag{15}$$

Allow Many-to-Many Mapping. Although it is assumed that the elements of the series are the smallest units in alignment, it is not always the case for the series treated in our study. Depending on the results of line detection, a sub-line may be further decomposed as shown in Fig. 3(right), or multiple sub-lines may be combined into one. In order to map such sub-lines correctly to GT texts in this situation, one-to-many and many-to-one mappings are needed. We chose to introduce many-to-many mapping. Specifically, sub-lines that are vertically close to each other are merged and registered as another element in the series. For example, i_1 in Fig. 3(right) also appears in the series as i_4. However, each time an element is mapped, all the elements intersecting with the mapped element should be removed. For example, if i_1 is mapped to j_1, then i_4 and i_6 are removed as well as i_1. j_5 is removed as well as j_1. If i_2 is mapped to j_6, then i_4, i_5 and i_6 are removed as well as i_2. j_2, j_3 and j_5 are removed as well as j_6. The elements to be removed with an element x can be easily obtained by the operation intersect_with(x) defined as follows:

$$\text{intersect\_with}(x) \overset{\text{def}}{=} \bigcup_{c\in\text{child}(x)\cup\{x\}} \text{parent}(c) \cup \{c\}, \tag{16}$$

where parent(x) is the set of elements that contain x, and child(x) is the set of elements that x contains. For example, for the series $\{j\}$ in Fig. 3(right), we have child(j_6) = $\{j_2,j_3\}$, parent(j_2) = $\{j_5,j_6\}$, parent(j_3) = $\{j_6\}$, parent(j_6) = $\{\}$. Then, intersect_with(j_6) = $\{j_2,j_3,j_5,j_6\}$ is correctly obtained. Introducing this idea, the objective function of alignment will be as follows:

$$\underset{align}{\text{argmax}} \ L_3(align) \tag{17}$$

$$s.t. \ L_3(align) = \prod_{(i_a,j_b)\in align} \text{sim}(i_a,j_b) \cdot \prod_{\substack{(i_a,j_b)\in align \\ (i_c,j_d)\in align \\ i_a \prec i_c}} \text{ord}(i_a,j_b,i_c,j_d) \tag{18}$$

$$(i_a,j_b) \in align \land (i_c,j_d) \in align \rightarrow i_a \neq i_c \land j_b \neq j_d \tag{19}$$

$$(i_a,j_b) \in align \land (i_c,j_d) \in align \rightarrow i_c \notin \text{intersect\_with}(i_a) \tag{20}$$

$$(i_a,j_b) \in align \land (i_c,j_d) \in align \rightarrow j_d \notin \text{intersect\_with}(j_b), \tag{21}$$

where $\{i\}$ and $\{j\}$ are expanded compared to the ones in Eq. (13). Equation (20) and (21) are to restrict the expanded part of the series. Moreover, $\log L_3(align)$ is used for the actual calculation instead of $L_3(align)$ itself.

4.2 Alignment Solver

Usually, the alignment problem is reduced to the shortest path problem and is solved by a standard method such as dynamic programming. However, in this research, we decided to solve it by *Beam Search* in order to realize our two ideas. The specific alignment algorithm is shown in Algorithm 1 and the overall algorithm is shown in Algorithm 2.

Algorithm 1 Alignment solved with Beam Search

---

procedure ALIGNMENT_SOLVER($\{gb_i, gt_i\}, \{pb_j, pt_j\}$)   ▷ gt_i and gb_i is GT text and bbox of i,
while pb_j and pt_j is bbox and predicted string of j extracted from autograph page
   create_similarity_matrix($\{gb_i, gt_i\}, \{pb_j, pt_j\}$)   ▷ calculated from eq.(1) to (6)
   $solutions$.push($initial\_solution$)   ▷ initial solution is $align = \{\}$
   repeat
      $s = solutions$.pop()   ▷ the most promising solution so far
      if s.alignment_done() **then**
         return s.all_aligned()
      $i_0 = s$.topmost_i()   ▷ rightmost text i_0 which is not aligned yet
      for all i in s.intersect_with(i_0) **do**   ▷ foreach text i related with i_0
         for all j in s.all_valid_j() **do**   ▷ foreach image j
            $s = s$.align(i, j)   ▷ align text i and image j, i.e., $align \mathrel{+}= (i, j)$
            for all i_1 in s.intersect_with(i) **do**
               $s = s$.unalign(i_1)   ▷ remove i_1 because of adopting i
            for all j_1 in s.intersect_with(j) **do**
               $s = s$.unalign(j_1)   ▷ remove j_1 because of adopting j
            $solutions$.push(s)   ▷ add s into the list of solutions
      $solutions$.push(s.unalign(i_0))   ▷ in case i_0 do not align to any j
      $solutions$.update_likelyhood_scores()   ▷ calculated from eq.(17) to (21)
      $solutions$.remove_inferior_solutions()
      $solutions$.truncate($beam\_width$)

Algorithm 2 Overall Algorithm

---

procedure OVERALL_TRAINING
   CRNN.pretrain()   ▷ pre-train CRNN on base data
   HRNet.pretrain()   ▷ pre-train HRNet on base data
   repeat until convergence
      $\{pb_j\}$ = HRNet.deploy(I)   ▷ predicted bboxes pb_j of page image I
      $\{l_j\}$ = apply($\{pb_j\}, I$)   ▷ crop I to get sub-line image l_j
      $\{pt_j\}$ = CRNN.deploy($\{l_j\}$)   ▷ pt_j is a predicted transcription of l_j
      $\{gb_i, gt_i\}$ = read_GT_transcription()   ▷ break down GT page transcription
                  ▷ into sub-line bbox gb_i and GT text gt_i
      $\{(i', j')\}$ = Alignment_solver($\{gb_i, gt_i\}, \{pb_j, pt_j\}$)
      $\{(i', j')\}$ = filter($\{(i', j')\}, \beta$)
      HRNet.refine($I, \{pb_{j'}\}$)   ▷ fine-tuning HRNet on target and base data
      CRNN.refine($\{l_{j'}\}, \{gt_{i'}\}$)   ▷ fine-tuning CRNN on target and base data

---

5 Experiments

5.1 Datasets

We used the following datasets. CODH44 is mostly based on printed documents with character-level annotation which we reconstructed into page-level, while the remaining six datasets are autograph documents with page-level annotation, of which we created the page-level GT transcriptions for the first four.

CODH44: Consists of 6151 images from 44 documents with annotation released by CODH [4]. It is used for pre-training as base data.

Arai: Vol.1 and 2 of a document called "Arai-kizaemon Yakuyou-Nikki" (新居喜左衛門役用日記), written in 1859, consists of 178 images.

Gaii: Vol.1, 2, 3, 5 and 6 of a document called "Gaii-raikan-shi" (外夷来艦誌), written about the year 1853 and 1854, consists of 254 images.

Soushuu1: A part of a document called "Soushuu O-ninzuu-dashi Kiroku" (相州御人数出記録), originally written from 1820 to 1821, consists of 220 images.

Soushuu2: A part of a document called "Soushuu Kiroku" (相州記録), originally written from 1824 to 1830, consists of 228 images.

Imizushi: A collection by a museum in Imizu-shi (射水市) consisting of 2354 images with page-level GT transcription[2].

Nakatsugawa: A collection by a city called Nakatsugawa (中津川) consisting of 2215 images with page-level GT transcription[3].

5.2 Implementation Details

We used $\alpha = 0.4$, $\sigma_x = \sigma_y = 0.15$, $sim_0 = 0.25$, $step\_num = 20$. β is gradually reduced in each iteration and kept at 0.01 after 0.4, 0.3, 0.2, 0.1. For pre-training, CODH44 reconstructed into page-level annotated data is used as base data. As the training data of fine-tuning, the same number of two types of data, that is, the base data of pre-training and the output of alignment, are randomly sampled and mixed. The output of alignment is also used as valid data. We used CRNN implementation by [10] with some ideas such as enlargement, random rotation and translation and RandomErasing [18].

5.3 Metrics

The metrics used for line detection is as follows:

$$\frac{1}{n} \sum_{j=1}^{n} \text{ite}(\text{IoU}(b_j, pb_j) \geq \gamma, 1, 0), \tag{22}$$

where $\text{ite}(\cdot, \cdot, \cdot)$ is the if-then-else clause and b_j is the ground truth region.

For line recognition, we use following two definitions of Accuracy Rate (AR):

$$\text{AR} = 1 - \frac{1}{\sum_j |t_j|} \sum_j \text{ed}(t_j, pt_j) \tag{23}$$

$$\tilde{\text{AR}} = 1 - \frac{1}{\sum_i |gt_i|} \sum_{(i,j)\in align} \text{ed}(gt_i, pt_j), \tag{24}$$

---

[2] https://trc-adeac.trc.co.jp/WJ11D0/WJJS05U/1621115100/1621115100100010.
[3] https://trc-adeac.trc.co.jp/Html/ImageView/2120605100/2120605100200010.

where t_j is the true ground truth text (if exists), gt_i is the GT text, pt_j is the predicted string. Actually, the true AR defined in Eq. (23) is the same metric as [17]. However, GT text gt_i in Eq. (24) is the text mapped as the result of alignment, which is not necessarily the ground truth in the true sense. Therefore, $\tilde{\text{AR}}$ is smaller than the true AR.

5.4   Experimental Results

Two experiments were performed: One to evaluate the accuracy of line detector and line recognizer on data with ground truth, and the other to test the effectiveness of learning with the autograph-derived data generated by the proposed algorithm.

Evaluation of Line Detector and Line Recognizer Using Ground Truth.
We chose "Hana-no-Shimadai" (花のしま台; HNSD) in CODH44 as the target document. Since HNSD has ground truth, we can accurately evaluate line detection and line recognition. In both pre-training and fine-tuning, 43 documents except the target document are used as page-level GT transcription.

Table 1. Accuracy of line detector on HNSD for various threshold γ in Eq. (22).

γ	0.5	0.55	0.6	0.65	0.7	0.75	0.8	0.85	0.9	0.95	[.5:.95]
IoU	0.980	0.980	0.977	0.972	0.926	0.789	0.535	0.263	0.121	0.056	0.660
IoUx	0.991	0.990	0.990	0.987	0.972	0.898	0.723	0.450	0.214	0.090	0.731
IoUy	0.984	0.984	0.983	0.982	0.981	0.980	0.978	0.976	0.973	0.840	0.966

Table 2. Accuracy of line recognizer on HNSD and various autograph documents. Metrics of [3] is F1-score, our result is $\tilde{\text{AR}}$ and all the rest are AR.

	HNSD	Arai	Gaii	Imizushi	Nakatsugawa	Soushuu1	Soushuu2
[13]	0.8712	–	–	–	–	–	–
[3]	0.8915	–	–	–	–	–	–
[8]	0.9558	–	–	–	–	–	–
[11]	0.9598	–	–	–	–	–	–
Ours	0.9524	0.8894	0.9298	0.6952	0.7796	0.9562	0.8301

Table 3. CRNN trained from scratch with the autograph-derived dataset and CODH. au: autograph, CO: CODH43, pau: part of autograph, and pCO: part of CODH43

Training/valid data	au	pCO	pau+pCO	CO	au+CO
# of lines	43,271	43,271	70,349	70,349	113,620
# of chars	506,307	666,453	952,806	1,083,284	1,589,591
[3]'s F1-score	–	–	–	0.5865	–
Our AR	0.5251	0.6940	0.7339	0.7193	0.7639

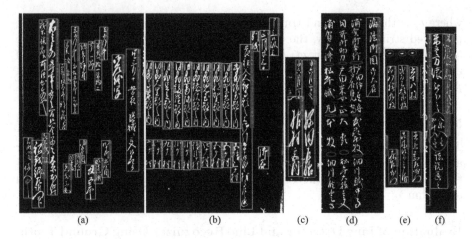

(a)   (b)   (c)   (d)   (e)   (f)

Fig. 4. (a) to (c) are successful samples. (d) to (f) are failed samples. (Color figure online)

Table 1 shows the accuracy of line detector; over 90% of b_j are properly predicted when the threshold γ is less or equal to 0.7. IoUx and IoUy are the one-dimensional version of IoU. They are introduced because IoU becomes inappropriate when the text region is elongated [7]. Practically, the accuracy of y-coordinates of the region proposal is critical for the subsequent line recognition. Table 1 shows that y-coordinate of the line detector is so accurate that over 97% of GT bboxes b_j are properly predicted even γ is 0.9.

Table 2 shows the accuracy of line recognizer. $\tilde{\text{AR}}$ for HNSD, which is a printed document, is high, but $\tilde{\text{AR}}$ for other datasets, which are autograph documents, is not necessarily high. This result shows that it is more difficult to recognize autograph documents than printed documents. In addition, our results achieve performance comparable to other state-of-the-art studies given line images, even though we are also learning line regions simultaneously.

Effect of Training with Generated Autograph-Derived Dataset. In order to examine the effect of training on autograph-derived data, we trained another CRNN model from scratch with the dataset generated by the proposed method.

We obtained 43,271 line images with GT text[4] from the six autograph documents executed separatedly. They are divided into 9 to 1 to obtain training and valid data. For test data, we selected the one from CODH44 called "Ryouri-kata Kokoroe-no-koto" (料理方心得之事; RKKK) which is an autograph document consisting of 373 lines. This is the case "au" (or "autograph") in Table 3. We also trained case "CO" (or "CODH43") consisting of the data derived from 43 documents in CODH44 excluding RKKK. For comparison, we trained case

---

[4] Downloadable from https://gadget.inf.gunma-u.ac.jp/dl/autograph.tar.gz.

"pau+pCO" (or "part of au" and "part of CO") which is obtained by sampling half of the data from "au" and "CO" respectively. "au+CO" is the case trained by the whole data of "au" and "CO". Table 3 shows that training only by autograph documents is worse than training only by printed documents. However, training with both autograph documents and printed documents becomes better than training only with printed documents. It tells us that, an efficient learning is to learn the basic writing style from printed documents and to learn various writing styles from autograph documents. Although no direct comparison can be made, the learned CRNN model outperforms previous study [3].

Successful Samples and Failed Samples. Figure 4(a) to (c) are successful samples. (c) has two lines; the right line is connected to one, while the left line is separated into two. However, both of them are correctly mapped to the appropriate GT texts. Figure 4(d) to (f) are failed samples. (d) is failed, since it is difficult to recognize the breaks between vertical sub-lines indicated by the red ovals. (e) fails to detect starting characters of itemization in the red ovals. (f) fails to detect where one line is split into two in the red oval.

6 Conclusion

We proposed a method that automatically generates a dataset consisting of a large amount of line images with GT texts. As a key technique in our algorithm, an alignment that maps line images to GT texts is also proposed. Two ideas are introduced in the alignment to cope with short sub-lines scattered in a page. That is, to allow out-of-order mapping for non-adjacent sub-lines and to allow many-to-many mapping. We applied our algorithm to Japanese historical autograph documents and obtained over 43,000 reliable line images with GT texts. As far as we know, this is the first autograph data of this scale. Experiment shows that HRNet is sensitive enough to learn the y-coordinates of the start and end points of the line. When learning with the obtained autograph-derived dataset, it is better to learn together with printed-derived dataset.

References

1. Aburai, H.: Historical documents are so interesting (in Japanese). Kashiwa Shobo (2005)
2. Aburai, H.: First step to the historical documents (in Japanese). Kashiwa Shobo (2008)
3. Clanuwat, T., Lamb, A., Kitamoto, A.: KuroNet: pre-modern Japanese Kuzushiji character recognition with deep learning. In: ICDAR, pp. 607–614 (2019)
4. CODH: Kuzushiji dataset. http://codh.rois.ac.jp/char-shape/
5. Jenckel, M., Bukhari, S., Dengel, A.: anyOCR: a sequence learning baesd OCR system for unlabeled historical documents. In: ICPR, pp. 4024–4029 (2016)
6. Kouta, K. (ed.): Cursive Examples Dictionary (in Japanese). Tokyo-do Press (1981)

7. Liao, M., Shi, B., Bai, X., Wang, X., Liu, W.: TextBoxes: a fast text detector with a single deep neural network. In: Proceedings of AAAI Conference on AI, pp. 4161–4167 (2017)
8. Ly, N.T., Nguyen, C.T., Nakagawa, M.: An attention-based end-to-end model for multiple text lines recognition in Japanese historical documents. In: ICDAR, pp. 629–634 (2019)
9. Masuda, T.: How to Read Historical Documents and Letters (in Japanese). Tokyo-do Press (2007)
10. Nagai, A.: On the improvement of recognizing single-line strings of Japanese historical cursive. In: ICDAR, pp. 621–628 (2019)
11. Nagai, A.: Recognizing Japanese historical cursive with pseudo-labeling-aided CRNN as an application of semi-supervised learning to sequence labeling. In: ICFHR, pp. 97–102 (2020)
12. Nakano, M.: Recommendation of Japanese Historical Books (in Japanese). Iwanami-Shinsho (2011)
13. Nguyen, H.T., Ly, N.T., Nguyen, K.C., Nguyen, C.T., Nakagawa, M.: Attempts to recognize anomalously deformed Kana in Japanese historical documents. In: Proceedings of 4th International Workshop on HIP2017, pp. 31–36 (2017)
14. Shi, B., Bai, X., Yao, C.: An end-to-end trainable neural network for image-based sequence recognition and its application to scene text recognition. IEEE Trans. PAMI **39**(11), 2298–2304 (2017)
15. Strecker, T., van Beusekom, J., Albayrak, S., Breuel, T.M.: Automated ground truth data generation for newspaper document images. In: ICDAR, pp. 1275–1279 (2009)
16. Sun, K., et al.: High-resolution representations for labeling pixels and regions. arXiv:1904.04514 (2019)
17. Yin, F., Wang, Q.F., Zhang, X.Y., Liu, C.L.: ICDAR 2013 Chinese handwriting recognition competition. In: ICDAR, pp. 1464–1470 (2013)
18. Zhong, Z., Zheng, L., Kang, G., Li, S., Yang, Y.: Random erasing data augmentation. In: Proceedings of AAAI Conference on AI, pp. 13001–13008 (2020)

DAP-BERT: Differentiable Architecture Pruning of BERT

Chung-Yiu Yau$^{(\boxtimes)}$, Haoli Bai, Irwin King, and Michael R. Lyu

The Chinese University of Hong Kong, Shatin, Hong Kong, China
1155109029@link.cuhk.edu.hk, {hlbai,king,lyu}@cse.cuhk.edu.hk

Abstract. The recent development of pre-trained language models (PLMs) like BERT suffers from increasing computational and memory overhead. In this paper, we focus on automatic pruning for efficient BERT architectures on natural language understanding tasks. Specifically, we propose differentiable architecture pruning (DAP) to prune redundant attention heads and hidden dimensions in BERT, which benefits both from network pruning and neural architecture search. Meanwhile, DAP can adjust itself to deploy the pruned BERT on various edge devices with different resource constraints. Empirical results show that the BERT$_{\text{BASE}}$ architecture pruned by DAP achieves 5× speed-up with only a minor performance drop. The code is available at https://github.com/OscarYau525/DAP-BERT.

Keywords: Natural language processing · Neural architecture search · Pruning · BERT

1 Introduction

In the study of natural language processing (NLP), pre-trained language models (PLMs) have shown strong generalization power on NLP tasks [7,15]. However, the high computational overhead and memory consumption of these PLMs prohibit the deployment of these models on resource-limited devices, and thus motivates various efforts towards network compression on PLMs. This area has been investigated by numerous studies, such as pruning [6,16,17], knowledge distillation [11,12], quantization [1,3,23,33].

Among these methods, network pruning starts from a pre-defined architecture and simplifies it by removing unimportant parameters in the network. However, most existing pruning methods rely on hand-crafted criteria to decide the sub-network structure, such as the magnitude of parameters [6] or its gradients [20]. On the other hand, neural architecture search (NAS) aims to automatically search for optimal network architectures and avoids human intervention at the stage of architecture design. Despite the success of NAS popularized in convolution neural networks and recurrent neural networks [8,14], little work has been put on applying NAS to attention-based Transformer networks such as BERT. This is due to the expensive pre-training of language models, which makes the searching process quite time-consuming. While there are some works

© Springer Nature Switzerland AG 2021
T. Mantoro et al. (Eds.): ICONIP 2021, LNCS 13108, pp. 367–378, 2021.
https://doi.org/10.1007/978-3-030-92185-9_30

that search architecture during the fine-tuning stage [5,16], they either suffer from computationally expensive CNN-based cells [5] or inaccurate control of model size through sparse regularization [16].

In this paper, we propose differentiable architecture pruning (DAP) for BERT, a novel approach that benefits from both BERT pruning and neural architecture search. Specifically, our proposed DAP can automatically discover the optimal head number for self-attention and the dimensionality for the feed-forward networks given the resource constraints from various edge devices. Inspired by [14], we assign each architecture choice with learnable parameters, which can be updated by end-to-end training. Meanwhile, to stabilize the searching algorithm, we further introduce rectified gradient update for architecture parameters, as well as progressive architecture constraint, such that the searching process can proceed smoothly. Finally, to find a sub-architecture that performs comparable to the original network, we apply knowledge distillation as a clue to architecture searching and model re-training so that the sub-network mimics the behaviours of its original network.

We conduct extensive experiments and discussions on the GLUE benchmark to verify the proposed approach. The empirical results show that the inference time of our pruned $\text{BERT}_{\text{BASE}}$ can be accelerated up to 5× with only a minor performance drop. Moreover, to the limit of compression, our pruned model can reach up to 27× inference speedup compared with the original $\text{BERT}_{\text{BASE}}$ model, while maintaining 95% of its average performance over GLUE tasks.

2   Related Work

Neural network pruning and neural architecture search are both rapidly growing fields with a large amount of literature. We summarize these two strands of research that are closely related to our proposed solution in the following sections.

2.1   Network Pruning for BERT

Network pruning aims to remove the unnecessary connections in the neural network [2,9,28,30]. Gordon et al. [10] investigate the effect of weight magnitude pruning during the pre-training stage on the transferability to downstream tasks. Prasanna et al. [20] apply gradient-informed structured pruning and unstructured weight magnitude pruning on BERT to verify the lottery ticket hypothesis [9], which indirectly points out the ineffectiveness of sensitivity-based structured pruning. McCarley et al. [16] incorporate distillation and structured pruning by L_0 regularization. Our proposed structured pruning method differs from the usual sensitivity-based approaches [16,20] or weight magnitude approaches [6,10] since we avoid hand-crafted criteria on pruning. Instead, we adopt a loss objective that accurately reflects the model FLOPs constraint and the prediction behaviour of the original model.

2.2 Neural Architecture Search

Neural architecture search (NAS) is to automatize the design of neural network architecture [14,27,35]. Expensive approaches such as reinforcement learning [35] and evolutionary algorithm [21] spend thousands of GPU days to obtain the optimal architecture for computer vision tasks. Recent efforts such as DARTS [14] follow differentiable architecture search, which adopt continuous relaxations over all possible operations in the search space for gradient-based optimization. Such NAS techniques can also be applied for network compression such as pruning [8,31] and quantization [13,29,31]. Especially, TAS [8] search for the optimal width and depth of CNN-based networks in a similar differentiable fashion and achieve effective network pruning. For BERT architecture, AdaBERT [5] performs differentiable cell-based NAS and searches for a novel small architecture by task-oriented knowledge distillation, and obtains efficient CNN-based models. Recently, NAS-BERT [32] performs block-wise NAS with knowledge distillation during the pre-training stage. However, these approaches search the architectures in the search space from scratch, which can be slow in practice. In this paper, instead of searching for a novel architecture from scratch, we start from existing BERT architectures. We follow the differentiable architecture search in a super-graph defined by the original BERT model to prune potentially ineffective connections.

3 Methodology

In this section, we present differentiable architecture pruning (DAP) for BERT, an automatic pruning solution that can be tailored for various edge devices with different resource constraints. The proposed algorithm is initialized with a trained PLM for the downstream task, which avoids the time-consuming pre-training of PLMs. An overview of our searching approach is illustrated in Fig. 1.

3.1 Definition of Search Space

We aim to find the best layer-wise configuration of heads in multi-head attention (MHA) and intermediate dimensionality in feed-forward network (FFN) of the transformer. The search space of MHA and FFN is designed as follows:

Multi-head Attention. Recall that with input X to the MHA, the i-th head H_i can be computed as

$$Q_i = XW_i^Q, \ K_i = XW_i^K, \ V_i = XW_i^V, \ H_i = \text{softmax}\left(\frac{Q_iK_i^T}{\sqrt{d}}\right)V_i, \quad (1)$$

where W_i^Q, W_i^K, W_i^V are projection matrices of query, key and value for the i-th head respectively, and d is the head size. To prune away unnecessary heads, we assign $\alpha^m \in \mathbb{R}^N$ as the architecture parameter for each head, where N is the

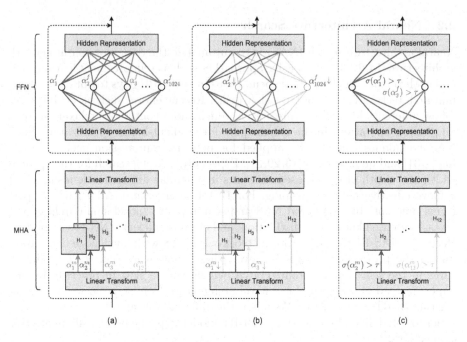

Fig. 1. An overview of the search method and search space of a BERT hidden layer, with multi-head attention block in the bottom and feed-forward block on the top. Dashed arrows represent skip connections. (a) The original network is initialized as the super-graph. An architecture parameter α_i is assigned to each group of weights of the same color in the diagram. (b) Learning the α_i w.r.t. the objective function. (c) Output the optimal sub-network by selecting α that exceeds the threshold τ.

total head numbers before pruning. We further assign a sigmoid function that ensures $\sigma(\alpha^m) \in [0,1]^N$ for head selection. Thus, the weighted MHA output can be written as

$$\text{MultiHead}(Q, K, V) = \text{Concat}(H_1 \odot \sigma(\alpha_1^m), \ ..., \ H_N \odot \sigma(\alpha_N^m)) \, W^O. \qquad (2)$$

Whenever $\sigma(\alpha_i^m) \to 0$, the i-th head can be safely pruned without affecting the output. Note that for each pruned head, the dimensionality of the projection matrices W^Q, W^K, W^V and W^O will be adjusted accordingly.

Feed-Forward Network. The feed-forward network is composed of an intermediate layer followed by an output layer. We aim at reducing the dimensionality of the intermediate representation, by introducing architecture parameters $\alpha^f \in \mathbb{R}^D$ for the D intermediate dimensions. The FFN output can thus be written as

$$\text{FFN}(X) = \max(0, XW_1 + b_1) \, \text{Diag}\big(\sigma(\alpha^f)\big) \, W_2 + b_2, \qquad (3)$$

where X is the input to FFN, $Diag(\cdot)$ represents the diagonal matrix, and W_1, W_2, b_1, b_2 are the parameters of the two linear layers.

3.2 Differentiable Architecture Pruning

Given the architecture parameters associated with the search space, we follow differentiable architecture search [14] to update α via end-to-end training. In order to find the best architecture on edge devices with different computational capacities, we also include the network FLOPs as part of the search objective. The overall searching problem can be formulated as a bi-level optimization problem as follows:

$$\min_{\alpha} \quad \mathcal{L}(w^*(\alpha), \alpha)$$
$$\text{subject to} \quad w^*(\alpha) = \operatorname{argmin}_w \mathcal{L}(w, \alpha), \tag{4}$$
$$\mathcal{F}(\alpha) \leq \mathcal{F}_{\text{target}},$$

where $\mathcal{F}(\alpha)$ denotes the number of FLOPs based on the architecture α, and $\mathcal{F}_{\text{target}}$ is the searching target FLOPs. To satisfy the FLOPs constraint in Eq. (4), we follow [8] to apply FLOPs penalty as a differentiable loss objective $\mathcal{L}_{\text{cost}}$ w.r.t. α as follows:

$$\mathcal{L}_{\text{cost}} = \begin{cases} \log(\mathbb{E}[\mathcal{F}(\alpha)]) & \text{if } \mathbb{E}[\mathcal{F}(\alpha)] > (1 + \delta) \times \mathcal{F}_{\text{curr}}(t), \\ 0 & \text{if } (1 - \delta) \times \mathcal{F}_{\text{curr}}(t) \leq \mathbb{E}[\mathcal{F}(\alpha)] \leq (1 + \delta) \times \mathcal{F}_{\text{curr}}(t), \\ -\log(\mathbb{E}[\mathcal{F}(\alpha)]) & \text{if } \mathbb{E}[\mathcal{F}(\alpha)] < (1 - \delta) \times \mathcal{F}_{\text{curr}}(t), \end{cases} \tag{5}$$

$$\mathbb{E}[\mathcal{F}(\alpha)] = \sum_{l=1}^{L} \left(\sum_{j=1}^{H} \sigma(\alpha_{lj}^m) \mathcal{F}_{\text{MHA}} + \sum_{j=1}^{D} \sigma(\alpha_{lj}^f) \mathcal{F}_{\text{FFN}} \right), \tag{6}$$

where $\mathbb{E}[\mathcal{F}(\alpha)]$ is the expected FLOPs of the current architecture summed over L hidden layers, $\mathcal{F}_{\text{MHA}}$ and $\mathcal{F}_{\text{FFN}}$ are the FLOPs of a single head and one intermediate dimension in FFN, and δ is a tolerance parameter. $\mathcal{F}_{\text{curr}}(t)$ denotes the target FLOPs, which is time-dependent as will be introduced in Eq. (8).

Rectified Update of α. During architecture search, we optimize both w and α by stochastic gradient descent. However, the update of α suffers from vanishing gradient as a result of the sigmoid activation $\sigma(\cdot)$. To solve this challenge, we introduce rectified update for α as follows. We adopt sign stochastic gradient descent (signSGD) [4] to enlarge the magnitude of gradients on α. It is known that signSGD avoids the problem of vanishing gradient since the magnitude of gradient is controlled [18]. Nevertheless, signSGD may bring oscillations that make the optimization unstable. These rapid oscillations to the architecture parameter α may lead to an immature solution. To smoothly stabilize the optimization process, we only backpropagate the top-10% gradient of α according to their magnitude within the search region, while the rest are masked as follows:

$$\hat{g}_i^{(t)} = \begin{cases} \text{sign}(g_i^{(t)}) & \text{if } g_i^{(t)} \text{ is top-10\% in magnitude}, \\ 0 & \text{otherwise}. \end{cases} \tag{7}$$

Progressive Architecture Constraint. Directly applying a FLOPs penalty with a fixed FLOPs target leads α to arrive at the target size early during

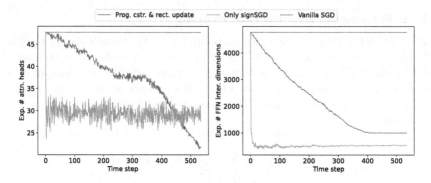

Fig. 2. The searching dynamics of different approaches. (**Left**) shows the expected number of attention heads. (**Right**) shows the FFN intermediate dimensions.

searching, while progressive pruning usually better identifies the less influential parameters [6]. To perform architecture search progressively, we adjust the FLOPs target $\mathcal{F}_{\mathrm{curr}}(t)$ at time step t as

$$\mathcal{F}_{\mathrm{curr}}(t) = \mathcal{F}_{\mathrm{original}} \exp\left(\frac{1}{T} \ln \frac{\mathcal{F}_{\mathrm{target}}}{\mathcal{F}_{\mathrm{original}}}\right)^{t}, \tag{8}$$

where T is the scheduled number of training steps, $\mathcal{F}_{\mathrm{original}}$ is the original architecture FLOPs, and $\mathcal{F}_{\mathrm{target}}$ is the desired FLOPs.

We visualize the effect of rectified update and progressive architecture constraint in Fig. 2. It is evident that vanilla SGD suffers from vanishing gradient, thus cannot achieve pruning. SignSGD directly arrives at the FLOPs target in the beginning, which may result in sub-optimal architectures. On the other hand, our progressive FLOPs constraint and rectified update enable smooth searching.

Loss Objective for Searching. Due to the intractability of the bi-level problem in Eq. (4), we simultaneously update w and α w.r.t. the objective function on the training set. The objective function involves two terms: the cross-entropy between the full-size model logits z^t and the searching model logits z^s (i.e., knowledge distillation); and the FLOPs penalty $\mathcal{L}_{\mathrm{cost}}$ as defined in Eq. (5). The searching objective is thus

$$\mathcal{L}_{\mathrm{search}} = \mathcal{L}_{\mathrm{ce}} + \lambda\mathcal{L}_{\mathrm{cost}}, \quad \text{where } \mathcal{L}_{\mathrm{ce}} = -\operatorname{softmax}(z^t) \cdot \operatorname{log\_softmax}(z^s). \tag{9}$$

After the training of α, pruning is achieved with a pre-set threshold τ, i.e., only keeping the connections that satisfy $\sigma(\alpha_i) > \tau$.

3.3   Fine-Tuning with Two-Stage Knowledge Distillation

After obtaining the slimmed BERT structure, we further fine-tune the network to recover from performance degradation due to pruning. The fine-tuning is based

---

Algorithm 1: Learning optimal sub-network of BERT

Initialize network weights w from a well-trained model;
Initialize architecture parameters α that satisfy $\sigma(\alpha_i) > \tau$;
▷ *Searching*;
for T *iterations* **do**
  | Forward pass to compute MHA and FFN by Eq. (2), (3);
  | Update $\mathcal{F}_{\mathrm{curr}}(t)$ according to Eq. (8);
  | Calculate the loss objective in Eq.(9);
  | Backpropagate to update w by Adam optimizer;
  | Backpropagate to update α with rectified gradient in Eq. (7);
end
▷ *Pruning*;
Prune by selecting $\sigma(\alpha_i) > \tau$;
▷ *Fine-tuning*;
Restore network parameter value from the original network;
Fine-tune the pruned network with two-stage knowledge distillation;

---

on two-stage knowledge distillation [12] given its previous success in model compression. The first stage aims at intermediate layer distillation, which minimizes the mean squared error (MSE) between the sub-network (student) and the original network (teacher) as follows:

$$\mathcal{L}_{\mathrm{int}} = \sum_{l=1}^{L} \|A_l^s - A_l^t\|_F^2 + \|F_l^s - F_l^t\|_F^2, \tag{10}$$

where $\|\cdot\|_F$ is the Frobenius norm, A_l^s, A_l^t and F_l^s, F_l^t denote the MHA attention maps and FFN output of l-th transformer layer from the student and teacher model, respectively. Note that as we do not modify the output shape of the student model, the MSE loss can be directly calculated without linear mapping to align the dimension as done in [12]. The second stage is prediction layer distillation, which similarly adopts the cross-entropy loss $\mathcal{L}_{ce}$ defined in Eq. (9).

In practice, we find that fine-tuning the network parameters immediately after searching usually lead to sub-optimal performance. Similar observations are also found in existing NAS literature [14,19,27], where instead they initialize the network parameters and train the architecture from scratch. Similarly, we only inherit the network structure (i.e., the configuration of heads and dimensions in MHA and FFN), and restore the parameters from the original model.

3.4 Summary of the Method

Algorithm 1 summarizes the overall workflow of our proposed method, which can be generally divided into three steps: searching, pruning and fine-tuning. Firstly, we initialize all architecture parameters by $\sigma(\alpha_i) > \tau$ such that all the prunable MHA heads and FFN dimensions are kept initially. We conduct searching by simultaneously minimizing the objective function in Eq. (9) w.r.t. architecture parameter α and network parameters w. To facilitate a smooth searching process, we rectify the update of α according to Eq. (7), and incorporate progressive architecture constraint in Eq. (8). After searching, we apply pruning based on α and restore the network parameters to their original states. Finally, we conduct two-stage knowledge distillation for fine-tuning.

4   Experiments

In this section, we empirically verify the proposed method on the GLUE benchmark [26]. We first introduce the experiment setup in Sect. 4.1. The main results are presented in Sect. 4.2, followed by comparisons with other state-of-the-art approaches in Sect. 4.3. Finally, we provide further discussions to better understand the proposed approach in Sect. 4.4.

4.1   Experiment Setup

Dataset and Metrics. The GLUE benchmark provides a variety of natural language understanding tasks. Unless specified, we report the metrics of each task as follows: Matthew's Correlation for CoLA, F1 score for MRPC and QQP, Spearman Correlation for STS-B, and accuracy for the remaining tasks. Following [12], We apply data augmentation to small datasets (RTE, MRPC, STS-B and CoLA) to improve fine-tuning of the pruned networks.

Implementation. The proposed method applies to any well-trained BERT models on downstream tasks. We take the $BERT_{BASE}$ [7][1] and TinyBERT [12][2] as the super-graph for searching. For each of the super-graphs, we experiment with different FLOPs constraint $\mathcal{F}_{target}$ and compare the performance drop across different models. For all the experiments, we initialize $\alpha_i = 5$ and use $\tau = 0.99$ as the pruning threshold. For small datasets, we search for 10 epochs and fine-tune for 10 epochs. For large datasets, we search for one or fewer epochs (i.e., using part of the training set) and fine-tune for 3 epochs. Then we fine-tune the sub-network using Adam optimizer with 5×10^{-5} learning rate.

4.2   Experiment Results

We evaluate DAP on $BERT_{BASE}$, $TinyBERT_4$, and $TinyBERT_6$, and results are shown in Table 1. We denote our results as +DAP-$p\%$, where $p\%$ denotes the pruning rate. It can be found that the accuracy drop depends on the original network size, where on the same scale of FLOPs reduction, small networks bear

---

[1] To obtain the task-specific parameters, we follow the standard fine-tuning pipeline in https://huggingface.co/bert-base-uncased.

[2] Task specific model parameters available at https://github.com/huawei-noah/Pretrained-Language-Model/tree/master/TinyBERT.

a larger percentage of accuracy drop than large networks. Notably, BERT$_{BASE}$ can be pruned to half without accuracy degradation.

Additionally, we also measure the practical inference speedup of the pruned networks in Table 2. It is shown that FLOPs reduction on the network architecture can bring up to 5.4× practical speed-up for BERT$_{BASE}$, and can even be 27.6× faster for TinyBERT$_4$ with DAP-30%.

Table 1. Experimental results of the proposed architecture searching algorithm, evaluated on the GLUE test set.

#	Models	FLOPs (B)	Param. (M)	MNLI-m 392k	QQP 363k	QNLI 108k	SST-2 67k	CoLA 8.5k	STS-B 5.7k	MRPC 3.5k	RTE 2.5k	Avg. (%↓)
1	BERT$_{BASE}$	22.3	109.5	84.0	70.7	91.1	92.7	55.3	82.5	86.6	65.5	78.6 (0.0)
2	+ DAP-50%	11.4	66.6	84.2	72.2	90.4	93.2	53.0	83.0	86.6	66.0	78.6 (0.0)
3	+ DAP-30%	7.1	51.6	83.6	71.6	89.9	91.9	49.9	82.6	86.5	65.2	77.7 (1.1)
4	+ DAP-10%	3.0	33.6	83.0	71.4	88.3	91.8	45.7	81.7	85.8	63.6	76.4 (2.7)
5	TinyBERT$_6$	11.1	67.0	84.6	71.6	90.4	93.1	51.1	83.7	87.3	70.0	79.0 (0.0)
6	+ DAP-30%	3.6	37.1	83.7	71.8	89.5	93.0	46.1	83.3	86.9	63.6	77.2 (2.2)
7	TinyBERT$_4$	1.2	14.5	82.5	71.3	87.7	92.6	44.1	80.4	86.4	66.6	76.5 (0.0)
8	+ DAP-30%	0.4	11.1	80.8	70.9	84.4	91.8	40.7	78.6	85.4	60.8	74.0 (3.2)

Table 2. Practical speedup of the sub-networks, presenting the inference time for a batch of 32 examples with 128 maximum sequence length on Intel(R) Xeon(R) CPU E5-2620 0 @ 2.00 GHz with 4 cores.

#	Models	FLOPs (B)	Speedup (×)	Time (s)	Practical Speedup (×)
1	BERT$_{BASE}$	22.3	1.0	7.28	1.0
2	+ DAP-50%	11.4	2.0	3.72	2.0
3	+ DAP-30%	7.1	3.1	2.49	2.9
4	+ DAP-10%	3.0	7.3	1.36	5.4
5	TinyBERT$_6$	11.1	2.0	3.69	2.0
6	+ DAP-30%	3.6	6.2	1.22	6.0
7	TinyBERT$_4$	1.2	17.9	0.62	11.8
8	+ DAP-30%	0.4	55.4	0.26	27.6

4.3 Comparison with State-of-the-arts

To further validate the proposed approach, we compare with several state-of-the-art compression baselines including vanilla BERT [25], DistilBERT [22], Mobile-BERT [24], NAS-BERT [32] and Mixed-vocab KD [34]. Evaluations on the GLUE test set and development set are shown in Table 3 and Table 4 respectively. The proposed DAP shows superior performance against the baselines. For instance, our DAP-BERT$_{12}$-10% achieves the averaged test score of 76.4 with only 3.0 FLOPs (B), which is just 2.2 score lower than the original BERT$_{BASE}$ model with more than 7× FLOPs reduction.

4.4   Discussion

Rectified Update and Progressive Architecture Pruning. The left side of Fig. 3 shows the ablation studies for our rectified update and progressive architecture constraint. It can be found that when armed with only progressive constraint, the searching algorithm fails to converge to the desired FLOPs. While pure signSGD can converge to network architectures with desired FLOPs, the performance is usually worse due to oscillating update of α, as previously discussed in Fig. 2. When combined with the rectified update, the performance is consistently improved at different FLOPs targets. Finally, when the rectified update is combined with progressive architecture constraint, the network performance is boosted since the searching dynamics is smooth and stabilized.

Table 3. Comparison with state-of-the-art compression approaches, evaluated on the GLUE test set.

#	Models	FLOPs (B)	Param. (M)	MNLI-m	QQP	QNLI	SST-2	CoLA	STS-B	MRPC	RTE	Avg.
1	BERT$_{BASE}$	22.3	109.5	84.0	70.7	91.1	92.7	55.3	82.5	86.6	65.5	78.6
2	BERT$_{SMALL}$	3.4	29.2	77.6	68.1	86.4	89.7	27.8	77.0	83.4	61.8	71.5
3	MobileBERT$_{TINY}$	3.1	15.1	81.5	68.9	**89.5**	91.7	**46.7**	80.1	**87.9**	**65.1**	76.4
4	DAP-BERT$_{12}$ – 10%	3.0	33.0	**83.0**	**71.4**	88.3	**91.8**	45.7	**81.7**	85.8	63.6	**76.4**
5	BERT$_{MINI}$	0.87	11.1	74.8	66.4	84.1	85.9	0.0	73.3	81.1	57.9	65.4
6	Mixed-vocab KD	-	10.9	80.7	-	-	90.6	-	-	87.2	-	-
7	DAP-BERT$_{4}$ – 30%	0.40	11.1	**80.8**	**70.9**	**84.4**	**91.8**	**39.6**	**78.6**	85.4	**60.8**	**74.0**

Table 4. Comparison with state-of-the-art compression approaches, evaluated on the GLUE development set.

#	Models	FLOPs (B)	Param. (M)	MNLI-m	QQP	QNLI	SST-2	CoLA	STS-B	MRPC	RTE	Avg.
1	NAS-BERT$_{10}$	2.3	10.0	76.4	88.5	86.3	88.6	34.0	84.8	79.1	66.6	75.5
2	DistilBERT$_{6}$	-	66.0	82.2	88.5	**89.2**	91.3	51.3	86.9	**87.5**	59.9	79.6
3	DAP-BERT$_{12}$ – 10%	3.0	33.0	**82.8**	**90.6**	88.9	**91.9**	**52.3**	**88.2**	85.3	**67.5**	**80.9**
4	NAS-BERT$_{5}$	0.86	5.0	74.4	85.8	84.9	87.3	19.8	83.0	79.6	**66.6**	72.7
5	DAP-BERT$_{4}$ – 30%	0.40	11.1	**81.2**	**90.6**	**86.2**	**92.2**	**45.8**	**85.8**	**86.0**	63.2	**78.9**

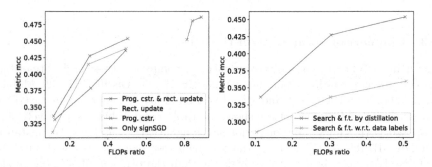

Fig. 3. (**Left**) shows the architecture accuracies under different approaches. (**Right**) shows the effect of knowledge distillation for searching and fine-tuning.

Distillation for Architecture Searching. We verify the advantage of knowledge distillation (soft labels from the original model) over data labels (hard labels from ground truth) by comparing the performance of the architectures found using these search objectives. The empirical result in the right of Fig. 3 shows that knowledge distillation using soft labels can generally find better architectures than using the ground truth data labels.

5   Conclusion

In this paper, we propose differentiable architecture pruning, an automatic neural architecture search for BERT pruning. Given the resource constraints from edge devices, the proposed approach can identify the best model architecture accordingly. Empirical results on the GLUE benchmark show that the pruned BERT model can perform on par with the original network, while enjoying significant inference speedup. Our work opens the door to deploying PLM to resource-limited edge devices and contributes to the various applications of NLP.

Acknowledgement. The work described in this paper was partially supported by the National Key Research and Development Program of China (No. 2018AAA0100204) and the Research Grants Council of the Hong Kong Special Administrative Region, China (No. CUHK 14210920 of the General Research Fund).

References

1. Bai, H., Hou, L., Shang, L., Jiang, X., King, I., Lyu, M.R.: Towards efficient post-training quantization of pre-trained language models. Preprint arXiv:2109.15082 (2021)
2. Bai, H., Wu, J., King, I., Lyu, M.: Few shot network compression via cross distillation. In: AAAI, vol. 34, pp. 3203–3210 (2020)
3. Bai, H., et al.: BinaryBERT: pushing the limit of BERT quantization. In: ACL (2020)
4. Bernstein, J., Wang, Y.X., Azizzadenesheli, K., Anandkumar, A.: signSGD: Compressed optimisation for non-convex problems. In: ICML (2018)
5. Chen, D., et al.: AdaBERT: task-adaptive BERT compression with differentiable neural architecture search. In: IJCAI (2021)
6. Chen, T., et al.: The lottery ticket hypothesis for pre-trained BERT networks. In: NeurIPS (2020)
7. Devlin, J., Chang, M.W., Lee, K., Toutanova, K.: BERT: pre-training of deep bidirectional transformers for language understanding. In: NAACL-HLT (2019)
8. Dong, X., Yang, Y.: Network pruning via transformable architecture search. In: NeurIPS (2019)
9. Frankle, J., Carbin, M.: The lottery ticket hypothesis: finding sparse, trainable neural networks. In: ICML (2018)
10. Gordon, M.A., Duh, K., Andrews, N.: Compressing BERT: studying the effects of weight pruning on transfer learning. In: ACL (2020)
11. Hou, L., Huang, Z., Shang, L., Jiang, X., Chen, X., Liu, Q.: DynaBERT: dynamic BERT with adaptive width and depth. In: NeurIPS (2020)

12. Jiao, X., et al.: TinyBERT: distilling BERT for natural language understanding. In: EMNLP (2020)
13. Li, Y., Wang, W., Bai, H., Gong, R., Dong, X., Yu, F.: Efficient bitwidth search for practical mixed precision neural network. Preprint arXiv:2003.07577 (2020)
14. Liu, H., Simonyan, K., Yang, Y.: DARTS: differentiable architecture search. In: ICLR (2019)
15. Liu, Y., et al.: RoBERTa: a robustly optimized BERT pretraining approach. Preprint arXiv:1907.11692 (2019)
16. McCarley, J.S., Chakravarti, R., Sil, A.: Structured pruning of a BERT-based question answering model. Preprint arXiv:1910.06360 (2021)
17. Michel, P., Levy, O., Neubig, G.: Are sixteen heads really better than one? In: NeurIPS (2019)
18. Pascanu, R., Mikolov, T., Bengio, Y.: On the difficulty of training recurrent neural networks. In: ICML (2013)
19. Pham, H., Guan, M.Y., Zoph, B., Le, Q.V., Dean, J.: Efficient neural architecture search via parameter sharing. In: ICML, pp. 4092–4101 (2018)
20. Prasanna, S., Rogers, A., Rumshisky, A.: When BERT plays the lottery, all tickets are winning. In: EMNLP (2020)
21. Real, E., Aggarwal, A., Huang, Y., Le, Q.V.: Regularized evolution for image classifier architecture search. In: AAAI (2019)
22. Sanh, V., Debut, L., Chaumond, J., Wolf, T.: DistilBERT, a distilled version of BERT: smaller, faster, cheaper and lighter. In: NeurIPS (2020)
23. Shen, S., et al.: Q-BERT: hessian based ultra low precision quantization of BERT. In: AAAI (2019)
24. Sun, Z., Yu, H., Song, X., Liu, R., Yang, Y., Zhou, D.: MobileBERT: a compact task-agnostic BERT for resource-limited devices. In: ACL (2020)
25. Turc, I., Chang, M.W., Lee, K., Toutanova, K.: Well-read students learn better: on the importance of pre-training compact models. Preprint arXiv:1908.08962v2 (2019)
26. Wang, A., Singh, A., Michael, J., Hill, F., Levy, O., Bowman, S.R.: GLUE: a multi-task benchmark and analysis platform for natural language understanding. In: ICLR (2019)
27. Wang, J., et al.: Revisiting parameter sharing for automatic neural channel number search. In: NeurIPS, vol. 33 (2020)
28. Wang, J., Bai, H., Wu, J., Cheng, J.: Bayesian automatic model compression. IEEE JSTSP **14**(4), 727–736 (2020)
29. Wang, K., Liu, Z., Lin, Y., Lin, J., Han, S.: HAQ: hardware-aware automated quantization with mixed precision. In: CVPR, pp. 8612–8620 (2019)
30. Wen, L., Zhang, X., Bai, H., Xu, Z.: Structured pruning of recurrent neural networks through neuron selection. NN **123**, 134–141 (2020)
31. Wu, J., et al.: PocketFlow: an automated framework for compressing and accelerating deep neural networks. In: NeurIPS, CDNNRIA workshop (2018)
32. Xu, J., et al.: Nas-BERT: task-agnostic and adaptive-size BERT compression with neural architecture search. In: KDD (2021)
33. Zhang, W., et al.: TernaryBERT: distillation-aware ultra-low bit BERT. In: EMNLP (2020)
34. Zhao, S., Gupta, R., Song, Y., Zhou, D.: Extremely small BERT models from mixed-vocabulary training. In: EACL (2021)
35. Zoph, B., Vasudevan, V., Shlens, J., Le, Q.V.: Learning transferable architectures for scalable image recognition. In: CVPR (2018)

Trash Detection on Water Channels

Mohbat Tharani[(✉)], Abdul Wahab Amin, Fezan Rasool, Mohammad Maaz,
Murtaza Taj, and Abubakar Muhammad

Lahore University of Management Sciences, Lahore, Pakistan
16060073@lums.edu.pk

Abstract. Rivers and canals flowing through cities are often used illegally for
dumping trash that contaminates freshwater channels, causes blockage in sewerage leading to urban flooding. The dumped trash is often found floating on
the water surface. We propose to automatically identify this trash through visual
inspection with the eventual goal of quantification, an early warning system to
avoid blockages and urban flooding. The trash could be disfigured, partially sub-
merged, or clumped together with other objects which obscure its shape and
appearance. Thus, we consider surface trash as a blob detection problem that
could either be solved as object detection or image segmentation or both. To
this extent, we evaluate and compare several deep-learning-based object detection
and segmentation algorithms. Unlike ocean trash, to the best of our knowledge,
there is no large dataset on urban trash on water channels. Thus, using IoT-based
camera nodes at multiple water channels, we collected a large dataset containing 48, 450 trash objects annotated for both bounding box and segmentation (the
dataset will be made publicly available (Dataset is available at https://cvlab.lums.
edu.pk/watertrash/)). In addition, we also propose modifications in state-of-the-
art detection and segmentation algorithms to cater to an issue such as partially
submerged, varying object sizes, and edge-based computing.

Keywords: Trash · Object detection · Segmentation

1 Introduction

Every year millions of tons of trash, particularly plastic, are discarded globally that pol-
lute our lands, rivers, and oceans. It causes environmental as well as economic repercus-
sions. According to the United Nations world water development report [5], annually
about 3.5 million people, mostly children, die from water-related infections. Almost
all research on plastic contamination in water systems focuses on oceans. However,
the source of this contamination, particularly in Asia and Africa, is the trash dumped
into fresh and sewerage water channels, especially those passing through urban centers.
These water channels transport this trash to rivers and finally to sea due to inadequate
disposal and handling of landfill, industrial and general wastes [12].

Thus it is essential to identify the trash objects such as papers, card-boards, food
residuals, plastic bottles, and bags, etc. which are present in freshwater ecosystems. In
this work, we propose to identify these floating trash objects through a camera-based

© Springer Nature Switzerland AG 2021
T. Mantoro et al. (Eds.): ICONIP 2021, LNCS 13108, pp. 379–389, 2021.
https://doi.org/10.1007/978-3-030-92185-9_31

Fig. 1. Sample camera views from the collected dataset. The variation in views, shadows of overpass bridge, reflection of buildings and presence of vegetation is clear in these images.

inspection system that will help in forming an index of water contamination by measuring the number of trash objects and water surface area covered by it. The identification and quantification of trash would then be used to remove trash via skimmer boats or by notifying concerned authorities for appropriate actions.

The existing work on vision-based approaches for the detection of trash could be divided into three categories i) Classification of trash in a controlled environment, applicable at waste recycling plants [22,29] ii) Detection of piles of trash, usually illegally dumped in cities [16,18] iii) Detection of sparse trash (street trash or marine litter) [7,10,11,15]. In this paper, we introduce a fourth category of detecting visual trash floating on the water channels, especially drainage canals. Different from the above-discussed studies, our problem focuses on surface trash present in canals running through dense urban areas.

There are many solutions to perform automatic waste classification in controlled environments such as recycling plants [2,22,29]. For comparatively less managed scenarios such as streets in the urban center, convolution neural network based object detectors are used [18]. This work identifies street trash by performing object detection on images obtained from an RGB camera mounted on a dashboard of a vehicle [18].

Image segmentation has also been used to classify trash pixels. Kelo *et al.* [11] uses unmanned aerial vehicles (UAV) to map the trash on beaches. This work first computes a point cloud using photogrammetry and then applies CNN based segmentation strategy for trash localization. Similarly, SpotGarbage [16] performs segmentation of trash on publicly available images such as those obtained from Bing Maps or through crowdsourcing (TACO Litter dataset [17]). Hyper-spectral images [31] and RGB-D images [28] obtained via depth sensor have also been used to address the problem of trash identification. Similarly, deep detection networks are used for localization of marine litter [7,10].

(a) Deformed object (plastic cup)   (b) Sub-merged trash objects   (c) Reflection of flying bird   (d) Reflection of buildings   (e) Micro-particles of trash

(f) Object in Reflection   (g) Color variation   (h) Pile of trash   (i) Sparse trash   (j) Air bubbles

Fig. 2. Sample cropped images from our dataset showing 10 challenging scenarios present in the fresh water and drainage canals.

Most of the existing methods are either focused on street litter or ocean trash, recently a solution is proposed to monitor plastic contamination in rivers [25]. This work first collects a dataset of 1272 images and then perform off the shelf image segmentation and classification for the identification of plastics. We instead study fresh and sewerage channels that are the source of contamination for rivers and oceans. Considering challenges; such as varying shape and appearance irregularities, trash localization can be considered as a *blob detection* problem and provide a comparative evaluation of various object detection and segmentation approaches. We also propose modifications in state-of-the-art detection and segmentation algorithms to cater for issues such as weak objects (such as partially submerged) and varying object sizes, we introduced log-attention, data imbalance is handled by replacing loss function with focal loss [30] and computational efficiency is improved through depth-wise separable convolution [9]. Furthermore, we present an IoT-based multi-camera setup deployed on these water channels for continuous data collection and monitoring. As a result, we also provide a comparatively large dataset containing 13, 500 images and 48, 450 object annotations (bounding boxes and segmentation mask).

The remainder of this paper is organized as follows. Section 2 introduces the dataset, comparative evaluation of object detection and segmentation approaches is discussed in Sect. 3 and Sect. 4 respectively. Finally, the paper is concluded in Sect. 5.

2 Dataset

2.1 Collection and Annotation

To obtain an extensive dataset containing wide-variety of realistic scenarios, we selected multiple sites across the city and performed video recording during different day times, weather conditions, and at different types of localities (market, dense urban, peri-urban, etc.). After removing noisy samples, a total of 1, 35, 00 images were extracted from these videos. These images were annotated for bounding boxes using LabelImg [24] and

Table 1. Distribution of small, medium and large instances (trash objects) in annotated images as per COCO [4] standard.

Size	No. of objects	Area (pixels)
Small	11,214	area $\leq 32^2$
Medium	32,078	$32^2 <$ area $\leq 96^2$
Large	5,158	area $> 96^2$
Total	48,450	area $> 7^2$

for segmentation using LabelMe [21]. Plastic bottles, plastic bags, papers, cardboards, and food residuals, etc. floating on the water surface either fully or partially visible (submerged) were regarded as garbage. We intentionally do not annotate micro-particles, plant leaves, and air bubbles. Thus, $4, 88, 98$ instances were opted as trash objects. We consider $1, 25, 00$ samples as train-validation set and the remaining $1, 000$ images are used as a test set. Apart from images from the same sites, the test set also contains images that are from a different distribution i.e. different view angles and sites. The example images are shown in Fig. 1, the dataset contains several challenging scenarios that are discussed next.

2.2   Challenges in Dataset

No Fixed Geometry: Objects in the trash do not have a defined geometrical shape and the shapes also change over time. For example, floating plastic bags can have any shape which also changes with time. The trash also contains objects which are damaged either before dumping or in the process such as plastic bottles or juice packets are collapsed and punctured and have wrinkled labels (see Fig. 2(a)).

Partial Visibility of Objects: Many of the objects are partially submerged in water and it has also been observed that due to the flow of water, objects sometimes sink and then resurface after a while. It has been observed that the texture of objects such as cardboards (see Fig. 2(b)), make them very difficult to be detected.

Reflections: Since the dataset was collected in dense urban areas, the reflections of both static and moving objects cause a significant change in the appearance of the water channel. Static objects include buildings and electricity poles, which depending upon the camera view-point and time of the day, may cover a significant portion of the water surface. Moving objects such as flying birds reflect over the water surface, which often exhibits very similar features to that of flowing garbage. Three samples of reflection are demonstrated in Fig. 2(c), (d) and (f).

Shadows: Similar to reflections, the shadows of buildings and poles, especially during morning and afternoon hours, create unfavorable conditions for deep detectors.

Table 2. Comparison with existing trash datasets. (Key: annot.: annotation, BBox: Bounding box for object detection, Mask: segmentation mask).

Dataset	Type	# images	# classes	# instances	Annot. type
TrashNet [29]	Indoor	2500	6	2500	Classification
VN-trash [26]	Indoor	5904	3	5904	Classification
Street [18]	Road	469	6	1421	BBox
TACO [17]	Multi	1500	60	4784	Mask
MJU-Waste [28]	Multi	2475	1	2524	Mask
Trash-det [7]	Marine	5720	3	–	BBox
TrashCan [10]	Marine	7212	22	12480	Mask
River Plastic [25]	Rivers	1272	1	14892	Mask
Our	Water channel	**13500**	1	48450	BBox, Mask

Water Opacity: The color of the water surface varies with the ratio of chemical discharge from industries that alters the opacity. It not only changes the texture of the water surface but also affects the reflection and refraction of sunlight.

Wide Density of Objects: As shown in Fig. 2(g)–(i), there are sparse objects as well as dense piles of trash. For sparse objects, each particle is considered as a different object, whereas the piles of trash are considered as a single object. Thus, a wide range of variable sizes of objects has been considered. But, the micro-particles as given in Fig. 2(g), are ignored since they are smaller in size and when the image is resized to less than 640 pixel, they become invisible.

Air Bubbles: Sometimes sewer gases are produced through the decomposition of organic household products or industrial waste, which results in air bubbles forming on the water surface as depicted in Fig. 2(j). The bubbles look similar to sub-merged objects or damaged cardboards and result in false detection from the networks.

2.3 Comparison with Existing Datasets

Table 2 compares existing datasets pertaining to trash detection in different scenarios. TrashNet [29] and VN-trash [26] contain images taken in controlled environments, therefore objects are clearly visible and each image contains only one object. TACO [17] is a cloud-sourced trash RGB images, MJU-Waste [28] contains point-cloud of trash objects, and Street [18] is collected through car dashboard camera. These datasets contain fewer image samples, therefore insufficient for generalization of deep learning models. Trash-det [7] and TrashCan [10] are underwater marine debris datasets. The recent River Plastic [25] dataset also contains very few images of specific locations and fixed lighting conditions. To the best of our knowledge, we have yet to find a large dataset that entails the study of trash in water channels. Our dataset is significantly larger than the existing datasets.

Table 3. Comparative evaluation of the state-of-the-art object detection techniques. (Key: AP: Average Precision, AP^S, AP^M and AP^L: AP small, medium and large size objects, M: Millions).

Model	#Param	AP^S	AP^M	AP^L	mAP
YOLO-v3 [19]	61.5M	2.7	11.4	33.7	35.1
YOLO-v3-Tiny	8.6M	0.4	2.8	14.5	9.0
RetinaNet [14]	55.4M	4.6	23.0	42.6	43.4
PeleeNet [27]	4.8M	3.5	10.9	30.1	29.9
M2Det (VGG) [32]	132M	5.2	19.8	35.2	45.8

3 Trash Object Identification

Most of the well-known object detectors [14, 19, 27, 32] are designed for general applications, especially for urban scenarios such as those related to surveillance and self-driving cars. Although they perform better on relevant benchmark datasets such as MS-COCO [4] and Pascal-VOC [6], however, the difficulty of trash detection is significantly higher due to the challenges present in the dataset.

We retrained five different detectors on our dataset namely YOLO-v3 [19], YOLO-v3-Tiny, RetinaNet [14], PeleeNet [27] and M2Det (VGG) [32] and demonstrate their results in Table 3. Overall, M2Det with VGG-16 as backbone outperforms all the other models whereas YOLO-v3-Tiny fails to learn complex information. The better performance of M2Det is due to its multi-level feature pyramid network (MLFPN). RetinaNet and PeleeNet have moderate performance. PeleeNet despite being a smaller network produces results comparable to RetinaNet and YOLO-v3. It can also be validated from Fig. 3(c)–(f), where both scenes contain different water textures and have a reflection from a pole and buildings, most of the objects are detected.

3.1 Analysis on Object Sizes

In order to find out the performance dependency on object sizes, we evaluated the trained networks at three scales of objects provided in Table 1. Table 4 demonstrates detection of smaller objects is difficult than medium and large objects. This is probably due to prior training of models for object detection task on the Pascal VOC dataset. Therefore, the information of objectness was retained for large objects whereas learned information about small objects is forgotten and learned again. As claimed by M2Det [32], it has outperformed on small as well as large objects. RetinaNet is second best as it is built upon a feature pyramid network (FPN). Although PeleeNet is a smaller model, yet its performance is comparable to YOLO-v3, even it has a better AP score for small objects. It would also be a better choice for deployment on an embedded computing platform.

3.2 Handling Smaller and Partially Visible Objects

To address the issue of poor performance on sub-merged, partially visible, and smaller objects by detection models, an attention layer is employed that tries to focus on weaker

(a) UNet   (b) UNet‡   (c) YOLO-v3   (d) RetinaNet   (e) Pelee   (f) M2Det

Fig. 3. Qualitative results of trained model: (a), (b) are segmentation output from vanilla UNet and UNet with separable convolution layer, respectively where green represents trash pixels, black and blue are land nad water pixels. (c)–(f) show the trash detected by with object detectors. (Color figure online)

activation. Attention based on *Sigmoid* and *Softmax* activation functions perform well for segmenting the pixels near boundaries of the objects but do not work in case of object detection. So, we amplify weaker activations through *Log* attention that can be defined as:

$$f_{i+1} = f_i \times log(ReLU(f_i) + 1),   \qquad (1)$$

where f_i is output of ith layer, and $i = 0, ..., N$ and N is number of layers. Here, *ReLU* discards the negative values in the activations and bias 1 shifts it one scale up, making it possible to compute *log*. The derivative of this attention layer would be:

$$\nabla f_{i+1} = \begin{cases} log(f_i + 1) + \frac{1}{f_i+1} & f_i \geq 0 \\ 1 & otherwise \end{cases} \qquad (2)$$

Table 4 demonstrates improvement in performance due to introduced log attention layer. YOLO-v3 has been improved for all objects, however, PeeleNet does not improve because it is a smaller model and has limited learning capacity.

4   Trash Pixel Identification

Trash may have blob-like irregular shape and appearance and it also breaks into smaller pieces as it floats through water channels. Instead of object-level analysis, pixel-level analysis via image segmentation results in better localization in such cases. Thus in this

Table 4. Effect of log attention layer on results of object detection. (Key: mAP: Mean Average Precision).

Model	AP^S	AP^M	AP^L	mAP
YOLO-v3+Attn	3.5	11.6	34.5	35.3
PeleeNet+Attn	3.4	10.3	31.5	29.4

work, we also provided an analysis of popular semantic segmentation models on our dataset. We first trained vanilla models (UNet [20] and SegNet [1]) that use standard convolution. From Table 5, it is clear that standard segmentation models show better performance as each pixel of trash is classified. However, these networks do have a larger number of parameters and are thus not suitable for continuous monitoring.

4.1   Reducing Model Size for Real Time Applications

We propose a modified UNet and SegNet architectures and improve their inference time by introducing depth-wise separable convolution [9], which can reduce the parameters in standard convolution [3,8,13]. As depicted in Table 5, when we replace standard convolution with separable convolution in alternate layers (Sep†), parameters of both models reduce to approximately half of the original and when all standard convolution layers are replaced with separable convolution layers (Sep‡), the models shrink by almost 10 times resulting 2x improvement in terms of frames per second. From Fig. 3, it is apparent that the segmentation map of before and after reducing the size of the UNet model are quite similar.

4.2   Addressing Class Imbalance

In segmentation, class distribution in each image instance varies. Sometimes one of the class has an extremely low number of pixels in the whole dataset, as in our case where the trash samples are almost 1% and water is more than 89%. To cater to this and to put specific emphasis on relatively difficult examples, we further modified the UNet and SegNet models and introduced focal loss [30] defined as:

$$CB_f(p, y) = \alpha \sum_{j=1}^{C} (1 - p_i) log(p_i), \tag{3}$$

where, p is prediction of model, y is ground truth label, C is number of classes and α is weighting factor. To address the problem of missing data, the inverse probability weighting (ipw) [23] is used which removes the bias by penalizing the estimator. It can be defined as:

Table 5. Performance of standard segmentation models along with their smaller versions achieved by employing separable convolution layers. Sep† is model with half of its layers employ separable convolution and Sep‡ has all layers as separable convolution. FPS = Frame per second, M = millions.

Measure	UNet			SegNet		
	Vanilla	Sep†	Sep‡	Vanilla	Sep†	Sep‡
#Param (M)	31	18	3.9	29	16.8	3.3
FPS	8	12	14	8	12	16
Acc (Trash)	87%	72%	57%	88%	70%	69%

$$CB_{ipw}(p, y) = \frac{\sum_{i=1}^{C} n_i}{n_c} \sum_{j=1}^{C} (1 - p_i)log(p_i), \qquad (4)$$

where, n_c number of pixels of cth classes in whole dataset. In our case, trash is seldom in the image instances, inverse probability weighting would be better class weighting scheme.

Table 6. Addressing class imbalance on UNet vanilla and its variants. CE is cross entropy loss, CB_f and CB_{ipw} are class balanced focal loss and our inverse probability loss functions.

Model	CE	CB_f	CB_{ipw}
UNet vanilla	87%	88%	92%
UNet Sep†	72%	78%	82%
UNet Sep‡	57%	75%	81%

These loss functions as depicted in Table 6 have significantly improved the accuracy of all models overcoming a drop in accuracy due to separable convolution. When UNet Sep‡ is trained using CB_{ipw} loss, despite the model is 800% times smaller there is only 10% drop in accuracy, the model applicable for real-time applications.

5 Conclusions and Future Work

This paper presents a new category of visual trash detection and provides a large benchmark trash dataset. A dataset of trash floating on canal surface in dense urban areas is collected and annotated. Due to various challenges in the dataset, both object detection and segmentation approaches were analyzed. Five recent and popular object detection models and six variants of segmentation models were trained. This paper studies the difficulty in detection of submerged, partially visible, and smaller objects and propose improvements through *log* attention, reduce model complexity through separable convolution layer for real-time deployment of models, and addresses the class imbalance in segmentation through focal loss functions. Overall, the detection of floating trash especially in water channels in urban areas is a challenging task and an emerging area of research. The dataset and comparative analysis provided in this work will serve as a stepping stone towards finding a solution to this problem. In the future, we plan to merge both object detection and segmentation approaches through architectural modifications such that it would force the model to learn better features of weak objects. We also plan to increase the dataset size and categorize trash into different classes such as organic, plastic, paper, leaves, metals, etc.

References

1. Badrinarayanan, V., Kendall, A., Cipolla, R.: SegNet: a deep convolutional encoder-decoder architecture for image segmentation. IEEE Trans. Pattern Anal. Mach. Intell. **39**, 2481–2495 (2017)

2. Bircanoglu, C., Atay, M., Beser, F., Genc, O., Kizrak, M.A.: RecycleNet: intelligent waste sorting using deep neural networks. In: IEEE Innovations in Intelligent Systems and Applications (2018)
3. Chen, L.C., Zhu, Y., Papandreou, G., Schroff, F., Adam, H.: Encoder-decoder with atrous separable convolution for semantic image segmentation. In: Proceedings of the European Conference on Computer Vision, pp. 801–818 (2018)
4. Chen, X., et al.: Microsoft COCO captions: data collection and evaluation server. arXiv preprint arXiv:1504.00325 (2015)
5. Engin, K., Tran, M., Connor, R., Uhlenbrook, S.: The united nations world water development report 2018: nature-based solutions for water; facts and figures. UNESCO (2018)
6. Everingham, M., Van Gool, L., Williams, C.K., Winn, J., Zisserman, A.: The pascal visual object classes (VOC) challenge. Int. J. Comput. Vis. **88**, 303–338 (2010)
7. Fulton, M., Hong, J., Islam, M.J., Sattar, J.: Robotic detection of marine litter using deep visual detection models. In: IEEE International Conference on Robotics and Automation (2019)
8. Gomez, A.N., Kaiser, L.M., Chollet, F.: Depthwise separable convolutions for neural machine translation (2020). US Patent App. 16/688,958
9. Hoang, V.T., Jo, K.H.: PydMobileNet: pyramid depthwise separable convolution networks for image classification. In: IEEE 28th International Symposium on Industrial Electronics (2019)
10. Hong, J., Fulton, M., Sattar, J.: TrashCan: a semantically-segmented dataset towards visual detection of marine debris (2020)
11. Kako, S., Morita, S., Taneda, T.: Estimation of plastic marine debris volumes on beaches using unmanned aerial vehicles and image processing based on deep learning. Mar. Pollut. Bull. **155**, 111127 (2020)
12. Kooi, M., Besseling, E., Kroeze, C., van Wenzel, A.P., Koelmans, A.A.: Modeling the fate and transport of plastic debris in freshwaters: review and guidance. In: Wagner, M., Lambert, S. (eds.) Freshwater Microplastics. THEC, vol. 58, pp. 125–152. Springer, Cham (2018). https://doi.org/10.1007/978-3-319-61615-5_7
13. Lei, C., Gong, W., Wang, Z.: Fusing recalibrated features and depthwise separable convolution for the mangrove bird sound classification. In: Proceedings of Machine Learning Research, vol. 101, pp. 924–939 (2019)
14. Lin, T.Y., Goyal, P., Girshick, R., He, K., Dollár, P.: Focal loss for dense object detection. In: IEEE International Conference on Computer Vision, pp. 2980–2988 (2017)
15. Liu, Y., Ge, Z., Lv, G., Wang, S.: Research on automatic garbage detection system based on deep learning and narrowband internet of things. In: Journal of Physics: Conference Series, vol. 1069, p. 012032. IOP Publishing (2018)
16. Mittal, G., Yagnik, K.B., Garg, M., Krishnan, N.C.: SpotGarbage: smartphone app to detect garbage using deep learning. In: ACM International Joint Conference on Pervasive and Ubiquitous Computing, pp. 940–945 (2016)
17. Proença, P.F., Simões, P.: TACO: trash annotations in context for litter detection. arXiv preprint arXiv:2003.06975 (2020)
18. Rad, M.S., et al.: A computer vision system to localize and classify wastes on the streets. In: Liu, M., Chen, H., Vincze, M. (eds.) ICVS 2017. LNCS, vol. 10528, pp. 195–204. Springer, Cham (2017). https://doi.org/10.1007/978-3-319-68345-4_18
19. Redmon, J., Farhadi, A.: YOLOv3: an incremental improvement. arXiv preprint arXiv:1804.02767 (2018)
20. Ronneberger, O., Fischer, P., Brox, T.: U-net: convolutional networks for biomedical image segmentation. In: Navab, N., Hornegger, J., Wells, W.M., Frangi, A.F. (eds.) MICCAI 2015. LNCS, vol. 9351, pp. 234–241. Springer, Cham (2015). https://doi.org/10.1007/978-3-319-24574-4_28

21. Russell, B., Torralba, A., Murphy, K., Freeman, W.: LabelMe: a database and web-based tool for image annotation. Int. J. Comput. Vis. **77**, 157–173 (2008)
22. Sakr, G.E., Mokbel, M., Darwich, A., Khneisser, M.N., Hadi, A.: Comparing deep learning and support vector machines for autonomous waste sorting. In: IEEE International Multidisciplinary Conference on Engineering Technology (2016)
23. Sherwood, B.: Variable selection for additive partial linear quantile regression with missing covariates. J. Multivar. Anal. **152**, 206–223 (2016)
24. Tzutalin: LabelImg (2015). https://github.com/tzutalin/labelImg. Accessed 21 Oct 2020
25. Van Lieshout, C., Van Oeveren, K., Van Emmerik, T., Postma, E.: Automated river plastic monitoring using deep learning and cameras. Earth Space Sci. **7**, e2019EA000960 (2020)
26. Vo, A.H., Vo, M.T., Le, T., et al.: A novel framework for trash classification using deep transfer learning. IEEE Access **7**, 178631–178639 (2019)
27. Wang, R.J., Li, X., Ling, C.X.: Pelee: a real-time object detection system on mobile devices. In: Advances in Neural Information Processing Systems, pp. 1963–1972 (2018)
28. Wang, T., Cai, Y., Liang, L., Ye, D.: A multi-level approach to waste object segmentation. Sensors **20**(14), 3816 (2020)
29. Yang, M., Thung, G.: Classification of trash for recyclability status. CS229 Project Report (2016)
30. Yun, P., Tai, L., Wang, Y., Liu, C., Liu, M.: Focal loss in 3D object detection. IEEE Robot. Autom. Lett. **4**, 1263–1270 (2019)
31. Zeng, D., Zhang, S., Chen, F., Wang, Y.: Multi-scale CNN based garbage detection of airborne hyperspectral data. Access **7**, 104514–104527 (2019)
32. Zhao, Q., et al.: M2Det: a single-shot object detector based on multi-level feature pyramid network. In: AAAI Conference on Artificial Intelligence (2019)

Tri-Transformer Hawkes Process: Three Heads are Better Than One

Zhi-yan Song, Jian-wei Liu$^{(\boxtimes)}$, Lu-ning Zhang , and Ya-nan Han

Department of Automation, College of Information Science and Engineering,
China University of Petroleum, Beijing (CUP), Beijing, China
`liujw@cup.edu.cn`

Abstract. Most of the real world data we encounter are asynchronous event sequence, so the last decades have been characterized by the implementation of various point process into the field of social networks, electronic medical records and financial transactions. At the beginning, Hawkes process and its variants which can simulate simultaneously the self-triggering and mutual triggering patterns between different events in complex sequences in a clear and quantitative way are more popular. Later on, with the advances of neural network, neural Hawkes process has been proposed one after another, and gradually become a research hotspot. The proposal of the transformer Hawkes process (THP) has gained a huge performance improvement, so a new upsurge of the neural Hawkes process based on transformer is set off. However, THP does not make full use of the information of occurrence time and type of event in the asynchronous event sequence. It simply adds the encoding of event type conversion and the location encoding of time conversion to the source encoding. At the same time, the learner built from a single transformer will result in an inescapable learning bias. In order to mitigate these problems, we propose a tri-transformer Hawkes process (Tri-THP) model, in which the event and time information are added to the dot-product attention as auxiliary information to form a new multi-head attention. The effectiveness of the Tri-THP is proved by a series of well-designed experiments on both real world and synthetic data.

Keywords: Hawkes process · Transformer Hawkes process · Encoding of event type · Encoding of time · Dot-product attention

1 Introduction

Modeling and predicting the asynchronous event sequences are extensively utilized in different domains, such as financial data [1], genome analysis [2], earthquake sequences [3], electronic medical records [4], social network [5], etc. In order to obtain effective information from these asynchronous event data, analyze the relationship between events and predict the events that may occur in

This work was supported by the Science Foundation of China University of Petroleum Beijing (No. 2462020YXZZ023).

the future, the most common and effective measure is the point process model [6]. And Hawkes process [7] and its variants can encapsulate the self-triggering and mutual triggering modes occurring among different complex event sequences in an unambiguous and quantitative fashion, so it is dominantly employed in a variety of application domain.

Hawkes process can be divided into parametric [8,9] and nonparametric Hawkes process [10–12]. For neural Hawkes process, Du et al. proposed the Recurrent Marked Temporal Point Processes (RMTPP) [13,14]. Xiao et al. introduced the Intensity RNN [15,16]. Mei et al. built the Neural Hawkes Process (NHP) [17].

In 2017, the transformer structure [17] has achieved good results, which has attracted extensive attention. This structure adopted by transformer completely gets rid of the other neural network structures, such as RNN, and CNN, and only puts attention mechanism to use for dealing with sequential tasks. In view of this, Zhang et al. devised Self Attention Hawkes Process (SAHP) [18], which takes advantage of the influence of historical events and finds the probability of the next event through self-attention mechanism. Zuo et al. presented the Transformer Hawkes Process (THP) [19], in which they integrated the transformer structure into the field of point process.

However, the previous Hawkes process model based on attention mechanism uses injudiciously the intrinsic event and time information existing in the asynchronous event sequence, but simply adds the encoding of event type conversion and the location encoding of time conversion to the encoder. Meanwhile, we speculate that the learner built from a single transformer may suffer from learning bias. Motivated by ensemble learning [20] and Transformer-XL [21], we generate a tri-transformer Hawkes process model (Tri-THP), in which event and time information are added to dot product attention as auxiliary information recurrently to form a new multi-head attention. The results show that compared with the existing models, the performance of our Tri-THP model has been greatly improved. The main contributions of our study are as follows:

1) The proposed three different THPs: event type embedding THP (ETE-THP), the primary THP (PRI-THP), and temporal embedding THP (TE-THP) are fused into a seamless organic whole, which are related and complemented with each other positively, and are utilized to distil event type and temporal embedding information underlying in the asynchronous event sequences.
2) We carry out two novel dot-product attention operations, so far as we know, there were not any research reports about this.
3) Finally, The experimental results on both real-life and synthetic asynchronous event sequence have verified empirically the effect of our Tri-THP algorithm.

2   Related Works

In this section, we briefly recapitulate Hawkes Process [7], Transformer [17], Transformer Hawkes Process [19] and ensemble learning [20].

Hawkes process [7], which is belong to self-exciting process, its intrinsic mechanism allows that previous historical events have stimulating effects on the occurrence of future ones, and the cumulative influence of historical events is intensity superposition. The intensity function for Hawkes process is formulated as follows:

$$\lambda(t) = \mu(t) + \sum_{i:t_i < t} \psi(t - t_i) \tag{1}$$

where $\mu(t)$ is the base intensity. For the sake of simplicity and convenient calculation, or only involving the trigger mode, i.e., causalities among events, $\mu(t)$ is usually set to constant. $\psi(\cdot)$ is a predefined decay function, also known as the influence function, in general, it is selected as the exponential decay function. From Eq. (1), we note that every event in the Hawkes process has a stimulating effect on the occurring current event, and the stimulating effect reduces with time. However, this Hawkes process model does not involve inhibition effect between events, which is obviously inconsistent with the actual situation and impairs the expressiveness of the model.

In 2017, Google launched the transformer structure [17], which has achieved promise results and become a current mainstream model. This structure replaces the RNN, CNN and other network structures with attention mechanism to deal with sequential tasks. However, each fragment in the model is divided according to the length, and some relevant information may be lost when cutting, which will lead to context fragmentation. Then, motivated by this, Dai et al. proposed transformer XL [26]. This model uses the segment level recurrence mechanism to solve the fragmentation problem. The attention calculation in each layer needs to be encoded by relative position, which is about 450% longer than the original transformer learning dependency.

Inspired by the encoder structure in transformer [17], Zou et al. design the Transformer Hawkes process (THP) [21]. THP extends the transformer structure to a continuous time domain. It uses the encoder in transformer to obtain the hidden representation of sequence data, and uses the hidden representation to represent the continuous conditional intensity function, and obtains very good performance. However, in theory, attention can be associated with two words at any distance, but in practice, due to the limited computing resources, it is still unable to deal with very long input. Therefore, inspired by [26], we introduce event and time auxiliary information into attention calculation to further enhance learning dependence.

The goal of machine learning is to learn a stable model with good performance in all aspects, but in most cases, we can only get more than one model with good performance in some aspects. Ensemble learning is to combine multiple weak supervised models to get a better and more comprehensive strong supervised model. The potential idea of ensemble learning is to improve the prediction accuracy of the final result by combining multiple weaker models. Therefore, in order to improve the robustness and prediction accuracy of the model, we design three heterogeneous learners to learn the hidden representation respectively, and then weighted the hidden representation to get the final hidden representation.

3 The Proposed Tri-THP Framework

In this section, we will elaborate our Tri-THP model in detail. Suppose that we have an event sequence $X = \{(t_i, k_i)\}_{i=1}^{N}$, in which each event has a type $k_i \in \{1, 2, \ldots, K\}$. Each pair of event and time stamp (t_i, k_i) corresponds to an event of type k_i which occurs at time t_i.

3.1 Tri-Transformer Hawkes Process

The core thought of our Tri-THP model is to build three THPs with different dot-product attention using different auxiliary information, and harvest diverse and heterogeneous side information from asynchronous event sequence. Before diving into the discussion of our Tri-THP model, we briefly present how to acquire the embedding of the occurrence time and type of the events, i.e., we need to transform the asynchronous event sequence to temporal and event encoding. Similar to [17] and [19], our temporal encoding expression is as follows:

$$[\mathbf{c}(t_i)]_j = \begin{cases} \cos\left(t_i/10000^{\frac{i-1}{Z}}\right), \text{if } j \text{ is odd,} \\ \sin\left(t_i/10000^{\frac{j}{Z}}\right), \text{if } j \text{ is even.} \end{cases} \tag{2}$$

where $\mathbf{c}$ is the temporal encoding, $\mathbf{c}(t_i) \in \mathbb{R}^Z$, Z is the preassigned dimension, here $j \in [0, \ldots, Z-1]$. That is to say, each component of temporal encoding denotes a sine curve. In terms of encoding event type, the unique one-hot encoding $\mathbf{k}_i \in \mathbb{R}^K$ of each event type is multiplied by the embedding matrix $\mathbf{M} \in \mathbb{R}^{Z \times K}$ to get the event encoding $\mathbf{M}\mathbf{k}_i \in \mathbb{R}^Z$, so the sequence encoding of event sequence $X = \{(t_i, k_i)\}_{i=1}^{N}$ is embedded as the sum of temporal encoding $\mathbf{C}^T$ and event encoding $(\mathbf{M}\mathbf{Y})^T$, where $\mathbf{C} = [\mathbf{c}(t_1), \ldots, \mathbf{c}(t_N)] \in \mathbb{R}^{Z \times N}$, and $\mathbf{Y} = [\mathbf{k}_1, \ldots, \mathbf{k}_N] \in \mathbb{R}^{K \times N}$. Therefore, each row in the final encoding corresponds to the temporal and event encoding in the sequence, respectively. Then, we feed the obtained sequence encoding into our Tri-THP model, and the schematic diagram of Tri-THP is shown in Fig. 1.

As shown in Fig. 1, we build three different learners: ETE-THP, PRI-THP, and TE-THP, which are complementary to each other, to carry out far more full-featured and diverse hidden state extraction. One that is shown on the left, we dub ETE-THP, is a learner that supplies with auxiliary information of event type embedding to multi-head attention module of the basic THP model; One that is depicted in the middle, we call PRI-THP, is the primary learner; One that is illustrated in the right, we call TE-THP, is a learner that equips with auxiliary information of temporal embedding to multi-head attention of the normal THP model. Three hidden states H_1, H_2, H_3 are learned by three different learners respectively, and then the final hidden state is obtained by weighting the learned three hidden states with the three weights $\lambda_1, \lambda_2, \lambda_3$ obtained by training. In order to prevent overfitting and improve the robustness and performance of the model, we also incorporate dropout [22], layer normalization [23] and residual connection [24]. The specific training process of our Tri-THP model is shown in Algorithm 1.

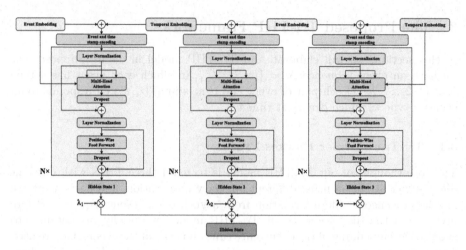

Fig. 1. Diagrammatic sketch for our proposed Tri-THP model. Our Tri-THP comprises three modules: the THP module with auxiliary information of event type (left), the THP module without auxiliary information (middle), and the THP module with auxiliary information of time (right).

Algorithm 1. Tri-Transformer Hawkes Process (Tri-THP)

Input: The number of encoding layers:n , event type encoding $(\mathbf{MY})^T$, temporal encoding $\mathbf{C}^T$.
Output: Hidden state of event sequence $\boldsymbol{H} \in \mathbb{R}^{Z \times N}$

1: Initialize state $\boldsymbol{H}_1, \boldsymbol{H}_2, \boldsymbol{H}_3 \leftarrow (\mathbf{MY})^T$
2: **for** i in n **do**
3:     $\boldsymbol{H}_1 \leftarrow \boldsymbol{H}_1 + \mathbf{C}^T$
4:     $\boldsymbol{H}_2 \leftarrow \boldsymbol{H}_2 + \mathbf{C}^T$
5:     $\boldsymbol{H}_3 \leftarrow \boldsymbol{H}_3 + \mathbf{C}^T$
6:     $\boldsymbol{H}_1 \leftarrow The\ i\text{-th Encoding\_Layer}(\boldsymbol{H}_1, (\mathbf{MY})^T)$
7:     $\boldsymbol{H}_2 \leftarrow The\ i\text{-th Encoding\_Layer}(\boldsymbol{H}_2)$
8:     $\boldsymbol{H}_3 \leftarrow The\ i\text{-th Encoding\_Layer}(\boldsymbol{H}_3, \mathbf{C}^T)$
9: **end for**
10: $\boldsymbol{H} = \lambda_1 \boldsymbol{H}_1 + \lambda_2 \boldsymbol{H}_2 + \lambda_3 \boldsymbol{H}_3$
11: **return** $\boldsymbol{H}$

As shown in Algorithm 1, in the initial stage, event encoding and temporal encoding are fed into the ETE-THP, PRI-THP, and TE-THP, respectively, and the input of each encoding layer is the output of the previous hidden layer plus temporal encoding. In our paradigm, we modify the conventional dot-product attention operation for multi-head attention in the encoding layer for different modules. In Fig. 1, the dot-product attention for the ETE-THP is written as follows:

$$A_1^s = Softmax \left[mask \left(\frac{(Q_1^s + b_1^{sq})(K_1^s)^T + (Q_1^s + b_1^{se})\big((\mathbf{MY})^T W_1^{events}\big)^T}{\sqrt{Z_K}} \right) \right] V_1^s \quad (3)$$

here we add the bias vector $\boldsymbol{b}_1^{sq}$ and $\boldsymbol{b}_1^{se}$ to the query matrix $\boldsymbol{Q}_1^s$,respectively. Incorporating the bias vector $\boldsymbol{b}_1^{sq}$ and $\boldsymbol{b}_1^{se}$ to the query matrix $\boldsymbol{Q}_1^s$ can strengthen model flexibility. The amended conventional dot-product attention operation is reflected in the item $(\boldsymbol{Q}_1^s + \boldsymbol{b}_1^{sq})(\boldsymbol{K}_1^s)^T$. The event information is introduced in the item $(\boldsymbol{Q}_1^s + \boldsymbol{b}_1^{se})\left((\mathbf{MY})^T \boldsymbol{W}_1^{event_s}\right)^T$ to heighten the attention to the event type. Event encoding $(\mathbf{MY})^T$ is linearly transformed into the item $(\mathbf{MY})^T \boldsymbol{W}_1^{event_s}$, and the s-th attention head is linearly translated into matrix $\boldsymbol{W}_1^{event_s} \in \mathbb{R}^{Z \times Z_K}$.

The dot-product attention of the PRI-THP is expressed as follows:

$$A_2^s = Softmax\left[mask\left(\frac{(\boldsymbol{Q}_2^s + \boldsymbol{b}_2^{sq})(\boldsymbol{K}_2^s)^T}{\sqrt{Z_K}}\right)\right] V_2^s \tag{4}$$

On the basis of the THP model, we add a bias vector to the query matrix of dot-product attention, and derive the dot-product attention for the TE-THP:

$$A_3^s = Softmax\left[mask\left(\frac{(\boldsymbol{Q}_3^s + \boldsymbol{b}_3^{sq})(\boldsymbol{K}_3^s)^T + (\boldsymbol{Q}_3^s + \boldsymbol{b}_3^{st})(\boldsymbol{C}^T \boldsymbol{W}_3^{tem_s})^T}{\sqrt{Z_K}}\right)\right] V_3^s \tag{5}$$

i.e., the event information added in Eq. (3) is replaced with the time information in Eq. (5).

For each module, e.g., ETE-THP, PRI-THP, and TE-THP, the query, key and value matrix for the s-th head attention are expressed as follows:

$$Q^s = HW_Q^s, K^s = HW_K^s, V^s = HW_V^s \tag{6}$$

here, H is the input of each encoding layer as described in algorithm 1. $W_Q^s \in \mathbb{R}^{Z \times Z_K}$, $W_K^s \in \mathbb{R}^{Z \times Z_K}$ and $W_V^s \in \mathbb{R}^{Z \times Z_V}$ are linear transformations of H respectively. The aim of mask operation is to ensure that future events in the matrix will not affect the attention weights of current events. In order to raise the expressiveness of the model, we utilize the multi-head attention of the following form:

$$A = [A_1, A_2, ..., A_S] W^{multi} \tag{7}$$

where W^{multi} is the aggregation matrix.

Then the attention weight matrix A is sent into position-wise feed-forward neural network to acquire the final hidden representation of the event sequence:

$$H = \text{ReLU}(AW_1^{FC} + b_1)W_2^{FC} + b_2 \tag{8}$$

The above is construction procedure of the multi-head attention and feed-forward network for each module. After learning three modules, we will get three hidden matrices H_1, H_2, H_3, and then multiply these three hidden matrices by the tradeoff coefficients λ_1, λ_2 and λ_3 obtained from training, and then calculate weighted sum to obtain the final hidden state.

$$H = \lambda_1 H_1 + \lambda_2 H_2 + \lambda_3 H_3$$
$$h(t_i) = H(i,:) \tag{9}$$

where $h(t_i)$ denotes the hidden state at time t_i.

3.2   Conditional Intensity Function

Conditional intensity function regulates the time point process. Similar to [19], we put the learned hidden state $\boldsymbol{h}(t_i)$ into the conditional intensity function:

$$\lambda_k(t\,|\mathcal{H}_t) = f(b_k + \alpha_k\frac{t-t_i}{t_i} + \boldsymbol{w}_k^T\boldsymbol{h}(t_i)) \tag{10}$$

Then the whole strength function condition on $\mathcal{H}_t$ for K types of sequences is formulated as follows:

$$\lambda(t\,|\mathcal{H}_t) = \sum_{k=1}^{K}\lambda_k(t\,|\mathcal{H}_t) \tag{11}$$

3.3   Loss Function for Forecasting Occurring Times and the Types of Events

The predictive value for occurring times and the types of event can be deduced as follows:

$$\hat{t}_{i+1} = \boldsymbol{W}^{time}\boldsymbol{h}(t_i)$$
$$\hat{\boldsymbol{p}}_{i+1} = Softmax(\boldsymbol{W}^{type}\boldsymbol{h}(t_i)) \tag{12}$$
$$\hat{k}_{i+1} = \arg\max_{k}\hat{\boldsymbol{p}}_{i+1}(k)$$

where $\boldsymbol{W}^{time} \in \mathbb{R}^{1\times Z}$ and $\boldsymbol{W}^{type} \in \mathbb{R}^{1\times Z}$ are the model parameters for occurring times and the types of event. More specifically, for sequence X, we utilize respectively the cross entropy loss of event type prediction and the square error of prediction occurring time of event to deduce the model parameters:

$$L_{time}(X) = \sum_{i=2}^{N}(t_i - \hat{t}_i)^2$$
$$L_{type}(X) = \sum_{i=2}^{N}-\mathbf{k}_i^T\log(\hat{\boldsymbol{p}}_i) \tag{13}$$

where the index i begins at 2 to keep away from predicting the first event.

3.4   Objective Function

We are now at the position to elaborate the objective function. The logarithm likelihood for a given sequence X can be inferred from Hawkes process theory:

$$L(X) = \sum_{i=1}^{N}\log\lambda(t_i|\mathcal{H}_i) - \int_{t_1}^{t_N}\lambda(t|\mathcal{H}_t)dt \tag{14}$$

Provided that there are L training sequences $X_1, X_2, \ldots, X_L$, according to the maximum likelihood estimation principle, we get

$$\max\sum_{i=1}^{L}L(X_i) \tag{15}$$

Unfortunately, directly using optimization approaches, such as the random gradient optimization algorithm Adam [25], to solve the problem (15) is unfeasible and unwise. The numeric expression for the second term $\Lambda = \int_{t_1}^{t_N} \lambda(t|\mathcal{H}_t)dt$ in (14) cannot be obtained because the function $\lambda(t|\mathcal{H}_t)$ is in the form of the deep neural network. Alternatively, the unbiased Monte Carlo integration method [26] and the biased numerical integration method [27] are incorporated to obtain the approximate value of Λ. For the first approach the approximate expression for Λ is written as follows:

$$\hat{\Lambda}_{MC} = \sum_{i=2}^{N} (t_i - t_{i-1})(\frac{1}{O}\sum_{o=1}^{O}\lambda(u_o)) \tag{16}$$

where u_o is sampled from uniform distribution $U(t_{i-1}, t_i)$. The second one obtains the approximate value of Λ by using trapezoidal rule:

$$\hat{\Lambda}_{NI} = \sum_{i=2}^{N} \frac{t_i - t_{i-1}}{2}(\lambda(t_i|\mathcal{H}_i) + \lambda(t_{i-1}|\mathcal{H}_{i-1})) \tag{17}$$

Now, recall the Eq. (13), and sum the three terms: $L(X_i)$, $L_{type}(X_i)$,and $L_{time}(X_i)$,we obtain the objective function for our Tri-THP model :

$$\min \sum_{i=1}^{X} -L(X_i) + L_{type}(X_i) + L_{time}(X_i) \tag{18}$$

4   Experimental Results

In this section, we first introduce the details of the dataset, and then compare our model with the existing baseline approaches on synthetic and real-life datasets. We evaluate the model through log-likelihood (in nats), event prediction accuracies and root mean square errors for occurring time of events.

4.1   Datasets

In this subsection, we use two artificial datasets:Synthetic and Neural Hawkes [17], and four real-world datasets of event sequences:Electrical Medical Records [28], StackOverflow [29], Financial Transactions [14] and Retweets [30], to carry out experiments. Table 1 describes the characteristics of each dataset.

Table 1. Characteristics of datasets used in experiments.

Dataset	C	Sequence length		
		Min	Aver.	Max
Synthetic	5	20	60	100
NeuralHawkes	5	20	60	100
Retweets	3	50	109	264
MIMIC-II	75	2	4	33
StackOverflow	22	41	72	736
Financial	2	829	2074	3319

4.2 Experimental Results and Comparison

We compare the performance of the Tri-THP model with that of the baseline model. First, we compare the log-likelihood value, which is the basic measure of the fitting degree of the model. Log-likelihood values of baseline and Tri-THP on different datasets are summed up in Table 2.

As shown in Table 2, the log-likelihood function values of Tri-THP on all test datasets are significantly better than the existing baselines. This shows that Tri-THP is more effective than the existing baselines in modeling event sequence. We think that this is because we use the idea of ensemble learning to design three learners. This way can effectively improve the stability and fitting similarity of the model, which is just the significance of the log-likelihood index. However, log-likelihood represents the similarity of the model, which has little effect on practical application.

Table 2. The value of log-likelihood function on the test datasets for different models.

Datasets	RMTPP	NHP	SAHP	THP	Tri-THP
Synthetic	\	−1.33	0.52	0.834	**6.036**
NeuralHawkes	\	−1.02	0.241	0.966	**6.601**
Retweets	−5.99	−5.06	−5.85	−4.69	**2.611**
StackOverflow	−2.6	−2.55	−1.86	−0.559	**−0.544**
MIMIC-II	−1.35	−1.38	−0.52	−0.143	**−0.081**
Financial	−3.89	−3.6	\	−1.388	**−0.651**

Therefore, we need to pay more attention to the prediction accuracies of event time and event type, which is of great practical significance. Through these two indicators, we can predict events in different fields and promote or prevent occurrences of unfavorable events. The prediction accuracies of event type are listed in Table 3.

Table 3. Predict accuracies of different models on various datasets.

Datasets	RMTPP	NHP	THP	Tri-THP
StackOverflow	45.9	46.3	46.79	**46.81**
MIMIC-II	81.2	83.2	83.2	**84.1**
Financial	61.95	62.2	62.23	**62.31**

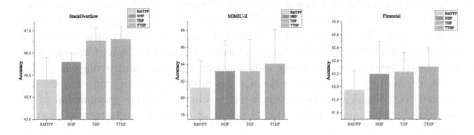

Fig. 2. Prediction accuracies of baselines and Tri-THP. On the basis of five train-dev-test partitions, five experiments are carried out on each dataset, and the mean and standard deviation of different models are obtained.

It can be seen from Table 3 and Fig. 2. that the event prediction accuracies of our Tri-THP model is better than that of the existing baseline models on complex datasets. We speculate that it is caused by the learner with event auxiliary information. By adding event auxiliary information to multi-head attention, the learner can pay more attention to event types, Therefore, it can effectively improve the accuracies of event type prediction.

For the prediction of the occurrence time of various events, we use root mean square error (RMSE) as the unified measurement and evaluation creteria. The RMSE on different datasets for the baselines and our Tri-THP model are compared in Table 4.

As shown in the Table 4 above, we can find that the RMSE of Tri-THP is higher than that of the baseline models. We guess that it is caused by the learner with time auxiliary information. By adding time auxiliary information to the multi-head attention, the learner can pay more attention to the time dependence relationships, so it can effectively reduce the root mean square error for occurrence time prediction. Our model shows some robustness on these datasets with obvious differences in sequence length and number of event types, which

Table 4. RMSE of different models on various datasets.

Datasets	RMTPP	NHP	THP	Tri-THP
StackOverflow	9.78	9.83	4.99	**3.89**
MIMIC-II	6.12	6.13	0.859	**0.858**
Financial	1.56	1.56	0.02575	**0.02550**

validates that the Tri-THP can effectively capture the short-term and long-term dependences between events.

5 Conclusions and Future Work

In this paper, we proposed a Tri-THP model. We introduce the event and time auxiliary information into the dot-product attention, and design three heterogeneous modules: ETE-THP, PRI-THP, and TE-THP. The experimental results verify that our Tri-THP model perform well on both real world and synthetic data, our Tri-THP model can effectively learn the complex short-term and long-term dependences. Future works can consider not only improving the performance of the model, but also reducing the model complexity and the computational cost. Finally, we hope this work encourages further exploration of the interplay between and relative strong points of ensemble and attention mechanism.

References

1. Bacry, E., Mastromatteo, I., Muzy, J.: Hawkes processes in finance. Mark. Microstruct. Liq. **1**, 1550005 (2015)
2. Reynaud-Bouret, P., Schbath, S.: Adaptive estimation for Hawkes processes; application to genome analysis. Ann. Stat. **38**, 2781–2822 (2010)
3. Ogata, Y.: Space-time point-process models for earthquake occurrences. Ann. Inst. Stat. Math. **50**, 379–402 (1998)
4. Wang, L., Zhang, W., He, X., Zha, H.: Supervised reinforcement learning with recurrent neural network for dynamic treatment recommendation. In: Proceedings of the 24th ACM SIGKDD International Conference on Knowledge Discovery & Data Mining, pp. 2447–2456 (2018)
5. Zhou, K., Zha, H., Song, L.: Learning social infectivity in sparse low-rank networks using multi-dimensional Hawkes processes. In: Artificial Intelligence and Statistics, pp. 641–649. PMLR (2013)
6. Vere-Jones, D., Daley, D.J.: An Introduction to the Theory of Point Processes. Springer Series in Statistics, Springer, New York (1988)
7. Hawkes, A.: Spectra of some self-exciting and mutually exciting point processes. Biometrika **58**, 83–90 (1971)
8. Ozaki, T.: Maximum likelihood estimation of Hawkes' self-exciting point processes. Ann. Inst. Stat. Math. **31**(1), 145–155 (1979). https://doi.org/10.1007/BF02480272
9. Veen, A., Schoenberg, F.P.: Estimation of space-time branching process models in seismology using an EM-type algorithm. J. Am. Stat. Assoc. **103**, 614–624 (2008)
10. Lewis, E., Mohler, G.: A nonparametric EM algorithm for multiscale Hawkes processes. J. Nonparametr. Stat. **1**, 1–20 (2011)
11. Marsan, D.O.L.: Extending earthquakes' reach through cascading. Science **319**, 1076–1079 (2008)
12. Xu, H., Farajtabar, M., Zha, H.: Learning granger causality for Hawkes processes. ArXiv abs/1602.04511 (2016)

13. Graves, A.: Supervised Sequence Labelling with Recurrent Neural Networks. Studies in Computational Intelligence (2008)

14. Du, N., Dai, H., Trivedi, R.S., Upadhyay, U., Gomez-Rodriguez, M., Song, L.: Recurrent marked temporal point processes: embedding event history to vector. In: Proceedings of the 22nd ACM SIGKDD International Conference on Knowledge Discovery and Data Mining (2016)

15. Hochreiter, S., Schmidhuber, J.: Long short-term memory. Neural Comput. **9**, 1735–1780 (1997)

16. Xiao, S., Yan, J., Yang, X., Zha, H., Chu, S.: Modeling the intensity function of point process via recurrent neural networks. In: AAAI (2017)

17. Mei, H., Eisner, J.: The neural hawkes process: a neurally self-modulating multivariate point process. In: NIPS (2017)

18. Zhang, Q., Lipani, A., Kirnap, O., Yilmaz, E.: Self-attentive Hawkes process. In: ICML, pp. 11183–11193 (2020)

19. Zuo, S., Jiang, H., Li, Z., Zhao, T., Zha, H.: Transformer Hawkes process. In: ICML, pp. 11692–11702 (2020)

20. Breiman, L.: Bagging predictors. Mach. Learn. **24**, 123–140 (1996)

21. Dai, Z., Yang, Z., Yang, Y., Carbonell, J., Le, Q.V., Salakhutdinov, R.: Transformer-XL: attentive language models beyond a fixed-length context. In: ACL (2019)

22. Krizhevsky, A., Sutskever, I., Hinton, G.E.: ImageNet classification with deep convolutional neural networks. Adv. Neural Inf. Process. Syst. **25**, 1097–1105 (2012)

23. Ba, J.L., Kiros, J.R., Hinton, G.E.: Layer normalization. arXiv preprint arXiv:1607.06450 (2016)

24. He, K., Zhang, X., Ren, S., Sun, J.: Deep residual learning for image recognition. In: Proceedings of the IEEE Conference on Computer Vision and Pattern Recognition, pp. 770–778 (2016)

25. Kingma, D.P., Ba, J.: Adam: a method for stochastic optimization. CoRR abs/1412.6980 (2015)

26. Robert, C., Casella, G.: Monte Carlo Statistical Methods. Springer Texts in Statistics, Springer, Heidelberg (2004). https://doi.org/10.1007/978-1-4757-4145-2

27. Neumaier, A.: Introduction to Numerical Analysis (2001)

28. Johnson, A.E.W., et al.: MIMIC-III, a freely accessible critical care database. Sci. Data **3**, 1–9 (2016)

29. Leskovec, J., Krevl, A.: SNAP datasets: Stanford large network dataset collection (2014)

30. Zhao, Q., Erdogdu, M., He, H.Y., Rajaraman, A., Leskovec, J.: SEISMIC: a self-exciting point process model for predicting tweet popularity. In: Proceedings of the 21th ACM SIGKDD International Conference on Knowledge Discovery and Data Mining (2015)

PhenoDeep: A Deep Learning-Based Approach for Detecting Reproductive Organs from Digitized Herbarium Specimen Images

Abdelaziz Triki[1]([⊠]) , Bassem Bouaziz[1]([⊠]) , Jitendra Gaikwad[2] ,
and Walid Mahdi[1]

[1] MIRACL Laboratory, University of Sfax, Sfax, Tunisia
`abdelaziz.triki@yahoo.fr`, `bassem.bouaziz@isims.usf.tn`
[2] Friedrich Schiller University, Jena, Germany

Abstract. Phenology is an important factor in studying climate change's effect on plant growth. Recent studies on herbarium specimens have afforded valuable information on plant phenology. The initiatives of herbaria to digitize their collections can extend plant phenological research rapidly by providing online access to significant collections of digitized specimen images. However, they present a major outstanding challenge when extracting reliable data from the specimen sheets. To effectively detect the presence/absence of the reproductive organs such as buds, flowers, and fruits from the specimen images, we developed PhenoDeep, a deep learning approach based on the refined Mask Scoring R-CNN approach. The Mask Scoring R-CNN backbone network was modified by exploiting the advantages of combining ResNet and DenseNet architectures. The experimental results indicate that PhenoDeep can segment the reproductive organs within different specimens, where the precision of PhenoDeep reached 94.1% and recall 94.3%.

Keywords: Reproductive organs · Mask scoring R-CNN · Digitized herbarium specimen images · ResNet · DenseNet

1 Introduction

In recent years, the effects of climate change on organisms have been widely studied by researchers. The changes in phenological events have been significant where species are failing to respond phenologically to climate change, causing severe implications for long-term diversity preservation [1].

Recent studies have shown that plant phenology information (such as flowers, buds, fruits, etc.) on herbarium specimens is of significant relevance [2]. In recent years, herbaria worldwide have gathered millions of specimens in physical form for species identification and cataloging. The collected specimen sheets contain relevant information for addressing scientific questions such as estimating the species' reproductive period and species detection [4].

© Springer Nature Switzerland AG 2021
T. Mantoro et al. (Eds.): ICONIP 2021, LNCS 13108, pp. 402–413, 2021.
https://doi.org/10.1007/978-3-030-92185-9_33

Several herbaria worldwide have recently started the massive digitization and online mobilization of herbarium specimens to facilitate access to millions of stored specimen images [5]. Researchers are using digitized herbarium specimen images (DHS) to study the phenological responses to climatic change through space and time [22].

The remainder of this paper is organized as follows. Section 2 shows the related work for reproductive organs detection and segmentation. The proposed approach for reproductive organs segmentation is detailed in Sect. 3. Section 4 describes the evaluation and the experimental results to examine the performance of our PhenoDeep approach. Section 5 presents the conclusion and future work.

2   Related Work

Over the past decade, several types of research applied to DHS images have concentrated on an individual phenological event, most generally the flowering time, to study climate change [23]. However, few researchers have simultaneously studied various reproductive organs (such as counting flower buds, young fruits, and mature fruits) and determining how different phenological events are related [24].

The approach proposed by Park et al. [7] is one of the first initiatives that manipulated the phenological data to determine the flowering time by joining the classification of the specimen images based on the presence/absence of flowers or fruits and the day of the year of gathering. However, they ignored other reproductive organs such as buds, seeds, and roots.

Recently, with the development of deep learning approaches, several solutions based on Convolutional Neural Networks (CNN) have been developed [8] to determine the reactiveness of various points in a particular phenophase. Lorieul et al. [9] demonstrated that a deep learning approach based on CNN could interpret fruits or flowers' presence within the specimen images with 90% accuracy. Also, Ellwood et al. [10] developed a model to categorize the phenological structures into two main categories: fertile and sterile specimens. However, their proposed solutions cannot localize or count the DHS images' reproductive organs.

In more recent work, Goëau et al. [11] utilized human-scored annotated images for training a model using Mask R-CNN [12]. The proposed model segments and counts the reproductive organs of a small dataset containing a single species (*Streptanthus tortuosus*). The experimental results demonstrated that the flowers are more reliably identified and counted than fruits within the DHS images. To deal with a larger number of species, Davis et al. [13] have trained a deep learning model based on Mask R-CNN to localize and count the reproductive organs of buds, flowers, and fruits within the DHS images. The annotated data is more than 3000 specimens of six common wildflower species of the eastern United States of America. Based on the experimental results, accurate counting was useful, but it varies based on the reproductive organs, where the flower counting is less accurate than buds and fruits due to its morphological variability on the scanned specimens.

In this paper, we developed PhenoDeep, a deep learning approach based on Mask Scoring R-CNN [14] network to segment and localize three reproductive organs, which are buds, flowers, and fruits on the herbarium sheets. PhenoDeep provides a bounding mask for each segmented object and classifies it. On the other hand, a collection of DHS images gathered from the herbarium Haussknecht of Jena (https://www.herbarium.uni-jena.de/) were used to train and evaluate our proposed approach.

3   Specimen Dataset Collection

The herbarium sheets used for training our PhenoDeep approach were selected from the herbarium Haussknecht of Jena, Germany, which comprises several species. This dataset consists of 4000 digitized specimens of various families' species with evident phenotypic variations among its reproductive phases to avoid singularities. Furthermore, the given dataset offers two main advantages. First, it facilitates accurate evaluation among human observers, resulting in a high-quality training dataset. Second, it promotes differentiation among different reproductive organs using deep learning algorithms.

The collected specimens were resized by adjusting their dimensions from 6400×3400 pixels to 2048×1024 pixels for height and width, respectively, including small objects such as the buds. This reduces excessive memory consumption and preserves the aspect ratio of the herbarium sheets. Moreover, each specimen was manually annotated using the VGG Image Annotator (VIA) annotation tool [15], which generates a JSON file representing each reproductive class's bounding mask within the scanned specimen images. Due to its irregular shape, each class is annotated by a bounding polygon used to represent its bounding mask. The speed of manual herbarium image annotation varied based on the complexity and number of different reproductive organs being annotated within the DHS images and was approximately 20 to 30 herbarium images per hour. With six masks per image, the overall number of annotated objects was 24,000 for the whole dataset. Furthermore, processing our data was sometimes challenging, where plant structures had difficulty being correctly labeled and localized. In specific scenarios, it is hard to distinguish between the reproductive organs such as buds, flowers, and fruits within the DHS images since they appear at various phases in the plant's life cycle. On the other hand, this paper randomly divides the dataset into two subsets: 70% of the dataset was used to train PhenoDeep and 30% for the testing process.

4   Methodology

The quantitative analysis of the reproductive organs within the DHS images, such as buds, flowers, and fruits, would significantly improve the use of specimen collections in climate change investigations. Given the enormous numbers of DHS images hosted in the herbarium Haussknecht of Jena and the variability of the reproductive organs presented within the herbarium scans (i.e., scales, orientations, aspect ratios, position, shape, and colors), we proposed a deep learning-based approach called PhenoDeep to detect the presence/absence of these reproductive organs and improve their detection performance. This approach is a refined version of the state-of-the-art instance segmentation approach, Mask Scoring RCNN [14] where the backbone model was modified by combining the ResNet [19], and DenseNet [18] architectures.

4.1 Mask Scoring RCNN

Mask Scoring RCNN is an improved version of Mask-RCNN [12] for object instance segmentation. Compared to Faster R-CNN [16], Mask R-CNN has better efficiency in terms of time and accuracy where the algorithm implements segmentation of pixel-level instances. Besides, Mask-RCNN uses the classification confidence as the consistency metric of the mask. However, the mask accuracy is commonly not well associated with the classification confidence, which leads to the poor precision and durability of the expected masks. Mask Scoring RCNN uses a particular network block for learning and predicting mask consistency. This block consists of the instance features and the expected mask. Furthermore, it measures the intersection-over-union (IoU) between the expected mask and the ground truth. The IoU is used as an indicator of the predicted mask. This allows Mask Scoring RCNN to gain better instance segmentation results compared to Mask-RCNN.

The Mask Scoring RCNN model consists mainly of three sections. The first section represents a typical convolutional layer, which is used to extract features of the image. Then, the generated feature maps will be used as the Region Proposal Network (RPN) input to produce the region proposals and correct them. The last section of the model is utilized for the adjustment of bounding boxes and the obtained masks.

Feature Extraction (ResNet + DenseNet). Designing deep convolutional networks of various weight layers is commonly used for image feature extraction. However, increasing the number of convolutional layers may increase the training error and decrease the classification accuracy. ResNet efficiently solved this problem by using several layers and parameters to learn the representation of residuals between inputs and outputs. Furthermore, ResNet considerably enhances the training speed and prediction accuracy by differentiating different classes of objects. Besides, it is based on a skip connection between the front and back layers, which helps the gradient's back-propagation during the training process to train a deeper CNN network.

The Mask Scoring RCNN model uses ResNet combined with Feature Pyramid Networks (FPN) [19] as the backbone to extract the comprehensive features. Based on the success of DenseNet and ResNet, the features reproducibility of PhenoDeep has been expanded by DenseNet since both ResNet and DenseNet networks come from the same dense topology [21]. The only difference is the form of connection, where the ResNet network uses addition and DenseNet uses concatenation. Additionally, the DenseNet connection increases the information flow between layers.

For each layer, the feature maps of all previous layers are used as input, and its feature map is used as input for all subsequent layers (Fig. 1). This combination significantly improves the feature diffusion and promotes feature reuse, so the reproductive organs with the tiny area (such as buds) are appropriately detected. Thus, DenseNet, in conjunction with ResNet, will be used as the backbone network to extract features. ResNet will help increase the training depth, while DenseNet will not entirely remove low-dimensional features. Moreover, a batch of convolution layers and ReLU are generated in the dense residual unit, while each unit's output is connected to the next unit's output.

Figure 1 represents a residual dense unit structure, which facilitates continuous information propagation. Convolutional networks are generally composed of L layers where x_0 represents the input image.

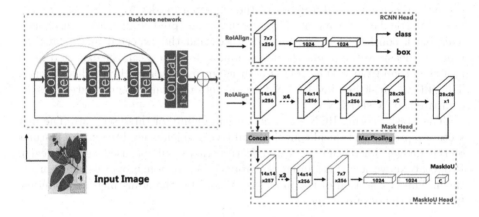

Fig. 1. Workflow of PhenoDeep model

Each layer applies a nonlinear transformation $H_i(.)$, while i denotes the ith layer, which requires as input the feature maps of all previous layers (Eq. 1).

$$x_i = H_i([x_0, x_1, ...x_{i-1}]) \tag{1}$$

where $[x_0, x_1, ...x_{i-1}]$ denotes a feature map cascade, and H_i is a composite function of two sequential processes: the ReLU function and the 3×3 convolution layers.

The input and output for the dth unit is specified as F_{d_1} and F_d, while the number of feature maps is G_0. The output of the cth convolution layer of the unit is represented as follows (Eq. 2):

$$F_{d,c} = H[F_{d-1}, F_{d,1}, ..., F_{d,c-1}] \tag{2}$$

The feature map at the end of the unit should be minimized due to the connection mode between the input unit and the convolutional layer. As a consequence, the number of feature maps dominated by 1×1 convolution can be described as (Eq. 3):

$$F_{d,LF} = H^d_{LFF}[F_{d-1}, F_{d,1}, ..., F_{d,c}] \tag{3}$$

where H^d_{LFF} denotes 1×1 convolution. The output unit is represented as follow (Eq. 4):

$$F_d = F_{d-1} + F_{d,LF} \tag{4}$$

To preserve the feed-forward propagation, the input of each layer is the aggregation of the mapping output of all the preceding layers, and its feature mapping result is often utilized as the input of the successive layers.

Region Proposal Neural Network and Region of Interest (RoI) Align. The proposed backbone produces the corresponding feature map of the reproductive organs within the DHS images. RPN first uses the generated feature maps to extract the candidate boxes likely to be reproductive organs. In addition, RPN generates nine anchors for each pixel by examining the feature map due to the reproductive organs' size variability and occlusion. The anchor will be monitored during the training phase by comparing various anchors using the IoU. To identify the candidate box anchors, IoU is calculated between multiple anchors and the ground truth. The Non-Maximum Suppression (NMS) approach is then used to select the best anchor. After determining the optimal anchors, the first proposal area is obtained using the bounding box regression approach.

PhenoDeep is an instance segmentation model. Therefore, RoI Align was often used to map Regions of Interest (RoI) generated by the previous step to feature maps of a standard size. To increase the pixel accuracy of the predicted masks, RoI Align employs bilinear interpolation rather than quantization. This process aligns the extracted features efficiently with the input.

Object Detection and Instance Segmentation. After completing the RoI Align phase, PhenoDeep adopts Fully Convolutional Networks (FCN) to perform the instance segmentation process. FCN precisely segments the reproductive organs within the DHS images. Each herbarium sheet is convolved to minimize the scale of its feature map. Therefore, the deconvolution process is performed, where the feature map resolution is increased steadily by applying the interpolation approach, and finally, each pixel is classified correctly.

On the other hand, the confidence of instance classification is utilized to indicate the consistency of the obtained masks. Nevertheless, the mask's segmentation accuracy is associated with the IoU between the expected mask and the ground truth. To tackle this disadvantage, Mask Scoring-RCNN proposes a novel MaskIoU head. It performs the IoU between the predicted mask and the corresponding ground truth by concatenating the mask head and the output feature maps of RoI Align.

5   Experiments and Discussion

The PhenoDeep model was implemented using Facebook's Open Source Maskrcnn-benchmark project (https://github.com/facebookresearch/maskrcnn-benchmark), and the experiments were conducted on Google's free cloud service named Google Collaboratory [21]. It is a free cloud service for machine learning and science education, providing 12 GB of RAM, and NVIDIA Tesla K80 carries out the training phase. Furthermore, PhenoDeep has been trained for 100 epochs with 200 steps for each epoch. As presented in Table 1, the stochastic gradient descent (SGD) was initialized with a learning rate of 0.001, a momentum of 0.9, and a weight decay of 0.0005.

During the test phase, 100 DHS images of six common species were selected, such as *Aplopappus stoloniferous*, DC., var; *Eupatorium scordonioides* A. Gray; *Penstemon pulchellus Lindl*; *Senecio Chapalensis*, Watson; *Russelia trachypleura*, Rob and, *Hirtella bracteosa Steud*. Each herbarium sheet image contains different reproductive organs with different sizes and orientations. For the purposes of this study, a total of 742 reproductive organs were identified and considered as ground-truth data, including 117 buds, 276 fruits, and 349 flowers (Table 3).

To assess the model's reliability and evaluate the segmentation accuracy, we measured the Precision (Eq. 5), Recall (Eq. 6), and Average Precision (AP) metrics for our PhenoDeep approach, the original Mask Scoring-RCNN and the Mask-RCNN using different backbone networks of ResNet-50/101 architectures.

$$Precision = \frac{TruePositives}{TruePositives + FalsePositives} \tag{5}$$

$$Recall = \frac{TruePositives}{TruePositives + FalseNegatives} \tag{6}$$

The test results presented in Table 2 demonstrated that the overall PhenoDeep precision and recall achieved 88.9%, and 90.4%, respectively. For example, only 12 buds were misclassified as flower samples due to the overlapping of different objects within the DHS in some cases.

As illustrated by the detection results in Table 2, PhenoDeep can detect the presence/absence of the reproductive organs accurately. Also, it can detect smaller objects with different occlusion degrees. Compared with the buds and fruits, PhenoDeep achieves higher results when detecting the presence/absence of the flowers, with a precision rate of 93% and a recall rate of 88.5%.

Table 1. Our proposed approach Hyper-parameters

Parameter	Value
Learning Rate	0.001
Momentum	0.9
Weight Decay	0.0005
Batch Size	4

Table 2. The precision/recall of PhenoDeep

Metrics	Buds	Fruits	Flowers	Overall
Precision	83.3%	90.5%	93%	88.9%
Recall	89.7%	93%	88.5%	90.4%

5.1 The Impact of Data Augmentation Approaches on Segmentation Results

Deep learning approaches have performed well in several areas of computer vision, such as object detection, classification, or segmentation. However, the lack of data is considered one of the main limitations since it requires huge amounts of labeled datasets to increase model accuracy and avoid overfitting. To address the lack of massive annotated datasets, we used the data augmentation technique. Furthermore, we aim in this paper to identify the impact of various image augmentation methods on the segmentation of the reproductive organs using PhenoDeep. The images were augmented using rotational transformation, translation transformation, scale transformation, and brightness [22].

To investigate PhenoDeep stability when detecting the presence/absence of the reproductive organs, we created the confusion matrix (Table 3), which visualizes the accuracy of PhenoDeep by comparing the ground truth and predicted reproductive organs after applying the data augmentation approaches.

Tables 3 and 4 present the PhenoDeep performance when applying the data augmentation techniques. By raising the volume and variety of training data, the overall precision of PhenoDeep is increased by 5.2%, while the overall recall is increased by 3.9%. This reveals that PhenoDeep with data augmentation can better boost the model's capability to perform the instance segmentation.

5.2 Experimental Results and Evaluation of Multiple Segmentation Approaches

To better evaluate the PhenoDeep efficiency for the segmentation of the reproductive organs, PhenoDeep was compared to other approaches with different backbones. The evaluation results are presented in Table 5. We adopted ResNet101 combined with DenseNet as our baseline, where our proposed approach's overall performance is better for finding the reproductive organs than Mask Scoring RCNN, and Mask-RCNN approaches.

Table 3. The confusion matrix of PhenoDeep when applying the data augmentation approaches using different species

Species Names	No. of DHS images	No. of Buds		No. of Fruits		No. of Flowers	
Aplopappus stoloniferous	19	Ground Truth	17	Ground Truth	43	Ground Truth	58
		Detected	14	Detected	30	Detected	55
Eupatorium scordonioides	12	Ground Truth	26	Ground Truth	39	Ground Truth	39
		Detected	24	Detected	29	Detected	37
Penstemon pulchellus	8	Ground Truth	8	Ground Truth	14	Ground Truth	34
		Detected	5	Detected	9	Detected	30
Senecio Chapalensis	22	Ground Truth	42	Ground Truth	58	Ground Truth	51
		Detected	40	Detected	45	Detected	47
Russelia trachypleura	26	Ground Truth	19	Ground Truth	63	Ground Truth	105
		Detected	18	Detected	53	Detected	102
Hirtella bracteosa Steud	13	Ground Truth	5	Ground Truth	59	Ground Truth	62
		Detected	4	Detected	48	Detected	60
Total	100	Ground Truth	117	Ground Truth	276	Ground Truth	349
		Detected	105	Detected	214	Detected	331

Table 4. The precision/recall of PhenoDeep when using the data augmentations

Metrics	Buds	Fruits	Flowers	Overall
Precision	92.5%	96.1%	93.8%	94.1%
Recall	95.7.7%	91%	96.2%	94.3%

Using the original Mask Scoring RCNN approach, several reproductive organs were misidentified, and the overall performance was decreased (the overall precision and recall achieved 87%, and 89%, respectively). Meanwhile, the accuracy curve is presented in Fig. 2 (left side) to verify the proposed model's efficiency. This curve gives additional insight into the proposed approach.

According to Fig. 2 (right side), the loss is almost unchanged after 100 epochs, while approximately 21 h are required to train the model. The overall loss of ResNet101 combined with DenseNet as the backbone is reduced significantly while accuracy increases during the training phase.

6   Discussion

PhenoDeep is designed to detect the presence/absence of buds, flowers, and fruits within the DHS images. To conduct this study, we manually annotated thousands of herbarium images containing several reproductive organs. However, PhenoDeep is heavily influenced by (1) the size of the training dataset and (2)

Table 5. Comparison results of PhenoDeep-50/101 with different other approaches

Models	Precison	Recall
Original Mask Scoring RCNN with ResNet50	86.3%	88%
Original Mask Scoring RCNN with ResNet101	87%	89%
Mask-RCNN with ResNet50	84%	82%
Mask-RCNN with ResNet101	85%	83%
PhenoDeep with ResNet50	90.6%	93%
PhenoDeep with ResNet101	**94.1%**	**94.3%**

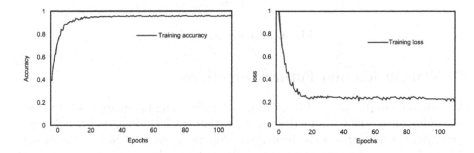

Fig. 2. Loss/Accuracy curves of PhenoDeep

the high variability of reproductive organs, which vary in size, shape, orientation, and color.

In the present paper, we consider organ size's impact on detecting reproductive organs within the herbarium sheets where smaller-sized objects like buds have lower performance compared to other objects. Various reasons are contributing to this poorer performance of buds. These include the limited number of buds organs within the DHS images, where around 70% of the samples had no buds. Furthermore, buds are smaller in size and have a lower visual distinctiveness than flowers or fruits.

In contrast, flowers appear to be the most detected object since they are (i) slightly more significant than buds and (ii) less isolated than fruits and buds (Fig. 3 and Table 4). Furthermore, the confusion matrix presented in Table 3 demonstrates that the total number of detected reproductive organs was very close to their ground truth values.

By comparing PhenoDeep with Mask Scoring-RCNN and Mask-RCNN, we have demonstrated that our approach worked well for segmenting and identifying the reproductive organs within the DHS images, where the precision and recall reached 94.1%, and 94.3%, respectively.

Fig. 3. PhenoDeep outputs

7   Conclusion and Future Directions

We proposed in this paper, PhenoDeep, a deep learning-based approach to local-ize and segment the reproductive organs within the DHS images. PhenoDeep is accurately segmenting buds, fruits, and flowers of different sizes and orienta-tions. Based on the obtained results, PhenoDeep achieves high precision and recall when identifying the reproductive organs. Also, it is essential to note that the PhenoDeep performance is based on a relatively small training dataset com-prising 4000 herbarium specimens. Thus, the performance could be considerably enhanced by enriching the dataset with specimens spanning different spatio-temporal scales. This can increase our model's reliability by using it in other tasks, such as reproductive organ counting and size measurements.

Acknowledgments. This work was part of the MAMUDS project (Management Multimedia Data for Science). It was supported by BMBF, Germany (Project No. 01D16009) and MHESR, Tunisia.

References

1. Piao, S., et al.: Plant phenology and global climate change: current progresses and challenges. Glob. Change Biol. **25**, 1922–1940 (2019)
2. Yalcin, H.: Plant phenology recognition using deep learning: Deep-Pheno. In: 2017 6th International Conference on Agro-Geoinformatics, pp. 1–5 (2017)
3. Milicevic, M., et al.: Application of deep learning architectures for accurate detec-tion of olive tree flowering phenophase. Remote Sen. **12**(13), 2120 (2020)
4. Borsch, T., et al.: A complete digitization of German herbaria is possible, sensible and should be started now. Res. Ideas Outcomes **6**, e50675 (2020)
5. Sweeney, P.W., et al.: Large-scale digitization of herbarium specimens: develop-ment and usage of an automated, high–throughput conveyor system. Taxon **67**, 165–178 (2018)

6. Hedrick, B.P., et al.: Digitization and the future of natural history collections. BioScience **70**(3), 243–251 (2020)
7. Park, D., et al.: Assessing Plant Phenological Patterns in the Eastern United States Over the Last 120 Years, Environmental Data Initiative (2018)
8. Pearson, K.D., et al.: Machine learning using digitized herbarium specimens to advance phenological research. BioScience **70**(7), 610–620 (2020)
9. Lorieul, T., et al.: Toward a large-scale and deep phenological stage annotation of herbarium specimens: case studies from temperate, tropical, and equatorial floras. Appl. Plant Sci. **7**(3), e01233 (2019). PMCID: PMC6426157
10. Ellwood, et al.: Emerging frontiers in phenological research. Appl. Plant Sci. **7**(3), e1234 (2019)
11. Goëau, H., et al.: A new fine-grained method for automated visual analysis of herbarium specimens: a case study for phenological data extraction. Appl. Plant Sci. **8**(6), e11368 (2020)
12. He, K., et al.: Mask R-CNN. In: 2017 IEEE International Conference on Computer Vision (ICCV), pp. 2980–2988 (2017)
13. Davis, C.C., et al.: A new method for counting reproductive structures in digitized herbarium specimens using mask R-CNN. Front. Plant Sci. **11**, 1129 (2020). PMID: 32849691; PMCID: PMC7411132
14. Huang, Z., et al.: Mask scoring R-CNN. In: IEEE/CVF Conference on Computer Vision and Pattern Recognition (CVPR), pp. 6402–6411 (2019)
15. Dutta, A., Zisserman, A.: The VIA annotation software for images, audio and video. In: Proceedings of the 27th ACM International Conference on Multimedia (MM 2019), p. 4, 21–25 October 2019, Nice, France. ACM, New York, NY, USA (2019)
16. Ren, S., et al.: Faster R-CNN: towards real-time object detection with region proposal networks. In: Proceedings of the 28th International Conference on Neural Information Processing Systems - Volume, pp. 91–99. MIT Press, Cambridge, MA, USA (2015)
17. Lin, T., et al.: Feature pyramid networks for object detection. In: IEEE Conference on Computer Vision and Pattern Recognition (CVPR), pp. 936–944 (2017)
18. Huang, G., et al.: Densely connected convolutional networks. In: 2017 IEEE Conference on Computer Vision and Pattern Recognition (CVPR), pp. 2261–2269 (2017)
19. He, K., Zhang, X., Ren, S., Sun, J.: Deep residual learning for image recognition. In: IEEE Conference on Computer Vision and Pattern Recognition (CVPR), pp. 770–778 (2016)
20. Francisco, M., Ross, G.: maskrcnn-benchmark: fast, modular reference implementation of instance segmentation and object detection algorithms. In: PyTorch (2018)
21. Shorten, C., Khoshgoftaar, T.M.: A survey on image data augmentation for deep learning. J. Big Data **6**, 60 (2019)
22. Lima, D.F., Mello, J.H.F., Lopes, I.T., Forzza, R.C., Goldenberg, R., et al.: Phenological responses to climate change based on a hundred years of herbarium collections of tropical Melastomataceae. PLOS ONE **16**(5), e0251360 (2021)
23. Lee, H.K., Lee, S.J., Kim, M.K., Lee, S.D.: Prediction of plant phenological shift under climate change in South Korea. Sustainability **12**, 9276 (2020)
24. Willis, C.G., et al.: Old plants, new tricks: phenological research using herbarium specimens. Trends Ecol. Evol. **32**(7), 531–546 (2017). ISSN 0169–53475

Document-Level Event Factuality Identification Using Negation and Speculation Scope

Heng Zhang, Zhong Qian$^{(\boxtimes)}$, Xiaoxu Zhu, and Peifeng Li

School of Computer Science and Technology, Soochow University, Suzhou, China
20204227055@stu.suda.edu.cn, {qianzhong,xiaoxzhu,pfli}@suda.edu.cn

Abstract. Document-level Event Factuality Identification (DEFI) is to identify the factuality of an event in document. DEFI is an important foundation for many Natural Language Understanding (NLU) tasks, such as information extraction and text understanding. The negation and speculation scope refers to the continuous segment of text, which is controlled by the words with negative or speculative semantics. In this paper, we explore the importance of negation and speculation scope for DEFI, and propose two methods to use the scope features. The model of detecting scope is trained from cross-domain corpus, and applied to the Document-Level Event Factuality (DLEF) corpus. Experimental results show that our DEFI model is superior to several baselines.

Keywords: Document-level event factuality identification · Negation and speculation scope · DLEF corpus

1 Introduction

Event Factuality Identification (EFI) is the task to point out the factuality of event. Depending on the granularity of the text to be processed, it can be divided into two subtasks: Sentence-level Event Factuality Identification (SEFI) and Document-level Event Factuality Identification (DEFI). SEFI is to identify the factuality of the event expressed within a sentence, while DEFI is to identify the factuality of the event in a document. The factuality values can be divided into five categories [1]: CerTain Positive (CT+), PoSsible Positive (PS+), CerTain Negative (CT-), PoSsible Negative (PS-), Underspecified (Uu).

In DEFI task, there are several sentences referring to the event of document, and the factuality values in these sentences are different, just as shown in Fig. 1. The current event is in **bold**, while the text with underline is the negation scope and the text with wavy line is the speculation scope.

The event of document is mentioned in (S1), S(3), S(4) and S(5), which can be represented by the event trigger "sale" in bold. In S(1), the factuality of event "sale" is PS+ according to the speculative cue "may"; the event's factuality is CT- in S(5) due to the negative cue "not"; in S(3) and S(4), "sale" is not

© Springer Nature Switzerland AG 2021
T. Mantoro et al. (Eds.): ICONIP 2021, LNCS 13108, pp. 414–425, 2021.
https://doi.org/10.1007/978-3-030-92185-9_34

(S1) <u>US may considers arms **sale**</u> to update Greece 's F_16 fleet.

(S2) US President Donald Trump welcomes visiting Greek Prime Minister Alexis Tsipras to the White House in Washington D.C., the United States, on Oct 17, 2017.

(S3) WASHINGTON _ US President Donald Trump said on Tuesday his administration had informed Congress of a arms **sale** to update Greece's F_16 aircrafts.

... ...

(S4) The proposed **sale** will require the assignment of approximately three to five additional US government or contractor representatives to Greece, the statement added.

(S5) <u>The arms **sale** is not concluded at the moment</u> as it still needs the approval of the US Congress.

Fig. 1. An example for document-level event factuality.

affected by any negative or speculative information, so the factuality is CT+. However, considering the whole document, the document-level factuality of the event "sale" is unique, i.e., PS+.

According to the analysis above, SEFI is the basis of DEFI, and since the sentences have different factuality values for the event of document, DEFI needs to take more global semantic information into account.

The factuality is closely related to the negative and speculative information in sentences, so the shortest syntactic paths from cue words to event triggers were commonly used in previous studies [1,2]. These syntactic features can address the problem of long distance on the relationship between cue words and event triggers, but the shortcoming is that there are some meaningless labels in syntactic paths, and they also lack semantic information compared with continuous text fragments. In addition, the process of obtaining syntactic paths is complex and depends heavily on the quality of cue words and syntactic trees generated by various NLP tools.

Negation and speculation scopes can help to identify the factuality from two aspects: On the one hand, scope is a direct reflection of the semantic distance between cue words and event triggers, so the judgment of factuality can be transformed into the judgment of whether the event is within the scope, i.e., when event trigger is in the negation/speculation scope, its factuality is CT-/PS+; while event trigger is not in any scopes, its factuality value is CT+. PS- can be regarded as the trigger in the scope of both types, while Uu is unknown or uncommitted and generally in a conditional sentence. PS- and Uu samples occupy the quite small proportion, so they are not considered in previous works [1,2] and this paper. On the other hand, the scope itself is the continuous natural language text, which contains rich semantic information. The concentrated information is valuable for DEFI. The major contributions of our work can be summarized as follows:

- We propose a model to detect the speculative and negative scopes of Chinese and English text, and use cross-domain scope corpora (i.e., BioScope and CNeSp) as the training set, which can be seen cross-domain data argumentation as well.
- We focus on using the scope features better in DEFI task, and propose two different methods. The experimental results show that our model can achieve excellent results on DLEF corpus.

2   Related Work

2.1   Event Factuality Identification

In the research of EFI, many scholars have made a long-term exploration in the sentence-level work, i.e., SEFI. SEFI started with rule-based methods [3,4], then many scholars used machine learning-based methods which relied on the annotated information [5–7], or combined the rules with machine learning [8]. The appearance of neural network model made the performance of SEFI further improved. For example, CNN-based model [9], hybrid neural network model combined BiLSTM and CNN [10], generative adversarial network model [11], graph based neural network model [12].

Compared with SEFI, the work on DEFI is still in the initial and exploratory stage. Qian et al. [1] constructed a Document-Level Event Factuality (DLEF) corpus and proposed a BiLSTM-based model, which encoded the sentences and syntactic paths, and then used the attention mechanism to fuse semantic and syntactic information together to obtain the representation of the whole document, and finally classified the factuality. Zhang et al. [2] proposed a Gated Convolution Neural Network (GCNN) model based on Qian's work, and got significant improvement on performance.

2.2   Negation and Speculation Scope Detection

With the wide application of neural networks in NLP, more and more deep learning models appeared in scope detection task to replace the heuristic rules methods and machine learning methods. Qian et al. [13] proposed a CNN-based model on BioScope dataset. Fancellu et al. [14,15] proposed a BiLSTM model, a Bidirectional Dependency LSTM (D-LSTM) model and a Graph Convolutional Network (GCN) model. Fabregat et al. [16] used the Recurrent Convolutional Neural Network (RCNN) for the negation scope detection. Fei et al. [17] used the recursive neural conditional random fields for scope detection. The evolution of these models was also inseparable from the development of relevant corpora, e.g., BioScope [18] corpus for English and CNeSp [19] corpus for Chinese.

3   DEFI with Scope

3.1   Overview

DLEF corpus does not annotate negation and speculation scopes, so the first step is that we need to learn a model from other corpora in order to obtain scopes. After that, we use the detected scopes as features for DEFI.

In Sect. 3.2, we propose a BERT-CRF model for scope detection. In Sect. 3.3 we explain how to apply the scopes to DEFI, and present the DEFI model structure.

3.2   Scope Detection

Task Definition. Negation and speculation scope detection can be regarded as a sequence labeling task. For each word in the input sentence, a label can be specified through the model to identify whether it falls within the scope. Our task doesn't take a fine-grained division as previous works [13,17], we just do a binary classification for each token [14,15]: Label "O" means the token is out of the scope, label "I" means the token is in the scope, a labeled sentence is shown as follows.

- *PMA/O treatment/O ,/O and/O not/I retinoic/I acid/I treatment/I of/I the/I U937/I cells/I acts/O in/O inducing/O NF-KB/O expression/O in/O the/O nuclei/O ./O*

BERT-CRF Model. The English corpus for training the model is BioScope and the Chinese one is CNeSp. Three sub-corpora with different language styles are labeled in both BioScope and CNeSp. Our goal is to obtain a good performance of scope detection on DLEF. Therefore, in order to enhance the generalization ability of the model, we combine all the sub-corpora in BioScope and CNeSp, respectively, and then divide the data uniformly for training and testing.

We use neither any annotated information, nor complex model structure to train. Our model is based on BERT [20], and a CRF layer is set up in the model, which is used to obtain a optimal prediction sequence through the relationship between adjacent tags.

3.3   DEFI

DEFI can be defined as a text classification problem, i.e., given a collection of tags and a document composed of multiple sentences, DEFI task is to map the document to a unique tag.

First of all, DEFI needs to consider the content of the whole document comprehensively, so the global feature which represent the whole semantic information is very important for factuality identification. In real-world applications, due to the excessive length of the document, information extraction from its content will be interfered with too much noise, and it is also difficult for the model to

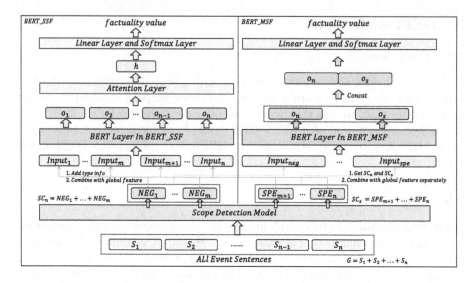

Fig. 2. Model structure for DEFI.

extract effective information from long sequence. Therefore, we consider the full text with all the event sentences (i.e., the sentences which refer to the current document event) as the global feature, and extract semantic information from them as global representation.

In Sect. 3.2, we have trained a BERT-CRF model. Now we propose two effective methods to use the scope features in DEFI as follows.

- We combine each scope with global feature to get the feature vector where the number of vectors is equal to the number of scopes detected in document. The vectors encode the semantic information of the scopes as well as the whole document, and the correlation between them. Then we learn a representation from multiple vectors for the final factuality classification.
- We divide the scopes into two categories, i.e., negation and speculation, and connect the two types of scopes separately in turn to get combination scopes. Both of scopes and global feature are combined as the input for further learning, respectively. The correlation between them is obtained in encoding vectors, which also contain rich semantic information. Finally, two vector representations are connected for the final identification.

3.4   Model Structure

The model structures of two methods are shown in Fig. 2.

Input and Global Feature. Firstly, we select event sentences from all the text in the document, and the set of event sentences is defined as $\{S_1, S_2, ..., S_n\}$.

All the event sentences are connected in turn as the global feature G, which represents the semantic information of the whole document:

$$G = S_1 + S_2 + ... + S_n \tag{1}$$

We denote method 1 as BERT_SSF (BERT_Single_Scope_Feature), and denote method 2 as BERT_MSF (BERT_Merge_Scope_Feature).

BERT_SSF. All the event sentences are taken as the input to BERT-CRF, we use SC to represent all the detected negation and speculation scopes in event sentences:

$$SC = \{NEG_1, ..., NEG_m, SPE_{m+1}, ..., SPE_n\} \tag{2}$$

After obtaining each scope, additional information is added in front of the scope to indicate the corresponding type as follows:

- For negation scope, add "The negation scope:" before the scope text.
- For speculation scope, add "The speculation scope:" before the scope text.
- If there is no scope in text, take "None scope in the document" as the scope text.

Then we combine these scopes with G as the input sequence I, whose main purpose is to learn the relationship between global feature and each scope:

$$I = \{[CLS], Type : Scope, [SEP], G, [SEP]\} \tag{3}$$

Since the number of detected scopes vary among documents, we treat each document as a batch separately. If the number of scopes for a document is n, then the BERT input in matrix form is $\boldsymbol{I} \in R^{n \times l}$, l is the length of input sequence. We take the last hidden layer of BERT, and take the first position (i.e., the $[CLS]$ token) in each sequence as output:

$$\boldsymbol{O} = \text{BERT}(\boldsymbol{I}) \tag{4}$$

where the output is $\boldsymbol{O} \in R^{n \times d}$, and d is the dimension of hidden states in BERT. Next, we use attention mechanism to extract the most important information from multiple sequences, and concatenate them into a vector $\boldsymbol{h} \in R^d$ to identify the factuality:

$$\boldsymbol{\alpha} = \text{softmax}(\boldsymbol{v}^T \tanh(\boldsymbol{O}^T)) \tag{5}$$

$$\boldsymbol{h} = \tanh(\boldsymbol{O}^T \boldsymbol{\alpha}^T) \tag{6}$$

where $\boldsymbol{v} \in R^d$ is the parameter. After getting the vector $\boldsymbol{h}$, we use a linear layer with a relu activation function to obtain a high-level abstract representation:

$$\boldsymbol{h_r} = \text{relu}(\boldsymbol{W_1}\boldsymbol{h} + \boldsymbol{b_1}) \tag{7}$$

where $\boldsymbol{W_1} \in R^{c_1 \times d}$ and $\boldsymbol{b_1} \in R^{c_1}$ are parameters, c_1 is the dimension of the hidden states in linear layer. The output $\boldsymbol{h_r} \in R^{c_1}$. The vector $\boldsymbol{h_r}$ is sent to the final softmax layer for classification, and we get the predication of the factuality according to the probability distribution computed as:

$$h_s = \text{softmax}(\boldsymbol{W_2}\boldsymbol{h} + \boldsymbol{b_2}) \qquad (8)$$

where $\boldsymbol{W_2} \in R^{c_2 \times c_1}$ and $\boldsymbol{b_2} \in R^{c_2}$ are parameters in softmax layer. c_2 is the number of factuality categories.

BERT_MSF. Different from BERT_SSF, detected scopes are separated by negation and speculation, then put scopes together which are in the same category to get SC_n and SC_s.

Each document can get two scopes (If there is no scope of negation/speculation, the content of SC_n / SC_s is empty). Similar to BERT_SSF, we combine scopes with global feature to get input I_n and I_s:

$$I_n = \{[CLS], SC_n, [SEP], G, [SEP]\} \qquad (9)$$
$$I_s = \{[CLS], SC_s, [SEP], G, [SEP]\} \qquad (10)$$

We can obtain the output $\boldsymbol{o_n} \in R^d$ and $\boldsymbol{o_s} \in R^d$, and then concatenate two vectors instead of using attention to get the representation:

$$\boldsymbol{o_f} = \text{concat}(\boldsymbol{o_n}, \boldsymbol{o_s}) \qquad (11)$$

Finally, $\boldsymbol{o_f}$ is fed to a linear layer and a softmax layer. The setups of two layers are the same as the previous one except for the input dimension. Both methods are trained following objective function:

$$L(\theta) = -\frac{1}{N} \sum_{i=0}^{N-1} \log P(y_i|x_i, \theta) \qquad (12)$$

where N is the number of training samples in the corpus, θ is the set of all the parameters in model that need gradient update. The term $P(y_i|x_i)$ is the probability of sample x_i to be predicted as the correct label y_i.

4   Experimentation

4.1   Dataset

DLEF corpus [1] contains 1727 English documents and 4649 Chinese documents. The English documents were selected from China Daily[1] website and Sina Bilingual News[2], and the Chinese documents were selected from Sina News[3]. There are five types of factuality values labeled in corpus, and the distribution of each type is shown in Table 1. Among them, the proportion of documents with the value of CT+ is largest, followed by CT- and PS+, while the number of Uu and PS- is very small.

---

[1] http://www.chinadaily.com.cn/.
[2] https://english.sina.com/.
[3] https://news.sina.com.cn/.

Table 1. Factuality distribution in DLEF.

Sub-corpus	CT+	CT-	PS+	PS-	Uu	Total
English	1150(66.59%)	279(16.16%)	274(15.87%)	12(0.69%)	12(0.69%)	1727
Chinese	2403(51.69%)	1342(28.87%)	848(18.24%)	36(0.77%)	20(0.43%)	4649

Table 2. Cross-domain performance of the BERT-CRF model.

Source	Target	Factuality	Total Documents	Scope Detected	Percent(%)
BioScope	DLEF English	CT-	279	263	94.27
		PS+	274	259	94.53
CNeSp	DLEF Chinese	CT-	1342	1250	93.14
		PS+	848	845	99.65

4.2 Implementation Details

In the experiments, we set $c_1 = 256$, $c_2 = 5$, and the dimension $d = 768$. In BERT_MSF, the batch size is set as 4. The learning rate is 10^{-5} and Adam algorithm is used for updating parameters in both methods. In order to avoid the contingency of results caused by random sample division, we perform 10-fold cross validation on both English and Chinese sub-corpora. It should be noted that there are few samples in categories of Uu and PS-, its identification needs the support of more training resources, so we focus on the performance of CT+, CT- and PS+.

4.3 Migration Application

We verify the performance of scope detection model on DLEF corpus, to prove the effectiveness of migration application. Since DLEF does not label the negation and speculation scopes, we take another way to verify: Select the documents which the factuality values are CT- or PS+, and the event sentences are extracted as the input of BERT-CRF. By observing the proportion of documents whose relevant scopes are detected (i.e., the negation scope for CT- and the speculation scope for PS+), we can prove whether the BERT-CRF model is effective in cross-domain detection, whose usefulness can be proven in Table 2.

4.4 Baselines

We use the following models as baselines for fair comparison with our model:

- **LSTM_Att** [1]: A BiLSTM model based on attention mechanism proposed by Qian.
- **GCNN** [2]: A gated convolution neural network for DEFI proposed by Zhang.
- **BERT_Global**: A strong baseline only consider global feature, i.e., use all the event sentences as input.

Table 3. Performance of each model on DLEF.

Models	English Sub-corpus					Chinese Sub-corpus				
	CT+	PS+	CT-	Micro	Macro	CT+	PS+	CT-	Micro	Macro
LSTM_Att	89.04	62.61	77.82	83.25	76.49	86.49	71.44	80.30	82.19	79.41
GCNN	91.19	70.76	80.28	86.37	80.74	89.60	76.81	85.38	86.03	83.93
BERT_Global	91.38	71.30	80.43	86.52	81.34	89.92	84.01	87.35	88.08	87.19
BERT_Syntax	91.89	72.93	81.71	87.35	82.37	90.45	83.69	87.99	88.53	87.50
BERT_SSF	92.37	76.15	**83.83**	88.34	**84.37**	90.94	85.43	88.53	89.20	88.37
BERT_MSF	**92.50**	**76.38**	83.71	**88.64**	84.24	**92.09**	**85.71**	**90.08**	**90.34**	**89.35**

- **BERT_Syntax:** Replace scope features with the shortest dependency syntactic paths of cue words and event triggers. The model structure is same as BERT_SSF (it performs better than BERT_MSF when using the syntactic path features). The cue words in English and Chinese are from BioScope and CNeSp corpus, respectively.

4.5 Results and Analysis

Table 3 show the performance of baseline models and proposed models. All the indicators in the table are F1-Measures, and all of them are in %.

According to the results in table, we can clearly observe the powerful effect of pre-training model in this downstream task: All the indicators comprehensively exceed the two baseline models of previous works, even if only global feature is considered. So we pay more attention to the improvement of proposed models compared to the BERT_Global.

In English sub-corpus, our works improve by 2.12% and 3.03% in Micro-F1 and Macro-F1, in terms of three main categories, i.e., CT+, PS+, CT-, the improvement is 1.12%, 5.08%, 3.40%, respectively. Our models also achieve excellent results in Chinese sub-corpus, the improvement of the five indicators is 2.26%, 2.16%, 2.17%, 1.70%, 2.73%, respectively. The above results prove that our models are effective.

In both English and Chinese corpora, BERT_Syntax which use syntactic path as the feature has a certain improvement compared with the baselines, but its performance is not ideal compared with the two methods using scopes. This is also a direct indication that the scope features are better than the syntactic path features in DEFI work.

4.6 Discussion

We enumerate the following three examples to illustrate the impact of different scope detection effects on the DEFI results. We replace the whole document with event sentences, use underline to indicate the detected negation scope, and wavy line to indicate the speculation one, the current event of document is shown in **bold**.

- *D1: (S1) White House Senior Advisor Kushner admits to four Russian meetings, <u>denies **collusion**</u>. (S2) WASHINGTON _ White House Senior Advisor Jared Kushner on Monday denied having **colluded** <u>with the Russian government</u> despite four meetings with Russian nationals during the campaign and the transition period. S(3) " <u>I did not **collude**, nor know of anyone else in the campaign who **colluded**, with any foreign government,</u>" Kushner said in a 11_page statement.*

- *D2: (S1) Lattice Semiconductor Corp will seek US President Donald Trump 's approval for its proposed $ 1.3 billion sale to China_backed Canyon Bridge Capital Partners, Lattice said on Friday, gambling that Trump will **approve** the tie_up against the advice of US national security officials. (S2) If Trump **approves** the transaction, it would be unprecedented.*

- *D3: (S1) Pensions system <u>not **fit** for</u> 21st century , says former government adviser. (S2)She called for an overhaul of defined contribution LRB DC RRB pensions which, she argued, " <u>are not **fit** for 21st century lives</u>".*

Obviously, the factuality of event "collude" in D1 is CT-. The model can also get the correct prediction through the correctly detected scopes, which contain negative information that is helpful for identifying factuality. When the negation and speculation scopes containing the event are correctly detected, even though both types of scopes can acquired simultaneously (e.g., the sample in introduction), model still has the ability to get the correct result.

When the negative or speculative information that can affect the event in the text is not detected in the form of scope, it may lead to identify the factuality mistakenly only by relying on the semantic information of the document. In D2, we can judge that the factuality of the event "approve" is PS+ acording to the word "gambling" in (S1) and "if" in (S2). However, the models could not find out any speculation scopes, fail to identify the event "approve" which is governed by the speculative cue "gambling", so it identify the wrong result of CT+. The reason for this situation is that the way of expressing speculative information in the sentence is too obscure, "gambling" can affect the factuality of the event in this context, but may not applicable in other contexts.

We observe all the detected scopes and find a phenomenon, i.e., the detection of negation scope is more accurate than that of the speculation one. For some sentences without speculative information, the models may mistakenly detect the existence of scopes. For the current event "fit" in D3, the models correctly detect the negation scopes, and mistakenly detect the speculation scopes even if the sentences don't contain speculative information. These speculation scopes lead to the error of DEFI.

Finally, we analyze the performance differences between the two proposed models: In English sub-corpus, the performance of two methods is very close, while in Chinese sub-corpus, the difference between them is more obvious. The reason is that scope detection errors happen more frequently in the Chinese sub-corpus. Since BERT_SSF pays more attention to the relationship between the single scope and global feature, error scopes have more obvious impacts on the result. However, BERT_MSF model encodes all the negative and speculative information respectively, so the individual error detection has a relatively smaller influence on the result, which means BERT_MSF is more robust.

5   Conclusion

In this paper, we construct a scope detection model by cross-domain learning, and propose two models using scopes for DEFI. Experimental results on DLEF show that our approaches are significantly better than the previous methods.

At present, we only conduct experiments in Chinese and English due to the limitations of corpus, so we could not determine the validity of the model in other languages. In addition, when the learned model for detecting scope is applied to other cross-domain corpus, although a good detection result has been achieved, problems such as language genre differences and labeling consistency still exist objectively. In the future work, we will try to construct a new corpus containing more other languages and train a better model for scope detection, which is more effective and can be used in different fields, so that it can be better beneficial to the EFI work.

Acknowledgments. The authors would like to thank the two anonymous reviewers for their comments on this paper. This research was supported by the National Natural Science Foundation of China (No. 62006167, 61836007 and 61772354.), and the Priority Academic Program Development of Jiangsu Higher Education Institutions (PAPD).

References

1. Qian, Z., Li, P., Zhu, Q., Zhou, G.: Document-level event factuality identification via adversarial neural network. In: Proceedings of the 2019 Conference of the North American Chapter of the Association for Computational Linguistics: Human Language Technologies, Volume 1 (Long and Short Papers), pp. 2799–2809 (2019)
2. Zhang, Y., Li, P., Zhu, Q.: Document-level event factuality identification method with gated convolution networks. Comput. Sci. **47**(3), 206–210 (2020)
3. Saurí, R.: A Factuality Profiler for Eventualities in Text. Brandeis University, Waltham (2008)
4. Lotan, A., Stern, A., Dagan, I.: Truthteller: annotating predicate truth. In: Proceedings of the 2013 Conference of the North American Chapter of the Association for Computational Linguistics: Human Language Technologies, pp. 752–757 (2013)
5. De Marneffe, M.C., Manning, C.D., Potts, C.: Did it happen? The pragmatic complexity of veridicality assessment. Comput. Linguist. **38**(2), 301–333 (2012)
6. Saurí, R., Pustejovsky, J.: Are you sure that this happened? Assessing the factuality degree of events in text. Comput. Linguist. **38**(2), 261–299 (2012)
7. Lee, K., Artzi, Y., Choi, Y., Zettlemoyer, L.: Event detection and factuality assessment with non-expert supervision. In: Proceedings of the 2015 Conference on Empirical Methods in Natural Language Processing, pp. 1643–1648 (2015)
8. Qian, Z., Li, P., Zhu, Q.: A two-step approach for event factuality identification. In: 2015 International Conference on Asian Language Processing (IALP), pp. 103–106. IEEE (2015)

9. Tianxiong, H., Peifeng, L., Qiaoming, Z.: Identifying Chinese Event Factuality with Convolutional Neural Networks. Presented at the (2018). https://doi.org/10.1007/978-3-319-73573-3_25

10. Qian, Z., Li, P., Zhou, G., Zhu, Q.: Event factuality identification via hybrid neural networks. In: Cheng, L., Leung, A., Ozawa, S. (eds.) Neural Information Processing. ICONIP 2018. LNCS, vol. 11305, pp. 335–347. Springer, Cham (2018). https://doi.org/10.1007/978-3-030-04221-9_30

11. Qian, Z., Li, P., Zhang, Y., Zhou, G., Zhu, Q.: Event factuality identification via generative adversarial networks with auxiliary classification. In: IJCAI, pp. 4293–4300 (2018)

12. Veyseh, A.P.B., Nguyen, T.H., Dou, D.: Graph based neural networks for event factuality prediction using syntactic and semantic structures. arXiv preprint arXiv:1907.03227 (2019)

13. Qian, Z., Li, P., Zhu, Q., Zhou, G., Luo, Z., Luo, W.: Speculation and negation scope detection via convolutional neural networks. In: Proceedings of the 2016 Conference on Empirical Methods in Natural Language Processing, pp. 815–825 (2016)

14. Fancellu, F., Lopez, A., Webber, B.: Neural networks for negation scope detection. In: Proceedings of the 54th annual meeting of the Association for Computational Linguistics (volume 1: long papers), pp. 495–504 (2016)

15. Fancellu, F., Lopez, A., Webber, B.: Neural networks for cross-lingual negation scope detection. arXiv preprint arXiv:1810.02156 (2018)

16. Fabregat, H., Araujo Serna, L., Martínez Romo, J.: Deep learning approach for negation trigger and scope recognition (2019)

17. Fei, H., Ren, Y., Ji, D.: Negation and speculation scope detection using recursive neural conditional random fields. Neurocomputing **374**, 22–29 (2020)

18. Vincze, V., Szarvas, G., Farkas, R., Móra, G., Csirik, J.: The bioscope corpus: biomedical texts annotated for uncertainty, negation and their scopes. BMC Bioinformatics **9**(11), 1–9 (2008)

19. Zou, B., Zhu, Q., Zhou, G.: Negation and speculation identification in Chinese language. In: Proceedings of the 53rd Annual Meeting of the Association for Computational Linguistics and the 7th International Joint Conference on Natural Language Processing (Volume 1: Long Papers), pp. 656–665 (2015)

20. Devlin, J., Chang, M.W., Lee, K., Toutanova, K.: Bert: Pre-training of deep bidirectional transformers for language understanding. arXiv preprint arXiv:1810.04805 (2018)

Dynamic Network Embedding
by Time-Relaxed Temporal Random Walk

Yifan Song[1], Darong Lai[1,2(✉)], Zhihong Chong[1,2], and Zeyuan Pan[1]

[1] School of Computer Science and Engineering, Southeast University, Nanjing, China
[2] MOE Key Laboratory of Computer Network and Information Integration,
Southeast University, Nanjing, China
{daronglai,chongzhihong}@seu.edu.cn

Abstract. Network embedding, also called network representation learning, aims at mapping high dimensional network information into low dimensional vectors. Previous studies mainly focus on static networks. In recent years, dynamic network embedding attracts much attention and methods specific to dynamic network are emerging. However, previous dynamic network embedding methods, such as CTDNE, still have drawbacks when using random walk to generate node sequences. Temporal random walk strictly requires that the time value of next edge be larger (i.e. later visiting) than that of the previous visited edge, which often leads to insufficient information obtained by random walk. In this article, a novel model named **T**ime-Relaxed **T**emporal Random **Walk(TxTWalk)** for dynamic network embedding is proposed. Firstly, a time-relaxed function is designed, which enables random walk to select the next edge in a time interval, not strictly larger than the time of previously visited edge. It can make the walking sequences obtained by **TxTWalk** contain a wider range of temporal information. Then the node sequences are put into the skip-gram model for training to generate embedding of nodes on dynamic networks. Experimental validations on various networks demonstrate that **TxTWalk** is more effective than other state-of-the-art methods in link prediction.

Keywords: Network embedding · Network representation learning · Link prediction · Random walk

1 Introduction

Networks arise naturally in applications that contain a great deal of useful information about the world around. In recent years, network embedding, also called network representation learning, which associates nodes with low-dimensional numerical vectors has become an effective way to process and use these information.

Existing studies mainly focus on static network by preserving as much as possible the structure of a network, such as **Deepwalk** [15], **LINE** [19] and **node2vec** [8]. Nonetheless, real-world networks always change over time, which

© Springer Nature Switzerland AG 2021
T. Mantoro et al. (Eds.): ICONIP 2021, LNCS 13108, pp. 426–437, 2021.
https://doi.org/10.1007/978-3-030-92185-9_35

provide a large amount of temporal information. For example, Fig. 1 shows a co-author network, whose nodes indicate authors and edges represent the collaboration between authors. The timestamp of an edge means the time of collaboration. In this network, the larger the timestamp of an edge is, the more similar the two connected nodes are (more recent collaboration). Therefore, node 6 and 8 in Fig. 1 will be closer to node 5 than node 4 after considering the timestamps of the edges, while in static network neglecting timestamps node 4, 6 and 8 are "equivalently close" to node 5.

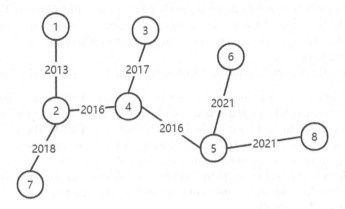

Fig. 1. An example of temporal network. Each edge is annotated with a timestamp denoting when the edge was created.

In order to generate better network embedding in dynamic networks, dynamic network embedding not only aims at preserving network structure but also takes into account temporal information [10]. Among previous dynamic network studies, one typical algorithm is **CTDNE** [13]. **CTDNE** takes into account the time information of each edge, that is the timestamp of an edge, in the random walk sampling of nodes, and requires all sampled edges in a random walk to be within a specific time interval. At the same time, the timestamp of each edge on the random walk path should be greater than that of the previous edge.

CTDNE [13] as well as other random walk based dynamic network representation learning methods perform well, however, they still have drawbacks when using random walk to generate node sequences. Temporal random walk strictly requires that the timestamp of next edge be larger (i.e. later visiting) than that of the previous visited edge. For instance, as shown in Fig. 1, when the edge between node 2 and 4 is selected, the next edge can not be the one connecting node 4 to node 5 since its timestamp is not larger than that of the edge between node 2 and 4. As a result, this type of methods may lead to insufficient information obtained by random walk when the timestamp of the initial edge is large.

For example, in the co-author network, as shown in Fig. 1, author 4 may first publish a paper with author 5 in 2016, and then published a paper with author

3 in 2017. It is obvious that node 3 should have a high degree of similarity to node 5. But in **CTDNE**, it is impossible to get a random walk from author 3 to author 5 because the timestamp of the edge from node 3 to node 4 is larger than the edge from node 4 to node 5. This situation is obviously unreasonable, and how to get temporal random walk to avoid this phenomenon becomes important.

In this article, a novel model called Time-Relaxed Temporal Random **Walk** (**TxTWalk**) has been designed for dynamic network representation learning. Firstly, a time relaxed function is designed, which enables the random walk to select the next edge in a time interval, not strictly larger than the timestamp of previous visited edge. It can make the walking sequences obtained by **TxTWalk** contain a wider range of temporal information. Then the node sequences are put into the skip-gram model for training to generate embedding of the nodes of dynamic networks.

In summary, the main contributions in this article are summarized as follows:

1) A new temporal random walk strategy is proposed by using time-relaxed function, which relaxes strict time restriction of temporal random walk.
2) A novel dynamic network embedding framework **TxTWalk** is proposed, which uses time-relaxed temporal random walk to sample node sequences and skip-gram to learn node representations of a network.
3) Experiments on various dynamic networks and the comparisons to state-of-the-art baseline network embedding methods are conducted to verify the effectiveness of **TxTWalk**.

2   Related Work

Most of the early classical methods to obtain network embedding are based on graph factorization [2–4]. With the introduction of word embedding technology, some methods based on skip-gram model are used to solve the problem of network embedding. Deepwalk [15] bridges the gap between network embedding and word embedding by treating nodes as words and generates short random walks as sentences. Skip-gram [12] can then be applied to these random walk sequences to obtain network embedding. LINE [19] defines the first-order and the second-order similarity for nodes. node2vec [8] introduces a biased random walk by adjusting two hyper parameters of the random walk.

For the past few years, researchers have focused on studying more intricate networks, such as heterogeneous network and attributed networks. HIN2Vec [6] defines metapaths in heterogeneous networks and proposes a model that use logistic classification to learn the embedding. GraphSAGE [9] uses neural network to generate embedding, which is useful for graphs that have rich node attribute information.

However, the great mass of study looked at dynamic networks [10], they study the dynamic network by dividing it into multiple snapshots. For example, DANE [11] considers situations in which attribute matrices of the networks evolve over time and learns the embedding by neural network. **CTDNE** [13] differs from other works that it uses continues-time dynamic network and can be served as a basis to popularize other deep learning methods based on random walk. Various other network embedding methods for dynamic networks are emerging, and details can be seen from the excellent surveys on dynamic network embedding [1,20].

3 Notations and Problem Definitions

In this section, the related notations and definitions of dynamic network embedding problem are given.

Definition 1. *(Continuous-Time Dynamic Network). A Continuous-Time Dynamic Network is a network with each edge being associated with a real value timestamp, which can be represented in the following way:*

$$G = \{V, E_T, T\}$$

where V is the set of nodes in the network, E_T is set of edges between nodes in V, and T is the set of all times of edge emergence. Each edge in E_T of the network is:

$$e_{u,v} = (u, v, t) \in E_T$$

where u and v are the two ending nodes of the connected edge $e_{u,v}$, respectively, and $t \in T$ is the timestamp of that edge.

Definition 2. *(Temporal Random Walk). A temporal walk [13] from v_1 to v_k in G is a sequence of nodes $(v_1, v_2, \cdots, v_k)$ such that $(v_i, v_{i+1}) \in E_t$ for $1 \le i < k$, as well as $T(v_i, v_{i+1}) \le T(v_{i+1}, v_{i+2})$ for $1 \le i < (k-1)$. When v_{i+1} is randomly selected after v_i in the node sequence $(v_1, v_2, \cdots, v_k)$, it is a temporal random walk. For two arbitrary nodes $u, v \in V$, we say that u is temporally (randomly) connected to v if there exists a temporal (random) walk from u to v.*

The link prediction in continuous-time dynamic network is to predict the existence of a link after time t on the basis of the edges generated until time t. Formally, the link prediction problem is

Definition 3. *(Link Prediction). A graph $G = (V, E_T^o, E_T^p, T)$ and two nodes $u, v \in V$ are given. Function Φ learning on the basis of E_T^o maps $\{u, v\}$ from E_T^p to 1 or 0:*

$$\Phi((u, v)) = \begin{cases} 1, & \text{then } e_{u,v} \in E_T^p \\ 0, & \text{otherwise} \end{cases}$$

where E_T^o is the set of the edges observed at time t, and E_T^p contains edges to be predicted after that time.

4   Time-Relaxed Temporal Random Walk(TxTWalk)

Given a temporal random walk (Definition 2), the temporal neighbourhood of node v is defined:

$$\Gamma_t(v) = \{(w, t') \mid e_{v,w} = (v, w, t') \in E_T \ \wedge \ t' \geq t\} \tag{1}$$

Here $\Gamma_t(v)$ is the set of node v's temporal neighbours after time t. $e_{v,w}$ represents the edge that starts at v and ends at w, generated at time t'. Temporal random walk only allows the selection of the next node within the temporal neighborhood. The rationale behind such neighborhood, as adopted in CTDNE [13], is that this definition can ensure the random walk respects the edge generation time and works logically in networks transmitting flows. A typical example is, in an email network, a person cannot forward an unreceived message to someone else in this network.

However, in a Users-Goods network as shown in Fig. 2, if the initial edge formed in a temporal random walk is from User 1 to Goods 1, then random walk will not be able to obtain the information of other nodes in Fig. 2, for the timestamp of the next edge in a walk needs to be strictly larger than that of the previous edge. But in our common sense, since User 1 and User 2 have bought the same goods, Goods 1 purchased by User 1 should be similar to Goods 3, which has been purchased by User 2.

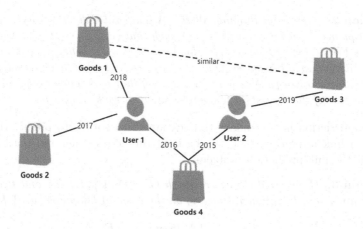

Fig. 2. An example of Users-Goods network.

To address the above-mentioned issue, **T**ime-Relaxed **T**emporal Random **W**alk(**TxTWalk**) has been proposed for continuous-time dynamic network embedding, which is inspired by the idea of image repairing [7]. In image repairing problem, to repair a given image, the input for training a neural network model are the original image together with all motion frames one second before the original image. Motivated by this, it is meaningful to use the edges whose timestamp is earlier than that of the current edge. Therefore, the timestamp restriction of selecting the next edge in temporal random walk can be relaxed by the following function:

$$\Gamma_t(v) = \{(w,t') \mid e = (v,w,t') \in E_T \ \wedge \ t' \geq \Theta(w,t)\} \tag{2}$$

$$\Theta(w,t) = t * max(\gamma, min(\sigma, \frac{t_{max} - t + 1}{t - t_{min} + 1})) \tag{3}$$

Here, $\Theta(w,t)$ is called time-relaxed function, and σ and γ are two parameters that can be manually adjusted based on the temporal information of a network. When the initial time(the timestamp of the first edge in random walk) is large and the time span in T is small, σ should be set to be large(but still smaller than 1) and γ be small(larger than 0). These two parameters are used to control maximum and minimum degree of relaxation. In relaxed function $\Theta(w,t)$, t_{max} and t_{min} represent the maximum and minimum timestamps of all edges starting with node w, respectively. $\frac{t_{max} - t + 1}{t - t_{min} + 1}$ ensures random walk can relax the time restriction of selecting the next edge when the timestamp of currently selected edge is large. It is not difficult to show that CTDNE [13] is in fact a special case of **TxTWalk** when $\sigma = 0$ and $\gamma = 1$.

Algorithm 1: Random Walk Sampling

Input: A continuous-time dynamic network $G = \{V, E_T, T\}$, initial edge (s,r), number of context windows X, maximum walk length l.

Output: Random Walk $S_t = \{v_i, v_{i+1}, ..., v_{i+l}\}$

1 $S_t \leftarrow (s,r)$;
2 $i \leftarrow r$;
3 **for** $k = 1 \rightarrow min(l, X)$ **do**
4 $\quad$ $\Theta(i,t) = t * min(\sigma, \frac{t_{max} - t + 1}{t - t_{min} + 1})$;
5 $\quad$ $\Gamma_t(i) = \{(w,t') \mid e = (i,w,t') \in E_T \ \wedge \ t' \geq \Theta(w,t)\}$;
6 $\quad$ **if** $|\Gamma_t(i)| > 0$; **then**
7 $\quad\quad$ $(j,t') \leftarrow$ *select a random element in* $\Gamma_t(i)$ *using equation 4*;
8 $\quad\quad$ Append j to S_t;
9 $\quad\quad$ $t \leftarrow t'$;
10 $\quad\quad$ $i \leftarrow j$;
11 $\quad$ **else**
12 $\quad\quad$ terminate random walk
13 $\quad$ **end**
14 **end**
15 **return** S_t;

During random walk sampling, the probability of each edge being selected is defined:

$$Pi(e) = \frac{exp(t_e - \Theta(w,t))}{\sum_{e' \in E_T} exp(t_{e'} - \Theta(w,t))} \tag{4}$$

Here Pi represents the probability that the edge e is selected. According to this equation, edges whose timestamp are larger can be more easily selected. As a result, multiple random walks can be sampled in a dynamic network by Eqs. 2 and 4.

The main advantage of **TxTWalk** over traditional static network embedding methods such as Deepwalk [15] is that it can effectively utilize the temporal information contained in dynamic networks. Compared with the dynamic network representation learning methods as CTDNE [13], **TxTWalk** has an advantage to avoid the insufficiency of random walk sampling.

Finally, the skip-gram model [12] is used to learn network embedding. Given a node sequence, the optimization of training skip-gram model is defined as follows:

$$minimize \; -logPr(\{v_{i-w}, ..., v_{i-1}, v_{i+1}, ..., v_{i+w}\} \mid \Phi(v_i)) \qquad (5)$$

here $\Phi(v_i)$ is the vector representation of node v_i. Node sequences sampled by random walk are fed into the skip-gram model to learn vector representations of nodes (network embedding).

Algorithm 2: TxTWalk Network Embedding

Input: A continuous-time dynamic network $G = \{V, E_T, T\}$, number of context windows X, minimum walk length ω, maximum walk length l, maximum number of walk μ, set of temporal random walk H_t, embedding dimension D.

Output: Node embedding matrix **Z**.

1 **while** $\mu - X > 0$ **do**
2   Sample an edge e = (u,v,t) by uniform distribution;
3      $S_t = RandomWalkSampling(G, e = (u, v), X = \omega + \mu - C - 1, l)$;
4   **if** $|S_t| > \mu$; **then**
5 | Append S_t to H_t; $X \leftarrow X + |S_t| - \omega + 1$;
6   **end**
7 **end**
8 $\mathbf{Z} = $ Skip-gram(μ, D, H_t);
9 **return Z**;

The learning algorithm is shown in Algorithm 1 and Algorithm 2, where ω is the minimum length of temporal random walk and is equivalent to the context window size for skip-gram [12].

5 Experiments

Experiments on five network datasets are conducted to demonstrate the effectiveness of the proposed **TxTWalk** model. Based on the experimental results, the following questions are to be answered.

- **Q1** Is **TxTWalk** model better than the baseline methods on the tested networks (especially better than **CTDNE** [13], an effective dynamic network embedding method)?
- **Q2** How different values of the parameters σ and γ can affect the performance of **TxTWalk**?

5.1 Experimental Setup

Datasets. The proposed **TxTWalk** model for dynamic network embedding is tested by applying to link prediction on five real-world network datasets of different scales. The datasets are listed with their topological information in Table 1 and detailed in the following.

Table 1. Topological information of the datasets

Data	$|V|$	$|E|$	Average degree
Enron	150	268	3.72
Forum	899	33720	74
FB-messages	1899	61734	63
CollegeMsg	1899	59835	112
ca-cit-HEPTH	22907	2673133	233

- **Enron** [18] is a email communication network. Nodes in this network represent employees while each edge represents a email between two employees.
- **Forum** [14] is a network that records users activities on Facebook. Nodes in this network indicate unique users and edges from u to v point out user u follows the user v.
- **FB-messages** [5] is a social network of students at University of California. The network includes the users that sent or received at least one message. Nodes in this network represent users while edges represents communications between users.
- **CollegeMsg** [16] is a network of messages sending at the University of California, Irvine. Nodes are users that could search for others to initiate conversations. An edge (u, v, t) means user u sent a private message to user v at time t.
- **ca-cit-HEPTH** [18] is from arXiv and covers all paper citations. Edge from paper u to paper v indicate that paper u cites another paper v.

Baseline Methods. For comparison, **TxTWalk** are compared to the following network embedding methods that either use temporal information or not.

- **Deepwalk** [15] is a static network embedding method which firstly treats nodes as words and generates short random walks as sentences to learn vector representations of nodes.

- **LINE** [19] defines the first-order and the second-order similarity for nodes, and optimizes the skip-gram model to learn nodes representations. It is also a static network embedding method.
- **node2vec** [8] employs a second order biased random walk controlled by two hyper-parameters to obtain node sequences for learning static network embedding.
- **CTDNE** [13] is a dynamic network embedding method which can make use of temporal information by temporal random walk.

Evaluation Metric. In the experiments, Area Under Curve (**AUC**) is used as the evaluation metric, which represents the area under the Receiver Operating Characteristic (**ROC**) curve [17]. The model performs better if **AUC** is closer to 1.

Parameter Settings. In all experiments, the parameters $\sigma = 0.85$ and $\gamma = 0.6$. The number of walks $\mu = 80$ and the length of walk $l = 10$. For all baseline methods, other parameters are set to the suggested values in the original papers, and the embedding dimension is fixed to 128.

5.2 Experimental Results

Link Prediction. To test the effectiveness of **TxTWalk** model, it was firstly trained to learn the vector embedding of each node. Then, the feature vector of an edge is computed by averaging the vectors of the two ending nodes of that edge. Logistic regression (LR) with hold-out validation of 25% edges is used on all datasets. Experiments are repeated for 10 random initialization and the averaged results compared with other baseline methods are shown in Table 2. From Table 2, **TxTWalk** outperforms almost all the baseline methods on all networks no matter whether the temporal information is used or not, which answers the question **Q1**. It is interesting to find that node2vec is slightly better than **TxTWalk** when tested on CollegeMsg network. By carefully examining the network, it is found that large portion of the edge associated times in CollegeMsg network falls in a narrow time interval, indicating that the temporal information in this network plays less important role.

Table 2. AUC score compared to baselines with 75% training links.

Data	Deepwalk	LINE	node2vec	CTDNE	TxTWalk
Enron	82.12 ± 0.58	83.04 ± 0.48	83.70 ± 0.57	84.12 ± 0.65	**87.02** $\pm$ **0.60**
Forum	71.30 ± 0.42	72.48 ± 0.43	71.10 ± 0.41	74.53 ± 0.61	**77.33** $\pm$ **0.50**
FB-messages	68.00 ± 0.38	67.74 ± 0.57	67.54 ± 0.62	83.88 ± 0.60	**87.75** $\pm$ **0.44**
CollegeMsg	90.22 ± 0.67	90.74 ± 0.68	**92.37** $\pm$ **0.71**	89.88 ± 0.54	91.05 ± 0.52
ca-cit-HEPTH	66.97 ± 0.65	67.51 ± 0.27	68.42 ± 0.48	83.51 ± 0.50	**84.41** $\pm$ **0.41**

When the temporal information is taken into account, it also shows that **TxTWalk** is more effective than CTDNE [13]. To fully reveal the superiority of **TxTWalk** to CTDNE on link prediction, different proportions of edges are used to form training set of FB-messages network. The results are shown in Table 3. From Table 3, **TxTWalk** works better in FB-messages than **CTDNE** under different proportions of training links, which shows that **TxTWalk** is more effective than **CTDNE**.

Table 3. The comparison between **TxTWalk** and **CTDNE** on link prediction with different proportion of training links of FB-messages network.

Data	30%	45%	60%	75%	90%
TxTWalk	**80.02** ± 0.48	**82.85** ± 0.50	**85.27** ± 0.42	**87.75** ±0.44	**88.12** ± 0.62
CTDNE	79.78 ± 0.49	80.81 ± 0.44	82.01 ± 0.46	83.88 ± 0.60	84.22 ± 0.52

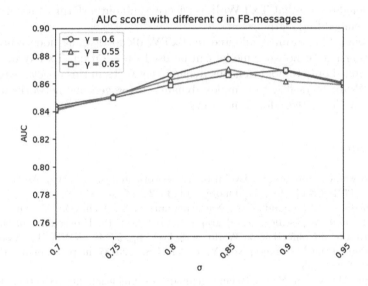

Fig. 3. Parameters sensitivity analysis

Parameters Sensitivity Analysis. For the question **Q2**, the impact of different values of the parameters is evaluated on FB-messages network. Different values of $\sigma = \{0.7, 0.75, 0.8, 0.85, 0.9, 0.95\}$ and $\gamma = \{0.55, 0.6, 0.65\}$ are searched respectively, and results of link prediction on FB-messages network with 75% training link are given. As shown in Fig. 3, with the increase of σ value, AUC first increases and then decreases, and the score is highest when $\gamma = 0.6$ and $\sigma = 0.85$. When σ is too large, as can be shown in Eq. 3, σ takes no effect. If σ is too small then $min(\sigma, \frac{t_{max}-t+1}{t-t_{min}+1}) = \sigma$, leading random walk to obtain useless information. Similarly, the value of γ can not be too large or too small, the reason of which is similar to that of σ. Therefore, $\gamma = 0.6$ and $\sigma = 0.85$ are the optimal setting of parameters, which were used in all experiments.

6   Conclusions and Discussions

In this article, a novel dynamic network embedding model **TxtWalk** is proposed. **TxtWalk** employs time-relaxed temporal random walk to sample node sequences that respect the times of edge emergence. In contrast to previous random walk based dynamic network embedding methods, such as CTDNE, requiring strict time respecting, **TxTWalk** relaxes such requirement to allow edges appear earlier within a small interval to be sampled to produce node sequences. **TxTWalk** first uses a time-relaxed temporal random walk to sample a series of node sequences starting from given nodes, and then employs skip-gram model to learn network embedding. Experiments are conducted on five networks and the results demonstrate that **TxTWalk** is effective and outperforms state-of-the-art network embedding methods.

Although Skip-gram is adopted in **TxTWalk** for leaning node representations, other neural network model can be used. In addition, **TxTWalk** is currently applied only to homogeneous dynamic networks in this article, extensions of **TxTWalk** to deal with complex dynamic heterogeneous networks are also feasible. We leave these for future work.

References

1. Aggarwal, C., Subbian, K.: Evolutionary network analysis: a survey. ACM Comput. Surv. **47**(1) (2014). https://doi.org/10.1145/2601412
2. Ahmed, A., Shervashidze, N., Narayanamurthy, S., Josifovski, V., Smola, A.J.: Distributed large-scale natural graph factorization. In: Proceedings of the 22nd International Conference on World Wide Web, pp. 37–48. WWW '13, Association for Computing Machinery, New York, NY, USA (2013). https://doi.org/10.1145/2488388.2488393
3. Belkin, M., Niyogi, P.: Laplacian eigenmaps for dimensionality reduction and data representation. Neural Comput. **15**(6), 1373–1396 (2003). https://doi.org/10.1162/089976603321780317
4. Cao, S., Lu, W., Xu, Q.: Grarep: learning graph representations with global structural information. In: Proceedings of the 24th ACM International on Conference on Information and Knowledge Management, pp. 891–900. CIKM 2015, Association for Computing Machinery, New York, NY, USA (2015). https://doi.org/10.1145/2806416.2806512
5. Opsahl, T., Panzarasa, P.: Clustering in weighted networks. Soc. Netw. **31**(2), 155–163 (2009)
6. Fu, T.Y., Lee, W.C., Lei, Z.: Hin2vec: explore meta-paths in heterogeneous information networks for representation learning. In: Proceedings of the 2017 ACM on Conference on Information and Knowledge Management. p. 1797–1806. CIKM '17, Association for Computing Machinery, New York, NY, USA (2017). https://doi.org/10.1145/3132847.3132953

7. Godard, C., Mac Aodha, O., Firman, M., Brostow, G.J.: Digging into self-supervised monocular depth estimation. In: Proceedings of the IEEE/CVF International Conference on Computer Vision (ICCV) (October 2019)
8. Grover, A., Leskovec, J.: node2vec: scalable feature learning for networks. In: Proceedings of the 22nd ACM SIGKDD International Conference on Knowledge Discovery and Data Mining, pp. 855–864 (2016)
9. Hamilton, W.L., Ying, R., Leskovec, J.: Inductive representation learning on large graphs. In: Proceedings of the 31st International Conference on Neural Information Processing Systems, pp. 1025–1035 (2017)
10. Hisano, R.: Semi-supervised graph embedding approach to dynamic link prediction. In: Cornelius, S., Coronges, K., Gonçalves, B., Sinatra, R., Vespignani, A. (eds.) Complex Networks IX. CompleNet 2018. Springer Proceedings in Complexity, pp. 109–121. Springer, Cham (2018). https://doi.org/10.1007/978-3-319-73198-8_10
11. Li, J., Dani, H., Hu, X., Tang, J., Chang, Y., Liu, H.: Attributed network embedding for learning in a dynamic environment. In: Proceedings of the 2017 ACM on Conference on Information and Knowledge Management, pp. 387–396. CIKM 20017, Association for Computing Machinery, New York, NY, USA (2017). https://doi.org/10.1145/3132847.3132919
12. Mikolov, T., Chen, K., Corrado, G., Dean, J.: Efficient estimation of word representations in vector space (2013)
13. Nguyen, G.H., Lee, J.B., Rossi, R.A., Ahmed, N.K., Koh, E., Kim, S.: Continuous-time dynamic network embeddings. In: Companion Proceedings of the The Web Conference 2018, pp. 969–976. WWW 2018, International World Wide Web Conferences Steering Committee, Republic and Canton of Geneva, CHE (2018)
14. Opsahl, T.: Triadic closure in two-mode networks: redefining the global and local clustering coefficients. Soc. Netw. (2011)
15. Perozzi, B., Al-Rfou, R., Skiena, S.: Deepwalk: online learning of social representations. In: Proceedings of the 20th ACM SIGKDD International Conference on Knowledge Discovery and Data Mining, pp. 701–710 (2014)
16. Panzarasa, P., Opsahl, T., Carley, K.M.: Patterns and dynamics of users' behavior and interaction: network analysis of an online community. J. Am. Soc. Inf. Sci. Technol. (2009)
17. Provost, F., Fawcett, T.: Analysis and visualization of classifier performance: comparison under imprecise class and cost distributions. In: Proceedings of the 3rd International Conference on Knowledge Discovery and Data Mining, pp. 43–48 (1997)
18. Rossi, R.A., Ahmed, N.K.: The network data repository with interactive graph analytics and visualization. In: AAAI (2015). http://networkrepository.com
19. Tang, J., Qu, M., Wang, M., Zhang, M., Yan, J., Mei, Q.: Line: large-scale information network embedding. In: Proceedings of the 24th International Conference on World Wide Web, pp. 1067–1077 (2015)
20. Xie, Y., Li, C., Yu, B., Zhang, C., Tang, Z.: A survey on dynamic network embedding. CoRR abs/2006.08093 (2020). https://arxiv.org/abs/2006.08093

Dual-Band Maritime Ship Classification Based on Multi-layer Convolutional Features and Bayesian Decision

Zhaoqing Wu[1][✉], Yancheng Cai[2], Xiaohua Qiu[1], Min Li[1][✉], Yujie He[1], Yu Song[1], and Weidong Du[1]

[1] Xi'an Research Institute of Hi-Tech, Xi'an 710025, China
[2] School of Information Science and Technology, Fudan University, Shanghai 201203, China

Abstract. There are some problems arising from the classification of visible and infrared maritime ship, for example, the small number of image annotated samples and the low classification accuracy of feature concatenation fusion. To solve the problems, this paper proposes a dual-band ship decision-level fusion classification method based on multi-layer features and naive Bayesian model. To avoid the occurrence of over-fitting caused by the small number of annotated samples, the proposed method is adopted. First of all, a convolutional neural network (CNN) which has been pre-trained on ImageNet dataset is used and fine-tuned to extract convolutional features of dual-band images. Then, principal component analysis is conducted to reduce the dimension of convolutional feature while L2 normalization is applied to normalize the features after dimensionality reduction. Meanwhile, multi-layer convolutional feature fusion is conducted through the period. In doing so, not only storage and computing resources is reduced, the information of feature representation is also enriched. Finally, a Bayesian decision model is constructed using support vector machine and naive Bayesian theory, for the subsequent dual-band ship fusion classification. According to the experiments results on the public maritime ship dataset, the classification accuracy of the proposed decision-level fusion method reaches 89.8%, which is higher not only than that of the dual-band feature-level fusion by 1.0%–2.0%, but also than that of the state-of-the-art method by 1.6%.

Keywords: Image classification · Principal component analysis · Naive Bayes model · Multi-layer convolutional features · Decision-level fusion

1 Introduction

In design, the maritime ship surveillance system is intended to assist the detection, tracking and classification of ship in the maritime environment. At present, it has played an increasingly important role in various civil and military fields, such as maritime traffic control and national security maintenance [1, 2]. Since infrared images provide supplementary information for those visible images of the same scene, visible (VIS) and

T. Mantoro et al. (Eds.): ICONIP 2021, LNCS 13108, pp. 438–449, 2021.
https://doi.org/10.1007/978-3-030-92185-9_36

infrared (IR) dual-band images have demonstrated significant advantages over single-band images. Therefore, the maritime surveillance system capable of integrating visible and infrared spectrum sensors performs in all weather and different lighting conditions [3].

Up to now, there has been remarkable progress made in the research on how to perform various computer vision tasks for visible images through deep learning, with the well-known convolutional neural network (CNN) proposed, such as VGGNet [4], GoogleNet [5] and ResNet [6]. Most recently, deep CNN has been applied in various fields related to VIS and IR dual-band images, for example, image classification [7], face recognition [8], pedestrian detection [9], object detection [10], and object tracking [11]. However, training deep CNN requires a large-scale annotated image dataset like ImageNet and it is challenging to collect and label large-scale data. In the absence of large-scale annotated data, it is difficult to apply deep CNN in specific fields. Fortunately, deep CNN shows hierarchical features from bottom to top, or in other words, from pixel-level features to semantic-level features [12]. Therefore, feature extraction based on convolution and model fine-tuning based on the well-known deep CNN models have been widely used at many computer vision tasks [13] and ship classification in maritime system [14–17].

Kanjir et al. [18] conducted review of the ship classification and detection methods using deep learning for optical remote sensing images. In comparison, our work focuses on dual-band maritime ship classification. After creating a maritime ship dataset containing VIS and IR images (shorted as VAIS), Zhang et al. [19] conducted fusion on classification results of the methods based on Gnostic area and VGG-16 model [4], to determine the baseline accuracy of ship classification. Combining the high-level features of VGG-19 model [4] with Multi-scale Complete Local Binary Patterns, Shi Q et al. [20] put forward a multi-feature learning ship classification algorithm for the VIS images in VAIS dataset. Based on this algorithm, Huang et al. [21] relied on extreme learning machine (ELM) to improve the efficiency of ship classification. Shi et al. [22] put forward a method for ship classification based on the multiple feature ensemble learning of convolutional neural networks (ME-CNN). With regard to dual-band ship classification, Aziz et al. [23] used a large-scale VIS images ship dataset to carry out training and VAIS dataset to fine-tune. Santos et al. [24] proposed a decision-level classification framework based on pre-trained VGG-19 model and probabilistic fusion model. Adopting the different pre-trained deep CNN models to extract convolutional features from dual-band images, Zhang et al. [7] proposed a multi-feature fusion classification method based on spectrum regression, structured fusion and discriminant analysis (SF-SRDA).

In practice, it is difficult to collect the ship dataset of the same magnitude as ImageNet dataset. Though the occurrence of over-fitting can be avoided by training deep network on a large visible ship dataset, it is unlikely to ensure that the classification result is as excellent as that of fine-tuning well-known CNN model which is pre-trained on ImageNet dataset. Furthermore, it is common for over-fitting to occur when fine-tuning is performed on dual-band ship dataset [19], which includes few annotated samples. Besides, convolutional features concatenation fusion provides low classification performance for taking well-known pre-trained deep CNN models as feature extractor, and it is difficult to learn common feature representation for convolutional features of dual-band images.

PCA is a classical method that can be adopted to drop the dimensionality of data for machine learning [25]. Decision-level fusion is often relied on for multiple classifiers fusion on single-band image in computer vision [26]. Recently, it has been further applied to assist dual-band image-based object recognition, pedestrian detection and other computer vision tasks. In order to solve the problems arising from dual-band ship classification as mentioned above, the information complementary in hierarchical convolutional features of the pre-trained CNN model is used. A method of decision-level fusion for dual-band surface ship classification based on multi-layer convolutional features and naive Bayesian theory is proposed. It is referred to as MLF-BDF in this study.

2   Decision-Level Fusion Classification Model

First of all, the proposed method (MLF-BDF) extracts the convolutional features of dual-band image which come from each layer of CNN model. Pre-trained weight on ImageNet dataset is loaded and fine-tuned to adapt ship dataset. Secondly, principal component analysis (PCA) and L2 normalization are relied on to reduce the dimension of convolutional features and to normalize the values of features, respectively. Thirdly, the normalized features of different layers are concatenated, and then are inputted into support vector machine (SVM) classifier for confusion matrix calculation. Finally, naïve Bayesian decision model is constructed to output the ship classification label of the dual-band image. Figure 1 shows the overall framework.

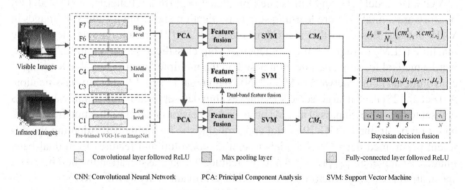

Fig. 1. An overview of decision-level fusion classification method

2.1   Feature Extraction

In our paper, the VGG-16 model is used to extract features from dual-band ship images, which is one of the most classic CNN. VGG network includes 8 layers, of which 5 layers named C1, C2, C3,C4 and C5 shown in Fig. 1 are convolution layers. The rest of it (F6–F8) are fully connected layers, which contains semantic information. The whole network architecture shows a bottom-up feature hierarchy. Layer C1 and C2 mainly extract low-level features, which are mostly visual features such as color, texture and edge. Since the low-level feature extraction is completed by calculating pixels, it is also called pixel-level feature. For higher layers (C3–C5), middle-level features are extracted from low-level features. Middle-level features mainly contain features such as circle, line and other basic geometry, which are more abstract than low-level features and include more semantic features. High-level features (F6–F8) which mainly comprised by semantic feature, are combined by middle-level features. High-level features can realize the understanding of image semantic information to a certain extent. VGG-16 model takes 3-channel image of 224 × 224 pixels input, and each ship image is put into convolution layers with the size of $r \times r \times K$, where $r \times r$ represents convolution kernel size, and K represents the number of convolution kernels. In this paper, all the convolution feature extracted from VGG-16 is flattened into a feature vector $\mathbf{f}_n \in \mathbb{R}^{r^2K} (n = 1, 2, \cdots N)$ for further processing, where N represents the number of samples.

2.2 Feature Dimensionality Reduction Using PCA

Containing both low, middle and high level feature, the extracted convolution features has high dimensionality, which not only occupies a lot of computing resources, but also contain noise as well as a lot of redundant information. In order to solve the problems, the convolution features are transformed by principal component analysis (PCA) from high-dimensional space into low-dimensional space. PCA aims to calculate the covariance matrix of high-dimensional features, and then decompose the eigenvalues to retain the eigenvectors corresponding to the first few eigenvalues. Thus, a low-dimensional feature space is created. There are two ways to obtain the low dimensional feature space. One is to set a fixed parameter for dimension, the other is to set a reconstruction threshold to control dimensionality reduction ratio.

The mathematical theory applied to set the reconstruction threshold is mainly based on the variance maximization theory. In image processing, image can be seen as a special type of signal. In most occasions, the useful signal which contains image information has a larger variance than noise. Therefore, the difference on variance can be used to dimensionality reduction. PCA can be seen as projection, which retain few eigenvectors with large variance and abandon eigenvectors with small variance. Meanwhile, the reconstruction threshold can be seen as the ratio of dimensions of large variance and small variance in essence.the reconstruction threshold can be expressed as Eq. (1).

$$\frac{\sum_{d=1}^{d_1} \lambda_d}{\sum_{d=1}^{d_h} \lambda_d} \geq t \tag{1}$$

Where λ_d represents the d-th eigenvalue in the covariance matrix, $d_h = r^2K$ indicates the size of high-dimensional space dimension, d_1 denotes the size of low-dimensional space dimension, and t refers to the reconstruction threshold.

Since the value range of convolution features of different layers various from each other in VGG-16 model for dual-band ship images, and meanwhile the number of images in VAIS dataset is significantly less than the size of convolutional feature dimension, it is important to set a reconstruction threshold for low-dimensional space.

2.3 Single-Band Multi-layer Features Fusion

The range of feature value is different from each other due to the difference of layer in VGG-16 network, causing difficulties for further feature fusion. Therefore, after PCA dimensionality reduction, it is necessary to do the normalization for each layer into a uniform value range. Furthermore, the normalization process after PCA is beneficial to improve the classification experiment performance. In our paper, the L2 norm is adopted to normalize features. First, calculate the L2 normalization for each sample. Then, to make the L2 normalization of each sample equal to 1, every element in the sample is divided by the L2 normalization. Set $\mathbf{f}'_n = \left(f'_1, f'_2, \cdots, f'_{d_1} \right)^{\mathrm{T}}$ as the n-th sample in the low-dimensional space, and $\mathbf{f}''_n = \left(f''_1, f''_2, \cdots, f''_{d_1} \right)^{\mathrm{T}}$ as the feature vector of the n-th sample normalized by the L2 normalization. The m-th element f''_m in the $\mathbf{f}''_n$ vector is calculated by Eq. (2).

$$f''_m = \frac{f'_m}{\left(\left| f'_1 \right|^2 + \left| f'_2 \right|^2 + \cdots + \left| f'_{d_1} \right|^2 \right)^{\frac{1}{2}}}, \quad m = 1, 2, \cdots d_1 \tag{2}$$

There are two classical feature fusion strategy, one of it is the serial strategy, the other one is the parallel strategy. Especially note that, the serial strategy requires the features having the same dimension, while the parallel strategy do not. Therefore, due to the difference on dimension between the PCA features from each layer of VGG, the parallel strategy is applied in this paper. Features of each sample are fused by concatenation based on the parallel strategy. The fusion processing can be calculated by Eq. (3).

$$f_f = (f''_{11}, f''_{12}, \cdots, f''_{1d_l}, \cdots, f''_{L1}, f''_{L2}, \cdots, f''_{Ld_l})^{\mathrm{T}} \tag{3}$$

Where, f_f represents feature vector after concatenation, f''_{Ld_l} indicates the d_l-th element in PCA feature vector from the L-th layer in CNN. Besides, in our method, L often set to 2 or 3.

2.4 Dual-Band Bayesian Decision Fusion Model Construction

Linear kernel-based SVM is adopted to calculate the posterior probability of each sample, while naive Bayes model is applied to perform fusion classification for obtaining the common classification label. Naive as a classification algorithm based on Bayesian theory and conditional independence assumption, Bayes model adopts a basic theory expressed as follows.

Assume that $\omega_k (k = 1, 2, \cdots c)$ is sample class list of dual-band image dataset, s_1 and s_2 are two independent SVM classifiers corresponding to each band image, which means $\mathbf{S} = \{s_1, s_2\}$. Denoted $P(s_m)$ the posterior probability of m-th SVM classifier labels the sample x in class ω_k. According to conditional independence assumption, the conditional probability $P(\mathbf{S}|\omega_k)$ can be calculated using Eq. (4).

$$P(\mathbf{S}|\omega_k) = P(s_1, s_2|\omega_k) = P(s_1|\omega_k)P(s_2|\omega_k) \tag{4}$$

Then, the posterior probability of the sample x in the proposed decision fusion model can be calculated using the prior probability and the conditional probability, as shown in Eq. (5).

$$P(\omega_k|S) = \frac{P(\omega_k)P(S|\omega_k)}{P(S)} = \frac{P(\omega_k)P(s_1|\omega_k)P(s_2|\omega_k)}{P(S)} \tag{5}$$

The denominator $P(S)$ does not depend on ω_k in Eq. (4) and can be ignored, while the support $\mu_k(x)$ of the sample x for class ω_k can be expressed as Eq. (6). The final class of the sample x is the class that maximum of $\mu = (\mu_1, \mu_2, \cdots \mu_c)$ corresponding to

$$\mu_k(x) \propto P(\omega_k)P(s_1|\omega_k)P(s_2|\omega_k) \tag{6}$$

The implementation of Bayesian decision fusion method on N samples is detailed as follows. The dual-band classifiers s_1 and s_2 are calculated on the basis of testing samples to obtain $c \times c$ confusion matrixes CM_1 and CM_2, respectively. The (k, s)-th entry of each CM, $cm_{k,s}^m$ is the number of elements of the testing samples whose true class is ω_k. It is assigned by the classifier to class ω_s. N_k denotes the total number of samples from class ω_k in the testing samples, $cm_{k,s}^m / N_k$ is treated as an estimate of the posterior probability, and N_k/N is viewed as an estimate of the prior probability. Then, the final support of the sample x for ω_k is obtained as Eq. (7). According to the maximum rule of $\mu = (\mu_1, \mu_2, \cdots \mu_c)$, the sample x will be labeled as class ω_k.

$$\mu_k(x) \propto \frac{1}{N_k}\left(cm_{k,s_1}^1 \times cm_{k,s_2}^2\right) \tag{7}$$

3   Experimental Results

3.1   Dataset

The only public VIS and IR dual-band maritime ship dataset referred to as VAIS is used to assess the performance of the proposed method. The dataset is comprised of 2865 ship images (1623 VIS images and 1242 IR images), including 1088 pairs of unregistered VIS-IR images. It has six ship categories, which are sailing, merchant, passenger, medium, small and tug. The dataset contains not only images taken during the day, but also images taken at night, dusk and dawn. Therefore, some samples in the dataset are blurred, setting difficulties for algorithms to recognize. According to reference [10], we separated the dataset into two parts, 549 pairs of VIS-IR images for testing and 539 pairs for training. The number for each category of testing and training samples is shown in Table 1. Some pairs of VIS and IR images is shown in Fig. 2.

3.2   Implementation Details

The implementation configuration is as follow: CPU i7-7700K 4.2 GHz processor, 31 Gb memory, GPU Geforce GTX1080 Ti, 64-bit operating system, and the experimental environment is Linux Ubuntu 16.04, Python3.6, Pycharm community condition 2019.3. Visual environment includes Keras, Tensorflow backend and the "sklearn" library which integrates SVM modules and PCA. Besides, we set the parameter t of PCA module to 0.99.

Table 1. Detail information for VAIS dataset testing and training

VAIS	Sailing	Merchant	Passenger	Medium	Small	Tug	Total
Testing	136	63	59	76	195	20	549
Training	148	83	58	62	158	30	539

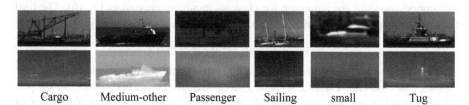

Cargo   Medium-other   Passenger   Sailing   small   Tug

Fig. 2. Examples of VAIS dataset

3.3 Analysis of Experimental Results

3.3.1 PCA Performance Analysis

Since the time complexity of linear SVM is linear with feature dimension of input samples, classification can be accelerated by reducing the feature dimensionality of input samples. In our experiments, the effect of PCA is assessed by analyzing and comparing the dimension and classification accuracy of PCA features and CNN features. Table 2 details the dimensional comparison between CNN features and PCA features of each layer on VGG-16 model. Figure 3 shows the comparison of experimental accuracy results between PCA features and CNN features of each layer on VGG-16 model. As shown in Table 2, after dimensionality reduction, the dimensions of PCA features are significantly smaller than that of CNN features. Meanwhile, as shown in Fig. 3(a), PCA method is generally effective in improving the accuracy results of CNN features for each band. There are two points revealed by comparing the classification accuracy of decision-level fusion (DF) and feature-level fusion (FF) based on the PCA features and CNN features in Fig. 3(b). On the one hand, the classification accuracy of fusion method based on PCA features is higher as compared to the fusion method based on CNN features. On the other hand, the classification result of decision-level fusion is higher than that of feature-level fusion, whether it is based on PCA features or CNN features. In summary, the PCA method is effective in reducing the dimension of CNN features.

Table 2. Dimensional comparison of PCA features and CNN features for each layer

Layer	C2	C3	C4	C5	F6	F7
CNN feature	401,408	200,704	100,352	25,088	4,096	4,096
PCA feature of VIS	469	478	477	463	427	388
PCA feature of IR	331	428	429	373	275	229

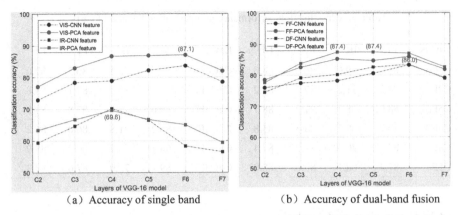

(a) Accuracy of single band   (b) Accuracy of dual-band fusion

Fig. 3. Classification accuracy comparison of PCA features and CNN features for each layer

3.3.2 Performance Analysis of Multi-layer Convolutional Feature Fusion

As for low-level features, they include the visual information of an image. Not only do middle-level features carry visual information, they also contain some semantic level information. In comparison, high-level features represent more semantic and abstractive information than middle-level features. Considering middle-level features maintain more semantic information compared to low-level features, we conduct concatenation on high-level features together with middle-level features in multi-layer feature fusion. Table 3 details the comparison between two layers and three layers of convolutional feature fusion on classification accuracy. In this table, C3F6 indicates that the convolutional features from C3 and F6 layers on VGG-16 model are concatenated, which is the same as other abbreviations.

As shown in Table 3 and Fig. 3(c), the classification accuracy of DF based on multi-layer features is higher as compared to the DF based on each single layer. For example, the classification accuracy reaches 83.8% and 82.5% for C3 and F7, respectively, despite the classification accuracy of C3F7 improved by 2.0% to 85.8%. Moreover, in comparison with two-layer feature fusion, three-layer feature fusion performs better in improving classification accuracy for DF. In addition, relative to the classification accuracy of FF and DF in Table 3, the classification accuracy of FF is lower than that of VIS images, while the classification accuracy result of DF is higher than that of FF by 1.0%-2.0%. In the experiments, the classification accuracy was not improved despite the attempted introduction of low-level features into the fusion of multi-layer features, suggesting that the middle-level features carry more abundant and abstractive visual information than features from low level. In summary, the multi-layer feature fusion conducted on middle-level as well as high-level features is conducive to improving the level of classification accuracy based on single convolutional layer in dual-band DF.

Table 3. Classification performance comparison of DF and FF(%)

Layers	VIS	IR	FF	DF	Layers	VIS	IR	FF	DF
C3F6	88.0	69.6	86.0	88.0	C3C4F6	87.4	72.1	86.5	87.6
C4F6	88.2	71.2	87.1	88.2	C3C5F6	89.4	70.3	87.1	89.8
C5F6	88.0	68.3	87.2	88.7	C4C5F6	89.6	71.8	87.8	89.8
C3F7	84.7	68.7	84.3	85.8	C3C4F7	86.2	71.8	86.0	86.2
C4F7	87.1	70.3	87.2	87.4	C3C5F7	88.7	70.7	86.2	88.9
C5F7	87.4	68.1	86.0	88.0	C4C5F7	89.1	72.9	87.6	89.3

3.3.3 Comparison and Analysis

The method proposed in this study is compared against seven existing methods. Table 4 shows the comparison of classification results with other classic methods in VAIS dataset. According to these methods, CNN + Gnostic Fields, Multimodal CNN, DyFusion and SF-SRDA are based on dual-band VIS and IR image, while CNN + Gabor + MS-CLBP, MFL (feature-level) + ELM and ME-CNN are premised on single-band VIS image. As shown in Table 4, the highest accuracy achieved the proposed MLF-BDF is 89.8%, which is significantly higher than other methods, including the state-of-the-art method by 1.6%. The SF-SRDA method achieved the best classification accuracy with 74.7% on single-band IR images. Figure 4 shows the decision-level fusion (C3C5F6) and (C4C5F6) normalized matrices with a classification accuracy of 89.8%. According to Fig. 4, the classification accuracy for cargo ships and sailing boats reaches 100%, while it is 89% for small boats. In comparison, the accuracy of other intermediate ships and tugs is basically lower, reaching 70% and 75%, respectively.

Table 4. Comparison of classification results on VAIS dataset (%)

Method	Year	VIS	IR	VIS + IR
CNN	2015	81.9	54.0	82.1
Gnostic Field	2015	82.4	58.7	82.4
CNN + Gnostic Field	2015	81.0	56.8	87.4
MFL (feature-level) + ELM	2018	87.6	–	–
CNN + Gabor + MS-CLBP	2018	88.0	–	–
Multimodal CNN	2018	–	–	86.7
DyFusion	2018	–	–	88.2
ME-CNN	2019	87.3	–	–
SF-SRDA	2019	87.6	**74.7**	88.0
Proposed MLF-BDF (C3C5F6)	2021	89.4	70.3	**89.8**
Proposed MLF-BDF (C4C5F6)	2021	**89.6**	71.8	**89.8**

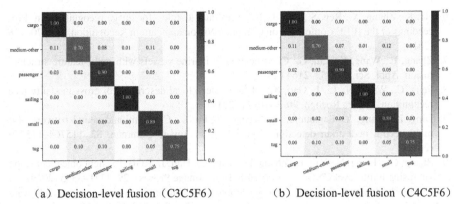

(a) Decision-level fusion (C3C5F6)   (b) Decision-level fusion (C4C5F6)

Fig. 4. Normalization confusion matrix for the best accuracy of decision-level fusion

4   Conclusion

Since dual-band images contain more object information than single-band images, dual-band ship classification is studied in this paper. With multi-layer convolution features extracted from the pre-trained well-known CNN models, not only is the occurrence of over-fitting caused by the small number of annotated samples of dual-band dataset avoided, the complementary information of hierarchical convolutional features is also leveraged. Besides, PCA and L2 normalization methods are applied to reduce storage and computing resources while improving classification performance. As for the decision-level fusion based on multi-layer features and naive, it solves the low classification accuracy for feature-level fusion in dual-band images and demonstrates its advantage over other existing methods. Finally, the above conclusions are substantiated by the experimental results obtained on the dataset VAIS. Moreover, the proposed fusion method is verified as effective. In the future, the focus of research will be placed on learning common feature representation for the convolutional features of dual-band images.

References

1. Oliveau, Q.: Ship classification for maritime surveillance. In: OCEANS 2019-Marseille, pp. 1–5. IEEE (2019)
2. Zhang, X., Lv, Y., Yao, L., Xiong, W., Fu, C.: A New benchmark and an attribute-guided multilevel feature representation network for fine-grained ship classification in optical remote sensing images. IEEE J. Sel. Topics Appl. Earth Obs. Remote Sens. **13**, 1271–1285 (2020)
3. Zhenzhen, L., Baojun, Z., Linbo, T., Zhen, L., Fan, F.: Ship classification based on convolutional neural networks. J. Eng. **2019**, 7343–7346 (2019)
4. Simonyan, K., Zisserman, A.: Very deep convolutional networks for large-scale image recognition (2014). arXiv preprint arXiv:1409.1556
5. Szegedy, C., Vanhoucke, V., Ioffe, S., Shlens, J., Wojna, Z.: Rethinking the inception architecture for computer vision. In: Proceedings of the IEEE Conference on Computer Vision and Pattern Recognition, pp. 2818–2826 (2016)

6. He, K., Zhang, X., Ren, S., Sun, J.: Deep residual learning for image recognition. In: Proceedings of the IEEE Conference on Computer Vision and Pattern Recognition, pp. 770–778 (2016)

7. Zhang, E., Wang, K., Lin, G.: Classification of marine vessels with multi-feature structure fusion. Appl. Sci. **9**, 2153–2164 (2019)

8. Peng, C., Wang, N., Li, J., Gao, X.: DLFace: Deep local descriptor for cross-modality face recognition. Pattern Recogn. **90**, 161–171 (2019)

9. Ding, L., Wang, Y., Laganiere, R., Huang, D., Fu, S.: Convolutional neural networks for multispectral pedestrian detection. Signal Process. Image Commun. **82**, 115764–115779 (2020)

10. Zhang, Q., Huang, N., Yao, L., Zhang, D., Shan, C., Han, J.: RGB-T salient object detection via fusing multi-level CNN features. IEEE Trans. Image Process. **29**, 3321–3335 (2019)

11. Zhang, H., Zhang, L., Zhuo, L., Zhang, J.: Object tracking in rgb-t videos using modal-aware attention network and competitive learning. Sensors **20**, 393–348 (2020)

12. Yosinski, J., Clune, J., Bengio, Y., Lipson, H.: How transferable are features in deep neural networks (2014). arXiv preprint arXiv:1411.1792

13. Ren, S., He, K., Girshick, R., Sun, J.: Faster R-CNN: towards real-time object detection with region proposal networks. Adv. Neural. Inf. Process. Syst. **28**, 91–99 (2015)

14. Dao-Duc, C., Xiaohui, H., Morère, O.: Maritime vessel images classification using deep convolutional neural networks. In: Proceedings of the Sixth International Symposium on Information and Communication Technology, pp. 276–281 (2015)

15. Gundogdu, E., Solmaz, B., Yücesoy, V., Koç, A.: Marvel: a large-scale image dataset for maritime vessels. In: Lai, S.-H., Lepetit, V., Nishino, K., Sato, Y. (eds.) ACCV 2016. LNCS, vol. 10115, pp. 165–180. Springer, Cham (2017). https://doi.org/10.1007/978-3-319-54193-8_11

16. Solmaz, B., Gundogdu, E., Yucesoy, V., Koc, A.: Generic and attribute-specific deep representations for maritime vessels. IPSJ Trans. Comput. Vision Appl. **9**(1), 1–18 (2017). https://doi.org/10.1186/s41074-017-0033-4

17. Milicevic, M., Zubrinic, K., Obradovic, I., Sjekavica, T.: Application of transfer learning for fine-grained vessel classification using a limited dataset. In: Ntalianis, K., Vachtsevanos, G., Borne, P., Croitoru, A. (eds.) APSAC 2018. LNEE, vol. 574, pp. 125–131. Springer, Cham (2019). https://doi.org/10.1007/978-3-030-21507-1_19

18. Liu, Y., Cui, H.-Y., Kuang, Z., Li, G.-Q.: Ship detection and classification on optical remote sensing images using deep learning. In: ITM Web of Conferences, pp. 5012–5025. EDP Sciences (2017)

19. Zhang, M.M., Choi, J., Daniilidis, K., Wolf, M.T., Kanan, C.: VAIS: a dataset for recognizing maritime imagery in the visible and infrared spectrums. In: Proceedings of the IEEE Conference on Computer Vision and Pattern Recognition Workshops, pp. 10–16 (2015)

20. Shi, Q., Li, W., Zhang, F., Hu, W., Sun, X., Gao, L.: Deep CNN with multi-scale rotation invariance features for ship classification. IEEE Access **6**, 38656–38668 (2018)

21. Huang, L., Li, W., Chen, C., Zhang, F., Lang, H.: Multiple features learning for ship classification in optical imagery. Multimedia Tools Appl. **77**(11), 13363–13389 (2017). https://doi.org/10.1007/s11042-017-4952-y

22. Shi, Q., Li, W., Tao, R., Sun, X., Gao, L.: Ship classification based on multifeature ensemble with convolutional neural network. Remote Sens. **11**, 419 (2019)

23. Aziz, K., Bouchara, F.: Multimodal deep learning for robust recognizing maritime imagery in the visible and infrared spectrums. In: Campilho, A., Karray, F., ter Haar Romeny, B. (eds.) ICIAR 2018. LNCS, vol. 10882, pp. 235–244. Springer, Cham (2018). https://doi.org/10.1007/978-3-319-93000-8_27

24. Santos, C.E., Bhanu, B.: Dyfusion: dynamic IR/RGB fusion for maritime vessel recognition. In: 2018 25th IEEE International Conference on Image Processing (ICIP), pp. 1328–1332. IEEE (2018)
25. Li, C., Ren, J., Huang, H., Wang, B., Zhu, Y., Hu, H.: PCA and deep learning based myoelectric grasping control of a prosthetic hand. Biomed. Eng. Online **17**, 1–18 (2018)
26. Sun, Y., et al.: Image classification base on PCA of multi-view deep representation. J. Vis. Commun. Image Represent. **62**, 253–258 (2019)

Context-Based Anomaly Detection via Spatial Attributed Graphs in Human Monitoring

Kang Zhang[(✉)], Muhammad Fikko Fadjrimiratno, and Einoshin Suzuki[(✉)]

Kyushu University, Fukuoka, Japan
`zhang.kang.737@s.kyushu-u.ac.jp, suzuki@inf.kyushu-u.ac.jp`

Abstract. Detecting anomalous regions focusing on interesting parts in surveillance images is a significant task in human monitoring. The most challenging problem is that a large number of anomalies in monitoring environments are context-based, which requires combined information between regions and contexts. Existing anomaly detection methods either capture normal patterns from individual regions or detect anomalies based on their defined relations among regions. Therefore, they cannot effectively capture the underlying relationships among regions to discriminate the context-based anomalies. In this paper, we propose to construct spatial attributed graphs to model the regions and contexts and devise a novel graph auto-encoder framework named Context-based Anomaly Detection Auto-Encoder (CADAE) to detect anomalous regions. Specifically, we assume that the context of a region is composed of its spatial relationships with other regions in the same image. The region and its context in an image are represented as a node and its neighborhoods in a spatial attributed graph, respectively. CADAE is first trained to learn normal patterns from nodes in graphs by compressing the representations of the nodes and their neighborhoods, and reconstructing the node attributes and structure information. CADAE then ranks the degrees of the abnormality of testing nodes by reconstruction errors. Unlike existing graph auto-encoders, by adopting a specified sum aggregator, CADAE can discriminate nodes whose neighbors have similar representations and thus capture the diversity of contexts although similar regions often exist in surveillance images. Experimental results on real-world datasets demonstrate the effectiveness of CADAE.

Keywords: Anomaly detection · Human monitoring · Graph modeling · Graph neural networks

1 Introduction

Anomaly detection in human monitoring aims to discover unexpected activities, including behaviors between humans and interactions between humans and objects, which deviate from normal patterns. In particular, detecting anomalous

T. Mantoro et al. (Eds.): ICONIP 2021, LNCS 13108, pp. 450–463, 2021.
https://doi.org/10.1007/978-3-030-92185-9_37

regions to track the most interesting parts in surveillance images is a critical task in this field. However, it is challenging to detect such a region-level anomaly due to its diverse contextual information and lack of supervision. The relationships among regions in the same image, which represent the underlying consistency between regions and contexts, are significant to understand the abnormality of regions, i.e., context-based anomalies.

Therefore, based on different contexts, similar regions can be defined as normal or abnormal instances. Such context-based anomalies widely exist in human monitoring since some activities are allowed in one area but forbidden in another area. For example, making phone calls (regions with red and green boxes in Fig. 4) is normal in the resting area but abnormal in the working area as it is not allowed and disturbing for other people.

Existing methods for region-level anomaly detection are mainly based on either mining normal patterns from individual image regions [10] or combining information of regions and their defined relationships among regions [6,16]. The former methods [10] only focus on learning variations of individual regions while neglecting the consistency between regions and their contexts. The latter methods exploit the combined information to detect anomalies in several ways, such as classifying defined relationships of regions [16] and only considering neighboring region pairs [6]. However, the data that they store are not informative enough for detecting context-based anomalies in human monitoring. Graph structure has shown its superiority in modeling regions and their relationships in many tasks, including object detection [15] and action recognition [20]. In addition, a Spatio-Temporal Context graph (STC graph) is built using the structural Recurrent Neural Networks (structural RNNs) for abnormal event detection in videos, where all regions in the same frame are connected for updating information in the structural RNNs [18].

In this paper, we propose to construct a spatial attributed graph (a.k.a. attributed network) to represent regions and contexts in a surveillance image and devise a novel graph auto-encoder framework named Context-based Anomaly Detection Auto-Encoder (CADAE) to detect anomalous regions. Different from the graph connecting all regions in [18], we consider spatial relationships of regions with our graph and assume building links between spatially adjacent regions can effectively capture the diversity of contexts in human monitoring. Therefore, the region, region feature and its spatial relations are represented as a node, node attribute and edges in our graph, respectively. The task of detecting anomalous regions is transformed to a task of detecting abnormal nodes in the graphs.

Several graph auto-encoders have been designed for graph anomaly detection tasks as unsupervised models. The main difference among them is their neighborhood aggregators, such as mean-pooling, max-pooling [5] and weighted average pooling by attention mechanism [7,17]. In human monitoring, as similar regions frequently appear in surveillance images, a spatial attributed graph usually includes nodes with similar attributes. However, nodes in a graph with such a structure cannot be well distinguished by an existing graph auto-encoder due to the limited discriminative power of its aggregator. Thus these models would fail to capture the diversity of contexts containing similar regions.

Instead, CADAE adopts a specified sum aggregator in Graph Isomorphism Network (GIN) [22] as components, which can compress the representations of a node with its neighbor nodes (the structure information) into injective embeddings to tackle this limitation. Moreover, two decoders in CADAE are devised to exploit the compressed embeddings for attribute and structure reconstructions. CADAE follows the unsupervised anomaly detection approach and trains a model on normal data only. Such a model can learn normal patterns and thus generates high reconstruction errors of abnormal data during testing. In summary, the main contributions of this paper are listed as follows.

1) We propose a spatial attributed graph for modeling the regions and their contexts in a surveillance image in human monitoring. The task of detecting anomalous regions is formalized as a graph anomaly detection problem.
2) We devise a deep graph auto-encoder framework CADAE for detecting region-level anomalies in human monitoring which can capture diverse contexts.
3) We utilize our autonomous mobile robot to collect about 100-hour surveillance videos in different areas of our laboratory and construct two surveillance datasets to evaluate our proposed method. Experimental results demonstrate the superiority of our method on other advanced anomaly detection methods.

2   Related Work

2.1   Anomaly Detection in Image Regions

Existing methods detect anomalous image regions in different scenarios. Several works focus on discovering pixel-wise or patch-level deviations by learning the consistency of the whole image [19,23] so that the abnormal regions (the sets of inconsistent pixels or patches) can be localized. These works are quite effective in tasks such as defect detection [23] which does not require considering contextual information. In the case of considering combined information of regions and contexts, Semantic Anomaly Detection (SAD) [16] defines several classes for relations, each of which is defined for a pair of regions, but the defined classes fail to cover several significant kinds of abnormal relations in human monitoring. Fadjrimiratno et al. [6] assume single-region anomalies and neighboring-region anomalies in human monitoring. The latter one is defined for a pair of overlapping regions, which does not cover anomalies on separate regions. Sun et al. [18] propose a context reasoning method which builds a STC graph for abnormal event detection. However, the STC graph connects all the regions in the same frame, and thus considers only their co-occurrence and neglects the spatial relationships. Therefore, these methods cannot represent the diverse contextual information for context-based anomaly detection in human monitoring.

2.2   Anomaly Detection in Attributed Networks

Graph Neural Networks (GNNs) are a family of deep learning models for graph embedding or network embedding [21]. GNNs for detecting abnormal nodes in attributed networks usually assume that anomalous nodes may induce high reconstruction errors when the networks are trained on graphs which consist of only normal nodes. Deep Anomaly Detection on Attributed Networks (DOMINANT) [5] constructs a graph auto-encoder framework with Graph Convolutional Network (GCN) [12] layers to reconstruct the node attributes and the structure information of the input graphs. Besides, Anomaly Dual Auto-Encoders (AnomalyDAE) [7] utilize two auto-encoders with graph attention layers to learn the underlying relations between the node attributes and the structure information in the graph. Similarly, by adopting graph attention layers to both an encoder and a decoder, Graph Attention Auto-Encoders (GATE) [17] exhibit superior performance in learning representations of nodes in attributed graphs.

However, these graph auto-encoders are not expressive to learn injective embeddings to discriminate nodes whose neighbor nodes are frequently similar [22], which may lead to sub-optimal performance on the reconstruction task. The main difference in these frameworks lies in their different aggregators to merge neighborhood information for updating the representations of nodes, such as mean pooling, max pooling, and weighted average pooling.

3   Target Problem

Our target problem is detecting anomalous regions in surveillance images [6,8]. Let the surveillance image be I^k with $k = 1, \ldots, K$, where K is the total number of images. I^k consists of n regions $\mathcal{R}^k = \{r_1^k, \ldots, r_n^k\}$ with captions extract by a pre-trained model for region captioning, such as DenseCap [11]. The region features consist of visual region features $\mathbf{B}^k = [\mathbf{b}_1^k, \ldots, \mathbf{b}_n^k]$ and region caption features $\mathbf{C}^k = [\mathbf{c}_1^k, \ldots, \mathbf{c}_n^k]$, both of which are extracted by pre-trained models, such as ResNet [9] and BERT [4].

It is unfeasible to collect all kinds of anomalous regions for our problem due to their rareness. Therefore, we tackle the problem as a one-class anomaly detection task [6,8], i.e., there are only normal data during training. We assume an anomalous region is highly dissimilar to the regions in normal data considering its content and context [6,8]. In the testing phase, the degrees of the abnormality of each r_i^k is computed by our method. The evaluation criteria for the target problem is ROC-AUC.

4   Context-Basded Anomaly Detection Auto-Encoder

In our method, we first construct a spatial attributed graph to represent regions and their contexts in a surveillance image. Then we devise a graph auto-encoder framework CADAE, which, given only normal data as inputs, aims to reconstruct node attributes and structure information of the graph in the training phase. During testing, an reconstruction error based anomaly score is designed to evaluate the degrees of the abnormality of each node in the graph, so that the corresponding anomalous regions in the image can be detected.

4.1   Constructing Spatial Attributed Graph

Extracting Regions from Images. We apply a pretrained dense-captioning model named DenseCap [11] to obtain bounding boxes and captions of salient regions in I^k, $k = 1, \ldots, K$. The top-n regions $\mathcal{R}^k = \{r_1^k, \ldots, r_n^k\}$ are selected in I^k, where $n = 10$ is the number of the most salient regions extracted by Densecap in our experiments.

Following the previous works [6,8], a pre-trained deep model named ResNet [9] is utilized to extract the visual feature of each region with the same dimension, denoted as $\mathbf{B}^k = [\mathbf{b}_1^k, \ldots, \mathbf{b}_n^k]$. BERT [4] is utilized to extract the semantic feature of each region caption with the same dimension, denoted as $\mathbf{C}^k = [\mathbf{c}_1^k, \ldots, \mathbf{c}_n^k]$. Then the visual and semantic features are concatenated as the whole region features, i.e., $\mathbf{f}_i^k = \mathrm{Concat}(\mathbf{b}_i^k, \mathbf{c}_i^k)$, where Concat denotes the concatenation operation. Thus all the region features in I^k are denoted as $\mathbf{F}^k = [\mathbf{f}_1^k, \ldots, \mathbf{f}_n^k]$.

Constructing Graphs. We construct a spatial attributed graph for I^k, i.e., $\mathcal{G}^k = \{\mathbf{A}^k, \mathbf{X}^k\}$. The regions, region features and their relationships in I^k are represented as nodes, node attributes (features) and structure information in $\mathcal{G}^k$, respectively. $\mathbf{X}^k \in \mathbb{R}^{n \times d}$ and $\mathbf{A}^k \in \mathbb{R}^{n \times n}$ denote the node attribute matrix and adjacent matrix, where d is the dimension of the node attributes. Node v_i $(i = 1, \ldots, n)$ in $\mathcal{G}^k$ represents r_i^k. The node attributes $\mathbf{x}_i^k$ in $\mathbf{X}^k$ represents $\mathbf{f}_i^k$ in $\mathbf{F}^k$. $\mathbf{A}_{ij}^k$ represents the element in $\mathbf{A}^k$, where $\mathbf{A}_{ij}^k = 1$ when the node v_i and v_j are connected, otherwise $\mathbf{A}_{ij}^k = 0$. We assume that spatially adjacent regions of r_i^k have strong relationships to express their contexts, which are represented by the neighboring nodes (structure information) of node v_i.

Figure 1 shows an example of constructing a spatial attributed graph from one surveillance image. The task of detecting anomalous regions in the surveillance images is transformed into a task of detecting abnormal nodes in the constructed spatial attributed graphs.

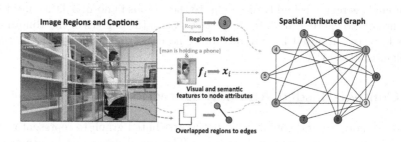

Fig. 1. Example of the spatial attributed graph of an image. The color of a node corresponds to the color of its image region.

4.2   The Proposed Model

CADAE is based on a graph auto-encoder framework, which consists of three modules: an attributed graph encoder, a structure decoder and an attribute decoder. Due to the diversity and rareness of anomalies, CADAE is trained in an unsupervised manner, i.e., it is trained on only normal nodes, the model can learn normal patterns and thus generates higher reconstruction errors of abnormal nodes than normal ones. Figure 2 illustrates the overall flow of CADAE: the attributed graph encoder updates the representations of nodes by aggregating the representations of their neighborhoods in a layer-wise way. Then the structure decoder and attribute decoder reconstruct the node attributes and structures with the learned latent representations, respectively.

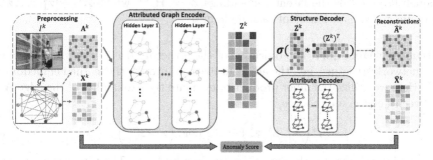

Fig. 2. The overall flow of CADAE. The surveillance image is firstly preprocessed into spatial graphs to represent the regions and contexts. Then CADAE recontructs the node attributes and structure information in the graph and evaluates the abnormality of nodes by reconstruction errors.

Motivation. In human monitoring, it is common that similar regions describing objects and human behaviors frequently appear in surveillance images, which means a large number of similar node features exist in the constructed graphs. As mentioned above, nodes in graphs with such structures are difficult to be distinguished by existing graph auto-encoder frameworks [5,7,17]. However, the sum aggregation strategy in GIN [22] is proved to generate explicitly injective representations of nodes to classify such graphs for graph classification tasks.

Considering the expressive power of the sum aggregation strategy in GIN and the typical structures in our spatial attributed graphs, CADAE adopts the specified sum aggregators to compress the representations of nodes with its neighborhoods into distinguised embeddings, which is beneficial to capture the diversity of contexts containing similar regions. In this way, CADAE addresses the aforementioned challenges by firstly devising the attributed graph encoder to precisely learn the distinguished embeddings of nodes. Then the structure decoder and attribute decoder to utilize the learned embeddings for reconstructions.

Attributed Graph Encoder. Trained on the spatial attributed graphs $\mathcal{G}^k = \{\mathbf{A}^k, \mathbf{X}^k\}_{k=1}^K$, the attributed graph encoder employs the sum aggregators to iteratively update the representation of each node $\mathbf{h}_i^{(l)}$ in l^{th} hidden layer by:

$$\mathbf{h}_i^{(l)} = \mathrm{MLP}_{\mathrm{Relu}}^{(l)} \left(\left(1 + \theta_i^{(l)} \right) \mathbf{h}_i^{(l-1)} + \sum\nolimits_{v_j \in \mathcal{N}(v_i)} \mathbf{h}_j^{(l-1)} \right), \tag{1}$$

where $\mathrm{MLP}_{\mathrm{Relu}}$ is a multi-layer perceptron with the ReLU activation funtion as the non-linear transformation module. $\theta_i^{(l)}$ is a learnable parameter in the overall learnable parameters $\boldsymbol{\theta}^{(l)}$ in l^{th} layer. $\mathcal{N}(v_i)$ is the set of the neighbor nodes adjacent to node v_i. We initialize $\mathbf{h}_i^{(0)} = \mathbf{x}_i^k$ in the input node attribute matrix $\mathbf{X}^k$. Therefore, the overall representations $\mathbf{H}^{(l)}$ in l^{th} layer is given by:

$$\mathbf{H}^{(l)} = \mathrm{MLP}_{\mathrm{Relu}}^{(l)} \left(\left(\mathbf{A}^k + \left(1 + \boldsymbol{\theta}^{(l)} \right) \cdot \mathbf{I} \right) \cdot \mathbf{H}^{(l-1)} \right). \tag{2}$$

Here $\mathbf{H}^{(0)} = \mathbf{X}^k$ is the input node attribute matrix. Thus the output of the last hidden layer L^{th} is the latent embeddings $\mathbf{Z}^k = \mathbf{H}^{(L)}$, where $\mathbf{Z}^k = \left[\mathbf{z}_1^k, \dots, \mathbf{z}_n^k \right]$ contains representation $\mathbf{z}_i^k$ of each node in $\mathcal{G}^k$.

Structure Decoder. The objective of the structure decoder is to reconstruct the network structure, i.e., the adjacent matrix, through the learned embeddings $\mathbf{Z}^k$. Here we utilize the inner product to estimate the probability of the edge $\widehat{\mathbf{A}}_{ij}^k$ between node v_i and node v_j:

$$P \left(\widehat{\mathbf{A}}_{ij}^k | \mathbf{z}_i^k, \mathbf{z}_j^k \right) = \mathrm{Sigmoid} \left(\mathbf{z}_i^k \cdot \mathbf{z}_j^{k^T} \right), \tag{3}$$

where Sigmoid denotes the sigmoid activation function. The total reconstructed adjacent matrix $\widehat{\mathbf{A}}^k$ is computed by the structure decoder as:

$$\widehat{\mathbf{A}}^k = \mathrm{Sigmoid} \left(\mathbf{Z}^k \cdot \mathbf{Z}^{k^T} \right). \tag{4}$$

Attribute Decoder. The attribute decoder is devised to decompress the embeddings $\mathbf{Z}^k$ for reconstructing the original node attributes. Similarly, the decoder also employs the specified sum aggregators in its layers, which helps to capture the combined information of nodes and their neighborhoods from the learned embeddings, for reconstructing the attribute matrix $\widehat{\mathbf{X}}^k$:

$$\widehat{\mathbf{X}}^k = \mathrm{MLP}_{\mathrm{Relu}} \left(\left(\mathbf{A}^k + (1 + \boldsymbol{\theta}) \cdot \mathbf{I} \right) \cdot \mathbf{Z}^k \right). \tag{5}$$

Optimization. As CADAE aims to reconstruct both the node attributes and the structure information in the input graphs, the model is optimized to jointly minimize both the structure and attribute reconstruction errors. Specifically, the objective function $\mathcal{L}$ of CADAE can be formulated as:

$$\mathcal{L} = \beta \mathbf{Rec}_{att} + (1 - \beta) \, \mathbf{Rec}_{str} \tag{6}$$

$$= \frac{1}{K} \sum\nolimits_K^{k=1} \left(\beta \| \widehat{\mathbf{X}}^k - \mathbf{X}^k \|_2 + (1 - \beta) \, \| \widehat{\mathbf{A}}^k - \mathbf{A}^k \|_2 \right), \tag{7}$$

where the ℓ^2-norm is used to measure the attribute reconstruction errors $\mathbf{Rec}_{att}$ and structure reconstruction errors $\mathbf{Rec}_{str}$. β is a hyper-parameter to adjust the influence of $\mathbf{Rec}_{att}$ and $\mathbf{Rec}_{str}$.

Anomaly Score. Trained on normal nodes from the spatial constructed graphs, CADAE can reconstruct both the attribute and structure information of normal nodes in high quality by optimizing the objective function. In the testing phase, the trained CADAE is supposed to generate high reconstruction errors of anomalous nodes. Therefore, the total reconstruction error of each node v_i can be defined as the anomaly score s_i to estimate its abnormality:

$$s_i = f(v_i) = \beta \|\widehat{\mathbf{x}}_i^k - \mathbf{x}_i^k\|_2 + (1 - \beta) \|\widehat{\mathbf{a}}_i^k - \mathbf{a}_i^k\|_2, \tag{8}$$

where $\mathbf{a}_i$ and $\widehat{\mathbf{a}}_i$ are the ith rows in $\mathbf{A}^k$ and $\widehat{\mathbf{A}}^k$, which are adjacent information and its reconstruction of the node v_i, respectively. Through ranking the anomaly scores, the anomalous degrees of the nodes and their corresponding image regions can be investigated.

The overall procedure of CADAE including the training and testing phases is illustrated in Algorithm 1.

Algorithm 1: The overall procedure of CADAE.

Input : Graph $\mathcal{G}_{train}^k = \{\mathbf{A}^k, \mathbf{X}^k\}_{k=1}^K$, $\mathcal{G}_{test}^r = \{\mathbf{A}^r, \mathbf{X}^r\}_{r=1}^R$;
Multi-layer perceptron with ReLU activation funtion
MLP$_{\text{Relu}}$; Learnable parameter $\boldsymbol{\theta}$; Hyper-parameter β;
Number L of the hidden layers of the encoder in CADAE;
Number T of the training epochs.

Output: Anomaly score function for each node $f(\cdot)$.

1 Randomly initialize $\boldsymbol{\theta}$ and the parameters in MLP$_{\text{Relu}}$;
   // Training phase.
2 for $t = 1, 2, \cdots, T$ do
3 $\quad$ for $k = 1, 2, \cdots, K$ do
4 $\quad\quad$ for $l = 1, 2, \cdots, L$ do
5 $\quad\quad\quad$ | Calculate $\mathbf{H}^{(l)}$ via Eq. (2);
6 $\quad\quad$ end
7 $\quad\quad$ $\mathbf{Z}^k = \mathbf{H}^{(L)}$;
8 $\quad\quad$ Calculate $\widehat{\mathbf{A}}^k, \widehat{\mathbf{X}}^k$ via Eq. (4-5);
9 $\quad\quad$ Calculate $\mathcal{L}$ via Eq. (7);
10 $\quad\quad$ Update $\boldsymbol{\theta}$ and the parameters in MLP$_{\text{Relu}}$ with the backpropagation algorithm.
11 $\quad$ end
12 end
   // Testing phase.
13 for $r = 1, 2, \cdots, R$ do
14 $\quad$ for $l = 1, 2, \cdots, L$ do
15 $\quad\quad$ | Calculate $\mathbf{H}^{(l)}$ via Eq. (2);
16 $\quad$ end
17 $\quad$ $\mathbf{Z}^r = \mathbf{H}^{(L)}$;
18 $\quad$ Calculate $\widehat{\mathbf{A}}^r, \widehat{\mathbf{X}}^r$ via Eq. (4-5);
19 $\quad$ Calculate the anomaly score of each node v_i in $\mathcal{G}_{test}^r$ via Eq. (8).
20 end

---

5   Experiments

5.1   Datasets

To evaluate our proposed method on human monitoring, two large-size surveillance video data are collected by our robot platform in a real laboratory environment. The two surveillance videos were already employed in several anomaly detection tasks for human monitoring [6,8]. To evaluate the performance of our method, we construct two human monitoring datasets from the surveillance videos as follows.

> **LabMonitoring** is selected from the first surveillance video frames collected by our autonomous mobile robot patrolling around the laboratory. LabMonitoring includes not only noticeable anomalies, such as a man holding a baseball bat, holding an umbrella in the lab., but also typical context-based anomalies, such as a man making a phone call in the working area (as making a phone call in the resting area is normal). It consists of 5146 images for training and 384 images for testing.
>
> **BehaviorMonitoring** is selected from the second video frames which were recorded almost 100 h by our robot to collect diverse and wide-ranging human behaviors in the environment. BehaviorMonitoring contains various context-based anomalies since similar behaviors in the working and resting areas are normal and abnormal, respectively, based on the diverse contexts, such as eating and sleeping in the working and resting areas. It consists of 5548 images for training and 696 images for testing.

Figure 3 shows some example images including normal and abnormal regions, such as someone making a phone call, playing with a ball, eating and sleeping in the working and resting areas, from two datasets. The two datasets are RGB images with the resolution 1366 × 768.

Fig. 3. Example images from the two datasets.

5.2   Experimental Settings

Preprocessing. We select top-10 salient regions and corresponding captions in the two datasets extracted by DenseCap. We annotate region-level anomalies when regions contain anomalous human behaviors or irregular co-occurrences between humans and objects. The region features include visual features extracted from the output of the penultimate layer whose dimension is 2048 in ResNet and semantic features of captions encoded by BERT, whose dimension is 768. The DenseCap, ResNet and BERT are in the default settings and pretrained on Visual Genome [14], ImageNet [3] and Wikipedia data [4], respectively.

Baselines. We adopt five popular anomaly detection methods and graph auto-encoders as our baselines.

> **AE** [2] is a classical unsupervised framework. Its auto-encoder is constructed by several non-linear fully-connected layers. AE reconstructs samples by handling them independently.
> **GANOMALY** [1] is a deep anomaly detection model with adversarial training. It is an effective reconstruction-based model for detecting anomalies without considering the relations among samples.
> **VGAE** [13] is a graph auto-encoder framework for unsupervised learning on graph-structured data based on the variational auto-encoder (VAE). It employs GCN layers to merge node features and reconstruct the structures of nodes in graphs.
> **DOMINANT** [5] is a popular deep graph convolutional auto-encoder for graph anomaly detection tasks. DOMINANT utilizes GCN layers to jointly learn the attribute and structure information and detect anomalies based on reconstruction errors.
> **GATE** [17] is also a graph auto-encoder framework with self-attention mechanisms. It generates the embeddings of nodes by merging the information of their neighbors using its graph attention layers.

Evaluation Metrics. To evaluate the performance of our method and the baselines, we adopt the ROC-AUC metrics. ROC curve is drawn by the true positive rate (TPR) against the false positive rate (FPR) with various threshold settings. AUC is the value of the area under the ROC curve. In our task, the anomalies are the positive class. A true positive is the result where our model correctly predicts the anomaly and a false positive is the result where our model incorrectly predicts the anomaly. For example, if the model correctly predicts the anomaly that a man is making calls in the working area, e.g., the red box in the lower image in Fig. 4, it is a true positive.

Parameter Settings. In the experiments, the number of epochs is 400 and Adam optimizer is adopted with the learning rate 0.002 and weight decay 8×10^{-5}

for all the methods. We devise 2 hidden layers in the attributed graph encoder and attribute decoder. The MLP module of the encoder in CADAE is 2-layer perceptrons with the dimension $(2816 - 256 - 256)$ in the first hidden layer and $(256 - 256 - 128)$ in the second hidden layer. Accordingly we reverse the dimensions of the MLP in the attribute decoder for reconstruction. For the sake of fairness, we keep the same dimension settings as CADAE for the five baselines and retain other parameter settings in their proposed papers. Hyper-parameter β in CADAE is set to 0.6 in LabMonitoring and 0.8 in BehaviorMonitoring.

5.3   Experimental Results and Analysis

Figure 4 shows the example results on how CADAE detects anomalous regions in human monitoring. We show ROC curves of all these methods in Fig. 5(a)-(b) and AUC values in Table 1. We also visualize the distributions of reconstruction error based anomaly scores of normal and abnormal regions by the boxplots ranging from the lower quartile to the upper quartile in Fig. 5(c)-(d). Through analyzing the experimental results, we can draw several conclusions.

Firstly, CADAE achieves the best performance in two datasets compared with the five baselines. This fact shows the effectiveness of CADAE for the task of detecting abnormal regions in human monitoring. Secondly, GANOMALY and AE perform well in LabMonitoring, but poorly in BehaviorMonitoring. It is because most anomalies in BehaviorMonitoring are context-based that cannot be distinguished by only learning normal patterns from the regions without considering their contexts. Lastly, the graph auto-encoders including DOMINANT, GATE and VGAE cannot achieve satisfying results, which proves that CADAE equipped with the specified sum aggregators gives accurate discriminative reconstruction errors of normal and abnormal nodes when similar node attributes frequently exist in the graphs.

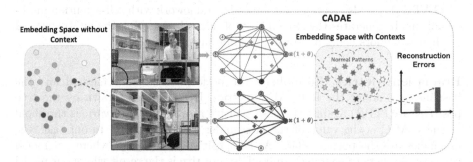

Fig. 4. Example of detecting context-based anomalous regions. The region of making phone calls is normal (green box) in the upper image while abnormal (red box) in the lower image due to their different contexts. They are represented as green and red nodes in the constructed graphs. Trained on normal data, our method compresses nodes with its neighborhoods (contexts) into distinguished embeddings. Then the abnormal nodes which deviate from the learned normal patterns (green dashed subspace) would exhibit high reconstruction errors during testing. (Color figure online)

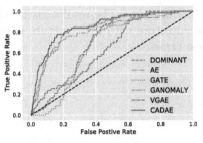
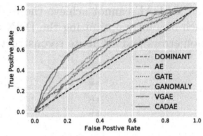

(a) ROC curve on LabMonitoring.   (b) ROC curve on BehaviorMonitoring.

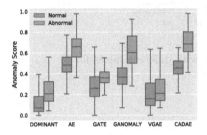
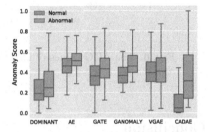

(c) Anomaly score on LabMonitoring. (d) Anomaly score on BehaviorMonitoring.

Fig. 5. The ROC curves and anomaly scores for all methods on two datasets. The boxplots in (c) and (d) show the distributions of anomaly scores ranging from the lower quartile to the upper quartile.

Table 1. AUC of all the methods on two datasets.

Methods	LabMonitoring	BahaviorMonitor
AE [2]	0.81	0.63
GANOMALY [1]	0.84	0.70
DOMINANT [5]	0.71	0.61
VGAE [13]	0.64	0.52
GATE [17]	0.66	0.64
CADAE	**0.86**	**0.75**

5.4 Parameter-Sensitivity Investigation

To further investigate the impact of the structure decoder, the attribute decoder and the hyper-parameter β on our task, we modify the value of β as $0, 0.1, 0.2, \ldots, 1$. In the extreme cases, we can verify the influence of only adopting the structure decoder when $\beta = 0$ and the attribute decoder when $\beta = 1$.

Figure 6 shows the AUC values of CADAE with different β on the two datasets. We see that the AUC value is not maximized when $\beta = 0$ and $\beta = 1$, which illustrates that by adopting only a structure decoder or an attribute decoder, CADAE cannot achieve the best performance. The results justify the approach of jointly optimizing the DNN architecture with the structure reconstruction errors and attribute reconstruction errors.

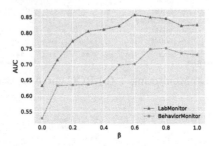

Fig. 6. The AUC values of CADAE with different values for β.

6 Conclusion

In this paper, we tackled the task of detecting anomalous regions in human monitoring by proposing spatial attributed graphs and developing a deep graph neural networks. In particular, through adopting the specified sum aggregators from GIN [22], CADAE can detect abnormal nodes when similar node attributes frequently appear in the constructed graphs, unlike existing graph auto-encoder frameworks. Experimental results and parameter-sensitivity investigation demonstrate the superiority of our CADAE over several popular anomaly detection methods.

We expect CADAE opens up a new perspective for the task of detecting anomalous regions in human monitoring. In future works, we will construct more informative attributed graphs to model the relations of regions in more complicated environments. Moreover, we will improve the proposed model to be more robust and effective for spotting complex anomalies in human monitoring.

Acknowledgements. This work was supported by China Scholarship Council (Grand No. 201906330075).

References

1. Akcay, S., Atapour-Abarghouei, A., Breckon, T.P.: GANomaly: semi-supervised anomaly detection via adversarial training. In: Proceedings of the ACCV, pp. 622–637 (2018)
2. Bengio, Y., Lamblin, P., Popovici, D., Larochelle, H.: Greedy layer-wise training of deep networks. In: Proceedings of the NeurIPS, pp. 153–160 (2007)
3. Deng, J., Dong, W., Socher, R., Li, L.J., Li, K., Fei-Fei, L.: ImageNet: a large-scale hierarchical image database. In: Proceedings of the CVPR, pp. 248–255 (2009)

4. Devlin, J., Chang, M.W., Lee, K., Toutanova, K.: BERT: Pre-training of Deep Bidirectional Transformers for Language Understanding. arXiv preprint arXiv:1810.04805 (2018)
5. Ding, K., Li, J., Bhanushali, R., Liu, H.: Deep anomaly detection on attributed networks. In: The Proceedings of the SIAM, pp. 594–602 (2019)
6. Fadjrimiratno, M.F., Hatae, Y., Matsukawa, T., Suzuki, E.: Detecting anomalies from human activities by an autonomous mobile robot based on fast and slow thinking. In: Proceedings of the VISIGRAPP, vol. 5, pp. 943–953 (2021)
7. Fan, H., Zhang, F., Li, Z.: AnomalyDAE: dual autoencoder for anomaly detection on attributed networks. In: Proceedings of the ICASSP, pp. 5685–5689 (2020)
8. Hatae, Y., Yang, Q., Fadjrimiratno, M.F., Li, Y., Matsukawa, T., Suzuki, E.: Detecting anomalous regions from an image based on deep captioning. In: Proceedings of the VISIGRAPP, vol. 5, pp. 326–335 (2020)
9. He, K., Zhang, X., Ren, S., Sun, J.: Deep residual learning for image recognition. In: Proceedings of the CVPR, pp. 770–778 (2016)
10. Ionescu, R.T., Khan, F.S., Georgescu, M.I., Shao, L.: Object-centric auto-encoders and dummy anomalies for abnormal event detection in video. In: Proceedings of the CVPR, pp. 7842–7851 (2019)
11. Johnson, J., Karpathy, A., Fei-Fei, L.: DenseCap: fully convolutional localization networks for dense captioning. In: Proceedings of the CVPR, pp. 4565–4574 (2016)
12. Kipf, T.N., Welling, M.: Semi-Supervised Classification with Graph Convolutional Networks. arXiv preprint arXiv:1609.02907 (2016)
13. Kipf, T.N., Welling, M.: Variational Graph Auto-Encoders. arXiv preprint arXiv:1611.07308 (2016)
14. Krishna, R., et al.: Visual genome: connecting language and vision using crowd-sourced dense image annotations. Int. J. Comput. Vis. **123**(1), 32–73 (2017)
15. Liu, Y., Wang, R., Shan, S., Chen, X.: Structure inference net: object detection using scene-level context and instance-level relationships. In: Proceedings of the CVPR, pp. 6985–6994 (2018)
16. Pasini, A., Baralis, E.: Detecting anomalies in image classification by means of semantic relationships. In: Proceedings of the AIKE, pp. 231–238 (2019)
17. Salehi, A., Davulcu, H.: Graph attention auto-encoders. In: Proceedings of the ICTAI, pp. 989–996 (2020)
18. Sun, C., Jia, Y., Hu, Y., Wu, Y.: Scene-aware context reasoning for unsupervised abnormal event detection in videos. In: Proceedings of the ACM MM, pp. 184–192 (2020)
19. Venkataramanan, S., Peng, K.C., Singh, R.V., Mahalanobis, A.: Attention guided anomaly localization in images. In: Proceedings of the ECCV, pp. 485–503 (2020)
20. Wang, X., Gupta, A.: Videos as space-time region graphs. In: Proceedings of the ECCV, pp. 399–417 (2018)
21. Wu, Z., Pan, S., Chen, F., Long, G., Zhang, C., Philip, S.Y.: A comprehensive survey on graph neural networks. IEEE Trans. Neural Netw. Learn. Syst. **32**(1), 4–24 (2020)
22. Xu, K., Hu, W., Leskovec, J., Jegelka, S.: How powerful are graph neural networks? In: Proceedings of the ICLR (2019)
23. Yi, J., Yoon, S.: Patch SVDD: patch-level SVDD for anomaly detection and segmentation. In: Proceedings of the ACCV (2020)

Domain-Adaptation Person Re-Identification via Style Translation and Clustering

Peiyi Wei[1], Canlong Zhang[1(✉)], Zhixin Li[1], Yanping Tang[2], and Zhiwen Wang[3]

[1] Guangxi Key Lab of Multi-source Information Mining and Security,
Guangxi Normal University, Guilin 541004, China
`zcltyp@163.com, lizx@mailbox.gxnu.edu.cn`
[2] School of Computer Science and Information Security, Guilin University
of Electronic Technology, Guilin 541004, China
[3] School of Electric, Electronic and Computer Science, Guangxi University of Science
and Technology, Liuzhou Guangxi, 545006, China

Abstract. To solve the two challenges of the high cost of manual labeling data and significant degradation of cross-domain performance in person re-identification (re-ID), we propose an unsupervised domain adaptation (UDA) person re-ID method combining style translation and unsupervised clustering method. Our model framework is divided into two stages: 1) In the style translation stage, we can get the labeled source image with the style of the target domain; 2) In the UDA person re-ID stage, we use Wasserstein distance as the evaluation index of the distribution difference between domains. In addition, to solve the problem of source domain ID labels information loss in the process of style translation, a feedback mechanism is designed to feedback the results of person re-ID to the style translation network, to improve the quality of image style translation and the accuracy of ID labels and make the style translation and person re-ID converge to the best state through closed-loop training. The test results on Market-1501, DukeMTMC, and MSMT17 show that the proposed method is more efficient and robust.

Keywords: Person re-identification · Style translation · Unsupervised clustering · Unsupervised domain adaptation

1 Introduction

Recently, the accuracy of person re-ID based on supervised learning has been greatly improved, thanks to labeled data sets for training. However, supervised person re-ID faces the following two challenges: 1) When the model obtained by using a labeled data set for learning is transferred to the real scene, the problem

---

Supported by Guangxi Key Lab of Multi-source Information Mining & Security, Guangxi Normal University.

T. Mantoro et al. (Eds.): ICONIP 2021, LNCS 13108, pp. 464–475, 2021.
https://doi.org/10.1007/978-3-030-92185-9_38

of domain mismatch often occurs due to the discrepancy between domains; 2) Labeling large-scale training data is costly, and manual labeling often brings great subjective discrepancy. UDA person re-ID is expected to overcome these problems. The main principle is to transfer the person re-ID model trained on the labeled source domain to the unlabeled target domain.

Recent research on UDA person re-ID is mainly divided into two categories: style-translation-based methods and pseudo-label-based methods. The former mainly uses a style translation model (such as Cyclegan [19]) to translate the style of tagged source domain image to unlabeled target domain image or translate the style of images captured by multiple cameras. The translated image will have the ID label and target domain style of the source domain to reduce the domain discrepancy. Then, the person re-ID model trained by the image after style translation can be generalized to a certain extent. The latter is to generate pseudo tags by clustering instance features or measuring the similarity with sample features, and then use pseudo tags for person re-ID learning. Compared with the former, the latter can achieve better performance, so it has become the mainstream research direction of UDA person re-ID. SPGAN [3] proposes to use the self-similarity before and after the image style translation to solve the image ID label loss during the translation process. However, it still can not solve this kind of problem very well. PTGAN [13] proposed a method of scene transfer combined with style translation to reduce the inter-domain differences caused by factors such as lighting and background. [17,18] mainly use the style translation between cameras to learn the characteristics of camera invariance, to solve the problem of intra-domain discrepancy caused by style changes in the images collected between different cameras, and then obtain a model of camera invariance and domain connectivity.[5] are through repeatedly mining the global information and local information of pedestrians to improve the accuracy of clustering to obtain a high-confidence training set with pseudo-labels for re-ID training. [14] designed an asymmetric collaborative teaching framework, which reduces the noise of pseudo-labels through cooperation and mutual iteration of two models, thereby improving the clustering effect.

However, whether based on style translation or pseudo-label clustering, the adaptation effect achieved is limited. Therefore, this article believes that combining image style translation and the clustering UDA person re-ID method will have greater potential. This paper proposes a UDA person re-ID method that combines style translation and clustering. Figure 1 shows the overall structure of the model. Specifically, cross-domain image translation (CIT) is on the left, which can transform the source domain image without a pair of samples. The converted image has the same identity label, and with the style of the target domain image, CIT can initially reduce the difference between the two domains. The Unsupervised Domain Adaptation Re-identification (UDAR) is on the right. In this process, the adversarial learning method is used. The data distribution of the two datasets is measured by a more stable Wasserstein distance so that the discrepancy between the domains can be further reduced. Then, we use the unsupervised clustering algorithm to assign pseudo labels to the target domain and then combine the source domain and the target domain into a unified dataset K for person re-ID training. Finally, to solve the ID labels in the style translation process loss and the quality of the generated image. We design a positive feedback mechanism, that is, the result of person re-ID is feedback to the image style translation module to ensure semantic consistency before and after the image style translation, that is, when an ideal pedestrian is obtained when recognizing the model, even if there is a visual difference between the image after the style translation and the original image, their ID labels are unchanged.

In summary, our key contributions are:

- Our proposed UDA person re-ID model of joint image style translation and clustering can make style translation and unsupervised clustering complementary. And better transfer the knowledge learned from the source domain to the target domain. To improve the generalization ability of the model.
- We design a positive feedback mechanism, which can feedback the person re-ID results of the second stage to the style translation of the first stage so that the image's visual quality after the style translation and the accuracy of the corresponding ID label information can be improved.
- In the unsupervised domain adaptation stage, we design a adversarial learning module and measure the distribution discrepancy between the two datasets by Wasserstein distance. The stability of the model is guaranteed when the domain-invariant representation is learned, the effect of to achieve reinforced domain adaptation.

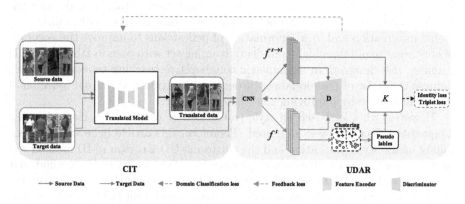

Fig. 1. The design of our UDA framework.

2   Proposed Method

Given a source domain sample S, it is represented as $S = \{(x_i^s, y_i^s)|_{i=1}^{N^s}\}$, where x_i^s and y_i^s are the i-th training sample of the source domain and its corresponding label, respectively. N^s is the number of images in the dataset. Given unlabeled target domain sample T, it is represented as $T = \{x_i^t|_{i=1}^{N^t}\}$, where x_i^t is the i-th training sample of the target domain. N^t is the number of images in the dataset. Define a feature encoder $F(\cdot|\theta_e)$ with θ_e as the parameter.

2.1 Cross-Domain Image Translation

CIT uses CycleGAN [19] network structure as the main body. The structure is mainly composed of two sets of generators-discriminators $\{G^{s \to t}, D_T\}$ and $\{G^{t \to s}, D_S\}$ to perform image style translation in two directions, forming a cyclic network structure to ensure consistent style translation. The source domain dataset S' after style translation can be obtained through CIT, where S' will keep the same label as S.

In CIT, GAN loss L_{GAN}, cycle reconstruction loss L_{cyc}, and appearance consistency loss L_{ac} [12] are mainly used. Due to the lack of paired training data, there may be infinitely many mapping functions to choose from. Therefore, CycleGan [19] introduced L_{cyc} to try to restore the original image after a cycle of translation and reverse translation, so as to reduce the possible mapping functions. L_{ac} can make the color composition of the image after the style translation similar to that before the translation. They are defined as follows.

$$
\begin{aligned}
L_{GAN}\left(G^{s \to t}, G^{t \to s}, D_S, D_T\right) = &\mathbb{E}_{x_i^s \sim S}\left[D_S\left(x_i^s\right)^2\right] \\
&+ \mathbb{E}_{x_i^t \sim T}\left[\left(D_S\left(G^{t \to s}\left(x_i^t\right)\right) - 1\right)^2\right] \\
&+ \mathbb{E}_{x_i^t \sim T}\left[D_T\left(x_i^t\right)^2\right] \\
&+ \mathbb{E}_{x_i^s \sim S}\left[\left(D_T\left(G^{s \to t}\left(x_i^s\right)\right) - 1\right)^2\right],
\end{aligned}
\tag{1}
$$

and

$$
\begin{aligned}
L_{cyc}\left(G^{s \to t}, G^{t \to s}\right) = &\mathbb{E}_{x_i^s \sim S}\left[\left\|G^{t \to s}\left(G^{s \to t}\left(x_i^s\right)\right) - x_i^s\right\|_1\right] \\
&+ \mathbb{E}_{x_i^t \sim T}\left[\left\|G^{s \to t}\left(G^{t \to s}\left(x_i^t\right)\right) - x_i^t\right\|_1\right],
\end{aligned}
\tag{2}
$$

and

$$
\begin{aligned}
L_{ac}\left(G^{s \to t}, G^{t \to s}\right) = &\mathbb{E}_{x_i^s \sim S}\left[\left\|G^{t \to s}\left(x_i^s\right) - x_i^s\right\|_1\right] \\
&+ \mathbb{E}_{x_i^t \sim T}\left[\left\|G^{s \to t}\left(x_i^t\right) - x_i^t\right\|_1\right],
\end{aligned}
\tag{3}
$$

In CIT, the source domain ID labels information is often lost. To solve this problem and further improve the quality of style translation, we design a positive feedback mechanism, which uses the person re-ID results of the second stage to feedback a loss L_{fb} to the first stage style transfer module. L_{fb} can make the two modules form a closed-loop, and the entire model framework is more closely connected. Here we use UDAR to calculate the relevant feature operation of the feedback loss, which is defined as follows.

$$L_{fb}\left(x_i^{s'}, x_i^t\right) = \lambda_{fb}\mathbb{E}_{x_i^s \sim S, x_i^{s'} \sim S'}\left[\left\|R\left(x_i^s\right) - R\left(x_i^{s'}\right)\right\|_1\right]$$
$$+\lambda_{fb\_recon}\mathbb{E}_{x_i^s \sim S, x_i^{s'} \sim S'}\left[\left\|R\left(G^{t\to s}\left(x_i^{s'}\right)\right) - R\left(x_i^s\right)\right\|_1\right]$$
$$+\lambda_{fb}\mathbb{E}_{x_i^t \sim T, x_i^{t'} \sim T'}\left[\left\|R\left(x_i^t\right) - R\left(x_i^{t'}\right)\right\|_1\right] \tag{4}$$
$$+\lambda_{fb\_recon}\mathbb{E}_{x_i^t \sim T, x_i^{t'} \sim T'}\left[\left\|R\left(G^{s\to t}\left(x_i^{t'}\right)\right) - R\left(x_i^t\right)\right\|_1\right],$$

where $R(\cdot)$ denotes the person re-ID calculation of UDAR. λ_{fb} and λ_{fb_recon} are equilibrium coefficients. By using the positive feedback mechanism, when an ideal person re-ID model is trained, even though S' and S or T' and T have visually existed differently, the translation image and the original image should have the same ID label information.

The overall loss for training the CIT is defined as

$$L_{CIT} = \lambda_{GAN}L_{GAN} + \lambda_{cyc}L_{cyc} + \lambda_{ac}L_{ac} + L_{fb}. \tag{5}$$

where λ_{GAN}, λ_{cyc} and λ_{ac} are weighting parameters.

2.2   Unsupervised Domain Adaptation for Person Re-ID

Domain-Invariant Representation Learning. In domain-invariant representation learning, when the marginal distributions of two domains are very different or even not overlapped, the gradient may disappear. Therefore, to prevent this problem, this paper uses Wasserstein distance [1] to measure the distribution discrepancy between the source domain and the target domain so that the model can better learn the invariant-feature representation of the domain while can keep the stability.

Specifically, we define a domain discriminator $F(\cdot|\theta_w)$ with θ_w as the parameter. $F(\cdot|\theta_w)$ can map the features $f^{s\to t} = F\left(x_i^{s'}|\theta_e\right)$ and $f^t = F(x_i^t|\theta_e)$ into the shared feature space. The Wasserstein distance between two different marginal data distributions $\mathbb{P}_{f^{s\to t}}$, $\mathbb{P}_{f^t}$ is

$$W_1\left(\mathbb{P}_{f^{s\to t}}, \mathbb{P}_{f^t}\right) = \sup_{\|F(\cdot|\theta_w)\|_L \leq 1} \mathbb{E}_{\mathbb{P}_{f^{s\to t}}}\left[F\left(f^{s\to t}|\theta_w\right)\right]$$
$$- \mathbb{E}_{\mathbb{P}_{f^t}}\left[F\left(f^t|\theta_w\right)\right]$$
$$= \sup_{\|F(\cdot|\theta_w)\|_L \leq 1} \mathbb{E}_{\mathbb{P}_{x_i^{s'}}}\left[F\left(F\left(x_i^{s'}|\theta_e\right)|\theta_w\right)\right] \tag{6}$$
$$- \mathbb{E}_{\mathbb{P}_{x_i^t}}\left[F\left(F\left(x_i^t|\theta_e\right)|\theta_w\right)\right],$$

where $\|\cdot\|_L$ denotes Lipschitz continuous [1,7], the Lipschitz constant of $F(\cdot|\theta_w)$ is set to 1 (i.e., 1-Lipschitz) for the convenience of calculation.

When the parameter of the function $F(\cdot|\theta_w)$ satisfies 1-Lipschitz, the critical loss L_{wd} of the parameter θ_w can be approximately estimated by maximizing the Wasserstein distance, as follows

$$L_{wd}\left(x_i^{s'}, x_i^t\right) = \frac{1}{N^s} \sum_{i=1}^{N^s} F\left(F\left(x_i^{s'}|\theta_e\right)|\theta_w\right)$$
$$- \frac{1}{N^t} \sum_{i=1}^{N^t} F\left(F\left(x_i^t|\theta_e\right)|\theta_w\right),$$

(7)

After each gradient update, the θ_w will be limited to the range of [-c, c] for clipping [1], which may cause the gradient to explode or disappear. Therefore, a gradient penalty L_{grad} [7] should be imposed on θ_w

$$L_{grad}\left(\hat{f}\right) = \mathbb{E}_{\hat{f} \sim \mathbb{P}_{\hat{f}}}\left[\left(\left\|\nabla_{\hat{f}} F\left(\hat{f}|\theta_w\right)\right\|_2 - 1\right)^2\right],$$

(8)

where $\hat{f}$ is the random point on the line between the feature distributions of two domains. Note that the gradient penalty here is for each individual input constraint, not for the whole batch. Therefore, layer normalization is used to replace batch normalization in the structure of domain discriminator.

Since the Wasserstein distance is continuous, we can first train the domain discriminator to the optimal result, then keep the parameters unchanged, minimize the Wasserstein distance, and finally make the feature encoder extract the feature representation with domain-invariance

$$\min_{\theta_e} \max_{\theta_w} \left\{L_{wd} + \lambda_{grad} L_{grad}\right\}.$$

(9)

where λ_{grad} is the balancing coefficient, which should be set to 0 in the process of minimization.

Unsupervised Person Re-ID. In this section, we use the unsupervised clustering algorithm (*e.g.*, *k*-means, DBSCAN [4]) to cluster the target domain dataset to obtain the target domain dataset $\hat{T} = \left\{(x_i^t, y_i^t)|_{i=1}^{N^t}\right\}$ with pseudo labels, where y_i^t denotes the pseudo label. And then merge the source domain and the target domain to create a unified dataset $K = S' \cup \hat{T}$, which is used for person re-ID training, where $K = \left\{(x_i, y_i)|_{i=1}^N\right\}$, $x_i = x_i^s \cup x_i^t$, $y_i = y_i^s \cup y_i^t$, $N = N^s + N^t$. Then, use an identity classifier C to predict the identity of the feature $f = f^{s \to t} \cup f^t$. Here, the classification loss and triple loss are adopted to training.

$$L_{id}\left(\theta_e\right) = \frac{1}{N} \sum_{i=1}^N L_{ce}\left(C\left(F\left(x_i|\theta_e\right)\right), y_i\right),$$

(10)

and

$$L_{tri}\left(\theta_e\right) = \frac{1}{N} \sum_{i=1}^N \max(0, m + \|F\left(x_i|\theta_e\right) - F\left(x_{i,p}|\theta_e\right)\|_2$$
$$- \|F\left(x_i|\theta_e\right) - F\left(x_{i,n}|\theta_e\right)\|_2),$$

(11)

where L_{ce} is the the cross-entropy loss, $\|\cdot\|$ denotes the L^2-norm distance, m is the margin of the triple distance. $x_{i,p}$ and $x_{i,n}$ is the positive and negative of the same sample.

Then the loss function of person re-ID is

$$L_R^1 = \lambda_{id} L_{id} + L_{tri}, \tag{12}$$

where λ_{id} is the weight coefficient of the two losses.

The overall loss for training the UDAR is defined as

$$L_R = \min_{\theta_e} \left\{ L_R^1 + \lambda_w \max_{\theta_w} \left(L_{wd} + \lambda_{\text{grad}} L_{\text{grad}} \right) \right\}. \tag{13}$$

where λ_w is the balance coefficient.

3   Experiments

3.1   Datasets and Evaluation Metrics

This section shows the experimental results, mainly including performance comparison with other network models and ablation studies. We used Market-1501 [15], DukeMTMC-reID [10] and MSMT17 [13] for experiments. Evaluation indicators mainly use average accuracy (mAP) and cumulative matching curve (CMC, rank-1) [15] to evaluate the performance of the model algorithm.

3.2   Implementation Details

The entire model framework is trained on two NVIDIA Tesla V100 GPUs (64g). The backbone uses ResNet-50 [8] trained on ImageNet [2].

Stage 1: The stage 1 training is set to 120 epochs, and there are 200 iterations in each epoch. The learning rate lr for the first 50 epochs is set to 0.0002, and the next 70 epochs will be gradually reduced to 0 according to the lr = lr $\times (1.0 - \max(0, \text{epoch} - 50)/50)$. Use Adam optimizer to optimize the network. The weighting coefficient are respectively set as $\lambda_{GAN} = 1$, $\lambda_{cyc} = 10$, $\lambda_{ac} = 1$, $\lambda_{fb} = 0.1$, $\lambda_{fb\_recon} = 10$.

Stage 2: The structure of the domain discriminator $F(\cdot|\theta_w)$ is similar to [9]. Except for the convolutional layer of the last layer, the other three layers all adopt the structure design of FC+LN+LeakyReLU. Stage 2 training defines 50 epochs. The learning rate is set to 10^{-4}, the Weighting coefficient $\lambda_w = 1$, $\lambda_{grad} = 10$.

Stage 3: In this stage, the unsupervised clustering algorithm k-means and DBSCAN [4] assign pseudo labels to the target domain. The stage 3 training is set to 120 epochs, and the initial learning rate is set to 3.5×10^{-4}, which is reduced to 3.5×10^{-5} after the 40th epoch and 3.5×10^{-6} after the 70th epoch, where each epoch has 200 iterations. Weighting coefficient $\lambda_{id} = 10$.

Table 1. Comparison with state-of-the-arts on DukeMTMC-reID, Market-1501 and MSMT17. Our proposed method has good performance for UDA person re-ID.

Method	D→M		M→D		M→MSMT		D→MSMT	
	mAP	R-1	mAP	R-1	mAP	R-1	mAP	R-1
PTGAN [13]	–	33.5	–	16.9	2.9	10.2	3.3	11.8
SPGAN [3]	22.8	51.5	22.3	41.1	–	–	–	–
CamStyle [18]	27.4	58.8	25.1	48.4	–	–	–	–
HHL [17]	31.4	62.2	27.2	46.9	–	–	–	–
ECN [16]	43.0	75.1	40.4	63.3	8.5	25.3	10.3	30.2
UDAP [11]	53.7	75.8	49.0	68.4	–	–	–	–
SSG [5]	58.3	80.0	53.4	73.0	13.2	31.6	13.3	32.2
ACT [14]	60.6	80.5	54.5	72.4	–	–	–	–
Ours(k-means)	61.3	82.7	53.9	73.0	15.9	41.7	18.7	45.0
Ours(DBSCAN)	**64.5**	**85.0**	**58.4**	**75.1**	**18.9**	**45.0**	**21.8**	**48.1**

3.3 Comparison with the State-of-the-art

We compare our proposed method with state-of-the-art methods on four domain adaptation person re-ID tasks. The experimental results are shown in Table 1, where D→M represents that the source domain is DukeMTMC-reID [10], the target domain is Market-1501 [15], and M→D is the opposite. It is not difficult to see that in D→M, M→D, M→MSMT, and D→MSMT, our method achieves 64.5%, 58.4%, 18.9%, and 21.8% accuracy, respectively. Experimental results show that the proposed method has better performance on UDA person re-ID tasks.

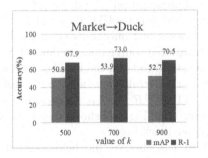

Fig. 2. The effect of different k values on model performance (w/k-means).

In the stage of unsupervised clustering to obtain pseudo-labels, we used k-means to unsupervised clustering of the target domain. Figure 2 shows the impact of different k values on the clustering effect during the clustering process. In the experiment of M→D, the performance is the best when k is set to 700, which shows that the performance of the model has a certain relationship with the number of cluster clusters set in the target domain. When the set value of k is closer to the actual number of identities in the target domain, the better the model performance.

3.4 Ablation Study

In this section, we mainly verify the effectiveness of the proposed method by comparing the results of ablation studies on D→M and M→D (see Table 2).

Fig. 3. Image style translation example. The first line represents the original image; The second line is the style translation image when L_{fb} is not used; The third line is to add the style translation image generated by L_{fb}.

For the convenience of presentation, $C^{(i)}$ denotes CIT ($i = 0, 1, 2$. Denotes the number of CIT training), $R^{(j)}$ denotes UDAR ($j = 0, 1, 2$. Denotes the number of UDAR training). $C^{(0)}$ denotes that only source domain data is used for training. $C^{(1)}$ denotes that the translated source domain dataset S' for training. $C^{(0)}R^{(1)}$ denotes that S and T are used to train the backbone network. $C^{(1)}R^{(1)}$ denotes that the backbone network is trained using the S' and T, only L_{GAN}, L_{cyc} and L_{ac} are applied to CIT. $C^{(2)}R^{(2)} + L_{fb}$ denotes that adding L_{fb} to the style translation module for training. Figure 3 shows the effect of the translated image before and after adding feedback loss.

As shown in Fig. 3, we can observe that before and after using the feedback loss, the quality of the generated image has been significantly improved in terms of color composition and clarity. Figure 4 shows some translated images. It can be seen intuitively that the image translation model can make the lighting and texture of the source domain image close to the style of the target domain image, so as to reduce the discrepancy between domains to a certain extent. The translation between Duke and Market datasets is not very obvious because visually, The image illumination and texture of the two datasets are not very

Table 2. Ablation studies for our proposed framework (w/k-means).

Method	D→M		M→D	
	mAP	R-1	mAP	R-1
$C^{(0)}$	25.2	56.3	16.3	31.1
$C^{(1)}$	30.5	60.5	23.6	45.1
$C^{(0)}R^{(1)}$	53.3	77.6	41.3	62.2
$C^{(1)}R^{(1)}$	57.1	79.3	47.8	68.5
$C^{(2)}R^{(2)} + L_{fb}$(full)	**61.3**	**82.7**	**53.9**	**73.0**

different. This can also explain that the accuracy of domain adaptive experiment between the two datasets is relatively high. Compared with the two datasets, the image discrepancy of the MSMT17 dataset is very obvious. Therefore, the visual discrepancy from the original image can be obviously seen after image translation. The translated image with better quality can be obtained through the method in this paper. At the same time, the semantic consistency of images before and after translation is maintained. The experimental data in Table 1 also shows that it can effectively improve the experimental accuracy.

Fig. 4. Part of the style translation images. The left column is the original images, and the right column is the translated images. The first line is the Market images, the second line is the Duke images, and the third line is the MSMT images.

Table 3 shows the experimental effects of domain adaptation using Wasserstein distance and MMD [6]. Experimental results show that the effect of using Wasserstein distance to measure the discrepancy in feature distribution between two domains is better than MMD [6] on complex tasks such as UDA person re-ID.

Table 3. Comparison of Wasserstein distance and MMD domain adaptation experiment effect (w/k-means).

Method	D→M		M→D	
	mAP	R-1	mAP	R-1
Ours +L_{MMD}	58.5	80.5	51.2	70.2
Ours +L_{wd}	61.3	82.7	53.9	73.0

4 Conclusion

In this paper, we propose a UDA person re-ID method based on image style translation and unsupervised clustering, which combines the labeled dataset with the unlabeled dataset. Among them, CIT can make the source domain image have the visual style of the target domain image and also can initially reduce the discrepancy between the domains. UDAR can further reduce the discrepancy between domains on the basis of CIT through adversarial training so that the backbone can better learn the domain-invariant feature representation. A positive feedback mechanism designed in this paper can combine style translation and unsupervised clustering person re-ID to form a closed-loop. Finally, a more generalized and robust model is obtained by closed-loop training of the two parts.

Acknowledgments. This work is supported by the National Natural Science Foundation of China (Nos. 61866004, 61966004, 61962007), the Guangxi Natural Science Foundation (Nos. 2018GXNSFDA281009, 2019GXNSFDA245018, 2018GXNSFDA294001), Research Fund of Guangxi Key Lab of Multi-source Information Mining & Security (No.20-A-03-01), Guangxi "Bagui Scholar" Teams for Innovation and Research Project, and Innovation Project of Guangxi Graduate Education(JXXYYJSCXXM-2021-007).

References

1. Arjovsky, M., Chintala, S., Bottou, L.: Wasserstein gan (2017)
2. Deng, J., Dong, W., Socher, R., Li, L.J., Li, K., Fei-Fei, L.: ImageNet: a large-scale hierarchical image database. In: 2009 IEEE Conference on Computer Vision and Pattern Recognition, pp. 248–255. IEEE (2009)
3. Deng, W., Zheng, L., Ye, Q., Kang, G., Yang, Y., Jiao, J.: Image-image domain adaptation with preserved self-similarity and domain-dissimilarity for person re-identification (2018)
4. Ester, M., Kriegel, H.P., Sander, J., Xu, X., et al.: A density-based algorithm for discovering clusters in large spatial databases with noise. In: KDD, vol. 96, pp. 226–231 (1996)
5. Fu, Y., Wei, Y., Wang, G., Zhou, Y., Shi, H., Huang, T.S.: Self-similarity grouping: A simple unsupervised cross domain adaptation approach for person re-identification. In: Proceedings of the IEEE/CVF International Conference on Computer Vision (ICCV) (2019)

6. Geng, B., Tao, D., Xu, C.: DAML: domain adaptation metric learning. IEEE Trans. Image Process. **20**(10), 2980–2989 (2011). https://doi.org/10.1109/TIP.2011.2134107
7. Gulrajani, I., Ahmed, F., Arjovsky, M., Dumoulin, V., Courville, A.: Improved training of Wasserstein GANs (2017)
8. He, K., Zhang, X., Ren, S., Sun, J.: Deep residual learning for image recognition. In: Proceedings of the IEEE Conference on Computer Vision and Pattern Recognition, pp. 770–778 (2016)
9. Radford, A., Metz, L., Chintala, S.: Unsupervised representation learning with deep convolutional generative adversarial networks (2016)
10. Ristani, E., Solera, F., Zou, R., Cucchiara, R., Tomasi, C.: Performance measures and a data set for multi-target, multi-camera tracking. In: Hua, G., Jégou, H. (eds.) ECCV 2016. LNCS, vol. 9914, pp. 17–35. Springer, Cham (2016). https://doi.org/10.1007/978-3-319-48881-3_2
11. Song, L., et al.: Unsupervised domain adaptive re-identification: theory and practice. Pattern Recognit. **102**, 107173 (2020)
12. Taigman, Y., Polyak, A., Wolf, L.: Unsupervised cross-domain image generation (2016)
13. Wei, L., Zhang, S., Gao, W., Tian, Q.: Person transfer GAN to bridge domain gap for person re-identification. In: Proceedings of the IEEE Conference on Computer Vision and Pattern Recognition (CVPR) (2018)
14. Yang, F., et al.: Asymmetric co-teaching for unsupervised cross domain person re-identification (2019)
15. Zheng, L., Shen, L., Tian, L., Wang, S., Wang, J., Tian, Q.: Scalable person re-identification: a benchmark. In: Proceedings of the IEEE International Conference on Computer Vision (ICCV) (2015)
16. Zhong, Z., Zheng, L., Luo, Z., Li, S., Yang, Y.: Invariance matters: exemplar memory for domain adaptive person re-identification. In: 2019 IEEE/CVF Conference on Computer Vision and Pattern Recognition (CVPR) (2019)
17. Zhong, Z., Zheng, L., Li, S., Yang, Y.: Generalizing a person retrieval model hetero- and homogeneously. In: Ferrari, V., Hebert, M., Sminchisescu, C., Weiss, Y. (eds.) ECCV 2018. LNCS, vol. 11217, pp. 176–192. Springer, Cham (2018). https://doi.org/10.1007/978-3-030-01261-8_11
18. Zhong, Z., Zheng, L., Zheng, Z., Li, S., Yang, Y.: Camstyle: a novel data augmentation method for person re-identification. IEEE Trans. Image Process. **28**(3), 1176–1190 (2019). https://doi.org/10.1109/TIP.2018.2874313
19. Zhu, J.Y., Park, T., Isola, P., Efros, A.A.: Unpaired image-to-image translation using cycle-consistent adversarial networks. In: Proceedings of the IEEE International Conference on Computer Vision (ICCV) (2017)

Multimodal Named Entity Recognition via Co-attention-Based Method with Dynamic Visual Concept Expansion

Xiaoyu Zhao and Buzhou Tang[✉]

Harbin Institute of Technology (Shenzhen), Shenzhen, China

Abstract. Multimodal named entity recognition (MNER) that recognizes named entities in text with the help of images has become a popular topic in recent years. Previous studies on MNER only utilize visual features or detected concepts from a given image directly without considering implicit knowledge among visual concepts. Taking the concepts not detected but relevant to those in the image into consideration provides rich prior knowledge, which has been proved effective on other multimodal tasks. This paper proposes a novel method to effectively take full advantage of external implicit knowledge, called Co-attention-based model with Dynamic Visual Concept Expansion (CDVCE). In CDVCE, we adopt the concept co-occurrence matrix in a large-scale annotated image database as implicit knowledge among visual concepts and dynamically expand detected visual concepts conditioned on the concept co-occurrence matrix and the input text. Experiments conducted on two public MNER datasets prove the effectiveness of our proposed method, which outperforms other state-of-the-art methods in most cases.

Keywords: Multimodal named entity recognition · External implicit knowledge · Co-occurrence matrix · Multimodal representation

1 Introduction

Named entity recognition plays a fundamental role in natural language processing field and serves as the cornerstone for many downstream tasks such as information extraction, question answering system, machine translating. Recent works argued that it is helpful to recognize textual named entities with the aid of visual clues [10,15,18]. With the emergence of social media platforms, a large amount of multimodal data generated by users has attracted much attention from researchers, and an increasing number of studies have been proposed for multimodal named entity recognition (MNER) in Twitter.

The current mainstream solution to MNER is to enhance the textual representation with visual information. Some researchers [9,10,14,20,22] employed image classification models such as ResNet [6] to extract region-level visual features as a supplement to text. However, there is an insuperable semantic gap

© Springer Nature Switzerland AG 2021
T. Mantoro et al. (Eds.): ICONIP 2021, LNCS 13108, pp. 476–487, 2021.
https://doi.org/10.1007/978-3-030-92185-9_39

Detected Concepts	Concepts with High Co-occurrence
sky	parasail, kite, moon, planes, tree branch, rainbow, clouds, sun, land, parachute, valley, mountain …
clouds	mountain, valley, sky, sun, moon, planes, range, horizon …
bush	tree branch, driveway, lawn, flower, yard, hedge, pathway, dirt road, grasses, twigs, shrubs, roses …
road	motorcycle, manhole, car, crosswalk, line, barricade, driveway, patch, town, tree branch …
rocks	town, mountain, valley, driveway, stream, polar bear, giraffes, rock wall, zebras, shore, enclosures …

@juliefowlis 1 mile from concert hall in [LOC green valley].

Fig. 1. An example of MNER which needs the help of the implicit knowledge among visual concepts. On the left, there is a tweet with an associated image. On the right, there are detected visual concepts with the concepts frequently co-occur with them according to the visual concepts co-occurrence correlations, where the underlined concepts co-occur with more than one detected concept, the concepts in red appear in the image but are not detected and "LOC" indicates the entity type "Location".

between text and the region-level features of image. Some researchers [18,21] adopted object detection models to reduce the semantic gap between text and image. However, these approaches did not explore the implicit knowledge among visual concepts, which may limit the representation ability of image.

The implicit knowledge among visual concepts can be depicted by the co-occurrence relationships among visual concepts, the effectiveness of which has been proved on many other multimodal tasks, such as image-text matching [12, 16], multi-label image recognition [2]. This idea is motivated by the fact that human beings are able to leverage outside-scene knowledge as supplements to comprehend images. For example, in Fig. 1, "sky", "clouds", "road", "bush" and "rocks" except "valley" are detected in the image due to the limitation of the object detection model [11], which could not provide enough visual information for MNER. According to our commonsense knowledge, "valley" co-occurs with "sky", "clouds" and "rocks" frequently. Therefore, we can employ the visual concept co-occurrence matrix to expand the visual concepts appropriately and make a correct inference. Introducing all co-occurred concepts of each detected concept may also introduce noises. For example, "planes" also appears frequently with "sky" and "clouds" but provides little information for MNER in Fig. 1 because they neither appears in the current scene and nor are related to the given text.

To take advantage of image effectively for MNER, we propose a novel method, called Co-attention-based method with Dynamic Visual Concept Expansion (CDVCE). CDVCE first adopts the concept co-occurrence matrix in a large-scale annotated image database as implicit knowledge among visual concepts and then dynamically expand detected visual concepts conditioned on the concept co-occurrence matrix and the input text. Given the detected concepts from a given image, we expand the concepts according to the co-occurrence matrix and regard the expanded ones as supplements. To avoid introducing too many noises from the expanded concepts, we design a Visual Concept Expansion algorithm to select concepts and a Text-guided Concept Self-Attention mechanism to extract the visual features conditioned on the associated words dynamically. The fusion of the expanded and original visual features are aggregated for each word through a multimodal co-attention mechanism.

In summary, our main contributions are listed as below.

- We make the first attempt to introduce external implicit knowledge in MNER task using the visual concepts co-occurrence matrix obtained from an additional large-scale annotated image database.
- We propose a novel Co-attention-based method with Dynamic Visual Concept Expansion (CDVCE) that effectively expands visual concepts.
- The experiments conducted on two public MNER datasets demonstrates that our method yields explainable predictions with better performance than other state-of-the-art methods in most cases.

2   Related Work

2.1   Multimodal NER

Recent years, lots of studies have shown that the performance of NER tasks can be improved by combining multimodal information such as textual and visual information. Early studies attempt to encode text through RNN-based methods and the image regions through CNN [9,10,22]. Moon et al. [10] proposed a multimodal attention module to selectively extract information from words, characters and images. Lu et al. [9] and Zhang et al. [22] incorporated the visual features into textual representation in the early stage and the late stage respectively. Despite the effectiveness of region-level visual information, Wu et al. [18] introduced object-level representation to extract entities precisely. With the great success of the pretrained model in natural language processing, Yu et al. [20] adopted BERT [3] as sentence encoder and ResNet [6] as image encoder, with a multimodal interaction module to capture alignments between words and image. Most previous works ignored the bias of unrelated image information, which causes the cross-modal attention mechanism to produce misleading cross-modality-aware representation. Yu et al. [20] leveraged text-based entity span detection as an auxiliary task to identify textual entities more precisely, which did not explore the relationship between text and image. Sun et al. [14,15] proposed a text-image relationship inference module which was based on a cross-modality transformer and trained on an labeled dataset to produce gated visual features. Zhang et al. [21] adopted text-guided object detection model to obtain text-related objects and introduced graph modeling in MNER tasks, which excluded those objects not relevant to text. However, the above methods only utilized the output of the visual feature extraction model while ignore the occluded or long tail concepts.

2.2 Multimodal Representation

A variety of multimodal fusion methods has been explored to generate multimodal representation.

Bilinear Fusion: To capture the correlations between words and image, bilinear pooling [5] produced multimodal features by computing their outer product, which resulted in generating n2-dimensional representation. To get higher efficiency, plenty of works were proposed and achieved better performance [1].

Cross-Attention-Based Methods: These approaches concatenated image and text as a sequence and fed it to the following feature extraction module. Such methods adopted a single-stream cross-modal transformer to learn deep interactions between two modalities and conducted pretraining tasks such as masked language modeling, masked region modeling, and image-text matching [13].

Co-attention-Based Methods: These approaches maintained sub-networks of all modalities to capture each single modality features and then aggregated features from other modalities [17]. Gao et al. [4] proposed a Dynamic Intra-modality Attention module to compute attention score within single modality conditioned on the information from the other modality. Zhang et al. [23] introduced semantics-based attention to capture latent intra-modal correlations.

3 Proposed Method

As most existing works in MNER, we formulate the problem as a sequence labeling task where given a sentence of n tokens $X = (w_1, w_2, \ldots, w_n)$ with an image V, the goal of the task is to identify the correct spans of entities as well as their types in the sentence, represented by named entity labels $Y = (y_1, y_2, \ldots, y_n)$.

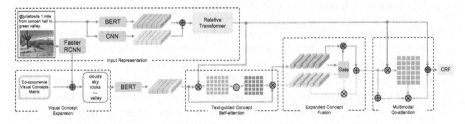

Fig. 2. Architecture of our proposed method CDVCE. It consists of six parts, Input Representation, Visual Concept Expansion, Text-guided Concept Self-attention, Expanded Concept Fusion, Multimodal Co-attention and CRF. The layer normalization and feed forward network are omitted for simplicity.

Figure 2 provides the detailed architecture of CDVCE for MNER. For sentence X with image V, we first use a BERT-based method to represent X and Faster R-CNN to detect objects in V respectively, and then expand the concepts about the detected objects dynamically via the Visual Concept Expansion, Text-guided Concept Self-Attention and Expanded Concept Fusion modules. Subsequently, the image-enhanced text representation is generated by the Multimodal Co-attention Module and fed to a CRF layer to make a prediction.

3.1   Input Representations

Textual Representation. We adopt pre-trained language model BERT[3] to encode the input sequence X, Meanwhile, to overcome the out-of-vocabulary (OOV) problem, CNN is adopted to capture the morphological information of each token. Next, we utilize relative transformer, a transformer-based encoder proposed by Yan et al. [19] with directional relative positional encoding and sharpened attention distribution, to obtain a direction- and distance-aware representation for each token. The representation of X is denoted as $\tilde{W} = [\tilde{w}_0, \tilde{w}_1, \cdots, \tilde{w}_n]$, where $\tilde{W} \in \mathbb{R}^{n \times d}$.

Visual Representation. In order to reduce the semantic gap between text and image, we first deploy Faster R-CNN [11] pretrained on Visual Genome [7], a fine-grained scene graph dataset with object, object attribute and object relationship annotations to detect objects from the input image V, and use the labels of the detected objects as the features of the image instead of the visual features generated by vision models in previous works for better interaction between words and visual concepts. Finally we utilize the embedding layer of BERT to encode these labels. Suppose that the objects detected from V by Faster R-CNN are denoted as $V = (v_0, v_1, \cdots, v_m)$ where m is the number of objects, and the visual representation of V is denoted as $\hat{V} = [\hat{v}_0, \hat{v}_1, \cdots, \hat{v}_m]$, where $\hat{V} \in \mathbb{R}^{m \times d}$.

3.2   Dynamic Visual Concept Expansion

Visual Concept Expansion. To explore knowledge among visual concepts, we expand the objects detected from image V according to the visual concept co-occurrence matrix based on relationships in Visual Genome. The co-occurrence matrix is constructed in the following way. When a relationship appears in an image, the corresponding objects are regarded as co-occurred concepts and the times of this object pair is increased by one. After scanning all images in Visual Genome, we regard the normalized times of all object pairs as the weights of all edges in the co-occurrence matrix. If two concepts appear frequently, the weight of the edge between them in the co-occurrence matrix is high, such as "sky" and "sun", "rocks" and "valley". We denote the visual concept co-occurrence matrix as C, where $C \in \mathbb{R}^{l \times l}$ and l is the size of visual concept vocabulary.

We design a simple rule to limit visual concept expansion to select concepts as close as possible to the detected concepts. Given the visual concepts detected by Faster R-CNN in image X, we retrieve their frequently co-occurrence concepts from C. For each detected concept vi , its frequently co-occurrence concept set is defined as $A_i = \{v_j | C_{ij} > t; v_i, v_j \in V_c, v_i \in V\}$ where t is the lower boundary of frequently co-occurrence weights and V_c is the concept vocabulary. If v_j appears in more than r frequently co-occurrence concept sets, we add it into the expanded visual concept set $V_e = (v_0, v_1, \cdots, v_m, v_{m+1}, v_{m+2}, \cdots, v_{m+e})$ where $(v_0, v_1, \cdots, v_m)$ is the original concepts and e is the number of expanded concepts. If we set r to 3, "planes", which is only related to "clouds" and "rocks" frequently, will not be added. The representation of the expanded visual concept set V_e is denoted as $\hat{V}_e = (\hat{v}_0, \hat{v}_1, \cdots, \hat{v}_{m+e})$ by BERT, where $\hat{V}_e \in \mathbb{R}^{(m+e) \times d}$.

Text-Guided Concept Self-attention. After the visual concept expansion, there is no doubt that we introduce potentially useful but undetected visual concepts into V_e as well as some useless concepts. To reduce helpless concepts, Text-guided Concept Self-attention is proposed to pay more attention on text-related visual concepts dynamically.

The naive self-attention mechanism only utilizes intra-modality information to estimate the object-to-object importance, ignoring the information from another modality. In MNER task, relations between the same visual concepts should have different weights conditioned on different associated text. So we modify the self-attention mechanism to pass the message from text to modeling the intra-modality interactions. First we calculate the similarities between visual concepts and textual tokens. The similarity matrices for V and V_e are:

$$
\begin{aligned}
S^{inter} &= \tilde{W}^T K \hat{V}, \\
S_e^{inter} &= \tilde{W}^T K_e \hat{V}_e,
\end{aligned}
\tag{1}
$$

where $S^{inter}, S_e^{inter} \in \mathbb{R}^{m \times n}$ and $K, K_e \in \mathbb{R}^{d \times d}$ are the weighted matrices. Each element S_{ij}^{inter} in the matrix represents the similarity between visual concept v_i and textual token w_j and the i-th row of the matrix indicates the relevance vector between the whole sentence and concept v_i. The visual concept similarities to text stimulates the intra-modality attention Mechanism to focus more on the concepts with stronger semantic relation with the text and less on those text-unrelated but included in the expansion procedure. Following [23], we assume that if two visual concepts have similar response to the text, they are semantically related to each other. Therefore, we calculate the similarity between the concept-to-sentence relevance vectors to measure the similarity of the concepts themselves. The intra-modality similarity matrix S^{intra} can be calculated by

$$
S^{intra} = S^{inter} K^{intra} S^{inter T},
\tag{2}
$$

where $K^{intra} \in \mathbb{R}^{\times}$ is the weighted matrix.

Then, we can update visual concept representation $\tilde{V}, \tilde{V}_e$ as follows:

$$
\begin{aligned}
\tilde{V} &= softmax(S^{intra})\hat{V}, \\
\tilde{V}_e &= softmax(S_e^{intra} \odot D_{V_e})\hat{V}_e.
\end{aligned}
\tag{3}
$$

Finally, we apply the residual connection operation, layer normalization and feed forward network to construct a transformer-based block.

Expanded Concept Fusion. We regard the original detected visual concepts and the expanded ones as two independent channels. To fuse the complementary information within the different channels, following [8], we apply a gate mechanism to calculate the ratio to maintain the information from each channel as follows:

$$
\begin{aligned}
M &= ReLU(FC([\tilde{V}, \tilde{V}_e, \tilde{V} - \tilde{V}_e, \tilde{V} \odot \tilde{V}_e])), \\
G &= Sigmoid(FC(M)),
\end{aligned}
\tag{4}
$$

where FC is fully-connected layer and $[\cdot, \cdot]$ denotes concatenation operation.

$$
\mathring{V} = G \odot \tilde{V} + (1 - G) \odot \tilde{V}_e.
\tag{5}
$$

$\mathring{V}$ is the final output of Dynamic Visual Concept Expansion and G is the gate value.

3.3 Multimodal Co-attention

This module models the interactions between words and visual concepts to aggregate visual information for each word. We employ co-attention mechanism to generate image-enhanced textual representation $\mathring{W}$ as follows:

$$
\begin{aligned}
CoATT(Q, K) &= softmax(\frac{QK^T}{\sqrt{d}})K, \\
\mathring{W}_{att} &= LN(CoATT(\tilde{W}, \mathring{V}) + \tilde{W}), \\
\mathring{W} &= LN(FFN(\mathring{W}_{att}) + \mathring{W}_{att}),
\end{aligned}
\tag{6}
$$

where $LN(\cdot)$ denotes layer normalization and $FFN(\cdot)$ represents feed forward network. Finally, a Conditional Random Fields (CRF) layer is applied to improve tagging accuracy considering the correlations between neighbour labels by

$$
\hat{y}_i = \arg\max_{y_i \in L} \frac{exp(E \cdot \mathring{W}_i + b)}{\sum_{y_{i-1}y_i} exp(E \cdot \mathring{W}_i + b)},
\tag{7}
$$

where $\hat{y}_i$ is the prediction for i-th token, W and b are trainable parameters to model the transition from y_{i-1} to y_i.

4 Experiments

We conduct experiments on two public MNER datasets (Twitter 2015 [22] and Twitter 2017 [9]) to evaluate the effectiveness of our CDVCE method by comparing it with other state-of-the-art methods.

4.1 Datasets and Experimental Settings

Datasets. The tweets with their associated images in 2015 and 2017 respectively collected by Zhang et al. [22] and Lu et al. [9] through Twitter's API, that is, Twitter 2015 and Twitter 2017, are used for experiments. In both two datasets, four types of named entities including Person, Location, Organization and Misc are annotated. The statistics of the datasets are listed in Table 1.

Experimental Settings. In this study, we use the BIO scheme to represent named entities. The co-occurrence visual concept matrix generated from Visual Genome contains a concept vocabulary of 1,600 concepts. For textual representation, we use BERT-base-cased to initialize BERT. We set the max length of sentences as 64, the sizes of kernels in CNN as 2, 3 and 4, the number of layers of relative transformer as 2. For visual representation, the amount of concepts in one image detected by Faster R-CNN is set as 8. The lower boundary of frequently co-occurrence weights t is set as 0.01. Only the concepts in more than 3 frequently co-occurrence concept sets of the detected objects are selected for expansion. We set the number of expanded concepts as 32, the number of layers of the Text-guided Concept Self-attention Module as 2, the learning rate for BERT as $3e-5$, the learning rate for the other modules as $1e-4$, the dropout rate as 0.3, the weight decay as 0.05, and the batch size as 16 respectively. Precision (P), recall (R) and F1-score are used as performance evaluation metrics.

Table 1. Statistics of Twitter 2015 and Twitter 2017

Entity type	Twitter 2015			Twitter 2017		
	Train	Dev	Test	Train	Dev	Test
Person	2217	552	1816	2943	626	621
Location	2091	522	1697	731	173	178
Organization	928	247	839	1674	375	395
Misc	940	225	726	701	150	157
Total	6176	1546	5078	6049	1324	1351
Tweets	4000	1000	3257	3373	723	723

4.2 Results

We start with three baselines, i.e., **BiLstm-CRF**, **BERT-CRF** and **BERT-CRF-Rel**, that are state-of-the-art methods only using textual information, and then compare **CDVCE** with other state-of-the-art methods, i.e., **OCSGA**, **UMT**, **UMGF** and **RpBERT**, using both textual and visual information in different settings. A brief introduction to these methods is shown in as follows:

BiLstm-CRF is the method that only uses BiLstm to encode input text and CRF for label prediction. The word embeddings of words are initialized as Glove embeddings. **BERT-CRF** is the method that only uses BERT to encode input text and CRF for label sequence prediction. **BERT-CRF-Rel** is an extension of BERT-CRF that utilizes CNN to capture character features and adopt a 2-layer relative transformer [19] to obtain a direction- and distance-aware representation for each token. **OCSGA** [18] is the first multi-modal method to explore object-level features for MNER, which utilized BiLstm as text encoder and initialize object embeddings with the Glove embeddings. **UMT** [20] is the first multi-modal method to adopt transformer to encode text and image with a span detection task to alleviate the bias of incorporating visual features. **UMGF** [21] is another multi-modal method based on transformer that utilizes a text-guided object detection model to detect objects in the image and a unified graph to model the interactions between text and the targeted visual nodes. **RpBERT** is a multi-task method for MNER and text-image relation classification, where an external Twitter annotation dataset with text-image relations is used for text-image relation classification. RpBERT-base and RpBERT-large are two versions of RpBERT that use basic BERT and BERT-large pretrained on external large-scale Twitter data respectively.

Table 2. Comparison of different methods. The results of the methods marked with † are obtained from published papers, while that marked with ‡ are obtained by rerunning experiments using the official implementation.

Approaches	Mechanism	Twitter 2015			Twitter 2017		
		P	R	F1	P	R	F1
BiLstm-CRF	None	66.54	68.12	67.32	79.29	78.34	78.81
OCSGA† [18]	BiLstm-based + object-level	74.71	71.21	72.92	–	–	–
OCSGA‡ [18]	BiLstm-based + object-level	73.44	**71.71**	72.56	81.78	80.43	81.10
BiLstm-CRF + CDVCE	BiLstm-based + object-level	**74.93**	71.21	**73.02**	83.52	80.68	82.07
BERT-CRF	None	70.55	74.82	72.63	85.18	82.09	83.60
BERT-CRF-Rel	Relative transformer	71.59	74.90	73.21	83.95	85.20	84.57
UMT† [20]	Transformer-based + region-level	71.67	75.23	73.41	85.28	85.34	85.31
UMGF† [21]	Transformer-based + object-level	74.49	75.21	74.85	86.54	84.50	85.51
RpBERT-base† [15]	BiLstm + transformer-based + region-level	–	–	74.40	–	–	87.40
RpBERT-large† [15]	BiLstm + transformer-based + region-level	–	–	74.90	–	–	**87.80**
CDVCE (ours)	Transformer-based + object-level + knowledge	72.94	**77.66**	**75.23**	**86.57**	**87.79**	87.17

In order to illustrate the effectiveness of CDVCE, we apply the similar structure to BiLstm-CRF, that is, BiLSTM-CRF+CDVCE. The comparison of different methods on Twitter 2015 and Twitter 2017 is shown in Table 2.

It can be observed that the multi-modal methods are superior to uni-modal methods in MNER and CDVCE outperforms all other state-of-the-art methods on Twitter 2015 and all other state-of-the-art methods except RpBERT-large on Twitter 2017. The reason why CDVCE does not achieve better performance than RpBERT-large on Twitter 2017 is that the relation propagation based on the text-image relation classification task in RpBERT brings an significant improvement. As reported in [15], the relation propagation brings an improvement of 1.2 in F1 for RpBERT-base. Therefore, it is unfair to compare CDVCE with RpBERT that uses an external unpublic annotation dataset for another joint task. In spite of this, CDVCE outperforms RpBERT on Twitter 2015. Compared to the RpBERT method that used basic BERT the same as CDVCE, CDVCE achieves much higher F1 by about 0.8% on Twitter 2015. On Twitter 2017, CDVCE outperforms RpBERT-base without using relation propagation by about 0.9% (87.17% vs 86.2%) according to the results reported in [15]. Introducing relation propagation into CDVCE may also bring improvement, which is one direction of our future work.

Compared to OCSGA, BiLstm-CRF+CDVCE achieves higher F1 on both Twitter 2015 and Twitter 2017. Compared to UMGF, CDVCE achieces much higher F1 by 0.38% and 1.66% on Twitter 2015 and Twitter 2017 respectively. These results indicate that the proposed dynamic visual concept expansion is useful for MNER by introducing external knowledge and brings more improvement than the mechanisms for multi-modal information fusion in other methods.

4.3 Ablation Studies and Case Studies

We also conduct ablation studies on CDVCE by removing each module of dynamic visual concept expansion. Table 3 shows the results, where "w/o" denotes "without", "Expansion" denotes the visual concept expansion module, "Text-guidance" denotes the text-guided concept self-attention module, and "Fusion" denotes the expanded concept fusion module. Each module is beneficial

Table 3. Ablation study of our CDVCE.

Approaches	Twitter 2015			Twitter 2017		
	P	R	F1	P	R	F1
CDVCE(ours)	**72.94**	**77.66**	**75.23**	**86.57**	87.79	**87.17**
- w/o Expansion	73.16	76.22	74.66	85.43	87.64	86.52
- w/o Text-guidance	72.91	76.67	74.74	85.83	**87.86**	86.83
- w/o Fusion	72.37	77.36	74.78	84.67	87.49	86.06

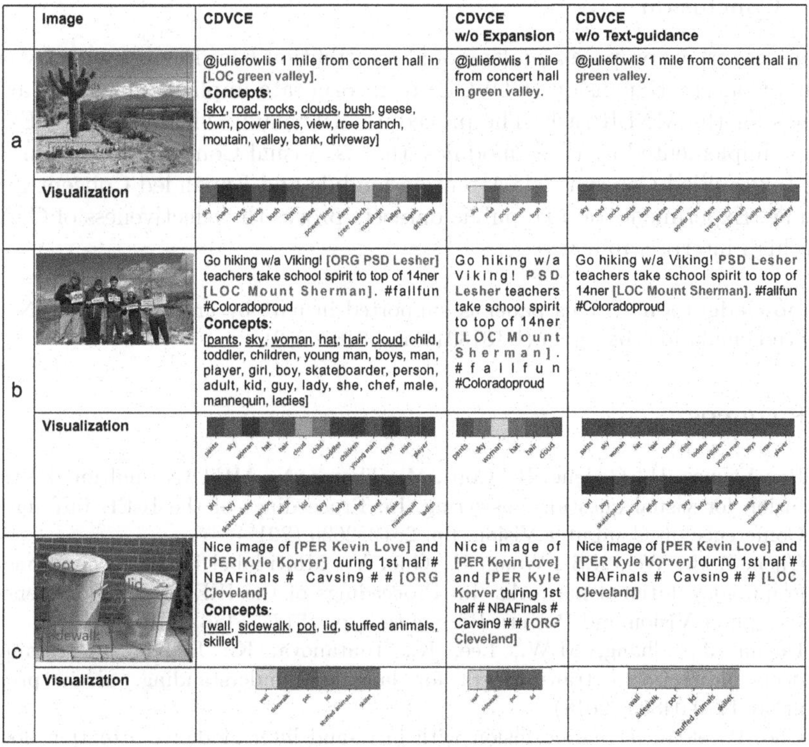

Fig. 3. Examples tested by CDVCE in different settings. The detected concepts are highlighted in underline, the text fragments in blue are named entities correctly recognized, the text fragments in red are named entities not or wrongly recognized, and the concepts with visualized attentions in brighter colors are more relevant to the sentences. (Color figure online)

for CDVCE. Among the three modules of dynamic visual concept expansion, the visual concept expansion module is the biggest influencing factor on CDVCE. It proves that external knowledge is very import for MNER and CDVCE provides an effective way to take full advantage of external knowledge again. We further conduct the visualization analysis on several examples to prove the interpretability of our proposed method. The visual concepts in images detected by Faster RCNN, the named entities recognized by CDVCE in different settings, the concepts related to the detected visual concepts and the visualized attentions of the related concepts to sentences about three samples from the two datasets (a and b from Twitter 2015, and c from Twitter 2017) are shown in Fig. 3.

It is clear that the visual concept expansion module can expand related concepts effectively, and the text-guided concept self-attention module can model implicit relationships between images and sentences. For samples a and b, more entities are correctly detected by CDVCE because of the visual concept expansion module. For sample c, the location entity are correctly detected by CDVCE because of the Text-Guided Self-Attention module.

4.4   Conclusion

In this paper, we propose a multi-modal co-attention-based model with Dynamic Visual Concept Expansion (CDVCE) to incorporate implicit knowledge among images for the MNER task. The procedure of Dynamic visual Concept Expansion is implemented by three modules, that is, Visual Concept Expansion module, Text-guided Concept Self-Attention module and Expanded Concept Fusion module. Experiments on two public datasets proves the effectiveness of CDVCE with high interpretability.

Acknowledgements. This paper is supported in part by grants: National Natural Science Foundations of China (U1813215 and 61876052).

References

1. Ben-Younes, H., Cadene, R., Cord, M., Thome, N.: MUTAN: multimodal tucker fusion for visual question answering. In: Proceedings of the IEEE International Conference on Computer Vision, pp. 2612–2620 (2017)
2. Chen, Z.M., Wei, X.S., Wang, P., Guo, Y.: Multi-label image recognition with graph convolutional networks. In: Proceedings of the IEEE/CVF Conference on Computer Vision and Pattern Recognition, pp. 5177–5186 (2019)
3. Devlin, J., Chang, M.W., Lee, K., Toutanova, K.: BERT: pre-training of deep bidirectional transformers for language understanding. arXiv preprint arXiv:1810.04805 (2018)
4. Gao, P., et al.: Dynamic fusion with intra-and inter-modality attention flow for visual question answering. In: Proceedings of the IEEE/CVF Conference on Computer Vision and Pattern Recognition, pp. 6639–6648 (2019)
5. Gao, Y., Beijbom, O., Zhang, N., Darrell, T.: Compact bilinear pooling. In: Proceedings of the IEEE Conference on Computer Vision and Pattern Recognition, pp. 317–326 (2016)

6. He, K., Zhang, X., Ren, S., Sun, J.: Deep residual learning for image recognition. In: Proceedings of the IEEE Conference on Computer Vision and Pattern Recognition, pp. 770–778 (2016)

7. Krishna, R., et al.: Visual genome: connecting language and vision using crowd-sourced dense image annotations. Int. J. Comput. Vis. **123**(1), 32–73 (2017)

8. Liu, L., Zhang, Z., Zhao, H., Zhou, X., Zhou, X.: Filling the gap of utterance-aware and speaker-aware representation for multi-turn dialogue. arXiv preprint arXiv:2009.06504 (2020)

9. Lu, D., Neves, L., Carvalho, V., Zhang, N., Ji, H.: Visual attention model for name tagging in multimodal social media. In: Proceedings of the 56th Annual Meeting of the Association for Computational Linguistics (Volume 1: Long Papers), pp. 1990–1999 (2018)

10. Moon, S., Neves, L., Carvalho, V.: Multimodal named entity recognition for short social media posts. arXiv preprint arXiv:1802.07862 (2018)

11. Ren, S., He, K., Girshick, R., Sun, J.: Faster R-CNN: towards real-time object detection with region proposal networks. arXiv preprint arXiv:1506.01497 (2015)

12. Shi, B., Ji, L., Lu, P., Niu, Z., Duan, N.: Knowledge aware semantic concept expansion for image-text matching. In: IJCAI, vol. 1, p. 2 (2019)

13. Su, W., et al.: Vl-BERT: pre-training of generic visual-linguistic representations. arXiv preprint arXiv:1908.08530 (2019)

14. Sun, L., et al.: RIVA: a pre-trained tweet multimodal model based on text-image relation for multimodal NER. In: Proceedings of the 28th International Conference on Computational Linguistics, pp. 1852–1862 (2020)

15. Sun, L., Wang, J., Zhang, K., Su, Y., Weng, F.: RpBERT: a text-image relation propagation-based BERT model for multimodal NER. arXiv preprint arXiv:2102.02967 (2021)

16. Wang, H., Zhang, Y., Ji, Z., Pang, Y., Ma, L.: Consensus-aware visual-semantic embedding for image-text matching. In: Vedaldi, A., Bischof, H., Brox, T., Frahm, J.-M. (eds.) ECCV 2020. LNCS, vol. 12369, pp. 18–34. Springer, Cham (2020). https://doi.org/10.1007/978-3-030-58586-0_2

17. Wang, Y., Huang, W., Sun, F., Xu, T., Rong, Y., Huang, J.: Deep multimodal fusion by channel exchanging. In: Advances in Neural Information Processing Systems, vol. 33 (2020)

18. Wu, Z., Zheng, C., Cai, Y., Chen, J., Leung, H.F., Li, Q.: Multimodal representation with embedded visual guiding objects for named entity recognition in social media posts. In: Proceedings of the 28th ACM International Conference on Multimedia, pp. 1038–1046 (2020)

19. Yan, H., Deng, B., Li, X., Qiu, X.: TENER: adapting transformer encoder for named entity recognition. arXiv preprint arXiv:1911.04474 (2019)

20. Yu, J., Jiang, J., Yang, L., Xia, R.: Improving multimodal named entity recognition via entity span detection with unified multimodal transformer. Association for Computational Linguistics (2020)

21. Zhang, D., Wei, S., Li, S., Wu, H., Zhu, Q., Zhou, G.: Multi-modal graph fusion for named entity recognition with targeted visual guidance (2021)

22. Zhang, Q., Fu, J., Liu, X., Huang, X.: Adaptive co-attention network for named entity recognition in tweets. In: Proceedings of the AAAI Conference on Artificial Intelligence, vol. 32 (2018)

23. Zhang, Q., Lei, Z., Zhang, Z., Li, S.Z.: Context-aware attention network for image-text retrieval. In: Proceedings of the IEEE/CVF Conference on Computer Vision and Pattern Recognition, pp. 3536–3545 (2020)

Ego Networks

Andrei Marin, Traian Rebedea, and Ionel Hosu[(✉)]

University Politehnica of Bucharest, Bucharest, Romania
andrei_mihai.marin@stud.etti.pub.ro,
{traian.rebedea,ionel.hosu}@cs.pub.ro

Abstract. Games on the Atari 2600 platform have served as a benchmark for reinforcement learning algorithms in recent years, and while deep reinforcement learning approaches make progress on most games, there are still some games that the majority of these algorithms struggle with. These are called hard exploration games. We introduce two new developments for the Random Network Distillation (RND) architecture. We apply self-attention and the mechanism of *ego motion* on the RND architecture and we evaluate them on three hard exploration tasks from the Atari platform. We find that the proposed ego network model improve the baseline of the RND architecture on these tasks.

Keywords: Deep reinforcement learning · Ego networks · Self-attention · Hard exploration · Atari

1 Introduction

Atari 2600 games have been used as a benchmark for reinforcement learning algorithms since the introduction of ALE [4]. While a large number of methods have made progress on most games on this platform, some still remain a challenge because of the sparsity of their rewards, meaning that a lot of actions have to be performed by the agent in the environment before a potential reward is received. Dealing with this type of games requires using either imitation learning [1], in which the agent learns to navigate its environment from demonstrations taken from a human expert, or by using intrinsic motivation, curiosity, and other methods that try to inform the agent's exploration in the absence of dense extrinsic rewards [3,5,7,16]. In Atari, 7 games have been classified as hard exploration games with sparse rewards: Montezuma's Revenge, Venture, Gravitar, Private Eye, Pitfall!, Solaris and Freeway. We test our method on 3 out of those 7 games, namely Montezuma's Revenge, Gravitar and Pitfall!.

In 2018 OpenAI released Random Network Distillation (RND) [7], which was one of the first methods to drastically improve the state of the art in Montezuma's Revenge, without the use of any human expert demonstrations. It uses intrinsic rewards generated from the prediction MSE error $\|\hat{f}(x;\theta) - f(x)\|^2$ between a fixed randomly initialized neural network, named the target network $f : \mathcal{O} \to \mathbb{R}^k$, and a predictor network $\hat{f} : \mathcal{O} \to \mathbb{R}^k$ that learns to predict the outputs of the target network given the same input to both networks. The intuition

T. Mantoro et al. (Eds.): ICONIP 2021, LNCS 13108, pp. 488–499, 2021.
https://doi.org/10.1007/978-3-030-92185-9_40

is that the predictor network will be able to correctly predict the output of the target network for inputs that are similar to previously seen ones, and it will fail to correctly predict for novel inputs, generating a higher error and consequently a higher intrinsic reward. They did experiments with both a CNN and a RNN policy. While on average they performed similar, they found that the RNN performed overall better and in one run managed to find a best return of 17500 and managed to explore all of the 24 rooms of the first level. We experimented only with the CNN policy, because we found better results with it early in our research, but in the future we should also experiment with the RNN policy.

Starting from RND as our baseline, we experimented with two approaches. In the first we implemented the self-attention mechanism [6] by applying it to two layers of the CNN policy in RND. In the second approach we extract a window centered on the agent in the 2D environment and add it as an input to the policy network. We call this approach the *Ego Network*. While both of these methods could be applied to any reinforcement learning network architecture, we chose to build it on top of RND because it already had good results on Montezuma's Revenge and other hard exploration games of interest.

Our main contributions are as follows:

1. We applied a self-attention mechanism on the last two convolutional layers of the CNN policy defined in RND [7] and found that the agent managed to explore more rooms, and yielded more rewards than the baseline RND.
2. We used a small window centered on the agent in order for what we call the Ego Network to have information about the agent's position in the environment and nearby objects. The results of this experiments show that the agent can outperform both the baseline RND and the RND with self-attention on Montezuma's Revenge.
3. We did an ablation study to find if the Ego Network can be used without the self-attention mechanism and found promising results.

2   Related Work

Attention based methods have been previously applied in RL. Some of these combine attention with LSTM networks to attend only to sections of the input images and make the models more interpretable [9,15,19].

Recently [21] applied self-attention on patches extracted from the input images in order to find the most relevant ones, and then these are converted to features using neuroevolution. They use this method because some operations that find the most relevant patches are not differentiable. Then these features are fed to a RNN controller to get the actions.

Methods similar to Ego Network include [11], in which an agent learns to interact with objects in its environment by being able to predict ego motion as part of a world model. The most notable similarity between this method and ours is the idea that learning the position of the agent is crucial to proper interaction and exploration of the environment. Another similar method is CoEX [10] which constructs grid cells in order to learn the localization of the agent using spatial attention. The self-localization mechanism that they construct is similar to ours, but the grid cells are used to create a discrete space in which an agent can be at a given time. While this helps with predicting the location of the agent in an automatic way, we benefit from having a continuously moving window centered on the agent at the expense of being feature engineered.

At the end we will make a comparison between Ego RND and multiple methods that perform well on Montezuma's Revenge and Gravitar. We will now briefly describe some of them.

Ape-X [12] introduced a distributed architecture for training agents that decouples the acting in the environment from the learning part. It uses a prioritized experience replay buffer in order to sample experience collected by a large number of agents. R2D2 [13] is very similar to Ape-X [12] in that it generates experiences in the environment using a lot of actors and then uses those batches to train a single centralized neural network. Some changes made by R2D2 to the Ape-X method include a LSTM layer after the convolution stack, a higher discount rate, not ending the episode when the agent died and others.

MuZero [18] is a model-based RL that builds a model that does not predict the whole environment state, like most other model-based reinforcement learning algorithms, but predicts elements that are relevant for planning, namely the policy, the value function and the immediate rewards. Never Give Up [3] tackles the problem of revisiting states, even in the absence of intrinsic rewards, by defining an intrinsic reward based on episodic and life-long novelty. It then learns an exploitation and an exploration policy that tries to maximize the score under the exploitation policy. Agent57 [2] is another successful method applied on Montezuma's Revenge, that unlike previous methods they outperform the human baseline on all Atari games. They based their work on NGU [3].

3   Method

We make two additions starting from RND. We first apply the self-attention mechanism by [6] on the last two convolution layers of the RND CNN policy.

The second addition extracts from the normal $210 \times 160 \times 3$ ALE frame, a smaller $51 \times 30 \times 3$ window frame centered on the agent which we call the ego window. The inspiration of this approach was driven mainly by [11] which places an agent in a 3D environment in which multiple objects of different shapes are present, and is then asked to learn a model of its environment by interacting with it through a first person perspective. The intuition behind the extraction of this ego window from the original frame was to help the policy network have easier access to the position of the agent in the environment and its nearby objects, such as enemies or objects with which it can interact. The normal and ego frames for Montezuma's Revenge are available in Fig. 1.

Instead of giving the policy network only the normal frame as input, we also give it the ego window. Finding the agent so that we could extract the ego window and setting the dimensions of the window was done manually, but in theory they could be automated. To manually extract the ego window we use

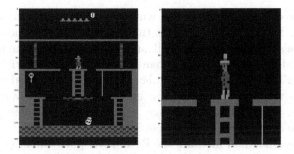

Fig. 1. Montezuma's Revenge frame. On the left side we can see the normal 210×160 frame, while on the right side we can see the ego window with size 51×30

different color matching algorithms depending on the colors representing the agent in every game that we considered. We tried to do template matching to find the agent, but we were unsuccessful because sometimes different parts of the agent would overlap with other parts. For example in Montezuma's Revenge we found frames where in the middle of the jump animation the agent's head would overlap with its torso. Finding the agent using its color was simpler because of the limited palette of colors present in Atari games in general and the fact that the agent had at least one color that was different from colors of the other objects in its environment.

3.1   Architecture

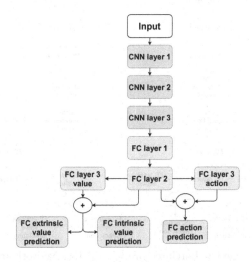

Fig. 2. Baseline RND CNN architecture

Setting the dimensions of the ego window was done by trial and error with different sizes. The general intuition is that because enemies and obstacles both in Montezuma's Revenge and Pitfall! are generally on the same level as the agent, the window should be wider than it is taller. After multiple experiments on Montezuma's Revenge training for about two hours we settled on using an ego window with size 51 × 30. We use this ego window size for every environment we trained on. Apart from this change, we use the exact same pre-processing pipeline as the original RND implementation.

Our proposed method modifies the RND CNN policy architecture [7] in two ways. First we replace the last convolution layer with two attention augmented convolutions [6] just before the first fully connected layer as seen in Fig. 3a. We then add the ego window mechanism on top of the CNN policy augmented with attention as seen in Fig. 3b and name this the Ego Network.

As a reference the baseline RND CNN policy architecture can be seen in Fig. 2. These figures represent a high level overview of the layers within the CNN policy neural network, more detailed information can be found in the RND paper.

The total trainable parameters of the baseline architecture amount to about 5 million, while the addition of the two convolution layers augmented with attention amount to about 15 million parameters. Each attention layer implements 8 attention heads and 256 dimensional query, key and value. Because of the limited time and hardware at our disposal, the way we found the best size of the attention layer, and the number of attention layers, is by training for about 5 h and then comparing the results between multiple settings.

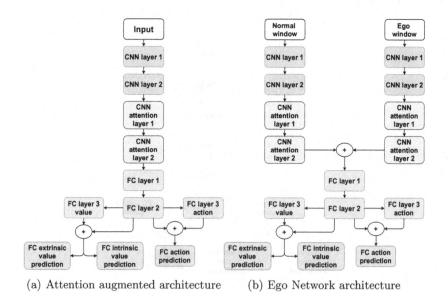

(a) Attention augmented architecture   (b) Ego Network architecture

Fig. 3. Proposed CNN policy neural network architectures

The Ego Network's only change is that it uses one additional input source. This input source is given by the ego window for every corresponding frame in order for the policy network to learn from both the normal 84 × 84 input frame, and from this ego window of size 51 × 30, which is processed the same as the input frame. The Ego Network has about 18.5 million learnable parameters.

4   Experiments

We experimented on three Atari games: Montezuma's Revenge, Gravitar and Pitfall. On each game we applied the attention augmented convolutions from [6] on the CNN policy of RND. We call this method the Attention Augmented RND (AA-RND). We then apply the Ego Network architecture from Fig. 3b on each game, and call this method the Ego RND. In order to validate the ego window contribution in the absence of the attention augmentation we also ran an agent without the attention augmented convolutional layers.

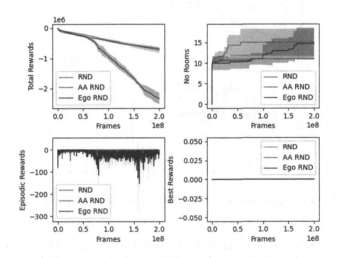

Fig. 4. Comparison of RND, AA RND and Ego RND on Pitfall

Almost all the hyperparameters that came by default with the RND implementation found in OpenAI's GitHub remained unchanged. What we did change is the extrinsic discount factor which we increased from 0.99 to 0.999, because it yielded better outcomes in all the experiments we performed, including the original approach. We also increased the intrinsic coefficient from 1 to 3 but only on our experiments with Ego RND. Increasing the intrinsic coefficient was necessary for the Ego RND agent because it became prone to only focus on finding extrinsic rewards, that it mostly stopped exploring. In order to balance this we found that increasing its intrinsic coefficient enabled the agent to continue exploring. We also introduced four other hyperparameters. On the attention augmentation side we introduced the number of attention heads and the number of dimensions of the query, key and value vectors. On the Ego Network side we introduce the ego window height and width. All games that we trained on used the same hyperparameters values, and all our experiments were trained on 200 million frames. For Montezuma's Revenge we averaged the results over 4 seeds, while for every other game we averaged the results over 3.

For the figures that we will present next, let's define the meaning of each plot. The *Total Rewards* plot represents the accumulated rewards from all the episodes that the agent was trained for. *No Rooms* represents the total number of rooms visited by the agent. *Best Rewards* represents the highest total reward gained by an agent in an episode and *Episodic Rewards* represent the average total reward between multiple episodes.

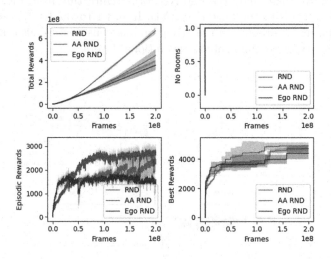

Fig. 5. Comparison of RND, AA RND and Ego RND on Gravitar

We will now start by comparing the results on Pitfall. These can be seen in Fig. 4 and they show that both AA RND and the baseline RND managed to get about the same number of rewards throughout training, but RND managed to explore a higher number of rooms. Ego RND performed by far the worse, and while higher variance in episodic rewards might be a sign of exploration, the Ego RND agent didn't manage to find higher rewards than the other approaches.

We next compare the methods on Gravitar. Figure 5 shows that AA RND performs better in every aspect compared to the other two approaches. Ego RND and the baseline RND were tied most of the training, with the Ego RND performing worse at the end. The number of rooms in this plot is irrelevant because none of the approaches managed to explore more than 1. Interestingly enough on our runs the baseline RND implementation trained for more than 100 million frames managed to achieve the highest rewards above 4000, whereas in [7] it didn't manage to break above 4000. While we didn't change any reward multiplier present in the RND implementation, it might be that they ran their experiments with a lower multiplier. Still the AA RND approach managed to find higher rewards than the baseline RND.

The main game on which we experimented and got the best performance is Montezuma's Revenge. These results can be seen in Fig. 6, and while arguably in terms of the total collected rewards, RND performed better than our approaches, AA RND and Ego RND managed to find higher rewards and explore more rooms than the baseline RND. On our repeated experiments on this environment, we found that AA RND generally explored 19 rooms, and its best rewards didn't go higher than 7900, while Ego RND generally manages to explore 20 rooms, and reaches the highest reward of 9500.

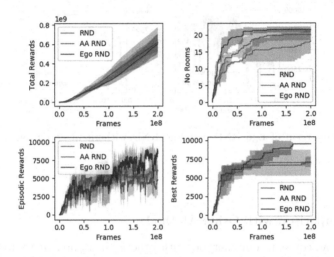

Fig. 6. Comparison of RND, AA RND and Ego RND on Montezuma's Revenge

4.1 Ablation Study

Because in all of our previous experiments, we built the ego network on top of the attention augmented network, and this network came with a lot more parameters than the baseline RND approach, we wished to find the actual contribution of the ego window to the results, and see if it can function on its own.

Compared to the 18.5 million parameters used in the Ego RND with attention, training Ego RND without attention only required 6.15 million parameters. About 1 million parameters more than the baseline RND method.

In Fig. 7 we can see the results from this ablation study, comparing the results of the Ego RND with and without the attention part. From our experiment we found that the results of the Ego RND without attention were pretty similar to the ones with attention. The Ego RND without attention managed to explore 1 more room and had a lower variance in episodic rewards, but also collected fewer rewards than the Ego RND with attention. Because of the high amount of time needed to train every experiment, we only managed to train this type of network once, so these are the results on only one seed.

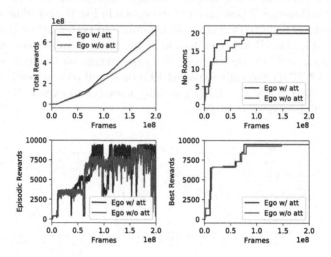

Fig. 7. Comparison of Ego RND and Ego RND without attention on Montezuma's Revenge

4.2   Comparison with Other Methods

We will now compare our method with others on Montezuma's Revenge and Gravitar. Because of the lack of results on Pitfall we have decided to skip such a comparison. Because there is a gap in performance between reinforcement learning algorithms that use imitation learning and those that use some form of intrinsic motivation and curiosity, we decided to restrict the comparison only to the latter type.

Table 1 shows the comparison between different methods applied on Montezuma's Revenge. The Score field represents the best mean episodic reward, and a timestep is equal to 4 frames. So for example our training of 50 million timesteps correspond to 200 million frames. Most of those methods only report the score and number of timesteps for which the algorithm was trained. Among those that did report were NGU with a maximum of 50 rooms when reaching the third level and a mean of 25 rooms explored and RND which explored a maximum of 24 rooms and a mean of 22. The methods present in this table were averaged over 3 seeds. Our best run on Montezuma's Revenge reached 21 rooms, and our mean is of 20.

Table 1. Table comparison between different methods on Montezuma's Revenge [10]

Method	Timesteps	Score
NGU [3]	8.75B	16,800
PPO+CoEX [10]	500M	11,618
Agent57 [2]	25B	9,352
Ego RND (ours)	50M	9,200
RND (original paper)	400M	7,570
A2C+CoEX [10]	100M	6,635
RND (our setup)	50M	6,600
DDQN+ [20]	25M	3,439
Sarsa-ϕ-EB [14]	25M	2,745
R2D2 [13]	–	2,666
DQN-PixelCNN [17]	37.5M	2,514
Curiosity-Driven [8]	25M	2,505

The RND in the table corresponds to a run on 32 parallel environments that trained for 400 million timesteps and got a mean reward of 7570 and the highest reward of 17500. They had other runs for which the mean reward was higher, but it was done on 1024 parallel environments and we thought that the one with 32 parallel environments was more comparable to our runs.

Table 2. Table comparison between different methods on Gravitar

Method	Timesteps	Best score
Agent57 [2]	25B	19,213
R2D2 [13]	–	15,680
NGU [3]	8.75B	14,200
MuZero [18]	4B	6,682
Ego RND (ours)	50M	4,900
RND (our setup)	50M	4200
RND (original paper)	500M	3,906
Ape-X [12]	5.7B	1,599

Table 2 shows the comparison between different methods on Gravitar. Here we used the Best Score field which represents the highest reward the methods have achieved during training.

5   Conclusion

Starting from [7] we added two new approaches, namely a self-attention mechanism [6] and the Ego Network. We tested these two methods on 3 hard exploration games from the Atari 2600 platform: Montezuma's Revenge, Gravitar and Pitfall.

All our experiments were done on 200 million frames collected from each environment. On Montezuma's Revenge we had the most performance increase with both our methods, while on Gravitar only AA RND managed to perform better than RND and Ego RND. The failure of Ego RND might be explained by the fact that the ego window did not have enough context to catch interactions between the enemies and the agent, so a bigger ego window might increase the performance of Ego RND. In Pitfall we had the worst performance, with RND performing better than AA RND and with Ego RND performing the worst.

In order to validate the performance of Ego RND without the self-attention modules we did an ablation study and found that the Ego RND with self-attention performed only slightly better on Montezuma's Revenge. This was especially important given that the self-attention mechanism tripled the number of parameters. The Ego RND without attention has about 1 million more parameters than the baseline RND, seeing that we could get such a similar result with a few more parameters is encouraging.

As a future work we would like to be to find the ego window automatically, without having to find the agent in the environment using color matching. We found one approach [10] that is capable of finding the agent in Montezuma's Revenge, Venture and others in an automatic way. But it has some limitations in the way it positions the agent inside the window, basically finding the agent inside the grid, instead of having a continuous moving window centered on the agent.

There are still problems inherent in the RND method such as the "Dancing with Skulls" identified in their paper, in that the agent seeks dangerous interactions with objects such as the skulls present in Montezuma's Revenge, and it generates intrinsic rewards because of the unpredictability of the outcome. This has further exacerbated the problem in the case of Ego RND because we increased the intrinsic reward coefficient.

In the future we should add the ego window to the RNN policy of the RND, and more importantly to other methods besides RND and to more games than those three.

References

1. Aytar, Y., Pfaff, T., Budden, D., Paine, T., Wang, Z., de Freitas, N.: Playing hard exploration games by watching YouTube. In: NeurIPS (2018)
2. Badia, A.P., et al.: Agent57: outperforming the Atari human benchmark. ArXiv abs/2003.13350 (2020)
3. Badia, A.P., et al.: Never give up: learning directed exploration strategies. ArXiv abs/2002.06038 (2020)

4. Bellemare, M.G., Naddaf, Y., Veness, J., Bowling, M.: The arcade learning environment: an evaluation platform for general agents (extended abstract). J. Artif. Intell. Res. **47**, 253–279 (2013)

5. Bellemare, M.G., Srinivasan, S., Ostrovski, G., Schaul, T., Saxton, D., Munos, R.: Unifying count-based exploration and intrinsic motivation. ArXiv abs/1606.01868 (2016)

6. Bello, I., Zoph, B., Vaswani, A., Shlens, J., Le, Q.V.: Attention Augmented Convolutional Networks. arXiv e-prints arXiv:1904.09925, April 2019

7. Burda, Y., Edwards, H., Storkey, A.J., Klimov, O.: Exploration by random network distillation. CoRR abs/1810.12894 (2018). http://arxiv.org/abs/1810.12894

8. Burda, Y., Edwards, H.A., Pathak, D., Storkey, A.J., Darrell, T., Efros, A.A.: Large-scale study of curiosity-driven learning. ArXiv abs/1808.04355 (2019)

9. Choi, J., Lee, B., Zhang, B.: Multi-focus attention network for efficient deep reinforcement learning. ArXiv abs/1712.04603 (2017)

10. Choi, J., et al.: Contingency-aware exploration in reinforcement learning. ArXiv abs/1811.01483 (2019)

11. Haber, N., Mrowca, D., Fei-Fei, L., Yamins, D.L.K.: Emergence of structured behaviors from curiosity-based intrinsic motivation. CoRR abs/1802.07461 (2018). http://arxiv.org/abs/1802.07461

12. Horgan, D., et al.: Distributed prioritized experience replay. ArXiv abs/1803.00933 (2018)

13. Kapturowski, S., Ostrovski, G., Dabney, W., Quan, J., Munos, R.: Recurrent experience replay in distributed reinforcement learning. In: International Conference on Learning Representations (2019). https://openreview.net/forum?id=r1lyTjAqYX

14. Martin, J., Sasikumar, S.N., Everitt, T., Hutter, M.: Count-based exploration in feature space for reinforcement learning. In: IJCAI (2017)

15. Mott, A., Zoran, D., Chrzanowski, M., Wierstra, D., Rezende, D.J.: Towards interpretable reinforcement learning using attention augmented agents. In: NeurIPS (2019)

16. Ostrovski, G., Bellemare, M.G., Oord, A., Munos, R.: Count-based exploration with neural density models. ArXiv abs/1703.01310 (2017)

17. Ostrovski, G., Bellemare, M.G., van den Oord, A., Munos, R.: Count-based exploration with neural density models. ArXiv abs/1703.01310 (2017)

18. Schrittwieser, J., et al.: Mastering Atari, go, chess and shogi by planning with a learned model. ArXiv abs/1911.08265 (2019)

19. Sorokin, I., Seleznev, A., Pavlov, M., Fedorov, A., Ignateva, A.: Deep attention recurrent Q-network. ArXiv abs/1512.01693 (2015)

20. Tang, H., et al.: Exploration: a study of count-based exploration for deep reinforcement learning. In: NIPS (2017)

21. Tang, Y., Nguyen, D., Ha, D.: Neuroevolution of self-interpretable agents. In: Proceedings of the 2020 Genetic and Evolutionary Computation Conference (2020)

Cross-Modal Based Person Re-identification via Channel Exchange and Adversarial Learning

Xiaohui Xu, Song Wu, Shan Liu, and Guoqiang Xiao[✉]

College of Computer and Information Science Southwest University,
Chongqing, China
gqxiao@swu.edu.cn

Abstract. The extensive Re-ID progress in the RGB modality has obtained encouraging performance. However, in practice, the usual surveillance system is automatically switched from visible modality to infrared modality at night. The task of infrared-visible modality-based cross-modal person Re-ID (IV-Re-ID) is required. However, the existing substantial semantic gap between the visible images and the infrared images results in the IV-Re-ID still challenging. In this paper, a Cross-modal Channel Exchange Network (CmCEN) is proposed. In the CmCEN, a channel exchange network is first designed. The magnitude of Batch-Normalization (BN) is calculated to adaptively and dynamically exchange discriminative information between two different modalities sub-networks. Then, a discriminator with adversarial loss is designed to guide the network to learn the similarity distribution in the latent domain space. The evaluation results on two popular benchmark datasets demonstrated the effectiveness of our proposed CmCEN, and it obtained higher performance than state-of-the-art methods on the task of IV-Re-ID. The source code of the proposed CmCEN is available at: https://github.com/SWU-CS-MediaLab/CmCEN.

Keywords: Person re-identification · Channel exchange · Cross-modality · Adversarial learning

1 Introduction

Person re-identification (Re-ID) can be regarded as an image retrieval or a classification task across non-overlapping cameras [1,29]. The wide application in the surveillance system has aroused widespread attention on the task of person re-identification. However, the performance of person Re-ID is significantly affected by the variations in time and location, such as changes in lighting and camera perspective, which makes it still a challenging task. Benefiting from the high abstraction capability of deep convolutional neural networks [23] (CNNs),

Supported by organization Southwest University.

T. Mantoro et al. (Eds.): ICONIP 2021, LNCS 13108, pp. 500–511, 2021.
https://doi.org/10.1007/978-3-030-92185-9_41

the recent CNNs based Re-ID methods have obtained encouraging performance in the visible spectrum [19], where the RGB modality images are captured by the visible cameras. Most of them focus on bridging the domain gap caused by the cross-view visible cameras. However, the person Re-ID in the dark situation is difficult for the visible cameras. Actually, in our daily life, most cameras of surveillance systems work in dual modes, where the RGB modality images are captured by the visible mode during the daytime, and the infrared modality images are captured by the infrared mode during the night. Thus, a new infrared-visible cross-modal person Re-ID has increased attention, aiming to effectively search a target person image in both infrared and visible domains.

The infrared-visible cross-modal person Re-ID is similar to text-image cross-modal retrieval [3,24]. This is mainly because a substantial semantic domain gap exists between heterogeneous data captured by the visible and infrared cameras. The RGB modality images have wavelengths in the spectrum between 0.4 and 0.7 microns, while infrared light has a spectrum from 0.7 to 1 mm, much broader than RGB modality images. As shown in Fig. 1, we can note that images captured by infrared cameras and visible cameras are essentially different, especially in the physical properties. It is clear that the detailed information of color and texture lacks in the infrared images. However, that information is very sensitive and important for the task of IV-Re-ID. Thus, compared to the Re-ID only relying on the RGB image modality, the lack of essential clues for the infrared image modality increases the difficulties of IV-Re-ID in practical application. Several studies have been proposed to improve the performance of IV-Re-ID. However, how to learn discriminative features to distinguish different person in different domains [17], and how to reduce the semantic gap between different modality domains are still challenging for the existing methods on the task of IV-Re-ID.

Actually, when people distinguish different objects among images in our daily life, the common way is to alternately look for discriminative features of the objects until they can determine whether the two objects are the same or not. Thus, in order to simulate the human-like way of objects matching, in this paper, a cross-modal channel exchange network (CmCEN) is proposed for the task of IV-Re-ID. As shown in Fig. 2, in our proposed CmCEN, firstly, the magnitude of Batch-Normalization [8] (BN) is calculated to adaptively focus on the discriminative information from channels and dynamically exchange them between two different modalities sub-networks; Secondly, a non-local attention mechanism is used in each sub-network, which can enhance the significance of important parts of the representations of two modalities in a holistic viewpoint; Thirdly, a semi-weight share mechanism is designed in the CmCEN, such that the low-level texture, color, shape, structure information of each modality can be effective preserved during the training, while the high-level semantic information is shared for the final reasoning. Moreover, since the visible images contains three color channels, the appearance and color information can be learned by deep network. Meanwhile, since the infrared images only contains one channel with infrared modality data, the structure and shape information can be learned by the deep network. Thus, in order to mitigate the semantic gap between different modalities, the cross-modal person re-identification should not only consider the difference of inter-modality, but also the difference of intra-modality. Inspired by the advantages of adversarial learning, a robust discriminator and an adversarial loss are designed to guide the network to learn the similarity distribution of two modalities.

The main contributions of our proposed CmCEN are summarized as follows:

- A channel exchange network is designed. Firstly, the magnitude of Batch-Normalization (BN) is calculated to adaptively and dynamically exchange discriminative information between two different modalities sub-networks; Secondly, a non-local attention mechanism is used in each sub-network to enhance the significance of essential parts of the representations of two modalities; Thirdly, a semi-weight share mechanism is designed to preserve the low-level information of each modality and share the high-level semantic information for the final reasoning.
- To mitigate the semantic gap between infrared and visible modality caused by physical properties, a discriminator and adversarial loss are designed to guide the network to learn the similarity distribution in the latent domain space, such that the feature distribution of two modality data with the same identity can be effectively preserved in the learned latent space.
- The evaluation results on two popular IV-Re-ID datasets show that the channel exchange network can effectively reduce the domain discrepancy between the infrared and visible data. Meanwhile, the proposed discriminator and adversarial loss can also learn discriminative features to distinguish different persons in each domain. Compared with the existing state-of-the-art baseline, our CmCEN achieves better gain in terms of rank-k accuracy and mAP.

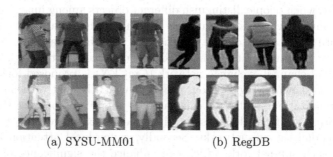

(a) SYSU-MM01   (b) RegDB

Fig. 1. This figure shows the difference between the two modals. (a) From the SYSU-MM01 dataset, we sample some images of two-modal. (b) Sampler from RegDB dataset, where images in each column from the same person.

2   Related Work

Most of the recent research concentrate on improving the performance of person matching in the RGB modality domain. There are two fundamental components in Person Re-ID: distance metric learning [12,16] and feature representations learning [11,30]. In the literature [1], it first summarized a review on the existing methods of person Re-ID based on RGB modality domain. Because of the existing large semantic domain gap between RGB domains and Infrared domains, these methods based on RGB domain cannot obtain accept performance on the infrared-visible based cross-modal person Re-ID. In order to address this limitation, several GANs based methods are proposed for IV-Re-ID [5,14,15,20]. The cmGAN [15] adopts GANs to learn distinctive features of two modalities. The D^2RL [21] adopts GANs to reduce the gap between two modalities. AlignGAN [18] proposed an unified GANs to accomplishes pixel level and feature alignment. In AlignGAN, GANs are adopted to translate RGB images into infrared images, demanding the generated infrared images could be mapped to the original RGB images. However, the GANs based deep models are sensitive to the parameters, difficult to converge, and computationally intensive during the training. In order to avoid the shortcomings of GANs, assistant modality is proposed in [9] to reformulate infrared-visible dual-mode cross-modal learning as an X-Infrared-Visible three-mode learning problem. A modality batch normalization method [10] based on the batch normalization is proposed to bridge the modality gap between RGB and infrared images.

3   The Proposed Method

The proposed Cross-modal Channel Exchange Network (CmCEN), as shown in Fig. 2, is introduced in this section, which mainly consists of two parts: a channel exchanging network and the a discriminator with adversarial loss.

3.1   The Exchange of Channel

Due to the popularity of low-cost sensors, various kind of modality data can be easily generated and captured. Thus, the analysis of multi-modality data has attracted increasing attentions, such as multi-modality based classification and regression [2]. Inspired by the network pruning [25,31], which calculates the BN scaling factor to estimate the importance of each channel, a channel exchange network is designed to achieve the information exchange and fusion of multi-modality data. In the framework of CmCEN, the process of information exchange and fusion of different modality is adaptively and dynamically, and it does not introduce extra parameters.

Since the magnitude of Batch-Normalization (BN) [8] scaling factor during training is used to estimate the importance of each channel. Here, the magnitude of the BN scaling factor is used to determine whether the two channels need to exchange. We first review the BN layer [?], which is used to estimate covariate shift and improve generalization. $x'_{l,c}$ denotes the $l-th$ layer feature maps. The batch normalization is formulated by:

$$x'_{l,c} = \gamma_{l,c} \frac{x_{l,c} - \mu_{l,c}}{\sqrt{\sigma^2_{l,c} + \epsilon}} + \beta_{l,c} \tag{1}$$

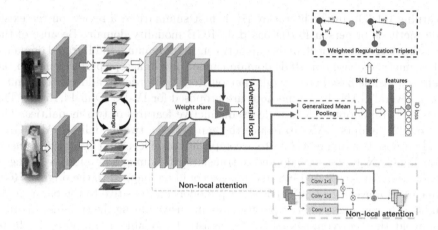

Fig. 2. Best view in color. The proposed architecture of Cross-modal Channel Exchange Network (CmCEN). The ResNet is used as the backbone to extract representations of two modalities, respectively. The yellow cube represents a non-local attention module. The orange trapezoidal represents a discriminator. The red and blue dashed arrow indicates that the scale factor of the exchanged channels is less than the threshold. The red dotted box indicates the last two layers are sharing weights during the CmCEN training.

where $\mu_c = \frac{1}{HW} \sum_{i=1}^{HW} x_i$ and $\sigma_c^2 = \frac{1}{HW} \sum_{i=1}^{HW} (x_i - \mu_c)^2$ compute the mean and standard deviation of all activation over all pixel level ($H and W$) of $c - th$ channel, respectively. $\gamma_{l,c}$ and $\beta_{l,c}$ are the learnable scaling and offset. ϵ is a small constant to avoid divisions by 0. And $x'_{l,c}$ is the input of the next layer. Inspired by the above mentioned network pruning, the scaling factor $\gamma_{l,c}$ indicates the degree of correlation between input $x_{l,c}$ and output $x'_{l,c}$. If $\gamma_{l,c}$ is close to zero, it means that $x_{l,c}$ will not be active on the final classification or prediction. This means that the channel becomes redundant, And if $\gamma_{l,c}$ near to zero, it will lead the $x_{l,c}$ becomes redundant in a certain training step, this process will last until the end of the training. Since these channels are useless, it prompts us to think of replacing them with other useful channels as shown in Fig. 3. Thus, it can achieve the interaction of both low-level and high-level features for cross-modality applications.

We define the condition of RGB modality replacing by infrared modality as Eq. 2. The replacement from infrared modality to RGB modality is in the similar way.

$$x'_{l,c} = \begin{cases} \gamma_{R,l,c} \frac{x_{R,l,c} - \mu_{R,l,c}}{\sqrt{\sigma_{R,l,c}^2 + \epsilon}} + \beta_{R,l,c}, & \text{if } \gamma_{R,l,c} > \theta \\ \gamma_{I,l,c} \frac{x_{I,l,c} - \mu_{I,l,c}}{\sqrt{\sigma_{I,l,c}^2 + \epsilon}} + \beta_{I,l,c}, & \text{else} \end{cases} \quad (2)$$

Where $\gamma_{R,l,c}$ is the scaling factor. θ is a certain threshold which is set to zero. if $\gamma > \theta$, the channel will remain the same. Otherwise, it means the channel of RGB modality has little impact on the re-identification, and we will replace it by the corresponding channel of infrared modality. Figure 3 depicts the replacement

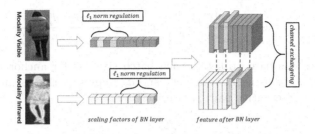

Fig. 3. This figure shows the process of channel exchange. $\ell_1 norm$ to perform sparsity constraints on the front and back parts of the two modal. Grey square in *scaling factors of BN layer* means the scaling factor less than threshold. Grey square in *feature after BN layer* represents the corresponding channel to be exchanged (Color figure online)

process. To normalize the distribution of $\gamma_{R,l,c}$ value, the $\ell_1 norm$ is used to perform sparsity constraints on the front and back parts of the two modality channels. This can ensure a smooth exchange between two sub-networks during model training.

3.2 The Discriminator and Adversarial Loss

In the task of cross-modal re-identification, due to the huge semantic domain gap, three-channel RGB images and single-channel infrared images contain different semantic information, resulting that their feature distribution is also different. A three-channel RGB image contains more discriminative information, such as color, than a single-channel infrared image. To make the feature representations of the two modalities consistent or as similar as possible, we propose a discriminator with adversarial loss as a constraint. The discriminator is composed of a three-layer network. It treats the features extracted from one modality as true and another modality as fake, with the constraint of the adversarial loss.

The F^R and F^I represent features extracted from RGB and infrared modality, respectively. We define the adversarial loss as L_{adv}, and it can be formulated as:

$$\mathcal{L}_{adv} = -\frac{1}{n} \sum_{i=1}^{n} \left(\log \left(D \left(\mathbf{f}_i^I; \theta_D \right) \right) + \log \left(1 - D \left(\mathbf{f}_i^R; \theta_D \right) \right) \right) \tag{3}$$

Where the $D(*)$ is the score of the input features of two modalities. F_i^R and F_i^I represent the features extracted from RGB and infrared modality in each batch size. The aim of the two learned modality features is to confuse the discriminator. When the score is larger, the feature is more similar to the infrared modality. θ_D is the parameter of the discriminator.

Finally we use the cross-entropy loss $\mathcal{L}_{id}$, weight regularization triplets loss $\mathcal{L}_{wrt}$ and adversarial loss $\mathcal{L}_{adv}$ as a constraint.

$$\mathcal{L}_{wrt}(i,j,k) = \log \left(1 + \exp \left(w_i^p d_{ij}^p - w_i^n d_{ik}^n \right) \right) \tag{4}$$

$$w_{ij}^p = \frac{\exp\left(d_{ij}^p\right)}{\sum_{d_{ij}^p \in \mathcal{P}_i} \exp\left(d_{ij}^p\right)}, w_{ik}^n = \frac{\exp\left(-d_{ik}^n\right)}{\sum_{d_{ik}^n \in \mathcal{N}_i} \exp\left(-d_{ik}^n\right)} \tag{5}$$

Where (i, j, k) is hard triplet in each batch size. As for anchor i_y, $\mathcal{P}_i$ is the positive set and $\mathcal{N}_i$ is negative set. d_{ij}^p and d_{ik}^n represent the distance of anchor and positive and negative pair, respectively.

$$\mathcal{L}_{id} = -\frac{1}{n} \sum_{i=1}^{n} \log\left(p\left(y_i \mid x_i\right)\right) \tag{6}$$

where n represents the number of training samples within each batch. Given an input image x_i with label y_i, the predicted probability of x_i being recognized as class y_i is encoded with a softmax function, represented by $p(y_i|x_i)$.

3.3   Optimization

By assembling the above five loss functions together, the final overall loss function is given as follows:

$$\min_{\gamma_I, \gamma_R, \gamma} \mathcal{L} = \lambda_1 \mathcal{L}_{id} + \lambda_2 \mathcal{L}_{wrt} + \lambda_3 \mathcal{L}_{adv} \tag{7}$$

where γ_I, γ_R are the parameters of the first two stages of ResNet50, and γ represents the parameters of embedded layers. λ_1, λ_2 and λ_3 are the weights of the loss function. A training strategy is employed to optimize Eq. 7. At the end of each epoch, some parameters will be optimized. The training details of CmCEN are given in Algorithm 1.

Optimize the Parameters γ_I, γ_R, and γ. The Stochastic Gradient Descent (SGD) algorithm is adopted to update parameters.

$$\gamma_I \leftarrow m \cdot \gamma_I - \eta \cdot \nabla_{\gamma_I} \frac{1}{n}\mathcal{L} \tag{8}$$

The parameters of sub-network γ_I is updated by SGD descending gradie of $\mathcal{L}$ in Eq. 8. The momentum parameter is set to 0.9. γ_R and γ are updated in the same way in Eq. 9 and 10. η is the hyperparameter of step size of gradient descent. n denotes the batch size of each epoch.

$$\gamma_R \leftarrow m \cdot \gamma_R - \eta \cdot \nabla_{\gamma_R} \frac{1}{n}\mathcal{L} \tag{9}$$

$$\gamma \leftarrow m \cdot \gamma - \eta \cdot \nabla_{\gamma} \frac{1}{n}\mathcal{L} \tag{10}$$

Algorithm 1. Optimization algorithm of CmCEN.

Input: RGB dataset R and infrared dataset I, pretrained backbone network $ResNet50$ and hyperparameter threshold θ;

Output: Optimization parameters γ_I, γ_R and γ;

Initialization: Initialize the parameters of network, set the mini-batch size to $n = 8$, setiteration number $iter$ and other hyper-parameters.

for $i = 1$ to $iter$ **do**

 Update γ_I, γ_R and γ by **SGD** to **descending** their gradients:

$$\gamma_I \leftarrow \gamma_I - \eta \cdot \nabla_{\gamma_I} \frac{1}{n} \mathcal{L}$$

$$\gamma_R \leftarrow \gamma_R - \eta \cdot \nabla_{\gamma_R} \frac{1}{n} \mathcal{L}$$

$$\gamma \leftarrow \gamma - \eta \cdot \nabla_{\gamma} \frac{1}{n} \mathcal{L}$$

 Update the paramters of pretrained backbone network ResNet50.

end

Until a fixed number of iterations or convergence;

4   Experiments

In this section, we compare our proposed method with several state-of-the-arts baselines and ablation experiment was carried out to validate the effectiveness of the key components of our proposed CmCEN.

4.1   Experimental Setting

We conducted experiments on two mainstream datasets: SYSU-MM01 [22] and RegDB [6]

- SYSU-MM01 is a large-scale visible-infrared person re-identification dataset. The dataset is split into the testing set with 96 identities and the training set with 395 identities.
- RegDB The RegDB data set contains 412 pedestrian identities. 10 RGB images and 10 infrared images are collected for each person. Among the 412 people, 156 were photographed from the front and 256 from the back. The dataset is split into two sets, one for training and the other for testing. Each set contains 2060 RGB images and 2060 infrared images.

Implementation Details. The ResNet50 pre-trained on ImageNet as our backbone network. We use the SGD optimizer with weight decay $5e^{-4}$, and the training epoch is 80. According to the experimental results, the threshold θ of Eq. 2 is set to $1.5e^{-2}$ and the weights of loss function $\lambda_1 = \lambda_2 = \lambda_3 = 1$. We adopt the widely used $rank - k$ accuracy, mean Average Precision (mAP), and a new evaluation standard mean Inverse Negative Penalty (mINP) as evaluation protocol.

4.2 Comparison with State-of-the-art Methods

The results are shown in Table 1 and Table 2, our method outperforms state-of-the-arts on both two datasets.

Table 1. Comparison with the state-of-the-arts on SYSU-MM01 dataset. Rank at r accuracy (%) and mAP (%) are reported.

Method	All search				Indoor search			
	$r=1$	$r=10$	$r=20$	mAP	$r=1$	$r=10$	$r=20$	mAP
D^2RL [21]	28.9	70.6	82.4	29.2	36.43	62.36	71.63	37.03
MAC [26]	33.26	79.04	90.09	36.22	36.43	62.36	71.63	37.03
MSR [7]	37.35	83.40	93.34	38.11	39.64	89.29	97.66	50.88
AlignGAN [18]	42.4	85.0	93.7	40.7	45.9	87.6	94.4	54.3
Hi-CMD [4]	34.9	77.6	–	35.9	–	–	–	–
cm-SSFT [13]	47.7	–	–	54.1	–	–	–	–
Baseline	47.54	84.39	92.14	47.65	54.17	91.14	95.98	62.97
Ours	**49.33**	**85.27**	**92.77**	**48.67**	**56.84**	**93.04**	**97.69**	**63.55**

Experimental Analysis. Table 1 shows the comparison results on the SYSU-MM01 dataset. Our method achieves significantly better performance than the comparison cross modal Re-ID method for every adopted evaluation metric. In the task of All Search, our method achieves an absolute gain of 1.02% in terms of mAP. As for Indoor Search, our method achieves a gain of 2.67% in terms of rank 1.

Table 2. Comparison with the state-of-the-arts on RegDB dataset on both query settings. Rank at r accuracy (%) and mAP (%) are reported. (*Both the visible to thermal and thermal to visible are evaluated.*)

Settings	Visible to thermal				Thermal to viaiable			
	$r=1$	$r=10$	$r=20$	mAP	$r=1$	$r=10$	$r=20$	mAP
D^2RL [21]	43.4	66.1	76.3	44.1		–	–	–
MAC [26]	36.43	62.36	71.63	37.03	36.20	61.68	70.99	36.63
MSR [7]	48.43	70.32	79.95	48.67	39.64	–	–	–
AlignGAN [18]	57.9	–	–	53.6	56.3	–	–	53.4
Hi-CMD [4]	70.93	86.39	–	66.04	–	–	–	–
cm-SSFT [13]	72.3	–	–	72.9	71.0	–	–	71.7
DDAG [27]	69.34	86.19	91.94	63.64	68.06	85.15	90.31	61.80
HAT [28]	71.83	87.16	92.16	67.56	70.02	86.45	91.61	66.30
Baseline	70.05	86.21	91.55	66.37	70.49	87.12	91.84	65.90
Ours	**74.03**	**88.25**	**92.38**	**67.52**	**74.22**	**87.22**	**91.99**	**67.36**

Both the visible to thermal and thermal to visible query settings are evaluated on the RegDB dataset. In the retrieval performance of visible-to-thermal and thermal-to-visible, we achieve a gain of 3.98% and 3.73% respectively in terms of rank 1.

Table 3. Ablation study on SYSU-MM01 dataset.

Method	$r = 1$	$r = 10$	$r = 20$	mAP	mINP
Baseline	47.54	84.39	92.14	47.65	35.30
Baseline + $CmCEN$	**48.99**	**86.43**	**93.55**	**48.22**	**35.96**
Baseline + Dis	**48.36**	**84.46**	**92.24**	**48.10**	**35.90**
$Ours$	**49.33**	**85.27**	**92.77**	**48.67**	**35.95**

Ablation Study. In order to verify the effectiveness of channel exchange and adversarial learning method, We performed ablation experiments on the large-scale SYSU-MM01 dataset and give the final results in Table 3. For the new evaluation index mINP we achieve an absolute gain. Firstly, we can observe that CmCEN can provide remarkable improvement in retrieval performance, this is because the channel exchange strategy can mutually share the information between two modalities, which can ensure a strong inter-modal correlation. Intuitively, the process of channel exchange is prone to the process of human's recognition on two images, which is an alternative observation on discriminative features that can distinguish or relate two images. Secondly, the improvement by using adversarial learning is significant, because the lack of adversarial learning in RGB and infrared features extract modules will lead to an inconsistent feature distribution between two modalities.

5 Conclusion

In this paper, we propose a Cross-modal Channel Exchange Network to exchange information between two heterogeneous modalities. The channels of the two modalities are exchanged according to the conditions of the specific human body, which is calculated by the magnitude of Batch-Normalization. The channel exchange method can make full use of the relevant information between two modalities. Intuitively, it is prone to the thinking habits of human beings when distinguishing different objects. At the same time, based on the idea of adversarial learning, a discriminator is proposed to make the learned distribution of the same identity of two modalities as similar as possible. The experimental results demonstrated that the CmCEN could learn discriminative features to distinguish different persons in different domains. Meanwhile, the discriminator with adversarial loss can reduce the semantic gap between different modality domains.

Acknowledgement. This work was supported by the National Natural Science Foundation of China (61806168), Fundamental Research Funds for the Central Universities (SWU117059), and Venture and Innovation Support Program for Chongqing Overseas Returnees (CX2018075).

References

1. Visible thermal person re-identification via dual-constrained top-ranking. In: Twenty-Seventh International Joint Conference on Artificial Intelligence IJCAI-18 (2018)
2. Baltrušaitis, T., Ahuja, C., Morency, L.-P.: Multimodal machine learning: a survey and taxonomy. IEEE Trans. Pattern Anal. Mach. Intell. **41**(2), 423–443 (2019)
3. Chen, S., Wu, S., Wang, L.: Hierarchical semantic interaction-based deep hashing network for cross-modal retrieval. PeerJ Comput. Sci. **7**(2), e552 (2020)
4. Choi, S., Lee, S., Kim, Y., Kim, T., Kim, C.: Hi-CMD: hierarchical cross-modality disentanglement for visible-infrared person re-identification. In: 2020 IEEE/CVF Conference on Computer Vision and Pattern Recognition, CVPR 2020, Seattle, WA, USA, 13–19 June 2020, pp. 10254–10263. IEEE (2020)
5. Dai, P., Ji, R., Wang, H., Wu, Q., Huang, Y.: Cross-modality person re-identification with generative adversarial training. In: Proceedings of the Twenty-Seventh International Joint Conference on Artificial Intelligence, IJCAI-18, pp. 677–683. International Joint Conferences on Artificial Intelligence Organization, July 2018
6. Dat, N., Hong, H., Ki, K., Kang, P.: Person recognition system based on a combination of body images from visible light and thermal cameras. Sensors **17**(3), 605 (2017)
7. Feng, Z.-X., Lai, J., Xie, X.: Learning modality-specific representations for visible-infrared person re-identification. IEEE Trans. Image Process. **29**, 579–590 (2020)
8. Ioffe, S., Szegedy, C.: Batch normalization: accelerating deep network training by reducing internal covariate shift. JMLR.org (2015)
9. Li, D., Wei, X., Hong, X., Gong, Y.: Infrared-visible cross-modal person re-identification with an x modality. In: Proceedings of the AAAI Conference on Artificial Intelligence, vol. 34, pp. 4610–4617, April 2020
10. Li, W., Ke, Q., Chen, W., Zhou, Y.: Bridging the distribution gap of visible-infrared person re-identification with modality batch batch normalization (2021)
11. Liao, S., Yang, H., Zhu, X., Li, S.Z.: Person re-identification by local maximal occurrence representation and metric learning. In: 2015 IEEE Conference on Computer Vision and Pattern Recognition (CVPR) (2015)
12. Liao, S., Li, S.Z.: Efficient PSD constrained asymmetric metric learning for person re-identification. In: 2015 IEEE International Conference on Computer Vision (ICCV), pp. 3685–3693 (2015)
13. Lu, Y., et al.: Cross-modality person re-identification with shared-specific feature transfer. In: 2020 IEEE/CVF Conference on Computer Vision and Pattern Recognition, CVPR2020, Seattle, WA, USA, 13–19 June 2020, pp. 13376–13386. IEEE (2020)
14. Mao, X., Li, Q., Xie, H.: AlignGAN: learning to align cross-domain images with conditional generative adversarial networks (2017)
15. Peng, Y., Qi, J., Yuan, Y.: CM-GANs: cross-modal generative adversarial networks for common representation learning. ACM Trans. Multimed. Comput. Commun. Appl. **15**(1), 1–24 (2017)

16. Shi, Z., Hospedales, T.M., Tao, X.: Transferring a semantic representation for person re-identification and search. IEEE (2015)

17. Wang, G., Yuan, Y., Chen, X., Li, J., Zhou, X.: Learning discriminative features with multiple granularities for person re-identification. ACM (2018)

18. Wang, G., Zhang, T., Cheng, J., Liu, S., Yang, Y., Hou, Z.: RGB-infrared cross-modality person re-identification via joint pixel and feature alignment. In2019 IEEE/CVF International Conference on Computer Vision, ICCV 2019, Seoul, Korea (South), 27 October–2 November 2019, pp. 3622–3631. IEEE (2019)

19. Wang, X., Zou, X., Bakker, E.M., Wu, S.: Self-constraining and attention-based hashing network for bit-scalable cross-modal retrieval. Neurocomputing **400**, 255–271 (2020)

20. Wang, Z., Wang, Z., Zheng, Y., Chuang, Y.Y., Satoh, S.: Learning to reduce dual-level discrepancy for infrared-visible person re-identification. In: 2019 IEEE/CVF Conference on Computer Vision and Pattern Recognition (CVPR) (2019)

21. Wang, Z., Wang, Z., Zheng, Y., Chuang, Y.-Y., Satoh, S.: Learning to reduce dual-level discrepancy for infrared-visible person re-identification. In: IEEE Conference on Computer Vision and Pattern Recognition, CVPR 2019, Long Beach, CA, USA, 16–20 June 2019, pp. 618–626. Computer Vision Foundation/IEEE (2019)

22. Wu, A., Zheng, W.-S., Yu, H.-X., Gong, S., Lai, J.: RGB-infrared cross-modality person re-identification. In: 2017 IEEE International Conference on Computer Vision (ICCV), pp. 5390–5399 (2017)

23. Wu, S., Oerlemans, A., Bakker, E.M., Lew, M.S.: Deep binary codes for large scale image retrieval. Neurocomputing **257**(sep.27), 5–15 (2017)

24. Xzab, D., Xw, A., Emb, C., Song, W.A.: Multi-label semantics preserving based deep cross-modal hashing. Signal Process.: Image Commun. **93**, 116131 (2021)

25. Ye, J., Xin, L., Zhe, L., Wang, J.Z.: Rethinking the smaller-norm-less-informative assumption in channel pruning of convolution layers (2018)

26. Ye, M., Lan, X., Leng, Q.: Modality-aware collaborative learning for visible thermal person re-identification. In: Amsaleg, L., et al. (eds.) Proceedings of the 27th ACM International Conference on Multimedia, MM 2019, Nice, France, 21–25 October 2019, pp. 347–355. ACM (2019)

27. Ye, M., Shen, J., J. Crandall, D., Shao, L., Luo, J.: Dynamic dual-attentive aggregation learning for visible-infrared person re-identification. In: Vedaldi, A., Bischof, H., Brox, T., Frahm, J.-M. (eds.) ECCV 2020. LNCS, vol. 12362, pp. 229–247. Springer, Cham (2020). https://doi.org/10.1007/978-3-030-58520-4_14

28. Ye, M., Shen, J., Shao, L.: Visible-infrared person re-identification via homogeneous augmented tri-modal learning. IEEE Trans. Inf. Forensics Secur. **16**, 728–739 (2021)

29. Chen, Y.-C., Zhu, X., Zheng, W.-S., Lai, J.-H.: Person re-identification by camera correlation aware feature augmentation. IEEE Trans. Pattern Anal. Mach. Intell. **40**, 392–408 (2017)

30. Zheng, L., Shen, L., Tian, L., Wang, S., Wang, J., Tian, Q.: Scalable personre-identification: a benchmark. IEEE (2016)

31. Zhuang, L., Li, J., Shen, Z., Gao, H., Zhang, C.: Learning efficient convolutional networks through network slimming. In: 2017 IEEE International Conference on Computer Vision (ICCV) (2017)

SPBERT: an Efficient Pre-training BERT on SPARQL Queries for Question Answering over Knowledge Graphs

Hieu Tran[1,2], Long Phan[3], James Anibal[4], Binh T. Nguyen[1,2], and Truong-Son Nguyen[1,2(✉)]

[1] University of Science, Ho Chi Minh City, Vietnam
ntson@fit.hcmus.edu.vn
[2] Vietnam National Univeristy, Ho Chi Minh City, Vietnam
[3] Case Western Reserve University, Cleveland, OH, USA
[4] National Cancer Institute, Bethesda, MD, USA

Abstract. Knowledge Graph is becoming increasingly popular and necessary during the past years. In order to address the lack of structural information of SPARQL query language, we propose SPBERT, a transformer-based language model pre-trained on massive SPARQL query logs. By incorporating masked language modeling objectives and the word structural objective, SPBERT can learn general-purpose representations in both natural language and SPARQL query language. We investigate how SPBERT and encoder-decoder architecture can be adapted for Knowledge-based QA corpora. We conduct exhaustive experiments on two additional tasks, including SPARQL Query Construction and Answer Verbalization Generation. The experimental results show that SPBERT can obtain promising results, achieving state-of-the-art BLEU scores on several of these tasks.

Keywords: Machine translation · SPARQL · Language models · Question answering

1 Introduction

During the last decade, pre-trained language models (LM) have been playing an essential role in many areas of natural language processing (NLP), including Question Answering (QA). Large pre-trained models like BERT [1], RoBERTa [5], and XLNet [17] derive contextualized word vector representations from immense text corpora. It can represent a significant deviation from traditional word embedding methods wherein each word is given a global representation. After training, one can usually fine-tune the corresponding models for downstream tasks. These pre-trained models have dramatically improved the state-of-the-art performance on multiple downstream NLP tasks. As such, this concept has been extended into various domains. One can retrain BERT models on corpora containing text specific to a particular area; for instance, Feng and colleagues used CodeBERT for tasks involving programming languages [2].

© Springer Nature Switzerland AG 2021
T. Mantoro et al. (Eds.): ICONIP 2021, LNCS 13108, pp. 512–523, 2021.
https://doi.org/10.1007/978-3-030-92185-9_42

There has also been widespread use of publicly available Knowledge Graph (KG) datasets, such as DBpedia[1], Wikidata[2], or Yago[3]. These datasets provide a valuable source of structured knowledge for NL relational tasks, including QA. These knowledge base (KB) endpoints use a query language called SPARQL, a standard graphing-matching query language to query graph data in the form of RDF triples [7] (an example format of a SPARQL query can be referred to Table 1). One of the key challenges in these SPAQRL-NL generation tasks is understanding the structured schemas between entities and relations in the KGs. In addition, it is necessary to correctly generate NL answers from SPARQL queries and generate SPARQL queries from NL descriptions. These challenges call for the development of pre-trained language models (LMs) that understand SPARQL schemas' structure in KGs and free-form text. While existing LMs (e.g., BERT) are only trained on encoding NL text, we present an unprecedented approach for developing an LM that supports NL and SPARQL query language (SL).

Previous works in this area [18] utilized the traditional word embedding method and built the vocabulary from tokens separated by spaces. The higher the number of entities and predicates in the dataset, the larger the vocabulary size. It is a notable drawback that needs to be improved to reduce time and cost complexity. We present SPBERT, a pre-trained model for SPARQL query language (SL) using a common vocabulary. SPBERT is a multi-layer transformer-based model [14], in which the architecture has been proven effective for various other large LMs models. We train SPBERT using standard masked language modeling [1]. To further advance the exhibit of the underlying solid structure of SPARQL schemas, we also incorporate the word structural objective [15]. These structural learning objectives can enable SPBERT to gain insights into the word-level structure of SL language during pre-training. We train SPBERT from large, processed SPARQL query logs. To show the effectiveness of our approach in creating a general-purpose pre-trained LMs for NL and SL, SPBERT is fine-tuned and evaluated on six popular datasets in SPARQL query construction and answer verbalization generation topics. In addition, we test various pre-training strategies, including different learning objectives and model checkpoints.

Our paper provides the following contributions: (1) We investigate how pre-trained language models can be applied to the sequence-to-sequence architecture in KBQA; (2) We introduce SPBERT, the first pre-trained language model for SPARQL Query Language that focuses on understanding the query structure; (3) We show that pre-training on large SPARQL query corpus and incorporating word-level ordering learning objectives lead to better performance. Our model achieves competitive results on Answer Verbalization Generation and SPARQL Query Construction; (4) We make our pre-trained checkpoints of SPBERT and related source code[4] for fine-tuning publicly available.

---

[1] https://dbpedia.org.

[2] https://wikidata.org.

[3] https://yago-knowledge.org.

[4] https://github.com/heraclex12/NLP2SPARQL.

2   Related Work

BERT [1] is a pre-trained contextualized word representation model which consists of the encoder block taken from the transformer method. [14]. BERT is trained with masked language modeling to learn word representation in both left and right contexts. The model incorporates information from bidirectional representations in a sequence - this has proved to be effective in learning a natural language. We hypothesize that this architecture will be effective in more domain-specific SPARQL query language contexts.

Multiple prior attempts use deep learning for enhancing performance on natural language-SPARQL query language-related tasks. For example, Yin et al. [18] presented different experiments with three types of neural machine translation models, including RNN-based, CNN-based, and Transformer-based models for translating natural language to SPARQL query tasks. The results showed a dominance of the convolutional sequence-to-sequence model over all of the proposed models across five datasets. This method is highly correlated with our approach, so we treat this work as our baseline later in this paper. Luz and Finger [6] proposed an LSTM encoder-decoder model that is capable of encoding natural language and decoding the corresponding SPARQL query. Furthermore, this work presented multiple methods for generating vector representation of Natural Language and SPARQL language. Finally, the paper introduced a novel approach to develop a lexicon representation vector of SPARQL. The results illustrated that this approach could achieve state-of-the-art results on natural language-SPARQL datasets.

There have also been attempts to develop a Knowledge-based question answering (KBQA) system that leverages multiple models for the various tasks. For example, Kapanipathi and colleagues introduced semantic parsing and reasoning-based Neuro-Symbolic Question Answering (NSQA) system [4]. The model consisted of an Abstract Meaning Representation (AMR) layer that could parse the input question, a path-based approach to transform the AMR into KB logical queries, and a Logical Neural Network (LNN) to filter out invalid queries. As a result, NSQA could achieve state-of-the-art performance on KBQA datasets.

3   SPBERT

3.1   Pre-training Data

To prepare a large-scale pre-training corpus, we leverage SPARQL queries from end-users, massive and highly diverse structures. These query logs can be obtained from the DBpedia endpoint[5] powered by a Virtuoso[6] instance. We only focus on valid DBpedia query logs spans from October 2015 to April 2016. These raw queries contain many duplicates, arbitrary variable names, and unnecessary components such as prefixes and comments.

---

[5] https://dbpedia.org/sparql.
[6] https://virtuoso.openlinksw.com.

Table 1. An example of SPARQL query and the corresponding encoding

Original query	SELECT DISTINCT ?uri WHERE { \<http://dbpedia.org/resource/Tom_Hanks\> \<http://dbpedia.org/ontology/spouse\> ?uri }
Encoded query	select distinct var_uri where brack_open \<dbr_Tom_Hanks\> \<dbo_spouse\> var_uri brack_close

To address these issues, we prepared a heuristic pipeline to clean the DBpedia SPARQL query logs and released this pipeline with the model (see footnote 4). This pipeline includes a cleaning process specific to SPARQL query language (i.e., removing comments, excluding prefix declarations, converting namespace, filtering unknown links, standardizing white space and indentations, etc.). In addition, the processed queries will go through an encoding process suggested by [18] to make these queries look more natural. We obtained approximately 6.8M queries. An example SPARQL query can be depicted in Table 1.

3.2 Input Representation

The input of SPBERT is a sequence of tokens of a single, encoded SPARQL query. We follow the style of input representation used in BERT. Every input sequence requires a special classification token, *[CLS]* as the first token, and a special end-of-sequence, *[SEP]*. *[CLS]* token contains representative information of the whole input sentence, while the *[SEP]* token is used to separate different sentences of an input. To alleviate the out-of-vocabulary problem in tokenization, we use WordPiece [16] to split the query sentence into subword units. In addition, we employ the same vocabulary with cased BERT for the following reasons: (i) SPARQL queries are almost made up of English words, and we aim to leverage existing pre-trained language models; (ii) these queries require strictly correct representations of entities and relations, which can contain either lowercase and uppercase characters. All tokens of the input sequence can be mapped to this vocabulary to pick up their corresponding indexes and then feed these indexes into the model.

3.3 Model Architecture

SPBERT uses the same architecture as BERT [1], which is based on a multi-layer bidirectional transformer [14]. BERT model is trained on two auxiliary tasks: masked language modeling and next sentence prediction. As our training corpus only contains independent SPARQL queries, we substitute the next sentence prediction with the word structural objective [15]. We illustrate how to combine these two tasks in Fig. 1.

Task #1: Masked Language Modeling (MLM) randomly replaces some percentage of tokens from a sequence (similar to [1], the rate of 15% of the

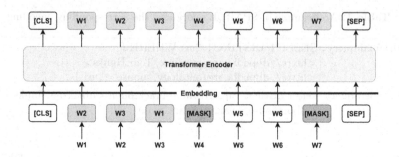

Fig. 1. An illustration about the combination of MLM and WSO

tokens will be covered) with a special token [MASK]. The model will then try to reconstruct this sequence by predicting the original token.

Task #2: Word Structural Objective (WSO) gives our model the ability to capture the sequential order dependency of a sentence. This objective corrupts the right order of words by randomly selecting n-grams from unmasked tokens and then permuting the elements. The model has to predict the original order of tokens to correct the sentence. Different from [15], our pre-training data are mostly short queries (less than 256 tokens). Therefore, we permute 10% of the n-grams rather than 5%.

We released three different versions of SPBERT. The first version begins training with randomly initialized weights, and the second starts with the weights from the pre-trained BERT model. Both these models use only the MLM objective, while we combine MLM and WSO to train the third SPBERT.

3.4 Pre-training Setup

In the pre-training step, we denote the number of Transformer encoder layers as L, the size of hidden vectors as H, and the number of self-attention heads as A. We followed the setting of $BERT_{BASE}$ (L = 12, H = 768, A = 12, total parameters = 110M) and continued to train 200K steps from cased $BERT_{BASE}$ checkpoint. The maximum sequence length was fixed to 512, and the batch size was set to 128. We used Adam with a learning rate of 2e-5 and epsilon of 1e-8 and employed cased $BERT_{BASE}$ vocabulary with 30K tokens.

3.5 Fine-Tuning SPBERT

Transformer-based sequence-to-sequence architecture [10] is an encoder-decoder architecture that employs multi-layer self-attention to efficiently parallelize long-term dependencies. This architecture has two components: an encoder and a decoder. The encoder can capture a contextualized representation of input sequences, and the decoder uses this information to generate target sequences.

Intuitively, this architecture is often used for generating a target sequence from a source sequence.

To fine-tune SPBERT on the natural language-SPARQL query language tasks required for an end-to-end KBQA, we apply pre-trained language models to the Transformer-based sequence-to-sequence architecture. BERT and SPBERT are encoder-only models developed for encoding language representations, so we can quickly assemble this model into the encoder. However, to adapt BERT-based models to the decoder, we must change the self-attention layers from bidirectional to left-context-only. We must also insert a cross-attention mechanism with random weights. For tasks with the input as a question and the output as a query, we initialize the encoder with BERT checkpoints and the decoder with SPBERT checkpoints. We denote this model as BERT2SPBERT. Conversely, when we use SPARQL queries as input and NL answers are the outputs, SPBERT will be initialized as the encoder. The decoder can be initialized at random (SPBERT2RND) or from BERT checkpoints (SPBERT2BERT). Similar to the pre-training phase, all SPARQL queries will be encoded before being put into the model.

We apply the weight typing method to reduce the number of parameters but still boost the performance. Same as [9], we tie the input embedding and the output embedding of the decoder that help the output embedding makes use of the weights learned from the input embedding instead of initializing randomly.

4   Experiments

This section first describes the datasets used to evaluate our proposed models and then explains the experimental setup and the results.

4.1   Datasets

An end-to-end KBQA system must perform two tasks sequentially. First, this system acquires a question and constructs a corresponding SPARQL query. We refer to this task as SPARQL Query Construction. The remaining task is Answer Verbalization Generation, which inputs the generated queries from the previous one and produces natural language answers. In these experiments, we only consider English datasets, and all queries can be encoded in the way described in Sect. 3.1. We represent both tasks below and summarize the evaluation datasets in Table 2.

SPARQL Query Construction. QALD-9 (Question Answering over Linked Data) [13] consists of a question and query pairs from real-world questions and query logs. LC-QuAD (Large-scale Complex Question Answering Dataset) [12] contains 5000 questions and corresponding queries from the DBPedia dataset. Each NL question in the LC-QuAD dataset is peer-reviewed by fluent English speakers to ensure the quality of the data.

Table 2. Data statistics about question answering datasets over knowledge graphs

Dataset	Amount			Tokens	Creation
	Training	Validation	Test		
QALD-9 [13]	408	–	150	1042	Manual
LC-QuAD [12]	4000	500	500	5035	Manual
Mon [11]	14588	100	100	2066	Automatic
Mon50 [18]	7394	1478	5916	2066	Automatic
Mon80 [18]	11830	1479	1479	2066	Automatic
VQuAnDa [3]	4000	500	500	5035	Manual

The Mon dataset was introduced by [11]. This dataset contains 38 pairs of handcrafted question and query templates. Each template is automatically inserted into one or two entities of the Monument ontology. In order to compare the performance between different number of training samples, [18] splits this dataset using two different ratios for training, test, and validation sets 80%-10%-10% (Mon80) and 50%-10%-40% (Mon50).

Answer Verbalization Generation. VQuAnDa (Verbalization QUestion ANswering DAtaset) [3] extends the LC-QuAD dataset by providing the query answers in natural language. These answers are created based on the questions and the corresponding SPARQL queries. An automatic framework generates the templates for the answers and uses a rule-based method to produce the first version. The final results are manually reviewed and rephrased to ensure grammatical correctness. We give a sample of this dataset in Table 3.

4.2   Experimental Setup

We verify the effectiveness of our proposed methods by comparing these models with three following types of network architectures from previous works [3,18].

– **RNN-based models:** These models are based on a standard sequence-to-sequence architecture and combine variants of RNNs (such as Long Short-Term Memory and Gated Recurrent Unit) with an attention mechanism.
– **CNN-based models:** CNN-based sequence to sequence (ConvS2S) models, which leverage an encoder-decoder architecture with an attention mechanism. In these cases, both the encoder and decoder consist of stacked convolutional layers.
– **Transformer models:** These models are based on [14], in which each layer of encoder-decoder architecture includes two major components: a multi-head self-attention layer and a feed-forward network. Since these models are initialized with random weights, they also are called RND2RND.

Table 3. An example of two KBQA tasks

Question	How many people play for the Dallas Cowboys?
With entites	
SPARQL	select distinct count(var_uri) where brack_open var_uri <dbo_team> <dbr_Dallas_Cowboys> brack_close
Answer	<ans> people play for the dallas cowboys
Covered entities	
SPARQL	select distinct count(var_uri) where brack_open var_uri <dbo_team> <ent> brack_close
Answer	<ans> people play for the <ent>

Performance Metrics. The most reliable evaluation method is to have appropriately qualified experts look at the translation and evaluate manually. However, this evaluation is costly and time-consuming. Therefore, we are using automatic metrics that are faster and cheaper, but still correlate with human judgments. We use corpus level Bilingual Evaluation Understudy (BLEU) [8] and Exact Match score (EM) as follows: (1) **BLEU** measures the differences in word choice and word order between candidate translations and reference translations by calculating the overlap of n-grams between a generated sequence and one or more reference sequences. (2) **EM** is the ratio of the number of generated sequences that perfectly match reference to the total number of input samples.

Experimental Settings. For each dataset, we fine-tuned our proposed models for a maximum of 150 epochs using Adam optimizer with a learning rate of 5e-5, a weight decay of 0.1, and a batch size of 16 or 32. We selected the best model based on the performance in the validation set. For SPARQL query language, we set the maximum input length as either 128 (QALD, LC-QuAD, VQuAnDa) or 256 (Mon, Mon50, Mon80). We fixed the maximum length of natural language as 64. In the decoding step, we used beam search of beam width 10 for all the experiments. All experiments were completed using Python 3.7.10, Pytorch 1.8.1, and Transformers 4.5.1.

4.3 Results

SPARQL Query Construction. Table 4 shows the BLEU and EM results on the five datasets. Our approaches perform pretty well compared to the previous works. On the LC-QuAD, BERT2SPBERT$_{MLM+WSO}$(B) achieves 69.03%, creating new state-of-the-art results on this dataset. Moreover, one can see that BERT2SPERT$_{MLM+WSO}$(B) outperforms the other models on the QALD dataset and improves by 3.81% over the baselines, indicating our model can perform well even with limited data. SPBERT also obtains competitive results on three simple Mon datasets, including Mon, Mon50, and Mon80. BERT-initialized

Table 4. Experiment results on SPARQL query construction

Model	QALD	Mon	Mon50	Mon80	LC-QuAD
RNN-Luong	31.77|5.33	91.67|76	94.75|85.38	96.12|89.93	51.06|1.20
Transformer	33.77|6.00	95.31|91	93.92|84.70	94.87|85.80	57.43|7.80
ConvS2S	31.81|5.33	97.12|95	**96.62**|90.91	96.47|90.87	59.54|8.20
BERT2RND	34.36|**6.67**	97.03|96	95.28|90.69	96.44|91.35	66.52|14.80
BERT2BERT	35.86|**6.67**	97.03|96	96.20|90.99	96.14|91.35	68.80|18.00
MLM					
BERT2SPBERT (S)	35.19|**6.67**	97.28|**97**	96.06|90.80	96.29|92.22	64.18|12.40
BERT2SPBERT (B)	35.95|**6.67**	96.78|95	96.45|**91.18**	**96.87|92.70**	68.08|**20.20**
MLM + WSO					
BERT2SPERT (B)	**37.58**|**6.67**	**97.33**|96	96.20|90.84	96.36|91.75	**69.03**|18.80

Notes: We train SPBERT (third group) from scratch (S) or initialized with the parameters of BERT (B), and we also use different learning objectives (only MLM, the combination of MLM and WSO). The left scores are BLEU and the right scores are EM. The best scores are in bold.

as an encoder can increase significant performance. However, the difference between BERT2RND and BERT2BERT is only apparent on QALD and LC-QuAD. They are almost equal on the three remaining datasets. According to Table 4, SPBERT outperforms the baselines in EM metric and is slightly better than BERT2BERT on some datasets. Specifically, BERT2SPERT$_{MLM}$(B) achieves state-of-the-art results by more than 1% on two out of five datasets. These results show that pre-training on SPARQL queries improves the construction of complete and valid queries.

Answer Verbalization Generation. In Table 5, the results show that our proposed models significantly outperform BERT-initialized as encoder and the baseline methods for both *With Entities* and *Covered Entities*. Fine-tuning SPBERT improves results on the test set by 17.69% (*With Entities*) and 8.18% (*Covered entities*). The results also indicate that combining MLM and WSO can perform better than MLM when the input data are the SPARQL queries. At the same time, BERT2BERT is much better than BERT2RND in *With Entities* setting, but BERT2BERT fails to predict in the sentences (*Covered Entities*) that have no context and are incomplete.

4.4   Discussion

The experimental results show that leveraging pre-trained models is highly effective. Our proposed models are superior to baseline models on both tasks. The main reason for this is that we employ robust architectures that can understand the language from a bidirectional perspective. These models train on massive corpora, learning a generalized representation of the language that can be fine-tuned on smaller evaluation datasets.

Table 5. BLEU score experiment results on answer verbalization generation

Model	With entities		Covered entities	
	Validation	Test	Validation	Test
RNN-Luong	22.29	21.33	34.34	30.78
Transformer	24.16	22.98	31.65	29.14
ConvS2S	26.02	25.95	32.61	32.39
BERT2RND	33.30	33.79	42.19	38.85
BERT2BERT	41.48	41.67	41.75	38.42
MLM				
- SPBERT2RND (S)	43.25	**43.64**	41.49	38.57
- SPBERT2RND (B)	42.21	41.59	**42.44**	39.63
- SPBERT2BERT (S)	42.58	41.44	41.39	38.56
- SPBERT2BERT (B)	41.75	40.77	42.01	39.59
MLM + WSO				
- SPBERT2RND (B)	**43.52**	42.39	41.74	38.48
- SPBERT2BERT (B)	43.23	41.70	41.97	**40.57**

The performance between the decoder initialized from the BERT checkpoint (BERT2BERT) and SPBERT checkpoint (BERT2SPBERT) is not significantly different in most cases. One possible reason is that we still need to randomly initialize the weights ($\sim$28M) for the attention mechanism between the encoder and the decoder, regardless of the decoder weights. This attention is used to align relevant information between the input and the output. Furthermore, we believe that this is why our proposed models underperform on SPARQL Query Construction compared to other tasks.

In Table 4, our proposed models performed very well with up to 97.33% BLEU score and 92.70% EM score on some datasets such as Mon, Mon50, Mon80. On the other hand, QALD and LC-QuAD only obtained 37.58% and 69.03% BLEU scores while achieving 6.67% and 20.20% EM scores. One possible reason for this significant disparity comes from creating the evaluation datasets and the number of samples in these datasets. As previously mentioned in Sect. 4.1, QALD and LC-QuAD contain many complex question-query pairs which are manually constructed by a human. Meanwhile, the Mon dataset is generated automatically and lacks a variety of entities even though this dataset holds many samples. In Table 5, our models drop the performance with *Covered Entities* setting when compared to *With Entities* setting. That is because SPBERT is trained on executable SPARQL queries that fully contain entities and their relationships.

Results in the SPARQL Query Construction task are much higher than results in the Answer Verbalization Generation. Although VQuAnDa is the extension of LC-QuAD by expanding verbalized answers, BERT2SPBERT$_{\text{MLM+WSO}}$(B) achieved 69.03% BLEU score in LC-QuAD and SPBERT2BERT$_{\text{MLM+WSO}}$(B) only obtained 40.57% BLEU score in VQuAnDa. The possible reason is SPARQL is a structured query language, which almost starts with **SELECT** keyword and ends up with `brack_close` (curly brackets). Meanwhile, natural answers are highly diverse, with many different sentences in the same meaning.

5   Conclusion

In this paper, we have employed pre-trained language models within a sequence-to-sequence architecture. We have also introduced SPBERT, which is the first structured pre-trained model for SPARQL query language. We conducted extensive experiments to investigate the effectiveness of SPBERT on two essential tasks of an end-to-end KBQA system. SPBERT obtains competitive results on several datasets of SPARQL Query Construction. The experimental results show that leveraging weights learned on large-scale corpora can outperform baseline methods in SPARQL Query Construction and Answer Verbalization Generation. SPBERT has also demonstrated that pre-training on SPARQL queries achieves a significant improvement in Answer Verbalization Generation task performance.

The number of our SPARQL queries is currently negligible compared to the number of triples (relationships) in the knowledge base. To improve the performance, we plan to expand the pre-training corpus. In addition, we can use an end-to-end architecture to improve efficiency. We plan to train a single BERT-SPBERT-BERT model on the two essential tasks rather than using BERT2SPBERT and SPBERT2BERT separately.

Acknowledgments. This research is partially supported by the research funding from the Faculty of Information Technology, University of Science, Ho Chi Minh city, Vietnam.

References

1. Devlin, J., Chang, M.W., Lee, K., Toutanova, K.: BERT: pre-training of deep bidirectional transformers for language understanding. In: Proceedings of the 2019 Conference of the North American Chapter of the Association for Computational Linguistics: Human Language Technologies, Volume 1 (Long and Short Papers), pp. 4171–4186, June 2019
2. Feng, Z., et al.: CodeBERT: A pre-trained model for programming and natural languages. In: Findings of the Association for Computational Linguistics: EMNLP 2020, pp. 1536–1547 (2020)
3. Kacupaj, E., Zafar, H., Lehmann, J., Maleshkova, M.: VQuAnDa: verbalization QUestion ANswering DAtaset. In: Harth, A., et al. (eds.) ESWC 2020. LNCS, vol. 12123, pp. 531–547. Springer, Cham (2020). https://doi.org/10.1007/978-3-030-49461-2_31
4. Kapanipathi, P., et al.: Question answering over knowledge bases by leveraging semantic parsing and neuro-symbolic reasoning (2020)
5. Liu, Y., et al.: RoBERTa: a robustly optimized BERT pretraining approach (2020)
6. Luz, F.F., Finger, M.: Semantic parsing natural language into SPARQL: improving target language representation with neural attention (2018). http://arxiv.org/abs/1803.04329

7. Manola, F., Miller, E.: RDF primer. W3C recommendation, February 2004. http://www.w3.org/TR/rdf-primer/

8. Papineni, K., Roukos, S., Ward, T., Zhu, W.J.: BLEU: a method for automatic evaluation of machine translation. In: Proceedings of the 40th Annual Meeting of the Association for Computational Linguistics, pp. 311–318, July 2002

9. Press, O., Wolf, L.: Using the output embedding to improve language models, pp. 157–163. Association for Computational Linguistics, Valencia, April 2017

10. Rothe, S., Narayan, S., Severyn, A.: Leveraging pre-trained checkpoints for sequence generation tasks. Trans. Assoc. Comput. Linguist. **8**, 264–280 (2020)

11. Soru, T., et al.: SPARQL as a foreign language (2020)

12. Trivedi, P., Maheshwari, G., Dubey, M., Lehmann, J.: LC-QuAD: a corpus for complex question answering over knowledge graphs. In: d'Amato, C., et al. (eds.) ISWC 2017. LNCS, vol. 10588, pp. 210–218. Springer, Cham (2017). https://doi.org/10.1007/978-3-319-68204-4_22

13. Usbeck, R., Gusmita, R.H., Ngomo, A.C.N., Saleem, M.: 9th challenge on question answering over linked data (QALD-9) (invited paper). In: Semdeep/NLIWoD@ISWC (2018)

14. Vaswani, A., et al.: Attention is all you need (2017). http://arxiv.org/abs/1706.03762

15. Wang, W., Bi, B., Yan, M., Wu, C., Bao, Z., Peng, L., Si, L.: StructBERT: incorporating language structures into pre-training for deep language understanding (2019). http://arxiv.org/abs/1908.04577

16. Wu, Y., et al.: Google's neural machine translation system: bridging the gap between human and machine translation (2016)

17. Yang, Z., Dai, Z., Yang, Y., Carbonell, J.G., Salakhutdinov, R., Le, Q.V.: XLNet: generalized autoregressive pretraining for language understanding (2019). http://arxiv.org/abs/1906.08237

18. Yin, X., Gromann, D., Rudolph, S.: Neural machine translating from natural language to SPARQL (2019). http://arxiv.org/abs/1906.09302

Deep Neuroevolution: Training Neural Networks Using a Matrix-Free Evolution Strategy

Dariusz Jagodziński[ID], Łukasz Neumann[ID], and Paweł Zawistowski[(✉)][ID]

Institute of Computer Science, Warsaw University of Technology,
Nowowiejska 15/19, Warsaw, Poland
d.jagodzinski@elka.pw.edu.pl,
{lukasz.neumann,pawel.zawistowski}@pw.edu.pl

Abstract. In this paper, we discuss an evolutionary method for training deep neural networks. The proposed solution is based on the Differential Evolution Strategy (DES) – an algorithm that is a crossover between Differential Evolution (DE) and the Covariance Matrix Adaptation Evolution Strategy (CMA-ES). We combine this approach with Xavier's coefficient-based population initialization, batch processing, and gradient-based mutations—the resulting weight optimizer is called neural Differential Evolution Strategy (nDES). Our algorithm yields results comparable to Adaptive Moment Estimation ADAM for a convolutional network training task (50K parameters) on the FashionMNIST dataset. We show that combining both methods results in better models than those obtained after training by either of these algorithms alone. Furthermore, nDES significantly outperforms ADAM on three classic toy recurrent neural network problems. The proposed solution is scalable in an embarrassingly parallel way. For reproducibility purposes, we provide a reference implementation written in Python.

Keywords: Neuroevolution · Neural network · Deep learning · Differential evolution · Genetic algorithm

1 Introduction

Deep artificial neural networks (DNNs) are among the most prominent breakthroughs in modern computer science. This success is largely founded on an effective way of establishing weights in DNNs. The backpropagation algorithm allows neural networks to be trained using a low computational cost gradient-based method. It enables all the weight updates to be simultaneously computed, using just one forward pass through the network, followed by a backward one.

Strong algorithmic efficiency orientation comes with several problems. The first-order methods used for weight optimization give no guarantees that the procedure will find the global minimum if the cost function is multimodal. Optimization may fall into the basin of attraction of one local optimum and prematurely

---

All authors contributed equally.

© Springer Nature Switzerland AG 2021
T. Mantoro et al. (Eds.): ICONIP 2021, LNCS 13108, pp. 524–536, 2021.
https://doi.org/10.1007/978-3-030-92185-9_43

converge. Computing the gradient one layer at a time and iterating backwards from the last may prevent the first layer's weights from changing the value of the gradient due to the vanishing gradient problem [11]. Gradient decreases exponentially with network depth, while the first layer's weight updates can be vanishingly small, or in some cases, the training process may completely stop. Backpropagation computes gradients by the chain rule, which leads to certain limitations of the neurons' activation functions [15, Chapter 7].

The restrictions imposed by the adopted learning method forced researchers to pursue new network layer types and connection schemes to address the vanishing gradient problem. This led to architectures like residual networks [10], or long short-term memory networks (LSTM) [12]. Also, standard activation functions like sigmoid and hyperbolic tangent ceased to be used in favour of the rectified linear unit (ReLU) [13], which has a constant derivative and has started to be the default activation when developing multi-layer perceptrons (MLPs).

Despite the progress, the root cause of the gradient optimization problems remains unchanged. Therefore, instead of struggling with the consequences of using a given optimization method, it might be more effective to use a different one altogether. One alternative lies in the field of neuroevolution [17], which uses evolutionary algorithms (EAs) to generate (or rather evolve) artificial neural networks.

This paper presents a neuroevolution method for deep neural network training based on an evolution strategy. We are motivated by previous research [1] that indicates the possibility of searching through the possible solution space with the contour fitting property without using any costly matrix algebra operations. Network parameters can be established according to the second-order optimization method based on the implicitly used Hessian matrix of DNNs' weight search space. The idea of this method is to use the Differential Evolution Strategy (DES [1]) algorithm, which is a crossover between Differential Evolution (DE) [18] and the Covariance Matrix Adaptation Evolution Strategy (CMA-ES) [9], with the use of batch processing and gradient-based mutation. We call the method the neural Differential Evolution Strategy—nDES. The two main contributions of this work are as follows:

- the paper introduces a new ES called nDES, tailored to high dimensional optimization tasks typical for neural network training,
- we explore the effectiveness of batch-processing and gradient mutations in the proposed method,
- we empirically analyze the properties of nDES by evaluating it against the well established gradient-based ADAM optimizer and show that our method achieves superior performance on synthetic tasks, which are considered hard for gradient-based methods while obtaining results comparable to ADAM in settings "easier" for the latter method,

The paper is organized in the following way. In Sect. 2 we provide relevant information regarding DES and finally introduce nDES. The method is then verified empirically in experiments described and discussed in Sect. 3. Section 4 concludes the paper and describes possible directions for future work.

2   Proposed Method

This Section introduces the (nDES) method, an evolutionary neural network training approach. Before formulating nDES, we provide some necessary background information that underlines both the desirable properties and limitations of evolutionary strategies from the deep learning perspective. These prompt the need for a new method.

2.1   Groundwork

Evolution strategies are popular randomized search heuristics in $\mathbb{R}^n$ that excel in the global optimization of continuous landscapes. Their mechanism is primarily characterized by the mutation operator, whose variance is drawn from a multivariate normal distribution using an evolving covariance matrix. It has been hypothesized that this adapted matrix approximates the inverse Hessian of the search landscape. This hypothesis has been proven for a static model relying on a quadratic approximation [16]. The covariance matrix over selected decision vectors in strategy $(1, \lambda)$ has the same eigenvectors as the Hessian matrix. When the population size is increased, the covariance becomes proportional to the inverse of the Hessian.

The process of generating points with multivariate normal distribution is relatively costly. Evolution strategies with a covariance matrix estimation have an internal computational complexity of at least $\mathcal{O}(n^2)$. CMA-ES is a popular evolution strategy within the field of global optimization, and it has a basic complexity $\mathcal{O}(n^3)$ [8]. The basic limitation is the number of degrees of freedom of the covariance matrix in the n-dimensional space. The $C^{(t)}$ matrix has $\frac{n^2+n}{2}$ parameters that must be updated to determine the $C^{(t+1)}$ matrix. So at least $\frac{n^2+n}{2}$ calculations in each generation are needed, just to determine the new covariance matrix. Generating individuals with a multivariate normal distribution with a given covariance matrix has a computational complexity of $\mathcal{O}(n^2)$. The implementation of a multivariate normal distribution generator $\mathcal{N}(m, C)$ with a given full covariance matrix C and expected value m is a computationally non-trivial task. However, the formula can be transformed as follows:

$$\mathcal{N}(m, C) \sim m + \mathcal{N}(0, C)$$
$$\sim m + C^{\frac{1}{2}}\mathcal{N}(0, I) \tag{1}$$

Spectral decomposition of a covariance matrix C allows $C^{\frac{1}{2}}$ to be factorized by using a series of transformations and substitutions [7]. The following formula can be obtained:

$$C^{\frac{1}{2}} = \mathbf{BDB}^T \tag{2}$$

where:

B is an orthogonal matrix $(B^{-1} = B^T$ and $BB^T = I)$, in which columns form an orthonormal basis of eigenvectors of matrix C

D is a diagonal matrix with square roots of eigenvalues of C as diagonal elements. $\mathbf{D}^2 = \mathrm{DD} = diag(d_1, \ldots, d_n)^2 = diag(d_1^2, \ldots, d_n^2)$, where d_n^2 is n-th eigenvalue of matrix C

From Eqs. (1) and (2), the following equation can be obtained [7]:

$$
\begin{aligned}
\mathcal{N}(m, C) &\sim m + C^{\frac{1}{2}}\mathcal{N}(0, I) \\
&\sim m + BD \underbrace{B^T \mathcal{N}(0, I)}_{\mathcal{N}(0,I)} \\
&\sim m + BD\mathcal{N}(0, I)
\end{aligned}
\tag{3}
$$

The formula (3) allows the generation of points with multivariate normal distribution to be simplified by using spherical (isotropic) normal distribution $\mathcal{N}(0, I)$. According to that formula, each generation of a single point requires the multiplication of the orthogonal matrix **B**, the diagonal matrix **D** and the n-dimensional result vector $\mathcal{N}(0, \mathbf{I})$. The process of matrix multiplication with a Coppersmith and Winograd algorithm [5] has a computational complexity of at least $\mathcal{O}(n^{2.376})$, but the complexity of matrix C factorization into a BDB^T is $\mathcal{O}(n^3)$.

Typically, evolution strategies are benchmarked on problems on a scale of up to hundreds of dimensions [2]. This is in stark contrast to the number of parameters of state-of-the-art neural networks. Here, the high computational complexity of evolution strategies makes their use virtually impossible. However, there is a possibility to generate new points using multivariate Gaussian distribution without an explicitly defined covariance matrix. (DES) [1] is an algorithm that is a crossover between DE and CMA-ES. It uses combinations of difference vectors between archived individuals and univariate Gaussian random vectors along directions of past shifts of population midpoints (Algorithm 1 Line 14). According to the experimental results, DES reveals a linear convergence rate for quadratic functions in a broad spectrum of Hessian matrix condition numbers. The authors of this method also experimentally verified that DES tends to perform contour fitting. The authors of DES have proved in their article that it is possible to achieve $w_i^{(t+1)} \sim \mathcal{N}(m, C)$ without an explicitly calculated or estimated covariance matrix.

Overcoming the computational difficulties connected with applying DES to larger dimensionalities is thus an auspicious direction for research. Potentially, this could lead to a method that has the desirable properties of very effective second-order approaches.

2.2   nDES Method Formulation

On a high level, the (nDES) method is a variant of DES specifically tailored for solving high dimensional optimization tasks typical for neural network training. Because of DES primary usability as evolutionary algorithm, it may manifest some numerical and memory complexity problems for high dimensional optimization. We introduce several crucial modifications to the original metaheuristic to make it applicable in such circumstances. To further exploit the fact that

1 $t \leftarrow 1$

2 initialize$\left(W^{(1)} = \left\{\mathbf{w}_{1,...,\lambda}^{(1)} : \mathbf{w}_i^{(1)} \sim \mathcal{N}(0, \frac{6}{f_{in}+f_{out}})\right\}\right)$

3 **while** *!stop* **do**

4 $\quad$ evaluate$\left(W^{(t)}; \hat{Q}\right)$

5 $\quad \mathbf{m}^{(t+1)} = \frac{1}{\mu}\sum_{i=1}^{\mu}\mathbf{w}_i^{(1)}$

6 $\quad \Delta^{(t)} \leftarrow \mathbf{m}^{(t+1)} - \mathbf{m}^{(t)}$

7 $\quad$ **if** $t = 1$ **then**

8 $\quad\quad \mid \mathbf{p}^{(t)} \leftarrow \Delta^{(t)}$

9 $\quad$ **else**

10 $\quad\quad \mathbf{p}^{(t)} \leftarrow (1 - c_c)\mathbf{p}^{(t-1)} + \sqrt{\mu c_c(2 - c_c)}\Delta^{(t)}$

11 $\quad$ **for** $i = 1, ..., \lambda$ **do**

12 $\quad\quad$ pick at random $\tau_1, \tau_2, \tau_3 \in \{1, ..., H\}$

13 $\quad\quad j, k \sim U(1, ..., \mu)$

14 $\quad\quad \mathbf{d}_i^{(t)} \leftarrow \sqrt{\frac{c_d}{2}}\left(\mathbf{w}_j^{(t-\tau_1)} - \mathbf{w}_k^{(t-\tau_1)}\right)$

15 $\quad\quad\quad + \sqrt{c_d}\Delta^{(t-\tau_2)} \cdot N(0,1) + \sqrt{1 - c_d}\mathbf{p}^{(t-\tau_3)} \cdot N(0,1)$

16 $\quad\quad \mathbf{w}_i^{(t+1)} \leftarrow \mathbf{m}^{(t+1)} + \mathbf{d}_i^{(t)}$

17 $\quad\quad$ **if** *gradientMutation* **then**

18 $\quad\quad\quad \mid \mathbf{w}_i^{(t+1)} \leftarrow \mathbf{w}_i^{(t+1)} + \epsilon\nabla\hat{Q}(w_i^{(t)})$

19 $\quad\quad$ **else if** *randomNoise* **then**

20 $\quad\quad\quad \mathbf{w}_i^{(t+1)} \leftarrow \mathbf{w}_i^{(t+1)} + (1 - c_{cov})^{t/2} \cdot N(\mathbf{0}, \mathbf{I})$

21 $\quad t \leftarrow t + 1$

22 **return** $\mathbf{w}_{best}$

Algorithm 1: Outline of the neural Differential Evolution Strategy

in such optimization scenarios, the gradient is usually available, we also propose ways to utilize this information in the algorithm.

Algorithm 1 presents the nDES outline and reveals a structure somewhat typical to evolution strategies: an initial population consisting of individuals is created (Line 2). This is later iteratively processed by the outer loop (Lines 3 to 21) and finally, after the stop criterion gets triggered, the best-found individual is returned (Line 22). When applied to neural networks, each individual in the population corresponds to the trained network's weights. We discuss the details of this process below.

Each individual in the first population is initialized (Line 2) according to a multivariate normal distribution, with zero mean and standard deviation equal to Xavier's coefficient for each layer of the underlying network [6]. In other words, standard deviation differs for each layer—it is proportional to the inverse of the total number of input (f_{in}) and output (f_{out}) connections to that layer. This approach takes into account the underlying structure of the optimized models and speeds up convergence.

A pass through the main loop of the algorithm (Lines 3 to 21) corresponds to a single iteration of the method. Each iteration begins with evaluating the individual networks according to the loss function $\hat{Q}$ (Line 4). If the training dataset is too big to calculate this function in a single forward pass, we introduce the *batching technique*, which involves multiple tweaks to the method. Before the optimization, the training dataset is split into batches, which stay constant throughout the training process. Next, an Exponentially Weighted Moving Average (EWMA) value is set to zero for each batch. Each individual in the population gets assigned to a data batch—the assignment is cyclical, in the order the batches were created. The individual's fitness is the difference between the loss function evaluated on the given batch and its EWMA value. After all points have been evaluated, the EWMA for each batch is updated with the new fitness values. The update follows the formula below:

$$\bar{Q}_t = \frac{\alpha_t}{n_t} \sum_i^{n_t} \hat{Q}_i + (1 - \alpha_t) \bar{Q}_{t-1} \tag{4}$$

$$\alpha_t = \frac{1}{\sqrt[3]{t}} \tag{5}$$

where $\bar{Q}_t$ is the EWMA value for the batch in the t-th iteration, $\hat{Q}_i$ is the fitness value for the i-th point that evaluated the batch, n_t is the total number the batch has been evaluated in the t-th iteration, and α_t is the multiplier value in the t-th iteration. Using EWMA-based fitness helps nDES to select the best points (i.e., ones that improve the current state) as opposed to favouring points evaluated on batches that were easy to classify. Note that this technique influences the procedure used to return the best individual found (Line 22), which is described below.

After evaluating the population, we use μ best individuals to calculate the center of the population and its shift from the previous iteration, which undergoes exponential smoothing (Lines 5 to 10). In the next steps, the new population gets initialized (Lines 11 to 20). The construction of each individual involves parameters drawn from three historical iterations (Line 12). These allow us to construct the difference vector (Line 14) and mutate the population center to form the new individual (Lines 17 to 20).

The nDES method uses one or two mutation operators. The first one is standard differential mutation, conducted by adding the difference vector to the population center (Line 16). The second one is optional and comes in two types: random noise perturbation (Line 20) or gradient-based rotation (Line 18), which we discuss below.

The rotation-based approach is dedicated to the tasks in which the gradient can easily be computed and does not cause numeric issues. Technically, these gradients are obtained as follows: after evaluating an individual (i.e., calculating the cost function), we calculate the backpropagation pass. The mutation then incorporates a classic SGD step, which is especially useful when the population size is smaller than the optimized network's number of parameters. Under such circumstances, the population spans a sub-hyperplane in the space of possible solutions (since $\lambda < N$). In such a case, one can think about the above technique as using the gradient information to rotate the sub-hyperplane, while nDES optimizes solutions within it.

To select the best solution ($\mathbf{w}_{best}$) for the optimization process (Line 22), different approaches are used depending on batching. Without batching, each point is evaluated on the entire dataset, so we select the point with the lowest fitness in the history as the best one. However, in the other case, the EWMA modification makes this approach unfeasible for nDES. The reason is that the modified fitness value would be the lowest for the point that showed the most improvement on a particular batch of data. Such a point would most probably not be the best solution, as the fitness values' improvements tend to drop near zero in the late stages of the training (i.e., there is relatively little progress as the optimization process starts to converge). Therefore, we evaluate all points in the final population on the validation set and return the best-performing one.

To better specialize nDES for various tasks, we introduce three different variants of the method:

- nDES: optimizing from scratch, with differential and random noise mutations,
- nDES-G: optimizing from scratch using differential and rotational mutations,
- nDES-B: bootstrapping the optimization with a gradient method and then optimizing this pre-trained network with nDES using only differential mutation.

In the nDES-B case, we first train the model using a gradient method (in our experiments, ADAM) with early stopping to prevent overfitting. Next, we use the weights of the model to initialize the population of our evolution strategy. Specifically, the mean of the sampling distribution for the initial population is set using the weights of the pre-trained model. Afterwards, normal nDES optimization is used. The motivation here is to check whether it is possible to make any improvements to a model for which a gradient optimizer has apparently converged. Note that using gradients within nDES-B would potentially harm the results in this case. The reason is that the gradient-based method uses early stopping - thus, the gradients calculated for the point at which the method ended could overfit the model rather than improve it.

3   Experiments

We conducted a series of experiments to evaluate the proposed method's performance on training neural networks. Below, we report the results obtained while training a CNN on one of the popular benchmark datasets to showcase how nDES handles highly dimensional optimization tasks. We also include a section dedicated to multiple RNN toy problems, which are very difficult for gradient-based methods.

All the experiments were conducted using an implementation created in PyTorch v. 1.5 framework [14] and a machine with an Intel Core i7-6850K processor, 128 GB RAM, and a single GTX 1080 Ti GPU card.

3.1   Training CNNs

The CNN experiments were conducted using the model summarized in Table 1 trained on the FashionMNIST [20] dataset. The neural network consisted of two convolutional layers with soft-sign activation and max pooling, and two linear layers – the first one with soft-sign and the second with soft-max activation functions. This architecture has 50,034 trainable parameters in total, which defines quite a challenging task in terms of metaheuristic optimization.

To thoroughly analyze the results, we will analyze multiple aspects connected with the training process and the models it produced. Firstly, we will report on the generated models' accuracy to get a high-level overview of their relative performance. Furthermore, to check whether there are any structural differences in their behaviour, we will compare their confusion matrices and analyze their robustness to adversarial noise. Finally, to focus on the stability of the training process, we will present learning curves from different training runs.

Table 1. Architecture of the neural network used to classify the FashionMNIST dataset

Layer	Layer's output dim	Param. #
Conv2d-1	$20 \times 24 \times 24$	520
Conv2d-2	$32 \times 8 \times 8$	16,032
Linear-3	64	32,832
Linear-4	10	650

Accuracies obtained for the FashionMNIST dataset are depicted in Table 2. nDES was run with the following hyperparameters: $c_{cum} = 0.96$; history $= 16$; batch size $= 64$; validation size $= 10000$. The dataset was normalized with respect to the mean and standard deviation prior to the optimization process, and no further augmentation techniques were used. The boundaries for weights were $[-2, 2]$.

To evaluate the training stability, in Fig. 1 (left) we also provide loss and accuracy learning curves obtained during multiple runs of nDES. From the plots, we may conclude that the training process seems stable and repeatable. In the case of nDES-B, the learning curves are presented in Fig. 1(right). At first sight,

Table 2. Experimental results obtained on the FashionMNIST dataset. We report accuracy values along with their standard deviations for each of the methods considered

Method	Test acc. mean	Test acc. σ	Test loss mean	Test loss σ
ADAM	89.89	0.17	0.2854	0.0031
nDES	86.44	0.24	0.3836	0.0048
nDES-G	89.15	0.37	0.3253	0.0064
nDES-B	90.67	0.17	0.2761	0.0032

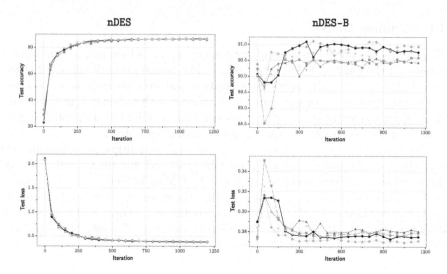

Fig. 1. Results of repeated training runs on the `FashionMNIST` dataset presented on individual curves. The left column depicts `nDES` results and the right one (`nDES-B`) the case in which the model is first trained with `ADAM` and then with `nDES`.

these might appear less stable; however, this is because the algorithm starts from a point already optimized using `ADAM`, and thus the plot's scale is different.

One measure in which `ADAM` and `nDES` differ significantly is the execution time. On average, training the model using `ADAM` in this experiment took about one minute, while using `nDES` took about 55 h. We do not present a precise performance comparison, as it would require more in-depth analysis. This difference stems from the fact that in order for `nDES` to work properly, it needs to make significantly more passes through the training dataset than `ADAM`. This problem is, however, embarrassingly parallel [19], and we discuss it in Sect. 4.

3.2   Training RNNs

RNN models are useful in tasks connected with sequence modelling. One crucial problem that escalates when these sequences become longer is vanishing/exploding gradients, which impair the performance of gradient-based methods. As previous research indicates [3], metaheuristics seem to be more robust against this problem. To evaluate whether this also holds in `nDES`, we conducted experiments on three synthetic datasets taken from the classic [3,12] papers— these are:

- the parity problem: given a sequence of 1's and −1, the model has to predict a class—1 if the sequence contains an odd number of 1's and −1 otherwise,
- the addition problem: given a sequence of real numbers (within [−1, 1]) and a binary mask, the task is to predict the sum of numbers for which the mask values are equal to 1,

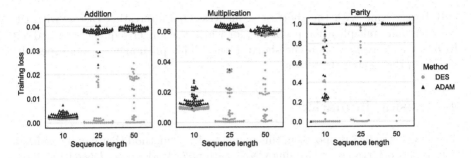

Fig. 2. Results obtained for all sequence modelling problems

- the multiplication problem: similar to the addition problem, the model has to predict the product of its masked input sequence.

During all three experiments, we used a neural network consisting of a single recurrent layer with four neurons, each having a hyperbolic tangent activation function and an output layer with a single, linearly activated neuron. All neurons in the recurrent layer have the same input, whose size depends on the task at hand. This architecture was chosen arbitrarily, as to our surprise, all of the classic papers describing these toy problems lack the precise descriptions of the model.

To compare nDES with a gradient method, we used the ADAM optimizer on all problems. We used mean squared error (MSE) loss for all problems and the results are presented in Fig. 2 and Table 3. Although these tasks are challenging for gradient-based methods, nDES successfully solved them. When analyzing the plots, it can be seen that the difficulty of the toy problems increased along with sequence size. Furthermore, not all optimization runs found the actual minimas (for which the loss equals zero). In all cases, nDES was more effective—ADAM could

Table 3. Results obtained for all sequence modelling problems—we report the obtained means and standard deviations. In each case, nDES obtained better results

Problem	Seq. length	Loss mean		Loss std dev.	
		ADAM	nDES	ADAM	nDES
Addition	10	0.0033	0.0018	0.0010	0.0002
	25	0.0379	0.0070	0.0024	0.0112
	50	0.0393	0.0179	0.0008	0.0148
Multiplication	10	0.0152	0.0093	0.0044	0.0007
	25	0.0625	0.0111	0.0048	0.0152
	50	0.0607	0.0320	0.0013	0.0237
Parity	10	0.7583	0.1894	0.3469	0.3019
	25	0.9994	0.7485	0.0021	0.4003
	50	1.0004	0.9494	0.0022	0.1881

only barely cope with the shortest sequences. One interesting case is with the addition and multiplication problems, for which the longer sequences turned out to be easier for nDES than the shortest ones. This phenomenon was especially visible in the case of the latter.

4   Closing Remarks

The changes to DES proposed in this paper lead to a method that can be useful in neural network training. Training CNNs with nDES leads to models with comparable accuracy to a modern gradient-based method, which experimentally proves that nDES can be a useful algorithm within the field of deep neuroevolution. From an optimization task perspective, using nDES during the conducted experiments, we were able to solve optimization tasks that had orders of magnitude more dimensions than the ones in previous literature concerning DES reports. Thus, our method resolves the computational complexity problems in evolution strategies, allowing them to be used on dimensionalities not manageable before.

Apart from experiments focused on pushing the dimensionality boundaries, the promising results obtained for RNNs also suggest an exciting research direction. Note that whereas CNN architectures seem to be a perfect match for gradient-based optimization methods, for RNNs, specialized architectures (like LSTM [12] or the more recent gated recurrent unit (GRU) [4]) have been developed to overcome the limitations of these methods and make training possible. The excellent performance obtained by nDES on challenging RNN tasks suggests that it is worth evaluating on a larger scale how well modern metaheuristic methods cope in such a setting. The possibility to effectively train RNN models without the need to use structural workarounds for gradient optimizers' deficiencies could significantly impact domains like natural language processing or recommender systems. Applying nDES to more real-world RNN architectures is thus also a direction worth pursuing.

The source code of nDES implementation and all experiments is available at https://github.com/fuine/nDES.

Further development of nDES will focus on solving the encountered problems. One particular source of problems, which limited the scope of experiments in this paper, is that nDES relies on the history of the population to perform well. Moreover, the population size should be multiple times larger than the problem's dimensionality, as suggested by the authors of DES ($\lambda = 4n$). Both these factors result in significant memory requirements and suggest that scaling nDES with the size of the architectures might pose a real difficulty, particularly in the case of deep convolutional models known for their sheer number of parameters. However, the parts of nDES that are the most computationally expensive (i.e., evaluation of the population) are embarrassingly parallel. Therefore, further increases in task dimensionality, and thus experiments on more complicated neural networks, seem to be only a technical hurdle—one resolvable by using multiple GPU cards and parallelizing the evaluation loop.

References

1. Arabas, J., Jagodziński, D.: Toward a matrix-free covariance matrix adaptation evolution strategy. IEEE Trans. Evol. Comput. **24**(1), 84–98 (2020)
2. Awad, N.H., Ali, M.Z., Liang, J.J., Qu, B.Y., Suganthan, P.N.: Problem definitions and evaluation criteria for the CEC 2017 special session and competition on single objective bound constrained real-parameter numerical optimization. Technical report, Nanyang Technological University, Singapore, November 2016
3. Bengio, Y., Simard, P., Frasconi, P.: Learning long-term dependencies with gradient descent is difficult. IEEE Trans. Neural Netw. **5**(2), 157–166 (1994)
4. Cho, K., et al.: Learning phrase representations using RNN encoder-decoder for statistical machine translation. arXiv preprint arXiv:1406.1078 (2014)
5. Coppersmith, D., Winograd, S.: Matrix multiplication via arithmetic progressions. J. Symb. Comput. **9**(3), 251–280 (1990). https://doi.org/10.1016/S0747-7171(08)80013-2. Computational algebraic complexity editorial
6. Glorot, X., Bengio, Y.: Understanding the difficulty of training deep feedforward neural networks. In: Teh, Y.W., Titterington, M. (eds.) Proceedings of the 13 International Conference on Artificial Intelligence and Statistics, vol. 9, pp. 249–256. PMLR, Chia Laguna Resort, Sardinia, 13–15 May 2010
7. Hansen, N.: The CMA evolution strategy: a tutorial (2005). https://hal.inria.fr/hal-01297037. arXiv e-prints, arXiv:1604.00772, pp. 1–39 (2016)
8. Hansen, N., Ostermeier, A.: Completely derandomized self-adaptation in evolution strategies. Evol. Comput. **9**, 159–195 (2001). https://doi.org/10.1162/106365601750190398
9. Hansen, N.: The CMA evolution strategy: a comparing review. In: Lozano, J.A., Larrañaga, P., Inza, I., Bengoetxea, E. (eds.) Towards a New Evolutionary Computation: Advances on Estimation of Distribution Algorithms. Studies in Fuzziness and Soft Computing, vol. 192, pp. 75–102. Springer, Heidelberg (2006). https://doi.org/10.1007/3-540-32494-1_4
10. He, K., Zhang, X., Ren, S., Sun, J.: Deep residual learning for image recognition. In: 2016 IEEE Conference on Computer Vision and Pattern Recognition (CVPR), pp. 770–778 (2016)
11. Hochreiter, S., Kolen, J.F., Kremer, S.C.: Gradient flow in recurrent nets: the difficulty of learning long term dependencies, pp. 237–243. Wiley-IEEE Press (2001)
12. Hochreiter, S., Schmidhuber, J.: Long short-term memory. Neural Comput. **9**, 1735–80 (1997). https://doi.org/10.1162/neco.1997.9.8.1735
13. Nair, V., Hinton, G.: Rectified linear units improve restricted Boltzmann machines Vinod Nair. In: Proceedings of ICML, vol. 27, pp. 807–814, June 2010
14. Paszke, A., et al.: PyTorch: an imperative style, high-performance deep learning library. In: Advances in Neural Information Processing Systems, vol. 32, pp. 8024–8035. Curran Associates, Inc. (2019)
15. Rojas, R.: Neural Networks: A Systematic Introduction. Springer, Heidelberg (1996). https://doi.org/10.1007/978-3-642-61068-4
16. Shir, O., Yehudayoff, A.: On the covariance-hessian relation in evolution strategies. Theor. Comput. Sci. **801**, 157–174 (2020)
17. Stanley, K., Clune, J., Lehman, J., Miikkulainen, R.: Designing neural networks through neuroevolution. Nat. Mach. Intell. **1**, 24–35 (2019). https://doi.org/10.1038/s42256-018-0006-z
18. Storn, R., Price, K.: Differential evolution: a simple and efficient adaptive scheme for global optimization over continuous spaces. J. Glob. Opt. **23** (1995)

19. Wikipedia contributors: Embarrassingly parallel – Wikipedia, the free encyclopedia (2020), https://en.wikipedia.org/wiki/Embarrassingly_parallel. Accessed 10 Sept 2020

20. Xiao, H., Rasul, K., Vollgraf, R.: Fashion-MNIST: a novel image dataset for benchmarking machine learning algorithms (2017)

Weighted P-Rank: a Weighted Article Ranking Algorithm Based on a Heterogeneous Scholarly Network

Jian Zhou[1], Shenglan Liu[1(✉)], Lin Feng[1], Jie Yang[2], and Ning Cai[3(✉)]

[1] School of Computer Science and Technology, Dalian University of Technology,
Dalian 116024, China
liusl@dlut.edu.cn
[2] Research Institute of Information Technology, Tsinghua University,
Beijing 100084, China
[3] School of Artificial Intelligence, Beijing University of Posts
and Telecommunications, Beijing 100876, China
caining91@tsinghua.org.cn

Abstract. The evaluation and ranking of scientific article have always been a very challenging task because of the dynamic change of citation networks. Over the past decades, plenty of studies have been conducted on this topic. However, most of the current methods do not consider the link weightings between different networks, which might lead to biased article ranking results. To tackle this issue, we develop a weighted P-Rank algorithm based on a heterogeneous scholarly network for article ranking evaluation. In this study, the corresponding link weightings in heterogeneous scholarly network can be updated by calculating citation relevance, authors' contribution, and journals' impact. To further boost the performance, we also employ the time information of each article as a personalized PageRank vector to balance the bias to earlier publications in the dynamic citation network. The experiments are conducted on three public datasets (arXiv, Cora, and MAG). The experimental results demonstrated that weighted P-Rank algorithm significantly outperforms other ranking algorithms on arXiv and MAG datasets, while it achieves competitive performance on Cora dataset. Under different network configuration conditions, it can be found that the best ranking result can be obtained by jointly utilizing all kinds of weighted information.

Keywords: Article ranking · Link weighting · Heterogeneous scholarly network · Weighted P-Rank algorithm

1 Introduction

Academic impact assessment and ranking have always been a hot issue, which plays an important role in the process of the dissemination and development of

This study was funded by National Natural Science Foundation of Peoples Republic of China(61672130, 61972064). The Fundamental Rearch Funds for the Central Universities(DUT19RC(3)01) and LiaoNing Revitalization Talents Program(XLYC1806006). The Fundamental Research Funds for the Central Universities, No. DUT20RC(5)010.

T. Mantoro et al. (Eds.): ICONIP 2021, LNCS 13108, pp. 537–548, 2021.
https://doi.org/10.1007/978-3-030-92185-9_44

academic research [1–3]. However, it is difficult to assess the real quality of academic articles due to the dynamic change of citation networks [4]. Furthermore, the evaluation result will be heavily influenced by utilizing different bibliometrics indicators or ranking methods [5]. As a traditional ranking method, PageRank [6] algorithm has already been widely and effectively used in various ranking tasks. Liu *et al.* [7], for instance, employed the PageRank algorithm to evaluate the academic influence of scientists in the co-authorship network. In [8], Bollen *et al.* utilized a weighted version of the PageRank to improve the calculation methodology of JIF. It is worth remarking that the vast majority of ranking algorithms such as PageRank and its variants deem the article (node) creation as a static citation network. In the real citation network, however, articles are published and cited in time sequence. Such approaches do not consider the dynamic nature of the network and are always biased to old publications. Therefore, the recent articles tend to be underestimated due to the lack of enough citations. To address this issue, Sayyadi and Getoor proposed a timeaware method, FutureRank [4], which calculates the future PageRank score of each article by jointly employing citation network, authorship network, and time information. In comparison to the other methods without time weight, FutureRank is practical and ranks academic articles more accurately. Furthermore, Walker *et al.* proposed a ranking model called CiteRank [9], which utilizes a simple network traffic model and calculates the future citations of each article by considering the publication time of articles. However, a main problem of the network traffic model is that it does not reveal the mechanism of how the article scores change. Moreover, although PageRank algorithm is advanced at exploring the global structure of the citation network, it neglects certain local factors that may influence the ranking results.

This paper aims to develop a weighted P-Rank algorithm based on a heterogeneous scholarly network and explore how the changes of the link weightings between different subnetworks influence the ranking result. To further boost the performance of weighted P-Rank algorithm, we utilize the time information of each article as a personalized PageRank vector to balance the bias to earlier publications in the dynamic citation network. The key contributions of this work can be summarized as follows:

- A weighted article ranking method based on P-Rank algorithm and heterogeneous graph is developed.
- The weighted P-Rank algorithm considers the influence of citation revelance, authors' contribution, journals' impact, and time information to the article ranking method comprehensively.
- We evaluate the performance of weighted P-Rank method under different conditions by manipulating the corresponding parameters that can be used to structure graph configurations and time settings.
- By introducing the corresponding link weightings in each heterogeneous graph, the performance of the weighted P-Rank algorithm significantly outperforms the original P-Rank algorithm on three public datasets.

2   Article Ranking Model

In this section, we introduce the proposed article ranking algorithm in detail. Specifically, we first define and describe a heterogeneous scholarly network that is composed of author layer, paper layer and journal layer, and how the different elements in the three layers are linked and interacted. Furthermore, a link weighting method based on P-Rank algorithm is developed to compute the article score in the heterogeneous scholarly network.

2.1   Heterogeneous Scholarly Network

A complete heterogeneous scholarly network consists of three subnetworks (i.e., author network, paper citation network, and journal network). There exist three types of edges in the network i.e., undirected edge between the authors and the papers, directed citation edge between the original paper and its citing papers, and undirected edge between the papers and the published journals. As stated in [10], the heterogeneous scholarly graph of papers, authors, and journals can be expressed as the following form:

$$G(V, E) = (V_P \cup V_A \cup V_J, E_P \cup E_{PA} \cup E_{PJ}) \qquad (1)$$

where V_P, V_A, and V_J are the paper nodes, author nodes, and journal nodes in the three layers respectively. E_P denotes the citation link in the paper layer, E_{PA} denotes the link between paper and author, and E_{PJ} denotes the link between paper and journal.

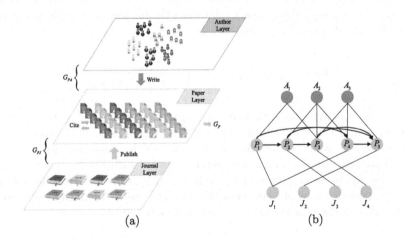

Fig. 1. Visualization of a heterogeneous scholarly network

As shown in Fig. 1a, the paper-author network and paper-journal network are two undirected graphs which can be represented as $G_{PA} = (V_P \cup V_A, E_{PA})$

and $G_{PJ} = (V_P \cup V_J, E_{PJ})$, respectively. In Fig. 1b, by contrast, the paper citation network is a directed graph $G_P = (V_P, E_P)$, the arrows point in the direction of paper citation: $P_4 \to P_5$ means P_4 cites P_5. In this work, we assign link weights to the corresponding subnetworks such that the three unweighted graphs can be updated as $G_P = (V_P, E_P, W_P)$, $G_{PA} = (V_P \cup V_A, E_{PA}, W_{PA})$, and $G_{PJ} = (V_P \cup V_J, E_{PJ}, W_{PJ})$, in which W_P, W_{PA}, and W_{PJ} refer to the link weight in the three graphs, respectively. With link weightings (W_P, W_{PA}, and W_{PJ}) defined in the corresponding G_P, G_{PA}, and G_{PJ}, the unweighted heterogeneous scholarly graph $G(V, E)$ becomes

$$G(V, E, W) = (V_P \cup V_A \cup V_J, E_P \cup E_{PA} \cup E_{PJ}, W_P \cup W_{PA} \cup W_{PJ}) \quad (2)$$

2.2 Link Weighting in Paper Citation Graph (G_P)

In this study, we develop a link weighting to assign weight in the paper citation graph (G_P) based on the citation relevance between two papers, which can be utilized to improve the reasonability of the article ranking. To be specific, the citation relevance (link weighting) between two different papers is mainly influenced by two factors, namely, text similarity (semantic-based) and citation network structure (structure-based). Supposing that the citation relevance between two papers is higher if the two papers are more likely to be similar in semantic and share mutual links and common nodes in the citation network.

In our work, the "slide" weighted overlap approach improved by ADW [11] is employed, which can be used to compute the semantic similarity between the abstracts T_i and T_j from papers i and j. Let S be the intersection of overlapping senses with non-zero probability in both signatures and r_i^j be the rank of sense $s_i \in S$ in signature j, where rank 1 represents the highest rank. The slide overlap Similarity$_1(P_i, P_j)$ can be computed using:

$$\text{Similarity}_1(P_i, P_j) = \tanh\left(\frac{\alpha \cdot \sum_{i=1}^{|S|} e^{(r_i^1 + r_i^2)^{-1}}}{\beta \cdot \sum_{i=1}^{|S|} e^{(2i)^{-1}}}\right) \quad (3)$$

where $\tanh(\cdot)$ is hyperbolic tangent function, and $\sum_{i=1}^{|S|}(2i)^{-1}$ is the maximum value to bound the similarity distributed over the interval [0,1]. Note that the maximum value would occur when each sense has the same rank in both signatures. Moreover, we normalize parameters α and β such that $\alpha + \beta = 1$.

In this work, we employ cosine similarity to measure the citation relevance of two papers in terms of network structure. The cosine similarity between two paper nodes in the citation network can be calculated by:

$$\text{Similarity}_2(P_i, P_j) = \cos(P_i, P_j) = \frac{\left|N_{P_i} \cap N_{P_j}\right|}{\sqrt{\left|N_{P_i}\right| \times \left|N_{P_j}\right|}} \quad (4)$$

where N_{P_i} denotes the neighborhood of node P_i, and $\left|N_{P_i} \cap N_{P_j}\right|$ denotes the number of nodes that link to both P_i and P_j.

Based on the Similarity$_1$ (semantic-based) and Similarity$_2$ (structure-based), the link weight between two paper nodes in the paper citation graph (G_P) can be represented as follows:

$$W_{i,j} = \lambda_1 \cdot \text{Similarity}_1(P_i, P_j) + \lambda_2 \cdot \text{Similarity}_2(P_i, P_j) \tag{5}$$

where $W_{i,j}$ is the weight from paper i to paper j in G_P, Similarity$_1$ and Similarity$_2$ are the semantic-based and structure-based similarities between two papers respectively. Parameters λ_1 and λ_2 are two corresponding coefficients, which can be defined as the following form:

$$\lambda_1 = e^{\mu[\text{Similarity}_1(P_i, P_j) - \varepsilon_1]} \tag{6}$$

$$\lambda_2 = e^{\mu[\text{Similarity}_2(P_i, P_j) - \varepsilon_2]} \tag{7}$$

with μ being a parameter shaping the exponential function, and ε_1 and ε_2 being the media values of Similarity$_1$ and Similarity$_2$ respectively. Here let $\mu = 6$ so that those similarity values that exceed the threshold can be constrained by the exponential curve. Parameters λ_1 and λ_2 are normalized as $\lambda_1 + \lambda_2 = 1$.

For a G_P with n papers, the adjacency matrix of the citation network can be denoted as an $n \times n$ matrix, where the link weight between two paper nodes can be calculated by:

$$M_{i,j} = \begin{cases} W_{i,j} & \text{if paper } i \text{ cites paper } j \\ 0 & \text{otherwise} \end{cases} \tag{8}$$

Let $\overline{M}$ be the fractionalized citation matrix where $\overline{M}_{i,j} = \frac{M_{i,j}}{\sum_{i=1}^n M_{i,j}}$. Let e be the n-dimensional vector whose elements are all 1 and v be an n-dimensional vector which can be viewed as a personalized vector [12]. Next let $x(v)_{\text{paper}}$ denote the PageRank vector corresponding to the vector $x(v)_{\text{paper}}$, and $x(v)$ can be calculated from $x = \overline{\overline{M}}x$ where $\overline{\overline{M}} = d\overline{M} + (1-d)ve^T$. Thus, PageRank vector x can be computed using:

$$x(v)_{\text{paper}} = (1-d)(I - d\overline{M})^{-1}v \tag{9}$$

where d (set at 0.85) is a damping factor. Let $Q = (1-d)(I - d\overline{M})^{-1}$, then $x = Qv$. For any given v, PageRank vector $x(v)$ can be obtained from Qv.

2.3   Link Weighting in Paper-Author Graph (G_{PA})

In the paper-author graph (G_{PA}), let $P = \{p_1, p_2, ..., p_n\}$ denote the set of n papers and $A = \{a_1, a_2, ..., a_m\}$ denote the set of m authors, then G_{PA} can be represented as an $n \times m$ adjacency matrix, where the link weight $A_{\text{author } i,j}$ from author j to paper i is:

$$A_{\text{author}}(i,j) = \begin{cases} 1 & \text{if author } j \text{ writes paper } i \\ 0 & \text{otherwise} \end{cases} \tag{10}$$

In this study, the link weights in G_{PA} can be deemed as the level of authors' contributions to their articles. Modified Raw Weight $(W_{R,j})$ [13] is adopted to assess the authors' contributions according to the relative rankings of authors in co-authored publications. For the author of rank j the Modified Raw Weight is:

$$W_{R,j} = \frac{n - \frac{j}{2} + 1}{\sum_{j=1}^{n} n_j} = \frac{2n - j + 2}{n \cdot (n+1)} \cdot \frac{2}{3} \qquad (11)$$

where $W_{R,j}$ is the Modified Raw Weight of author j, j is the position of author j in the author list, n is the total number of authors in the paper, and $\sum_{j=1}^{n} n_j$ is the sum of author positions. Hence, the unweighted G_{PA} can be updated by:

$$A_{\text{author}}(i,j) = \begin{cases} W_{R,j} & \text{if author } j \text{ writes paper } i \\ 0 & \text{otherwise} \end{cases} \qquad (12)$$

2.4   Link Weighting in Paper-Journal Graph (G_{PJ})

In the initial P-Rank algorithm, the paper-journal graph (G_{PJ}) can be represented as an $n \times q$ adjacency matrix, where n and q are the number of papers and journals, respectively:

$$A_{\text{journal}}(i,j) = \begin{cases} 1 & \text{if paper } i \text{ is published on journal } j \\ 0 & \text{otherwise} \end{cases} \qquad (13)$$

Here we develop a weighted G_{PJ} in which the corresponding link weight can be updated by the journal impact factors [14,15]. Similar to G_{PA}, the link weights in G_{PJ} can be regarded as the level of journals' impact to the published articles. Here, the "mapminmax" function defined in MATLAB R2018b version is used to normalize the JIF list, the range distributed over the interval [0.1,1]. The formula 13 can thus be rewritten as below:

$$A_{\text{journal}}(i,j) = \begin{cases} \text{Normalize}[\text{JIF}_j] & \text{if paper } i \text{ is published on } j \\ 0 & \text{otherwise} \end{cases} \qquad (14)$$

2.5   Weighted P-Rank Algorithm

The weighted P-Rank score of papers can be expressed as $x(v)_{\text{paper}}$ in Eq. 9, where the personalized vector is

$$v = (\varphi_1((\frac{x(v)_{\text{author}}}{n_{\text{p\_author}}})^T \times A_{\text{author}}^T) + \varphi_2((\frac{x(v)_{\text{journal}}}{n_{\text{p\_journal}}})^T \times A_{\text{journal}}^T))^T \qquad (15)$$

where $n_{\text{p-author}}$ represents a vector with the number of publications for each author, and $n_{\text{p-journal}}$ represents a vector with the number of publications for each journal. The mutual dependence (intra-class and inter-class walks) of papers, authors, and journals is coupled by the parameters φ_1 and φ_2, which are set at 0.5 as default. The weighted P-Rank scores of author and journal can be expressed as:

$$x(v)_{\text{author}} = A_{\text{author}}^T \times x(v)_{\text{paper}} \tag{16}$$

$$x(v)_{\text{journal}} = A_{\text{journal}}^T \times x(v)_{\text{paper}} \tag{17}$$

In this study, we adopt a time weight T_i to eliminate the bias to earlier publications, which can be regarded as a personalized PageRank vector. Here according to the time-aware method proposed in FutureRank [4], the function T_i is defined as:

$$T_i = e^{-\rho \times (T_{\text{current}} - T_{\text{publish}})} \tag{18}$$

where T_{publish} denotes the publication time of paper i, and $T_{\text{current}} - T_{\text{publish}}$ denotes the number of years since the paper i was published. ρ is a constant value set to be 0.62 based on FutureRank [4]. The sum of T_i for all the articles is normalized to 1.

Taken together, the weighted P-Rank score of a paper can be calculated by:

$$x(v)_{\text{paper}} = \gamma \cdot Pagerank(\overline{M}, v) + \delta \cdot T + (1 - \gamma - \delta) \cdot \frac{1}{n_p} \tag{19}$$

with parameters γ and δ being constants of the algorithm. $(1 - \gamma - \delta) \cdot \frac{1}{n_p}$ represents the probability of random jump, where n_p is the number of paper samples.

In the proposed algorithm, the initial score of each paper is set to be $\frac{1}{n_p}$. For articles which do not cite any other papers, we suppose that they hold links to all the other papers. Hence, the sum of $x(v)_{\text{paper}}$ for all the papers will keep to be 1 in each iteration. The steps above are recursively conducted until convergence (threshold is set at 0.0001). The pseudocode of the weighted P-Rank algorithm is given in Algorithm 1.

3   Experiments

3.1   Datasets and Experimental Settings

Three public datasets are used in this study, i.e. arXiv (hep-th), Cora, and MAG. The summary statistics of three datasets are listed in Table 1. It is worth remarking that the A_{journal} values of all conference articles were sampled from the average JIF of all journals calculated in the corresponding dataset.

All experiments are conducted on a computer with 3.30 GHz Intel i9-7900X processor and 64 GB RAM under Linux 4.15.0 operating system. The program codes of data preprocessing and graphs modeling are written by Python 3.6.9, which is available on https://github.com/pjzj/Weighted-P-Rank.

Algorithm 1: Weighted P-Rank Algorithm Based on Heterogeneous Network

---

Input	: G_P, G_{PA}, G_{PJ}, JIF list of all journals, and time list of all papers

Output   : Weighted P-Rank score of paper $x(v)_{\text{paper}}$
Parameters: α, β, γ, δ, ρ, λ_1, λ_2, μ, ε_1, ε_2, d, φ_1, φ_2
Steps   :

1 Initialize all the scores of papers: $x(v)_{\text{paper}} = \frac{ones(n_p,1)}{n_p}$, where n_p is the number of paper samples
2 Normalize JIF of each journal in dataset: mapminmax$[J] \leftarrow$ JIF list (J)
3 Compute and normalize time score of each paper based on Eq. 18:
   $$T_i = \text{Normalize}[e^{-\rho \times (T_{\text{current}} - T_{\text{publish}})}]$$
4 Update G_P by Eqs. 5 and 8: $C_w \leftarrow C$
5 Update G_{PA} by Eqs. 11 and 12: $A_w \leftarrow A$
6 Update G_{PJ} with Eq. 14: $J_w \leftarrow J$
7 **while** *not converging* **do**
8 | Eq. 17: $x(v)_{\text{journal}} = A_{\text{journal}}^T \times x(v)_{\text{paper}}$
9 | Eq. 16: $x(v)_{\text{author}} = A_{\text{author}}^T \times x(v)_{\text{paper}}$
10 | Eq. 15: $v = (\varphi_1((\frac{x(v)_{\text{author}}}{n_{\text{p\_author}}})^T \times A_{\text{author}}^T) + \varphi_2((\frac{x(v)_{\text{journal}}}{n_{\text{p\_journal}}})^T \times A_{\text{journal}}^T))^T$
11 | Calculate $Pagerank(\overline{M}, v)$
12 | Update the score of each paper based on time information (Eq. 19):
    | $x(v)_{\text{paper}} = \gamma \cdot Pagerank(\overline{M}, v) + \delta \cdot T + (1 - \gamma - \delta) \cdot \frac{1}{n_p}$
13 **end**
14 **return** $x(v)_{\text{paper}}$, $x(v)_{\text{author}}$, and $x(v)_{\text{journal}}$

Table 1. The datasets utilized in experiments

Dataset	Articles	Citations	Authors	Journals
arXiv	28,500	350,000	14,500	410
Cora	16,252	43,850	12,348	8156
MAG	15,640	200,483	26,430	9575

3.2   Evaluation Metrics

Spearman's Rank Correlation

In this paper, Spearman's rank correlation coefficient is used to assess the performance of proposed algorithm under different conditions. For a dataset $\mathbf{X} = [\mathbf{x}_1, \mathbf{x}_2, ..., \mathbf{x}_N] \in \mathbb{R}^{D \times N}$ with N samples, N original data are converted into grade data, and the correlation coefficient ρ can be calculated by:

$$\rho = \frac{\sum_{i=1}^{n} (R_1(P_i) - \overline{R}_1)(R_2(P_i) - \overline{R}_2)}{\sqrt{\sum_{i=1}^{n} (R_1(P_i) - \overline{R}_1)^2 \sum_{i=1}^{n} (R_2(P_i) - \overline{R}_2)^2}} \tag{20}$$

where $R_1(P_i)$ denotes the position of paper P_i in the first rank list, $R_2(P_i)$ denotes the position of paper P_i in the second rank list, and $\overline{R}_1$ and $\overline{R}_2$ denote the average rank positions of all papers in the two rank lists respectively.

Robustness

Here according to the corresponding historical time point on three datasets, the whole time on each dataset can be divided into two periods. The time period before the historical time point can be denoted as T_1, while the whole period can be denoted as T_2. The robustness of algorithm can thus be measured by calculating the correlation of ranking scores in T_1 and T_2.

3.3 Experimental Results

Graph Configurations

Two parameters can be set in graph configurations: φ_1 and φ_2. By using various combinations of graphs, we compare and assess four different cases of P-Rank algorithm with previous works. The cases and the associated parameters are listed below:

- G_P ($\varphi_1 = 0$, $\varphi_2 = 0$): which is the traditional PageRank algorithm for rank calculation.
- $G_P + G_{PA}$ ($\varphi_1 = 1$, $\varphi_2 = 0$): A new graph (G_{PA}) is introduced into the heterogeneous network which only utilizes citation and authorship.
- $G_P + G_{PJ}$ ($\varphi_1 = 0$, $\varphi_2 = 1$): A new graph (G_{PJ}) is introduced into the heterogeneous network which only utilizes citation and journal information.
- $G_P + G_{PA} + G_{PJ}$ ($\varphi_1 = 0.5$, $\varphi_2 = 0.5$): Two new graphs (G_{PA} and G_{PJ}) are introduced into the heterogeneous network which uses citation, authorship, and journal information simultaneously.

Time Settings

Based on whether to use time information, there exist two kinds of settings:

- No-Time ($\delta = 0$): which does not utilize article time information to enhance the effect of the recent published articles.
- Time-Weighted (see Eq. 19): which can be used to balance the bias to earlier published articles in the citation network.

With these assumptions, we are now ready to verify Spearman's ranking correlation of different cases on three datasets, as shown in Tables 2, 3, 4 and 5. From an analysis of Table 2, it can be found that the best performance (arXiv: 0.5449; Cora: 0.3352; MAG: 0.4994) of proposed algorithm is all achieved from the weighted graph configurations as follows: $G_P + G_{PA} + G_{PJ}$. In addition, we note that under the four graph configuration conditions (G_P; $G_P + G_{PA}$; $G_P + G_{PJ}$; $G_P + G_{PA} + G_{PJ}$), an important observation from the experimental results is that weighted graphs significantly outperform unweighted graphs.

Table 2. Spearman's ranking correlation of different graph configurations on three datasets.

Graph configurations	arXiv		Cora		MAG	
	Unweighted	Weighted	Unweighted	Weighted	Unweighted	Weighted
G_P	0.4153	0.4339	0.2607	0.2793	0.3521	0.3764
$G_P + G_{PA}$	0.4133	0.4490	0.2879	0.3096	0.4125	0.4530
$G_P + G_{PJ}$	0.4082	0.4273	0.2730	0.2894	0.4049	0.4254
$G_P + G_{PA} + G_{PJ}$	0.4915	**0.5449**	0.3135	**0.3352**	0.4748	**0.4994**

Table 3. Spearman's ranking correlation of two time settings on arXiv dataset.

Time settings	G_P		$G_P + G_{PA}$		$G_P + G_{PJ}$		$G_P + G_{PA} + G_{PJ}$	
	Unweighted	Weighted	Unweighted	Weighted	Unweighted	Weighted	Unweighted	Weighted
No-time	0.4153	0.4339	0.4133	0.4490	0.4082	0.4273	0.4915	0.5449
Time-weighted	0.5880	**0.6228**	0.5616	**0.6496**	0.5800	**0.6574**	0.6753	**0.7115**

Table 4. Spearman's ranking correlation of two time settings on Cora dataset.

Time settings	G_P		$G_P + G_{PA}$		$G_P + G_{PJ}$		$G_P + G_{PA} + G_{PJ}$	
	Unweighted	Weighted	Unweighted	Weighted	Unweighted	Weighted	Unweighted	Weighted
No-time	0.2607	0.2793	0.2879	0.3096	0.2730	0.2894	0.3135	0.3352
Time-weighted	0.3120	**0.3490**	0.3593	**0.3848**	0.3116	**0.3729**	0.3772	**0.3962**

Table 5. Spearman's ranking correlation of two time settings on MAG dataset.

Time settings	G_P		$G_P + G_{PA}$		$G_P + G_{PJ}$		$G_P + G_{PA} + G_{PJ}$	
	Unweighted	Weighted	Unweighted	Weighted	Unweighted	Weighted	Unweighted	Weighted
No-time	0.3521	0.3764	0.4125	0.4530	0.4049	0.4254	0.4778	0.4994
Time-weighted	0.4245	**0.5051**	0.4693	**0.5474**	0.4500	**0.5139**	0.5548	**0.5933**

The best performance is highlighted in bold.

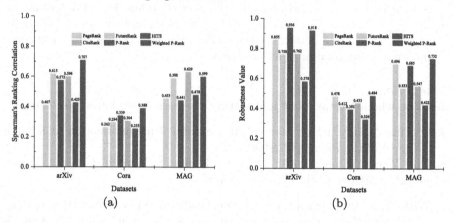

Fig. 2. Spearman's ranking correlation and robustness of six algorithms on three datasets.

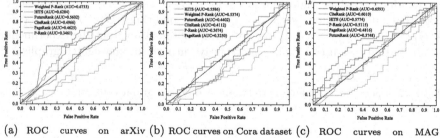

(a) ROC curves on arXiv dataset (b) ROC curves on Cora dataset (c) ROC curves on MAG dataset

Fig. 3. ROC curves obtained by 6 ranking algorithms (Weighted P-Rank, P-Rank, PageRank, FutureRank, HITS, and CiteRank) on three different datasets.

By comparing and analyzing the data from Tables 3, 4 and 5, under the conditions of two time settings (No-Time and Time-Weighted), it can be seen that the performance of Time-Weighted configurations always outperform the results of corresponding No-Time configurations, and the best performance (arXiv: 0.7115; Cora: 0.3962; MAG: 0.5933) is obtained by jointly utilizing all kinds of configurations as follows: $G_P + G_{PA} + G_{PJ}$+Time-Weighted.

For better comparison, we also measure the performance of the weighted P-Rank and five famous algorithms (PageRank, FutureRank, HITS, CiteRank, and P-Rank) on three datasets by using Spearman's rank correlation and robustness. We see from Fig. 2 that weighted P-Rank achieved superior rank correlation (arXiv: 0.707; Cora: 0.388; MAG: 0.599) and robustness performance (arXiv: 0.918; Cora: 0.484; MAG: 0.732), in particular compared to the initial P-Rank algorithm.

It can be seen from Fig. 3 that weighted P-Rank algorithm (as plotted by red curve) significantly outperforms other ranking algorithms on arXiv and MAG datasets, while it achieves competitive performance on Cora dataset. The AUC vales obtained by weighted P-Rank on arXiv, Cora, and MAG datasets are 0.6733, 0.5586, and 0.6593 respectively. By a sharp contrast, the AUC values achieved by initial P-Rank algorithm are unsatisfactory, especially on arXiv dataset (only 0.3461). This result indicates that link weighting plays an important role in heterogeneous graphs, which will be very helpful to improve the performance of the article ranking algorithm.

4   Conclusion

This paper developed a weighted P-Rank algorithm based on a heterogeneous scholarly network for article ranking evaluation. The study is dedicated to assigning weight to the corresponding links in G_P, G_{PA}, and G_{PJ} by calculating citation relevance (G_P), authors' contribution (G_{PA}), and journals' contribution (G_{PJ}). Under conditions of two weighting combinations (Unweighted and Weighted) and four graph configurations (G_P, $G_P + G_{PA}$, $G_P + G_{PJ}$, and $G_P + G_{PA} + G_{PJ}$), the performance of weighted P-Rank algorithm is further evaluated and analyzed. The experimental results showed that the weighted P-Rank method achieved promising performance on three different datasets, and the best ranking result can be achieved by jointly employing all kinds of weighting information. Additionally, we note that the article ranking result can be further improved by utilizing time-weighting information.

In the future, a series of meaningful studies can be conducted subsequently, combining network topology and link weighting. For instance, we would test the effect of link weighting on more ranking methods and verify how the parameters influence the performance of the algorithms.

References

1. Cai, L., et al.: Scholarly impact assessment: a survey of citation weighting solutions. Scientometrics **118**(2), 453–478 (2019)

2. Zhou, J., Cai, N., Tan, Z.-Y., Khan, M.J.: Analysis of effects to journal impact factors based on citation networks generated via social computing. IEEE Access **7**, 19775–19781 (2019)

3. Zhou, J., Feng, L., Cai, N., Yang, J.: Modeling and simulation analysis of journal impact factor dynamics based on submission and citation rules. Complexity, no. 3154619 (2020)

4. Sayyadi, H., Getoor, L.: FutureRank: ranking scientific articles by predicting their future PageRank. In: Proceedings of the 2009 SIAM International Conference on Data Mining, pp. 533–544. SIAM (2009)

5. Feng, L., Zhou, J., Liu, S.-L., Cai, N., Yang, J.: Analysis of journal evaluation indicators: an experimental study based on unsupervised Laplacian Score. Scientometrics **124**, 233–254 (2020)

6. Page, L.: The PageRank citation ranking: bringing order to the web. Technical report. Stanford Digital Library Technologies Project, 1998 (1998)

7. Liu, X., Bollen, J., Nelson, M.L., Van de Sompel, H.: Co-authorship networks in the digital library research community. Inf. Pocess. Manag. **41**(6), 1462–1480 (2005)

8. Bollen, J., Rodriquez, M.A., Van de Sompel, H.: Journal status. Scientometrics **69**(3), 669–687 (2006)

9. Walker, D., Xie, H., Yan, K.K., Maslov, S.: Ranking scientific publications using a model of network traffic. J. Stat. Mech: Theory Exp. **2007**(6), 1–5 (2007)

10. Yan, E., Ding, Y., Sugimoto, C.R.: P-Rank: an indicator measuring prestige in heterogeneous scholarly networks. J. Am. Soc. Inform. Sci. Technol. **62**(3), 467–477 (2011)

11. Pilehvar, M.T., Jurgens, D., Navigli, R.: Align, disambiguate and walk: a unified approach for measuring semantic similarity. In: Proceedings of the 51st Annual Meeting of the Association for Computational Linguistics (Volume 1: Long Papers), vol. 1, pp. 1341–1351 (2013)

12. Haveliwala, T., Kamvar, S., Jeh, G.: An analytical comparison of approaches to personalizing PageRank. Technical report, Stanford (2003)

13. Trueba, F.J., Guerrero, H.: A robust formula to credit authors for their publications. Scientometrics **60**(2), 181–204 (2004)

14. Garfield, E.: Citation analysis as a tool in journal evaluation. Science **178**(4060), 471–479 (1972)

15. Garfield, E.: The history and meaning of the journal impact factor. JAMA **295**(1), 90–93 (2006)

Clustering Friendly Dictionary Learning

Anurag Goel[1,2] and Angshul Majumdar[1(✉)]

[1] Indraprastha Institute of Information Technology, New Delhi 110020, India
{anuragg,angshul}@iiitd.ac.in
[2] Delhi Technological University, New Delhi 110042, India

Abstract. In this work we propose a dictionary learning based clustering app-roach. We regularize dictionary learning with a clustering loss; in particular, we have used sparse subspace clustering and K-means clustering. The basic idea is to use the coefficients from dictionary learning as inputs for clustering. Compari-son with state-of-the-art deep learning based techniques shows that our proposed method improves upon them.

Keywords: Dictionary learning · k-means clustering · Sparse subspace clustering

1 Introduction

This work introduces a new formulation for clustering based on the paradigm of dictionary learning. There have been a few studies that use dictionary learning itself as a clustering algorithm [1, 2]. Such studies are a logical extension to non-negative matrix factorization based clustering [3–5]. Such clustering techniques were popular at the turn of the century and still are being used in document clustering.

We follow the current trend of representation learning based clustering [6–13]. In these studies, autoencoders are used for learning the representation, and the learnt representations are fed into a popular clustering algorithm. Such a piecewise solution was initially proposed in [6]. However, later studies, embed the clustering loss inside the autoencoder cost and jointly solve the ensuing problem.

The problem with autoencoders is that one needs to learn both the encoder and the decoder; the representation from the encoder is used for clustering. The filters learnt by the decoder do not play any role in clustering. When samples are limited, learning the redundant decoder filters may lead to over-fitting. In dictionary learning, only a single set of filters/basis need to be learnt. This will reduce the chance of over-fitting.

A concrete example may give better clarity. Say the dimension of the samples is 100 and we want the features to be of size 50. Then an autoencoder will learn 100×50 values for the encoder and another 100×50 values for the decoder filters. In dictionary learning, only one dictionary of size 100×50 needs to be learnt. This is our motivation for having a dictionary based clustering approach.

The rest of the paper will be organized into several sections. A brief literature review on representation learning based clustering will be discussed in Sect. 2. The proposed formulations are given in Sect. 3. The experimental results will be shown in Sect. 4. The conclusions of the work will be discussed in Sect. 5.

© Springer Nature Switzerland AG 2021
T. Mantoro et al. (Eds.): ICONIP 2021, LNCS 13108, pp. 549–557, 2021.
https://doi.org/10.1007/978-3-030-92185-9_45

2   Literature Review

One of the first studies in deep learning based clustering is [6]; in there stacked autoencoder is learnt and the representation from the deepest layer is fed into a separate clustering algorithm like k-means or spectral clustering. A later study [7], embedded the (sparse subspace) clustering algorithm into the stacked autoencoder formulation. It was found that the jointly learnt formulation [7] yielded better results than the piecemeal technique [6].

Other studies like [8–10] were also based on ideas similar to [7]; unlike the latter which embedded sparse subspace clustering into the stacked autoencoder formulation, [8–10] incorporated the K-means algorithm. The difference between [8] and [9, 10] lies in the definition of the distance metric used in the K-means; while [9, 10] uses the standard Euclidean distance, [8] used the Student's t-distribution kernel.

While all the prior studies [6–9] were based on the autoencoder formulation, [11] proposed a convolutional autoencoder based clustering technique. As in [8] Student's t-distribution based K-means clustering loss was embedded in the deepest layer of the convolutional autoencoder for segmenting the samples.

The initial work on deep clustering [6] proposed a piecemeal solution back in 2014. Over the years [7–11] it was found that jointly learnt solutions that incorporate the clustering loss into the network always improved the results. Following this observation, [12] embedded spectral clustering loss into the autoencoder based formulation.

An interesting deep clustering formulation was put forth in [13]. It proposed deep matrix factorization, which is deep dictionary learning with ReLU activation. The relationship between matrix factorization and clustering is well known [4, 5]; especially in document clustering [3]. The aforesaid study [13] leveraged this relationship and argued that different layers corresponded to different 'concepts' in the data. For example, if the task was face clustering, the first layer probably corresponded to clustering gender, the second layer to age, the third layer to ethnicity, and so on.

Recently a transform learning based clustering approach has also been proposed. Transform learning is the analysis counterpart of dictionary learning [14]. In [15] the subspace clustering losses were embedded in the transform learning framework. An extended deeper version has also been proposed [16].

We have reviewed the major studies in deep learning based clustering. Given that it is a concise paper, we may have omitted papers that are application specific.

3   Proposed Approach

In this section, the technical formulation of our work will be discussed We put forth two variants. In the first, K-means clustering is incorporated into dictionary learning. Next, we propose to regularize dictionary learning with sparse subspace clustering.

3.1 K-means Friendly Dictionary Learning

The popular way to express K-means clustering is via the following formulation:

$$\sum_{i=1}^{k}\sum_{j=1}^{n} h_{ij}\left\| z_j - \mu_i \right\|_2^2$$

$$h_{ij} = \begin{cases} 1 & \text{if } x_j \in \text{Cluster } i \\ 0 & \text{otherwise} \end{cases}$$

(1)

where z_j denotes the j_{th} sample and μ_i the ith cluster.

In [17], it was shown that (1) can be alternately represented in the form of matrix factorization.

$$\left\| Z - ZH^T \left(HH^T \right)^{-1} H \right\|_F^2$$

(2)

where Z is the data matrix formed by stacking z_j's as columns and H is the matrix of binary indicator variables h_{ij}.

In our formulation, the input to the K-means (Z) is not the raw data but the coefficients from dictionary learning. Incorporating the K-means cost into the said formulation we get,

$$\min_{D,Z,H} \underbrace{\|X - DZ\|_F^2}_{\text{Dictionary Learning}} + \mu \underbrace{\left\| Z - ZH^T \left(HH^T \right)^{-1} H \right\|_F^2}_{\text{K - means}}$$

(3)

In (3) μ controls the relative importance of the dictionary learning and K-means costs. We do not see a reason to give one cost more importance than the other, hence we keep $\mu = 1$.

$$\min_{D,Z,H} \|X - DZ\|_F^2 + \left\| Z - ZH^T \left(HH^T \right)^{-1} H \right\|_F^2$$

(4)

Note that we have dropped the sparsity promoting term on the coefficients Z; it is usually there in dictionary learning. The sparsity penalty is important for solving inverse problems but plays a significant role in such representation learning problems apart from being a regularizer. Furthermore, the sparsity promoting term complicates the solution by requiring iterative updates.

One can solve (4) using alternating minimization, i.e. by updating each of the variables assuming the others to be constant. This leads to the following sub-problems.

$$\min_{D} \|X - DZ\|_F^2$$

(5)

$$\min_{Z} \|X - DZ\|_F^2 + \left\| Z - ZH^T \left(HH^T \right)^{-1} H \right\|_F^2$$

(6)

$$\min_{H} \left\| Z - ZH^T \left(HH^T\right)^{-1} H \right\|_F^2 \tag{7}$$

The closed from update for D is given by –

$$D_k = XZ_k^{\dagger} \tag{8}$$

To solve Z we have to take the gradient of the expression and equate it to 0.

$$\nabla \left(\|X - DZ\|_F^2 + \left\| Z - ZH^T \left(HH^T\right)^{-1} H \right\|_F^2 \right) = 0$$

$$\Rightarrow D^T X - D^T DZ - Z\left(I - H^T \left(HH^T\right)^{-1} H\right) = 0$$

$$\Rightarrow D^T X = D^T DZ + Z\left(I - H^T \left(HH^T\right)^{-1} H\right)$$

The last step implies that Z in the form of Sylvester's equation of the form $AZ + ZB = C$, where $C = D^T X$, $A = D^T D$ and $B = \left(I - H^T \left(HH^T\right)^{-1} H\right)$.

The last step is to update H. This is obtained by solving,

$$H^k \leftarrow \min_{H} \left\| Z - ZH^T \left(HH^T\right)^{-1} H \right\|_F^2 \tag{9}$$

This is nothing but K-means algorithm applied on Z. We apply the simple Llyod's algorithm to solve the K-means clustering.

The computational complexity of updating the dictionary is $O(n^3)$ this follows from a matrix product one of which is a pseudo-inverse. The solution of Sylvester's equation (assuming $m \sim n$) by Bartels-Stewart[1] algorithm is also $O(n^3)$. The complexity of solving K-means clustering by Lyod's algorithm is $O(Nkn)$ where N is the number of samples, k is the number of clusters and n the dimensionality of Z.

The algorithm is shown in a succinct fashion below. Once the convergence is reached, the clusters can be found from H. Since the problem is a non-convex function, we do not have any guarantees for convergence. We stop the iterations when the H does not change significantly in subsequent iterations.

Algorithm: DL+K-means

Initialize: D_0, Z_0, H_0
Repeat till convergence
Update D_k using (8)
Update Z_k by solving Sylvester's eqn
Update H_k by K-means clustering
End

---

[1] https://en.wikipedia.org/wiki/Bartels%E2%80%93Stewart_algorithm.

3.2 Sparse Subspace Clustering Friendly Dictionary Learning

In sparse subspace clustering (SSC) [18] it is assumed that the samples belonging to the same cluster lie in the same subspace. The formulation for SSC is as follows.

$$\sum_i \|z_i - Z_{i^c} c_i\|_2^2 + \lambda \|c_i\|_1, \forall i \text{ in } \{1, ..., m\} \tag{10}$$

Here z_i is the ith data point, Z_{i^c} represents all the data points barring the ith one and $c_i(\in \mathbb{R}^{m-1})$ corresponds to the sparse linear weights that represent samples in Z_{i^c} belonging to the same cluster as z_i. The l_1-norm imposes sparsity. Once all the c_i's are solved, 0's are imputed in appropriate (i^{th}) locations to make them vectors of length m; the vectors are then stacked into a matrix $C_{m \times m}$. The affinity matrix is computed from C using,

$$A = |C| + |C|^T \tag{11}$$

The affinity matrix is segmented using normalized cuts algorithm as in [19].

As we did for K-means, we will embed the SSC formulation (10) in dictionary learning leading to the joint optimization problem,

$$\min_{D,Z,H} \underbrace{\|X - DZ\|_F^2}_{\text{Dictionary Learning}} + \underbrace{\sum_i \|z_i - Z_{i^c} c_i\|_2^2 + \lambda \|c_i\|_1}_{\text{SSC}}, \forall i \text{ in } \{1, ..., m\} \tag{12}$$

Here we are giving equal importance to the dictionary learning and clustering at the onset.

As we did before, we solve (12) using alternating minimization. The updates for the dictionary remain the same as (8) hence we do not repeat it. The update for Z is given by –

$$Z \leftarrow \min_Z \|X - D_1 D_2 D_3 Z\|_F^2 + \|Z - ZC\|_F^2 \tag{13}$$

Here Z is formed by stacking the z_i's as columns. To solve Z, we take the gradient of the expression in (13) and equate it to 0.

$$\nabla \left(\|X - DZ\|_F^2 + \|Z - ZC\|_F^2 \right) = 0$$
$$\Rightarrow D^T DZ + Z(I - C) - D^T X = 0$$
$$\Rightarrow D^T DZ + Z(I - C) = D^T X$$

The solution to Z turns out to be via Sylvester's equation as was the case in K-means. The final step is to update the c_i's

$$c_i^k \leftarrow \min_{c_i} \|z_i - Z_{i^c} c_i\|_2^2 + \lambda \|c_i\|_1, \forall i \tag{14}$$

This is solved using SPGL1[2].

_____
[2] https://www.cs.ubc.ca/~mpf/spgl1/index.html.

The algorithm proceeds by iterative solving for the dictionaries using (8), updating the representation by solving Sylvester's equation and updating the coefficients c_i's by SPGL1. As in the case of K-means, (15) is non-convex. Hence, we can only expect to reach a local minimum; however we do not have any theoretical guarantees regarding convergence. In practice, we stop the iterations when the values of c_i's do not change significantly with iterations. We emphasize on c_i's since it has a direct consequence on the clustering performance.

Once the c_i's are obtained the actual clustering proceeds the same as in SSC, i.e. the affinity matrix is computed from the c_i's which then is segmented using normalized cuts. The complete algorithm is shown in a succinct fashion below.

Algorithm: DL+SSC

Initialize: D_0, Z_0, H_0
Repeat till convergence
Update D_k using (8)
Update Z_k by solving Sylvester's eqn
Update Solve c_i's using SPGL1
End
Compute affinity matrix: $A = \lvert C \rvert + \lvert C \rvert^T$
Use N-cuts to segment A

The updation of D is of complexity $O(n^3)$. As before, the updation of Z has a complexity of $O(n^3)$. The solution for l_1-minimization is also $O(n^3)$. Normalized cuts can be solved efficiently[3] in $O(n)$.

4   Experimental Results

We have compared with several state-of-the-art benchmark techniques in deep learning based clustering. They are – deep convolutional embedded clustering (DCEC) [11], deep k-means (DKM) [10], deep clustering network (DCN) [9] and deeply transformed subspace clustering (DTSC) [10].

We have experimented on three popular databases. They are ARFaces[4] (4000 samples, 126 clusters), COIL20[5] (1440 samples, 20 clusters) and Extended Yale B (EYaleB)[6] (2432 samples, 38 clusters). Owing to limitations in space, we request the reader to peruse the references if interested.

We have proposed two variants in this work – with K-means (DLK) and with sparse subspace clustering (DLS). For DLK there is no parameter to tune. In DLS there is only

---

[3] https://en.wikipedia.org/wiki/Segmentation-based_object_categorization#Computational_C omplexity.
[4] http://www2.ece.ohio-state.edu/~aleix/ARdatabase.html.
[5] https://www.cs.columbia.edu/CAVE/software/softlib/coil-20.php.
[6] http://vision.ucsd.edu/~leekc/ExtYaleDatabase/ExtYaleB.html.

one parameter to specify, i.e. λ; we have kept it constant at $\lambda = 0.1$. One needs to fix the number of dictionary atoms in each case. Here we have used the number of atoms as half the dimensionality of the input.

For the experiments, we have assumed that the number of clusters is known. We have used Normalized Mutual Information (NMI) and Adjusted Rand Index (ARI) as the metrics to evaluate the performance of various deep clustering algorithms. The comparative results are shown in Table 1. For the AR Faces and the EYale B, our methods performs considerably better than the others. For the COIl20, the deep methods perform better than others; but our method is slightly worse than the best benchmarks on this dataset.

Table 1. Comparison of clustering performance

Algorithms →	DCEC		DKM		DCN		DTSC		DLK		DLS	
Dataset ↓	NMI	ARI	NMI	ARI	NMI	ARI	NMI	ARI	NMI	ARI	NMI	ARI
ARFaces	0.26	0.02	0.45	0.04	0.45	0.05	0.46	0.15	0.56	0.23	**0.64**	**0.26**
COIL20	**0.79**	**0.64**	0.79	0.60	0.78	0.60	0.78	0.60	0.67	0.56	0.72	0.59
EYaleB	0.20	0.04	0.17	0.02	0.22	0.04	0.42	0.14	**0.36**	**0.11**	0.29	0.10

Table 2. Comparison of run-time (in seconds)

Dataset ↓	DCEC	DKM	DCN	DTSC	DDLK	DDLS
ARFaces	1013	1548	1495	996	95	389
COIL20	959	517	921	502	29	201
EYaleB	1180	2054	2065	1252	73	564

The experiments were carried out on a 64 bit Intel Core i5-8265U CPU @ 1.60 GHz, 16 GB RAM running Ubuntu. The run-times for different algorithms are shown in Table 2. Between DLK and DLS, the former is faster by almost an order of magnitude. This is because it does not require solving a computationally costly l_1-norm minimization problem like the latter in every iteration. Our algorithms are faster than all the benchmarks. This is expected since it is a shallow technique.

The empirical convergence plots for the proposed techniques are given in Fig. 1. We observe that although our problems are non-convex our algorithms converge, at least to a local minimum.

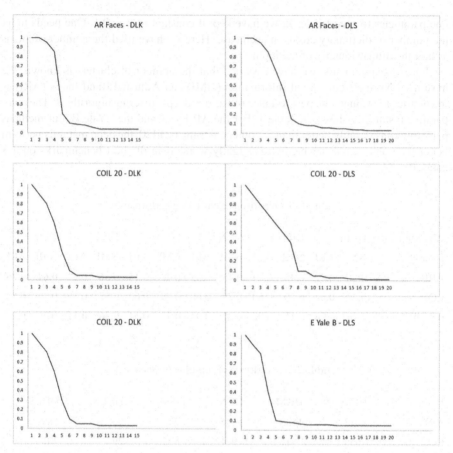

Fig. 1. Empirical convergence for all datasets. Left – DLK. Right – DLS. X-axis: number of iterations. Y-axis: normalized cost

5 Conclusion

This work proposes to regularize dictionary learning by a clustering loss. Although there are many studies on supervising dictionary learning via various losses such as label consistency [20] or fisher discrimination [21]. To the best of our knowledge this is the first work to embed clustering loss into the dictionary learning framework. In particular we have incorporated two clustering losses in this paper – K-means and sparse subspace.

We have carried out thorough experimentation on three popular datasets and four state-of-the-art deep learning based clustering techniques. On two datasets our algorithms perform better than the benchmarks; on the third dataset our results are slightly worse. In terms of speed our proposed algorithms are much faster than the rest.

Acknowledgement. This work is supported by Infosys Center for Artificial Intelligence at IIIT Delhi.

References

1. Ramirez, I., Sprechmann, P., Sapiro, G.: Classification and clustering via dictionary learning with structured incoherence and shared features. In: IEEE CVPR, pp. 3501–3508 (2010)
2. Sprechmann, P., Sapiro, G.: Dictionary learning and sparse coding for unsupervised clustering. In: IEEE ICASSP, pp. 2042–2045 (2010)
3. Xu, W., Liu, X., Gong, Y.: Document clustering based on non-negative matrix factorization. In: ACM SIGIR, pp. 267–273 (2003)
4. Ding, C., He, X., Simon, H.D.: On the equivalence of nonnegative matrix factorization and spectral clustering. In: SIAM SDM, pp. 606–610 (2005)
5. Li, T., Ding, C.: The relationships among various nonnegative matrix factorization methods for clustering. In: IEEE ICDM, pp. 362–371 (2006)
6. Tian, F., Gao, B., Cui, Q., Chen, E., Liu, T.Y.: Learning deep representations for graph clustering. In: AAAI, pp. 1293–1299 (2014)
7. Peng, X., Xiao, S., Feng, J., Yau, W.Y., Yi, Z.: Deep sub-space clustering with sparsity prior. In: IJCAI, pp. 1925–1931 (2016)
8. Xie, J., Girshick, R., Farhadi, A.: Unsupervised deep embedding for clustering analysis. In: ICML, pp. 478–487 (2016)
9. Yang, B., Fu, X., Sidiropoulos, N.D., Hong, M.: Towards k-means-friendly spaces: Simultaneous deep learning and clustering. In: ICML, pp. 3861–3870 (2017)
10. Fard, M.M., Thonet, T., Gaussier, E.: Deep k-means: Jointly clustering with k-means and learning representations. Pattern Recogn. Lett. **138**, 185–192 (2020)
11. Guo, X., Liu, X., Zhu, E., Yin, J.: Deep clustering with convolutional autoencoders. In: ICONIP, pp. 373–382 (2017)
12. Yang, X., Deng, C., Zheng, F., Yan, J., Liu, W.: Deep spectral clustering using dual autoencoder network. In: IEEE CVPR, pp. 4061–4070 (2019)
13. Trigeorgis, G., Bousmalis, K., Zafeiriou, S., Schuller, B.W.: A deep matrix factorization method for learning attribute representations. IEEE Trans. Pattern Anal. Mach. Intell. **39**(3), 417–429 (2017)
14. Ravishankar, S., Bresler, Y.: Learning sparsifying transforms. IEEE Trans. Signal Process. **61**(5), 1072–1086 (2013)
15. Maggu, J., Majumdar, A., Chouzenoux, E.: Transformed Subspace Clustering. IEEE Trans. Knowl. Data Eng. **33**, 1796–1801 (2020)
16. Maggu, J., Majumdar, A., Chouzenoux, E., Chierchia, G.: Deeply transformed subspace clustering. Signal Process. **174**, 107628 (2020)
17. Bauckhage, C.: K-means clustering is matrix factorization (2015). arXiv preprint arXiv:1512.07548,
18. Elhamifar, E., Vidal, R.: Sparse Subspace clustering: algorithm, theory, and applications. IEEE Trans. Pattern Anal. Mach. Intell. **35**(11), 2765–2781 (2013)
19. Shi, J., Malik, J.: Normalized cuts and image segmentation. IEEE Trans. Pattern Anal. Mach. Intell. **22**(8), 888–905 (2000)

Understanding Test-Time Augmentation

Masanari Kimura[(✉)]

Ridge-i Inc., Tokyo, Japan
mkimura@ridge-i.com

Abstract. Test-Time Augmentation (TTA) is a very powerful heuristic that takes advantage of data augmentation during testing to produce averaged output. Despite the experimental effectiveness of TTA, there is insufficient discussion of its theoretical aspects. In this paper, we aim to give theoretical guarantees for TTA and clarify its behavior.

Keywords: Data augmentation · Ensemble learning · Machine learning

1 Introduction

The effectiveness of machine learning has been reported for a great variety of tasks [3,11,14,15,22]. However, satisfactory performance during testing is often not achieved due to the lack of training data or the complexity of the model.

One important concept to tackle such problems is data augmentation. The basic idea of data augmentation is to increase the training data by transforming the input data in some way to generate new data that resembles the original instance. Many data augmentations have been proposed [13,25,29,37], ranging from simple ones, such as flipping input images [20,26], to more complex ones, such as leveraging Generative Adversarial Networks (GANs) to automatically generate data [7,8]. In addition, there are several studies on automatic data augmentation in the framework of AutoML [9,18].

Another approach to improve the performance of machine learning models is ensemble learning [4,27]. Ensemble learning generates multiple models from a single training dataset and combines their outputs, hoping to outperform a single model. The effectiveness of ensemble learning has also been reported in a number of domains [5,6,17].

Influenced by these approaches, a new paradigm called Test-Time Augmentation (TTA) [23,34,35] has been gaining attention in recent years. TTA is a very powerful heuristic that takes advantage of data augmentation during testing to produce averaged output. Despite the experimental effectiveness of TTA, there is insufficient discussion of its theoretical aspects. In this paper, we aim to give theoretical guarantees for TTA and clarify its behavior. Our contributions are summarized as follows:

- We prove that the expected error of the TTA is less than or equal to the average error of an original model. Furthermore, under some assumptions, the expected error of the TTA is strictly less than the average error of an original model;

T. Mantoro et al. (Eds.): ICONIP 2021, LNCS 13108, pp. 558–569, 2021.
https://doi.org/10.1007/978-3-030-92185-9_46

- We introduce the generalized version of the TTA, and the optimal weights of it are given by the closed-form;
- We prove that the error of the TTA depends on the ambiguity term.

2 Preliminaries

Here, we first introduce the notations and problem formulation.

2.1 Problem Formulation

Let $\mathcal{X} \in \mathbb{R}^d$ be the d-dimensional input space, $\mathcal{Y} \in \mathbb{R}$ be the output space, and $\mathcal{H} = \{h(\boldsymbol{x}; \boldsymbol{\theta}) : \mathcal{X} \to \mathcal{Y} \mid \boldsymbol{\theta} \in \Theta\}$ be a hypothesis class, where $\Theta \subset \mathbb{R}^p$ is the p-dimensional parameter space. In supervised learning, our goal is to obtain $h^* \in \mathcal{H} : \mathcal{X} \to \mathcal{Y}$ such that

$$h^* = \arg\min_{h \in \mathcal{H}} \mathcal{R}^\ell(h) = \arg\min_{h \in \mathcal{H}} \mathbb{E}\Big[\ell(y, h(\boldsymbol{x}; \boldsymbol{\theta}))\Big], \tag{1}$$

where

$$\mathcal{R}^\ell(h) := \mathbb{E}\Big[\ell(y, h(\boldsymbol{x}; \boldsymbol{\theta}))\Big] \tag{2}$$

is the expected error and $\ell : \mathcal{Y} \times \mathcal{Y} \to \mathbb{R}_+$ is some loss function. Since we can not access $\mathcal{R}^\ell(h)$ directly, we try to approximate $\mathcal{R}^\ell(h)$ from the limited sample $S = \{(y_i, \boldsymbol{x}_i)\}_{i=1}^N$ of size $N \in \mathbb{N}$. It is the ordinal empirical risk minimization (ERM) problem, and the minimizer of the empirical error $\hat{\mathcal{R}}_S^\ell := \frac{1}{N}\sum_{i=1}^N \ell(y_i, h(\boldsymbol{x}_i))$ can be calculated as

$$\hat{h} = \arg\min_{h \in \mathcal{H}} \hat{\mathcal{R}}_S^\ell(h) = \arg\min_{h \in \mathcal{H}} \frac{1}{N}\sum_{i=1}^N \ell(y_i, h(\boldsymbol{x}_i; \boldsymbol{\theta})). \tag{3}$$

It is known that when the hypothesis class is complex (e.g., a class of neural networks), learning by ERM can lead to overlearning [10]. To tackle this problem, many approaches have been proposed, such as data augmentation [26,31,36] and ensemble learning [4,5,27]. Among such methods, Test-Time Augmentation (TTA) [23,34,35] is an innovative paradigm that has attracted a great deal of attention in recent years.

2.2 TTA: Test-Time Augmentation

The TTA framework is generally described as follows: let $\boldsymbol{x} \in \mathcal{X}$ be the new input variable at test time. We now consider multiple data augmentations $\{\tilde{\boldsymbol{x}}_i\}_{i=1}^m$ for $\boldsymbol{x}$, where $\tilde{\boldsymbol{x}}_i \in \mathbb{R}^d$ is the i-th augmented data where $\boldsymbol{x}$ is transformed and m is the number of strategies for data augmentation. Finally, we compute the output $\tilde{y}$ for the original input $\boldsymbol{x}$ as $\tilde{y} = \sum_{i=1}^m h(\tilde{\boldsymbol{x}}_i)$. Thus, intuitively, one would expect $\tilde{y}$ to be a better predictor than y. TTA is a very powerful heuristic, and its effectiveness has been reported for many tasks [2,23,30,34,35]. Despite its experimental usefulness, the theoretical analysis of TTA is insufficient. In this paper, we aim to theoretically analyze the behavior of TTA. In addition, at the end of the manuscript, we provide directions for future works [28] on the theoretical analysis of TTA in light of the empirical observations given in existing studies.

3   Theoretical Results for the Test-Time Augmentation

In this section, we give several theoretical results for the TTA procedure.

3.1   Re-formalization of TTA

First of all, we reformulate the TTA procedure as follows.

Definition 1. *(Augmented input space) For the transformation class $\mathcal{G}$, we define the augmented input space $\bar{\mathcal{X}}$ as*

$$\bar{\mathcal{X}} := \mathcal{X} \cup \left(\bigcup_{i=1}^{\infty} g(\mathcal{X}; \boldsymbol{\xi}_i) \right) = \mathcal{X} \cup \left(\bigcup_{i=1}^{\infty} \bigcup_{j=1}^{\infty} g(\boldsymbol{x}_j; \boldsymbol{\xi}_i) \right). \tag{4}$$

Definition 2. *(TTA as the function composition) Let $\mathcal{F} = \{f(\boldsymbol{x}; \boldsymbol{\theta}_{\mathcal{F}}) \mid \boldsymbol{\theta}_{\mathcal{F}} \in \Theta_{\mathcal{F}} \subset \Theta\} \subset \mathcal{H}$ be a subset of the hypothesis class and $\mathcal{G} = \{g(\boldsymbol{x}; \boldsymbol{\xi}) : \mathcal{X} \to \bar{\mathcal{X}} \mid \boldsymbol{\xi} \in \Xi\}$ be the transformation class. We assume that $\{g_i = g(\boldsymbol{x}; \boldsymbol{\xi}_i)\}_{i=1}^{m}$ is a set of the data augmentation strategies, and the TTA output $\tilde{y}$ for the input $\boldsymbol{x}$ is calculated as*

$$\tilde{y}(\boldsymbol{x}, \{\boldsymbol{\xi}_{i=1}^{m}\}) := \sum_{i=1}^{m} f \circ g_i(\boldsymbol{x}) = \frac{1}{m} \sum_{i=1}^{m} f(g(\boldsymbol{x}; \boldsymbol{\xi}_i); \boldsymbol{\theta}_{\mathcal{F}}). \tag{5}$$

From these definitions, we have the expected error for the TTA procedure as follows.

Definition 3. *(Expected error with TTA) The empirical error $\mathcal{R}^{\ell, \mathcal{G}}$ of the hypothesis $h \in \mathcal{H}$ obtained by the TTA with transformation class $\mathcal{G}$ is calculated as follows:*

$$\mathcal{R}^{\ell, \mathcal{G}}(h) := \int_{\mathcal{X} \times \mathcal{Y}} \ell(y, \tilde{y}(\boldsymbol{x}, \{\boldsymbol{\xi}_{i=1}^{m}\})) p(\boldsymbol{x}, y) d\boldsymbol{x} dy. \tag{6}$$

The next question is, whether $\mathcal{R}^{\ell, \mathcal{G}}(h)$ is less than $\mathcal{R}^{\ell}$ or not. In addition, if $\mathcal{R}^{\ell, \mathcal{G}}(h)$ is strictly less than $\mathcal{R}^{\ell}$, it is interesting to show the required assumptions.

3.2   Upper Bounds for the TTA

Next we derive the upper bounds for the TTA. For the sake of argument, we assume that $\ell(a, b) = (a - b)^2$ and we decompose the output of the hypothesis for $(\boldsymbol{x}, y)$ as follows:

$$h(\boldsymbol{x}; \boldsymbol{\theta}) = y + \epsilon(\boldsymbol{x}, y; \boldsymbol{\theta}) \quad (\forall h \in \mathcal{H}). \tag{7}$$

Then, the following theorem holds.

Theorem 1. *Assume that $f \circ g \in \mathcal{H}$ for all $f \in \mathcal{F}$ and $g \in \mathcal{G}$, and $\mathcal{G}$ contains the identity transformation $g : x \mapsto x$. Then, the expected error obtained by TTA is bounded from above by the average error of single hypothesises:*

$$\mathcal{R}^{\ell,\mathcal{G}}(h) \leq \bar{\mathcal{R}}^{\ell}(h) := \mathbb{E}\left[\frac{1}{m}\sum_{i=1}^{m}\ell(y, h(x; \theta_i))\right]. \tag{8}$$

Proof. From the definition, the ordinal expected average error is calculated as

$$\bar{\mathcal{R}}^{\ell}(h) = \int_{\mathcal{X} \times \mathcal{Y}} \frac{1}{m}\sum_{i=1}^{m}(y - h(x; \theta_i))^2 p(x, y) dx dy \tag{9}$$

$$= \int_{\mathcal{X} \times \mathcal{Y}} \frac{1}{m}\sum_{i=1}^{m} \epsilon(x, y; \theta_i)^2 p(x, y) dx dy. \tag{10}$$

On the other hand, the expected error of TTA is

$$\mathcal{R}^{\ell,\mathcal{G}}(h) = \int_{\mathcal{X} \times \mathcal{Y}} \left(y - \frac{1}{m}\sum_{i=1}^{m} f \circ g_i(x)\right)^2 p(x, y) dx dy$$

$$= \int_{\mathcal{X} \times \mathcal{Y}} \left(\frac{1}{m}\sum_{i=1}^{m}(y - f \circ g_i(x))\right)^2 p(x, y) dx dy \tag{11}$$

$$= \int_{\mathcal{X} \times \mathcal{Y}} \left(\frac{1}{m}\sum_{i=1}^{m} \epsilon(x, y; \theta_i)\right)^2 p(x, y) dx dy. \tag{12}$$

Then, from Eq. (10) and (12), we have the proof of the theorem.

By making further assumptions, we also have the following theorem.

Theorem 2. *Assume that $f \circ g \in \mathcal{H}$ for all $f \in \mathcal{F}$ and $g \in \mathcal{G}$, and $\mathcal{G}$ contains the identity transformation $g : x \mapsto x$. Assume also that each ϵ has mean zero and is uncorrelated with each other:*

$$\int_{\mathcal{X} \times \mathcal{Y}} \epsilon(x, y; \theta_i) p(x, y) dx dy = 0 \quad (\forall i \in \{1, \ldots, m\}), \tag{13}$$

$$\int_{\mathcal{X} \times \mathcal{Y}} \epsilon(x, y; \theta_i)\epsilon(x, y; \theta_j) p(x, y) dx dy = 0 \quad (i \neq j). \tag{14}$$

In this case, the following relationship holds

$$\mathcal{R}^{\ell,\mathcal{G}}(h) = \frac{1}{m}\bar{\mathcal{R}}^{\ell}(h) < \bar{\mathcal{R}}^{\ell}(h). \tag{15}$$

Proof. From the assumptions (13), (14) and Eq. (10) and (12), the proof of the theorem can be obtained immediately.

3.3 Weighted Averaging for the TTA

We consider the generalization of TTA as follows.

Definition 4. *(Weighted averaging for the TTA) Let* $\mathcal{F} = \{f(x; \theta_{\mathcal{F}}) \mid \theta_{\mathcal{F}} \in \Theta_{\mathcal{F}} \subset \Theta\} \subset \mathcal{H}$ *be a subset of the hypothesis class and* $\mathcal{G} = \{g(x; \xi) : \mathcal{X} \to \bar{\mathcal{X}} \mid \xi \in \Xi\}$ *be the transformation class. We assume that* $\{g_i = g(x; \xi_i)\}_{i=1}^m$ *is a set of the data augmentation strategies, and the TTA output* $\tilde{y}$ *for the input* x *is calculated as*

$$\tilde{y}_w(x, \{\xi_{i=1}^m\}) := \sum_{i=1}^m w_i f \circ g_i(x) = \sum_{i=1}^m w(\xi_i) f(g(x; \xi_i); \theta_{\mathcal{F}}), \qquad (16)$$

where $w_i = w(\xi_i) : \Xi \to \mathbb{R}_+$ *is the weighting function:*

$$w_i \geq 0 \ (\forall i \in \{1, \dots, m\}), \quad \sum_{i=1}^m w_i = 1 \qquad (17)$$

Then, we can obtain the expected error of Eq. (16) as follows.

Proposition 1. *The expected error of the weighted TTA is*

$$\mathcal{R}^{\ell, \mathcal{G}, w}(h) = \sum_{i=1}^m \sum_{j=i}^m w_i w_j \Gamma_{ij}, \qquad (18)$$

where

$$\Gamma_{ij} = \int_{\mathcal{X} \times \mathcal{Y}} \Big(y - f \circ g_i(x) \Big) \Big(y - f \circ g_j(x) \Big) p(x, y) dx dy. \qquad (19)$$

Proof. We can calculate as

$$\mathcal{R}^{\ell, \mathcal{G}, w}(h) = \int_{\mathcal{X} \times \mathcal{Y}} \left(y - \sum_{i=1}^m w_i f \circ g_i(x) \right)^2 p(x, y) dx dy$$

$$= \int_{\mathcal{X} \times \mathcal{Y}} \left(y - \sum_{i=1}^m w_i f \circ g_i(x) \right) \left(y - \sum_{j=1}^m w_j f \circ g_j(x) \right) p(x, y) dx dy$$

$$= \sum_{i=1}^m \sum_{j=i}^m w_i w_j \Gamma_{ij}. \qquad (20)$$

Proposition 18 implies that the expected error of the weighted TTA is highly depending on the correlations of $\{g_1, \dots, g_m\}_{i=1}^m$.

Theorem 3. *(Optimal weights for the weighted TTA) We can obtain the optimal weights* $w = \{w_i, \dots, w_j\}$ *for the weighted TTA as follows:*

$$w_i = \frac{\sum_{j=1}^m \Gamma_{ij}^{-1}}{\sum_{k=1}^m \sum_{j=1}^m \Gamma_{kj}^{-1}}, \qquad (21)$$

where Γ_{ij}^{-1} *is the* (i, j)*-element of the inverse matrix of* (Γ_{ij})*.*

Proof. The optimal weights can be obtained by solving

$$\boldsymbol{w} = \arg\min_{\boldsymbol{w}} \sum_{i=1}^{m} \sum_{j=1}^{m} w_i w_j \Gamma_{ij}. \tag{22}$$

Then, from the method of Lagrange multiplier,

$$\frac{\partial}{\partial w_k} \left\{ \sum_{i=1}^{m} \sum_{j=1}^{m} w_i w_j \Gamma_{ij} - 2\lambda \left(\sum_{i=1}^{m} w_i - 1 \right) \right\} = 0 \tag{23}$$

$$2 \sum_{j=1}^{m} w_k \Gamma_{kj} = 2\lambda \tag{24}$$

$$\sum_{j=1}^{m} w_k \Gamma_{kj} = \lambda. \tag{25}$$

From the condition (17), since $\sum_{i=1}^{m} w_i = 1$ and then, we have

$$w_i = \frac{\sum_{j=1}^{m} \Gamma_{ij}^{-1}}{\sum_{k=1}^{m} \sum_{j=1}^{m} \Gamma_{kj}^{-1}}. \tag{26}$$

From Theorem 3, we obtain a closed-form expression for the optimal weights of the weighted TTA. Furthermore, we see that this solution requires an invertible correlation matrix Γ. However, in TTA we consider the set of $\{f \circ g_i\}_{i=1}^{m}$, and all elements depend on $f \in \mathcal{F}$ in common. This means that the correlations among $\{f \circ g_i\}_{i=1}^{m}$ will be very high, and such correlation matrix is generally known to be singular or ill-conditioned.

3.4 Existence of the Unnecessary Transformation Functions

To simplify the discussion, we assume that all weights are equal. Then, from Eq. (20), we have

$$\mathcal{R}^{\ell,\mathcal{G},w}(h) = \sum_{i=1}^{m} \sum_{j=1}^{m} \Gamma_{ij}/m^2. \tag{27}$$

If we remove g_k from $\{g_1, \ldots, g_m\}$, the error $\tilde{\mathcal{R}}^{\ell,\mathcal{G},w}(h)$ is recomputed as follows.

$$\tilde{\mathcal{R}}^{\ell,\mathcal{G},w}(h) = \sum_{\substack{i=1 \\ i \neq k}}^{m} \sum_{\substack{j=1 \\ j \neq k}}^{m} \Gamma_{ij}/(m-1)^2. \tag{28}$$

Here we consider how the error of the TTA changes when we remove the k-th data augmentation. If we assume that $\mathcal{R}^{\ell,\mathcal{G},w}(h)$ is greater than or equal to $\tilde{\mathcal{R}}^{\ell,\mathcal{G},w}(h)$, then

$$(2m-1) \sum_{i=1}^{m} \sum_{j=1}^{m} \Gamma_{ij} \leq 2m^2 \sum_{\substack{i=1 \\ i \neq k}}^{m} \Gamma_{ik} + m^2 \Gamma_{kk}. \tag{29}$$

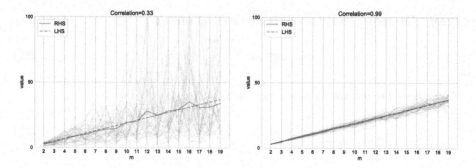

Fig. 1. $(2m-1)\sum_{i=1}^{m}\sum_{j=1}^{m}\Gamma_{ij}$ = LHS vs RHS = $2m^2\sum_{i\neq k}^{m}\Gamma_{ik} + m^2\Gamma_{kk}$ (Eq. (29)). When the correlation is 0.33, the numerical calculation yields $\Pr(RHS \geq LHS) \approx 0.38$. On the other hand, when the correlation is 0.99, we yields $\Pr(RHS \geq LHS) \approx 0.49$.

From the above equation, we can see that the group of data augmentations with very high correlation is redundant, except for some of them. Figure 1 shows an example of a numerical experiment to get the probability that Eq. (29) holds. In this numerical experiment, we generated a sequence of random values with the specified correlation to obtain a pseudo (Γ_{ij}), and calculated the probability that Eq. (29) holds out of 100 trials. From this plot, we can see that (Γ_{ij}) with high correlation is likely to have redundancy. In the following, we introduce ambiguity as another measure of redundancy and show that this measure is highly related to the error of TTA.

3.5   Error Decomposition for the TTA

Knowing what elements the error can be broken down into is one important way to understand the behavior of TTA. For this purpose, we introduce the following notion of ambiguity.

Definition 5. *(Ambiguity of the hypothesis set [16]) For some $x \in \mathcal{X}$, the ambiguity $\varsigma(h_i|x)$ of the hypothesis set $h = \{h_i\}^m$ is defined as*

$$\varsigma(h_i|x) := \left(h_i(x) - \sum_{i=1}^{m} w_i h_i(x) \right)^2 \quad (\forall i \in \{1,\ldots,m\}). \tag{30}$$

Let $\bar{\varsigma}(h|x)$ be the average ambiguity: $\bar{\varsigma}(h|x) = \sum_{i=1}^{m} w_i \varsigma(h_i|x)$. From Definition 5, the ambiguity term can be regarded as a measure of the discrepancy between individual hypotheses for input x. Then, we have

$$\bar{\varsigma}(h|x) = \sum_{i=1}^{m} w_i (y - f \circ g_i(x))^2 - (y - \sum_{i=1}^{m} w_i f \circ g_i(x))^2. \tag{31}$$

Since Eq. (31) holds for all $\boldsymbol{x} \in \mathcal{X}$,

$$\sum_{i=1}^{m} w_i \int_{\mathcal{X} \times \mathcal{Y}} \varsigma(h_i | \boldsymbol{x}) p(\boldsymbol{x}, y) d\boldsymbol{x} dy$$

$$= \sum_{i=1}^{m} w_i \int (y - f \circ g_i(\boldsymbol{x}))^2 p(\boldsymbol{x}, y) d\boldsymbol{x} dy - \int \left(y - \sum_{i=1}^{m} w_i f \circ g_i(\boldsymbol{x}) \right)^2 p(\boldsymbol{x}, y) d\boldsymbol{x} dy.$$

Let

$$err(f \circ g_i) = \mathbb{E}\left[(y - f \circ g_i(\boldsymbol{x}))^2 \right] = \int (y - f \circ g_i(\boldsymbol{x}))^2 p(\boldsymbol{x}) d\boldsymbol{x} dy, \qquad (32)$$

and

$$\varsigma(f \circ g_i) = \mathbb{E}\left[\varsigma(f \circ g_i | \boldsymbol{x}) \right] = \int \varsigma(f \circ g_i | \boldsymbol{x}) p(\boldsymbol{x}) d\boldsymbol{x}. \qquad (33)$$

Then, we have

$$\mathcal{R}^{\ell, \mathcal{G}, w}(h) = \sum_{i=1}^{m} w_i \cdot err(f \circ g_i) - \sum_{i=1}^{m} w_i \cdot \varsigma(f \circ g_i)), \qquad (34)$$

where the first term corresponds to the error, and the second term corresponds to the ambiguity. From this equation, it can be seen that TTA yields significant benefits when each $f \circ g_i$ is more accurate and more diverse than the other.

To summarize, we have the following proposition.

Proposition 2. *The error of the TTA can be decomposed as*

$$\mathcal{R}^{\ell, \mathcal{G}, w}(h) = \Big[errors \ of \ f \circ g_i \Big] + \Big[ambiguities \ of \ f \circ g_i \Big]. \qquad (35)$$

3.6   Statistical Consistency

Finally, we discuss the statistical consistency for the TTA procedure.

Definition 6. *The ERM is the strictly consistent if for any non-empty subset $\mathcal{H}(c) = \{ h \in \mathcal{H} : \mathcal{R}^\ell(h) \geq c \}$ with $c \in (-\infty, +\infty)$ the following convergence holds:*

$$\inf_{h \in \mathcal{H}(c)} \hat{\mathcal{R}}_S^\ell(h) \xrightarrow{p} \inf_{h \in \mathcal{H}(c)} \mathcal{R}^\ell(h) \quad (N \to \infty). \qquad (36)$$

Necessary and sufficient conditions for strict consistency are provided by the following theorem [32, 33].

Theorem 4. *If two real constants $a \in \mathbb{R}$ and $A \in \mathbb{R}$ can be found such that for every $h \in \mathcal{H}$ the inequalities $a \leq \mathcal{R}^\ell(h) \leq A$ hold, then the following two statements are equivalent:*

1. *The empirical risk minimization is strictly consistent on the set of functions* $\{\ell(y, h(\boldsymbol{x})) \mid h \in \mathcal{H}\}$.
2. *The uniform one-sided convergence of the mean to their expectation takes place over the set of functions* $\{\ell(y, h(\boldsymbol{x})) \mid h \in \mathcal{H}\}$, *i.e.*,

$$\lim_{N \to \infty} \Pr\left[\sup_{h \in \mathcal{H}} \left\{ \mathcal{R}^\ell(h) - \hat{\mathcal{R}}_\mathcal{S}^\ell(h) \right\} > \epsilon \right] = 0 \quad (\forall \epsilon > 0). \tag{37}$$

Using these concepts, we can derive the following lemma.

Lemma 1. *The empirical risk* $\frac{1}{m} \sum_{i=1}^m \hat{\mathcal{R}}_\mathcal{S}^{\ell,\mathcal{G}}(h)$ *obtained by ERM with data augmentations* $\{g_1, \ldots, g_m\}_{i=1}^m$ *is the consistent estimator of* $\mathbb{E}_{\bar{\mathcal{X}} \times \mathcal{Y}} \left[\ell(y, f(\boldsymbol{x})) \right]$, *i.e.*,

$$\inf_{h \in \mathcal{H}(c)} \hat{\mathcal{R}}_\mathcal{S}^{\ell,\mathcal{G}}(h) \xrightarrow{p} \mathbb{E}_{\bar{\mathcal{X}} \times \mathcal{Y}} \left[\ell(y, f(\boldsymbol{x})) \right] \quad (N \to \infty). \tag{38}$$

Proof. Let $\bar{\mathcal{X}}$ be the augmented input space with $\{g_1, \ldots, g_m\}_{i=1}^m$. Then, we have

$$\mathbb{E}_{\mathcal{X} \times \mathcal{Y}} \left[\hat{\mathcal{R}}_\mathcal{S}^{\ell,\mathcal{G}}(h) \right] = \int_{\mathcal{X} \times \mathcal{Y}} \left\{ \frac{1}{N} \sum_{i=1}^N \frac{1}{m} \sum_{j=1}^m \ell(y, f \circ g_j(\boldsymbol{x})) \right\} p(\boldsymbol{x}, y) d\boldsymbol{x} dy \tag{39}$$

$$= \int_{\mathcal{X} \times \mathcal{Y}} \left\{ \frac{1}{Nm} \sum_{i=1}^N \sum_{j=1}^m \ell(y, f \circ g_j(\boldsymbol{x})) \right\} p(\boldsymbol{x}, y) d\boldsymbol{x} dy \tag{40}$$

$$= \int_{\bar{\mathcal{X}} \times \mathcal{Y}} \left\{ \frac{1}{Nm} \sum_{i=1}^{Nm} \ell(y, f(\boldsymbol{x})) \right\} p(\boldsymbol{x}, y) d\boldsymbol{x} dy \tag{41}$$

$$= \int_{\bar{\mathcal{X}} \times \mathcal{Y}} \ell(y, f(\boldsymbol{x})) p(\boldsymbol{x}, y) d\boldsymbol{x} dy = \mathbb{E}_{\bar{\mathcal{X}} \times \mathcal{Y}} \left[\ell(y, f(\boldsymbol{x})) \right]. \tag{42}$$

From Lemma 1, we can confirm that the ERM with data augmentation is also minimizing the TTA error. This means that the data augmentation strategies used in TTA should also be used during training.

4   Related Works

Although there is no existing research that discusses the theoretical analysis of the TTA, there are some papers that experimentally investigate the behavior of the TTA [28]. In those papers, the following results are reported:

- the benefit of TTA depends upon the model's lack of invariance to the given Test-Time Augmentations;
- as the training sample size increases, the benefit of TTA decreases;
- when TTA was applied to two datasets, ImageNet [15] and Flowers-102 [24], the performance improvement on the Flowers-102 dataset was small.

Because of the simplicity of the concept, several variants of TTA have also been proposed [12,19,21]. It is also a critical research direction to consider whether a theoretical analysis of these variants is possible using the same procedure as discussed in this paper.

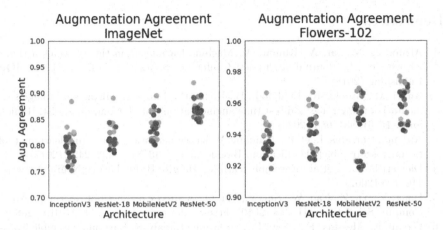

Fig. 2. Architectures that benefit least from standard TTA are also the least sensitive to the augmentations. Note that this figure is created by [28], and see their paper for more details.

5 Conclusion and Discussion

In this paper, we theoretically investigate the behavior of TTA. Our discussion shows that TTA has several theoretically desirable properties. Furthermore, we showed that the error of TTA depends on the ambiguity of the output.

5.1 Future Works

In the previous work, some empirical observations are reported [28]. The future of research is to construct a theory consistent with these observations.

- When TTA was applied to two datasets, ImageNet [15] and Flowers-102 [24], the performance improvement on the Flowers-102 dataset was small. This may be because the instances in Flower-102 are more similar to each other than in the case of ImageNet, and thus are less likely to benefit from TTA. Figure 2 shows the relationship between the model architectures and the TTA ambiguity for each dataset [28]. This can be seen as an analogous consideration to our discussion of ambiguity.
- The benefit of TTA varies depending on the model. Complex models have a smaller performance improvement from TTA than simple models. It is expected that the derivation of the generalization bound considering the complexity of the model such as VC-dimension and Rademacher complexity [1,22] will provide theoretical support for this experiment.
- The effect of TTA is larger in the case of the small amount of data. It is expected to be theorized by deriving inequalities depending on the sample size.

References

1. Alzubi, J., Nayyar, A., Kumar, A.: Machine learning from theory to algorithms: an overview. In: Journal of Physics: Conference Series, vol. 1142, p. 012012. IOP Publishing (2018)
2. Amiri, M., Brooks, R., Behboodi, B., Rivaz, H.: Two-stage ultrasound image segmentation using u-net and test time augmentation. Int. J. Comput. Assist. Radiol. Surg. 15(6), 981–988 (2020)
3. Arganda-Carreras, I., et al.: Trainable Weka segmentation: a machine learning tool for microscopy pixel classification. Bioinformatics 33(15), 2424–2426 (2017)
4. Dietterich, T.G., et al.: Ensemble learning. Handb. Brain Theory Neural Netw. 2, 110–125 (2002)
5. Dong, X., Yu, Z., Cao, W., Shi, Y., Ma, Q.: A survey on ensemble learning. Front. Comput. Sci. 14(2), 241–258 (2019). https://doi.org/10.1007/s11704-019-8208-z
6. Fersini, E., Messina, E., Pozzi, F.A.: Sentiment analysis: Bayesian ensemble learning. Decis. Support Syst. 68, 26–38 (2014)
7. Frid-Adar, M., Diamant, I., Klang, E., Amitai, M., Goldberger, J., Greenspan, H.: Gan-based synthetic medical image augmentation for increased CNN performance in liver lesion classification. Neurocomputing 321, 321–331 (2018)
8. Frid-Adar, M., Klang, E., Amitai, M., Goldberger, J., Greenspan, H.: Synthetic data augmentation using GAN for improved liver lesion classification. In: 2018 IEEE 15th International Symposium on Biomedical Imaging (ISBI 2018), pp. 289–293. IEEE (2018)
9. Hataya, R., Zdenek, J., Yoshizoe, K., Nakayama, H.: Faster AutoAugment: learning augmentation strategies using backpropagation. In: Vedaldi, A., Bischof, H., Brox, T., Frahm, J.-M. (eds.) ECCV 2020. LNCS, vol. 12370, pp. 1–16. Springer, Cham (2020). https://doi.org/10.1007/978-3-030-58595-2_1
10. Hawkins, D.M.: The problem of overfitting. J. Chem. Inf. Comput. Sci. 44(1), 1–12 (2004)
11. Indurkhya, N., Damerau, F.J.: Handbook of Natural Language Processing, vol. 2. CRC Press (2010)
12. Kim, I., Kim, Y., Kim, S.: Learning loss for test-time augmentation. arXiv preprint arXiv:2010.11422 (2020)
13. Kimura, M.: Why mixup improves the model performance, June 2020
14. Kotsiantis, S.B., Zaharakis, I., Pintelas, P.: Supervised machine learning: a review of classification techniques. Emerg. Artif. Intell. Appl. Comput. Eng. 160(1), 3–24 (2007)
15. Krizhevsky, A., Sutskever, I., Hinton, G.E.: ImageNet classification with deep convolutional neural networks. In: Advance Neural Informational Processing System, vol. 25, pp. 1097–1105 (2012)
16. Krogh, A., Vedelsby, J.: Validation, and active learning. In: Advances in Neural Information Processing Systems, vol. 7, p. 231 (1995)
17. Li, Y., Hu, G., Wang, Y., Hospedales, T., Robertson, N.M., Yang, Y.: Differentiable automatic data augmentation. In: Vedaldi, A., Bischof, H., Brox, T., Frahm, J.-M. (eds.) ECCV 2020. LNCS, vol. 12367, pp. 580–595. Springer, Cham (2020). https://doi.org/10.1007/978-3-030-58542-6_35
18. Lim, S., Kim, I., Kim, T., Kim, C., Kim, S.: Fast AutoAugment, May 2019
19. Lyzhov, A., Molchanova, Y., Ashukha, A., Molchanov, D., Vetrov, D.: Greedy policy search: a simple baseline for learnable test-time augmentation. In: Peters, J., Sontag, D. (eds.) Proceedings of the 36th Conference on Uncertainty in Artificial

Intelligence (UAI). Proceedings of Machine Learning Research, vol. 124, pp. 1308–1317. PMLR, 03–06 August 2020. http://proceedings.mlr.press/v124/lyzhov20a.html

20. Mikołajczyk, A., Grochowski, M.: Data augmentation for improving deep learning in image classification problem. In: 2018 International Interdisciplinary PhD Workshop (IIPhDW), pp. 117–122. IEEE (2018)

21. Mocerino, L., Rizzo, R.G., Peluso, V., Calimera, A., Macii, E.: Adaptive test-time augmentation for low-power CPU. CoRR abs/2105.06183 (2021). https://arxiv.org/abs/2105.06183

22. Mohri, M., Rostamizadeh, A., Talwalkar, A.: Foundations of Machine Learning. MIT Press (2018)

23. Moshkov, N., Mathe, B., Kertesz-Farkas, A., Hollandi, R., Horvath, P.: Test-time augmentation for deep learning-based cell segmentation on microscopy images. Sci. Rep. 10(1), 1–7 (2020)

24. Nilsback, M.E., Zisserman, A.: Automated flower classification over a large number of classes. In: Indian Conference on Computer Vision, Graphics and Image Processing, December 2008

25. Park, D.S., et al.: SpecAugment: a simple data augmentation method for automatic speech recognition, April 2019

26. Perez, L., Wang, J.: The effectiveness of data augmentation in image classification using deep learning. arXiv preprint arXiv:1712.04621 (2017)

27. Polikar, R.: Ensemble learning. In: Zhang, C., Ma, Y. (eds.) Ensemble Machine Learning, pp. 1–34. Springer, Heidelberg (2012). https://doi.org/10.1007/978-1-4419-9326-7_1

28. Shanmugam, D., Blalock, D., Balakrishnan, G., Guttag, J.: When and why test-time augmentation works. arXiv preprint arXiv:2011.11156 (2020)

29. Shorten, C., Khoshgoftaar, T.M.: A survey on image data augmentation for deep learning. J. Big Data 6(1), 1–48 (2019)

30. Tian, K., Lin, C., Sun, M., Zhou, L., Yan, J., Ouyang, W.: Improving auto-augment via augmentation-wise weight sharing, September 2020

31. Van Dyk, D.A., Meng, X.L.: The art of data augmentation. J. Comput. Graph. Stat. 10(1), 1–50 (2001)

32. Vapnik, V.: The Nature of Statistical Learning Theory. Springer, Heidelberg (2013)

33. Vapnik, V.N.: An overview of statistical learning theory. IEEE Trans. Neural Netw. 10(5), 988–999 (1999)

34. Wang, G., Li, W., Aertsen, M., Deprest, J., Ourselin, S., Vercauteren, T.: Aleatoric uncertainty estimation with test-time augmentation for medical image segmentation with convolutional neural networks. Neurocomputing 338, 34–45 (2019)

35. Wang, G., Li, W., Ourselin, S., Vercauteren, T.: Automatic brain tumor segmentation using convolutional neural networks with test-time augmentation. In: Crimi, A., Bakas, S., Kuijf, H., Keyvan, F., Reyes, M., van Walsum, T. (eds.) BrainLes 2018. LNCS, vol. 11384, pp. 61–72. Springer, Cham (2019). https://doi.org/10.1007/978-3-030-11726-9_6

36. Zhang, H., Cisse, M., Dauphin, Y.N., Lopez-Paz, D.: Mixup: beyond empirical risk minimization. arXiv preprint arXiv:1710.09412 (2017)

37. Zhong, Z., Zheng, L., Kang, G., Li, S., Yang, Y.: Random erasing data augmentation. In: AAAI, vol. 34, no. 07, pp. 13001–13008 (2020)

SphereCF: Sphere Embedding for Collaborative Filtering

Haozhuang Liu[1,2], Mingchao Li[1,2], Yang Wang[1,2], Wang Chen[1,2],
and Hai-Tao Zheng[1,2(✉)]

[1] Department of Computer Science and Technology, Tsinghua University,
Beijing, China
zheng.haitao@sz.tsinghua.edu.cn
[2] Tsinghua Shenzhen International Graduate School, Tsinghua University,
Shenzhen, China

Abstract. Recently, metric learning has shown its advantage in Collaborative Filtering (CF). Most works capture data relationships based on the measure of Euclidean distance. However, in the high-dimensional space of the data embedding, the directionality between data, that is, the angular relationship, also reflects the important relevance among data. In this paper, we propose Sphere Embedding for Collaborative Filtering (SphereCF) which learns the relationship between cosine metric learning and collaborative filtering. SphereCF maps the user and item latent vectors into the hypersphere manifold and predicts by learning the cosine similarity between the user and item latent vector. At the same time, we propose a hybrid loss that combines triplet loss and logistic loss. The triplet loss makes the inter-class angle between the positive and negative samples of a user as large as possible, and the logistic loss makes the intra-class angle of the positive user-item pairs as small as possible. We consider the user and item latent vector as a point on the hypersphere, which makes the margin in the triplet loss only depend on the angle, thus improving the performance of the model. Extensive experiments show that our model significantly outperforms state-of-the-art methods on four real-world datasets.

Keywords: Metric learning · Collaborative filtering · Data mining

1 Introduction

In this era of information overload, we have to deal with a huge amount of information every day. The time that users spend browsing information and items is limited. Therefore, it makes sense to construct an effective recommendation system by utilizing the user's past interaction among the item, such as the score of the movie, the label of the book, and the music. Thus recommendation model picks more suitable items for the user in a large number of items [17].

H. Liu, M. Li, Y. Wang and W. Chen—Equal contribution.

T. Mantoro et al. (Eds.): ICONIP 2021, LNCS 13108, pp. 570–583, 2021.
https://doi.org/10.1007/978-3-030-92185-9_47

Collaborative filtering (CF) is a common recommendation system. CF uses matrix factorization (MF) to model the user's explicit feedback by mapping users and items to latent factor spaces [13]. So that user-item interactions can be captured by the dot product of their latent factors. Its essence is to calculate the similarity of user and item implicit vectors for scoring.

Collaborative metric learning (CML) combines the characteristics of metric learning and collaborative filtering [7]. CML looks for the relationship between user and item latent vectors from the perspective of distance. More specifically, it minimizes the Euclidean distance of latent vector for positive user-item pairs and uses negative sampling to maximize the distance of latent vector for negative user-item pairs [21].

The current methods of applying metric learning to collaborative filtering, which capture the data relationships based on the measure of Euclidean distance. However, in the high-dimensional space of the data embedding, the angular relationship also reflects the important relationship among data. In addition, the measure of Euclidean distance will cause that the length of embeddings is positively correlated with the frequency of occurrence. Because the longer the vector, the larger the score will get when doing the inner product. This phenomenon will cause the long tail problem. Therefore, this paper proposes Sphere Embedding for Collaborative Filtering (SphereCF), which combines angle metric learning with collaborative filtering. SphereCF maps users and items into a geometric space of angular metrics and ranks items by calculating the cosine similarity between user and item vectors.

Moreover, We utilize triplet loss to optimize our model. Triplet loss is a widely used loss function in deep learning [19]. It is used to train less sensitive samples and applied to tasks such as face recognition [16]. For the recommendation task, we have a lot of items and users. The difference between a considerable number of items and users is relatively small, so the triplet loss is also suitable for the recommendation.

At the same time, we explore the limitations of cosine distance application in triplet loss. The goal of our model is to predict the similarity of user-item pairs by fitting the cosine distance of the angle between user and item vectors. However, the margin in triplet loss will be affected by both vector length and angle. This results in the model not being able to rely solely on angles for prediction, which reduces the performance of the model. To overcome this problem, we map all user and item latent vectors to a hypersphere manifold, so that the margin between classes depends only on the angular difference between the user and the item vectors. At the same time, the hypersphere manifold also gives our model a new geometric interpretation. We treat both the item and the user as points above the hypersphere. As the model trains, the angle between positive user-item pairs will be reduced, and the angle between negative user-item pairs will be increased. To achieve this constraint, we regard CF as a binary classification task and use logistic loss [22] to minimize the angular distance between the user vector and the positive vectors.

Finally, we combine the triplet loss and logistic loss in metric learning and propose a novel hybrid loss. A hyperparameter a controls the weight between two loss functions. The triplet loss makes the inter-class angle between the positive and negative samples of a user as large as possible, and the logistic loss makes the intra-class angle of the positive user-item pairs as small as possible. We analyze the effects of the hybrid loss qualitatively and quantitatively on real-world datasets. In summary, our major contributions can be summarized as follows:

- We propose a novel angle-based collaborative metric learning model (SphereCF). SphereCF maps all user and item vectors to a hypersphere manifold, treating users and items as points on the hypersphere, which gives our model a clear and novel geometric interpretation.
- We innovatively propose a novel hybrid loss that combines triplet loss function with logistic loss function. The former loss function increases the angle between the positive and negative user-item pairs, and the latter decreases the angle between positive user-item pairs.
- We are the first to introduce angular metric learning into CF and show the effectiveness of sphere embedding. Extensive analysis and experiment show that our model obtains significant improvements over the state-of-the-art methods.

2   Related Work

2.1   Metric Learning

Metric learning aims to learn a distance (similarity) function, which captures important relationships between data by distance metrics [22]. Metric learning has been widely used in computer vision [8,16], knowledge graph embedding [14], etc. Most traditional metric learning usually learns a positive semi-definite matrix A for a distance metric $\|x_1 - x_2\|_A = \sqrt{(x_1 - x_2)^{\mathrm{T}} A(x_1 - x_2)}$ upon the given features x_1, x_2 [22]. Recent approaches [11] attempt to use nonlinear transformation functions, such as kernel trick or neural network, to improve the accuracy of metric learning. In deep learning, the most widely used loss functions are contrastive loss, triplet loss [19] and center loss [23].

2.2   Collaborative Filtering

Traditional collaborative filtering implements item recommendations by calculating a large amount of structured data to find other users' items similar to the given user. In the past, Matrix Factorization (MF) was the most widely used method in CF because of its excellent performance [5,6].

In particular, CML combines metric learning with collaborative filtering and uses pairwise hinge loss to train the model [7]. On the basis of CML, LRML [20] uses the attention mechanism [15] to capture the hidden relationship of user-item pairs and alleviate the problem of CML's geometric inflexibility.

2.3   Collaborative Metric Learning (CML)

CML is a competitive method for CF, which was proposed recently and focuses on implicit feedback [7]. Totally different from explicit feedback, implicit feedback methods represent the user similarity based on some implicit interaction

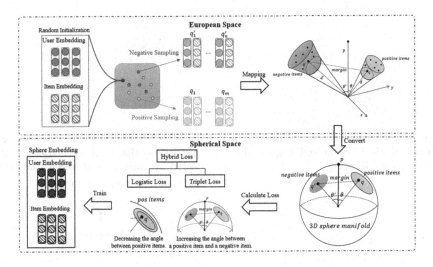

Fig. 1. The whole structure of SphereCF.

data such as likes, click-through, and bookmarks. The model can map user and item latent vector to metric space and learns via a pairwise hinge loss. The main contribution of CML is that CML captures users-items relationships in an Euclidean space, which aims to minimizes the distance between each user-item pair. The Euclidean distance of user vector and item vector can be defined as:

$$s(p,q) = \|p - q\|_2 \tag{1}$$

and the loss of CML is a pairwise hinge loss:

$$\mathcal{L}_m(d) = \sum_{(i,j)\in\mathcal{S}} \sum_{(i,k)\notin\mathcal{S}} w_{ij} \left[m + d(i,j)^2 - d(i,k)^2 \right]_+ \tag{2}$$

where j is an item user i liked, k is an item he did not like. CML is one of the state-of-the-art methods which obeys the triangle inequality and is a prerequisite for fine-grained fitting of users and items in vector space.

3   Our Model

The whole structure of our model can be seen as Fig. 1. We use user embedding and item embedding as input and sample to get the positive and negative vectors as users p, items q (positive), and q' (negative). Then, in order to better optimize our model in angular space, we convert the vector into a hypersphere manifold and calculate cosine distance for each user-item pair $<p,q>$ and $<p,q'>$. At the same time, this conversion solves the problem that the margin in triplet loss is affected by both angle and vector length. Then we use triplet loss to help our model to distinguish positive and negative samples. Besides, in order to make

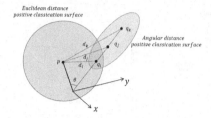

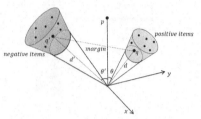

(a) Comparison between Euclidean distance metric learning and angular distance metric learning.

(b) The situation with no hypersphere restriction. the value of margin is affected by both angle (θ and θ') and the norm of item vectors (d and d')

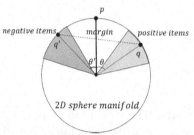

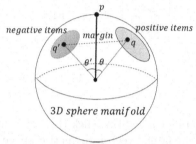

(c) The spherefold in 2D. All the users and items are on the circle. The margin between positive item q and negative item q' is only affected by angle (θ and θ')

(d) The spherefold in 3D. All the users and items are on the sphere. The margin between positive item q and negative item q' is only affected by angle (θ and θ')

Fig. 2. Explanation of hypersphere manifold

the positive sample as close as possible with users, we introduce a logistic loss in our model. Finally, we propose a novel hybrid loss that combines triplet loss with logistic loss and use a hyperparameter a to balance the weight of these two losses.

3.1 Cosine Metric Learning

Recently, some works, like CML [7] and LRML [20], have shown a promising way of introducing metric learning in CF. However, those methods only consider the Euclidean distance between users and items, which ignore that the angle is also an important factor to model the relationship between users and items. For example, as Fig. 2(a) has shown, the q_i, q_j and q_k are all the positive items which user p like. These three items have different Euclidean distance with user p as d_i, d_j and d_k ($d_i < d_j < d_k$), if we only consider Euclidean distance, the positive classification surface based on Euclidean distance may just contain q_i. However, by taking the angular relationship into account, we can classify them correctly into the angular distance classification surface.

In order to achieve this goal, we introduce cosine metric learning into our model and use cosine distance to represent the similarity of user vectors and item vectors. For each user-item pair (p, q), we defined their cosine distance as:

$$d(p, q) = -cos(\theta_{p,q}) \tag{3}$$

$\theta_{p,q}$ is the angle between user vector p and item vector q. By adding a negative sign, we turn $d(p, q)$ into a monotonically increasing function when $\theta_{p,q}$ is between 0 and π, so that an item that user likes can have smaller cosine distance than other items that user did not like.

3.2 Triplet Loss

By introducing cosine metric learning into CF, we define a novelty geometric space to model the relationship between users and items. In this space, in order to better distinguishing positive and negative samples, we need to make the cosine distance of the positive and negative samples as large as possible. Inspired by other works in metric learning [5,7], we find that triplet loss is a great loss function for this purpose. In our model, we use the following triplet loss:

$$loss_{triplet}(d) = - \sum_{(p,q)\in\mathcal{S}} \sum_{(p,q')\notin\mathcal{S}} \left[m + d(p, q) - d(p, q') \right]_{+} \tag{4}$$

where $\mathcal{S}$ is the set of positive user-item pairs, p is the user vector, q is the positive item vector, and q' is the negative item vector. Especially, m is a hyperparameter called "margin" and is an important parameter for our model. During the training step, $d(p, q)$ is converging to -1 and $d(p, q')$ is converging to 1. Because triplet loss only takes the positive value, so the margin can control whether it is far enough for the cosine distance between positive and negative samples.

3.3 Hypersphere Manifold

Triplet loss helps to distinguish the positive and negative items. However, in cosine metric space, the margin value in triplet loss is affected by both angles and its length. We can see an example in the Fig. 2(b), the margin between positive items q and negative items q' is not only affected by angles θ and θ', but also vector length d and d'. It may lead to that the model optimizes the triplet loss by changing vector length rather than angle, which makes the triplet loss function fail to achieve the goal of optimizing the model in a way. To solve this problem, we convert the user vectors and item vectors into a hypersphere manifold, so that these vectors can have the same length.

For each user vector p and item vector q, we convert them to p^* and q^* as:

$$p^* = r\frac{p}{\|p\|}, q^* = r\frac{q}{\|q\|} \tag{5}$$

where $\|p\|$ and $\|q\|$ is the second-order norm of p and q, and r is the radius of the hypersphere. Then, we change the calculation of triplet loss in Eq. 4 as:

$$loss_{triplet}(d) = - \sum_{(p^*,q^*) \in S} \sum_{(p^*,q^{*'}) \notin S} \left[m + d(p^*,q^*) - d(p^*,q^{*'}) \right]_+ \qquad (6)$$

We show the sphere manifold for 2D and 3D respectively in the Fig. 2(c) and Fig. 2(d), by doing this conversion, the margin between user vector p and item vector q, $q^{'}$ only depends on their angle. Meanwhile, by limiting user and item vectors to the contain hypersphere, each user or item vector can be seen as a point on a hypersphere, which makes our model have a novel geometric interpretation.

3.4   Hybrid Loss

Most metric learning models pay attention to increase the distance between positive and negative items to improve the effect. But it is also a natural and effective way to make the positive items more closely cluster around the users vectors. Meanwhile, we can regard CF as a binary classification and use logistic loss in our model to achieve this goal as:

$$s_i = \text{sigmoid}\left(-d\left(p,q\right)\right) \qquad (7)$$

$$loss_{logistic} = -\frac{1}{n} \sum \left[y_i * \ln s_i + (1 - y_i) \ln (1 - s_i) \right] \qquad (8)$$

where $d\left(p,q\right)$ is cosine distance of user p and positive item q, y_i is the category label of item q. We consider the classification of positive items in the logistic loss. During the training, the $-d\left(p,q\right) = cos(\theta_{p,q})$ approaches 1, and it means the angle of user and positive item approaches 0, which allows our model to better aggregate positive items. Then, combining with the triplet loss, a novel hybrid loss is created as:

$$loss_{hybrid} = \alpha \cdot loss_{triplet} + (1 - \alpha) \cdot loss_{logistic} \qquad (9)$$

the α is a hyperparameter to balance the influence of triplet loss and logistic loss. By combining two types of loss, our model can aggregate positive items as well as separating positive and negative items.

4   Experiments

In this section, we evaluate our proposed model SphereCF against other state-of-the-art algorithms. Our experimental evaluation is designed to answer the following research questions.

RQ1. Does our proposed model SphereCF outperforms other baselines and state-of-the-art methods for collaborative filtering?

RQ2. What role does the triplet loss in SphereCF play in the model for the recommendation? How does the value of margin m affect the performance of the model?

RQ3. How does the proposed combine loss in SphereCF affect model performance? What is the performance of the model under different α values?

4.1 Experimental Settings

Datasets. In the spirit of rigorous experimentation, we conducted our experiments on four publicly accessible datasets of different sizes.

Netflix Prize. This dataset includes over 480,000 randomly selected anonymous users on Netflix, scoring more than 1 trillion movies for more than 17,000 movies. Since the entire Netflix Prize dataset is extremely large, we construct a subset of the Netflix Prize dataset. Specifically, we selected user-item ratings from August to December 2015 and filtered out the users who have less than 50 interactions.

MovieLens. This movie rating dataset is widely used to evaluate the performance of collaborative filtering algorithms. We used a version with 1 million user-item ratings (MovieLens-1M).

LastFM. LastFM is the recommended dataset for music. For each user in the dataset, including a list of their most popular artists and the number of plays. It also includes user application tags that can be used to build content vectors.

Delicious. Delicious is the user's tag information for bookmarks. Specifically, it includes the relationship between users, the title and URL of the bookmark, and the tag used by the user for the bookmark.

The four datasets are summarized in Table 1.

Table 1. Characteristics of the datasets.

Dataset	Interactions	# Users	# Items	%Density
Netflix Prize	21,005,149	246,458	17770	0.47
MovieLens	1,000,209	6040	3952	4.19
LastFM	92,834	1,892	17,632	0.28
Delicious	437,593	1,867	69,223	0.34

Evaluation Metrics and Protocol. Our evaluation metrics and protocol follow [5,20]. For each dataset, we adopt *leave-one-out* evaluation consistent with [1,2,18]. The development set consists of the last item for each user. For datasets that do not contain timestamp information, items for the development set are randomly sampled. At the same time, because ranking all items is very time-consuming, we follow the common strategy [3,12] that randomly samples 100 negative samples for each user and ranking the test item among 100 items. These negative samples are those items that do not have interactions with the user. Since our problem is the Top-N ranking for recommendation, the performance of the model is evaluated by two metrics, *Hit Ratio* (HR) and *Normalized Discounted Cumulative Gain* (NDCG) [4]. We rank the sampled 100 items and the test item and calculate the number of occurrences and rankings of the test item in the Top-10. HR indicates the frequency at which the test item appears in the Top-10, and NDCG reflects the position of the test item.

Baselines. In this section, we introduce the baselines compared with our proposed SphereCF method.

Bayesian Personalised Ranking (BPR). BPR [18] is a Bayesian personalized ranking learning model using a pairwise ranking loss for implicit preference data.

Matrix Factorization (MF). MF [17] is a standard baseline for CF that learning the relationship between user and item using inner products.

Multi-layered Perceptron (MLP). MLP is a baseline that uses multiple non-linear layers to fit the relationship between the user and the item [5].

Neural Matrix Factorization (NeuMF). NeuMF [5] is the state-of-the-art unified model combining MF with MLP.

Collaborative Metric Learning (CML). CML [7] can be considered as our baseline model which models user and item vectors using metric learning with euclidean distance.

Latent Relational Metric Learning (LRML). LRML [20] are proposed to use the attention mechanism to model the hidden vector.

Multiplex Memory Network for Collaborative Filtering (MMCF). MMCF [10] leverages multiplex memory layers to jointly capture user-item interactions and the co-occurrence contexts simultaneously in one framework.

Dual Channel Hypergraph Collaborative Filtering (DHCF). DHCF [9] introduces a dual-channel learning strategy to learn the representation of users and items and uses the hypergraph structure to model users and items with explicit hybrid high-order correlations.

Implementation Details. All hyperparameters are tuned to perform the best on the development set based on the NDCG metric. All models are trained until convergence,i.e., until the performance of NDCG metric does not improve after 50 epochs. The dimensionality of user and item embeddings d is tuned of $[20, 50, 100]$. The learning rate for all models is tuned of $[0.01, 0.005, 0.001]$. For models that minimize the hinge loss or triplet loss, the margin m is tuned of $[0.1, 0.3, 0.5, 0.7, 0.9]$. For NeuMF and MLP models, we follow the configuration and architecture proposed in [5]. In addition, for a fair comparison, we do not pretrain the MLP and MF models in NeuMF since it acts as an ensemble classifier [20]. For SphereCF, the parameter α of hybrid loss is tuned of $[0.1, 0.3, 0.5, 0.7, 0.9]$.

4.2   Performance Comparison (RQ1)

The experimental results on all datasets are shown in Table 2, and our proposed model SphereCF shows strong competitiveness among many strong baselines. The result of evaluation metrics of SphereCF surpasses other baselines in all

Table 2. Model performance under Top-10 item ranking

	Netflix		MovieLens		LastFM		Delicious	
	NDCG@10	HR@10	NDCG@10	HR@10	NDCG@10	HR@10	NDCG@10	HR@10
BPR	72.82	94.72	39.28	65.11	61.36	77.58	39.03	44.63
MF	71.42	94.26	37.07	64.21	61.19	77.93	29.25	41.57
MLP	72.66	94.59	39.58	67.35	59.62	77.40	34.60	44.12
NeuMF	73.84	95.13	40.81	68.53	61.89	78.60	35.61	46.17
CML	73.97	95.52	41.59	67.58	<u>64.76</u>	75.37	37.53	45.42
LRML	74.23	95.34	40.12	<u>68.36</u>	63.32	77.68	40.37	45.38
MMCF	<u>74.96</u>	**95.59**	41.71	67.58	62.94	78.97	<u>41.01</u>	46.37
DHCF	74.47	95.51	<u>42.67</u>	64.49	63.58	<u>79.90</u>	40.83	<u>46.65</u>
SphereCF	**76.19**	<u>95.57</u>	**43.96**	**70.35**	**68.13**	**80.70**	**45.05**	**49.15**

datasets except HR@10 on Netflix Prize. This also answers RQ1: our proposed model outperforms the other methods on different public datasets. Not only that, Table 2 shows boldly the second-best performance in other models, we found that these values belong to different models.

Comparison with Different Top-N. To demonstrate our model performance in more detail. We conduct extensive experiments on Top-3,5,10 and 20 on two datasets, MovieLens and Delicious respectively. Table 3 gives a comparison of the best results of all the other baselines with the results of SphereCF on these two datasets. Observing the experimental results, except that DHCF achieves HR@20 performance of 83.59% on MovieLens, which is higher than our proposed model's 82.28%. SphereCF achieved the best performance in all experiments.

Table 3. Performance under different Top-N

Dataset	MovieLens				Delicious			
Model	Others(best)		SphereCF		Others(best)		SphereCF	
Top-N	NDCG	HR	NDCG	HR	NDCG	HR	NDCG	HR
3	33.45	42.01	**33.99**	**43.09**	39.18	42.11	**42.09**	**43.62**
5	37.58	52.46	**38.72**	**54.34**	40.56	44.01	**41.01**	**46.56**
10	42.67	68.36	**43.96**	**70.35**	41.01	46.65	**45.05**	**49.15**
20	46.17	**83.59**	**46.85**	82.28	42.49	53.46	**46.60**	**54.26**

4.3 What Role Does the Triplet Loss in SphereCF Play in the Model for Recommendation (RQ2)

To answer *RQ2*, we explored our model's performance under different values of margin m on the MovieLens and LastFM datasets. The experimental results

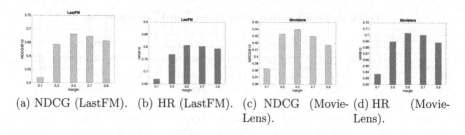

(a) NDCG (LastFM).   (b) HR (LastFM).   (c) NDCG   (Movie-   (d) HR   (Movie-
                                          Lens).   Lens).

Fig. 3. Model performance under different margin m.

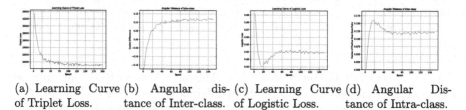

(a) Learning Curve (b) Angular   dis- (c) Learning Curve (d) Angular   Dis-
of Triplet Loss.   tance of Inter-class. of Logistic Loss.   tance of Intra-class.

Fig. 4. Angular distance and loss curves with epoch

are shown in Fig. 3. When margin m ranges from 0.1 to 0.9, the metric value of HR@10 and NDCG@10 rise first and decrease on both datasets. This also means that the margin m needs to take a suitable range of values to distinguish the cosine angle difference between positive and negative user-item pairs. When the margin value is too large, the triplet loss will separate the positive and negative samples, but it might also cause over-fitting problems, which will degrade the performance of SphereCF.

Figure 4(a) gives the learning curve of our model's triplet loss during the training process on MovieLens. Similarly, Fig. 4(b) gives the difference curve between positive and negative user-item pairs versus the cosine value. This difference can be considered as the angular distance of inter-class. We find that when the triplet loss gradually decreases and converges as our model trains, the angular distance of inter-class increases and stabilizes simultaneously. Both of them are exactly negatively related.

Therefore, the conclusions of *RQ2* are summarized as follows: the margin value will affect the performance of the model to some extent. The excessive margin might bring the model over-fitting problem, thus lower the performance of our model. Triplet loss takes the role of distinguishing the inter-class cosine angle between positive and negative user-item pairs. As the model is trained, the angular separation between positive and negative samples is better distinguished.

4.4   How Does the Proposed Combine Loss in SphereCF Affect Model Performance (RQ3)

To answer *RQ3*, we explored our model's performance under different values of α on the MovieLens and LastFM datasets. The experimental results are shown in Fig. 5. When α ranges form 0.1 to 0.9, the metric value of HR@10 and NDCG@10 rise first and then decrease on both dataset. From Eq. 9, we know that when α is equal to zero, the loss of our model will degenerate into logistic loss. Similarly, when α is equal to one, the hybrid loss of our model degenerates into triplet loss. α acts as an adjustment parameter for two different losses, balancing the training weight of angle interval between inter-class and intra-class.

Figure 4(c) gives the learning curve of our model's logistic loss during the training process on MovieLens. Similarly, Fig. 4(d) gives the cosine curve for the user and positive items. The cosine value can be considered as the angular distance of intra-class. As shown in Fig. 4(c) and Fig. 4(d), both of them are exact negatively related similar to Fig. 4(a) and Fig. 4(b) in *RQ2*. When SphereCF was trained to 20 epochs, the logistic loss was minimized, as was the intra-class angle, but then the logistic loss rose and converged to a certain value. This is because, at 20 epochs, although the angular distance of inter-class angle is small, the angular distance of intra-class is still relatively large. So the best performance is not obtained at this time. Under the adjustment of α, the model distinguishes between positive and negative items and gathers the positive items for the user in a suitable range of angles. Finally, after 140 epochs of training, SphereCF converges and achieves its best performance.

Therefore, the conclusions of *RQ3* are summarized as follows: α controls the proportion of intra-class angle and inter-class angle for user-item pairs during the model training process. We not only want to distinguish positive and negative items for given users but also hope that our model can gather positive items for given users to a certain extent. Experiments have also shown that with the training of the model, logistic loss gradually gathers the positive items for the anchor, i.e., the given user. The hybrid of triplet loss and logistic loss does improve the performance of our proposed model.

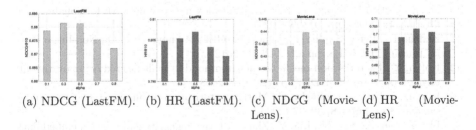

(a) NDCG (LastFM).   (b) HR (LastFM).   (c) NDCG (Movie-Lens).   (d) HR (Movie-Lens).

Fig. 5. Model performance under different α.

5   Conclusion

In this paper, we introduce cosine metric learning into the Collaborative Filtering task and propose Sphere Embedding for Collaborative Filtering (SphereCF). SphereCF learns the relationship between the user and item latent vector according to their cosine distance.

Acknowledgement. This research is supported by National Natural Science Foundation of China (Grant No. 61773229 and 6201101015), Alibaba Innovation Research (AIR) programme, Natural Science Foundation of Guangdong Province (Grant No. 2021A1515012640), the Basic Research Fund of Shenzhen City (Grand No. JCYJ20210324120012033 and JCYJ20190813165003837), and Overseas Cooperation Research Fund of Tsinghua Shenzhen International Graduate School (Grant No. HW2021008).

References

1. Bayer, I., He, X., Kanagal, B., Rendle, S.: A generic coordinate descent framework for learning from implicit feedback. In: WWW, pp. 1341–1350 (2017)
2. Cheng, H.-T., et al.: Wide & deep learning for recommender systems. In: Proceedings of the Recsys, pp. 7–10 (2016)
3. Elkahky, A.M., Song, Y., He, X.: A multi-view deep learning approach for cross domain user modeling in recommendation systems. In: WWW, pp. 278–288 (2015)
4. He, X., Chen, T., Kan, M.-Y., Chen, X.: TriRank: review-aware explainable recommendation by modeling aspects. In: CIKM, pp. 1661–1670 (2015)
5. He, X., Liao, L., Zhang, H., Nie, L., Hu, X., Chua, T.-S.: Neural collaborative filtering. In: WWW, pp. 173–182 (2017)
6. He, X., Zhang, H., Kan, M.-Y., Chua, T.-S.: Fast matrix factorization for online recommendation with implicit feedback. In: SIGIR, pp. 549–558 (2016)
7. Hsieh, C.-K., Yang, L., Cui, Y., Lin, T.-Y., Belongie, S., Estrin, D.: Collaborative metric learning. In: WWW, pp. 193–201 (2017)
8. Hu, J., Lu, J., Tan, Y.-P.: Discriminative deep metric learning for face verification in the wild. In: CVPR, pp. 1875–1882 (2014)
9. Ji, S., Feng, Y., Ji, R., Zhao, X., Tang, W., Gao, Y.: Dual channel hypergraph collaborative filtering. In: SIGKDD, pp. 2020–2029 (2020)
10. Jiang, X., Hu, B., Fang, Y., Shi, C.: Multiplex memory network for collaborative filtering. In: SDM, pp. 91–99 (2020)
11. Kedem, D., Tyree, S., Sha, F., Lanckriet, G.R., Weinberger, K.Q.: Non-linear metric learning. In: NIPS, pp. 2573–2581 (2012)
12. Koren, Y.: Factorization meets the neighborhood: a multifaceted collaborative filtering model. In: SIGKDD, pp. 426–434 (2008)
13. Koren, Y., Bell, R., Volinsky, C.: Matrix factorization techniques for recommender systems. Computer **8**, 30–37 (2009)
14. Lin, Y., Liu, Z., Sun, M., Liu, Y., Zhu, X.: Learning entity and relation embeddings for knowledge graph completion. In: AAAI (2015)
15. Mnih, V., Heess, N., Graves, A., et al.: Recurrent models of visual attention. In: NIPS, pp. 2204–2212 (2014)
16. Song, H.O., Xiang, Y., Jegelka, S., Savarese, S.: Deep metric learning via lifted structured feature embedding. In: CVPR, pp. 4004–4012 (2016)

17. Rendle, S.: Factorization machines. In: ICDM, pp. 995–1000 (2010)
18. Rendle, S., Freudenthaler, C., Gantner, Z., Schmidt-Thieme, L.: BPR: Bayesian personalized ranking from implicit feedback. In: Proceedings of the Twenty-Fifth Conference on Uncertainty in Artificial Intelligence, pp. 452–461 (2009)
19. Schroff, F., Kalenichenko, D., Philbin, J.: FaceNet: a unified embedding for face recognition and clustering. In: CVPR, pp. 815–823 (2015)
20. Tay, Y., Tuan, L.A., Hui, S.C.: Latent relational metric learning via memory based attention for collaborative ranking. In: WWW, pp. 729–739 (2018)
21. Tran, V.-A., Hennequin, R., Royo-Letelier, J., Moussallam, M.: Improving collaborative metric learning with efficient negative sampling. In: SIGIR, pp. 1201–1204 (2019)
22. Xing, E.P., Jordan, M.I., Russell, S.J., Ng, A.Y.: Distance metric learning with application to clustering with side-information. In: NIPS, pp. 521–528 (2003)
23. Zabihzadeh, D., Monsefi, R., Yazdi, H.S.: Sparse Bayesian approach for metric learning in latent space. Knowl.-Based Syst. **178**, 11–24 (2019)

Concordant Contrastive Learning for Semi-supervised Node Classification on Graph

Daqing Wu[1,2], Xiao Luo[1,2], Xiangyang Guo[1,2], Chong Chen[2], Minghua Deng[1], and Jinwen Ma[1(✉)]

[1] School of Mathematical Sciences, Peking University, Beijing, China
{wudq,xiaoluo,guoxy}@pku.edu.cn, {dengmh,jwma}@math.pku.edu.cn
[2] Damo Academy, Alibaba Group, Hangzhou, China
cheung.cc@alibaba-inc.com

Abstract. Semi-supervised object classification has been a fundamental problem in relational data modeling recently. The problem has been extensively studied in the literature of graph neural networks (GNNs). Based on the homophily assumption, GNNs smooth the features of the adjacent nodes, resulting in hybrid class distributions in the feature space when the labeled nodes are scarce. Besides, the existing methods inherently suffer from the non-robustness, due to the deterministic propagation. To address the above two limitations, we propose a novel method Concordant Contrastive Learning (CCL) for semi-supervised node classification on graph. Specifically, we generate two group data augmentations by randomly masking node features and separately perform node feature propagation with low- and high-order graph topology information. Further, we design two granularity regularization losses. The coarse-grained regularization loss (i.e., center-level contrastive loss) preserves the identity of each class against the rest, which benefits to guide the discriminative class distributions. The fine-grained regularization loss (i.e., instance-level contrastive loss) enforces consistency between soft assignments for different augmentations of the same node. Extensive experiments on different benchmark datasets imply that CCL significantly outperforms a wide range of state-of-the-art baselines on the task of semi-supervised node classification.

Keywords: Graph convolution networks · Semi-supervised learning · Contrastive learning

1 Introduction

Graphs have shown great importance for modeling structured and relational data in many fields. It can benefit numerous real-world problems and applications to mine graph information. Recently, graph neural networks (GNNs) [15,27] have attracted increasing attention for graph-structured data. Based on the homophily assumption that the connected nodes squint towards the similar

© Springer Nature Switzerland AG 2021
T. Mantoro et al. (Eds.): ICONIP 2021, LNCS 13108, pp. 584–595, 2021.
https://doi.org/10.1007/978-3-030-92185-9_48

features and labels, GNNs combine node features within local neighborhoods on the graph to learn node representations for various graph-based tasks, such as node classification, link prediction, and etc. In this study, we focus on the task of semi-supervised node classification on graph, which seeks to predict the label of unlabeled nodes few labeled nodes. For this task, it usually requires a large number of labeled data to train GNNs in a supervised manner. However, data annotations are often unaffordable in practice resulting in scarce labeled nodes. Fortunately, contrastive learning (CL) has been slathered in semi-supervised and unsupervised tasks in many fields, e.g., computer vision [4,7,12,19]. Generally, CL is based on the classical information maximization principle and usually maximizes the mutual information between the input and its deep representations by contrasting positive pairs with negative-sampled counterparts. Inspired by popular CL methods, DGI [28] incorporates CL into GNN models. Specifically, DGI augments the original graph by simply shuffling node features and then proposes a contrastive objective to maximize the mutual information between node embeddings and the global graph embedding. CSSL [32] incorporates MoCo [12] into graph representation learning and propose two frameworks based on pre-training and data-dependent regularization. Qiu et al. proposed a pre-training framework GCC [21], which conducts subgraph instance discrimination in and across networks with InfoNCE [20] objective.

Despite the prevalence of the aforementioned CL techniques on graphs, the existing GNNs methods have two key limitations for semi-supervised node classification task: **First**, GNNs inherently suffer from the non-robustness owing to the deterministic node feature propagation. By changing only a few links or perturbing node attributes, the performance of GNNs would be drastically degraded [34]. In the other words, GNNs is vulnerable for graph attack. **Second**, GNNs usually produce an ambiguous discrimination pattern with few labeled nodes, especially performing high-order node feature propagation. To be specific, GNNs are based on the homophily assumption to perform the node feature propagation by smoothing the features of the adjacent nodes. However, since nearby nodes may belong to distinct classes in practice, the supervised learning objective is hard to optimize a remarkable discrimination pattern with scare labeled nodes. The previous graph CL methods aim to learn the discriminative node representation to retard the smoothness between any two nodes (For any node, its positive object is the augmentation of itself). But this constrain is contradictory with the scenario in which two nodes have the same class semantic information, which would degrade the performance of the supervised learner.

To address the above two limitations, we propose a novel graph method named Concordant Contrastive Learning (CCL) for semi-supervised node classification on graph. In CCL, we first produce two perturbed feature matrices by randomly mask each node's features either partially. Then CCL separately performs node feature propagation over the graph by incorporating multiple adjacency matrices with low- and high- orders. Furthermore, we design two granularity regularization losses. The coarse-grained regularization loss (i.e., center-level contrastive loss) preserves the identity of each class against the rest, which benefits to guide the discriminative class distributions. The fine-grained regularization loss (i.e., instance-level contrastive loss) enforces consistency between soft assignments for different augmentations of the same node. With the perturbed propagation and contrastive learning objectives, CCL can increase the robustness of GNN. To summarize, our main contributions can be summarized as follows: (1) We propose a novel graph method CCL for semi-supervised node classification on graph. (2) To improve the robustness of GNNs, we introduce the random propagation and contrastive learning. (3) To enhance the discrimination pattern learning, we present the prototypical prediction and center-level regularization loss. (4) We conduct extensive experiments on a range of well-known benchmark datasets to evaluate the CCL. Experimental results demonstrate that CCL achieves impressive performance compared with the state-of-the-arts.

2   Related Works

2.1   Graph Neural Networks

Recently, a large number of works have been devoted toward applying deep learning into graph data learning in the form of GNNs. Graph neural networks are divided into two types: spectral-based techniques and spatial-based methods. Graph spectral theory is used to learn node representations in spectral-based techniques. Bruna et al. [3] generalize convolution operation from spectral domain via the graph Laplacian matrix. To simplify the convolution operation, ChebNet [8] approximates the graph Fourier coefficients with Chebyshev polynomials. GCN [15] sets the order of the polynomial as 1 to simply the ChebNet. Note that GCN can also be regarded as a spatial-based method. From a node perspective, GCN aggregates information from its neighbors to update node representations. Many spatial-based techniques for aggregating and transforming neighborhood data have been suggested recently, including GraphSAGE [10], GAT [27].

2.2   Contrastive Learning for Graph Neural Networks

Contrastive learning (CL) is based on the use of self-supervised information between contrastive pairs that are created by applying random perturbation on original data. CL develops high-quality representations for both images [2,26] and texts [5,18,23] by regularizing disturbed representations through pulling the positive pairs and pushing the negative pairs. However, GNNs only compute the loss on few labeled nodes, which restricts their modeling capabilities. CL has recently been used to graph-based semi-supervised learning, since it takes a large number of unlabeled nodes into consideration and learns pairwise node embeddings distribution. These researches now focus on designing augmentation techniques on graph: [29] assigns different graph views with different encoders and proposes the graph generative learning target. [11] leverages contrasting encodings with graph diffusion to generate various views of graph. [31] considers different combination of graph augmentations and analyzes its influence.

[33] designs some heuristic graph augmentations from the view of graph node centrality. [9] performs graph data augmentation with a random propagation strategy.

3   Method

In this section, we first give the definition of semi-supervised node classification, then present our proposed model Concordant Contrastive Learning (CCL). CCL develops two granularity contrastive learning: center-level contrastive learning, which endows the distribution of the features with discriminative information (i.e., class semantics), and instance-level contrastive learning, which aims to preserve the consistency between different data augmentations. With the enhancement of center- and instance-level contrastive learning, CCL can improve the generalization capacity.

3.1   Problem Definition

For a graph $\mathcal{G} = (\mathcal{V}, \boldsymbol{X}, \mathcal{E}, \boldsymbol{A})$, $\mathcal{V}$ is a set of N nodes in the graph, $\boldsymbol{X} = [\boldsymbol{x}_1, \boldsymbol{x}_2, \cdots, \boldsymbol{x}_N]^T$ is the attribute features of nodes where $\boldsymbol{x}_i \in \mathbb{R}^F$ and F is the number of attributes, $\mathcal{E} \subseteq \mathcal{V} \times \mathcal{V}$ is the edge set of graph and $\boldsymbol{A} \in \{0, 1\}^{N \times N}$ is the adjacency matrix of graph. For $\forall v_i, v_j \in \mathcal{V}$, $A_{ij} = 1$ if there exists an edge between v_i and v_j, otherwise, $A_{ij} = 0$. In this study, we focus on the task of semi-supervised node classification on graph. In this task, each node v_i is associated with a feature vector $\boldsymbol{x}_i \in \mathbb{R}^F$ and an one-hot label vector $\boldsymbol{y}_i \in \{0, 1\}^K$ where K represents the number of classes. In semi-supervised settings, $M(M < N)$ nodes have labels $\mathcal{Y}^L$ and the labels of the remaining $N - M$ nodes are missing. The goal is to learn a predictive function $f : \mathcal{G}, \mathcal{Y}^L \to \mathcal{Y}^U$ to predict the missing labels $\mathcal{Y}^U$ for unlabeled nodes.

3.2   Graph Convolutional Networks with High-Order Propagation

Given an input graph $\mathcal{G} = (\mathcal{V}, \boldsymbol{X}, \mathcal{E}, \boldsymbol{A})$, let $\boldsymbol{S}$ denotes the normalized adjacency matrix with added self-loops, i.e., $\boldsymbol{S} = \widetilde{\boldsymbol{D}}^{-\frac{1}{2}} \widetilde{\boldsymbol{A}} \widetilde{\boldsymbol{D}}^{-\frac{1}{2}}$, where $\widetilde{\boldsymbol{A}} = \boldsymbol{A} + \boldsymbol{I}$ and $\widetilde{\boldsymbol{D}}$ is the degree matrix of $\widetilde{\boldsymbol{A}}$. Denote l as the order of the normalized adjacency matrix $\boldsymbol{S}$, $\boldsymbol{S}^l$ is defined as $\boldsymbol{S}^l = \underbrace{\boldsymbol{S} \cdot \boldsymbol{S} \cdots \boldsymbol{S}}_{l}$. The value of l determines the global and local topology information of the graph $\mathcal{G}$. A large l means $\boldsymbol{S}^l$ contains more global topology information. A small l means $\boldsymbol{S}^l$ contains more local topology information. In other words, $\boldsymbol{S}^l$ is equal to the information of k-step random walk in the graph $\mathcal{G}$. In graph convolutional networks, the output of each layer are the average of neighbors representations. Therefore, $\boldsymbol{S}^l \boldsymbol{X}$ is the node representations which aggregate the l-hop features for each node. In order to make full use of local and global topology information, we consider $1, 2, \cdots, L-$hop propagation, i.e.,

$$\overline{\boldsymbol{X}} = \alpha_0 \boldsymbol{X} + \alpha_1 \boldsymbol{S} \boldsymbol{X} + \alpha_2 \boldsymbol{S}^2 \boldsymbol{X} + \cdots + \alpha_K \boldsymbol{S}^L \boldsymbol{X} \tag{1}$$

where $\alpha_l \geq 0$ denotes the importance of the l-hop node representations in constituting the final embedding. To keep model simplicity, we find that setting α_l uniformly as $\frac{1}{L+1}$ leads to good performance in our experiments.

It's worth noting that the homophily assumption indicates that the connected nodes squint towards the similar features and labels, and the Eq. (1) conducts node feature propagation by smoothing the adjacent nodes representations. As a result, the target node feature information could be compensated by its neighbors. Based on this fact, we employ the random drop node input features strategy as the graph data augmentations. Specifically, we first sample two binary mask vectors $\boldsymbol{m}^{(a)} \in \{0,1\}^N, a = 1,2$ where each element is independently drawn from a Bernoulli distribution, i.e., $m_i^{(a)} \sim Bernoulli(1 - \delta)$ (δ is the keep probability and generally set to 0.5). Then, we multiply each node input features $\boldsymbol{x}_i$ with its corresponding mask, i.e., $\boldsymbol{x}_i^{(a)} = m_i^{(a)}\boldsymbol{x}_i$. Finally, we conduct node feature propagation to obtain two augmented node embeddings according to

$$\boldsymbol{H}^{(a)} = \text{ReLU}(\overline{\boldsymbol{X}}^{(a)}\boldsymbol{W}), \qquad a = 1,2 \tag{2}$$

where $\overline{\boldsymbol{X}}^{(a)} = \frac{1}{L+1}\sum_{l=0}^{L} \boldsymbol{S}^l \boldsymbol{X}^{(a)}$ and $\boldsymbol{W} \in \mathbb{R}^{F \times d}$ is the feature transformation parameters to transfer input feature space $\mathbb{R}^F$ into a low-dimensional feature space $\mathbb{R}^d$ ($d \ll F$).

3.3   Prototypical Prediction

Inspire with the nearest-neighbor algorithms which classify unseen cases by finding the class labels of instances which are closest to representative prototypes set [17], we introduce trainable class prototypes $\boldsymbol{P} = \{\boldsymbol{p}_1, \boldsymbol{p}_2, \cdots, \boldsymbol{p}_K\}^T$ and categorize nodes by a softmax operation in the similarities between the node embeddings and prototypes, i.e.,

$$q^{(a)}(k|v_i) = \frac{\exp\left(\boldsymbol{p}_k^T \cdot \boldsymbol{h}_i^{(a)}/t_p\right)}{\sum_{k'=1}^{K} \exp\left(\boldsymbol{p}_{k'}^T \cdot \boldsymbol{h}_i^{(a)}/t_p\right)}, \qquad a = 1,2 \tag{3}$$

where t_p is the temperature parameter. All prototypes and node embeddings are l_2-normalized in the $\mathbb{R}^d$ space. Hence, the prediction function performs a soft partitioning of the graph nodes based on the cosine similarity between the node embeddings and the class prototypes. We can view it as a prototypical discriminative classification module.

3.4   Concordant Contrastive Learning

After performing node feature propagation and prototypical prediction, all nodes have embeddings $\boldsymbol{h}_i^{(a)}$ and soft assignments $\boldsymbol{q}_i^{(a)}$. To improve the generalization capacity of semi-supervised node classification on graph, we not only leverage supervised loss but also two granularities regularization losses.

Supervised Loss. In the graph semi-supervised node classification task, M labeled nodes have accurate class semantic information. For these labeled nodes, we can construct the average cross-entropy loss over two augmentations, i.e.,

$$\mathcal{L}_{\text{sup}} = -\frac{1}{2}\Big(\sum_{i \in \mathcal{Y}^L} \boldsymbol{y}_i^T \log \boldsymbol{q}_i^{(1)} + \sum_{i \in \mathcal{Y}^L} \boldsymbol{y}_i^T \log \boldsymbol{q}_i^{(2)} \Big) \tag{4}$$

Center-level Contrastive Loss. Note that all node embeddings are located in the $\mathbb{R}^d$ unit hypersphere. Any two distinct classes should have a wide border in terms of feature distribution, which means that all class centers in the unit hypersphere should be discriminative. Therefore, we first compute the class centers and then push away them. Denote $\boldsymbol{r}_k^{(a)}$ as the k-th class center representation under augmentation a, it can be generated by

$$\boldsymbol{r}_k^{(a)} = \Big(\sum_{i=1}^{N} q^{(a)}(k|v_i)\boldsymbol{h}_i^{(a)} \Big) \Big/ \Big\| \sum_{i=1}^{N} q^{(a)}(k|v_i)\boldsymbol{h}_i^{(a)} \Big\|_2 \tag{5}$$

where $\|\cdot\|_2$ refers to the l_2-norm. Now, we define the similarity between $\boldsymbol{r}_k^{(a)}$ and $\boldsymbol{r}_{k'}^{(a)}$ as $S_{k,k'}^{a,a'} = \exp(\boldsymbol{r}_k^{(a)T}\cdot\boldsymbol{r}_{k'}^{(a')}/t_c)$ where t_c is the temperature parameter. A large $S_{k,k'}^{a,a'}$ indicates that $\boldsymbol{r}_k^{(a)}$ and $\boldsymbol{r}_{k'}^{(a)}$ have the similar class semantic information. Therefore, we design a contrastive objective that preserves the identity of each class center against the rest, i.e.,

$$\mathcal{L}_{\text{cen}} = -\frac{1}{K} \sum_{k=1}^{K} \Big(\log \frac{S_{k,k}^{1,2}}{\sum_{k'\neq k} S_{k,k'}^{1,1} + S_{k,k'}^{2,2} + S_{k,k'}^{1,2} + S_{k,k'}^{2,1}} \Big) \tag{6}$$

Instance-level Contrastive Loss. We conduct two augmentations and produce two soft assignments $\boldsymbol{q}_i^{(1)}$ and $\boldsymbol{q}_i^{(2)}$ for each node v_i. To preserve the consistency of two augmentations, we minimize the distance between $\boldsymbol{q}_i^{(1)}$ and $\boldsymbol{q}_i^{(2)}$, i.e.,

$$\mathcal{L}_{\text{ins}} = \sum_{i=1}^{N} \|\boldsymbol{q}_i^{(1)} - \boldsymbol{q}_i^{(2)}\|_2 \tag{7}$$

Training and Inference. The overall learning objective consists of two main parts, the supervised loss and regularization loss. Thus, the final loss of CCL is

$$\mathcal{L} = \mathcal{L}_{\text{sup}} + \underbrace{\lambda\,\mathcal{L}_{\text{cen}} + \mu\,\mathcal{L}_{\text{ins}}}_{\text{regularization}} \tag{8}$$

where λ and μ are hyper-parameters that control the balance among these losses. During inference, we directly utilize the original attribute features $\boldsymbol{X}$ to perform node feature propagation and conduct prototype-based prediction with trained class prototypes.

4   Experiment

In this section, we first give a brief introduction of datasets used in experiments and baseline methods for comparison. Details of experimental settings are provided. Then, the effectiveness of CCL is shown. Further ablation studies are carried out to investigate the contribution of two granularity regularization losses in CCL. Following that, we plot the loss curves to demonstrate the generalizability and conduct random attack to verify the robustness. Last, we study the influence of hyper-parameters.

4.1   Datasets

To evaluate our CCL approach, we consider three standard citation network benchmark datasets: Cora, Citeseer and Pubmed [25]. In all of these datasets, nodes represent documents and edges indicate citation relations. Node attributes are the bag-of-words representation of corresponding document. We utilize only 20 nodes per class for training, evaluate the trained model performance on 1000 test nodes, and 500 additional nodes are considered as validation. The detailed description of these datasets are provided in Table 1.

Table 1. Statistical details of the evaluation datasets.

Dataset	# Nodes	# Edges	# Features	# Classes	# Train	# Valid	# Test
Cora	2,708	5,429	1,433	7	140	500	1000
Citeseer	3,327	4,732	37,03	6	120	500	1000
Pubmed	19,717	44,338	500	3	60	500	1000

4.2   Experimental Settings

In experiments, we implement our proposed model CCL based on PyTorch. The embedding size d is fixed to 32 for all datasets, which is suitable for the model to learn a strong representation embedding. We optimize CCL with Adam optimizer [14] having learning rate 0.01 and weight decay $5e^{-4}$ for Cora and Citeseer, 0.1 and $1e^{-3}$ for Pubmed. Normal initialization is used here to initialize the parameters. To detect over-fitting, we use an early stopping strategy on the accuracy on the validation nodes with a patience of 300 epochs for Cora and Citeseer, 100 epochs for Pubmed. On the stage of node feature propagation, the orders are $8, 2, 5$ for Cora, Citeseer and Pubmed, respectively. In terms of hyper-parameters, we apply a grid search method on loss balance coefficients λ and μ are searched in $[0, 0.1, 0.2, 0.3, 0.4, 0.5, 0.6, 0.7, 0.8, 0.9, 1.0]$.

4.3   Baseline Methods

We compare CCL with the following baseline methods: Graph convolutions networks: GCN [15], GAT [27], SGC [30], MixHop [1], GMNN [22], APPNP [16] and GCNII [6]. Self-supervised based methods: DGI [28] and SimP-GCN [13]. Augmentation based method: DropEdge [24].

Table 2. Test nodes classification accuracy (%).

Models	Cora	Citeseer	Pubmed
GCN [15]	81.5	70.3	79.0
GAT [27]	83.0 ± 0.7	72.5 ± 0.7	79.0 ± 0.3
SGC [30]	81.0 ± 0.0	71.9 ± 0.1	78.9 ± 0.0
MixHop [1]	81.9 ± 0.4	71.4 ± 0.8	80.8 ± 0.6
GMNN [22]	83.7	72.9	81.8
APPNP [16]	83.8 ± 0.3	71.6 ± 0.5	79.7 ± 0.3
GCNII [6]	82.3	68.9	78.2
DGI [28]	82.3 ± 0.6	71.8 ± 0.7	76.8 ± 0.6
SimP-GCN [13]	82.8	72.6	81.1
DropEdge [24]	82.8	72.3	79.6
CCL	$\mathbf{84.1 \pm 0.5}$	$\mathbf{72.9 \pm 0.6}$	$\mathbf{82.4 \pm 0.5}$

4.4 Overall Performance Comparison

Over 100 runs with random seeds, the averaged results of CCL are recorded, and we remove the outliers (70% confidence intervals) when calculating their statistics due to high variance. Table 2 shows the overall results. We can find that:

- Methods with self-supervised modules (SimP-GCN, DGI, CCL) generally outperform that without self-supervised modules, since it can excavate the semantics implied on the unlabeled nodes and generate more discriminative node embeddings in the feature space.
- For the recent augmentation-based model DropEdge, CCL achieves 1.5%, 0.8% and 3.5% improvements. Due to that the labeled nodes are scarce, DropEdge may intercept the feature propagation from labeled nodes to unlabeled nodes. CCL conducts perturbations by masking node features that would not harm propagation.
- CCL improves upon the inchoate GCN by a margin of 3.2%, 3.7% and 4.3%. Moreover, when compared with a very recent model SimP-GCN, CCL outperforms by 1.5%, 0.4% and 1.6%, demonstrating that the designed regularization losses are of great significance.

Table 3. Ablation study of two granularities regularization losses.

Models	Cora	Citeseer	Pubmed
w/o CR and IR	81.9 ± 0.3	70.8 ± 0.5	79.8 ± 0.5
w/o CR	82.4 ± 0.3	71.2 ± 0.4	81.2 ± 0.4
w/o IR	82.9 ± 0.5	71.7 ± 0.6	81.4 ± 0.5
CCL	$\mathbf{84.1 \pm 0.5}$	$\mathbf{72.9 \pm 0.6}$	$\mathbf{82.4 \pm 0.5}$

4.5   Ablation Study

We conduct an ablation study to investigate the contributions of two granularity regularization losses in CCL. Without center-level regularization (abbreviated as w/o CR): we exclusively examine instance-level contrastive loss, i.e., $\lambda = 0$. Without instance-level regularization (w/o IR): we only utilize center-level contrastive loss, i.e., $\mu = 0$. Without any regularization constraints (w/o CR and IR), we set $\lambda = 0$ and $\mu = 0$. The results of the ablation study is shown in Table 3. Compared to the non-regularization loss model, it is worthwhile to note that w/o CR and w/o IR can produce acceptable performance, but they cannot beat the full model. This suggests that learning with our designed two granularity regularization losses is essential.

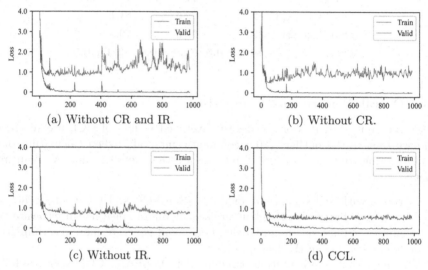

Fig. 1. Loss curves for variants of CCL on Pubmed. (x-axis represents epoch.)

4.6   Generalization Analysis

To verify the generalization capacity of the proposed two granularity regularization losses. We conduct analysis on the supervised loss on both training and validation sets on Pubmed. From Fig. 1(a), without CR and IR, there is a huge difference between validation and training losses, suggesting an apparent overfitting issue. From Fig. 1 (b) and (c), we can find that this gap may be narrowed somewhat using CR or IR. For the full model CCL in Fig. 1 (d), It's clear that this gap is remarkably diminished. This experiment indicates that our designed regularization operations can significantly improve generalization capability.

4.7   Robustness Analysis

We also examine the robustness of CCL by randomly adding edges between two nodes. Figure 2 presents the classification accuracies of GCN, GAT and CCL with

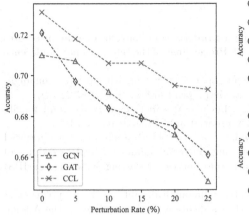
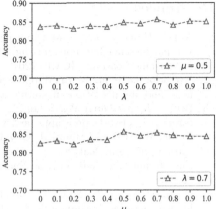

Fig. 2. Robustness analysis on citeseer.   **Fig. 3.** Sensitivity of λ and μ on cora.

the perturbation rates from 0% to 25% on Citeseer. Apparently, CCL reliably enhances GCN performance under various random attack perturbation rates. When the perturbation rate is higher, CCL enhances GCN by a larger margin. These observations suggest that CCL can be robust to random attacks.

4.8   Parameter Sensitivity

We further study the influence of hyper-parameters λ and μ. Figure 3 shows the effect of these two hyper-parameters on Cora. We first fix μ to 0.5 and evaluate the accuracy by λ from 0 to 1. The performance of the model is not sensitive to the λ in the range of $[0,1]$. Similarly, we show the accuracy by varying μ from 0 to 1 with λ fixed to 0.7. The result is also insensitive to μ in the range of $[0,1]$.

5   Conclusion

In this study, we present the Concorant Contrastive Learning (CCL) to tackle the task of semi-supervised node classification on graph. In CCL, we stochastically generate two graph data augmentations by random masking strategy, then we utilize concordant contrastive regularization from both the center- and instance- level to improve the model generalization and robustness. On benchmark datasets, we demonstrate CCL has incredible performance improvement over a range of state-of-the-art baselines. In the future, we will try to expand CCL for heterogeneous graph data.

Acknowledgments. This work was partially supported by the National Key Research and Development Program of China under grant 2018AAA0100205.

References

1. Abu-El-Haija, S., et al.: MixHop: higher-order graph convolutional architectures via sparsified neighborhood mixing. In: Proceedings of the International Conference on Machine Learning (ICML) (2019)
2. Berthelot, D., et al.: ReMixMatch: semi-supervised learning with distribution alignment and augmentation anchoring. arXiv preprint arXiv:1911.09785 (2019)
3. Bruna, J., Zaremba, W., Szlam, A., LeCun, Y.: Spectral networks and locally connected networks on graphs. arXiv preprint arXiv:1312.6203 (2013)
4. Caron, M., Misra, I., Mairal, J., Goyal, P., Bojanowski, P., Joulin, A.: Unsupervised learning of visual features by contrasting cluster assignments. In: Proceedings of the International Conference on Neural Information Processing Systems (NeurIPS) (2020)
5. Chen, J., Yang, Z., Yang, D.: MixText: linguistically-informed interpolation of hidden space for semi-supervised text classification. In: Proceedings of the Annual Meeting of the Association for Computational Linguistics (ACL) (2020)
6. Chen, M., Wei, Z., Huang, Z., Ding, B., Li, Y.: Simple and deep graph convolutional networks. In: Proceedings of the International Conference on Machine Learning (ICML) (2020)
7. Chen, T., Kornblith, S., Norouzi, M., Hinton, G.: A simple framework for contrastive learning of visual representations. In: Proceedings of the International Conference on Machine Learning (ICML) (2020)
8. Defferrard, M., Bresson, X., Vandergheynst, P.: Convolutional neural networks on graphs with fast localized spectral filtering. In: Proceedings of the International Conference on Neural Information Processing Systems (NeurIPS) (2016)
9. Feng, W., et al.: Graph random neural networks for semi-supervised learning on graphs. In: Proceedings of the International Conference on Neural Information Processing Systems (NeurIPS) (2020)
10. Hamilton, W.L., Ying, Z., Leskovec, J.: Inductive representation learning on large graphs. In: Proceedings of the International Conference on Neural Information Processing Systems (NeurIPS) (2017)
11. Hassani, K., Khasahmadi, A.H.: Contrastive multi-view representation learning on graphs. In: Proceedings of the International Conference on Machine Learning (ICML) (2020)
12. He, K., Fan, H., Wu, Y., Xie, S., Girshick, R.: Momentum contrast for unsupervised visual representation learning. In: Proceedings of the IEEE/CVF Conference on Computer Vision and Pattern Recognition (CVPR) (2020)
13. Jin, W., Derr, T., Wang, Y., Ma, Y., Liu, Z., Tang, J.: Node similarity preserving graph convolutional networks. In: Proceedings of the ACM International Conference on Web Search and Data Mining (WSDM) (2021)
14. Kingma, D.P., Ba, J.: Adam: a method for stochastic optimization. arXiv preprint arXiv:1412.6980 (2014)
15. Kipf, T.N., Welling, M.: Semi-supervised classification with graph convolutional networks. In: Proceedings of the International Conference on Learning Representations (ICLR) (2017)
16. Klicpera, J., Bojchevski, A., Günnemann, S.: Predict then propagate: graph neural networks meet personalized PageRank. In: Proceedings of the International Conference on Learning Representations (ICLR) (2019)
17. Lam, W., Keung, C.K., Liu, D.: Discovering useful concept prototypes for classification based on filtering and abstraction. IEEE Trans. Pattern Anal. Mach. Intell. **24**(8), 1075–1090 (2002)

18. Luo, F., Yang, P., Li, S., Ren, X., Sun, X.: CAPT: contrastive pre-training for learning denoised sequence representations. arXiv preprint arXiv:2010.06351 (2020)
19. Luo, X., et al.: CIMON: towards high-quality hash codes. In: Proceedings of the International Joint Conference on Artificial Intelligence (IJCAI) (2021)
20. van den Oord, A., Li, Y., Vinyals, O.: Representation learning with contrastive predictive coding. arXiv preprint arXiv:1807.03748 (2018)
21. Qiu, J., et al.: GCC: graph contrastive coding for graph neural network pre-training. In: Proceedings of the ACM SIGKDD International Conference on Knowledge Discovery and Data Mining (2020)
22. Qu, M., Bengio, Y., Tang, J.: GMNN: graph Markov neural networks. In: Proceedings of the International Conference on Machine Learning (ICML) (2019)
23. Qu, Y., Shen, D., Shen, Y., Sajeev, S., Han, J., Chen, W.: CoDA: contrast-enhanced and diversity-promoting data augmentation for natural language understanding. arXiv preprint arXiv:2010.08670 (2020)
24. Rong, Y., Huang, W., Xu, T., Huang, J.: DropEdge: towards deep graph convolutional networks on node classification. In: Proceedings of the International Conference on Learning Representations (ICLR) (2019)
25. Sen, P., Namata, G., Bilgic, M., Getoor, L., Galligher, B., Eliassi-Rad, T.: Collective classification in network data. AI Mag. **29**(3), 93–93 (2008)
26. Sohn, K., et al.: FixMatch: simplifying semi-supervised learning with consistency and confidence. In: Proceedings of the International Conference on Neural Information Processing Systems (NeurIPS) (2020)
27. Veličković, P., Cucurull, G., Casanova, A., Romero, A., Lio, P., Bengio, Y.: Graph attention networks. In: Proceedings of the International Conference on Learning Representations (ICLR) (2018)
28. Velickovic, P., Fedus, W., Hamilton, W.L., Liò, P., Bengio, Y., Hjelm, R.D.: Deep graph infomax. In: Proceedings of the International Conference on Learning Representations (ICLR) (2019)
29. Wan, S., Pan, S., Yang, J., Gong, C.: Contrastive and generative graph convolutional networks for graph-based semi-supervised learning. arXiv preprint arXiv:2009.07111 (2020)
30. Wu, F., Souza, A., Zhang, T., Fifty, C., Yu, T., Weinberger, K.: Simplifying graph convolutional networks. In: Proceedings of the International Conference on Machine Learning (ICML) (2019)
31. You, Y., Chen, T., Sui, Y., Chen, T., Wang, Z., Shen, Y.: Graph contrastive learning with augmentations. In: Proceedings of the International Conference on Neural Information Processing Systems (NeurIPS) (2020)
32. Zeng, J., Xie, P.: Contrastive self-supervised learning for graph classification. In: Proceedings of the AAAI Conference on Artificial Intelligence (2021)
33. Zhu, Y., Xu, Y., Yu, F., Liu, Q., Wu, S., Wang, L.: Graph contrastive learning with adaptive augmentation. In: Proceedings of the World Wide Web Conference (WWW) (2021)
34. Zügner, D., Günnemann, S.: Adversarial attacks on graph neural networks via meta learning. In: Proceedings of the International Conference on Learning Representations (ICLR) (2018)

Improving Shallow Neural Networks via Local and Global Normalization

Ning Jiang[1]([✉]), Jialiang Tang[1], Xiaoyan Yang[1], Wenxin Yu[1], and Peng Zhang[2]

[1] School of Computer Science and Technology, Southwest University of Science and Technology, Mianyang, China
`jiangning@swust.edu.cn`
[2] School of Science, Southwest University of Science and Technology, Mianyang, China

Abstract. Convolutional neural networks (CNNs) have achieved great success in computer vision. In general, CNNs can achieve superior performance depend on the deep network structure. However, when the network layers are fewer, the ability of CNNs is hugely degraded. Moreover, deep neural networks often with great memory and calculations burdens, which are not suitable for practical applications. In this paper, we propose Local Feature Normalization (LFN) to enhance the local competition of features, which can effectively improve the shallow CNNs. LFN can highlight the expressive local regions while repressing the unobvious local areas. We further compose LFN with Batch Normalization (BN) to construct an LFBN by two ways of concatenating and adding. LFN and BN are excel at handle local features and global features, respectively. Therefore LFBN can significantly improve the shallow CNNs. We also construct a shallow LFBN-Net by stacking the LFBN and conduct extensive experiments to validate it. LFBN-Net has achieved superior ability with fewer layers on various benchmark datasets. And we also insert the LFBN to exiting CNNs. These CNNs with the LFBN all achieve considerable performance improvement.

Keywords: Convolutional neural networks · Local feature normalization · Batch normalization · LFBN · LFBN-Net

1 Introduction

In recent years, convolutional neural networks (CNNs) have promoted the development of many fields, such as image classification [10,16], semantic segmentation [13,20], and object detection [8,27]. Generally speaking, deep neural networks (DNNs) with more parameters and calculations can achieve significant performance than shallow neural networks (SNNs). For example, the 152 layers ResNet152 [10] can achieve 21.43% top-1 error rates and 5.71% top-5 error rates on the large-scale image dataset ImageNet [6]. However, ResNet152 has up to 60 million (M) parameters, requires more than 20 Giga (G) floating-point operations

© Springer Nature Switzerland AG 2021
T. Mantoro et al. (Eds.): ICONIP 2021, LNCS 13108, pp. 596–608, 2021.
https://doi.org/10.1007/978-3-030-92185-9_49

(FLOPs) to process a 224×224 pixels image. So huge memory and computational burdens of DNNs are unaffordable by existing smart devices, which leads these excellent DNNs unable to apply in real life.

Many researchers have worked to compress complicated CNNs, mainly include lightweight networks [3,12] and model compression algorithms [11,35]. The lightweight network designs an optimized network artificially that requires fewer parameters to achieve a comparative performance with complicated DNNs. But the design of a lightweight network is time-consuming and laborious. Model compression algorithms mainly include knowledge distillation [11] and network pruning [23]. Knowledge distillation transfers knowledge from a large teacher network to a small student network to promote student network performance. However, knowledge distillation requires the student network and the teacher network with an identical structure. Network pruning finds and prunes channels or filters that contribute less to network capability to effectively compress the network. But the capability of the compressed model still reduced to a certain extent. Unlike the above methods, we design a Local Feature Normalization (LFN) to improve the performance of SNNs almost without additional computational cost. Extensive experiments demonstrate that SNNs with LFN can achieve similar, even superior performance than complicated DNNs (see details in Sect. 5).

Batch Normalization (BN) is widely used in existing neural networks to accelerate model convergence and improve its performance. BN calculates the mean and variance across the whole features to normalize the features. However, BN only normalizes features from the global perspective and ignores the importance of the local features. Therefore, we propose LFN to process the local features. The LFN can highlight the more expressive local areas while suppressing the other areas. The normalization procedure of LFN is shown in Fig. 1. Instead of normalizing the global features, LFN normalizes the identical local positions of multiple channel maps. LFN is beneficial for handling local features, and BN is good at handling global features. Therefore, we combine LFN and BN to propose a new network structure LFBN. There are two ways to compose the LFN and BN. In one way, the output features of LFN F_l and BN F_b are concatenated together along the channel dimension to obtain the final features F. In the other way, the output features of LFN F_l and BN F_b are added pixel by pixel under the adjustment of two learnable parameters as the final features F. Experiments show that both two ways are effective on various tasks. Based on LFBN, we built a shallow LFBN-Net, which has achieved superior performance on multiple image classification benchmark datasets. And extensive experiments demonstrate that LFBN can deploy to existing CNNs, significantly improving the performance of these networks. The contributions of this paper are concluded as follows:

1. We analyze the problem of CNNs that need a deeper structure to achieve satisfactory performance and unable to apply to real life. Moreover, we propose Local Feature Normalization (LFN) to solve this problem.

2. We propose two ways to combine LFN and BN with building LFBN and constructing a high-performance shallow LFBN-Net.
3. We have applied LFBN into popular CNNs and successfully improved these CNNs on benchmark datasets include CIFAR-10 [15], CIFAR-100 [15], STL-10 [4] and CINIC-10 [5].

The rest of the paper is shown as follows: Sect. 2 introduces the related works of this paper. Section 3 shows the details of the LFN and how to compose LFBN. Section 4 exhibits the structure of LFBN-Net. Section 5 conducts extensive experiments to demonstrate the effectiveness of our LFBN. Section 6 concludes this paper.

2   Related Works

The related works include network normalization, model compression and convolutional neural networks.

2.1   Network Normalization

Normalization methods normalize the input data and intermediate features to faster the training and improve the performance of CNNs. In the previous works, Local Response Normalization (LRN) [22] normalized each pixel with a small neighborhood, which was used as a component of AlexNet [16]. But LRN's effect is unsatisfactory and is almost replaced by dropout operations [30].

Batch Normalization (BN) [14] is a landmark technology of global normalization. BN calculates the mean and variance at the batch dimension to regularize the input features. Even though BN is well normalizing CNN's features, BN still has two disadvantage. First, BN requires a larger batch of data, the effect of BN drop drastically when the batch is smaller. Second, BN only performs normalization from a global perspective and ignores local feature normalization, limiting the further improvement of CNN's performance. There are a series of works to solve the first inadequate of BN. Group Normalization (GN) [33] divides the channels into groups and normalizes the feature in each group independently, successfully improve the performance of CNNs with a small batch input data. Ba et al. [1] proposes a batch-independent Layer Normalization (LN), which normalizes different channels in the same sample. Instance Normalization (IN) [32] is also a normalization method independent of the batch. Unlike LN, IN normalizes the feature of a single channel. Luo et al. [21] combine BN, LN, and IN to propose a differentiable normalization layer named Switchable Normalization (SN). Each layer can adaptively select various normalization methods according to the input data and the outputs of SN is the weighted sum of the outputs of the three normalization methods. Positional Normalization (PN) [18] regularizes the entire features along the channel. Still, PN neglect the local competition of the feature and only applicable to generative adversarial networks [24]. Filter Response Normalization (FRN) [29] operates each activation map of each batch sample independently to eliminate the batch dependence, but it needs to be used with Thresholded Linear Unit (TLU). Pan et al. [25] are similar to us, which combined IN and BN to propose IBN, but IBN only combines existing algorithms and neglects to process the local feature. The above methods all focus on solving the batch dependency of BN, ignoring the normalization of local features. This paper proposes to perform feature normalization from a local perspective, which effectively improves the performance of CNNs.

2.2   Model Compression

Model compression algorithms mainly include lightweight network, knowledge distillation, and neural network pruning. The lightweight network constructs an optimized neural network through manual design and experimental verification. Thanks to the depthwise separable convolution and the group convolution, lightweight CNNs ShuffleNet [37], MobileNet [12], etc., use only a small calculation and memory consumption to achieve performance similar to that of complicated CNNs. Instead of using a more optimal network structure, Adder-Net [3] replaces multiplication operation with the addition operation to reduce CNN's size hugely. Knowledge distillation uses a large teacher network to train a small student network. Hinton et al. [11] propose to use the softmax function with a soften parameter to extract the output of CNNs as the soft target for knowledge transfer, which contains more information than the original output. Recently, correlation congruence knowledge distillation (CCKD) [26] takes the sample correlation consistency as knowledge, and dual attention knowledge distillation (DualKD) [31] uses the spatial and channel attention feature maps as the knowledge. These methods successfully train the student network to achieve better performance than the teacher network. Neural network pruning is based on various criteria to find channels/filters that are unimportant to the network. The early network pruning Optimal Brain Damage [17] removes unimportant weights in the fully connected layers. Recent network pruning cuts the convolutional layer and the fully connected layer together to compress the model size further. Molchanov et al. [23] propose the Taylor expansion-based criterion to calculate the channel importance and remove unimportant channels. Network slimming [19] adds a scale factor to each channel to determine its importance. After training, the low-scale factor channels are pruned almost no loss of network ability.

2.3   Convolutional Neural Networks

Since Alex et al. [16] propose that AlexNet won the 2012 ImageNet competition, CNNs have continuously proved its outstanding capability. VGGNet [28] builds a deeper CNN by stacking 3×3 convolutional layers, which significantly improves the performance of CNNs. ResNet [10] creatively proposes the residual connection to solve the problem of gradient disappearance and further deepen the layer of CNNs. Many researchers work to update classic CNNs. Xie et al.

[34] propose a wider residual block to construct ResNeXt, each residual block contains multiple convolutional layer branches, and the output of each branch is gathered together as the final outputs. Ding et al. [7] propose a VGG-like CNN called RepVGG, which adds a 1×1 convolution branches into VGG blocks.

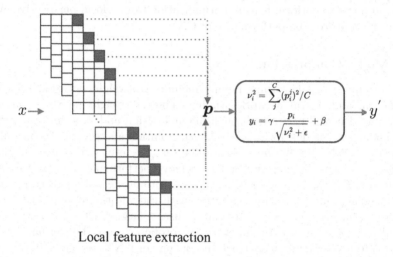

Local feature extraction

Fig. 1. The calculate process of our Local Feature Normalization.

3   Approch

In this section, we first introduce the principle of BN, then describe how to implement LFN, and finally combine LFN and BN to construct LFBN.

3.1   Batch Normalization

BN [14] is a fundamental component that normalizes the global features on the batch dimension to speed up network training while improving network performance. To simplify the description, we define the input features as $x = \{x_1, x_2, \cdots, x_n\}$, and the normalized features as $y = \{y_1, y_2, \cdots, y_n\}$, n represents the batchsize. First, the mean and variance of the input features are calculated as:

$$\mu_\mathcal{B} = \frac{1}{n} \sum_{i=1}^{n} x_i, \quad \sigma_\mathcal{B}^2 = \frac{1}{n} \sum_{i=1}^{n} (x_i - \mu_\mathcal{B})^2 \tag{1}$$

Then, the input features x is normalized via the mean $\mu_\mathcal{B}$ and variance $\sigma_\mathcal{B}^2$:

$$\widehat{x}_i = \frac{x_i - \mu_\mathcal{B}}{\sqrt{\sigma_\mathcal{B}^2 + \epsilon}} \tag{2}$$

In the end, the normalized features are scaled and shifted by two learnable parameters γ and β:

$$y_i = \gamma \widehat{x}_i + \beta \tag{3}$$

After normalized by BN, the output features y conform to the Gaussian distribution.

3.2 Local Feature Normalization

Even though BN is effective, it ignores the processing of local features, which limits the further improvement of CNNs. In this section, we describe LFN how to handle the local features. The calculate process of LFN is shown as Fig. 1. Specifically, we first extract the same local pixel of various channels as the local feature $p_i = \{p_i^1, p_i^2, \cdots, p_i^C\}$ and take p_i to show the normalization of LFN. For the input features $x \in R^{C \times H \times W}$ (to simplify the description, we omit the batch dimension in subsequent description, the C, H, and W represent the channel number, height and width of x). There are $N = H \times W$ number of local features, and the set of local features is defined as $p = \{p_1, p_2, \cdots, p_N\}$. Then, we calculate the mean square of the local feature p_i as:

$$\nu_i^2 = \sum_{j=1}^{C} (p_i^j)^2 / C \tag{4}$$

The mean square ν_i^2 is effective to highlight the competition of local feature and the p_i is normalized as follow:

$$y_i = \frac{p_i}{\sqrt{\nu_i^2 + \epsilon}} \tag{5}$$

ϵ is a small non-zero number to prevent division by zero errors. After proceed by Eq. (5), if the local region of the features normalized by LFN is more expressive, its value will be bigger, and conversely, the value will be smaller. Similar to BN, we also use two learnable parameters γ and β to scale and shift each normalized feature in $y = \{y_1, y_2, \cdots, y_N\}$ as:

$$y_i = \gamma \frac{p_i}{\sqrt{\nu_i^2 + \epsilon}} + \beta \tag{6}$$

In the end, y is sequentially passed through the subsequent Rectified Linear Unit (ReLU) [9] layer and convolutional layer.

3.3 How to Construct LFBN

BN is powerful in handling global features, and LFN is effective in dealing with local features. In this section, we combine BN and LFN as LFBN in two ways, the features of LFN and BN are added or concatenated together across the channel dimension. In the way of the add operation, we named the block as LFBN-A,

Table 1. The configuration of LFBN-Net. The conv3×3-$N \times M$ represents the 3×3 convolutional layer with N channels, and LFBN-module contains M convolutional layers. The total number of layers of the network is equal to the sum of the number of convolutional layers and fully connected layers, each convolutional layer with BN, LFBN, and ReLU.

LFBN-Net configuration		
LFBN-Net9	LFBN-Net11	LFBN-Net14
9 weight layers	11 weight layers	14 weight layers
Input (32×32 RGB image)		
conv3×3-64 ×1	conv3×3-64 ×2	conv3×3-64 ×2
Maxpool		
conv3×3-64 ×1	conv3×3-64 ×2	conv3×3-64 ×2
Maxpool		
conv3×3-64 ×2	conv3×3-64 ×2	conv3×3-64 ×3
Maxpool		
conv3×3-64 ×2	conv3×3-64 ×2	conv3×3-64 ×3
Maxpool		
conv3×3-64 ×2	conv3×3-64 ×2	conv3×3-64 ×3
Averagepool		
Fully connected layer-512		
Soft-max		

the input features $x \in R^{C \times H \times W}$ are first input into LFN and BN, then the output features of LFN $F_l \in R^{C \times H \times W}$ and BN $F_b \in R^{C \times H \times W}$ are added with two learnable parameters α_1 and α_2 as:

$$F_A = \alpha_1 \cdot F_b + \alpha_2 \cdot F_l \tag{7}$$

$F_A \in R^{C \times H \times W}$ are the outputs of LFBN-A. In the way of the concatenate operation, we named this way as LFBN-C, the input features $x \in R^{C \times H \times W}$ are first divided as two set features of $x_b \in R^{C/2 \times H \times W}$ and $x_l \in R^{C/2 \times H \times W}$, then the x_b and x_l are input to LFN and BN respectively to obtain the output features of LFN $F_l \in R^{C/2 \times H \times W}$ and BN $F_b \in R^{C/2 \times H \times W}$. The two obtained features are concatenate together along the channel dimension to get the final outputs of LFBN-C $F_C \in R^{C \times H \times W}$:

$$F_C = \mathcal{C}(F_b, F_l) \tag{8}$$

$\mathcal{C}$ is the concatenate operation, subsequently, all F_A and F_C are input to ReLU to maintain its nonlinearity.

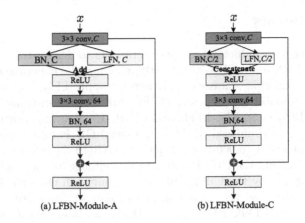

Fig. 2. The two ways to construct LFBN-module.

Table 2. Results of various LFBN-Net on CIFAR dataset.

Model	CIFAR-10	CIFAR-100	STL-10	CINIC-10
LFBN-Net9-A	93.81%	74.87%	86.29%	84.42%
LFBN-Net9-C	93.64%	74.53%	86.24%	84.34%
LFBN-Net11-A	94.40%	75.35%	87.68%	85.73%
LFBN-Net11-C	94.23%	75.36%	87.63%	85.55%
LFBN-Net14-A	94.44%	75.80%	86.38%	85.59%
LFBN-Net14-C	95.42%	75.49%	86.43%	85.48%

4 Construct the LFBN-Net

In this section, we construct two types LFBN-Net based on LFBN, Fig. 2 shows the two types LFBN-module and LFBN-Net is composed of multiple LFBN-modules. For the LFBN-Module-A (the subfigure (a) in Fig. 2) based on LFBN-A, the features of BN and LFN are added together same as described in Subsect. 3.3. For the types of LFBN-Module-C (the subfigure (b) in Fig. 2) based on LFBN-C, the features of BN and LFN are concatenated together. There is a residual connection in each LFBN-module, for the input x, the final output of LFBN-module is $y = \mathcal{F}(x) + x$, $\mathcal{F}$ represents the convolutional layer, LFN, BN and ReLU contains in LFBN-module. The residual function is learned to align x in the identity path with the $\mathcal{F}(x)$. LFN will increase the local competitiveness of features, which may harm the identity mapping of residual connections. So we only use LFBN after the first convolutional layer of the LFBN-module. The final LFBN-Net (from 9 to 14 layers) is composed of five LFBN-modules, the network setups are shown in Table 1.

5   Experiments

In this section, we conduct extensive experiments to demonstrate the effectiveness of our LFBN and LFBN-Net. Firstly, we show the experimental results of LFBN-Net on multiple datasets, and then we apply LFBN to various CNNs. Finally, we compare the proposed LFBN with other normalization methods. We choose CIFAR [15], STL-10 [4], and CINIC-10 [5] as experimental datasets. CIFAR dataset is composed of 32×32 pixel RGB images, including 50,000 training samples and 10,000 test samples. CIFAR-10 dataset contains 10 categories of images, and the CIFAR-100 dataset contains 100 categories of images. STL-10 is inspired by the CIFAR-10 dataset, but each class has fewer labeled training examples, and the resolution of this dataset (96×96) is higher than that of CIFAR-10. CINIC-10 comprises CIFAR and ImageNet [6] images, which contains 270,000 32×32 pixel images in ten categories.

Table 3. Results of LFBN applied to selected CNNs on the benchmark datasets. Acc means the accuracy of CNNs with LFBN and $Acc^\dagger$ means the accuracy improvement of this CNNs compared to original CNNs.

Model	CIFAR-10		CIFAR-100		STL-10		CINIC-10	
	Acc	$Acc^\dagger$	Acc	$Acc^\dagger$	Acc	$Acc^\dagger$	Acc	$Acc^\dagger$
VGGNet11 [28]	93.70%	1.34%	74.56%	2.66%	84.20%	0.54%	83.21%	0.93%
VGGNet13 [28]	94.21%	1.03%	74.73%	1.63%	85.60%	0.98%	84.72%	0.53%
VGGNet16 [28]	94.10%	0.79%	74.34%	0.98%	84.71%	0.14%	84.67%	0.33%
ResNet18 [10]	95.04%	1.12%	77.07%	0.54%	88.75%	1.14%	86.74%	0.39%
ResNet34 [10]	95.58%	0.73%	78.42%	1.08%	89.13%	0.59%	87.44%	0.36%
WRS16_2 [36]	93.91%	0.72%	73.17%	1.55%	85.92%	0.55%	84.64%	0.61%
WRS28_2 [36]	94.69%	0.44%	74.97%	0.89%	87.26%	0.29%	85.97%	0.40%

5.1   Experiments of LFBN-Net

In this part, we use LFBN-Net (9–14 layers) to conduct experiments on CIFAR-10, CIFAR-100, STL-10, and CINIC-10 datasets, each LFBN-Net built by stacking five LFBN-modules as shown in Sect. 4. All LFBN-Net use the stochastic gradient descent (SGD) [2] as the optimizer. For the training setups of CIFAR and CINIC-10 datasets, the learning rate is 0.1 (divide by 10 at 80th, 120th, and 160th epoch), the weight-decay and momentum are set as 1e-4 and 0.9. For the STL-10, SGD is used as the optimizer and the learning rate is set to 0.01, other setups are the same as that of CIFAR. The batchsize is set as 128, and each LFBN-Net trains 200 epochs. Table 2 exhibitions the results of LFBN-Net train on benchmark datasets. In general, the effect of LFBN-Net-A is better than LFBN-Net-C, which proves that the LFBN constructed by the add operation is better than the LFBN builted by the concatenate operation. It is worth noting that the 11 layers LFBN-Net11-A have achieved better performance than the deeper 14 layers LFBN-Net14-A and LFBN-Net14-C on STL-10 and CINIC-10, reach 87.68% accuracy on the STL-10, and 85.73% accuracy on the CINIC-10.

5.2   Applied LFBN to Various CNNs

To further prove the effectiveness of LFBN, we use LFBN in various CNNs. We select widely used VGGNet13 [28], VGGNet16 [28], ResNet18 [10], WideRes-Net16_2 (abbreviated WideResNet as WRS) [36] as experimental CNNs to train on CIFAR, STL-10, CINIC-10. According to the experimental results in the Subsect. 5.1, all selected CNNs use LFBN-A. The training setups are same as described in Subsect. 5.1.

Table 3 shows the results of selected CNNs with our LFBN on benchmark datasets. These selected CNNs used LFBN all achieve considerable performance improvement. It is noteworthy that the shallow models have gained a remarkable improvement over the deep models. The 11 layers VGGNet11 has achieved accuracy improvements of 1.34%, 2.66%, 0.54%, and 0.93% on the CIFAR-10, CIFAR-100, STL-10, and CINIC-10 datasets, respectively, significantly better than that of 16 layers VGGNet16. The shallow ResNet18 and WRS16_2 all achieve significant improvements than the deep ResNet34 and WRS28_2. These experimental results demonstrate that our LFBN can be widely used to improve the performance of CNNs. And LFBN is more effective in enhancing shallow networks.

Table 4. Results of VGGNet with various normalization on CIFAR dataset.

Model	Dataset	BN [14]	GN [33]	SN [21]	FRN [29]	IBN [25]	LFBN-A
VGGNet16 [28]	CIFAR-10	93.31%	92.98%	93.50%	93.45%	93.87%	**94.10%**
	CIFAR-100	73.36%	73.12%	73.61%	73.74%	73.95%	**74.34%**

5.3   Compared to Others Normalization Algorithms

In this subsection, we compare our LFBN (LFBN-A) with other normalization algorithms. We select VGGNet16 to train on CIFAR dataset, the training setups are the same as Subsects. 5.1. VGGNet16 with LFBN-A are compared with VGGNet16 with BN [14], GN [33], SN [21], FRN [29], IBN [25]. Table 4 depicts the results of VGGNet16 with various normalization methods. We can find that VGGNet16 with our LFBN achieved 94.10% accuracy on the CIFAR-10 dataset and 74.34% accuracy on the CIFAR-100 dataset. The performance of VGGNet16 with LFBN is significantly better than that of VGGNet16 with BN, GN, SN, and FRN. Even compared with IBN, which composes the IN and BN together to normalize the features, the accuracy of VGGNet16 with LFBN are also 0.23% and 0.39% higher than VGGNet16 with IBN.

6 Conclusion

This paper proposes LFBN based on LFN and BN, which can process local and global features together. We have conducted extensive experiments on multiple benchmark datasets to prove that LFBN can effectively improve the performance of CNNs, especially shallow neural networks. We also elaborate on how to construct LFBN-Net based on LFBN in two ways. Experimental results demonstrate that LFBN-Net only with fewer layers achieves a similar or even better performance than deep neural networks. In future work, we will explore the LFBN to apply to other fields, such as object detection and image segmentation tasks.

Acknowledgement. This work was supported in part by the Sichuan Science and Technology Program under Grant 2020YFS0307, Mianyang Science and Technology Program 2020YFZJ016, SWUST Doctoral Foundation under Grant 19zx7102, 21zx7114.

References

1. Ba, J.L., Kiros, J.R., Hinton, G.E.: Layer normalization. arXiv preprint arXiv:1607.06450 (2016)
2. Bottou, L.: Stochastic gradient descent tricks. In: Montavon, G., Orr, G.B., Müller, K.-R. (eds.) Neural Networks: Tricks of the Trade. LNCS, vol. 7700, pp. 421–436. Springer, Heidelberg (2012). https://doi.org/10.1007/978-3-642-35289-8_25
3. Chen, H., et al.: AdderNet: do we really need multiplications in deep learning? In: Proceedings of the IEEE/CVF Conference on Computer Vision and Pattern Recognition, pp. 1468–1477 (2020)
4. Coates, A., Ng, A., Lee, H.: An analysis of single-layer networks in unsupervised feature learning. In: Proceedings of the Fourteenth International Conference on Artificial Intelligence and Statistics, pp. 215–223. JMLR Workshop and Conference Proceedings (2011)
5. Darlow, L.N., Crowley, E.J., Antoniou, A., Storkey, A.J.: CINIC-10 is not ImageNet or CIFAR-10. arXiv preprint arXiv:1810.03505 (2018)
6. Deng, J., Dong, W., Socher, R., Li, L.J., Li, K., Fei-Fei, L.: ImageNet: a large-scale hierarchical image database. In: 2009 IEEE Conference on Computer Vision and Pattern Recognition, pp. 248–255. IEEE (2009)
7. Ding, X., Zhang, X., Ma, N., Han, J., Ding, G., Sun, J.: RepVGG: Mmaking VGG-style convnets great again. arXiv preprint arXiv:2101.03697 (2021)
8. Girshick, R., Donahue, J., Darrell, T., Malik, J.: Rich feature hierarchies for accurate object detection and semantic segmentation. In: Proceedings of the IEEE Conference on Computer Vision and Pattern Recognition, pp. 580–587 (2014)
9. Glorot, X., Bordes, A., Bengio, Y.: Deep sparse rectifier neural networks. In: Proceedings of the Fourteenth International Conference on Artificial Intelligence and Statistics, pp. 315–323 (2011)
10. He, K., Zhang, X., Ren, S., Jian, S.: Deep residual learning for image recognition. In: 2016 IEEE Conference on Computer Vision and Pattern Recognition (CVPR) (2016)
11. Hinton, G., Vinyals, O., Dean, J.: Distilling the knowledge in a neural network. Comput. Sci. **14**(7), 38–39 (2015)

12. Howard, A.G., et al.: MobileNets: efficient convolutional neural networks for mobile vision applications. arXiv preprint arXiv:1704.04861 (2017)
13. Huang, Z., Huang, L., Gong, Y., Huang, C., Wang, X.: Mask scoring R-CNN. In: Proceedings of the IEEE/CVF Conference on Computer Vision and Pattern Recognition, pp. 6409–6418 (2019)
14. Ioffe, S., Szegedy, C.: Batch normalization: accelerating deep network training by reducing internal covariate shift. In: International Conference on Machine Learning, pp. 448–456. PMLR (2015)
15. Krizhevsky, A., Hinton, G., et al.: Learning multiple layers of features from tiny images (2009)
16. Krizhevsky, A., Sutskever, I., Hinton, G.E.: ImageNet classification with deep convolutional neural networks. In: Advances in Neural Information Processing Systems, pp. 1097–1105 (2012)
17. LeCun, Y., Denker, J.S., Solla, S.A.: Optimal brain damage. In: Advances in Neural Information Processing Systems, pp. 598–605 (1990)
18. Li, B., Wu, F., Weinberger, K.Q., Belongie, S.: Positional normalization. arXiv preprint arXiv:1907.04312 (2019)
19. Liu, Z., Li, J., Shen, Z., Huang, G., Yan, S., Zhang, C.: Learning efficient convolutional networks through network slimming. In: Proceedings of the IEEE International Conference on Computer Vision, pp. 2736–2744 (2017)
20. Long, J., Shelhamer, E., Darrell, T.: Fully convolutional networks for semantic segmentation. In: Proceedings of the IEEE Conference on Computer Vision and Pattern Recognition, pp. 3431–3440 (2015)
21. Luo, P., Ren, J., Peng, Z., Zhang, R., Li, J.: Differentiable learning-to-normalize via switchable normalization. arXiv preprint arXiv:1806.10779 (2018)
22. Lyu, S., Simoncelli, E.P.: Nonlinear image representation using divisive normalization. In: 2008 IEEE Conference on Computer Vision and Pattern Recognition, pp. 1–8. IEEE (2008)
23. Molchanov, P., Tyree, S., Karras, T., Aila, T., Kautz, J.: Pruning convolutional neural networks for resource efficient inference. arXiv preprint arXiv:1611.06440 (2016)
24. Odena, A.: Semi-supervised learning with generative adversarial networks. arXiv preprint arXiv:1606.01583 (2016)
25. Pan, X., Luo, P., Shi, J., Tang, X.: Two at once: enhancing learning and generalization capacities via IBN-Net. In: Proceedings of the European Conference on Computer Vision (ECCV), pp. 464–479 (2018)
26. Peng, B., et al.: Correlation congruence for knowledge distillation. In: Proceedings of the IEEE International Conference on Computer Vision, pp. 5007–5016 (2019)
27. Ren, S., He, K., Girshick, R., Sun, J.: Faster R-CNN: towards real-time object detection with region proposal networks. In: Advances in Neural Information Processing Systems, pp. 91–99 (2015)
28. Simonyan, K., Zisserman, A.: Very deep convolutional networks for large-scale image recognition. Computer Science (2014)
29. Singh, S., Krishnan, S.: Filter response normalization layer: eliminating batch dependence in the training of deep neural networks. In: Proceedings of the IEEE/CVF Conference on Computer Vision and Pattern Recognition, pp. 11237–11246 (2020)
30. Srivastava, N., Hinton, G., Krizhevsky, A., Sutskever, I., Salakhutdinov, R.: Dropout: a simple way to prevent neural networks from overfitting. J. Mach. Learn. Res. 15(1), 1929–1958 (2014)

31. Tang, J., Liu, M., Jiang, N., Yu, W., Yang, C.: Spatial and channel dimensions attention feature transfer for better convolutional neural networks. In: 2021 IEEE International Symposium on Circuits and Systems (ISCAS), pp. 1–5. IEEE (2021)
32. Ulyanov, D., Vedaldi, A., Lempitsky, V.: Instance normalization: the missing ingredient for fast stylization. arXiv preprint arXiv:1607.08022 (2016)
33. Wu, Y., He, K.: Group normalization. In: Proceedings of the European Conference on Computer Vision (ECCV), pp. 3–19 (2018)
34. Xie, S., Girshick, R., Dollár, P., Tu, Z., He, K.: Aggregated residual transformations for deep neural networks. In: Proceedings of the IEEE Conference on Computer Vision and Pattern Recognition, pp. 1492–1500 (2017)
35. Xu, S., Ren, X., Ma, S., Wang, H.: meProp: sparsified back propagation for accelerated deep learning with reduced overfitting. In: ICML 2017 (2017)
36. Zagoruyko, S., Komodakis, N.: Wide residual networks. arXiv preprint arXiv:1605.07146 (2016)
37. Zhang, X., Zhou, X., Lin, M., Sun, J.: ShuffleNet: an extremely efficient convolutional neural network for mobile devices. In: Proceedings of the IEEE Conference on Computer Vision and Pattern Recognition, pp. 6848–6856 (2018)

Underwater Acoustic Target Recognition with Fusion Feature

Pengyuan Qi, Jianguo Sun, Yunfei Long, Liguo Zhang$^{(\boxtimes)}$, and Tianye

College of Computer Science and Technology, Harbin Engineering University,
Harbin 150001, China
zhangliguo@hrbeu.edu.cn

Abstract. For the underwater acoustic targets recognition, it is a challenging task to provide good classification accuracy for underwater acoustic target using radiated acoustic signals. Generally, due to the complex and changeable underwater environment, when the difference between the two types of targets is not large in some sensitive characteristics, the classifier based on single feature training cannot output correct classification. In addition, the complex background noise of target will also lead to the degradation of feature data quality. Here, we present a feature fusion strategy to identify underwater acoustic targets with one-dimensional Convolutional Neural Network. This method mainly consists of three steps. Firstly, considering the phase spectrum information is usually ignored, the Long and Short-Term Memory (LSTM) network is adopted to extract phase features and frequency features of the acoustic signal in the real marine environment. Secondly, for leveraging the frequency-based features and phase-based features in a single model, we introduce a feature fusion method to fuse the different features. Finally, the newly formed fusion features are used as input data to train and validate the model. The results show the superiority of our algorithm, as compared with the only single feature data, which meets the intelligent requirements of underwater acoustic target recognition to a certain extent.

Keywords: Acoustic signal · Target recognition · Feature fusion

1 Introduction

Passive sonar is widely used in the underwater target recognition because of its excellent concealment and long working distance. It is usually designed to detect and identify targets from the ubiquitous clutter, a typical scenario shown in Fig. 1. During the passive sonar detection, pattern classification method is used to detect the underlying pattern or structures in the acoustic signal received by the front end. The sonar target classification system recognition process is illustrated in Fig. 2. The methods for underwater acoustic targets classification

Supported by Natural Science Foundation of Heilongjiang Province No. F2018006.

T. Mantoro et al. (Eds.): ICONIP 2021, LNCS 13108, pp. 609–620, 2021.
https://doi.org/10.1007/978-3-030-92185-9_50

are far from practical application, especially in a real-ocean environment. The reasons include the acoustic characteristics of different types of targets overlap, the complex and changeable ocean environment, the low signal-to-noise ratio of the receiving signal, high-quality data is rare and costly to obtain. These factors make the process of object classification a complicated problem. So far, the recognition of underwater acoustic signals has attracted widespread attention from scholars.

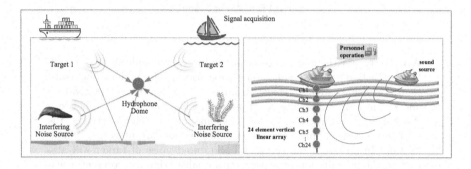

Fig. 1. Sonar working schematic diagram. An object can use the sonar equipment to analyze the underwater signals. The figure on the left shows the array element acquiring target information. The figure on the right shows a vertical line array sonar, the hydrophone array used is composed of 24 array elements, namely Ch1 to Ch24, and the array elements are equally spaced.

Aiming at the problem of underwater acoustic signal identification, a variety of identification methods are proposed. The characteristic parameters of time-domain waveforms and time-frequency analysis, nonlinear characteristic parameters [1,2] and spectrum analysis with the line spectrum characteristics, Low-Frequency Analysis and Recording(LOFAR), high-order spectra, Detection of Envelope Modulation on Noise(DEMON) are used commonly. The extracted auditory characteristic parameters commonly include Mel Frequency Cepstrum Coefficient(MFCC), Linear Predictive Cepstral Coefficient(LPCC) [3–5].

With the development of big data technology and the improvement of computer computing power, Machine Learning(ML), especially Deep Learning(DL) has been widely used in related application fields, e.g., Support Vector Machines, Back Propagation Neural Networks, K-Nearest Neighbor is employed for underwater acoustic signal recognition [6–10]. However, with the increase of the amount of data, ML can hardly meet the needs of existing recognition tasks. DL showed strong data processing and feature learning capabilities by the commonly used of Denoising Auto-Encoder (DAE) [11], LSTM [12], Deep Convolutional Neural Networks (DCNN) [13]. More and more scholars have begun to apply DL to the underwater acoustic target recognition, e.g., Z. Xiangyang et al. proposed a method of transforming the original one-dimensional underwater acoustic signal into a multi-dimensional acoustic spectrogram [14]. H. Yang et al. proposed an LSTM-based DAE collaborative network [15]. H. Yang et al. proposed an end-to-end deep neural network based on auditory perception-Deep Convolutional Neural Network (ADCNN) [16], Y. Gao et al. proposed the combination of Deep Convolutional Generative Adversarial Network (DCGAN) and Densely Connected Convolutional Networks (DenseNet), which extracts deep features for underwater acoustic targets [17]. J. Chen et al. proposed a LOFAR spectrum enhancement (LSE)-based underwater target recognition scheme [18]. Considering the relative scarcity of underwater acoustic data sets for training, G. Jin et al. presented a novel framework that applied the LOFAR spectrum for preprocessing to retain key features and utilized Generative Adversarial Networks (GAN) for the expansion of samples to improve the performance classification [19]. The above works show that deep network has powerful modeling ability for complex functions with high dimensional input.

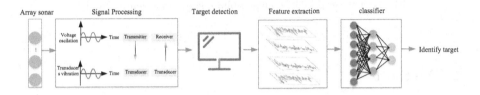

Fig. 2. The sonar target recognition process. Sonar equipment can detect objects through electro-acoustic conversion and information processing.

The deep network models rely only on a single spectral feature, such as the STFT feature [20] and the LPS feature [21], some important characteristics of radiated noise from underwater targets may be lost. In this paper, by extracting these two kinds of features, the advantages of LSTM system in complex underwater acoustic signal modeling are further studied. In addition, the audio signal has timing characteristics, the LSTM network is usually more excellent than other networks for processing the timing information. Influenced by this, we do this by exploring two different properties of the radiated noise training data set: the frequency spectrum and the phase spectrum in low-frequency band. The framework of the proposed underwater target classification model is described in Fig. 3. Our experimental results show that the proposed method performs significantly better than the single feature in terms of recognition accuracy. The contributions of this paper are summarised as follows:

(1) The model is used to automatically learn the effective feature representation of complex target signals, and it can greatly improve the performance of pattern recognition system compared with the previous manual feature extraction.
(2) We construct a joint feature for the depth model based on the spectrum and phase spectrum information, and make full use of the advantages of the depth structure to achieve feature complementarity and reduce the impact of the inherent defects of a single feature.

(3) Our method is tested on the underwater acoustic signals which is different from the previous work under simulation conditions and achieves outstanding performance compared with the single method.

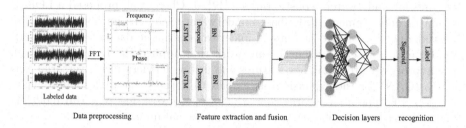

Fig. 3. The proposed frequency-phase spectrum identify model.

2   Method

2.1   Model Overview

In the first phase, we need to extract the low-level features of different domains based on LSTM network and the multi-domain feature vectors are spliced into joint feature inputs suitable for model training. The joint feature is composed of the frequency spectrum feature and phase spectrum feature. In this paper, the feature subsets of frequency and phase are fused directly in the series form to form multi-category fusion features. In the classification stage, CNN was used to classify and identify the targets. The design of the framework based on CNN is described in Fig. 4. In the prediction classification stage, the above process was repeated to obtain the fusion feature subset of the test samples, and the trained classifier was used to identify the target category.

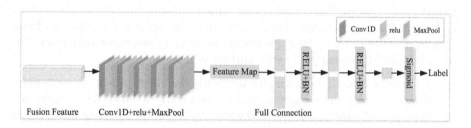

Fig. 4. The framework based on Convolutional Neural Network.

2.2   Frequency-Phase Spectrum Analysis

In the actual marine environment, the underwater acoustic signal is commonly affected by the following two aspects: 1) environmental noise; 2) experimental platform noise. Figure 5 displays the time-domain waveform of the original underwater acoustic signal, which is part of the underwater target in the data set. The strong background noise caused that the time-domain waveform of the original underwater acoustic signal shows noise-like characteristics. In order to verify the effectiveness of multidimensional feature fusion method proposed in this paper, we choose to analyze the frequency spectrum and the phase spectrum of signals by Fourier Transform on the time-domain waveform, Fourier Transform is shown in Eq. (1). Figure 6(a) and Fig. 6(b) are respectively the frequency spectrum and phase spectrum of the underwater acoustic signal, in which the red line represents the signal with sailing ship target and the blue line represents the signal without sailing ship target (background noise).

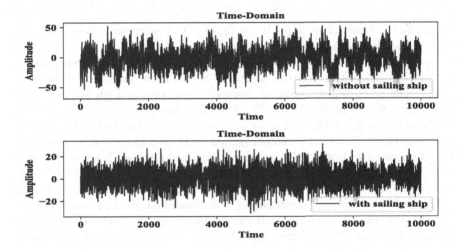

Fig. 5. Time-domain waveform.

Continuous spectrum and line spectrum make up the frequency spectrum of the underwater acoustic signal commonly. The ship-radiated noise when railing includes three kinds of low-frequency line spectrum, which all in 100 Hz–600 Hz. Therefore, in Fig. 6(a) the frequency spectrum comparison chart, the peak-to-peak value of the signal with the sailing ship is significantly higher than the background noise 420 Hz–460 Hz and 520 Hz–560 Hz because of line spectrum. In the phase spectrum comparison chart, the difference of peak-to-peak value is equally obvious within the aforementioned frequency range. If the features of the underwater acoustic signal are analyzed only from the frequency spectrum, that will lose part information of the signal. Taking the frequency spectrum and phase spectrum of the underwater acoustic signal as the input of the recognition model can effectively compensate for the lack of underwater acoustic signal characteristics.

$$F(\omega) = \mathcal{F}[f(t)] = \int_{-\infty}^{\infty} f(t)e^{-iwt}dt \tag{1}$$

Where, $f(t)$ refers to the time-domain data of original underwater acoustic signal.

2.3 Frequency-Phase Feature Fusion Recognition

Figure 3 describes the process of expressing feature extraction. The frequency feature and phase feature can be obtained from the spectrogram. The process can finally extract two-dimensional features and form new feature vectors. The new feature vectors $\vec{N}$ can be expressed as:

$$\vec{N}_i = \{F_i(t), P_i(t)\} \tag{2}$$

where t is time series, $F_i(t)$ is the frequency characteristic value at time i, $P_i(t)$ is the phase characteristic value at time i.

In this paper, the joint feature input N is build for deep learning network to identify underwater acoustic signal. In Sect. 2.2, we analyze feature of the signal. The peak-to-peak value of frequency spectrum and phase spectrum is obviously different in 100 Hz–600 Hz. Therefore, when preprocessing the underwater acoustic signal, we need to obtain the frequency spectrum and the phase spectrum of the underwater acoustic signal by Fourier transform in 100 Hz–600 Hz and normalize them by the Deviation Standard method. Take processed frequency spectrum and phase spectrum as the input of model, and feature learning from them through LSTM network. The dimensions of the feature are 50. Then, obtain fusion feature by concatenating. The dimensions of the fused feature are 100. Finally, the recognition result can be implemented by the FC layers and Sigmoid function. The specific Algorithm 1 of the proposed multi-dimensional fusion feature is as follows:

Algorithm 1. Multi-dimensional Fusion Feature.

Require: original underwater acoustic signals $s(t)$;
Ensure: fusion feature vector;
 1: Initialization: LSTM, Dropout(rate=0.25),BatchNormalization(BN),P,F,N;
 2: Let $x(t) \leftarrow s(t)$;
 3: Calculate x(k) using Equation (1);
 4: Decompose $x(k)$ into $freq(k)$ and $phase(k)$;
 5: Calculate Frequency_ feature F using BN(Dropout(LSTM($freq(k)$)));
 6: Calculate Phase_ feature P using BN(Dropout(LSTM($phase(k)$)));
 7: Calculate output_feature $N \leftarrow F + P$;
 8: **return** N;

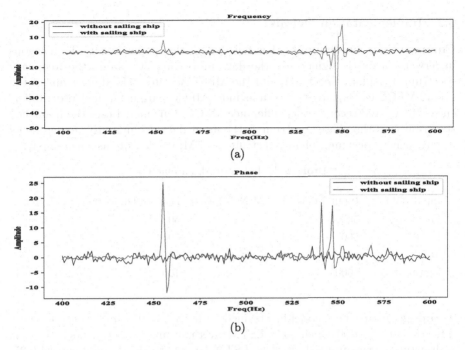

Fig. 6. The frequency and phase comparative chart by Fourier transform of the time-domain waveform, where the first row shows a frequency spectrum of underwater target. Wherein, the ordinate is the spectrum value. The second row shows a phase spectrum of underwater target. Wherein, the ordinate is the phase value.

3   Experiments

In this section, we introduce the implementation details and quality assessment criteria. Finally, the experimental results are given which proves the superiority of the method proposed in this paper.

3.1   Dataset and Experiment Platform

The method is verified by two kinds of signal data: is there a sailing ship. Each signal in this paper comes from a passive sonar in the marine and a sampling rate of 25600 Hz. The total number of samples of the model is 1.8×10^4, in which the number of the signal with a sailing ship is 1.0×10^4, the number of the signal without a sailing ship is 0.8×10^4. In order to ensure the validity of the verification results, the paper randomly selected 1.72×10^4 from the sample library to form a number of training sets, the remaining samples as the test set. We train our model on the NVIDIA TITAN XP by CUDA 9.0 and Keras.

3.2   Implementation Details

Comparative Experiments. For fairness, in this paper, we use homologous underwater acoustic signals. By the data processing, we obtain the frequency spectrum, the phase spectrum, and the MFCC feature. The dimensions of the overall MFCC feature are 96, which includes MFCC parameter, first-order difference MFCC, and second-order difference MFCC. Our model uses the frequency spectrum, and the phase spectrum as the input data. Comparative experiments use frequency spectrum, phase spectrum, and MFCC feature as single input.

Table 1. LSTM network parameters

Input data	Input Nodes	LSTM NO	Layers	Feature Dimension	FC Layer
Frequency	500	1	1	50	2
Phase	500	1	1	50	2
MFCC	96	1	1	50	2
Feature fusion	1000	2	1	100	2

Training Setup. The model has trained with 1.72×10^4 samples, the learning rate is 5×10^{-6} by 700 epochs. The LSTM layer params of experiments in Table 1. A dropout layer is inserted after the LSTM layer with the dropout rate of 0.25.

Quality Assessment Criteria. This paper mainly focuses on the classification problem of the 2 types of acoustic signals. In order to evaluate the proposed method, we used the data samples from real marine data, F1 score and accuracy rate as evaluation indexes. We used the true negative rate, true positive rate, false positive rate, and false negative rate from ML and use F1-score which is the harmonic mean recall of Recall rate (R) and Precision rate (P) to evaluate the model's recognition effect of underwater targets in the test set. The F1-score calculation formula is shown in Eq. (3), and the Accuracy Rate is shown in Eq. (4).

$$F1 = 2 \times \frac{P \times R}{P + R} \tag{3}$$

Where P is Precision rate, R is Recall rate.

$$\text{Accuracy Rate} = \frac{A_{\text{acc}}}{T_{\text{total}}} \times 100\% \tag{4}$$

Where A is the total number of objects that can be correctly identified, T is the total number of the two targets.

3.3   Experimental Results

In this paper, we compare the performance of the proposed method with other methods, i.e., MFCC, frequency, and phase on the validation set. In this paper,

Table 2. Confusion matrix of experimental results

Recognition result	Method		Prediction real label	Prediction goalless label	In total
Real targeted lable	Sigle feature	Frequency	341	77	418
		Phase	302	116	
		MFCC	349	69	
	Fusion feature	Our method	381	37	
Real goalless label	Sigle feature	Frequency	307	75	382
		Phase	322	60	
		MFCC	329	53	
	Fusion feature	Our method	337	45	

the F1 score and accuracy are adopted as the evaluation indexes. As can be seen from Table 2, the F1-score in frequency, phase and MFCC is 64%, 57.9% and 63.1% respectively, the calculated F1-Score is 72.1% with our method by Eq. (3), compared with frequency feature and phase feature, the fusion feature can improve the performance of recognition precision.

To simulate practical applications of recognition for ship-radiated noise, the classification accuracy of each acoustic event is used to measure the classification performance of the model, which is defined as the percentage of all acoustic events that are correctly classified. The classification accuracy of the proposed method and the comparison method is shown in Table 3. As we can see from

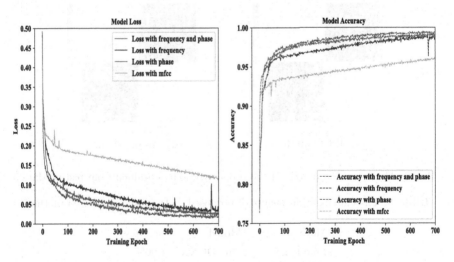

Fig. 7. Loss and accuracy of the model, where the left shows the classification loss of the proposed method with frequency, phase and MFCC, the right shows the classification accuracy of the proposed method with frequency, phase and MFCC.

Table 3, compared with a single feature of underwater acoustic target recognition methods, the proposed fusion method effectively improves the classification accuracy of the underwater acoustic target.

To further show the effectiveness of our proposed model, the recognition performance results on the validation set are illustrated in Fig. 7, which details the classification accuracy improvement of feature fusion relative to phase spectrum, frequency spectrum, and MFCC in each class. As shown in Fig. 7, in the process of model training, there is no over-fitting or under-fitting phenomenon, and there is no gradient disappearance or gradient explosion. By testing the model with measured data, with the number of model training steps increases, the proposed method can achieve a higher recognition accuracy on the validation set. We provide a confusion matrix for the recognition result of the proposed model, as shown in Fig. 8. Each row of the confusion matrix correspond to the real label and each column corresponds to the predicted label.

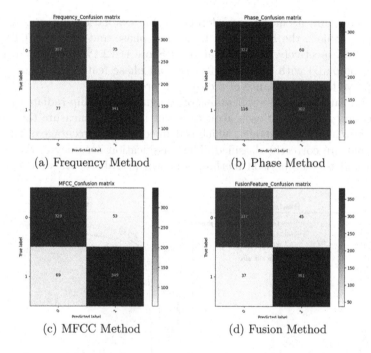

(a) Frequency Method   (b) Phase Method

(c) MFCC Method   (d) Fusion Method

Fig. 8. The confusion matrix of the proposed model obtained from testing data.

Table 3. The classification results of proposed method and compared methods

Method	Model	Accuracy
Frequency	One-DCNN	81.00%
Phase	One-DCNN	78.00%
MFCC	One-DCNN	84.75%
Fusion-Feature	One-DCNN	**89.75%**

4 Conclusion

This paper focuses on how to introduce the acoustic feature of the frequency-phase spectrum and two types of feature fusion model into the passive recognition problem. In order to alleviate the identify difficulty in the actual marine, the recognition method based on frequency-phase spectrum analysis is proposed. In this method, the LSTM is used for multi-class feature extract. By analyzing the target and background noise, two kinds of target data are obtained. Experiments show that the frequency-phase spectrum recognition method proposed can effectively distinguish the above two types of target, and the recognition results of the two types of feature fusion are better than other cases. In addition, our method strengthens the interpretability of the features extracted compared to deep learning technology.

In this paper, only the two discriminant methods were studied and introduced. However, due to the lack of relevant research on optimization selection, the comparison of the discriminant effects after optimization between the two methods needs to be further studied and discussed in the future.

References

1. Oswald, J.N., Au, W.W., Duennebier, F.: Minke whale (balaenoptera acutorostrata) boings detected at the station aloha cabled observatory. J. Acoust. Soc. Am. **129**(5), 3353–3360 (2011)
2. Esfahanian, M., Zhuang, H., Erdol, N.: Using local binary patterns as features for classification of dolphin calls. J. Acoust. Soc. Am. **134**(1), EL105–EL111 (2013)
3. Chinchu, M., Supriya, M.: Real time target recognition using labview. In: International Symposium on Ocean Electronics (SYMPOL), pp. 1–9. IEEE (2015)
4. Wang, W., Li, S., Yang, J., Liu, Z., Zhou, W.: Feature extraction of underwater target in auditory sensation area based on MFCC. In: IEEE/OES China Ocean Acoustics (COA), pp. 1–6 (2016)
5. Zhang, L., Wu, D., Han, X., Zhu, Z.: Feature extraction of underwater target signal using mel frequency cepstrum coefficients based on acoustic vector sensor. J. Sens. **11–17** (2016)
6. Liu, H., Wang, W., Yang, J.-A., Zhen, L.: A novel research on feature extraction of acoustic targets based on manifold learning. In: International Conference on Computer Science and Applications (CSA), pp. 227–231. IEEE (2015)
7. Sun, L., Kudo, M., Kimura, K.: Reader: robust semi-supervised multi-label dimension reduction. IEICE Trans. Inf. Syst. **100**(10), 2597–2604 (2017)
8. Sherin, B., Supriya, M.: Sos based selection and parameter optimization for underwater target classification. In: OCEANS MTS/IEEE Monterey, pp. 1–4. IEEE (2016)
9. Li, H., Cheng, Y., Dai, W., Li, Z.: A method based on wavelet packets-fractal and SVM for underwater acoustic signals recognition. In: International Conference on Signal Processing (ICSP), pp. 2169–2173. IEEE (2014)
10. Barngrover, C., Althoff, A., DeGuzman, P., Kastner, R.: A brain-computer interface (BCI) for the detection of mine-like objects in sidescan sonar imagery. IEEE J. Oceanic Eng. **41**(1), 123–138 (2015)

11. Verwimp, L., Pelemans, J., Wambacq, P., et al.: Character-word LSTM language models, arXiv preprint arXiv:1704.02813 (2017)
12. Mimura, M., Sakai, S., Kawahara, T.: Speech dereverberation using long short-term memory. In: Sixteenth Annual Conference of the International Speech Communication Association (2015)
13. Wang, P., Peng, Y.: Research on feature extraction and recognition method of underwater acoustic target based on deep convolutional network. In: International Conference on Advances in Electrical Engineering and Computer Applications (AEECA), pp. 863–868. IEEE (2020)
14. Xiangyang, Z., Jiaruo, H., Lixiang, M.: Image representation of acoustic features for the automatic recognition of underwater noise targets. In: Third Global Congress on Intelligent Systems, pp. 144–147. IEEE (2012)
15. Yang, H., Xu, G., Yi, S., Li, Y.: A new cooperative deep learning method for underwater acoustic target recognition. In: OCEANS 2019-Marseille, pp. 1–4. IEEE (2019)
16. Yang, H., Li, J., Shen, S., Xu, G.: A deep convolutional neural network inspired by auditory perception for underwater acoustic target recognition. Sensors **19**(5), 1104 (2019)
17. Gao, Y., Chen, Y., Wang, F., He, Y.: Recognition method for underwater acoustic target based on DCGAN and DenseNet. In: 2020 IEEE 5th International Conference on Image, Vision and Computing (ICIVC), pp. 215–221. IEEE (2020)
18. Chen, J., Liu, J., Liu, C., Zhang, J., Han, B.: Underwater target recognition based on multi-decision lofar spectrum enhancement: a deep learning approach. arXiv preprint arXiv:2104.12362 (2021)
19. Jin, G., Liu, F., Wu, H., Song, Q.: Deep learning-based framework for expansion, recognition and classification of underwater acoustic signal. J. Exp. Theoret. Artif. Intell. **32**(2), 205–218 (2020)
20. Kamal, S., Mohammed, S.K., Pillai, P.S., Supriya, M.: Deep learning architectures for underwater target recognition. In: 2013 Ocean Electronics (SYMPOL), pp. 48–54. IEEE (2013)
21. Cao, X., Zhang, X., Yu, Y., Niu, L.: Deep learning-based recognition of underwater target. In: 2016 IEEE International Conference on Digital Signal Processing (DSP), pp. 89–93. IEEE (2016)

Evaluating Data Characterization Measures for Clustering Problems in Meta-learning

Luiz Henrique dos S. Fernandes[1]([⊠]) [iD], Marcilio C. P. de Souto[2] [iD],
and Ana C. Lorena[1] [iD]

[1] Instituto Tecnológico de Aeronáutica, São José dos Campos, SP 12228-900, Brazil
{lhsf,aclorena}@ita.br
[2] Université d'Orléans, Orléans 45100, France
marcilio.desouto@univ-orleans.fr

Abstract. An accurate data characterization is essential for a reliable selection of clustering algorithms via meta-learning. This work evaluates a set of measures for characterizing clustering problems using beta regression and two well-known machine learning regression techniques as meta-models. We have observed a subset of meta-features which demonstrates greater resourcefulness to characterize the clustering datasets. In addition, secondary findings made it possible to verify the direction and magnitude of the influence and the importance of such measures in predicting the performance of the algorithms under analysis.

Keywords: Machine learning · Meta-learning · Clustering

1 Introduction

One of the main challenges of building meta-learning (MtL) systems in Machine Learning (ML) is how to properly characterize the datasets under analysis [26]. When clustering datasets are concerned, this issue is amplified. Whilst there are multiple definitions of what constitutes a cluster in the literature [14], the lack of an expected output also hinders the definition of measures which might capture the underlying clustering structure more effectively. In this work we gather a set of measures following different clustering criteria in an attempt to characterize better the variety of structures a clustering dataset may contain. The objective is to be capable of adequately and consistently characterizing varied clustering datasets. These meta-features are then evaluated in a MtL setup aimed to predict the expected Adjusted Rand Index (ARI) of eighth different clustering algorithms for datasets of known structure.

Three regression techniques are employed as meta-models: beta-regression, k-nearest neighbours (k-NN) and random forest (RF). Beta regression is a parametric model which accommodates well asymmetries in the distribution of the responses and non-constant variances of the residuals, aka heterocedasticity [11].

© Springer Nature Switzerland AG 2021
T. Mantoro et al. (Eds.): ICONIP 2021, LNCS 13108, pp. 621–632, 2021.
https://doi.org/10.1007/978-3-030-92185-9_51

It was employed to perform a preliminary analysis of meta-features in predicting the ARI values. The objective was to identify the direction and magnitude of influence of the data characterization measures, as well as to objectively raise the statistical significance of each measure under analysis. On the other hand, k-NN and RF are two well known and widely disseminated non-parametric ML models. Their cross-validation performance in ARI estimation was computed and the importance of the meta-features according to the RF models was assessed. As a side result, we identify a group of measures that are consistent for the characterization of various clustering datasets and which can support better future MtL studies for clustering problems.

The remainder of this paper is organized as follows: In Sect. 2, we present a theoretical summary and previous studies on meta-learning and clustering. Section 3 describes the data characterization measures applied to clustering problems evaluated in this paper. In Sect. 4 we describe the experimental methodology adopted. Section 5 presents and discusses the experimental results of the meta-learners. Conclusions and future work are presented in Sect. 6.

2   Meta-learning and Clustering

The characterization of ML problems has been widely studied in a branch known as meta-learning (MtL) [26, 28]. In this paper, MtL refers to the exploration of meta-knowledge from previous known problems in order to obtain more efficient models and solutions for new problems [3]. Figure 1 illustrates this process as it is usually applied in ML. First, multiple datasets for which the ML solutions can be evaluated are gathered. A meta-dataset is then generated by extracting data characterization measures (aka meta-features) from the datasets along with algorithmic performance results for a pool of candidate ML algorithms. A meta-model can then be generated by a meta-learner using the meta-dataset as input, which can support different algorithm selection fronts. Regarding clustering problems, one may estimate the number of clusters a dataset has [23], the best clustering algorithm to be applied to a given dataset [10] or a ranking of algorithms to be recommended [22]. Here we are concerned on evaluating the ability and importance of different sets of meta-features for the proper characterization of clustering problems.

One of the first and seminal MtL contributions from the literature was developed by Rice in 1976 [24]. The author deals with the algorithm selection problem (ASP), i.e., the choice of an algorithm, among a given portfolio, that is more appropriate to solve a particular problem instance. The meta-features play a primordial role in the algorithm selection process and must be able to reveal the underlying structure and hardness level of the problem's instances [26].

The use of MtL in ASP for clustering has been explored in works such as [6, 10, 22, 23, 27]. De Souto et al. [6] was the first work to highlight the potentialities of MtL for clustering algorithm recommendation. Ferrari and Nunes [9, 10] have expanded MtL in clustering algorithm recommendation, using additional meta-features as input. More recently, Pimentel and Carvalho [22] proposed new data characterization meta-features for selecting clustering algorithms based on the

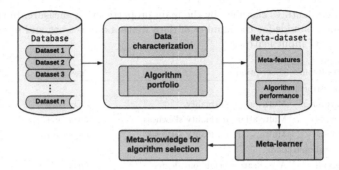

Fig. 1. Acquisition of meta-knowledge for algorithm selection.

similarities of the observations from a dataset. In a subsequent work, they used the same meta-features for designing a meta-model to recommend the number of clusters a dataset has [23]. Saez et al. [25] have also employed a MtL model to determine the number of clusters a dataset has based on clustering validation indices. In a recent paper, Fernandes et al. [8] applied a methodology known as Instance Space Analysis derived from MtL to visually evaluate the performance of different clustering algorithms across a benchmark of more than 500 datasets.

All previous work have considered different sets of meta-features for characterizing the clustering datasets. But a question remains on whether they are able to characterize the variety of structures different clustering datasets have. In this paper we carefully organize a set of meta-features according to different clustering criteria, so as to extract diverse views of the datasets and highlight the possible clustering structures they might have. Next, we evaluate their effectiveness in a MtL setting designed in order to predict the expected performance of different clustering algorithms for a set of 599 datasets.

3 Data Characterization Measures for Clustering Problems

Twenty five data characterization measures in total were selected and evaluated in this work. These candidate meta-features were divided into seven distinct categories, according to the main properties they regard in a dataset: distribution, neighbourhood, density, dimensionality, network centrality, distance and entropy. These categories reflect distinct existent clustering criteria from the literature. Table 1 presents, for each meta-feature, its acronym, description, asymptotic complexity and reference from which the meta-feature was extracted from. It is important to notice that multiple measures can be extracted at once, reducing the overall time complexity for extracting the meta-features values for a same dataset. Most of the measures were used in previous work or adapted here for the clustering scenario, with exception of the entropy measures, whose use is introduced here. All measures receive as input a dataset D containing n observations with m input features each.

Table 1. Meta-features classified by the main properties they measure. In the Asymptotic column, n stands for the number of data items a dataset has and m corresponds to its number of input features.

Meta-feature	Description	Asymptotic	Ref.
1) Distribution			
multi_norm	Multivariate normality	$O(m \cdot n + n^2)$	[21]
skewness	Multivariate normality skewness	$O(m \cdot n + n^2)$	[21]
kurtosis	Multivariate normality kurtosis	$O(m \cdot n + n^2)$	[21]
2) Neighbourhood			
avg_nnd	Avg. nearest neighbour degree	$O(m \cdot n^2)$	[1]
contrast	Contrast	$O(n^2)$	[12]
3) Density			
clust_coef	Clustering coefficient	$O(m \cdot n^2)$	[19]
net_dens	Network density	$O(m \cdot n^2)$	[19]
perc_out	Percentage of outliers	$O(n^2)$	[6]
4) Dimensionality			
number_ex	$\log_{10}$ number of examples	$O(n)$	[6]
number_ftr	$\log_{10}$ number of attributes	$O(m)$	[27]
ratio_ftr_ex	Ratio number of attributes to examples	$O(m + n)$	[15]
avg_abs_cor	Avg. absolute correlation	$O(n)$	[9]
intr_dim	Intrinsic dimensionality	$O(n^2)$	[12]
avg_pca	Avg. number of points per PCA dimension	$O(m^2 \cdot n + m^3)$	[19]
ratio_pca	Ratio PCA to the original dimension	$O(m^2 \cdot n + m^3)$	[19]
5) Network Centrality			
power_cent	Bonacich's power centrality	$O(n^3)$	[2]
eigen_cent	Eigenvalue centrality of MST	$O(n^2)$	[2]
hub_score	Kleinberg's hub centrality	$O(n^3)$	[17]
6) Distance			
mean_dist	Mean distance	$O(n^2)$	[10]
var_dist	Variance of distances	$O(n^2)$	[10]
sd_dist	Standard deviation of distances	$O(n^2)$	[10]
high_dist	Percentage of points of high distance	$O(n^2)$	[10]
low_dist	Percentage of points of low distance	$O(n^2)$	[10]
7) Entropy			
cop_entropy	Copula entropy	$O(n^2 + n \cdot \log n + n \cdot \sqrt{k \cdot n})$	[20]
knn_entropy	k-NN method entropy	$O(n^2 + n \cdot \sqrt{k \cdot n})$	[18]

Distribution-based measures quantify if the data distribution roughly approximates to a normal distribution. If this is the case, the dataset might have compact hyper-spherical clusters.

Neighbourhood-based measures quantify the local nearest neighbour influence in clustering, being a rough indicator of connectivity. The avg_nnd measure involves finding for each example the distance to its nearest neighbour and averaging the values found for all the observations in the dataset. The measure involves first building an ϵ-nearest neighbour graph from the data (in this work ϵ corresponds to 15% of the number of examples). The contrast of a point is defined as the relative difference in the distances to its nearest and farthest neighbour.

Density measures quantify whether there are dense regions of data in the input space, a surrogate for the presence of dense clusters. As avg_nnd, clust_coef and net_dens involve first building an ϵ-nearest neighbor graph from the data and then computing the average density and clustering coefficient of the vertices.

Dimensionality measures regard on the dimensions of the dataset concerning the number of examples and input features. Some quantify whether the data is sparsely distributed. intr_dim is the squared average distance between all points divided by the variance of those distances. avg_pca and ratio_pca first involve reducing data dimensionality by Principal Component Analysis (PCA) and retrieving the number of remaining features after retaining 95% of data variability.

Network centrality measures quantify whether there are connected structures on data. Eigenvalue centrality scores correspond to the values of the first eigenvector of the graph adjacency matrix of a Minimum Spanning Tree (MST) built from data. power_cent and hub_score involve first building an ϵ-nearest neighbor graph from the data and then computing the Bonacich's power centrality and Kleinberg's hub centrality measures, respectively. Bonacich's power measure corresponds to the notion that the power of a vertex is recursively defined by the sum of the power of its alters.

Distance-based measures quantify the relative differences of distances between the dataset observations. They resemble the neighborhood-based measures, but whilst the former considers all distances in the dataset, the later takes local distance-based information only. For a dataset, we take the average over all observations it has. high_dist and low_dist involve computing the pairwise distances between all examples and counting the number of examples with a high distance (higher than the average distance plus standard deviation) and with a low distance (lower than the average distance minus the standard deviation).

Entropy measures quantify the statistical dependence among random variables. cop_entropy is a mathematical concept for multivariate statistical independence measuring and testing, and proved to be equivalent to mutual information [20]. knn_entropy involves estimating entropy from data by k-NN [18].

4   Experimental Methodology

In this work, we evaluate the behavior of the previous meta-features in the characterization of clustering datasets for a MtL setup designed to predict the Adjusted Rand Index (ARI) of different clustering algorithms. We have chosen eight clustering algorithms of different biases widely used in the literature to compose the algorithm portfolio: K-Means (KME), Fuzzy C-Means (FCM), Hierarchical Agglomerative Single Linkage (SLK), Hierarchical Agglomerative Complete Linkage (CLK), High Dimensional Gaussian Mixture Model (GMM),

Bagged Clustering (BAG), Spectral Clustering (SPC) and Hierarquical Density Based Clustering of Applications with Noise (HDB). Therefore, our meta-dataset results in eight regression problems, each predicting the ARI of one of the previous clustering techniques for a set of 599 different datasets, described next.

4.1 Building the Meta-dataset

A preliminary and relevant step of this work is the consolidation of a meta-dataset. It is composed by the mentioned meta-features for a set of datasets and performance measures of the algorithms under analysis. In clustering, some specific aspects had to be observed. As in [6], the number of clusters was extracted from the ground truth of the datasets and the ARI external validation index was chosen for evaluating the results of the clustering algorithms.

The meta-dataset was composed by 599 datasets. 380 of them are synthetic clustering datasets and collected from diverse repositories, with number of examples ranging from 100 to 5000, number of attributes ranging from 2 to 1024 and number of embedded clusters ranging from 1 to 40. Among them are 80 Gaussian datasets of low dimensionality and 80 ellipsoidal datasets of high dimensionality [13]. Another 220 datasets have different shapes, number of examples, attributes and clusters. The remaining 219 real datasets were collected from the OpenML repository [22]. In those datasets, the number of examples ranges from 100 to 5000, the number of attributes ranges from 2 to 168, and the number of embedded clusters ranges from 2 to 30. Figure 2 presents the distribution of meta-features values for all 599 datasets. Some measures present very concentrated values (five meta-features in the left corner of the plot), whilst others have a larger variation (mainly ratio_pca, number_ex and number_ftr). There are many outlier values for most of the measures, except from ratio_pca, clust_coef, avg_abs_cor, number_ex and number_ftr.

The eight algorithms composing our portfolio were run using the implementations contained in the following R packages: e1071, dbscan, HDclassif, Spectrum and stats. The non-deterministic algorithms were run 10 times. Default

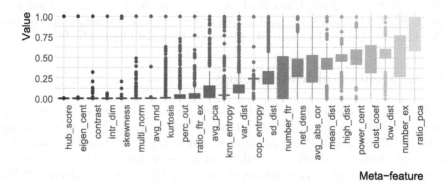

Fig. 2. Meta-features distribution. All values have been scaled to the $[0, 1]$ range.

parameters were used, with the exception of the HDBSCAN minimum number of neighbours (minPts), which was set to 4 [5]. As previously mentioned, ARI [16] was chosen to quantify performance. The higher the ARI value, the larger the agreement between the clustering results and the reference ground truth partition. Figure 3 illustrates the ARI distribution for each algorithm under analysis. SLK and GMM presented a significant peak of low ARI performance. SLK had a lower average ARI performance compared to the other clustering algorithms. SLK is very sensitive to noise and outliers, whereas GMM is a model-based algorithm that is sensitive to parameter tuning.

4.2 Preliminary Analysis of Meta-dataset with Beta Regression

Beta regression is a class of supervised learning tools for regression problems with univariate and limited response [7]. According to [11], beta regression is commonly used to model dependent variables that assume values in the $(0, 1)$ interval, such as rates, proportions and the ARI. The technique is based on the density function of the beta distribution in terms of two parameters and accommodates naturally non-constant variances of residuals (heteroscedasticity) and skewness. Figure 3 shows that the ARI distributions tend to be asymmetric. According to [11], beta density function is expressed by

$$f(y; \mu, \phi) = \frac{\Gamma(\phi)}{\Gamma(\mu\phi)\Gamma((1-\mu)\phi)} y^{\mu\phi-1}(1-y)^{(1-\mu)\phi-1}, \tag{1}$$

with $0 < y < 1$, $0 < \mu < 1$ and $\phi > 0$. The ϕ parameter is known as the precision parameter, for fixed μ. In beta regression there is an additional flexibility, as the researcher can choose a link function that best fits the model. In this work we use the logit function. The estimates generated by beta regression allow to assess the significance and influence of the predictors, which in this case correspond to the meta-features. Beta regression models were implemented using the betareg package from R.

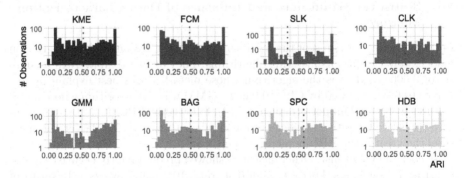

Fig. 3. ARI distribution for each clustering algorithm in the 599 datasets considered. Number of observations have been $\log_{10}$-scaled. The red dashed line represents the mean of the ARI. (Color figure online)

4.3 ML Meta-learners

Besides beta regression models, two other well-known ML regression techniques were used as meta-learners: k-nearest neighbours (k-NN) and random forest (RF). Given an observation, k-NN selects the k closest observations in the training dataset and averages their outputs. RF, on the other hand, combines multiple tree models induced from bootstrap samples of the dataset and random subsets of input features [4].

The ML regression models were trained using the caret package from R. We have used 10-fold cross validation, repeated 30 times across the meta-dataset. Coefficient of determination (Pseudo-R^2) and Root Mean Square Error (RMSE) are used as performance measures of the models. R^2 is a statistic generated in ordinary least squares regression (OLS), whereas Pseudo-R^2 is used with similar purpose to evaluate the goodness-of-fit of models estimated from maximum likelihood or ML techniques. During the cross-validation process, RMSE was used to select the number of neighbours in k-NN (k) and the number of variables randomly sampled as candidates at each split of the RF models (mtry). k was tested for odd values in the range [5, 23] during training. mtry was tested assuming the following values: {2, 7, 13, 19, 25}.

5 Experimental Results

This section summarises the meta-models results. Firstly, the statistical significance and influence of data characterization measures on the ARI performance of each clustering algorithm were assessed with use of beta regression. Next, the performance in ARI prediction was examined by means of two well-known ML techniques. All code was executed on an Intel Core i7-7500U CPU with 2.70 GHz, 16 GB RAM, and the Microsoft Windows 10 operating system. Code and data are available in an open access repository (see https://github.com/ml-research-clustering/meta-features).

5.1 Statistical Significance and Influence of Data Characterization Measures

In this subsection we present the results of the beta regression models. Regression models were run for all clustering algorithms. They achieved an average pseudo-R^2 of 0.600. KME was the algorithm whose meta-features best explain performance (0.683), followed by CLK (0.0.662). GMM was the algorithm whose meta-features worst explain performance (0.414), followed by SPC (0.552). Some measures presented positive coefficients for all algorithms, i.e., the larger the measure, the greater the ARI: number_ex, perc_out, avg_abs_cor, low_dist, net_dens, clust_coef and mean_dist. Other measures showed predominantly negative trends: multi_norm, skewness, kurtosis, high_dist, ratio_ftr_ex and avg_pca. In terms of influence, eigen_cent showed, on average, a high negative impact on ARI scores and power_cent had the smallest impact on the performance measure. Figure 4 illustrates the influence of all meta-features on the ARI scores.

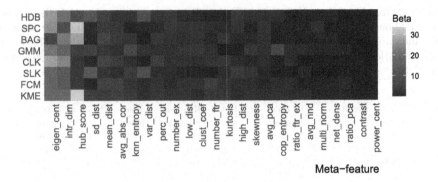

Fig. 4. Influence of meta-features on the ARI scores, where *Beta* represents the absolute value of estimated coefficients in beta regression models.

Despite the verified influence, not all meta-features had statistical significance. Four meta-features were not statistically significant for any algorithm: eigen_cent, contrast, hub_score and power_cent. We can highlight a group of 10 meta-features that were statistically significant for at least 75% of the clustering algorithms in the portfolio when predicting their ARI scores: kurtosis (distribution); clust_coef and perc_out (density); number_ex, number_ftr, avg_abs_cor and avg_pca (dimensionality); mean_dist, high_dist and low_dist (distance).

5.2   Performance Prediction and Importance of Meta-Features

In this subsection we proceed to examine the meta-models with k-NN and RF. Table 2 summarises the results of the ML regression models. A naive regressor that uses the mean of the training set ARI scores to predict the ARI values for the test set was considered as the baseline.

According to Table 2, RF and k-NN models were consistently better in terms of RMSE than the baseline in predicting the ARI of the clustering algorithms considered. They also obtained better overall performance in terms of Pseudo-R^2 when compared to parametric models implemented with beta regression. RF models obtained better results in terms of Pseudo-R^2 and RMSE for all algorithms when compared to k-NN models. The RMSE values observed were lower for the clustering algorithms CLK, KME and HDB, while they were the highest for the algorithms SLK, GMM and BAG, corresponding to a worst performance. The low RMSE standard deviation values verified for the RF meta-model is a strong indicator of robustness of the meta-models.

Besides their superior predictive performance in our meta-learning setting, the RF models allow an objective assessment of the importance of the meta-features, by means of a measure known as node purity increase. Figure 5 shows the node purity increase results of the RF models. The higher the node purity increase, the greater the importance of the meta-feature. A ranking was created based on the results obtained for all algorithms. It is clear that the importance of meta-features for the case of the SLK and GMM algorithms, which

Table 2. Results of Naive, k-NN and Random Forest regressors, where k is the number of neighbours for k-NN models, mtry is the number of variables randomly sampled as candidates at each split for Random Forest models, Pseudo-R^2 is the coefficient of determination, RMSE is the Root Mean Square Error and SD is the standard deviation of RMSE.

Alg.	Naive		k-NN					Random forest			
	RMSE	SD	k	Pseudo-R^2	RMSE	SD	mtry	Pseudo-R^2	RMSE	SD
KME	0.338	0.016	5	0.733	0.175	0.004	7	0.788	0.155	0.002
FCM	0.349	0.009	5	0.744	0.177	0.008	13	0.791	0.160	0.002
SLK	0.371	0.027	5	0.705	0.201	0.011	7	0.751	0.186	0.002
CLK	0.351	0.012	5	0.770	0.168	0.008	19	0.814	0.151	0.002
GMM	0.381	0.010	5	0.700	0.209	0.008	7	0.802	0.170	0.002
BAG	0.380	0.011	5	0.726	0.199	0.008	7	0.779	0.179	0.002
SPC	0.367	0.016	5	0.736	0.189	0.004	7	0.790	0.168	0.002
HDB	0.367	0.014	5	0.772	0.175	0.013	13	0.826	0.153	0.001
Mean	0.363	0.014	5	0.736	0.187	0,008	10	0.793	0.165	0.002

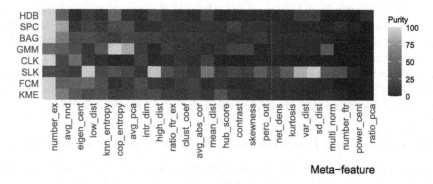

Fig. 5. Increases in node purities of random forest prediction models.

are the worse performing algorithms in the portfolio as ARI is regarded, differs from the other algorithms. And, for most of the algorithms, one very important meta-feature was number_ex, followed by avg_nnd. A relevant finding to highlight from the results of the beta regression and RF models is that a subset of 10 meta-features constitutes a kind of base core in the characterization of the examined datasets for the algorithm portfolio. According to Table 2, this quantity corresponds exactly to the average number of variables randomly sampled as candidates at each split for RF models (mtry). These measures belong to the following categories: distribution (kurtosis), density (clust_coef and perc_out), dimensionality (number_ex, number_ftr, avg_abs_cor and avg_pca) and distance (mean_dist, high_dist and low_dist). Among them, there are no neighbourhood, network centrality or entropy measures.

6   Conclusion

In this paper we evaluated 25 data characterization measures applied to clustering problems using a meta-learning methodology, grouped into different categories according to the aspect they emphasize. Preliminarily, beta regression was used to evaluate the statistical significance and influence of the meta-features. This technique fills a gap existing in several ML models related to the comprehensive objective analysis of the influence of input feature selection on the response variable. Then, k-NN and RF were used as meta-models and their cross-validation predictive performance in estimating the ARI of the clustering algorithms was assessed. RF was the best performing meta-regressor. It also allowed to highlight a ranking of the importance of meta-features used as inputs. These two joint approaches allowed us to identify a subset of consistent meta-features which stood out in the performance prediction of clustering algorithms. These categories reflect the distinct cluster structures inherent in the data. Future work will include methodologies that allow visualizing datasets in an instance space, in order to obtain better insights about performance of clustering algorithms and their relationship to the evidenced data characterization measures.

Acknowledgements. To the Brazilian research agency CNPq.

References

1. Barrat, A., Barthelemy, M., Pastor-Satorras, R., Vespignani, A.: The architecture of complex weighted networks. Proc. Natl. Acad. Sci. **101**(11), 3747–3752 (2004)
2. Bonacich, P.: Power and centrality: a family of measures. Am. J. Sociol. **92**(5), 1170–1182 (1987)
3. Brazdil, P., Carrier, C.G., Soares, C., Vilalta, R.: Metalearning: Applications to Data Mining. Springer, Heidelberg (2008). https://doi.org/10.1007/978-3-540-73263-1
4. Breiman, L.: Random forests. Mach. Learn. **45**(1), 5–32 (2001)
5. Campello, R.J.G.B., Moulavi, D., Sander, J.: Density-based clustering based on hierarchical density estimates. In: Pei, J., Tseng, V.S., Cao, L., Motoda, H., Xu, G. (eds.) PAKDD 2013. LNCS (LNAI), vol. 7819, pp. 160–172. Springer, Heidelberg (2013). https://doi.org/10.1007/978-3-642-37456-2_14
6. De Souto, M.C., et al.: Ranking and selecting clustering algorithms using a meta-learning approach. In: 2008 IEEE International Joint Conference on Neural Networks, pp. 3729–3735 (2008)
7. Espinheira, P.L., da Silva, L.C.M., Silva, A.D.O., Ospina, R.: Model selection criteria on beta regression for machine learning. Mach. Learn. Knowl. Extract. **1**(1), 427–449 (2019)
8. Fernandes, L.H.D.S., Lorena, A.C., Smith-Miles, K.: Towards understanding clustering problems and algorithms: an instance space analysis. Algorithms **14**(3), 95 (2021)
9. Ferrari, D.G., de Castro, L.N.: Clustering algorithm recommendation: a meta-learning approach. In: Panigrahi, B.K., Das, S., Suganthan, P.N., Nanda, P.K. (eds.) SEMCCO 2012. LNCS, vol. 7677, pp. 143–150. Springer, Heidelberg (2012). https://doi.org/10.1007/978-3-642-35380-2_18

10. Ferrari, D.G., De Castro, L.N.: Clustering algorithm selection by meta-learning systems: a new distance-based problem characterization and ranking combination methods. Inf. Sci. **301**, 181–194 (2015)
11. Ferrari, S., Cribari-Neto, F.: Beta regression for modelling rates and proportions. J. Appl. Stat. **31**(7), 799–815 (2004)
12. Fränti, P., Sieranoja, S.: K-means properties on six clustering benchmark datasets. Appl. Intell. **48**(12), 4743–4759 (2018). https://doi.org/10.1007/s10489-018-1238-7
13. Handl, J., Knowles, J.: Cluster generators for large high-dimensional data sets with large numbers of clusters (2005). https://personalpages.manchester.ac.uk/staff/Julia.Handl/generators.html. Accessed 5 Aug 2021
14. Handl, J., Knowles, J., Kell, D.B.: Computational cluster validation in post-genomic data analysis. Bioinformatics **21**(15), 3201–3212 (2005)
15. Ho, T.K., Basu, M.: Complexity measures of supervised classification problems. IEEE Trans. Pattern Anal. Mach. Intell. **24**(3), 289–300 (2002)
16. Hubert, L., Arabie, P.: Comparing partitions. J. Classif. **2**(1), 193–218 (1985). https://doi.org/10.1007/BF01908075
17. Kleinberg, J.M.: Authoritative sources in a hyperlinked environment. J. ACM **46**(5), 604–632 (1999)
18. Kraskov, A., Stögbauer, H., Grassberger, P.: Estimating mutual information. Phys. Rev. E **69**(6), 066138 (2004)
19. Lorena, A.C., Garcia, L.P., Lehmann, J., Souto, M.C., Ho, T.K.: How complex is your classification problem? A survey on measuring classification complexity. ACM Comput. Surv. (CSUR) **52**(5), 1–34 (2019)
20. Ma, J.: Estimating transfer entropy via copula entropy. arXiv preprint. arXiv:1910.04375 (2019)
21. Mardia, K.V.: Measures of multivariate skewness and kurtosis with applications. Biometrika **57**(3), 519–530 (1970)
22. Pimentel, B.A., de Carvalho, A.C.: A new data characterization for selecting clustering algorithms using meta-learning. Inf. Sci. **477**, 203–219 (2019)
23. Pimentel, B.A., de Carvalho, A.C.: A meta-learning approach for recommending the number of clusters for clustering algorithms. Knowl.-Based Syst. **195**, 105682 (2020)
24. Rice, J.R.: The algorithm selection problem. In: Advances in Computers, vol. 15, pp. 65–118. Elsevier (1976)
25. Sáez, J.A., Corchado, E.: A meta-learning recommendation system for characterizing unsupervised problems: on using quality indices to describe data conformations. IEEE Access **7**, 63247–63263 (2019)
26. Smith-Miles, K.A.: Cross-disciplinary perspectives on meta-learning for algorithm selection. ACM Comput. Surv. (CSUR) **41**(1), 6 (2009)
27. Soares, R.G.F., Ludermir, T.B., De Carvalho, F.A.T.: An analysis of meta-learning techniques for ranking clustering algorithms applied to artificial data. In: Alippi, C., Polycarpou, M., Panayiotou, C., Ellinas, G. (eds.) ICANN 2009. LNCS, vol. 5768, pp. 131–140. Springer, Heidelberg (2009). https://doi.org/10.1007/978-3-642-04274-4_14
28. Vanschoren, J.: Meta-learning: a survey. arXiv preprint arXiv:1810.03548 (2018)

ShallowNet: An Efficient Lightweight Text Detection Network Based on Instance Count-Aware Supervision Information

Xingfei Hu, Deyang Wu, Haiyan Li, Fei Jiang, and Hongtao Lu[✉]

Department of Computer Science and Engineering, Shanghai Jiao Tong University,
Shanghai, China
{huxingfei,wudeyang,lihaiyan_2016,jiangf,htlu}@sjtu.edu.cn

Abstract. Efficient models for text detection have a wide range of applications. However, existing real-time text detection methods mainly focus on the inference speed without considering the number of parameters. In this paper, we propose an efficient, lightweight network called ShallowNet with fewer layers, which significantly reduces the number of parameters and improves the inference speed. Fewer layers usually incur small receptive fields that degrade the performance of large-size text detection. To address such an issue, we utilize dilated convolution with various receptive fields that satisfies the need for different sizes of text detection. Nevertheless, the design of dilated convolution ignores the continuity property of text and results in stickiness and fragment in the text detection tasks. To tackle this challenge, we introduce instance count-aware supervision information that guides the network focus on text instances and preserves the boundaries. The instance count-aware supervision is used as the weight of the loss function. Extensive experiments demonstrate that the proposed method achieves a desirable trade-off between the accuracy, size of the model, and inference speed.

Keywords: Dilated convolution · Instance count-aware supervision · Lightweight · Text detection

1 Introduction

Scene text detection is an active research direction in computer vision because of its various practical applications, such as image search, ancient book retrieval, and autonomous driving [1].

This paper is supported by NSFC (No. 62176155, 61772330, 61876109), Shanghai Municipal Science and Technology Major Project (2021SHZDZX0102), the Shanghai Key Laboratory of Crime Scene Evidence (no. 2017XCWZK01), and the Interdisciplinary Program of Shanghai Jiao Tong University (no. YG2019QNA09).
H. Lu—Also with the MoE Key Lab of Artificial Intelligence, AI Institute, Shanghai Jiao Tong University.

© Springer Nature Switzerland AG 2021
T. Mantoro et al. (Eds.): ICONIP 2021, LNCS 13108, pp. 633–644, 2021.
https://doi.org/10.1007/978-3-030-92185-9_52

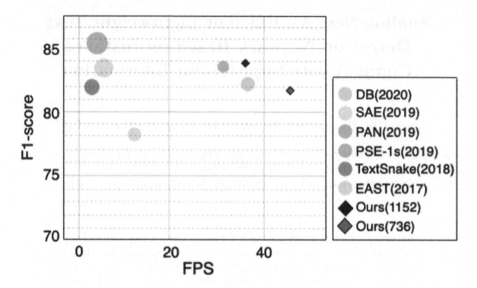

Fig. 1. Comparison with some scene text detection methods in state-of-the-art on ICDAR2015 dataset. 'FPS' means the number of images processed per second. Except for F1-score and FPS, the size of the circle represents the number of parameters. The larger the number of parameters is, the larger the circle will be. Our method considers parameters, accuracy, and speed, and achieves best trade-off.

As a special task of general object detection, text detection has many unique characteristics. The most significant difference is that the scene text has various shapes, especially curved, and often has a huge aspect ratio. For different types of text detection, regression-based methods and segmentation-based methods are derived. Regression-based text detection methods [2–6] can directly get the bounding box, so they are relatively faster. However, they cannot handle complex shapes very well. Segmentation-based scene text detection methods [7–14] can detect any text shapes, including curved and rectangular text, because they perform predictions at the pixel level. On the downside, segmentation-based methods consist of complex post-processing operations, which make the inference process slower. Besides, to increase the receptive field to maintain higher accuracy, regression and segmentation-based methods usually adopt deeper networks, such as ResNet50 [15], which slows down the inference process. To make the model more efficient, several existing methods [9,16] first utilize a downsampling operation to reduce the size of images to alleviate the computational burden during the training process. Afterward, they use upsampling operation to expand to the original image size. However, such sampling operation significantly loses the spatial information and impedes dense prediction tasks [17].

With the above analysis, we propose an efficient, lightweight network, namely **ShallowNet**. The structure is shown in Fig. 2, and the contributions are threefold. First, we explore a new neural network structure and build a shallow network with fewer layers to increase the inference speed and reduce the parame-

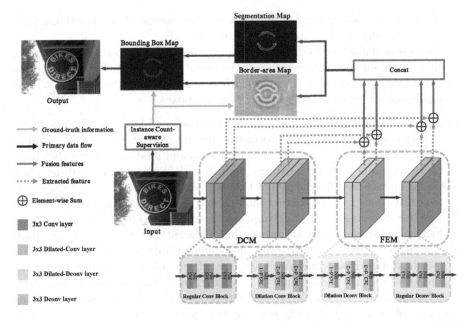

Fig. 2. The overall pipeline of our framework. DCM is used to extract feature. FEM could help to enhance the feature.

ters. Instead of using sampling operations, ShallowNet adopts dilated convolution to extract advanced features that alleviate the smaller receptive field issue. Second, an auxiliary instance counting supervision information is further introduced to learn more comprehensive representations, avoiding continuous feature loss caused by dilated convolution. The instance counting task is modeled as a quantity matching problem, forcing the network to predict the number of text boxes as close as possible to the number in ground-truth. Third, Our model is on par with state-of-art algorithms but with having fewer parameters and faster speed.

2 Related Work

There are two different types of methods in text detection: regression-based methods and segmentation-based methods.

Regression-based methods directly get the bounding boxes from the text instances. [2] was the first method to use the anchor mechanism to solve the problem of text detection. To better adapt to text detection, [6] adjusted the size of the anchor. [9] further improved the representation of bounding boxes to predict multi-oriented Text. [4] suppressed the background by introducing attention, which makes the regression effect better. In addition, some methods were not based on anchor. [3] made regression at the pixel level. [5] used the dimension decomposition region proposal network instead of the traditional RPN to match the object width and height independently.

Regression-based methods usually require simple post-processing operations, so they are often speedy. Nevertheless, in recent years, the task has become to detect arbitrary shape text. The curved text is a representative type that these methods cannot regress the accurate bounding boxes.

Segmentation-based methods predicte at the pixel level and then generate bounding boxes by some post-processing operations. [10] used the method of instance segmentation to get the text line region and relationship and then finds the corresponding text line's external rectangle. [12] achieved a good detection effect for curved text by the global text segmentation branch and reduces the false-positive rate by text context module and re-score mechanism. [13] used a lighter backbone based on [7] to improve speed and designed Feature Pyramid Enhancement Module (FPEM) to introduce different scale features. [1] simplified the process of segmentation by proposing differentiable binarization modules.

Segmentation-based methods can get more accurate results than the regression-based method because of pixel-level prediction. It can be more suitable for the detection of arbitrary shape text. However, complex post-processing steps are often needed after the segmentation, such as pixel aggregation [13], or cluster processing [14]. These operations are often very time-consuming, resulting in slow inference speed.

Efficient text detection methods pay more attention to the inference speed, the number of parameters and simultaneously achieve higher accuracy. Many methods [6,9] aimed at regular text and achieved some results in inference speed by trying various network structures. Based on these methods, [1] proposed an end-to-end trainable differentiable binarization method, which further improves the speed and can well process the text of any shape.

However, although these methods adopt different structures to improve the reasoning speed on every dataset, the number of model parameters is ignored. As shown in Fig. 1, The number of parameters for these methods is more than 10M, and some approaches have even 20M, limiting their applications in different platforms. To solve this problem, we propose the ShallowNet, which reduces the depth of the network with a very small number of parameters and still achieves good results.

3   Proposed Method

The overall architecture of the proposed ShallowNet is shown in Fig. 2. The whole model consists of three parts: DCM (Dilated Convolution Module), FEM (Feature Enhancement Module), and the post-processing operations based on instance count-aware supervision information. DCM consists of five blocks, and each block includes three convolution layers. Then, the features processed by DCM are fed into FEM to extract more semantic information and concatenated to produce fused feature F. Afterward, the segmentation map (SM) and the border-area map (BM) are predicted according to F. The bounding box map

Table 1. Comparison of different network structures. 'SD', 'MD', 'LD' means different rates of dilation, and set as (1,2,3),(1,3,4),(2,3,5), respectively. 'R0', 'R1', 'R2' represent that we use 0, 1, and 2 regular convolution layers. 'F1', 'F2', 'F3' indicates that how many layers of de-convolution are used in FEM. The 'SC' showed that shortcut architecture is used in FEM. 'ICAS' means we use instance count-aware supervision information.

Structures	ICDAR 2015			Total-text		
	R	P	F	R	P	F
ShallowNet(SD-R0-F2-ICAS)	74.0	83.4	78.4	72.7	84.5	78.1
ShallowNet(MD-R0-F2-ICAS)	71.8	80.8	76.0	71.6	83.6	77.2
ShallowNet(LD-R0-F2-ICAS)	66.9	81.0	73.3	70.7	84.3	76.9
ShallowNet(SD-R1-F2-ICAS)	75.6	85.0	80.1	72.4	83.4	77.6
ShallowNet(SD-R2-F2-ICAS)	76.1	**86.6**	81.0	75.0	**85.6**	80.1
ShallowNet(SD-R3-F2-ICAS)	76.2	85.9	80.8	72.8	84.1	78.0
ShallowNet(SD-R2-F3-ICAS)	75.8	84.3	79.8	74.9	82.4	78.5
ShallowNet(SD-R2-F1-ICAS)	76.0	85.8	80.6	73.7	85.1	79.0
ShallowNet(SD-R2-F2-ICAS-SC)	76.9	85.9	**81.2**	80.6	84.8	**82.6**
ShallowNet(SD-R2-F2-SC)	**78.4**	82.6	80.4	75.2	84.5	79.6

(BBM) is obtained after the binarization of SM with BM. During training, the results from SM and BBM are used as instance count-aware supervision. For the label generation, we adopt the same method as [1]. In the inference period, the bounding boxes can be obtained directly from the SM.

3.1 Network Architecture

Dilated Convolution Module (DCM). Generally, deeper network architectures lead to higher model performance. Therefore, in the OCR domain, ResNet50 is widely used for higher accuracy. However, the deeper networks come at the cost of a huge amount of parameters and slow the inference speed. Moreover, numerous upsampling and downsampling operations in deep networks degrades the precision of detection [17].

To overcome the above issues, we construct a DCM module without sampling operations. The DCM consists of five blocks, and each block contains three convolution layers, as shown in Fig. 2. The first two blocks use regular convolution to preserve the detailed features. The sizes of the convolution kernel are 3×3, and the dilation rate is all 1. The rest of the blocks use dilated convolution to enlarge the receptive field. The size of each convolution kernel is also 3×3, but the dilation rates are 1, 2, and 3, respectively. Moreover, to improve the training process, we add a shortcut structure to each block, similar to ResNet [15]. It is to be noted that sampling operations are almost not utilized, which means that the feature maps have the same sizes after the input image is downsampled.

Table 2. Comparison with MobileNetV3 and ResNet on ICDAR2015. Two different sizes are used for inference. '736' and '1152' means the size of image is 736×1280 and 1152×2048, respectively.

Method	ICDAR2015				Params
	R	P	F	FPS	
MobileNetV3	83.1	57.7	68.1	**54**	1.7M
ResNet-18	77.1	85.3	81.0	31	14M
ResNet-50	79.6	**86.1**	82.7	25	28M
ShallowNet(736)	76.9	85.9	81.2	47	1.1M
ShallowNet(1152)	**81.4**	85.7	**83.5**	33	**1.1M**

Feature Enhancement Module (FEM). To combine low-resolution with high-resolution and enhance semantic information [18], we utilize a top-down architecture called FEM. FEM has four blocks, and each block is composed of three de-convolution layers. Corresponding to DCM, the first two blocks adopt a dilated de-convolution kernel, while the remaining blocks adopt regular de-convolution. The dilation rate is set 1 for the first two blocks and 3,2,1 for the remaining blocks. These blocks also use the shortcut structure.

3.2 Optimization

Loss Function. To train the ShallowNet, we construct a weighted loss function L, which contains the segmentation loss L_p, the bounding box regression loss L_b, and the border-area regression loss L_t:

$$L = \alpha L_t + w_p L_p + w_b L_b \tag{1}$$

where α is empirically set to 10. The values of w_p and w_b are calculated dynamically according to the predicted results during the training.

We apply a binary cross-entropy (BCE) loss for L_p. To overcome the unbalance problem of text segmentation, hard negative mining is used in the BCE loss by sampling the hard negatives:

$$L_p = \frac{1}{n} \left(\sum_{i \in S_l} (y_i \log(x_i) + (1 - y_i) \log(1 - x_i)) \right) \tag{2}$$

where y_i and x_i are ground-truth and predicted values, respectively; S_l is the sampled set where the ratio of positives and negatives is 1 : 3; n represents the total number of pixels in S_l.

The border-area loss L_t is computed as the sum of L_1 distances between the predictions and labels:

$$L_t = \sum_{i \in R_d} |y_i - x_i| \tag{3}$$

Table 3. Experiment on instance count-aware supervision. ShallowNet(W) means training without the instance count-aware supervision information.

Method	ICDAR 2015			Total-text		
	R	P	F	R	P	F
ShallowNet(W)	**78.4**	82.6	80.4	78.9	84.2	81.4
ShallowNet	76.9	**85.9**	**81.2**	**80.6**	**84.8**	**82.6**

where R_d is a set of indexes of the pixels inside the ground-truth; y_i is the label for the border-area map; x_i is the predicted value of the corresponding position.

For the bounding box loss L_b, we use dice loss since the probability values in the bounding box map are very close to 0 and 1.

$$L_b = 1 - \frac{2\,|Y \cap X|}{|Y| + |X|} \tag{4}$$

where Y is the feature map generated from ground truth, and X is the predicted probability value.

Instance Count-Aware Supervision Information. For text detection, the locations and number of text boxes are two main factors to affect the prediction results. In our numerical investigation, we found that the predicted text boxes were almost always correct. That is, we should pay more attention to the prediction of the number of text boxes. Therefore, we propose a kind of supervision information based on instance counting. The difference between the number of predicted bounding boxes and ground-truth is utilized as the weights on L_t and L_b, which is calculated as follows:

$$w = \min(\frac{2 \times |g - p|}{g + \delta} + 1, W) \tag{5}$$

where g and p represent the number of text boxes in ground-truths and the predicted boxes per image, respectively; δ is a tiny number, say $1e-6$, to avoid having a zero in the denominator. The W is a constant(3), which prevents the calculated weight value from being too large.

To be specific, for the probability map, we first filter through a threshold (0.3) to obtain the corresponding connected regions and then calculate the average probability value of all pixel points on the connected region. If it is greater than 0.5, it is considered a valid prediction box. We believe that this method of calculation avoids the prediction of broken text boxes. For the binary map, we use the threshold (0.6) for binarization, and the result is directly used to count the prediction boxes. This calculation method can help us to separate the text better close to each other.

In the inference period, considering that the model has been sufficiently trained and to speed up its operation, we no longer generate the threshold map

Fig. 3. Qualitative results on multi-oriented text, curved text, and text-line. The red line represents the ground-truth, and the green line is the result we generated. (Color figure online)

but directly use the probability map to generate the predicted box. The box process involves three steps: (1) a constant threshold (0.3) is used to binary the probability map to get the binary map; (2) the connected regions are obtained from the binary map; (3) the regions are dilated to get the final detection results.

4   Experiments

4.1   Datasets

In the experiment, we use four datasets, Syntext150K [19], ICDAR2015 [20], Total-Text [21], and CTW1500 [22]. Syntext150K is used for pre-train only. ICDAR 2015 consists of 1,500 images with lots of horizontal and slanted texts. Total-Text includes the text of various shapes, especially curved text. CTW1500 is also made up of curved text, but the text instances are annotated by fourteen coordinate points in the text-line level.

4.2   Implementation Details

For all the models, we first pre-train them with the SynText150k for 10 epochs. Then, we train the models on the corresponding real-world datasets for 600 epochs. We follow a "poly" learning rate policy where the learning rate at the current iteration equals the initial learning rate multiplying $(1 - \frac{power}{max\_epoch})^{power}$. According to [1], the initial learning rate and the power are set to 0.007 and 0.9, respectively. We use a weight decay of 0.001 and a momentum of 0.9.

Data augmentation methods include (1) Random rotation with an angle range of $(-10°, 10°)$; (2) Random cropping; (3) Random Flipping. All the images are adjusted to 640×640 during the training.

Table 4. Detection results on Total-Text, CTW1500. The results with '*' indicates we inference with official code in our platform. '+' means the parameters are inferred by the backbone. The ones with no data are represented as '−'. The 640 and 900 in our method refer to inference with different image sizes.

Method	Total-Text				CTW500				Params
	R	P	F	FPS	R	P	F	FPS	
TextSnake(2018)	74.5	82.7	78.4	−	85.3	67.9	75.6	1.1	19.1M
PSE-1s(2018)	84.8	79.7	82.2	3.9	84.8	79.7	82.2	3.9	28.6M
PAN-320(2019)	71.3	84.0	77.1	**82.4**	72.6	82.2	77.1	84.2	14M+
PAN-640(2019)	79.4	**88.0**	**83.5**	39.6	77.7	**84.6**	81.0	39.8	14M+
TextField(2019)	79.9	81.2	80.6	−	79.8	83.0	81.4	−	17M+
SAE(2019)	−	−	−	−	77.8	82.7	80.1	3	27.8M
MASK TTD(2020)	74.5	79.1	76.7	−	79.0	79.7	79.4	−	28M+
DB-Res18(2020)	**86.8**	78.4	82.3	33*	**88.3**	77.9	**82.8**	37*	14M*
Ours(640)	76.7	81.8	79.2	72	75.9	81.3	78.5	**93**	**1.1M**
Ours(900)	80.6	84.8	82.6	48	80.3	83.2	81.7	65	**1.1M**

4.3   Ablation Study

Network Architecture. We firstly demonstrate the function of dilation convolution in ShallowNet. The proposed ShallowNet containing fewer layers encounters small receptive fields, which degrades the detection performance of large-size texts. Thus, we enlarge the receptive fields of features by introducing dilated convolution. As shown in Table 1, where 'R', 'P' and 'F' are abbreviations of 'Recall', 'Precision' and 'F1-score', respectively. It is observed that too large dilation rates and too many dilated convolution layers reduce the detection accuracy. This is because the valuable features of the text are more sparse than those of the generic target, while the dilated convolution does not use all the pixels to calculate and lose several features. The experimental results in Table 1 show that the dilation rates of each block are set to 1, 2, and 3, and the use of the dilated convolution only in the last three blocks of DCM achieves better results.

For FEM, we first explore the use of dilated de-convolution to extract semantic features. Table 1 shows that the best result is achieved by using the dilated de-convolution layer in the last two blocks of FEM. Then, we verify that adding a shortcut structure in FEM is effective, also shown in Table 1. Same experimental results are obtained on Total-text. To sum up, the final FEM is designed as follows: the first two blocks use conventional de-convolution, and the second two blocks use dilated de-convolution, with convolution rates of 3, 2, and 1, respectively.

To show the effectiveness and efficiency of ShallowNet, we compare it with ResNet18 and ResNet50 on ICDAR2015. Table 2 indicates that with a small input size 736 × 1280, ShallowNet achieves comparable results with ResNet50 and outperforms ResNet18, with large input size 1152×2048, ShallowNet exceeds

ResNet50. What's more, the number of parameters for ResNet18 and ResNet50 is 14M and 28M, respectively, far more than ShallowNet of 1.1M. Due to fewer parameters, the inference speed of ShallowNet is much faster than those of ResNet18 and ResNet50, even with the large size input. We also compared MobileNetv3, another well-known lightweight network, which is faster than ours but far less accurate than our model.

Instance Count-Aware Supervision. In Table 3, we conduct an ablation study on the ICDAR2015 and Total-Text to show the effectiveness of our proposed instance count-aware supervision information. The supervision information represents the quantity matching relationship between the predicted boxes and the number of text boxes in ground-truths. It is attached to the loss function as a dynamic weight. The weight focuses on the continuity of features and forces the network to learn better at the instance level. The weight can bring about 0.8% and 1.2% for ICDAR 2015 and Total-text, respectively.

Table 5. Detection results ICDAR2015. The results with '*' indicates we inference by our platform. '+' means the parameters are inferred by the backbone. The ones with no data are represented as '−'.

Method	ICDAR2015				Params
	R	P	F	FPS	
TextSnake(2018)	80.4	84.9	82.6	1.1	19.1M
PSE-1s(1)(2018)	84.5	86.9	**85.7**	1.6	28.6M
PAN(2019)	81.9	84.0	82.9	26.1	14M+
TextField(2019)	80.5	84.3	82.4	5.5	17M+
SAE(2019)	**84.5**	85.1	84.8	3	27.8M
DB-res18(2020)	78.4	**86.8**	82.3	31*	14M*
Ours	81.4	85.7	83.5	**33**	**1.1M**

4.4   Comparisons with Previous Methods

We compare our method with previous methods on three standard benchmarks, including two benchmarks for curved text, one benchmark for multi-oriented text. Our method achieves faster speed with only 1.1M parameters and also performs well on accuracy.

As shown in Table 5, our method achieves higher FPS than the previous real-time method [1] and outperforms it by 1% on the ICDAR2015. We also demonstrate the generalization of our method on two curved text benchmarks (Total-Text and CTW1500). According to Table 4, although the speed of our method only ranks second, the fastest method is at the cost of accuracy, and its f1-score is 4% - 5% lower than ours. As we reduced the size of the image, our speed increased. We tried to increase the speed as much as possible while ensuring accuracy. Some qualitative results are visualized in Fig. 3.

5 Conclusion

In this paper, we have proposed a novel framework (ShallowNet) to detect text of any shape efficiently. Besides, we introduce instance count-aware supervision information to improve the performance of the model. Experimental results on multiple datasets demonstrate the performance of our model. We achieve competitive performance on three datasets with the fewest parameters and faster speed.

References

1. Liao, M., Wan, Z., Yao, C., Chen, K., Bai, X.: Real-time scene text detection with differentiable binarization. In: Proceedings of AAAI, Artificial Intelligence, pp. 11474–11481 (2020)
2. Tian, Z., Huang, W., He, T., He, P., Qiao, Y.: Detecting text in natural image with connectionist text proposal network. In: Leibe, B., Matas, J., Sebe, N., Welling, M. (eds.) ECCV 2016. LNCS, vol. 9912, pp. 56–72. Springer, Cham (2016). https://doi.org/10.1007/978-3-319-46484-8_4
3. Zhou, X., et al.: EAST: an efficient and accurate scene text detector. In: Proceedings of the IEEE Conference on Computer Vision and Pattern Recognition, pp. 2642–2651 (2017)
4. He, P., Huang, W., He, T., Zhu, Q., Qiao, Y., Li, X.: Single shot text detector with regional attention. In: Proceedings of the IEEE International Conference on Computer Vision (ICCV), pp. 3066–3074 (2017)
5. Xie, L., Liu, Y., Jin, L., Xie, Z.: DeRPN: taking a further step toward more general object detection. In: Proceedings of AAAI, Artificial Intelligence, pp. 9046–9053 (2019)
6. Liao, M., Shi, B., Bai, X., Wang, X., Liu, W.: Textboxes: a fast text detector with a single deep neural network. In: Proceedings of AAAI, Artificial Intelligence, pp. 4161–4167 (2017)
7. Wang, W., et al.: Shape robust text detection with progressive scale expansion network. In: Proceedings of CVPR, Computer Vision and Pattern Recognition, pp. 9336–9345 (2019)
8. Wang, H., et al.: All you need is boundary: toward arbitrary-shaped text spotting (2019). http://arxiv.org/abs/1911.09550
9. Liao, M., Shi, B., Bai, X.: TextBoxes++: a single-shot oriented scene text detector. IEEE Trans. Image Process **27**(8), 3676–3690 (2018)
10. Deng, D., Liu, H., Li, X., Cai, D.: Pixellink: detecting scene text via instance segmentation. In: Proceedings of AAAI, Artificial Intelligence, pp. 6773–6780 (2018)
11. Xu, Y., Wang, Y., Zhou, W., Wang, Y., Yang, Z., Bai, X.: Textfield: learning a deep direction field for irregular scene text detection. IEEE Trans. Image Process. **28**(11), 5566–5579 (2019)
12. Xie, E., Zang, Y., Shao, S., Yu, G., Yao, C., Li, G.: Scene text detection with supervised pyramid context network. In: Proceedings of AAAI, Artificial Intelligence, pp. 9038–9045 (2019)
13. Wang, W., et al.: Efficient and accurate arbitrary-shaped text detection with pixel aggregation network. In: Proceedings of ICCV Conference on Computer Vision, pp. 8439–8448 (2019)

14. Tian, Z., et al.: Learning shape-aware embedding for scene text detection. In: Proceedings of CVPR, Computer Vision, pp. 4234–4243 (2019)

15. He, K., Zhang, X., Ren, S., Sun, J.: Deep residual learning for image recognition. In: Proceedings of CVPR, Computer Vision, pp. 770–778 (2016)

16. Liu, Y., Jin, L., Fang, C.: Arbitrarily shaped scene text detection with a mask tightness text detector. IEEE Trans. Image Process. **29**, 2918–2930 (2020)

17. Chen, L.-C., Papandreou, G., Schroff, F., Adam, H.: Rethinking atrous convolution for semantic image segmentation (2017). https://doi.org/10.1109/TIP.2019.2900589

18. Lin, T., Dollár, P., Girshick, R.B., He, K., Hariharan, B., Belongie, S.J.: Feature pyramid networks for object detection. In: Proceedings of CVPR, Computer Vision, pp. 936–944 (2017)

19. Liu, Y., Chen, H., Shen, C., He, T., Jin, L., Wang, L.: ABCNet: real-time scene text spotting with adaptive Bezier-curve network. In: Proceedings of CVPR, Computer Vision, pp. 9806–9815 (2020)

20. Karatzas, D., et al.: ICDAR 2015 competition on robust reading. In: Proceedings of ICDAR, Computer Society, pp. 1156–1160 (2015)

21. Chng, C.K., Chan, C.S.: Total-text: a comprehensive dataset for scene text detection and recognition. In: Proceedings of ICDAR, Computer Society, pp. 935–942 (2017)

22. Liu, Y., Jin, L., Zhang, S., Luo, C., Zhang, S.: Curved scene text detection via transverse and longitudinal sequence connection. Pattern Recognit. **90**, 337–345 (2019)

Image Periodization for Convolutional Neural Networks

Kailai Zhang[✉], Zheng Cao, and Ji Wu

Department of Electronic Engineering, Tsinghua University, Beijing, China
zhangkl17@mails.tsinghua.edu.cn

Abstract. In last few years, convolutional neural networks (CNNs) have achieved great success on many image classification tasks, which shows the effectiveness of deep learning methods. The training process of CNN is usually based on a large number of training data, and the CNN can learn local image translation invariance property by using the convolution operation and pooling operation, which is very important for image classification. However, the performance of CNN is limited when the location of the key object varies greatly in the image or the number of data is insufficient. Addressing this problem, in this paper, we propose a novel method named image periodization for CNN on image classification tasks. We extend the original image periodically and resample it to generate new images, while we design a circular convolutional layer to replace the original convolutional layer. Our method can be used as a data augmentation method, and it can provide complete translation invariance property for CNNs. Our method can be easily plugged into common CNNs, and the experiment results show consistent improvement on different CNN-based models.

Keywords: Image periodization · Circular convolution · Convolutional neural network

1 Introduction

Image classification is an important task in computer vision field. Given an input image, image classification aims to give the image a label from the pre-defined label set. There are a large number of classical works [14–17] for image classification tasks. In recent years, deep learning methods especially the convolutional neural networks (CNNs), have been developed rapidly, and there are a large number of works for image classification tasks based on CNN methods. Some famous CNN-based models such as VGG [1], Resnet [2,3], Resnext [4] and Densenet [5] have been proposed, and they show great success in many image classification tasks [6,7]. These models are usually trained based on a large number of training data. In most CNNs, the convolutional layers and pooling layers are the main components. The convolutional layer can extract features while the pooling layers can select features from feature maps and reduce the size of feature map. By stacking these layers, the CNN can get different level features and achieve great performance on final classification.

© Springer Nature Switzerland AG 2021
T. Mantoro et al. (Eds.): ICONIP 2021, LNCS 13108, pp. 645–656, 2021.
https://doi.org/10.1007/978-3-030-92185-9_53

In CNN-based models, the image translation invariance property, which means that the classification result of network should be the same as the input image shifts, is very important for classification. An example is shown in Fig. 1. By using convolutional layers and pooling layers, the CNN can learn the local image translation invariance property. For instance, when the key object moves a little in the input image, the max-pooling layers can still find the same maximum value in a small range by using the sliding window, and output the same feature map for latter layers, which can provide the local translation invariance property. However, when the location of the key object varies greatly in the image, the translation invariance property can not be learned sufficiently. This case happens especially when the number of training data is insufficient. Addressing this problem, to provide CNN with complete translation invariance property, we propose a novel method named image periodization, which is based on a data augmentation method with convolutional layer design. There are a lot of data augmentation methods for data pre-processing. The rotation, flip and crop are the basic operations which are commonly used as a combination in image classification tasks. While these basic operations achieve better performance in some tasks, they may lose key features by cropping the image, or decrease the resolution of images by rotating the image. Although the flip operation will not lose the information in images, it can only provide two variants by horizontal flip or vertical flip. Circular shift [11] is also proposed as a basic operation, which divides the image to pieces to generate new images. Besides, some advanced methods are proposed to utilize the combination of different images. Cutout [8] uses a randomly generated mask image to cover the original image. Mixup [9] directly combines two images and their labels in a proportion to generate the synthetic image. Cutmix [10] replaces a part of the original image with the relevant part of another randomly selected image. The Generative Adversarial Networks (GAN) [12,13] can also generate new data by sending a random noise to the network, and the network can output a generated image. The performance of these methods depends on the quality of datasets. Some of these data augmentation methods can also be regarded as network regularization methods, which can provide their own property for the network.

For our proposed image periodization method, it has two components: on the one hand, for each training epoch, we extend the original image periodically in a plane, and randomly resample a new image from the plane with the same size, which can generate a large number of images and can be used as a data augmentation method. It can be regarded as a basic operation since this operation is only based on a single original image. On the other hand, we design the circular convolutional layer to replace the standard convolutional layer. In convolutional layers, the padding operation is used to extend the margin of feature map, which can keep the size of output feature map after the convolution. Usually zero value is used in the padding operation, which actually has no meaning for CNNs. Addressing this case, our circular convolutional layer can change the padding value according to the input image so that the padding value can also provide information for CNNs. The combination of image exten-

(a)   (b)

Fig. 1. An example of image translation. The location of the key object changes in the image as the image shifts. The network should output the same classification results for these two images.

sion and resampling, and circular convolution can provide complete translation invariance property for CNNs. In our experiments, three datasets are used for evaluation, and our proposed method achieves consistent improvement on four commonly used CNN-based models. The visualized results of CNN also show that the CNN can indeed learn the translation invariance property based on our image periodization method.

The main contributions of our method[1] are: (1) We propose a novel data augmentation operation by using periodical image extension and resampling for CNNs. (2) We design a novel circular convolutional layer by revising the padding operation, which can provide complete translation invariance property by combining with image extension and resampling, and it can be easily plugged into common CNN-based models. (3) The experiment results show consistent improvement on different CNN-based models, and the visualized results show the effectiveness of our method.

2 Methods

2.1 Image Extension and Resampling

For image classification based on CNN methods, the input image $\mathbf{x}$ is sent to the network, and it outputs a n-dimensional probability vector $\mathbf{p}$ where n is the number of classes. During network training, the groundtruth label $\mathbf{y}$ is required. It is a n-dimensional one-hot vector, and the commonly used cross entropy loss is defined as

$$loss = -\mathbf{y} \cdot \log(\mathbf{p}) \tag{1}$$

[1] This work is sponsored by the National Natural Science Foundation of China (Grant No.61571266), Beijing Municipal Natural Science Foundation (No. L192026), and Tsinghua-Foshan Innovation Special Fund (TFISF) (No. 2020THFS0111).

Fig. 2. An example of our image periodization. In the left figure, the image is extended periodically and a random patch (red bounding box) with the same size is chosen. In the right figure, during convolution, we use the pixel in the extended feature map (assume the image is a feature map) as the padding value (outside the red bounding box). (Color figure online)

The back propagation algorithm is applied to update the network parameters according to the loss function. On the test stage, the CNN outputs $\mathbf{p}$ according to the input image $\mathbf{x}$, so that we can get the final classification result by choosing the class index that has the maximum value in $\mathbf{p}$.

On the training stage, our image periodization method has two steps. In the first step of our image periodization, we try to generate more images based on the original image. We consider the input image $\mathbf{x}$ which has width w and height h, so that $x_{i,j}(0 \leqslant i < w, 0 \leqslant j < h)$ represents a pixel of $\mathbf{x}$. We extend $\mathbf{x}$ periodically so that it does not have width and height explicitly, and we have $x_{i,j} = x_{i-w,j}$ and $x_{i,j} = x_{i,j-h}$. After the periodical extension, we randomly resample a patch $\mathbf{x}^{\mathbf{new}}$ which has the same size with the original image. In the resampled image, $x_{i,j}(t \leqslant i < t + w, s \leqslant j < s + h)$ represents a pixel of $\mathbf{x}^{\mathbf{new}}$ where t and s are randomly chosen. An example of the whole process is shown in Fig. 2. According to this operation, we can generate different variants from the original image in each training epoch. It is worth mentioning that all the pixels in the original image still remain in the generated image, while their absolute coordinates have been changed, so that the location of relevant feature also changes. In our methods, $w * h$ variations of the original image can be generated, which can provide sufficient variants for network training, so that this operation can be used as a data augmentation method.

2.2 Circular Convolution

By using the image extension and resampling, we can get a lot of generated images. However, this operation also breaks the space relation of some pixels, so that the key feature may be divided to several parts. To handle this situation, in the second step of image periodization, we consider the convolutional layer itself. For an input feature map $\mathbf{F}$ with m channels, the convolution operation is usually defined as

$$G_{k,l,n} = \sum_{i,j,m} K_{i,j,m,n} F_{k+i-1,l+j-1,m} \qquad (2)$$

where n is the number of channels of output feature map $\mathbf{G}$, and k, l is the relevant pixel in $\mathbf{G}$. i, j, m will traverse the width, height and channel of convolutional kernel $\mathbf{K}$. In convolutional layers, the convolution kernel scans the original feature map to generate new feature map, and the padding operation is commonly applied in the margin of original feature map, which is used to keep the size of feature map after convolution. The zero value is usually used as the padding value. When the size of feature map reduces, the proportion of padding value increases, especially in the feature map output by the last several convolutional layers. However, this value has no meaning and may influence the network training. Addressing this problem, we propose the circular convolution to replace the standard convolution, and thus design the circular convolutional layer. Instead of using zero value, we consider pixels in the extended feature map, which is shown in Fig. 2. That is, for padding operation, we also extend the feature map periodically, and the neighbour pixels outside the original feature map are chosen as the padding value if needed, so that the padding value is also relevant to the feature map, which can provide more information for the network. Assuming a feature map $\mathbf{F}$ has width w and height h, the padding operation in our circular convolutional layer can be described as

$$F_{0:w-1,-j} = F_{0:w-1,h-j} \quad F_{0:w-1,h+j-1} = F_{0:w-1,j-1} \qquad (3)$$

$$F_{-i,0:h-1} = F_{w-i,0:h-1} \quad F_{w+i-1,0:h-1} = F_{i-1,0:h-1} \qquad (4)$$

where $i, j > 0$ and $i, j < padding\ range$, and the : means the range of index. For example, during the circular convolution, we use the value of the right margin of feature map as the padding value outside the left margin of feature map. We ignore the channel index in the above formulas, but we use this operation for each channel of the feature map.

The image extension and resampling, and circular convolution can be applied simultaneously for CNN. It is worth mentioning that we can provide the network with complete translation invariance property by combining these two steps. For an input image, the image extension and resampling can generate a large number of variations. While all the pixels are still remained in the generated image, the absolute coordinate of each pixel varies in different generated images. However, the circular convolution can keep the relative coordinate between pixels. In all the circular convolutional layers, the pixels in the feature maps still use the same neighbour for convolution because of our padding operation, which is just the same as the original image. This property forces the network to learn the feature of image when the absolute location of pixels varies randomly but the information between neighbour pixels does not lose, so that the CNN can learn complete translation invariance property in the global range.

In summary, our image periodization includes two steps: extend the original image and resample it, and use the circular convolutional layer to replace the standard convolutional layer. We use both of them during network training. On test stage, only circular convolutional layer is remained in CNN, and the original image is input to the network to get the classification result.

3 Experiments

3.1 Datasets Description

For evaluation, three commonly used image datasets are chosen for the evaluation of our method.

CIFAR: The CIFAR10 and CIFAR100 dataset, both of which are available in [18], consist of 50000 training images and 10000 test images with 32×32 pixels. CIFAR10 has 10 classes and CIFAR100 has 100 classes. For the two datasets, the number of images in each class are equal in both training set and test set.

CALTECH101: This dataset consists 101 classes, which is available in [19, 20]. It has 9145 images in total. Most of classes have about 50 images. The width and length of most images is not fixed, which are commonly between 200 and 300 pixels.

Table 1. The classification results on CIFAR10 and CIFAR100 dataset by using different CNN-based models. O means the original model. R means the image extension and resampling. C means the circular convolution.

Methods		Top1 error (CIFAR10)	Top1 error (CIFAR100)	Top5 error (CIFAR100)
VGG16	O	11.47 ± 0.26	36.64 ± 0.14	15.50 ± 0.24
	O+R	8.45 ± 0.27	29.63 ± 0.15	10.21 ± 0.16
	O+C	11.38 ± 0.23	36.22 ± 0.12	15.15 ± 0.24
	O+R+C	$\mathbf{7.92 \pm 0.21}$	$\mathbf{29.14 \pm 0.14}$	$\mathbf{9.47 \pm 0.25}$
Resnet50	O	10.85 ± 0.16	33.33 ± 0.17	10.77 ± 0.16
	O+R	5.76 ± 0.19	22.30 ± 0.09	5.41 ± 0.17
	O+C	10.66 ± 0.19	32.92 ± 0.22	10.58 ± 0.20
	O+R+C	$\mathbf{4.79 \pm 0.12}$	$\mathbf{22.04 \pm 0.10}$	$\mathbf{5.36 \pm 0.15}$
Resnext50	O	8.70 ± 0.18	29.38 ± 0.14	9.60 ± 0.24
	O+R	5.05 ± 0.11	21.60 ± 0.18	5.88 ± 0.18
	O+C	8.61 ± 0.19	29.14 ± 0.17	9.42 ± 0.24
	O+R+C	$\mathbf{5.03 \pm 0.14}$	$\mathbf{21.48 \pm 0.15}$	$\mathbf{5.61 \pm 0.12}$
Densenet121	O	8.21 ± 0.24	26.42 ± 0.24	7.40 ± 0.14
	O+R	5.17 ± 0.14	22.44 ± 0.28	$\mathbf{5.67 \pm 0.11}$
	O+C	7.98 ± 0.23	26.38 ± 0.25	7.35 ± 0.16
	O+R+C	$\mathbf{4.66 \pm 0.12}$	$\mathbf{21.31 \pm 0.24}$	5.74 ± 0.18

3.2   Parameters Setting

For experimental setting, the test environment of our experiments is running on ubuntu 14.04 with gpu 1080Ti, and the software we used is python with pytorch package. In all the experiments, we apply the stochastic gradient descent(SGD) optimizer with momentum. We set the initial learning rate and momentum to 0.1 and 0.9 respectively, with the learning rate decay of 0.2 for each 45 epoches. The number of training epoches is 180. To provide robust results, we use the average performance by repeating the training process five times. For CIFAR10 and CIFAR100 dataset, the batch size is set to 128, and the 32×32 images are input to CNNs. For CALTECH101 dataset, due to the large image size, the batch size is set to 32, and the images are randomly divided to the training set, validation set and test set, with the proportion of 7:1:2. All the images are resized to 224×224 pixels before input to the network.

3.3   Evaluation and Comparison

In our experiments, we consider some famous CNN structures as baselines. Four commonly used CNN-based models including VGG, Resnet, Resnext and Densenet are used for the evaluation. For evaluation metrics, we calculate the top1 error for CIFAR10 dataset, while we calculate both top1 error and top5 error for CIFAR100 and CALTECH101 dataset because of the large number of

Table 2. The classification results on CALTECH101 dataset. O means the original model. R means the image extension and resampling. C means the circular convolution.

Methods		Top1 error	Top5 error
VGG16	O	23.45 ± 0.25	10.65 ± 0.23
	O+R	13.93 ± 0.10	5.49 ± 0.31
	O+C	22.96 ± 0.36	10.02 ± 0.32
	O+R+C	$\mathbf{13.50 \pm 0.16}$	$\mathbf{5.01 \pm 0.26}$
Resnet50	O	16.50 ± 0.26	5.85 ± 0.16
	O+R	14.07 ± 0.42	4.98 ± 0.16
	O+C	15.98 ± 0.29	5.50 ± 0.27
	O+R+C	$\mathbf{13.70 \pm 0.27}$	$\mathbf{4.38 \pm 0.15}$
Resnext50	O	17.50 ± 0.16	7.70 ± 0.37
	O+R	11.86 ± 0.25	$\mathbf{4.64 \pm 0.21}$
	O+C	17.52 ± 0.20	7.63 ± 0.35
	O+R+C	$\mathbf{11.33 \pm 0.11}$	4.80 ± 0.42
Densenet121	O	14.60 ± 0.21	5.69 ± 0.18
	O+R	12.65 ± 0.28	4.06 ± 0.21
	O+C	14.30 ± 0.41	5.42 ± 0.18
	O+R+C	$\mathbf{11.23 \pm 0.31}$	$\mathbf{3.58 \pm 0.15}$

classes. We first consider the performance of original models, models with image extension and resampling, and models with the whole image periodization. The results for three datasets are shown in Table 1 and Table 2.

According to the results, on three datasets, we achieve consistent performance improvement for different CNN-based models by using image extension and resampling, which means that this data augmentation method can indeed provide variations for datasets with large images or small images. While the image extension and resampling can provide variations for performance improvement, by plugging the circular convolutional layers into the network, the performance can be further increased in most cases, because we keep the connection between pixels during convolution. The relative position of pixels stays the same, which can prevent key feature from dividing into different parts, so that the complete translation invariance property can be learned by the network. We also find that by using the circular convolutional layer alone without image extension and resampling, the performance shows little difference with the baseline, because the original images do not require this operation to keep the space structure. The results indicate that in our image periodization, the image extension and resampling makes the main contribution in the performance improvement, and the circular convolutional layer can be used as a supplement when the network can not learn sufficiently from the pieces of images.

For further evaluation, we also compare our image periodization with other data augmentation methods. We consider the combination of basic operation including crop, flip and rotation, which are commonly used for image classification tasks. Some advanced data augmentation methods such as mixup [9], cutmix [10] and circular shift [11] are also chosen for comparison. Since that our method is easy to plug into CNN-based models and it can also be combined with other methods, we also combine our proposed method with the listed methods

Table 3. The top1 error on three dataset by using Resnet50 as the baseline. O means the original model. I means our image periodization. C means the cutmix method. M means the mixup method. S means the circular shift method. H means the combination of flip, crop and rotation.

Methods	CIFAR10	CIFAR100	CALTECH101
O	10.85 ± 0.16	33.33 ± 0.17	16.50 ± 0.26
O+S	8.59 ± 0.14	26.60 ± 0.19	14.23 ± 0.48
O+H	7.88 ± 0.19	25.07 ± 0.24	16.96 ± 0.31
O+M	4.48 ± 0.17	22.17 ± 0.33	14.47 ± 0.25
O+C	4.12 ± 0.15	21.61 ± 0.26	13.43 ± 0.32
O+I	4.79 ± 0.12	22.04 ± 0.10	13.70 ± 0.27
O+H+I	4.36 ± 0.18	22.10 ± 0.23	13.18 ± 0.19
O+M+I	4.01 ± 0.22	21.89 ± 0.27	13.14 ± 0.22
O+C+I	$\mathbf{3.73 \pm 0.12}$	$\mathbf{21.08 \pm 0.31}$	$\mathbf{12.95 \pm 0.26}$

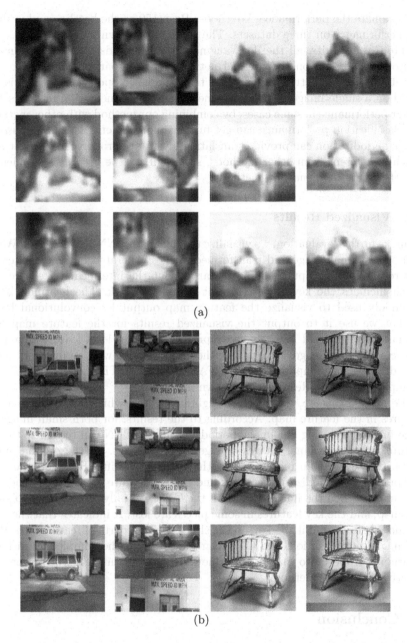

Fig. 3. Some visualized results from CIFAR10 dataset(a) and CALTECH101 dataset(b). The images in first column and third column of the first row are original images while the images in second column and forth column are transformed images. The second row and third row are the relevant heatmaps output by CNN-based models without image periodization and with image periodization respectively. All the images has been resized for display. In these heatmaps, the network will lighten the key areas.

and evaluate the performance. We choose Resnet50 as the baseline and evaluate the performance on three datasets. The results are shown in Table 3.

From the results, all the data augmentation methods can get performance improvement based on the original network. By comparison, our image periodization achieves better performance than the basic operations, which is also based on a single image. Although some advanced augmentation methods have better performance in some cases, by combining our method with other methods, the classification performance can get further improvement. This is because our image periodization can provide translation invariance property for the network, which is different from other methods, so that it can be used as a supplement for other data augmentation methods.

3.4   Visualized Results

To make further evaluation, we visualize the output of CNN-based model. We use both the original image and transformed images generated by image extension and resampling as input. The CNN can output the intermediate feature map which includes the key features learnt by the network. The heatmap [22] is commonly used to visualize the feature map output by convolutional layers, so that we use it to output the visualized results for the feature map from the last convolutional layer. It can represent the high level feature that CNN extracts from the image. We evaluate the model trained with and without image periodization for comparison. For CIFAR10 and CALTECH101 dataset, some examples output by Resnet50 are shown in Fig. 3.

From the heatmap, the CNN will lighten the area which it recognizes as key features in the feature map. According to the results, for both small images or large images, by using our image periodization, the CNN can recognize the key feature for both original images and generated images. As the original image shifts, the key feature recognized by the CNN also shifts. However, the CNN without image periodization can not find the key feature accurately in the original image. When the image is shifted, the CNN can only output an ambiguous heatmap, and the key features in some pieces are lost. It indicates that the CNN can not perform well if the key object shifts or some parts of feature lose. With our image periodization, the CNN can indeed learn the complete translation invariance property, so that it can achieve better performance, which shows the effectiveness of our method.

4   Conclusion

In this paper, we propose a novel method named image periodization for CNN-based models on image classification tasks. The translation invariance property is very important for image classification. To provide CNN with complete translation invariance property, we first apply the image extension and resampling as a data augmentation method, then we use the circular convolutional layer to replace the standard convolutional layer, which can generate a large number of variants and keep the relative position for the network. Our method can be easily plugged into different CNN-based models. In experiments, we achieve consistent performance improvement on three datasets for different CNN-based models, and the visualized results show that the CNN with our image periodization can indeed learn the complete translation invariance property, which proves the effectiveness of our method. In the future work, we will make further improvement based on our current method.

References

1. Simonyan, K., Zisserman, A.: Very deep convolutional networks for large-scale image recognition. Computer Science (2014)
2. He, K., et al.: Deep residual learning for image recognition. In: 2016 IEEE Conference on Computer Vision and Pattern Recognition (CVPR). IEEE Computer Society (2016)
3. Mahmood, A., Bennamoun, M., An, S., et al.: ResFeats: residual network based features for image classification (2016)
4. Xie, S., Girshick, R., Dollár, P., et al.: Aggregated residual transformations for deep neural networks. In: 2017 IEEE Conference on Computer Vision and Pattern Recognition (CVPR). IEEE (2017)
5. Huang, G., et al.: Densely connected convolutional networks. In: Proceedings of the IEEE Conference on Computer Vision and Pattern Recognition (2017)
6. Deng, J., Dong, W., Socher, R., et al.: ImageNet: a large-scale hierarchical image database. In: IEEE Conference on Computer Vision and Pattern Recognition (2009)
7. Russakovsky, O., Deng, J., Su, H., et al.: ImageNet large scale visual recognition challenge. Int. J. Comput. Vision **115**(3), 211–252 (2015). https://doi.org/10.1007/s11263-015-0816-y
8. DeVries, T., Taylor, G.W.: Improved regularization of convolutional neural networks with cutout. arXiv preprint arXiv:1708.04552 (2017)
9. Zhang, H., et al.: mixup: beyond empirical risk minimization. arXiv preprint arXiv:1710.09412 (2017)
10. Yun, S., et al.: Cutmix: regularization strategy to train strong classifiers with localizable features. In: Proceedings of the IEEE International Conference on Computer Vision (2019)
11. Zhang, K., Cao, Z., Wu, J.: Circular shift: an effective data augmentation method for convolutional neural network on image classification. In: 2020 IEEE International Conference on Image Processing (ICIP), pp. 1676–1680. IEEE (2020)
12. Goodfellow, I., et al.: Generative adversarial nets. In: Advances in Neural Information Processing Systems (2014)
13. Radford, A., Metz, L., Chintala, S.: Unsupervised representation learning with deep convolutional generative adversarial networks. arXiv preprint arXiv:1511.06434 (2015)
14. Bosch, A., Zisserman, A., Muñoz, X.: Image classification using random forests and ferns. In: 2007 IEEE 11th International Conference on Computer Vision. IEEE (2007)
15. Chapelle, O., Haffner, P.: SVMs for histogram-based image classification. IEEE Trans. Neural Netw. **10**(5), 1055–1064 (1999)

16. Lu, D., Weng, Q.: A survey of image classification methods and techniques for improving classification performance. Int. J. Remote Sens. **28**(5/6), 823–870 (2007)
17. Perronnin, F., Sánchez, J., Mensink, T.: Improving the fisher kernel for large-scale image classification. In: Daniilidis, K., Maragos, P., Paragios, N. (eds.) ECCV 2010. LNCS, vol. 6314, pp. 143–156. Springer, Heidelberg (2010). https://doi.org/10.1007/978-3-642-15561-1_11
18. Krizhevsky, A., Hinton, G.: Learning multiple layers of features from tiny images, vol. 1. no. 4. Technical report, University of Toronto (2009)
19. Fei-Fei, L., et al.: Learning generative visual models from few training examples: an incremental Bayesian approach tested on 101 object categories. In: Computer Vision and Image Understanding (2007)
20. Feifei, L., Fergus, R., Perona, P.: One-shot learning of object categories. IEEE Trans. Pattern Anal. Mach. Intell. **28**(4), 594–611 (2006)
21. Iandola, F.N., Han, S., Moskewicz, M.W., et al.: SqueezeNet: AlexNet-level accuracy with 50x fewer parameters and $< 0.5MB$ model size (2016)
22. Selvaraju, R.R., et al.: Grad-CAM: visual explanations from deep networks via gradient-based localization. Int. J. Comput. Vision 1–24 (2019)

BCN-GCN: A Novel Brain Connectivity Network Classification Method via Graph Convolution Neural Network for Alzheimer's Disease

Peiyi Gu[1] , Xiaowen Xu[2], Ye Luo[1(✉)], Peijun Wang[2(✉)], and Jianwei Lu[1]

[1] Tongji University, Shanghai, China
{pierespy,yeluo,jwlu}@tongji.edu.cn
[2] Department of Medical Imaging, Tongji Hospital,
Tongji University School of Medicine, Tongji University, Shanghai, China
1710451@tongji.edu.cn, tongjipjwang@vip.sina.com

Abstract. Brain connectivity network (BCN) is a non-directed graph which represents the relationship between brain regions. Traditional feature extractor methods has been applied to identify diseases related to brains such as Alzheimer's disease (AD) with BCN. However, the feature extraction capabilities of these methods are insufficient to collect various topological properties of non-directed graphs. They can only take into consider the local topological properties or frequent sub-graph topological properties. To explicitly exploit these properties, this paper proposes a BCN based fully-supervised Graph Convolution Network (GCN) to select features and classify disease automatically, skipping the step of manual feature selection and keeping all information in the brain connectivity network. Extensive experiments show that we outperform other method in classifying different stages of AD. Furthermore, our method can find the most relevant brain regions to classify different stages of AD, showing better interpretation ability compared to traditional methods.

Keywords: Brain connectivity network · Graph convolutional neural network · Alzheimer's disease

1 Introduction

Alzheimer's disease (AD) is a progressive neurodegenerative disease. There is still no effective way to cure AD patients. Fortunately, AD has an early stage, mild cognitive impairment (MCI), which can be blocked in the disease progression. Thus the early diagnosis of MCI and AD has become increasingly critical.

P. Gu and X. Xu–These authors contributed equally to this work.
This work was supported by the General Program of National Natural Science Foundation of China (NSFC) (Grant No. 61806147).

T. Mantoro et al. (Eds.): ICONIP 2021, LNCS 13108, pp. 657–668, 2021.
https://doi.org/10.1007/978-3-030-92185-9_54

In the past few decades, many different methods have been put forward to distinguish AD from MCI [14]. According to different ways of medical imaging, AD and MCI can be distinguished by structural magnetic resonance imaging (sMRI) [14] or functional magnetic resonance imaging (fMRI) [6]. sMRI is a commonly used method to diagnose AD by detecting the atrophy of the brain [14]. Meanwhile, fMRI is a series of four dimensional brain scanning data that can demonstrate the connectivity strength of people's brain and has been used to diagnose AD patients from MCI and normal controls (NCs) [7]. Here we put emphasis on AD classification on the fMRI while other modalities are not our focus in this paper.

As for fMRI, most of the existing AD classification methods are based on people's brain connectivity network (BCN) which is constructed from fMRI. The motivation behind is that fMRI provides a way to map the patterns of functional interaction of brain regions, and BCN can characterizes the interaction patterns of brain regions [4]. By exploring connectivity in brain networks, we can better understand the pathological underpinnings of neurological disorder, thus diseases like AD/MCI can be well distinguished by BCNs [3]. Details to construct BCNs from fMRI can be referred to Sect. 3.2.

The key step of BCN based AD classification is to handle the BCN data which is a non-directed graph. In this non-directed graph, each node denotes a particular brain region and each edge represents the connectivity between two nodes. Unlike feature vectors, BCN has special topological structure, thus some graph kernel based methods are proposed to compute the similarity of BCNs [2,3]. The graph kernel can project the graph data onto a proper feature space such that AD and MCI can be better distinguished by some classifiers. However, most of the graph kernels are defined via comparing small sub-graphs such as walks [2], paths, or graphlets [3], which are all manually extracted features and dependent on human experiences. Moreover, even though the sub-graph like features are extracted, it still need to design a graph kernel to compute the similarity, and it is still not clear what kind of kernel can project the extracted features onto the most proper feature space for classification.

Meanwhile, as the remarkable success of the graph convolutional neural network (GCN) applied into various domains, GCN based methods for graph data classification achieve much attention recently. In [5], a semi-supervised GCN method is proposed to perform the node classification in a graph. With the GCN, nodes with no class label can be assigned a label after the information propagation. The semi-supervised GCN has been used to diagnose AD in [7] via building the patient graph, and the patient as the graph node is assigned the class label via the network learning. However, GCN based method has rarely been proposed to diagnose AD based on BCNs which are non-directed graphs.

In this paper, we propose a BCN based fully-supervised Graph Convolution Network for AD classification. By the proposed GCN, features can be automatically extracted from BCNs. Moreover, these features can keep the topological structures of BCNs and discriminate themselves from other classes at the same time. In order to obtain features like that, the convolution kernel of our GCN is derived from BCNs' adjacent matrices. Thus, as the information propagation, the feature values of the nodes change according to the topological structure of the kernel, leading to the fact that patients with similar BCN structures will have similar features and therefore are more likely to be grouped into the same class. For the BCNs whose structures are with slight difference but belonging to the same class, their features can be tuned by adjusting the weights of GCN so that they can still have the same class label. Experimental results on ADNI dataset further validate the superiority of our method against graph kernel based methods. Our method can also be compatible with GAT module [13] and get further improvement.

Our main contributions are three-fold: (1) A novel GCN is proposed to automatically extract topological graph features from BCNs for AD classification. To the best of our knowledge, it is among the first attempts to build fully-supervised GCN for AD classification directly on graph data. (2) The important brain regions found by our method to discriminate AD from MCI are in line with previous researches' conclusions that AD has close relationship to the default mode network. (3) Experimental results on ADNI dataset validate the effectiveness of the proposed method on AD classification and get further improvement after combining with GAT module [13].

2   Related Work

2.1   fMRI Image Based Methods

Handcraft Feature Based. Handcrafted features have been applied to diagnose Alzheimer's disease directly using MRI images. They manually extract texture features from fMRI. After recognizing some region of interests, they gather texture information of pixels at different frequencies such as Hippocampal volumetric integrity, gray scale, neuropsychological scores and others. Then, different classifiers are used to classify these features, such as SVM in [1] and random forest in [9]. These methods are limited to manually extracted features and some dynamic features like brain connectivity [10] are therefore ignored.

Convolution Neural Network Based. Recently, some deep neural network based methods that automatically extract features from images are proposed for AD diagnosis. Convolution neural network (CNN) is mostly used for two-dimensional and three-dimensional image feature extraction. It's typically used to extract features from sMRI, and rarely used for fMRI which is a four-dimension image. GCN is another type of CNN, in which the convolution kernel is designed for graphical data. It was proposed for classification task in a semi-supervised way to extract the features of graphical data [15]. Specifically, a non-directed graph is generated to represent the relationships among the node samples. Each node represents the raw features of the data sample and the edge between two nodes describes the relationship between them. Only a part of the nodes are assigned with labels. Some samples with no labels can get their labels after the training procedure as shown in Fig. 1. The semi-supervised GCN has been used to diagnose AD in [7], in which each node represents a patient and the feature of the node is extracted from the intensity of the patient's MRI image directly. Some text information such as age and sex is encoded as the weight of edges. Hence, their graph is to build the connectivity among different patients and assign the label to the node via information propagation among labeled patients within one graph. However, this kind of method needs lots of text information to generate edges and it's difficult to get these information on public databases so we are not able to reproduce this method. Different from them, in our method, for each patient, we build the BCN graph to model the connectivity between two particular brain regions, and each graph is given a class label after passing the learned GCN.

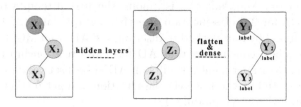

Fig. 1. The main idea of the semi-supervised GCN for AD classification. Each X_k is a patient, and each Y_k represents the corresponding label after the GCN. Z_k is the intermediate feature learned by the GCN.

2.2   Brain Connectivity Network Based Methods

Directly using fMRI image data couldn't capture the functional connectivity features between brain regions. Thus most of the conventional methods employ BCNs for AD classification task. The key of the conventional BCN based AD classification method is the graph kernel which is used to map the BCNs to a feature space such that the classifiers can be used successfully. Different graph kernels have been proposed to classify different stages of AD. One of them is the shortest path kernel [2] which converts the original BCN graph to the shortest path graph by connecting each pair of nodes through the shortest path. The ego-network graph kernel [12] aims to calculate node-centered indicators and use these indicators as the characteristics of the nodes. It then compares the distances between these characteristics as its kernel functions. The Weisfeiler-Lehman-subtree method [11] encodes the node labels of the graph structure. While the original nodes of the encoded label core retain the mapping relationship. It decompose each graph into a tree structure containing all its neighbors based on graph nodes. The Weisfeiler-Lehman edge kernel [11] and Weisfeiler-Lehman shortest path kernel [2] are other examples of the Weisfeiler-Lehman kernel framework. The sub-network graph kernel algorithm (SKL) [3] attempts to compare the similarity between these sub-networks by constructing a sub-network for each node instead of comparing the entire network structure, because the comparison of the entire network structure is a NP puzzle. By generating sub-graphs and then calculating the similarity between them, a graph kernel function can be formed.

Although graph kernel based methods achieved performance gain on AD classification task, it still depends on the manually design of the kernel. And it's still very challenging for graph kernel to extract the features of BCN (i.e. non-directed graph). We propose a BCN-GCN based method which can automatically extract crucial information from BCN for AD classification via a novel GCN network.

3   The Proposed Method

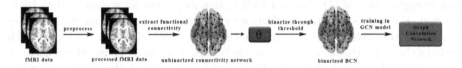

Fig. 2. The flowchart of the proposed method by classifying brain connectivity networks via graph convolution neural networks.

In this paper, we propose a AD classification method which utilizes a novel fully-supervised graph convolutional neural network to distinguish the brain connectivity networks (BCNs). Figure 2 shows the flowchart of the proposed method. fMRI images are first pre-processed such that the same position in different fMRI slices can be registered and also denoised. Then, the connectivity network of the whole brain is constructed on the processed fMRI images. After removing the edges with small values by some thresholds, we binarize the whole brain BCN into a binary adjacent matrix which only manifests the connection or non-connection relationship of different brain regions. The binarized BCNs are as inputs and sent into the GCN for training. Finally, given a test sample, binarized BCNs can be similarly generated and then can be labeled with the trained GCN.

3.1   Subjects and Image Pre-processing

Subject Description. We use the public Alzheimer's disease Neuroimaging Initiative (ADNI) data set to validate the proposed method. We use 311 subjects in total, which include 110 AD subjects, 95 MCI subjects and 106 NC subjects. Since our method is based on fMRI, all data employed is resting state fMRI.

Image Pre-processing.   To pre-process the fMRI images, we use Resting-State fMRI Data Analysis Toolkit plus V1.22. The pre-processing procedure is a standardized operation. It includes removing the first n time points from fMRI considering they may have noises caused by the adjusts of patients or machines, time slicing correction, realigning to the first volume(motion correction), space normalizing, spatial smoothing, detrending to eliminate the impact of the offset generated by the sensor, filtering between 0.01 Hz–0.1 Hz and regressing out nuisance covariates into gray matter, white matter, and cerebrospinal fuid. After pre-processing all fMRI, we are ready to construct brain connectivity networks for fMRI.

3.2   Brain Connectivity Network Construction

Given a pre-precessed fMRI which is a time series data of 3D brain volume, to build the brain connectivity network, we parcellate each 3D brain volume into 90 brain regions (i.e. region of interest) by warping the automated anatomical labeling (AAL) template to it. For each brain region in each volume, we calculate an average value of GM-masked BOLD signals among all voxels in that brain region. For a specific brain region, we concatenate all these calculated values along the time axis and then calculate the Pearson correlation coefficient of each pair of them. By this way, we obtain a fully connected brain connectivity network represented by a 90×90 matrix C, in which the value of element c_{ij} describes the connectivity between the ith and the jth brain regions. The larger c_{ij} is, the stronger connectivity between the ith and the jth brain regions is, and vice versa.

To outline the topological properties of discriminative networks, we emphasize the relationship of two brain regions with strong connectivity value by thresholding their edge weight with some pre-defined values. Thus, a binary matrix A with an element a_{ij} is obtained via the following formula:

$$a_{ij} = \begin{cases} 1 & if \quad c_{ij} \geq \theta \\ 0 & otherwise \end{cases}, \tag{1}$$

where θ is a pre-defined threshold which is empirically set and A is also the adjacent matrix of the sample. Detailed analysis of this parameter is provided in our experimental Section.

3.3   BCN Classification by the Proposed GCN

As we all know that even samples are with different topological structures (i.e. BCNs), some similar features should exist to imply that they belong to the same category. Motivated by this, we propose a novel GCN network to automatically extract features from BCNs, by which the topological structure of BCNs are maintained while the category related information are extracted. Details of the network architecture is shown in Fig. 3. Given a sample, its binarized BCN (i.e. adjacent matrix) and the initial feature of this BCN are passed into the network. After multiple propagation in the GCN, features of all nodes at the last graph convolution layer are stacked together and then passed to a flatten layer. And the label of the BCN is the output after the dense layer and the softmax activation layer.

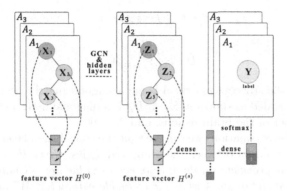

Fig. 3. Illustration the framework of the proposed BCN-GCN based AD classification network. A_k is the adjacent matrix of BCN for the kth sample. X_n is the nth node of the BCN, and Z_n is the intermediate feature learned by the GCN for X_n. Node features of X_n and Z_n are stacked separately to be the initial feature $H^{(0)}$ and the learned feature $H^{(s)}$. Y_k is the output label of A_k after processing $H^{(s)}$ with the learned GCN.

The Input of the BCN-GCN Network. The inputs of the GCN are closely related to the adjacent matrix A including the adjacent matrix A which maintains the topological structure of the BCN and the initial feature of the BCN computed from A. Specifically, given a sample, its adjacent matrix can be obtained as the above Section, and the initialized feature of the BCN can be denoted by $F \in \mathbb{R}^{N \times D}$, where N is the number of graph nodes and D is the dimension of the feature. We calculate the ith element of F as the degree of this node in the BCN associated with A: $F_i = \sum_j A_{ij}$. In other words, the initial feature of the GCN is the stack of the node features from the 1th brain region to the Nth brain region. For simplicity, we set $D = 1$ which means the feature dimension of each node is set to one.

Features of BCNs Learned by GCN. The ideal features automatically learned from BCNs should characterize the topological structures of the BCNs and could discriminate themselves from different classes at the same time. In our proposed GCN network, we use the graph convolution kernel to maintain the topological structure and enhance the discrimination abilities of features via the trainable network parameters. Mathematically, $H^{(l+1)}$ denotes the output feature of a graph convolution at the layer $l + 1$ and it can be written as:

$$H^{(l+1)} = \sigma(LH^{(l)}W^{(l)}), \tag{2}$$

where $H^{(l)}$ is the output feature learned at the lth layer, $W^{(l)}$ is the trainable weight matrix in the lth layer, and $\sigma(.)$ denotes an activation function. Here we use $H^{(0)} = F$ and activation function as $ReLU(x) = max(0, x)$. L is the kernel of the GCN and derived from the adjacent matrix A. Mathematically, it can be written as:

$$L = \tilde{D}^{-\frac{1}{2}} \tilde{A} \tilde{D}^{-\frac{1}{2}}, \tag{3}$$

$$\tilde{A} = A + I_N, \tag{4}$$

$$\tilde{D}_{ii} = \sum_{j=0}^{i} \tilde{A}_{ij}. \tag{5}$$

Here, to avoid the adjacent matrix A being a singular matrix, an identity matrix $I_N \in \mathbb{R}^{N \times N}$ is added as shown in Eq. 4. $\tilde{D}_{ii}$ is a diagonal matrix of $\tilde{A}$ (i.e. Eq. 5). N is the number of the brain regions, and $N = 90$.

From the above equations we can see that the convolution layer can propagate the feature information between the connected nodes. *Given the adjacent matrix A, after multiple propagation, the feature values of the nodes change according to the topological structure of A while the trainable parameters W make the learned features discriminate themselves from different class categories.*

Training and Testing. Training and testing of our proposed GCN is different from that of the previous methods [7,15] because all their samples (i.e. patients or nodes) construct only one graph. Only this graph together with labeled and unlabeled nodes are sent to the GCN. After information propagation, the labeled and the unlabeled nodes sharing similar features tend to have the same category labels. In other words, it is a semi-supervise classification method. While in our method, each graph (i.e. BCN) stands for a patient and each node stands for a brain region. After training the GCN with labeled graphs, unlabeled graphs can classified by GCN. In other words, our proposed GCN is to perform the fully supervised classification on multiple BCN graphs.

For different group of classification, the training and the testing datasets are split according to the ratio of 9 to 1. With the learned GCN, given a test sample, the BCN can be assigned to a specific class via the following equation:

$$Y = softmax(W^{(s+1)}\sigma(W^{(s)}H^{(s)})), \tag{6}$$

where $W^{(s)} \in \mathbb{R}^{h \times N}$ and $W^{(s+1)} \in \mathbb{R}^{2 \times h}$ are the trainable weight of the dense layers s and $s + 1$, h means the output feature dimension of dense layer, $s + 1$ is the last layer of the GCN and $Y \in \mathbb{R}^{2 \times 1}$ is the output label of the BCN associated with the adjacent matrix A.

Table 1. Classification performance comparisons with different methods under three control groups.

Method	AD/MCI			MCI/NC			AD/NC		
	Accuracy(%)	bACC(%)	AUC	Accuracy(%)	bACC(%)	AUC	Accuracy(%)	bACC(%)	AUC
Shortest-path [2]	71.4	61.5	0.6	69.8	62.4	0.60	81.0	80.2	0.84
Ego-net [12]	75.9	60.7	0.59	71.8	59.0	0.60	85.7	84.2	0.83
WL-edge [11]	76.7	55.4	0.54	73.2	67.0	0.72	79.8	79.7	0.83
WL-subtree [11]	79.7	61.3	0.58	76.5	65.5	0.72	85.7	85.2	0.88
WL-shortestpath [2]	76.7	55.4	0.55	73.2	67.4	0.70	83.3	82.7	0.82
SKL [3]	82	68.6	0.71	82.6	74.5	0.80	90.5	89.6	0.93
Our Method	92.9	93.7	0.93	86.3	86.2	0.84	88.9	89.9	0.91
Our Method + GAT	94.7	95.0	0.95	88.9	88.9	0.83	89.9	89.4	0.90

4   Experimental Results and Analysis

We first compare the state-of-the-art AD classification methods with the proposed BCN-GCN method. Then, the sensitivity of the hyper-parameter θ is analyzed. To validate our classification results being objective and fitting to the employed dataset characteristic, the heat maps showing the difference of BCNs between the target class against the control group are provided. At last, the importance of various brain regions to distinguish AD from MCI is analyzed.

Comparisons with State-of-the-arts. To show the superority of our BCN-GCN based method for AD classification, we conduct three groups of classifications: AD/MCI, MCI/NC and AD/NC, and six graph kernel based methods are compared: the shortest-path graph kernel method [2], the Ego-net graph kernel method [12], the WL-edge graph kernel method [11], the WL-subtree graph kernel method [11], the WL-shortestpath graph kernel method [2], and the SKL graph kernel method [3]. Experimental results in Table 1 show that our method achieved the best results in accuracy, AUC and balanced accuracy (bACC) which is the average between TPR and TNR.

Classification Results v.s. Dataset Characteristics. From Table 1 we get the best result for AD/MCI classification, followed by AD/NC, and MCI/NC is the worst, and here we provide the analysis to show that this conclusion is fit to the characteristics of the employed data. In other words, if the two classes (i.e. AD/MCI) achieve the best classification accuracy, they should have larger BCN difference than that of the other two (i.e. AD/NC, MCI/NC). In order to show that the difference of the connectivity of a brain region from another brain region (i.e. $A_{i,j}$) of the first class between that of the second class, we accumulate the connectivity $A_{i,j}$ for all the samples within the first class, then take the absolute difference value between the two classes as: $D_{ij} = \left| \frac{1}{N_1} \sum_{k=0}^{N_1} A_{ij}^k - \frac{1}{N_2} \sum_{k=0}^{N_2} M_{ij}^k \right|$, where N_1 is the number of samples of the first class, N_2 is the number of samples of the second class. A_{ij}^k is the value of the adjacent matrix at (i, j) for the kth sample in the first class. M_{ij} is the counterpart of A_{ij} for the second class.

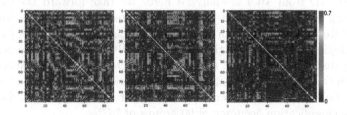

Fig. 4. Heat maps of different class against its control group by averaging the difference between corresponding BCNs. Hot color means large difference and vice verse. (AD/MCI left, AD/NC middle, and MCI/NC right).

We plot the heat map of $D \in R^{N \times N}$ for AD/MCI, AD/NC and MCI/NC respectively and show them in Fig. 4. From Fig. 4, we can see that most elements of D for AD/MCI are larger than that of the other two groups. On the contrary, MCI/NC comes to the most difficult one, and this is in line with our classification results above.

Sensitivity of the Threshold θ. Considering that there is no golden standard to select the optimal threshold θ, and different threshold will lead to different BCNs, we conduct the classification by setting θ with a range of $[0.1, 1.3]$ and 0.05 as the interval. As shown in Fig. 2, for three groups of classifications, thresholds 0.45 perform best in AD/MCI classification, 0.3 perform best in AD/NC classification and 0.35 perform best in MCI/NC classification (Fig. 5).

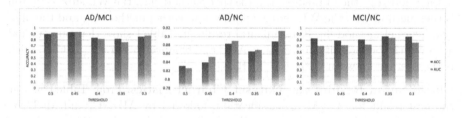

Fig. 5. Accuracy results by different thresholds to binarize the BCN.

Compatibility with Other Methods. Our method can also be combined with other modules for further performance exploration. For instance, our network can be equipped with GAT (Graph Attention Network) [13]. To fit GAT with our method, we combined the GAT module after our BCN-GCN main network. The results in Table 1 empirically validates this idea. Through BCN-GCN method with GAT module, it extracts more important category related information and still maintains the topological structure of BCNs. This can reflect the power of graph attention brought by GAT module.

Brain Region Importance Analysis. The importance of brain regions to distinguish AD from MCI is analyzed here. We take forward the experiment of removing each brain region at a time and testing the results with BCNs generated from the left 89 brain regions. If the classification accuracy of the left 89 brain region is much lower than that of the BCN with 90 brain regions, we can claim that the brain region removed is a crucial one to distinguish AD from MCI. In total there are 90 brain region for check and 90 groups of BCNs for classification. We run a 5 fold cross validation for each group of BCNs. Figure 6 shows the classification accuracy of 90 groups of BCNs.

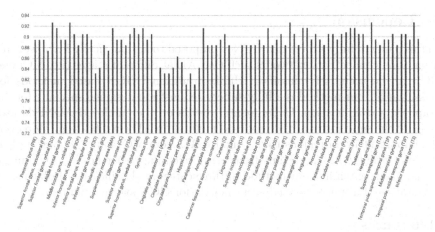

Fig. 6. Classification accuracy of AD/MCI for 90 groups of BCNs, each of which contain 89 brain regions. The one with lower accuracy means the removed brain region plays a more important role to classify AD from MCI.

We rank these brain regions according to their classification accuracy and select the 14 brain regions with the lowest classification accuracy. Figure 7 is the visualizations of those selected brain regions according to their normalized accuracy: $nAcc = \frac{\max(Acc)-acc}{\max(Acc)-\min(Acc)}$, where Acc means the accuracy vector of 90 brain regions and acc means the accuracy result of classification using the group of BCNs without a particular brain region. We can see that most of the important brain regions to distinguish AD from MCI found by our method are used to control the human memories and emotions, and this discovery is also in line with the claims from other scientific researches [8].

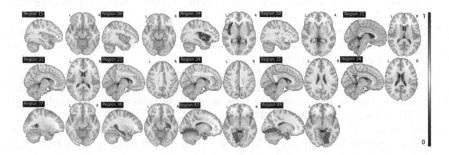

Fig. 7. Visualizations of the top 14 important brain regions according to their normalized classification accuracy: "Inferior frontal gyrus, orbital", "Insula", "Cingulate gyrus, anterior part", "Cingulate gyrus, mid part", "Cingulate gyurs, posterior part", "Hippocampus" and "Parahippocampus".

5   Conclusion

In this paper, a fully-supervised GCN is proposed to automatically learn features from brain connectivity networks such that AD at different stages (i.e. MCI and NC) can be accurately classified. Unlike traditional GCN based methods for classification, our method is a fully-supervised method designed specially for the graph data classification. Experimental results on the employed ADNI dataset validate the superiority of the proposed method on AD classification. Moreover, the important brain regions to distinguish AD from MCI found by our method are also in line with the claims from other scientific researches [8]. Our future work include extending our method to three category classification problems, an exploration of a non-binary adjacent matrix for GCN learning, and etc.

References

1. Bi, X.A., Shu, Q., Sun, Q., et al.: Random support vector machine cluster analysis of resting-state fMRI in Alzheimer's disease. PloS One **13**(3), e0194479 (2018)
2. Borgwardt, K.M., Kriegel, H.P.: Shortest-path kernels on graphs. In: ICDM, pp. 8-pp. IEEE (2005)
3. Jie, B., Liu, M., Zhang, et al.: Sub-network kernels for measuring similarity of brain connectivity networks in disease diagnosis. IEEE TIP **27**(5), 2340–2353 (2018)
4. Kaiser, M.: A tutorial in connectome analysis: topological and spatial features of brain networks. Neuroimage **57**(3), 892–907 (2011)
5. Kipf, T.N., Welling, M.: Semi-supervised classification with graph convolutional networks. preprint arXiv:1609.02907 (2016)
6. Machulda, M.M., Ward, H., Borowski, B., et al.: Comparison of memory fMRI response among normal, MCI, and Alzheimer's patients. Neurology **61**(4), 500–506 (2003)
7. Parisot, S., Ktena, S.I., Ferrante, E., et al.: Disease prediction using graph convolutional networks: application to autism spectrum disorder and Alzheimer's disease. Med. Image Anal. **48**, 117–130 (2018)
8. Raichle, M.E.: The brain's default mode network. Annu. Rev. Neurosci. **38**, 433–447 (2015)
9. Sarica, A., Cerasa, A., Quattrone, A.: Random forest algorithm for the classification of neuroimaging data in Alzheimer's disease: a systematic review. Front. Aging Neurosci. **9**, 329 (2017)
10. Sarraf, S., Tofighi, G., Initiative, A.D.N., et al.: DeepAD: Alzheimer's disease classification via deep convolutional neural networks using MRI and fMRI. BioRxiv (2016)
11. Shervashidze, N., Schweitzer, P., Leeuwen, et al.: Weisfeiler-Lehman graph kernels. J. Mach. Learn. Res. **12**(Sep), 2539–2561 (2011)
12. Shrivastava, A., Li, P.: A new space for comparing graphs. In: International Conference on Advances in Social Networks Analysis and Mining, pp. 62–71. IEEE Press (2014)
13. Veličković, P., Cucurull, G., Casanova, A., et al.: Graph attention networks. arXiv preprint arXiv:1710.10903 (2017)
14. Vemuri, P., Jack, C.R.: Role of structural MRI in Alzheimer's disease. Alzheimer's Res. Ther. **2**(4), 23 (2010)
15. Zhu, X., Ghahramani, Z., Lafferty, et al.: Semi-supervised learning using gaussian fields and harmonic functions. In: Proceedings of the 20th International conference on Machine learning (ICML-03), pp. 912–919 (2003)

Triplet Mapping for Continuously Knowledge Distillation

Xiaoyan Yang[1], Jialiang Tang[1], Ning Jiang[1（✉）], Wenxin Yu[1], and Peng Zhang[2]

[1] School of Computer Science and Technology, Southwest University of Science
and Technology, Mianyang, China
jiangning@swust.edu.cn
[2] School of Science, Southwest University of Science and Technology,
Mianyang, China

Abstract. Knowledge distillation is regarded as a widely used technology that trains a small high-performance network and knowledge
is transferred from a large network (teacher) to a miniature network
(student). However, the student network simulates the teacher network
quickly during training, which leads the effect of knowledge transmission
is extremely reduced. In this paper, we propose a triplet mapping knowledge distillation (TMKD) for continuing and efficient knowledge transfer,
which contains a new structure, an assistant network. The teacher and
the assistant are employed to supervise the training of the student, helping it to gain expressive knowledge from the teacher and discard the
useless knowledge of the assistant. To maintain effectiveness of knowledge transfer, we map the original features of the above three networks
by the multilayer perceptron (MLP). The metric function is used to
calculate the difference between the original features and the mapped
features that have a large difference. The experimental results demonstrate that the TMKD can obviously enhance the capability of the student. The student VGGNet13 achieves 1.67%, 3.73%, 0.42%, and 2.89%
performance improvements on the CIFAR-10, CIFAR-100, SVHN, and
STL-10 datasets, reaching 94.72%, 76.22%, 95.55%, and 88.18% accuracies respectively.

Keywords: Convolutional neural networks · Model compression ·
Triplet mapping knowledge distillation · Multilayer perceptron

1 Introduction

Convolutional neural networks (CNNs) have promoted the progress of many
fields of artificial intelligence broadly [5,11,18] with their outstanding performance. In general, a large size of CNNs often results in higher performance.
But limited by real-life requirements and hardware (such as smart headphones,
smartwatch, etc.), the size of CNNs cannot increase arbitrarily. Knowledge distillation is an effective model compression technique that learns a small network

© Springer Nature Switzerland AG 2021
T. Mantoro et al. (Eds.): ICONIP 2021, LNCS 13108, pp. 669–680, 2021.
https://doi.org/10.1007/978-3-030-92185-9_55

called the student from a large pretrained network called the teacher. The layer-wise or soft labels features of the teacher are considered as knowledge by extant knowledge distillation algorithms [9,24]. The student can achieve similar performance to the teacher by learning from the extracted knowledge. The teacher is only introduced in training, and only the student predicts the data during the testing. Therefore, it is effective for knowledge distillation to compress the student model and apply it into practice.

However, knowledge distillation still has two problems. The first is that there is a high similarity between the soft labels of the teacher and hard labels. The second is that the student can quickly simulate the teacher to produce features similar to the teacher. The above two problems huge drop the efficiency of knowledge transmission during training, limiting the further improvement of the student performance. To solve those problems, we propose a triplet mapping knowledge distillation (named TMKD). The overview of the TMKD structure is shown in Fig. 1, TMKD contains four networks named teacher, student, assistant, and multilayer perceptron (MLP). The size and performance of the assistant are smaller than the student. During training, the student learns to close to the teacher while keeping away from the assistant. To maintain the efficiency of knowledge transfer throughout the training, we map the features of the teacher, student, and assistant by MLP. Formally, we define the original features of teacher, student, and assistant as F_t, F_s, and F_a, respectively. Firstly, we input these features into the MLP to obtain the corresponding mapped features M_t, M_s, and M_a. Then we measure the discrepancies between the original features and the mapped features by metric function as knowledge for transfer (the details are described in Sect. 4). The student can produce more accurate predictions by learning from the teacher and avoid false predictions by maximizing the distance from the assistant. Since the mapping operation, the discrepancies between the features are continuous, which means that the student is efficiently learned from the knowledge throughout the training. Experimental evaluations demonstrate that the student trained with TMKD has achieved better performance than the teacher on multiple benchmark datasets. The contributions of this paper are as follows:

1. We analyze problems of knowledge distillation and propose a triplet mapping knowledge distillation framework (TMKD) to solve those problems.
2. We conduct extensive experiments on popular CNNs and successfully train small CNNs to achieve superior performance than that of large CNNs on benchmark datasets such as CIFAR-10 [13], CIFAR-100 [13], SVHN [14], and STL-10 [4].
3. To the best of our knowledge, compared with other knowledge distillation algorithms, TMKD trains the student to reach the state-of-the-art (SOTA) performance.

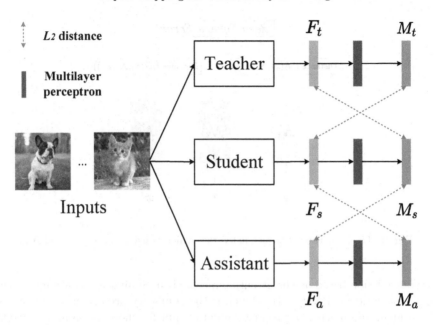

Fig. 1. The total structure of our proposed triplet mapping knowledge distillation, which consists of a teacher, student, assistant and multilayer perception.

2   Related Work

The related works of TMKD include knowledge distillation and triplet networks.

2.1   Knowledge Distillation

Knowledge distillation aims to train a lightweight student that can produce a comparable performance to a complicated teacher. The concept of knowledge distillation was first elaborated by Hinton et al. [9]. They named the large network as the teacher and the small network as the student, and the outputs of the teacher are input into the softmax function to obtain the soft labels as knowledge. FitNets [19] first use the middle layer features to train a deeper student successfully. Paying more attention (PMAT) [24] proposes that it's inefficient to use the feature of the network for knowledge transmission directly. Therefore, PMAT extracts the attention maps for knowledge transmission, which effectively improves the performance of the student. Xu et al. [23] implement knowledge distillation by simulating the Generative Adversarial Networks [6,15,26], which introduce a discriminator to determine whether a feature comes from the teacher or student. Instead of transferring the features of samples, the relational knowledge distillation (RKD) [16] transmits the interrelation between the samples from the teacher to student. Wang et al. [22] propose attentive feature distillation and selection (AFDS), which dynamically determines important features

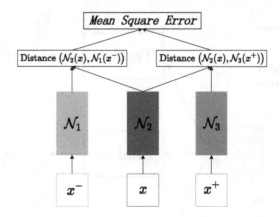

Fig. 2. The structure of triplet networks, which consist of $\mathcal{N}_1$, $\mathcal{N}_2$, and $\mathcal{N}_3$.

for transmission based on the attention mechanism. Similarity-preserving knowledge distillation (SPKD) [21] calculates the similarity-preserving to encourage the student and teacher to product similar output for the same sample. Correlation congruence knowledge distillation (CCKD) [17] not only transmits information at the instance-level but also the relationship between instances from the teacher to the student. Although above knowledge distillation methods are effective to improve the performance of students, these transmission effects continue to decrease with training.

2.2   Triplet Networks

Triplet networks [10] are widely used to determine whether a sample belongs to the same category as another sample, which comprises three shared weights networks with the same structure. Triplet networks are developed from the siamese networks [1,28], which consist of two networks that share parameters. Bromley et al. [3] first propose siamese networks for checking signature. In subsequent studies, siamese networks have been widely used in target tracking due to their ability to compare sample similarities. Hoffer et al. [10] propose the triplet networks improved by the siamese networks. The inputs of the triplet networks include positive sample, normal sample, and negative sample. During training, the triplet networks strive to promote the normal sample to be closer to the positive sample and stay away from the negative sample. Inspired by the triplet networks, our TMKD introduces an assistant into the knowledge distillation framework.

3   Triplet Networks

Triplet networks [10] consist of three networks that shared weights as Fig. 2 shows. To simplify the description, we define the three networks contained in triplet networks as $\mathcal{N}_1$, $\mathcal{N}_2$, and $\mathcal{N}_3$. And we define the inputs as positive sample x^+, normal sample x, and negative sample x^-. During the training of triplet networks, positive sample x^+, normal sample x, and negative sample x^- are input into $\mathcal{N}_1$, $\mathcal{N}_2$, and $\mathcal{N}_3$, respectively, and the corresponding outputs are $\mathcal{N}_1(x^-)$, $\mathcal{N}_2(x)$, and $\mathcal{N}_3(x^+)$. Based on the obtained outputs, the L_2 distance function is used to calculate the similarity between samples. The distance d^+ between normal sample and positive sample and the distance d^- between normal sample and negative sample are calculated as follows:

$$d^+ = \frac{e^{\left\|\mathcal{N}_2(x)-\mathcal{N}_3\left(x^+\right)\right\|_2}}{e^{\left\|\mathcal{N}_2(x)-\mathcal{N}_3(x^+)\right\|_2}+e^{\left\|\mathcal{N}_2(x)-\mathcal{N}_1\left(x^-\right)\right\|_2}} \tag{1}$$

$$d^- = \frac{e^{\left\|\mathcal{N}_2(x)-\mathcal{N}_1\left(x^-\right)\right\|_2}}{e^{\left\|\mathcal{N}_2(x)-\mathcal{N}_3(x^+)\right\|_2}+e^{\left\|\mathcal{N}_2(x)-\mathcal{N}_1\left(x^-\right)\right\|_2}} \tag{2}$$

The softmax function is applied to the calculated L_2 distance to better measure the sample similarity. To minimize the d^+ while maximizing the d^-, follow the work [10], use the Mean Squared Error (MSE) to measure as:

$$\mathcal{L}_{mse}\left(d^+, d^-\right) = \left\|\left(d^+, d^- - 1\right)\right\|_2^2 \tag{3}$$

Processed by softmax function, the values in d^+ and $d^- \in (0,1)$, therefore the values in $(d^- - 1) \in (-1,0)$. Therefore, minimizing the $\mathcal{L}_{mse}$ is equivalent to reduce the distance between the normal sample and the positive sample while enlarging the distance between the normal sample and the negative sample. It means that triplet networks can precisely determine whether two samples belong to the same category.

4   Triplet Mapping Knowledge Distillation

The overall structure of TMKD is shown in Fig. 1, TMKD is composed of three networks that are defined as the teacher $\mathcal{N}_t$, student $\mathcal{N}_s$, assistant $\mathcal{N}_a$, and MLP $\mathcal{M}$, which are described as follows:

Teacher is a large pre-trained CNN with the largest network structure, and its performance is better than the student and assistant.

Student is a smaller initialized CNN, the size and performance of the student are between the teacher and the assistant.

Assistant is a pre-trained network with the smallest structure, besides, it's also the lowest performance among the three networks.

MLP consists of multiple linear layers and the teacher supervises its training process.

In this section, we introduce the details of how to implement TMKD. We show how to train MLP, describe how to train the student, and propose the total objective function.

4.1   Training the Multilayer Perceptron

In this subsection, for mapping the features of CNNs, we train the MLP with the teacher. We define the inputs of teacher as $x = \{x^1, x^2, \cdots, x^n\}$, n is the batch-size. We first input x into the teacher $\mathcal{N}_t$ to get the features after the penultimate layer as $F_t = \mathcal{N}_t(x) = \{F_t^1, F_t^2, \cdots, F_t^n\}$. Then, we map F_t by MLP as $M_t = \mathcal{M}(F_t)$. We expect that the mapped features are different from the original features at local values and obey the same distribution in general. Therefore, we calculate the L_2 distance to measure the discrepancy between original features F_t and mapped features M_t:

$$\mathcal{L}_{mlp} = \|F_t - M_t\|_2^2 \tag{4}$$

By minimizing the $\mathcal{L}_{mlp}$ through backpropagation until the MLP convergence, the mapped features which are similar to but slightly different from original features can be obtained.

4.2   Triplet Mapping Knowledge Distillation

For simulating triplet networks, an assistant is introduced into the teacher-student structure by us. During training, the student learns to output features similar to the teacher's features while keeping different from the assistant's features. Instead of inputting various samples to the three networks, all networks in TMKD use the same input. The teacher's performance is the highest among the three networks, we assume that the features of the teacher are more informative. Therefore, we take the features of the teacher F_t as a positive sample. Meanwhile, we use the features of the assistant F_a and student F_s as a negative sample and normal sample. All mentioned features are extracted after the penultimate layer. To carry out continuous and effective knowledge transfer, we use MLP to map the features of the three networks to maintain the differences between the features. The mapped features are obtained as:

$$M_t = \mathcal{M}(F_t), \quad M_s = \mathcal{M}(F_s), \quad M_a = \mathcal{M}(F_a) \tag{5}$$

M_t, M_s, M_a mean the mapped features of the teacher, student, and assistant, respectively. As described in Sect. 3, based on the original features F_t, F_s, F_a and mapped features M_t, M_s, M_a, the discrepancies between the teacher and student, and the discrepancies between the assistant and student are calculated as follows:

$$D(F_t, M_s) = \frac{e^{\|M_s - F_t\|_2}}{e^{\|M_s - F_t\|_2} + e^{\|M_s - F_a\|_2}}, \quad D(M_t, F_s) = \frac{e^{\|F_s - M_t\|_2}}{e^{\|F_s - M_t\|_2} + e^{\|F_s - M_a\|_2}}$$

$$D(F_a, M_s) = \frac{e^{\|M_s - F_a\|_2}}{e^{\|M_s - F_t\|_2} + e^{\|M_s - F_a\|_2}}, \quad D(M_a, F_s) = \frac{e^{\|F_s - M_a\|_2}}{e^{\|F_s - M_t\|_2} + e^{\|F_s - M_a\|_2}} \tag{6}$$

Where the function $D(a, b)$ denotes the distance between a and b. Then, we add $D(F_t, M_s)$ and $D(M_t, F_s)$ as d^+ and add $D(F_a, M_s)$ and $D(M_a, F_s)$ as d^-. In the end, we calculate the MSE of d^+ and d^- as the loss for knowledge transfer:

Table 1. The results of the student trained by TMKD on the benchmark datasets. The Student represents the original student, the Student* represents the student trained by TMKD, WRS is the abbreviation of WideResNet.

Model	Algorithm	Parameters	FLOPs	CIFAR-10	CIFAR-100	SVHN	STL-10
Teacher	VGGNet16	~14.72M	~0.31G	93.31%	73.36%	95.38%	85.73%
Student	VGGNet13	~9.41M	~0.23G	93.05%	72.49%	95.13%	85.29%
Student*	VGGNet13	~9.41M	~0.23G	**94.72%**	**76.22%**	**95.55%**	**88.18%**
Assistant	VGGNet11	~9.23M	~0.15G	91.75%	71.25%	94.71%	84.98%
Teacher	ResNet34	~21.28M	~1.16G	94.85%	77.34%	95.73%	89.05%
Student	ResNet18	~11.17M	~0.56G	93.92%	76.53%	95.44%	88.47%
Student*	ResNet18	~11.17M	~0.56G	**95.46%**	**78.12%**	**95.97%**	**89.36%**
Assistant	VGGNet11	~9.23M	~0.15G	91.75%	71.25%	94.71%	84.98%
Teacher	WRS40_2	~2.25M	~0.36G	93.99%	73.84%	95.55%	85.26%
Student	WRS28_2	~1.47M	~0.24G	93.51%	73.43%	95.07%	84.87%
Student*	WRS28_2	~1.47M	~0.24G	**94.95%**	**75.05%**	**95.79%**	**86.79%**
Assistant	WRS16_2	~0.69M	~0.13G	92.73%	72.46%	94.72%	84.56%

$$\mathcal{L}_{tmkd}\left(d^+, d^-\right) = \left\| \left(d^+, d^- - 1\right) \right\|_2^2 \tag{7}$$

By minimizing $\mathcal{L}_{tmkd}$, the distance between the features of the student and teacher will be smaller, and the distance between the features of the student and assistant will be larger. This means that the student could learn much knowledge from the teacher while avoiding false predictions through the assistant.

4.3 Trained by Hard Labels

To facilitate the student classification more precisely, we train the student by the hard labels. In detail, the input images $x = \left\{x^1, x^2, \cdots, x^n\right\}$ are input to the student to get the classification results y_s. And the cross-entropy loss function between the classification results $y_s = \left\{y_s^1, y_s^2, \cdots, y_s^n\right\}$ and the hard labels $l = \left\{l^1, l^2, \cdots, l^n\right\}$ is calculated as:

$$\mathcal{L}_{cls} = \frac{1}{n}\sum_{i=1}^{n}\mathcal{H}_{cross}\left(l^i, y_s^i\right) = \frac{1}{n}\sum_{i=1}^{n}l^i\log(y_s^i) \tag{8}$$

By minimizing $\mathcal{L}_{cls}$, the student can generate predictions that are more consistent with the hard labels.

4.4 Total Objective Function

To encourage the student to learn from the teacher, assistant, and hard labels together, we combine these loss functions proposed above, the total objective function of TMKD can be written as:

$$\mathcal{L}_{total} = \alpha \cdot \mathcal{L}_{tmkd} + \beta \cdot \mathcal{L}_{cls} \tag{9}$$

The α and β are hyper-parameters to balance the student learning process.

5   Experiments

In this section, we select the popular VGGNet-family networks [20], ResNet-family networks [7], WideResNet-family networks [25] as the experimental CNNs to train on benchmark datasets CIFAR [13], SVHN [14], STL-10 [4] datasets for empirical evaluation. We first trained the chosen CNNs on all datasets to demonstrate that our TMKD can be used to improve CNN's performance extensively. Then, we compare our TMKD with advanced knowledge distillation algorithms to show that the TMKD can achieve SOTA performance. Finally, we conduct ablation studies to understand the effectiveness of the teacher and assistant in training the student. The experimental setups are shown as follows:

CIFAR is composed of the 10 categories CIFAR-10 and 100 categories CIFAR-100, which contain 60,000 32×32 sized RGB images. Specifically, CIFAR-10 contains 5,000 images in training set and 1,000 images in testing set. CIFAR-100 contains 500 images for training and 100 images for testing. For the training setups of CIFAR, Stochastic Gradient Descent (SGD) [2] is used as the student optimizer, the learning rate (divided by 10 at 40%, 60%, 80% of the total epochs), momentum, weight-decay are set as 0.1, 0.9, 5e−4. The hyper-parameters in Eq. (9) are set as $\alpha=0.9$, $\beta=0.1$.

SVHN consists of street house numbers from 0 to 9, the training set contains 73,257 digits and the test set contains 26,032 digits. Adam [12] is chosen as the optimizer, the learning rate is 0.01, other settings are the same as that of CIFAR.

STL-10 is inspired by the CIFAR-10, there are 10 categories of 96 × 96 sized RGB images in it. Compared with CIFAR-10, STL-10 contains fewer training samples, and the image size of STL-10 is three times that of CIFAR-10. The training setups of STL-10 are the same as that of CIFAR.

5.1   Experiments on Benchmark Datasets

To extensively prove the effectiveness of our TMKD, we use TMKD to train all selected CNNs. For the VGGNet-family, VGGNet16, VGGNet13, and VGGNet11 are selected as the teacher, student, and assistant. For the ResNet-family, ResNet34, ResNet18, and VGGNet11 are set as the teacher, student, and assistant. For the WideResNet-family (WRS), WideResNet40_2, WideResNet28_2 and WideResNet16_2 are set as teacher, student, and assistant. Table 1 shows the experiment results of selected CNNs training on benchmark datasets.

CIFAR is a popular image recognition dataset. Table 1 reports that the performance of all students trained by TMKD has been significantly improved, surpassing the teacher. The student VGGNet13 achieves 94.72% and 76.22% accuracies on the CIFAR-10 and CIFAR-100, which obtain the highest improvement, 1.67% and 3.73% on CIFAR-10 and CIFAR-100, respectively. The student ResNet18 has 11.17M parameters and 0.56G FLOPs, which are only about half those of teacher ResNet34. It is surprising that the student has achieved better performance than the teacher, 95.46% accuracy on the CIFAR-10 and 78.12% accuracy on the CIFAR-100. Similar to the results of ResNet, the student WRS28_2 achieves better performance of 94.95% and 75.05% accuracies on

Table 2. The results of the student trained by knowledge distillation algorithms and our TMKD on the CIFAR dataset. Accuracy† means the accuracy improvement.

Algorithm	Parameters	FLOPs	CIFAR-10		CIFAR-100	
			Accuracy	Accuracy†	Accuracy	Accuracy†
Teacher	~21.28M	~1.16G	94.85%	\	77.34%	\
Student	~11.17M	~0.56G	93.92%	\	76.53%	\
Assistant	~9.23M	~0.15G	91.75%	\	71.25%	\
DML [27]	~11.17M	~0.56G	94.95%	1.03%	77.44%	0.91%
PMAT [24]	~11.17M	~0.56G	95.12%	1.20%	77.47%	0.94%
KDGAN [23]	~11.17M	~0.56G	95.05%	1.13%	77.52%	0.99%
RKD [16]	~11.17M	~0.56G	95.13%	1.21%	77.64%	1.11%
AB [8]	~11.17M	~0.56G	95.04%	1.12%	77.53%	1.00%
SPKD [21]	~11.17M	~0.56G	95.08%	1.16%	77.59%	1.06%
CCKD [17]	~11.17M	~0.56G	95.21%	1.29%	77.72%	1.19%
TMKD (ours)	~11.17M	~0.56G	**95.46%**	**1.54%**	**78.12%**	**1.59%**

CIFAR-10 and CIFAR-100, although the memory and computing consumptions are much smaller than those of the teacher WRS40_2.

SVHN is also an image classification dataset. Selected baseline CNNs have achieved high accuracy on the SVHN dataset (over 95% accuracy), which means that the student is difficult to improve further. However, the student trained by TMKD has achieved considerable performance improvement, surpassing the teacher. The student VGGNet13, ResNet18, and WRS28_2 achieve 95.55%, 95.97% and 95.79% on the SVHN, improved by 0.42%, 0.53% and 0.24 respectively.

STL-10 is similar to CIFAR-10, but there are fewer training samples in STL-10, which means that CNNs are more difficult to train on STL-10 to achieve good performance. Nevertheless, all students have improved obviously, after trained by TMKD. In particular, VGGNet13 achieves a 2.89% improvement on the STL-10 dataset, reaching an accuracy of 88.18%.

These experimental results prove that TMKD can be widely used in various structure CNNs to improve the performance of CNNs on various datasets.

5.2 Compared to Other Methods

In the subsection, we compare our TMKD with the advanced knowledge distillation algorithms: DML [27], PMAT [24], KDGAN [23], RKD [16], AB [8], SPKD [21], CCKD [17]. We select the ResNet34, ResNet18 and VGGNet11 as the teacher, student, assistant respectively, all CNNs are trained on CIFAR dataset.

Table 2 exhibits the experimental results of CNNs trained by various knowledge distillation algorithms. Through learning from the teacher, all students achieve considerable improvement. The performance of the student trained by

Table 3. The results of ablation experiments.

Dataset	Student	TMKD-$\mathcal{N}_t$	TMKD-$\mathcal{N}_a$	TMKD
CIFAR-10	93.05%	94.56%	94.42%	**94.72%**
CIFAR-100	72.49%	75.93%	75.80%	**76.22%**

knowledge distillation algorithms all surpasses the teacher. It shows that knowledge transfer can effectively train a high-capability network. Moreover, we can find that our proposed TMKD achieves the best accuracy among all methods. The students trained by TMKD obtains 95.46% accuracy and 78.12% accuracy on CIFAR-10 and CIFAR-100, respectively. Compared to the original student, the student trained by TMKD improves 1.54% and 1.90% accuracies on CIFAR-10 and CIFAR-100, even compared to the teacher, the accuracy of the student trained by TMKD also improves 0.61% and 0.78%. The experimental results demonstrate that our TMKD can achieve SOTA performance among the existing knowledge distillation algorithms.

5.3   Ablation Experiments

To further understand the influence of the teacher and assistant on the student. In this subsection, we train the student by the teacher (named TMKD-$\mathcal{N}_t$) and assistant (named TMKD-$\mathcal{N}_a$) individually. TMKD-$\mathcal{N}_t$ is equivalent to minimize the distance the d^+ in Eq. (7), while TMKD-$\mathcal{N}_a$ is equivalent to maximize the d^- in Eq. (7). We choose the CIFAR dataset for ablation experiments. We used the VGGNet16, VGGNet13, and VGGNet11 as the teacher, student, and assistant. Table 3 reports the results of various experimental designs. The student trained by TMKD-$\mathcal{N}_t$ and TMKD-$\mathcal{N}_a$ all achieve better performance than the original student. TMKD-$\mathcal{N}_t$ gets 94.56% and 75.93% accuracies on the selected data sets CIFAR-10 and CIFAR-100, and TMKD-$\mathcal{N}_a$ obtains 94.42% and 75.80% accuracies on CIFAR-10 and CIFAR-100. This proves that both the teacher and assistant contribute to the training of the student. And then, the student trained by the teacher and assistant together achieves the best accuracies of 94.72% and 76.22% on the CIFAR-10 and CIFAR-100. This suggests that the combination of the teacher and assistant can train students more effectively.

6   Conclusion

This work introduces a novel knowledge distillation framework (TMKD) based on the triplet networks and multilayer perception (MLP). TMKD maps original features with MLP and explores feature discrepancies between original features and mapped features for continuous knowledge transfer. We take the discrepancies as knowledge to encourage the student to obtain more information from the superior teacher and keep the student away from the shallow assistant at the same time to avoid the error predictions. Students trained by our TMKD achieve state-of-the-art performance among exiting knowledge distillation algorithms. Moreover, it is noteworthy that the compact student trained by TMKD can reach a better performance than the complicated teacher.

Acknowledgement. The Sichuan Science and Technology Program under Grant 2020YFS0307 gave this work partial support, Mianyang Science and Technology Program 2020YFZJ016, SWUST Doctoral Foundation under Grant 19zx7102, 21zx7114.

References

1. Bertinetto, L., Valmadre, J., Henriques, J.F., Vedaldi, A., Torr, P.H.S.: Fully-convolutional siamese networks for object tracking. In: Hua, G., Jégou, H. (eds.) ECCV 2016. LNCS, vol. 9914, pp. 850–865. Springer, Cham (2016). https://doi.org/10.1007/978-3-319-48881-3_56
2. Bottou, L.: Stochastic gradient descent tricks. In: Montavon, G., Orr, G.B., Müller, K.-R. (eds.) Neural Networks: Tricks of the Trade. LNCS, vol. 7700, pp. 421–436. Springer, Heidelberg (2012). https://doi.org/10.1007/978-3-642-35289-8_25
3. Bromley, J., Guyon, I., LeCun, Y., Säckinger, E., Shah, R.: Signature verification using a "siamese" time delay neural network. Adv. Neural Inf. Process. Syst. **6**, 737–744 (1993)
4. Coates, A., Ng, A., Lee, H.: An analysis of single-layer networks in unsupervised feature learning. In: Proceedings of the Fourteenth International Conference on Artificial Intelligence and Statistics, pp. 215–223. JMLR Workshop and Conference Proceedings (2011)
5. Girshick, R., Donahue, J., Darrell, T., Malik, J.: Rich feature hierarchies for accurate object detection and semantic segmentation. In: Proceedings of the IEEE Conference on Computer Vision and Pattern Recognition, pp. 580–587 (2014)
6. Goodfellow, I.: Nips 2016 tutorial: Generative adversarial networks. arXiv preprint arXiv:1701.00160 (2016)
7. He, K., Zhang, X., Ren, S., Jian, S.: Deep residual learning for image recognition. In: 2016 IEEE Conference on Computer Vision and Pattern Recognition (CVPR) (2016)
8. Heo, B., Lee, M., Yun, S., Choi, J.Y.: Knowledge transfer via distillation of activation boundaries formed by hidden neurons. In: Proceedings of the AAAI Conference on Artificial Intelligence, vol. 33, pp. 3779–3787 (2019)
9. Hinton, G., Vinyals, O., Dean, J.: Distilling the knowledge in a neural network. Comput. Sci. **14**(7), 38–39 (2015)
10. Hoffer, E., Ailon, N.: Deep metric learning using triplet network. In: Feragen, A., Pelillo, M., Loog, M. (eds.) SIMBAD 2015. LNCS, vol. 9370, pp. 84–92. Springer, Cham (2015). https://doi.org/10.1007/978-3-319-24261-3_7
11. Huang, Z., Huang, L., Gong, Y., Huang, C., Wang, X.: Mask scoring R-CNN. In: Proceedings of the IEEE/CVF Conference on Computer Vision and Pattern Recognition, pp. 6409–6418 (2019)
12. Kingma, D.P., Ba, J.: Adam: a method for stochastic optimization. arXiv preprint arXiv:1412.6980 (2014)
13. Krizhevsky, A., Hinton, G., et al.: Learning multiple layers of features from tiny images (2009)
14. Netzer, Y., Wang, T., Coates, A., Bissacco, A., Wu, B., Ng, A.Y.: Reading digits in natural images with unsupervised feature learning (2011)

15. Odena, A.: Semi-supervised learning with generative adversarial networks. arXiv preprint arXiv:1606.01583 (2016)
16. Park, W., Kim, D., Lu, Y., Cho, M.: Relational knowledge distillation. In: Proceedings of the IEEE/CVF Conference on Computer Vision and Pattern Recognition, pp. 3967–3976 (2019)
17. Peng, B., et al.: Correlation congruence for knowledge distillation. In: Proceedings of the IEEE International Conference on Computer Vision, pp. 5007–5016 (2019)
18. Ren, S., He, K., Girshick, R., Sun, J.: Faster R-CNN: towards real-time object detection with region proposal networks. In: Advances in Neural Information Processing Systems, pp. 91–99 (2015)
19. Romero, A., Ballas, N., Kahou, S.E., Chassang, A., Gatta, C., Bengio, Y.: FitNets: hints for thin deep nets. arXiv preprint arXiv:1412.6550 (2014)
20. Simonyan, K., Zisserman, A.: Very deep convolutional networks for large-scale image recognition. arXiv preprint arXiv:1409.1556 (2014)
21. Tung, F., Mori, G.: Similarity-preserving knowledge distillation. In: Proceedings of the IEEE/CVF International Conference on Computer Vision, pp. 1365–1374 (2019)
22. Wang, K., Gao, X., Zhao, Y., Li, X., Dou, D., Xu, C.Z.: Pay attention to features, transfer learn faster CNNs. In: International Conference on Learning Representations (2020)
23. Xu, Z., Hsu, Y.C., Huang, J.: Training shallow and thin networks for acceleration via knowledge distillation with conditional adversarial networks. arXiv preprint arXiv:1709.00513 (2017)
24. Zagoruyko, S., Komodakis, N.: Paying more attention to attention: Improving the performance of convolutional neural networks via attention transfer. arXiv preprint arXiv:1612.03928 (2016)
25. Zagoruyko, S., Komodakis, N.: Wide residual networks. arXiv preprint arXiv:1605.07146 (2016)
26. Zhang, H., et al.: StackGAN: text to photo-realistic image synthesis with stacked generative adversarial networks. In: Proceedings of the IEEE International Conference on Computer Vision, pp. 5907–5915 (2017)
27. Zhang, Y., Xiang, T., Hospedales, T.M., Lu, H.: Deep mutual learning. In: Proceedings of the IEEE Conference on Computer Vision and Pattern Recognition, pp. 4320–4328 (2018)
28. Zhang, Z., Peng, H.: Deeper and wider siamese networks for real-time visual tracking. In: Proceedings of the IEEE/CVF Conference on Computer Vision and Pattern Recognition, pp. 4591–4600 (2019)

A Prediction-Augmented AutoEncoder for Multivariate Time Series Anomaly Detection

Sawenbo Gong[1,2], Zhihao Wu[1,2,3], Yunxiao Liu[1,2,3], Youfang Lin[1,2,3,4], and Jing Wang[1,2,3(✉)]

[1] School of Computer and Information Technology, Beijing Jiaotong University, Beijing, China
{gongsa,zhwu,lyxiao,yflin,wj}@bjtu.edu.cn
[2] Beijing Key Laboratory of Traffic Data Analysis and Mining, Beijing, China
[3] CAAC Key Laboratory of Intelligent Passenger Service of Civil Aviation, Beijing, China
[4] Key Laboratory of Transport Industry of Big Data Application Technologies for Comprehensive Transport, Beijing, China

Abstract. To ensure the normal running of IT systems, multivariate time series changing constantly related to system states need to be monitored to detect unexpected events and anomalies and further prevent adverse effects on the systems. With the rapid increase of complexity and scales of the systems, a more automated and accurate anomaly detection method becomes crucial. In this paper, we propose a new framework called Prediction-Augmented AutoEncoder (PAAE) for multivariate time series anomaly detection, which learns a better representation of normal data from the perspective of reconstruction and prediction. Predictive augmentation is introduced to constrain the latent space to improve the ability of the model to recognize anomalies. And a novel anomaly score is developed considering both the reconstruction errors and prediction errors to reduce false negatives. Extensive experiments prove that the introduction of prediction is effective and show the superiority and robustness of PAAE. Particularly, PAAE obtains at best 7.9% performance improvement on the SMAP dataset.

Keywords: Anomaly detection · Time series · Autoencoder · Reconstruction · Prediction

1 Introduction

IT systems monitor a large number of metrics changing over time, also known as multivariate time series, to identify the state of systems. Some anomalies and unexpected events could be reflected by these metrics. Identifying these anomalies and unexpected events is essential due to anomalies may cause unexpected disasters. The process of detecting unexpected behaviors through metrics monitored is

© Springer Nature Switzerland AG 2021
T. Mantoro et al. (Eds.): ICONIP 2021, LNCS 13108, pp. 681–692, 2021.
https://doi.org/10.1007/978-3-030-92185-9_56

typically formalized as the multivariate time series anomaly detection. Due to the heavy workload of data labeling, data is usually unlabeled, so approaches we used for anomaly detection are usually in an unsupervised way. The early approaches are based on expert knowledge. In this type of approach, domain experts set thresholds for different observation metrics according to the knowledge derived from historical data. If the observed metrics exceed the threshold, we consider that there may be an anomaly or event in the system. However, as IT systems dynamic changes frequently, expert-based approaches are no longer applicable. Thus, automated monitoring of IT systems becomes inevitable.

A number of machine learning based approaches have emerged, in which the most common approaches are clustering-based, such as DBSCAN [5], distance-based, such as K-NN [3] and iForest [4], density-based GMM [14]. But, as systems become more complex and larger, these approaches are inapplicable. Because of the increase of dimensions, they often result in suboptimal performance due to the curse of dimensionality. Recently, methods based on deep learning have been shining in various fields, among which there are many methods applied in anomaly detection of time series [1,10,16,18].

Since systems are usually running normally, many methods will learn the distribution of normal data to implement reconstructing data samples. A well-trained reconstruction-based method can reconstruct the normal samples well, while the abnormal samples have large reconstruction error. Under such a premise, autoencoders (AE) [13] is natively suitable for anomaly detection based on reconstruction. AE maps samples to a low-dimensional latent space and then reconstruct it back into the original space, and there are many works based on AE, such as Deep Autoencoding Gaussian Mixture Model (DAGMM) [19], Multi-Scale Convolutional Recurrent Encoder-Decoder (MSCRED) [18], Omni-Anomaly [16], UnSupervised Anomaly Detection (USAD) [1].

DAGMM jointly learns deep autoencoder and gaussian mixture model. MSCRED, a RNN-based autoencoder with an attention mechanism based on Convolutional Long-Short Term Memory (ConvLSTM) [15], uses the signature matrix as the object of model reconstruction. Both OmniAnomaly and USAD are currently the best performing methods for multivariate time series anomaly detection. The former learns robust representations with a stochastic variable connection and a planar normalizing flow to detect anomalies by the reconstruction probabilities, and the latter builds a stable model under an adversarial training strategy.

Regardless of whether the input is abnormal or not, these methods and other variants of AE provide a most likely reconstruction, which may cause the detection results to contain a large of false negatives to degree model performance. To address this challenge, we propose a new framework called Prediction-Augmented AutoEncoder (PAAE) for multivariate time series anomaly detection. Behind the inspiration of that a good latent representation of normal data can not only reconstruct itself well but also predict the future to some extent, while anomalies can not, we utilize the latent space representation to predict assisting the model with reconstruction to improve the ability of the model to recognize anomalies. We employ a CNN-based encoder and decoder and an attention module to form the skipped connection between the encoder and the decoder to reconstruct and use a prediction module to constrain the latent space. Thus our model is trained by minimizing the reconstruction errors and prediction errors both of which enable the model to learn a better representation for anomaly detection. We also introduce a novel anomaly score considering both reconstruction and prediction to reduce false negatives. The main contributions of this paper are summarized as follows:

- We propose a new framework PAAE for multivariate time series anomaly detection, which learns a better representation of normal data from the perspective of reconstruction and prediction.
- We introduce predictive augmentation to constrain the latent space into multivariate time series anomaly detection to improve the ability of the model to recognize anomalies.
- We also developed a novel anomaly score considering both the reconstruction errors and prediction errors to reduce false negatives.
- Experiments performed on four public datasets demonstrated that our model obtained a superior performance in comparison with several most advanced models. Parameter sensitivity experiments further proved the robustness and stability of our model.

2 Method

2.1 Problem Formulation

Multivariate time series is made up of multiple univariate time series. Each univariate time series records one metric to form a sequence of observed data points. In other words, multivariate time series contains more than one variable at each time instant. In this paper, for multivariate time series with m variables, each of which has T time steps, we denote it as

$$\Gamma = \{X_1, X_2, \ldots, X_T\}, \ X_t = \{x_t^1, x_t^2, \ldots, x_t^m\}, \ t \in \{1, 2, \ldots, T\}$$

where X_t contains m variables at given time t. We assume we obtained an unsupervised model using Γ_{train}, in which contains only normal points, as the training set, so the target of anomaly detection is to determine whether an unknown observation $\hat{X}_{T+1}$ is conformed to the distribution of Γ_{train}. To measure the dissimilarity distance between $\hat{X}_{T+1}$ and the distribution of Γ_{train}, we define an anomaly score and a threshold to label whether it is abnormal.

To capture the temporal dependence, we use sliding windows as the input of models, which we denote as W_t, a sliding window sample of length L at given time t:

$$W_t = \{X_{t-L+1}, X_{t-L+1}, \ldots, X_t\}.$$

Therefore we get a sequence of sliding windows $W = \{W_L, \ldots, W_T\}$. For a sliding window sample of the test dataset, which is denoted as $\hat{W}_{T+1}$, we set a label y_{T+1}

to indicate whether it is an anomaly at time $T + 1$. If its anomaly score is greater than a given threshold, it is regarded as an anomaly, $y_{T+1} = 1$. Otherwise, it is normal and $y_{T+1} = 0$.

2.2   Model Framework

Based on the idea of that a reconstruction model which is well trained with normal data can easily reconstruct normal samples, while it is hard to reconstruct an unseen or abnormal sample, we use an autoencoder (AE), consists of an encoder E and a decoder D, to reconstruct and further we utilize the latent space representation to predict considering that the latent representation of normal data should have the ability to predict to some extent. Thus, the training objective is to minimize the reconstruction error $\mathcal{L}_{rec}$ and prediction error $\mathcal{L}_{pre}$.

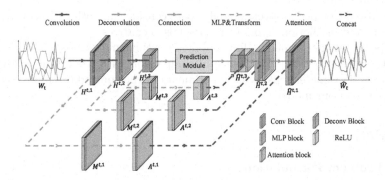

Fig. 1. Model Framework. The framework consists of AE with solid lines and the attention module with dotted lines. The solid arrows go through the E, the prediction module and D, where the three Conv Blocks forms the E and the three Deconv Blocks forms the D. And the dotted arrows go through the attention module which forms the skipped connection between E and D. Different colored arrows indicate different operations displayed at the top of the figure and the meanings of the different colored blocks are shown in the lower right of the figure. (Color figure online)

Our model PAAE is shown in Fig. 1. It is an improvement on the raw AE with an encoder E made up of three convolutional layers, a decoder D made up of three deconvolutional layers, a prediction module P in latent space and an attention module A, which forms the skipped connection of E and D. For a sliding window sample W, we first get the representation H^l of each convolution layer though E. After that, we use each hidden layer to obtain corresponding A^l through attention module, and use H^3 to obtain the latent representation Z to predict. Finally, we utilize the latent representation Z and A^l to get reconstruction $\hat{W}$ through D. Next we will introduce the details of the autoencoder, prediction module, training objective and anomaly score respectively.

Autoencoder with Attention Module. We employ three convolution layers as the encoder to encode spatial patterns of input window W_t. Assuming $H^{t,l} \in \mathbb{R}^{l_w * l_h * l_c}$ denotes the l-th hidden layer with l_c channels representation obtained after $(l-1)$-th convolution at given time t, the hidden layer representation is given by:

$$H^{t,l} = conv_{l-1}(H^{t,l-1}),$$

where $conv_{l-1}(\cdot)$ denotes the $(l-1)$-th convolution operation and $H^{t,0}$ equivalent to W_t.

In order to model the temporal information, we introduced an attention module based Multi-Layer Perceptron (MLP) with two linear layers. Given the different scale of feature maps $H^{t,l}$, we capture the temporal information through a MLP, which is formulated as:

$$M_{t,l} = transform\ to\ 2D(MLP_l(transform\ to\ 1D(H^{t,l}))),$$

where the $transform\ to\ 2D(\cdot)$ and the $transform\ to\ 1D(\cdot)$ denote the shape transformation of feature map, and the MLP_l denotes the l-th MLP. And then, we adopt an attention mechanism [2] to adaptively select the step most relevant to the current step and aggregate it into a feature representation $A_{t,l}$, which is given by:

$$A_{t,l} = \sum_{i \in (t-L,t)} H^{i,l} * \frac{exp\{\frac{H^{i,l} \cdot H^{t,l}}{bs}\}}{\sum_{j \in (t-L,t)} exp\{\frac{H^{j,l} \cdot H^{t,l}}{bs}\}},$$

where bs is the batch size of each training epoch. After getting the l-th representation $A_{t,l}$ through attention module, we feed it into the corresponding layer of the decoder D. As l increases, the information $A_{t,l}$ contains is more global and vice versa.

We employ three deconvolution layers as the decoder to decode the latent representation Z_t which will introduce in next section. Assuming $\hat{H}^{t,l} \in \mathbb{R}^{l_w * l_h * 2l_c}$ denotes the l-th hidden layer with $2l_c$ channels representation obtained after deconvolution and $A_{t,l}$ is the l-th representation through the attention module, the hidden layer representation is given by:

$$\hat{H}^{t,l-1} = deconv_l(concat(\hat{H}^{t,l}, A_{t,l})),$$

where $deconv_l(\cdot)$ denotes the $(l-1)$-th deconvolution operation and $concat(\cdot)$ denotes concatenating operation along the direction of channel. While the $\hat{H}^{t,3}$ is obtained by the latent representation Z_t through a linear layer and transformation. In this work, we use Rectified Linear Unit (ReLU) as the activation function for the encoder E and decoder D.

Prediction Module. We believe that the latent representation of normal data should have the ability to predict to some extent, while anomalies can not. Thus, we constrain the latent space with predictive augmentation to further assist the model with learning a valuable latent representation. The process of prediction module we describe later can be seen at Fig. 2. The latent representation Z_t is obtained after the $H^{t,3}$ going through a linear layer and transformation, which we formulate as:

$$Z_t = f(transform\ to\ 1D(H^{t,3})),$$

where the $f(\cdot)$ denotes MLP with one linear layer. And then, we employ MLP with one layer to predict utilizing the latent representation Z_t, which can be formulated as:

$$\hat{W}_{[t+1:t+p]} = transform\ to\ 2D(f(Z_t)),$$

where $\hat{W}_{[t+1:t+p]} = \{X_{t+1}, \ldots, X_{t+p}\}$ is the prediction with p steps after the window W_t. Thus, the prediction $\hat{W}_{[t+1:t+p]}$ should be as close as possible with the real $W_{[t+1:t+p]}$.

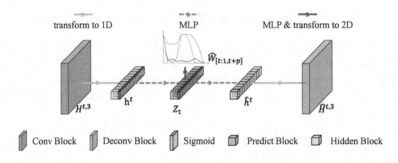

Fig. 2. Prediction Module. This figure is the detail of the prediction module in Fig. 1. The prediction $W_{[t+1:t+p]}$ is obtained by the latent space representation Z_t, while the latent space representation Z_t by the last hidden layer representation $H^{t,3}$ of E. Then, the representation $\hat{H}^{t,3}$ obtained through the MLP layer by latent space representation Z_t is fed to D. Different colored arrows indicate different operations displayed at the top of the figure and the meanings of the different colored blocks are shown at the bottom of the figure. (Color figure online)

Training Objective. For PAAE, we minimize the reconstruction error and the prediction error simultaneously. Thus, the objective $\mathcal{L}$ is consists of two parts, namely reconstruction error $\mathcal{L}_{rec}$ and prediction error $\mathcal{L}_{pre}$, which can be formulated as:

$$\mathcal{L} = \alpha * \mathcal{L}_{rec} + (1 - \alpha) * \mathcal{L}_{pre}$$
$$= \alpha * \sum(||W_t - \hat{W}_t||_2) + (1 - \alpha) * \sum(||W_{[t+1:t+p]} - \hat{W}_{[t+1:t+p]}||_2),$$

where α, $\alpha \in [0, 1]$ is a hyperparameter that trades off between reconstruction errors and prediction errors and $||\cdot||_2$ means the L2-norm. The choice and effect of α will be discussed at Sect. 4.

Anomaly Score. After sufficient training, the model PAAE is utilized for reconstructing and predicting the testing data. Combining with reconstruction error and prediction error, we developed a novel anomaly score to reduce the false negatives. We define it as:

$$s(W_t) = \lambda * ||W_t - \hat{W}_t||_2 + (1 - \lambda) * ||W_{[t+1:t+p]} - \hat{W}_{[t+1:t+p]}||_2,$$

where the $\lambda \in [0, 1]$ is the balance between reconstruction and prediction. The choice and effect of λ will also be discussed at Sect. 4.

3   Experimental Setup

This section describes datasets, the hyperparameters configurations and evaluation metrics.

3.1   Datasets

In this paper, we use four public datasets for experiments, including Mars Science Laboratory (MSL) rover Datasets, Soil Moisture Active Passive (SMAP) satellite datasets, Server Machine Dataset (SMD), and Secure Water Treatment (SWaT) Dataset. The basic information of datasets is shown in Table 1 and the description is introduced respectively as follows.

Table 1. Dataset information

Dataset	Train	Test	Dimensions	Anomalies(%)
MSL	58317	73729	27*55	10.72
SMAP	135183	427617	55*25	13.13
SMD	708405	708420	28*38	4.16
SWaT	496800	449919	51	11.98

MSL and SMAP. These two datasets are real-world public datasets collected from NASA [7]. Each dataset has been split into training and testing subsets and the testing subset contains anomalies that are all expert-labeled. MSL has 27 entities, each of which has 55 metrics, while SMAP has 55 entities, each of which has 25 metrics.

SMD[1]. This dataset is collected by a large Internet company [16]. It records a 5-week-long data from 28 server machines with monitored 33 metrics. This dataset is divided into two equally sized subsets in which the first half is the training subset and the left is the testing subset. Like MSL and SMAP, anomalies in the testing subset were labeled by the domain experts.

---

[1] https://github.com/smallcowbaby/OmniAnomaly.

SWaT[2]. This dataset is recorded by simulating the production of filtered water from industrial water plants in the real world [6,11]. The whole dataset is with a duration of 11-day, in which the first 7-day is recorded under normal working conditions, used as the training subset, and the left 4-day is injected into cyber-attacks, used as the testing subset.

3.2   Hyperparameters Configurations

For the training and testing data, the length of sliding windows L is set to 60 in our all experiments. In the phase of model training, the batch size is set to 64 and the number of epochs is set to 100. The Adam [8] optimiter is used to update the parameters of the network. The learning rate is set to $1e^{-5}$ for the encoder and $1e^{-4}$ for the decoder.

3.3   Evaluation Metrics

To evaluate the anomaly detection performance of different models, we use Precision (P), Recall (R), F1-Score (denoted as F1).

$$P = \frac{TP}{TP + FP}, \ R = \frac{TP}{TP + FN}, \ F1 = 2 \cdot \frac{P \cdot R}{P + R},$$

where TP is the True Positives, FP is the False Positives, and FN is the False Negatives. In this paper, our work is not focused on choosing an appropriate threshold, so we traverse over all possible thresholds to get optimal performance in terms of F1-Score.

In many previous works [9,16,17], as long as at least one anomaly is observed in an anomalous segment, we assume that all anomalies in that segment have been detected, even if they were not. Thus, [16] proposed a point-adjust approach to improve the detecting results. As this approach describes, whenever at least an anomaly is detected in an abnormal segment, all the point in this segment is considered to be marked as anomalies. The results shown in Sect. 4 has been adjusted by point-adjust, a more realistic criterion.

4   Experiments and Results

In this section, we demonstrate the performance and characteristics of our proposed model PAAE through three experiments: overall performance, ablation study, and parameters sensitivity.

4.1   Overall Performance

Six unsupervised anomaly detection methods including AE, IF [4], DAGMM [19], LSTM-VAE [12], Omnianomaly (Omni) [16], and USAD [1], are adopted to compare with our model PAAE. The results are shown in Table 2. Our model

---

[2] https://github.com/JulienAu/Anomaly_Detection_Tuto.

obtained the best performance on three datasets, MSL, SMAP and SMD, and the second-best performance on SWaT.

IF and DAGMM, performed poorly on the four datasets, take m variables of each time step as the input of the model, so the temporal information is not considered. While AE, LSTM-VAE, Omnianomaly and USAD take the sequential inputs and could capture the temporal information. But the first three methods, regardless of whether the input is abnormal or not, perform the most likely reconstruction. The last method USAD is built through adversarial training, which makes it robust. Our model PAAE also uses sequential input, the difference is that we use the prediction of latent space variables as an augmentation to make the reconstruction more stable. Compared with USAD, PAAE is over 1.6% on MSL, 7.9% on SMAP, and 3.4% on SMD, and PAAE is on par with USAD on SWaT in terms of F1-score. PAAE does not perform well on SWaT is due to a large number of contextual anomalies in this dataset.

Table 2. Overall performance

Methods		AE	IF	LSTM-VAE	DAGMM	Omni	USAD	**PAAE**
MSL	P	0.8534	0.5681	0.8599	0.7562	0.9140	0.8810	0.9345
	R	0.9748	0.6740	0.9756	0.9803	0.8891	0.9786	0.9173
	F1	0.8792	0.5984	0.8537	0.8112	0.8952	_0.9109_	**0.9259**
SMAP	P	0.7216	0.4423	0.7164	0.6334	0.7585	0.7697	0.8425
	R	0.9795	0.5105	0.9875	0.9984	0.9756	0.9831	0.9279
	F1	0.7776	0.4671	0.7555	0.7124	0.8054	_0.8186_	**0.8831**
SMD	P	0.8825	0.5938	0.8698	0.6730	0.9809	0.9314	0.9583
	R	0.8037	0.8532	0.7879	0.8450	0.9438	0.9617	0.9816
	F1	0.8280	0.5866	0.8083	0.7231	_0.9441_	0.9382	**0.9698**
SWaT	P	0.9913	0.9620	0.7123	0.8292	0.7223	0.9870	0.9894
	R	0.7040	0.7315	0.9258	0.7674	0.9832	0.7402	0.7374
	F1	0.8233	0.8311	0.8051	0.7971	0.8328	**0.8460**	_0.8450_

4.2   Ablation Study

Employing MSL, SMAP, and SWaT, we evaluated the role of each module of PAAE. Figure 3 shows performance comparison with PAAE in terms of F1-score using PAAE without attention module (w/o attention module), PAAE without prediction module (w/o prediction module), PAAE without attention module and prediction module (w/o both). The frame of PAAE without attention module is consists of a symmetric encoder and decoder without skip connection between each layer. To train PAAE without a prediction module, we set the parameter α and λ as 1. PAAE without attention module and prediction module uses both of the former two configurations.

From Fig. 3, we can find that both the attention module and prediction module contribute greatly to the performance improvement of the model, especially the prediction module with 18.7% enhancement by comparing w/o attention module and w/o both. The reason is that it constrains the latent space representations and allows a broader view of model observation extending to the future. This also proves that a good latent space representation should have the ability to predict and reconstruct. Moreover, adopting an attention module means can further improve the performance of anomaly detection.

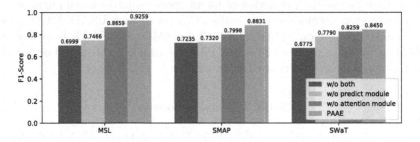

Fig. 3. Results of ablation study

4.3   Parameters Sensitivity

We study the effect of three parameters, length p of prediction, α of training objective, and λ of anomaly score, on the performance of our method using MSL and SMAP. Figure 4 a) responds how to choose an appropriate choice of predictive length p. As the length of the prediction window increases, the performance of the model decreases slightly due to more redundant future information incorporated. This also implies that the choice of this hyperparameter arrange from 5 to 25 has little effect on model performance.

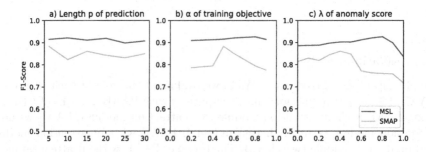

Fig. 4. Effect of parameters.

Figure 4 b) shows the performance of PAAE at different values of α while λ fixed as 0.5 and Fig. 4 c) shows the performance of PAAE at different values of λ while α fixed as 0.5. We can see that the optimal parameter selection is different for different datasets, but for the same dataset, the optimal selection of two parameters, α and λ, is similar. So we can set the two parameters same, i.e. $\alpha = \lambda = 0.8$ for MSL and $\alpha = \lambda = 0.5$ for SMAP.

5   Conclusion and Outlook

In this paper, a novel framework named PAAE is proposed for multivariate time series anomaly detection. PAAE learns a better reconstruction representation with predictive augmentation and utilizes an attention module to form the skipped connection of the encoder and the decoder to further improve the anomaly detection performance. The introduction of predictive augmentation to constrain the latent space into multivariate time series anomaly detection enhance the ability of the model to recognize anomalies. Four open datasets are employed to evaluate the performance of the proposed method. Extensive experiments demonstrated that our proposed method is superior to state-of-the-art techniques. The parameter sensitivity experiments also showed the robustness and stability of the model. Nevertheless, our model has the drawback of failing to contextual threshold occasionally and the execution efficiency is not optimal. In future, we plan to introduce a graph structure to capture the relationship among variables and optimize the model structure to further address these problems.

Acknowledgments. This work was supported by the Fundamental Research Funds for the Central Universities (2021JBM007) and the National Natural Science Foundation of China under Grant 61603029.

References

1. Audibert, J., Michiardi, P., Guyard, F., Marti, S., Zuluaga, M.A.: USAD: unsupervised anomaly detection on multivariate time series. In: Proceedings of the 26th ACM SIGKDD International Conference on Knowledge Discovery & Data Mining, pp. 3395–3404 (2020)
2. Bahdanau, D., Cho, K., Bengio, Y.: Neural machine translation by jointly learning to align and translate. arXiv preprint arXiv:1409.0473 (2014)
3. Chaovalitwongse, W.A., Fan, Y.J., Sachdeo, R.C.: On the time series k-nearest neighbor classification of abnormal brain activity. IEEE Trans. Syst. Man Cybern.-Part A: Syst. Hum. **37**(6), 1005–1016 (2007)
4. Ding, Z., Fei, M.: An anomaly detection approach based on isolation forest algorithm for streaming data using sliding window. IFAC Proc. Vols. **46**(20), 12–17 (2013)
5. Ester, M., Kriegel, H.P., Sander, J., Xu, X., et al.: A density-based algorithm for discovering clusters in large spatial databases with noise. In: KDD, vol. 96, pp. 226–231 (1996)
6. Goh, J., Adepu, S., Junejo, K.N., Mathur, A.: A dataset to support research in the design of secure water treatment systems. In: Havarneanu, G., Setola, R., Nassopoulos, H., Wolthusen, S. (eds.) CRITIS 2016. LNCS, vol. 10242, pp. 88–99. Springer, Cham (2017). https://doi.org/10.1007/978-3-319-71368-7_8

7. Hundman, K., Constantinou, V., Laporte, C., Colwell, I., Soderstrom, T.: Detecting spacecraft anomalies using LSTMs and nonparametric dynamic thresholding. In: Proceedings of the 24th ACM SIGKDD International Conference on Knowledge Discovery & Data Mining, pp. 387–395 (2018)

8. Kingma, D., Ba, J.: Adam: a method for stochastic optimization. Comput. Sci. (2014)

9. Li, Z., Chen, W., Pei, D.: Robust and unsupervised KPI anomaly detection based on conditional variational autoencoder. In: 2018 IEEE 37th International Performance Computing and Communications Conference (IPCCC), pp. 1–9. IEEE (2018)

10. Malhotra, P., Ramakrishnan, A., Anand, G., Vig, L., Agarwal, P., Shroff, G.: LSTM-based encoder-decoder for multi-sensor anomaly detection. arXiv preprint arXiv:1607.00148 (2016)

11. Mathur, A.P., Tippenhauer, N.O.: SWaT: a water treatment testbed for research and training on ICS security. In: 2016 International Workshop on Cyber-physical Systems for Smart Water Networks (CySWater), pp. 31–36. IEEE (2016)

12. Park, D., Hoshi, Y., Kemp, C.C.: A multimodal anomaly detector for robot-assisted feeding using an LSTM-based variational autoencoder. IEEE Robot. Autom. Lett. **3**, 1544–1551 (2018)

13. Rumelhart, D.E.: Learning internal representations by error propagations. In: Parallel Distributed Processing (1986)

14. Schnell, T., et al.: Robot health estimation through unsupervised anomaly detection using Gaussian mixture models. In: 2020 IEEE 16th International Conference on Automation Science and Engineering (CASE), pp. 1037–1042. IEEE (2020)

15. Shi, X., Chen, Z., Wang, H., Yeung, D.Y., Wong, W.K., Woo, W.C.: Convolutional LSTM network: a machine learning approach for precipitation nowcasting. arXiv preprint arXiv:1506.04214 (2015)

16. Su, Y., Zhao, Y., Niu, C., Liu, R., Sun, W., Pei, D.: Robust anomaly detection for multivariate time series through stochastic recurrent neural network. In: Proceedings of the 25th ACM SIGKDD International Conference on Knowledge Discovery & Data Mining, pp. 2828–2837 (2019)

17. Xu, H., et al.: Unsupervised anomaly detection via variational auto-encoder for seasonal KPIs in web applications. In: Proceedings of the 2018 World Wide Web Conference, pp. 187–196 (2018)

18. Zhang, C., et al.: A deep neural network for unsupervised anomaly detection and diagnosis in multivariate time series data. In: Proceedings of the AAAI Conference on Artificial Intelligence, vol. 33, pp. 1409–1416 (2019)

19. Zong, B., et al.: Deep autoencoding gaussian mixture model for unsupervised anomaly detection. In: International Conference on Learning Representations (2018)

Author Index

Agyapong, Dorothy Araba Yakoba   77
Amin, Abdul Wahab   379
Anibal, James   512

Bai, Haoli   367
Barucca, Paolo   14
Bouaziz, Bassem   402

Cai, Ning   537
Cai, Yancheng   438
Cao, Guitao   177
Cao, Zheng   645
Challa, Aravind   165
Chen, Chong   584
Chen, Wang   570
Chen, Xinyu   152
Chong, Zhihong   426

Dai, Pei   65
Dai, Zhongjian   317
de Souto, Marcilio C. P.   621
Deng, Minghua   584
Ding, Xiang   227
Ding, Xiaotian   317
Du, Weidong   438

Fadjrimiratno, Muhammad Fikko   450
Feng, Lin   537
Fernandes, Luiz Henrique dos S.   621
Frempong, Lady Nadia   77
Fu, Yeu-Shin   91

Gaikwad, Jitendra   402
Gao, Min   305
Goel, Anurag   549
Goel, Srishti   40
Gong, Sawenbo   681
Gong, Yuchen   190
Gu, Peiyi   657
Gu, Zhihao   190
Guo, Junjie   27
Guo, Xiangyang   584

Han, Ya-nan   390
He, Baoyin   65
He, Penghao   141
He, Yujie   438
Hosu, Ionel   488
Hu, Xing   103
Hu, Xingfei   633
Huang, Lei   3
Huang, Yinqiu   305

Jagodziński, Dariusz   524
Jiang, Fei   633
Jiang, Hualiang   266
Jiang, Linhua   103
Jiang, Ning   342, 596, 669

Kang, Guixia   116
Kimura, Masanari   558
King, Irwin   367
Kumar, Neeraj   40

Lai, Darong   426
lall, Brejesh   40
Li, Haiyan   633
Li, Min   438
Li, Mingchao   570
Li, Peifeng   152, 414
Li, Xiaoyong   27
Li, Xinyi   254
Li, Zhixin   464
Lin, Jia-Xiang   129
Lin, Youfang   681
Liu, Cheng-Lin   279
Liu, Gongshen   27, 203
Liu, Haozhuang   570
Liu, Jian-wei   215, 390
Liu, Jing   103
Liu, Mingjin   342
Liu, Shan   500
Liu, Shenglan   537
Liu, Yan   317
Liu, Yang   103, 227
Liu, Yunxiao   681
Long, Yunfei   609

Lorena, Ana C.   621
Lu, Hongtao   633
Lu, Jianwei   657
Luo, Xiao   584
Luo, Ye   657
Luo, Zhigang   54
Lyu, Michael R.   367

Ma, Jinwen   584
Ma, Lizhuang   190
Ma, Xiaohu   266
Ma, Xurui   54
Ma, Zhiyuan   27
Maaz, Mohammad   379
Mahdi, Walid   402
Majumdar, Angshul   549
Marin, Andrei   488
Meng, Kui   203
Muhammad, Abubakar   379

Nagai, Ayumu   354
Narang, Ankur   40
Nartey, Obed Tettey   77
Neumann, Łukasz   524
Nguyen, Binh T.   512
Nguyen, Truong-Son   512

Pan, Zeyuan   426
Phan, Long   512
Piczak, Karol J.   330

Qi, Pengyuan   609
Qian, Zhong   414
Qin, Zhenyue   227
Qiu, Xiaohua   438

Rao, Sathwik   165
Rasool, Fezan   379
Rebedea, Traian   488

Sadowski, Michał   330
Sanodiya, Rakesh Kumar   165
Sarpong, Asare K.   77
Satwik, Sai   165
Shao, Wei   3
Sharma, Chinmay   165
Shen, Lian   129
Shen, Yao   266
Shu, Kai   305
Song, Liang   103

Song, Yifan   426
Song, Yu   438
Song, Zhi-yan   390
Spurek, Przemysław   330
Suleiman, Basem   292
Sun, Dongting   54
Sun, Jianguo   609
Suzuki, Einoshin   240, 450

Taj, Murtaza   379
Tang, Buzhou   476
Tang, Jialiang   342, 596, 669
Tang, Yanping   464
Tharani, Mohbat   379
Tianye   609
Tran, Hieu   512
Triki, Abdelaziz   402
Trzciński, Tomasz   330

Wang, Chang-Ying   129
Wang, Fuzhou   3
Wang, Hanqi   103
Wang, Jia   305
Wang, Jiangtao   177
Wang, Jing   681
Wang, Mengzhu   54
Wang, Peijun   657
Wang, Yang   570
Wang, Zhiwen   464
Wei, Peiyi   464
Wong, Ka-Chun   3
Wu, Daqing   584
Wu, Deyang   633
Wu, Ji   645
Wu, JinZhao   77
Wu, Song   254, 500
Wu, Zhaoqing   438
Wu, Zhihao   681

Xiao, Bing-biao   215
Xiao, Guoqiang   254, 500
Xie, Weidun   3
Xiong, Yangyang   317
Xiong, Yongyu   279
Xu, Chaoyue   141
Xu, Sheng   152
Xu, Tianyu   65
Xu, Xiaohui   500
Xu, Xiaowen   657
Xu, Xinyue   227

Yang, Dongdong   266
Yang, Guowu   77
Yang, Hao   141
Yang, Jie   537
Yang, Peipei   279
Yang, Xiaoyan   596, 669
Yao, Leehter   165
Yaqub, Waheeb   292
Yau, Chung-Yiu   367
Yin, Nan   54
Yin, Zexuan   14
Yu, Wei   54
Yu, Wenxin   342, 596, 669
Yu, Ying   141

Zawistowski, Paweł   524
Zhang, Canlong   464
Zhang, Guanhua   103
Zhang, Heng   414
Zhang, Kailai   645
Zhang, Kang   450
Zhang, Liguo   609

Zhang, Linwei   91
Zhang, Lu-ning   390
Zhang, Manli   116
Zhang, Mei   65
Zhang, Peng   596, 669
Zhang, Tianming   54
Zhang, Yang   266
Zhang, Yueling   177
Zhang, Zhenghao   190
Zhao, Haodong   27
Zhao, Jiaxin   266
Zhao, Xiaoyu   476
Zhao, Yichun   203
Zheng, Hai-Tao   570
Zheng, Zhong   116
Zhou, Bei   292
Zhou, Jian   537
Zhou, Shunkai   177
Zhu, Jinhui   65
Zhu, Qiaoming   152
Zhu, Xiaoxu   414
Zou, Qiming   240